PROBIOTICS

Edited by **Everlon Cid Rigobelo**

Probiotics
http://dx.doi.org/10.5772/3444
Edited by Everlon Cid Rigobelo

Contributors
Danfeng Song, Salam Ibrahim, Saeed Hayek, Alice Maayan Elad, Uri Lesmes, Carina Paola Van Nieuwenhove, Victoria Terán, Silvia Nelina González, Z. Denkova, A. Krastanov, Fariborz Akbarzadeh, Aziz Homayouni, Emiliane Andrade Araújo, Ana Clarissa dos Santos Pires, Maximiliano Soares Pinto, Gwénaël Jan, Antônio Fernandes de Carvalho, Roel J. Vonk, Gerlof A.R. Reckman, Hermie J.M. Harmsen, Marion G. Priebe, R. Nyanzi, P.J. Jooste, Maedeh Alizadeh, Hossein Alikhah, Vahid Zijah, Esteban Boza-Méndez, Rebeca López-Calvo, Marianela Cortés-Muñoz, Gabriel-Danut Mocanu, Elisabeta Botez, Dorota Żyżelewicz, Ilona Motyl, Ewa Nebesny, Grażyna Budryn, Wiesława Krysiak, Justyna Rosicka-Kaczmarek, Zdzisława Libudzisz, Antigoni Mavroudi, Hani Al-Salami, Rima Caccetta, Svetlana Golocorbin-Kon, Momir Mikov, Giselle Nobre Costa, Lucia Helena S. Miglioranza, Marcin Łukaszewicz, Rosa Helena Luchese, Mikhail Lakhtin, Vladimir Lakhtin, Alexandra Bajrakova, Andrey Aleshkin, Stanislav Afanasiev, Vladimir Aleshkin, Fatma Nur Sari, Ugur Dilmen, Parvin Bastani, Violet Gasemnezhad Tabrizian, Somayeh Ziyadi, Marie-José Butel, Anne-Judith Waligora-Dupriet, Julio Aires, Petar Nikolov, María Chávarri, Izaskun Marañón, María Carmen Villarán, Kamila Goderska, Saddam S. Awaisheh, Andrea Carolina Aguirre Rodríguez, Jorge Hernán Moreno Cardozo, I.E. Luis-Villaseñor, A.I. Campa-Córdova, F.J. Ascencio-Valle, Mariella Rivas, Carlos Riquelme

Published by InTech
Janeza Trdine 9, 51000 Rijeka, Croatia

Publishing Process Manager Petra Nenadic
Typesetting InTech DTP Team
Cover InTech Design Team

Additional hard copies can be obtained from orders@intechopen.com

Probiotics, Edited by Everlon Cid Rigobelo
p. cm.
ISBN 978-953-51-0776-7

For EU product safety concerns:
IN TECH d.o.o., Prolaz Marije Krucifikse Kozulić 3, 51000 Rijeka, Croatia,
info@intechopen.com or visit our website at intechopen.com.

Meet the editor

Everlon Cid Rigobelo has graduated from Agronomy School, Universidade Estadual Paulista, Brazil, in 2000. He received his M.S. degree in Animal Science Microbiology from same University in 2002. He obtained the Ph.D from same University. Rigobelo has experience in genetics, epidemiology, acting on the following subjects: microbial biotechnology, molecular genetics, and bacterial genomics. He works with the probiotic strains against colonization caused by Escherichia coli STEC, Pasteurella multocida, Leptospirosis. Previously he worked with E. coli shiga like toxin isolated from cattle and beef carcass.

Contents

Preface

Probiotics are specific strains of microorganisms, which when served to human in proper amount, have a beneficial effect, improving health or reducing risk of get sick. They are used of functional foods and pharmaceutical products and play an important role in promoting and maintaining human health. This book comprehensively reviews and compiles information on probiotics strains in 30 chapters which cover the use of probiotics the editor has tried arrange the book chapters in a issue order to make it easier for the readers to find what they need.

Section 1 – Use of Probiotics in food, which includes chapters 1-12 is showed issues related with the use of probiotics in food on different approaches such as lactose intolerance and functional foods development

Section 2 – Probiotics in Health, which includes chapters 13-22 is showed issues related with the use of probiotics in human′s health such as application in inflammatory diseases, interaction in the gut and prevention of necrotizing.

Section 3 – Probiotics in Biotechnological Aspects, which includes chapters 23-26 is showed issues related with the Biotechnological Aspects such as probiotics stabilization, types and specifications.

Section 4 – Probiotics in Aquaculture, which includes chapters 27-28, chapters related with probiotics in shrimp larvae and biofilms.

This book is written by authors from America, Europe, Asia and Africa, yet, the editor has tried arrange the book chapters in a issue order to make it easier for the readers to find what they need.

The scientists selected to publishing of this book were guests due to their recognized expertise and important contributions on fields in which they are acting. Without these scientists, their dedication and enthusiasm the publishing this book would have not been possible. I recognize the efforts them in the attempt of contribute to animals production contributing thus to the developing Human and I′m very gratefully for that.

This book will hopefully be of help to many scientists, doctors, pharmacists, chemicals and other experts in a variety of disciplines, both academic and industrial. It may not only support research and development, but also be suitable for teaching.

I would like to thank Professor Fernando Antonio de Ávila by his life lessons and also by he to be my scientific mentor.

Finally, I would like to thank my daughter Maria Eduarda and my wife Fernanda for their patience and also my son that is coming and in this moment is inside of comfortable womb from Mom. I extend my apologies for many hours spent on the preparation of my chapter and the editing of this book, which kept me away from them.

Prof. Dr. Everlon Cid Rigobelo
Laboratory of Microbiology & Hygiene,
UNESP Univ Estadual Paulista
Animal Science Course
Dracena
Brazil

Section 1

Use of Probiotic in Food

Chapter 1

Recent Application of Probiotics in Food and Agricultural Science

Danfeng Song, Salam Ibrahim and Saeed Hayek

Additional information is available at the end of the chapter

http://dx.doi.org/10.5772/50121

1. Introduction

Probiotic foods are a group of functional foods with growing market shares and large commercial interest [1]. Probiotics are live microorganisms which when administered in adequate amounts confer a beneficial health benefit on the host [2]. Probiotics have been used for centuries in fermented dairy products. However, the potential applications of probiotics in nondairy food products and agriculture have not received formal recognition. In recent times, there has been an increased interest to food and agricultural applications of probiotics, the selection of new probiotic strains and the development of new application has gained much importance. The uses of probiotics have been shown to turn many health benefits to the human and to play a key role in normal digestive processes and in maintaining the animal's health. The agricultural applications of probiotics with regard to animal, fish, and plants production have increased gradually. However, a number of uncertainties concerning technological, microbiological, and regulatory aspects exist [3].

1.1. Definition of probiotics

Probiotics are live microbes that can be formulated into many different types of products, including foods, drugs, and dietary supplements. Probiotic is a relatively new word that is used to name the bacteria associated with the beneficial effects for the humans and animals. The term probiotic means "for life" and it was defined by an Expert Committee as "live microorganisms which upon ingestion in certain numbers exert health benefits beyond inherent general nutrition" [4]. FAO/WHO Expert Consultation believes that general guidelines need to provide to how these microorganisms can be tested and proven for safety and potential health benefits when administered to humans.

Lactobacillus and *Bifidobacterium* are most commonly used probiotics in food and feed (Table 1). Other microorganisms such as yeast *Saccharomyces cerevisiae* and some *Escherichia coli* and *Bacillus* species are also used as probiotics. Lactic acid bacteria (LAB) which have been used for food fermentation since the ancient time, can serve a dual function by acting as food fermenting agent and potentially health benefits provider. LAB are GRAS (general recognized as safe) with no pathogenic, or virulence properties have been reported. For the use of LAB as probiotics, some desirable characteristics such as low cost, maintaining its viability during the processing and storage, facility of the application in the products, resistance to the physicochemical processing must be considered.

Lactobacillus species	Bifidobacterium species	Others
L. acidophilus	*B. adolescentis*	*Bacillus cereus*
L. amylovorus	*B. animalis*	*Clostridium botyricum*
L. brevis	*B. breve*	*Enterococcus faecalis*[a]
L. casei	*B. bifidum*	*Enterococcus faecium*[a]
L. rhamnosus	*B. infantis*	*Escherichia coli*
L. crispatus	*B. lactis*	*Lactococcus lactis* subsp. *cremoriss*
L. delbrueckii subsp. *bulgaricus*	*B. longum*	*Lactococcus lactis* subsp. *lactis*
L. fermentum		*Leuconostoc mesenteroides* subsp. *dextranicum*
L. gasseri		*Pediococcus acidilactici*
L. helveticus		*Propionibacterium freudenreichii*[a]
L. johnsonii		*Saccharomyces boulardii*
L. lactis		*Streptococcus salivarius* subsp. *thermophilus*
L. paracasei		*Sporolactobacillus inulinus*[a]
L. plantarum		
L. reuteri		
L. salivarius		
L. gallinarum[a]		

[a] mainly applied in animals

Table 1. Probiotic microorganisms. Adapted from [5, 6]

1.2. Characteristics of probiotics

Characteristics of probiotics will determine their ability to survive the upper digestive tract and to colonize in the intestinal lumen and colon for an undefined time period. Probiotics are safe for human consumption and no reports have found on any harmfulness or production of any specific toxins by these strains [7, 8]. In addition, some probiotics could produce antimicrobial substances like bacteriocins. Therefore, the potential health benefit will depend on the characteristic profile of the probiotics. Some probiotic strains can reduce intestinal transit time, improve the quality of migrating motor complexes [9], and temporarily increase the rate of mitosis in enterocytes [10, 11].

The most common probiotics are *Lactobacillus* and *Bifidobacterium*. In general most probiotics are gram-positive, usually catalase-negative, rods with rounded ends, and occur in pairs, short, or long chains [7]. They are non-flagellated, non-motile and non-spore-forming, and are intolerant to salt. Optimum growth temperature for most probiotics is 37°C but some strains such as *L. casei* prefer 30 °C and the optimum pH for initial growth is 6.5-7.0 [7]. *L. acidophilus* is microaerophilic with anaerobic referencing and capability of aerobic growth. *Bifidobacterium* are anaerobic but some species are aero-tolerant. Most probiotics bacteria are fastidious in their nutritional requirements [12, 13]. With regard to fermentation probiotics are either obligate homofermentative (ex. *L. acidophilus, L. helvelicas*), obligate heterofermentative (ex. *L. brevis, L. reuteri*), or facultative heterofermentative (ex. *L. casei, L. plantarum*) [14]. Additionally, probiotics produce a variety of beneficial compounds such as antimicrobials, lactic acid, hydrogen peroxide, and a variety of bacteriocins [15, 16] . Probiotics should have the ability to interact with the host microflora and competitive with microbial pathogens, bacterial, viral, and fungal [16].

2. Probiotics health benefits

Probiotic research suggests a range of potential health benefits to the host organism. The potential effects can only be attributed to tested strains but not to the whole group of probiotics. Probiotics have shown to provide a diverse variety of health benefits to human, animal, and plans. However, viability of the microorganisms throughout the processing and storage play an important role in transferring the claimed health effects. Therefore, the health benefits must be documented with the specific strain and specific dosage [17].

2.1. Human health

Probiotics display numerous health benefits beyond providing basic nutritional value [4]. These evidences have been established by the scientific testing in the humans or animals, performed by the legitimate research groups and published in peer-reviewed journals [16, 18]. Some of these benefits have been well documented and established while the others have shown a promising potential in animal models, with human studies required to substantiate these claims [18]. Health benefits of probiotic bacteria are very strain specific; therefore, there is no universal strain that would provide all proposed benefits and not all strains of the same species are effective against defined health conditions [18].

Probiotics have been used in fermented food products for centuries. However, nowadays it has been claimed that probiotics can serve a dual function by their potentially importing health benefits. The health benefit of fermented foods may be further enhanced by supplementation of *Lactobacillus* and *Bifidobacterium* species [19]. *L. acidophilus, Bifidobacterium* spp. and *L. casei* species are the most used probiotic cultures with established human health in dairy products, whereas the yeast *Saccharomyces cerevisiae* and some *E. coli* and *Bacillus* species are also used as probiotics [20].

Several studies have documented probiotic effects on a variety of gastrointestinal and extraintestinal disorders, including prevention and alleviation symptoms of traveler's diarrhea and antibiotic associated diarrhea [21], inflammatory bowel disease [21], lactose intolerance [22], protection against intestinal infections [23], and irritable bowel syndrome. Some probiotics have also been investigated in relation to reducing prevalence of atopic eczema later in life [24], vaginal infections, and immune enhancement [25], contributing to the inactivation of pathogens in the gut, rheumatoid arthritis, improving the immune response of in healthy elderly people [26], and liver cirrhosis.

In addition, probiotics are intended to assist the body's naturally occurring gut microbiota. Some probiotic preparations have been used to prevent diarrhea caused by antibiotics, or as part of the treatment for antibiotic-related dysbiosis. Although there is some clinical evidence for the role of probiotics in lowering cholesterol but the results are conflicting. Probiotics have a promising inhibitory effect on oral pathogens especially in childhood but this may not necessarily lead to improved oral health [27]. Antigenotoxicity, antimutagenicity and anticarcinogenicity are important potential functional properties of probiotics, which have been reported recently. Observational data suggest that consumption of fermented dairy products is associated with a lower prevalence of colon cancer, which is suggested that probiotics are capable of decreasing the risk of cancer by inhibition of carcinogens and pro-carcinogens, inhibition of bacteria capable of converting pro-carcinogens to carcinogens [18].

2.2. Animal health

Probiotics which are traditional idea in the human food have been extended to animals by developing fortified feed with intestinal microbiota to benefit the animals. The microflora in the gastrointestinal tracts of animals plays a key role in normal digestive processes and in maintaining the animal's health. Probiotics can beneficially improve the intestinal microbial balance in host animal. Commercial probiotics for animal use are claimed to improve animal performance by increasing daily gain and feed efficiency in feedlot cattle, enhance milk production in dairy cows, and improve health and performance of young calves [28] and in improving growth performance of chickens [29]. Probiotics can attach the mucosal wall, adjust to immune responses [30], and compete the pathogenic bacteria for attachment to mucus [31, 32]. Probiotics provide the animal with additional source of nutrients and digestive enzymes [33, 34]. They can stimulate synthesis vitamins of the B-group and enhancement of growth of nonpathogenic facultative anaerobic and gram positive bacteria by producing inhibitory compounds like volatile fatty acids and hydrogen peroxide that inhibit the growth of harmful bacteria enhancing the host's resistance to enteric pathogens [32, 35]. Probiotics stimulate the direct uptake of dissolved organic material mediated by the bacteria, and enhance the immune response against pathogenic microorganisms [36, 37]. Finally, probiotics can inhibit pathogens by competition for a colonization sites or nutritional sources and production of toxic compounds, or stimulation of the immune system.

2.3. Plant health

The more beneficial the bacteria and fungi are, the more "fertile" the soil is. These microorganisms break down organic matter in the soil into small, usable parts that plants can uptake through their roots. The healthier the soil, the lower the need for synthetic herb/pesticides and fertilizers.The concept that certain microorganisms 'probiotics' may confer direct benefits to the plant acting as biocontrol agents for plants. The plant probiotic bacteria have been isolated and commercially developed for use in the biological control of plant diseases or biofertilization [38]. These microorganisms have fulfilled important functions for plant as they antagonize various plant pathogens, induce immunity, or promote growth [38-40]. The interaction between bacteria and fungi with their host plants has shown their ability to promote plant growth and to suppress plant pathogens in several studies [41-44].

3. Food applications of probiotics

Today an increase in knowledge of functional foods has led to develop foods with health benefits beyond adequate nutrition. The last 20 years have shown an increased interest among consumers in functional food including those containing probiotics. The presence of probiotics in commercial food products has been claimed for certain health benefits. This has led to industries focusing on different applications of probiotics in food products and creating a new generation of 'probiotic health' foods. This section will summarize the common applications of probiotics in food products.

3.1. Dairy-based probiotic foods

Milk and its products is good vehicle of probiotic strains due to its inherent properties and due to the fact that most milk and milk products are stored at refrigerated temperatures. Probiotics can be found in a wide variety of commercial dairy products including sour and fresh milk, yogurt, cheese, etc. Dairy products play important role in delivering probiotic bacteria to human, as these products provide a suitable environment for probiotic bacteria that support their growth and viability [45-48]. Several factors need to be addressed for applying probiotics in dairy products such as viability of probiotics in dairy [19, 48], the physical, chemical and organoleptic properties of final products [49-51], the probiotic health effect [52, 53], and the regulations and labeling issues [4, 54].

3.1.1. Drinkable fresh milk and fermented milks

Among probiotics carrier food products, dairy drinks were the first commercialized products that are still consumed in larger quantities than other probiotic beverages. Functional dairy beverages can be grouped into two categories: fortified dairy beverages (including probiotics, prebiotics, fibers, polyphenols, peptides, sterol, stanols, minerals, vitamins and fish oil), and whey-based beverages [55]. Among the probiotic bacteria used in the manufacture of dairy beverages, *L. rhamnosus* GG is the most widely used. Owing to *L. rhamnosus* GG acid and bile resistance [56], this probiotic is very suitable for industrial applications. Özer and Avnikirmaci have reported several examples of commercial probiotic dairy beverages showing that *L. acidophilus, L. casei, L. rhamnosus,* and *L. plantarum* as most applied probiotics [55].

Several factors have been reported to affect the viability of probiotic cultures in fermented milks. Acidity, pH, dissolved oxygen content, redox potential, hydrogen peroxide, starter microbes, potential presence of flavoring compounds and various additives (including preservatives) affect the viability of probiotic bacteria and have been identified as having an effect during the manufacture and storage of fermented milks [19, 48, 57]. Today, a wide range of dairy beverages that contain probiotic bacteria is available for consumers in the market including: Acidophilus milk, Sweet acidophilus milk, Nu-Trish A/B, Bifidus milk, Acidophilus buttermilk, Yakult, Procult drink, Actimel, Gaio, ProViva, and others [55].

Probioticts such as *Lactobacillus* and *Bifidobacterium* strains grow weakly in milk due to their low proteolytic activity and inability to utilize lactose [47, 57]. These bacteria also need certain compounds for their growth which is missing in milk [19, 58, 59]. To improve growth and viability of probiotics in dairy beverages various substances have been tested in milk. Citrus fiber presence in fermented milks was found to enhance bacterial growth and survival of probiotic bacteria in fermented milks [60]. Addition of soygerm powder has shown certain positive effects on producing fermented milk with *L. reuteri*. Soygerm powder may release important bioactive isoflavones during fermentation that could protect *L. reuteri* from bile salt toxicity in the small intestine [61]. Other substances include fructooligosaccahrides (FOS), aseinomacropeptides (CMP), whey protein concentrate (WPC), tryptone, yeast extracts, certain amino acids, nucleotide precursors and an iron source were also documented [59, 63, 64]. Additionally, the selection of probiotic strains and optimization of the manufacturing conditions (both formulation properties and storage conditions) are of utmost importance in the viability of probiotic bacteria in fermented milk [47, 65].

3.1.2. Yogurt

Yogurt is one of the original sources of probiotics and continues to remain a popular probiotic product today. Yogurt is known for its nutritional value and health benefits. Yogurt is produced using a culture of *L. delbrueckii* subsp. *bulgaricus* and

Streptococcus salivarius subsp. *thermophilus* bacteria. In addition, other lactobacilli and bifidobacteria are also sometimes added during or after culturing yogurt. The probiotic characteristics of these bacterial strains that form the yogurt culture are still debatable. The viability of probiotics and their proteolytic activities in yoghurt must be considered. Numerous factors may affect the survival of *Lactobacillus* and *Bifidobacterium* spp. in yogurt. These include strains of probiotic bacteria, pH, presence of hydrogen peroxide and dissolved oxygen, concentration of metabolites such as lactic acid and acetic acids, buffering capacity of the media as well as the storage temperature [19, 66, 67].

Although yogurt has been widely used as probiotics vehicle, most commercial yogurt products have low viable cells at the consumption time [19, 68]. Viability of probiotics in yogurt depends on the availability of nutrients, growth promoters and inhibitors, concentration of solutes, inoculation level, incubation temperature, fermentation time and storage temperature. Survival and viability of probiotic in yogurt was found to be strain dependant. The main factors for loss of viability of probiotic organisms have been attributed to the decrease in the pH of the medium and accumulation of organic acids as a result of growth and fermentation. Among the factors, ultimate pH reached at the end of yogurt fermentation appears to be the most important factor affecting the growth and viability of probiotics. Metabolic products of organic acids during storage may further affect cell viability of probiotics [66]. The addition of fruit in yogurt may have negative effect on the viability of probiotics, since fruit and berries might have antimicrobial activities. Inoculation with very high level of probiotics with attempts to compensate the potential viability loss, might result in an inferior quality of the product. The present of probiotic was found to affect some characteristics of yogurt including: acidity, texture, flavor, and appearance [69]. However, encapsulation in plain alginate beads, in chitosancoated alginate, alginate-starch, alginate-prebiotic, alginate-pectin, in whey protein-based matrix, or by adding prebiotics or cysteine into yogurt, could improve the viability and stability of probiotics in yogurt [70-79].

3.1.3. Cheese

Yogurt and milk are the most common vehicles of probiotics among dairy products. However, alternative carriers such as cheese seem to be well suited. Cheeses have a number of advantages over yogurt and fermented milks because they have higher pH and buffering capacity, highly nutritious, high energy, more solid consistency, relatively higher fat content, and longer shelf life [80, 81]. Several studies have demonstrated a high survival rate of probiotics in cheese at the end of shelf life and high viable cells [45, 48, 82, 83]. Probiotics in cheese were found to survive the passage through the simulated human gastrointestinal tract and significantly increase the numbers of probiotic cells in the gut [82]. However, comparing the serving size of yogurt to that of cheese, cheese needs to have higher density of probiotic cells and higher viability to provide the same health benefits. Cheese was introduced to probiotic industry in 2006 when Danisco decided to test the growth and survival of probiotic strains in cheese [84]. At that time, only few probiotic cheese products were found on the market. The test showed that less than 10% of the bacteria were lost in the cheese whey. Based on the process, a commercial probiotic cheese was first developed by the Mills DA, Oslo, Norway. Nowadays, there are over 200 commercial probiotic cheeses in various forms, such as fresh, semi-hard, hard cheese in the marketplaces. Semi-hard and hard cheese, compared to yogurt as a carrier for probiotics, has relatively low recommended daily intake and need relatively high inoculation level of probiotics (about 4 to 5 times). Fresh cheese like cottage cheese has high recommended daily intake, limited shelf life with refrigerated storage temperature. It may, thus, serve as a food with a high potential to be applied as a carrier for probiotics.

3.1.4. Other dairy based products

Other dairy products including quark, chocolate mousse, frozen fermented dairy desserts, sour cream, and ice cream can be good vehicles of probiotics. Quark was tested with two probiotic cultures to improve its nutrition characteristics and the results showed that probiotics can ensure the highest level of utilization of fat, protein, lactose, and phosphorus partially in skimmed milk [85]. Chocolate mousse with probiotic and prebiotic ingredients were developed [86]. Probiotic chocolate mousse was supplemented with *L. paracasei* subsp. *paracasei* LBC 82, solely or together with inulin and the results showed that chocolate mousse is good vehicle for *L. paracasei* [86]. Sour cream was investigated as probiotic vehicle and the results showed that using sour cream as a probiotic carrier is proved feasible [87]. Ice creams are among the food products with high potential for use as probiotic vehicles. Cruz and others have reviewed the technological parameters involved in the production of probiotic ice creams [88]. They have pointed several factors that need to be controlled, including the appropriate selection of cultures, inoculums concentration, the appropriate processing stage for the cultures to be added, and the processing procedures and transport and storage temperatures. They concluded that probiotic cultures do not modify the sensory characteristics of the ice-creams and frozen desserts also these products hold good viability for probiotics during the product storage period.

3.2. Non dairy based probiotic products

Dairy products are the main carriers of probiotic bacteria to human, as these products provide a suitable environment for probiotic bacteria that support their growth and viability. However, with an increase in the consumer vegetarianism throughout the developed countries, there is also a demand for the vegetarian probiotic products. Nondairy probiotic products have shown a big interest among vegetarians and lactose intolerance customers. According to the National Institute of Diabetes and Digestive and Kidney Diseases (NIDDK) of the U.S. National Institutes of Health, about 75% of the world population is lactose intolerant. The development of new nondairy probiotic food products is very much challenging, as it has to meet the consumer's expectancy for healthy benefits [89, 90]. Granato and others have overview of functional food development, emphasizing nondairy foods that contain probiotic bacteria strains [91]. From their review, some nondairy probiotic products recently developed are shown in Table 2.

3.2.1. Vegetable-based probiotic products

Fermentation of vegetables has been known since ancient time. Fermented vegetables can offer a suitable media to deliver probiotics. However, it shows that the low incubation temperature of vegetable fermentation is a problem for the introduction of the traditional *L. acidophilus* and *Bifidobacterium* probiotic bacteria. Probiotic of *L. rhamnosus*, *L. casei* and *L. plantarum* are better adapted to the vegetable during fermentation [94]. Nevertheless, when the temperature is adjusted at 37ºC, probiotic bacteria grow quite rapidly in plant-based substrates [95].

Category	Product
Fruit and vegetable based	Vegetable-based drinks
	Fermented banana pulp
	Fermented banana
	Beets-based drink
	Tomato-based drink
	Many dried fruits
	Green coconut water
	Peanut milk
	Cranberry, pineapple, and orange juices
	Ginger juice
	Grape and passion fruit juices
	Cabbage juice
	Carrot juice
	Noni juice
	Onion
	Probiotic banana puree
	Nonfermented fruit juice beverages
	Blackcurrant juice
Soy based	Nonfermented soy-based frozen desserts
	Fermented soymilk drink
	Soy-based stirred yogurt-like drinks
Cereal based	Cereal-based puddings
	Rice-based yogurt
	Oat-based drink
	Oat-based products
	Yosa (oat-bran pudding)
	Mahewu (fermented maize beverage)
	Maize-based beverage
	Wheat, rye, millet, maize, and other cereals fermented probiotic beverages
	Malt-based drink
	Boza (fermented cereals)
	Millet or sorghum flour fermented probiotic beverage
Other nondairy foods	Starch-saccharified probiotic drink
	Probiotic cassava-flour product
	Meat products
	Dosa (rice and Bengal gram)

Table 2. Some nondairy probiotic products recently developed. Adapted from [91]

To develop new probiotic vegetable products, many studies have been carried out. The suitability of carrot juice as a raw material for the production of probiotic food with Bifidobacterium strains was investigated [96]. Kun and others have found that Bifidobacteria were capable of having biochemical activities in carrot juice without any nutrient supplementation [96]. Yoon and others studied the suitability of tomato juice for the production of a probiotic product by *L. acidophilus, L. plantarum, L. casei* and *L. delbrueckii.* They reported that the four LAB were capable of rapidly utilizing tomato juice for cell synthesis and lactic acid production without nutrient supplementation and pH adjustment [109]. Yoon and others also tested the suitability of cabbage to produce probiotic cabbage juice and suggested that fermented cabbage juice support the viability of probiotics and serve as a healthy beverage [97]. The viability of various bifidobacteria in kimchi was investigated under various conditions and the results show the acceptable levels of probiotics in kimchi [98]. In addition, sauerkraut-type products such as fermented cabbage, carrots, onions, and cucumbers based on a lactic fermentation by *L. plantarum* could be good probiotic carrier. Yoon and others have evaluated the potential of red beets as substrate for the production of probiotic beet juice by four strains of lactic acid bacteria and all strains were capable of rapidly utilizing the beet juice for the cell synthesis and lactic acid production [99]. However, traditional methods of production might result in inactivation of the probiotic cultures and the use of probiotics in fermented vegetables would require low temperature storage of the products [94].

Moreover, soybean has received attention from the researchers due to its high protein and quality. Soymilk is suitable for the growth of LAB and bifidobacteria [100, 101]. Several studies have focused on developing fermented soymilk with different strains of LAB and Bifidobacteria to produce a soymilk product with improved health benefits [62, 101-103]. Soymilk is now known for their health benefits such as prevention of chronic diseases such as menopausal disorder, cancer, atherosclerosis, and osteoporosis, therefore, soymilk fermented with bifidobacteria may be a unique functional food [62, 104]. In probiotic soy products, fermentation by probiotics has the potential to (1) reduce the levels of some carbohydrates possibly responsible for gas production in the intestinal system, (2) increase the levels of free isoflavones, which has many beneficial effects on human health, and (3) favor desirable changes in bacterial populations in the gastrointestinal tract. Supplementing soymilk with prebiotics such as, fructooligosaccharides (FOS), mannitol, maltodextrin and pectin, was found to be a suitable medium for the viability of probiotic bacteria [105].

3.2.2. Fruit-based probiotic products

Nowadays, there is increasing interest in the development of fruit-juice based probiotic products. The fruit juices contain beneficial nutrients that can be an ideal medium for probiotics [106, 107]. Fruit juices have pleasing taste profiles to all age groups and they are perceived as being healthy and refreshing. The fruits are rich in several nutrients such as minerals, vitamins, dietary fibers, antioxidants, and do not contain any dairy allergens that might prevent usage by certain segments of the population [107, 108]. Those characteristics allow the selection of appropriate strains of probiotics to manufacture enjoyable healthy fruit juice. However, the sensory impact of probiotic cultures would have different taste profiles compared to the conventional, nonfunctional products. The different aroma and flavors have been reported when *L. plantarum* was added to orange juices which consumers do not prefer. But if their health benefits information is provided the preference increases over the conventional orange juices. Different attempts have been made to reduce the sensations of unpleasant aromas and flavors in probiotic fruit juice. Luckow and others reported that the perceptible off-flavors caused by probiotics that often contribute to consumer dissatisfaction may be masked by adding 10% (v/v) of tropical fruit juices, mainly pineapple, but also mango or passion fruit [108].

To develop probiotic fruits, many studies have been carried out. The suitability of noni juice as a raw material for the production of probiotics was studied by Wang and others and found that *B. longum* and *L. plantarum* can be optimal probiotics for fermented noni juice [109]. Suitability of fermented pomegranate juice was tested using *L. plantarum, L. delbruekii, L. paracasei, L. acidophilus.* Pomegranate juice was proved to be a suitable probiotic drink as results have shown desirable microbial growth and viability for *L. plantarum* and *L. delbruekii* [110]. Optimized growth conditions of *L. casei* in cashew apple juice were studied. *L. casei* has shown suitable survival ability in cashew apple juice during 42 days of refrigerated storage. It was observed that *L. casei* grew during the refrigerated storage and cashew apple juice showed to be suitable probiotic product [111]. Tsen and others reported that *L. acidophilus* immobilized in Ca-alginate can carry out a fermentation of banana puree, resulting in a novel probiotic banana product with higher number of viable cells [112]. Kourkoutas and others reported that *L. casei* immobilized on apple and quince pieces survived for extended storage time periods and adapted to the acidic environment, which usually has an inhibitory effect on survival during lactic acid production [113].

3.2.3. Cereal-based probiotic products

Cereal-based probiotic products have health-benefiting microbes and potentially prebiotic fibers. The development of new functional foods which combine the beneficial effects of cereals and health promoting bacteria is a challenging issue. Nevertheless, cereal-based products offer many possibilities. Indeed, numerous cereal-based products in the world require a lactic fermentation, often in association with yeast or molds. Cereals are good substrates for the growth of probiotic strains and due to the presence of non-digestible components of the cereal matrix may also serve as prebiotics [114, 115]. Due to the complexity of cereals, a systematic approach is required to identify the factors that enhance the growth of probiotic in cereals [116]. Champagne has listed number of cereal-based products that require a lactic fermentation, often in association with yeast or molds. We have found it useful to include part of these products in Table 3.

Food	Country	Ingredients	Microorganisms
Adai	India	Cereal, legume	*Pediococcus* spp., *Streptococcus* spp., *Leuconostoc* spp.
Anarshe	India	Rice	Lactic acid bacteria
Aya-bisbaya	Mexico	Rice	Lactic acid bacteria
Bhatura	India	Wheat	Lactic acid bacteria, yeasts
Burukutu	Nigeria	Sorghum, cassava	Lactic acid bacteria, *Candida* spp., *S. cerevisiae*
Fermented oatmeal (ProViva)	Sweden	Oatmeal	*L. plantarum*
Llambazi, lakubilisa	Zimbabwe	Maize	Lactic acid bacteria, yeasts, molds
Injera	Ethiopia	Sorghum, tef, corn, millet, barley, wheat	*L. plantarum*, *Aspergillus* spp., *Penicillium* spp., *Rhodotorula* spp., *Candida* spp.
Kishk, kushuk, trahanas	Egypt, Syria, Lebanon	Milk (yoghurt), wheat	*L. casei*, *L. plantarum*, *L. brevis*, *B. subtilis*, *B. licheniformis*, *B. megaterium*, yeasts
Kisra	Sudan, Irak, Arabian Gulf	Sorghum, millet	*Lactobacillu.* spp., *L. brevis*, *L. fermentum*, *E. faecium*, *Acetobacter* spp., *S. cerevisiae*
Togwa	Tanzania	Maize, sorghum	*L. plantarum*, *L. brevis*, *L. fermentum*, *L. cellobiosus* *P. pentosaceus*, *W. confusa*, *S. cerevisiae*, *C. tropicalis*

Table 3. Fermented cereal products that carry a lactic fermentation [94]

A multitude of fermented cereal products have been created, but only recently probiotic microorganisms involved in traditional fermented cereal foods have been reported. Strains of *L. plantarum,Candida rugosa* and *Candida lambica* isolated from a traditional Bulgarian cereal-based fermented beverage exhibited probiotic properties, being resistant up to 2% bile concentration, which enables them to survive bile toxicity during their passage through the gastrointestinal system [117]. More studies are being done to demonstrate that cereals are suitable substrates for the growth of some probiotic bacteria. Rozada-Sa´nchez and others have studied the growth and metabolic activity of four different *Bifidobacterium* spp. in a malt hydrolisate using four *Bifidobacterium* strains with the aim of producing a potentially probiotic beverage [92]. The study has reported potential use for malt hydrolysate as probiotic beverage with the addition of a growth and yeast extract. Angelov and others have used a whole-grain oat substrate to obtain a drink with probiotics and oat prebiotic beta-glucan. They have found that viable cell counts reached at the end of the process were about 7.5×10^{10} cfu/ ml. Also the addition of sweeteners aspartame, sodium cyclamate, saccharine and Huxol (12% cyclamate and 1.2% saccharine) had no effect on the dynamics of the fermentation process and on the viability of the starter culture during product storage [93]. Charalapompoulos and others have done experiments with different cereals to determine the main parameters that need to be considered in the growth of probiotic microorganisms, defining them as follows: the composition and processing of cereal grains, the substrate formulation, the growth capability and productivity of the starter culture, the stability of the probiotic strain during storage, the organoleptic properties and the nutritional value of the final product [114]. They reported that many cereals supported the growth of probiotics with some differences. Malt medium supported the growth of all examined strains (*L. plantarum, L. fermentum, L. acidophilus* and *L. reuteri*) better than barley and wheat media due to its chemical composition. Also, wheat and barley extracts were found to exhibit a significant protective effect on the viability of *L. plantarum, L. acidophilis* and *L. reuteri* under acidic conditions (pH 2.5).

Oat is often used in studies of cereal fermented by probiotic bacteria. Several studies have evaluated the potential of oat as substrates for the development of a probiotic product. Kedia and others have explored the potential of using mixed culture fermentation to produce cereal-based foods with high numbers of probiotic bacteria. In this study, LAB growth was enhanced by the introduction of yeast and the production of lactic acid and ethanol were increased in comparison against pure LAB culture. They have fermented whole oat flour with *L. plantarum* along with white flour and bran in order to compare the suitability of these substrates for the production of a probiotic beverage. Those substrates were found to enhance probiotic viability at the end of fermentation above the minimum required in a probiotic product [118]. Martensson and others have studied the development of nondairy fermented product based on oat [119]. Yosa is a snack food made from oat bran pudding cooked in water and fermented with LAB and Bifidobacteria. It is mainly consumed in Finland and other Scandinavian countries. It has a texture and a flavor similar to yogurt but it is totally free from milk or other animal products. It is lactose-free, low in fat, contains beta-glucan and it is suitable for vegetarians [120]. Yosa is therefore considered a healthy food due to its content of oat fiber and probiotic LAB, which combine the effect of beta-glucan for cholesterol reduction and the effect of LAB benefits to maintain and improve the intestinal microbiota balance of the consumer.

Other cereals and cereal components that can be used as fermentation substrates for probiotics have been studied. Survival of probiotics in a corn-based fermented substrate was reported [121]. Autoclaved maize porridge was fermented with probiotic strains (grown separately): *L. reuteri, L. acidophilus* and *L. rhamnosus* for 24h at 37 ◦C. All strains examined showed good growth in maize porridge with added barley malt. Probiotic fermented maize products could have a good world-wide acceptance, since maize fermentation induces fruity flavors in traditional Mexican foods. Prado and others have summarized some of the international cereal based probiotic beverages including: *Boza* made from wheat, rye, millet and other cereals in Bulgaria, Albania, Turkey and Romania, *Bushera* made from sorghum, or millet flour in Western highlands of Uganda, *Mahewu* (amahewu) made from corn meal in Africa and some Arabian Gulf countries, *Pozol* made from maize in the Southeastern Mexico, and *Togwa* made from maize flour and millet malt in Africa [5].

Normally sourdoughs are the cereal products fermented by LAB cultures. However, baking will kills most probiotic bacteria and only probiotics which synthesize a thermostable bioactive compound during leavening can be of use in bread making. Different studies have shown the ability of human derived strains of *L. reuteri* to resist simulated gastric acidity and bile acid, and also to grow well in a number of cereal substrates [89, 116]. In this perspective, *L. reuteri* has potential use in bread making due to reuterin synthesis [122]. The *L. reureri* cells might be inactivated by heating, but the bioactive compound might remain active. Probiotic *Bacillus* strains could better adapt to bread making due to their spore-forming characteristics.

3.2.4. Meat-based probiotic foods

Probiotic applications are restricted to fermented meats, such as dry sausages. The idea of using probiotic bacteria in fermenting meat products has introduced the idea of using antimicrobial peptides, i.e. bacteriocins, or other antimicrobial compounds as an extra hurdle for meat products. Meat starter culture was defined as preparations which contain living or resting microorganisms that develop the desired metabolic activity in the meat [123]. LAB are the most common used starter culture in meat which produce lactic acid from glucose or lactose. As meat content of these sugars are low, sugar is added at 0.4–0.7% (w/w) for glucose and 0.5–1.0% (w/w) for lactose to the sausage matrix [124]. Some LAB strains such as *L. rhamnosus* GG are not able to utilize lactose, therefore, the starter culture properties have to be taken into account for successful applications. From pentoses, such as arabinose and xylose, meat starter LAB produce both lactic acid and acetic acid [125]. As indicated in commercial catalogues LAB strains currently most employed in meat starter cultures are *L. casei, L. curvatus, L. pentosus, L. plantarum, L. sakei, Pediococcus acidilactici* and *Pediococcus pentosaceus* [124].

LAB have been used for dry sausage manufacturing process since 1950s in order to ensure the safety and quality of the end product. Dry sausages are non heated meat products, which may be suitable carriers for probiotics into the human gastrointestinal tract [124]. Dry sausage is made from a mixture of frozen pork, beef and pork fat with the addition of sugars, salt, nitrite, and nitrate, ascorbates and spices. The raw sausage material is stuffed into casing material of variable diameters and hung vertically in fermentation and ripening chambers for several weeks. Salt, nitrite, and added spices are the main contributors in the inhibition of different bacteria on the surface of the sausages. Lactic acid bacteria and staphylococci used as starter cultures to ferment the sausage. Salt decreases the initial water activity inhibiting or at least delaying the growth of many bacteria while favoring the growth of starter LAB and starter staphylococci. During the first day of fermentation the growth of microbes in sausage material uses up all the oxygen mixed in the sausage matrix during the chopping. After few days of fermentation, LAB decrease the pH to about 5.0 which acts as a hurdle for several Gram-negative bacterial species [126, 127]. The presence LAB in the food suggests that bacteriocins may be active in the human small intestine against food pathogens as long as they are able to survive the environment of gastrointestinal tract [27]. Likewise, probiotic strains with antimicrobial effects on food act similarly and therefore might be more successful than commonly used food fermenting bacteria. It could be concluded that dry sausage is suitable carrier for probiotics. However, human clinical studies are needed before the final answer concerning the health promoting effects of probiotic dry sausage.

Some traditional Indian fermented fish products such as Ngari, Hentak and Tungtap have been analyzed for microbial load [128]. LAB were identified as *Lactococcus lactis* subsp. *cremoris, Lactococcus plantarum, Enterococcus faecium, L. fructosus, L. amylophilus, L. coryniformis* subsp. *torquens*, and *L. plantarum*. Most strains of LAB had a high degree of hydrophobicity, indicating that these microorganisms have a probiotic potential.

4. Agricultural applications of probiotics

Probiotics applications have been extended from human applications to diversity of agricultural application. Agricultural applications include animal and plants.

4.1. Animal

Probiotics, with regard to animal applications, were defined as live microbial feed supplements beneficially improve the intestinal microbial balance in host animal [26]. They have been approved to provide many benefits to the host animal and animal products production. They are used as animal feed to improve the animal health and to improve food safety with examples of the application in poultry, ruminant, pig and aquaculture.

The microflora in the gastrointestinal tracts of poultry plays a key role in normal digestive processes and in maintaining the animal's health. Some feed additives can substantially affect this microbial population and their health promoting effects. Recently, concerns about some unwanted harmful side effects caused by antibiotics [129] has grown in many countries, so that there is an increasing interest in finding alternatives to antibiotics in poultry production. Probiotic has provided a possible natural alternative to antibiotics in poultry production to produce foods of reliable quality and safety [130]. In addition, the application of probiotic to chicken feed was shown to increase the internal and external quality of eggs. Addition of probiotic to chicken feed increased egg weight shell thickness, shell weight, albumen weight, and specific gravity and decreased shape index [131]. Farm animals are often subjected to environmental stresses which can cause imbalance in the intestinal ecosystem and could be a risk factor for pathogen infections. Applications of probiotics in feed have decreased the pathogen load in the farm animals. Feeding probiotic LAB and yeast to calve was found to promote the growth and suppress diarrhea in Holstein calve [132]. Gaggia and others have reviewed the applications of probiotics and prebiotics in animal feeding that can introduce to safe food production [133]. Probiotics has been used to intervene in decreasing pathogen load and in ameliorating gastrointestinal disease symptoms in pigs. Beside the in vitro test to identify the best potential probiotics, several studies are conducted in vivo utilizing different probiotic microorganisms. Most of the studies showed a beneficial role of improving the number of beneficial bacteria, decreasing the load of pathogens, stimulating the immune cell response towards pathogens in comparison to control, and increasing defensive tools against pathogenic invasion. In contrast, some authors reported an enhancement of the course of infection or a partial alleviation of diarrhea.

Applications of probiotics in aquaculture generally depend on producing antimicrobial metabolites and their ability to attach to intestinal mucus. *Aeromonas hydrophila* and *Vibrio alginolyticus* are common pathogens in fish, however, addition of probiotics strains (isolated from the clownfish, *Amphiprion percula*) were found capable to prevent the adhesion of these microbes to fish intestinal mucus and to compete with the pathogens [31]. Feeding probiotics to shrimp was found to reduce disease caused by *Vibrio parahaemolyticus* in shrimp [36]. Balcazar and others have reviewed the use of probiotics for prevention of bacterial diseases in aquaculture [134].

4.2. Plant

A strong growing market for plant probiotics for the use in agricultural biotechnology has been shown worldwide with an annual growth rate of approximately 10%. Based on the mode of action and effects, the plant probiotics products can be used as biofertilizers, plant strengtheners, phytostimulators, and biopesticides [38]. Berg has reported several advantages of using plant probiotics over chemical pesticides and fertilizers including: more safe, reduced environmental damage, less risk to human health, much more targeted activity, effective in small quantities, multiply themselves but are controlled by the plant as well as by the indigenous microbial populations, decompose more quickly than conventional chemical pesticides, reduced resistance development due to several mechanisms, and can be also used in conventional or integrated pest management systems [38]. Plant growth promotion can be achieved by the direct interaction between beneficial microbes and their host plant and also indirectly due to their antagonistic activity against plant pathogens. Several model organisms for plant growth promotion and plant disease inhibition are well-studied including: the bacterial genera *Azospirillum* [44, 135], *Rhizobium* [136], *Serratia* [137], *Bacillus* [138, 139], *Pseudomonas* [140, 141], *Stenotrophomonas* [142], and *Streptomyces* [143] and the fungal genera *Ampelomyces*, *Coniothyrium*, and *Trichoderma* [144]. Some examples of commercial products that have plant probiotics are listed in Table 4.

Several mechanisms are involved in the probiotics-plant interaction. It is important to specify the mechanism and to colonize plant habitats for successful application. Steps of colonization include recognition, adherence, invasion, colonization and growth, and several strategies to establish interactions. Plant roots initiate crosstalk with soil microbes by producing signals that are recognized by the microbes, which in turn produce signals that initiate colonization [43, 51]. Colonizing bacteria can penetrate the plant roots or move to aerial plant parts causing a decreasing in bacterial density in comparison to rhizosphere or root colonizing populations [43]. Furthermore, in the processes of plant growth, probiotic bacteria can influence the hormonal balance of the plant whereas phytohormones can be synthesized by the plant themselves and also by their associated microorganisms [38].

Microorganism	Name of the product	Plant pathogens, or pathosystem	Company
Ampelomyces quisqualis M-10	AQ10 Biofungicide	Powdery mildew on apples, cucurbits, grapes, omamentals, strawberries, and tomatoes.	Ecogen
Azospirillum spp.	Biopromoter	Paddy, millets, oilseeds, fruits, vegetables, sugarcane, banana	Manidharma Biotech
Bacillus subtilis GB03	Kodiak	Growth promotion; *Rhizoctonia* and *Fusarium* spp.	(Gustafson); Bayer CropScience
Bradyrhizobium japonicum	Soil implant	Soy bean	Nitragin
Bacillus pumilus GB34	YiedShield	Soil-born fungal pathogens	(Gustafson); Bayer CropScience
Coniothyrium minitans	Contans WG, Intercept WG	*Sclerotinia sclerotiorum, S. minor*	Prophyta Biologischer Pflanzenschutz
Delftia acidovorans	BioBoost	Canola	Brett-Young Seeds Limited
Phlebiopsis gigantea	Rotex	*Heterobasidium annosum*	E~nema Biologischer Pflanzenschutz
Pseudomonas chlororaphis	Cedomon	Leaf stripe, net blotch, *Fusarium* sp., sot blotch, leaf spot, etc. on barley and oats	BioAgri AB
Streptomyces griseoviridis K61	Mycostop	*Phomopsis* spp., *Botrytis* spp., *Pythium* spp.,*Phythophora* spp.	Kemira Agro Oy
Trichoderma harzianum T22	RootShield, PlantShield T22, Planter box	*Pythium spp., Rhizoctonia solani, Fusarium* spp	Bioworks
Pseudomonas spp.	Proradix	*Rhizoctonia solani*	Sourcon Padena

Table 4. Examples of commercial products that have plant probiotics. Adapted from [38]

Besides these mechanisms, probiotic bacteria can supply macronutrients and micronutrients. They metabolize root exudates and release various carbohydrates, amino acids, organic acids, and other compounds in the rhizosphere [43]. Bacteria may contribute to plant nutrition by liberating phosphorous from organic compounds such as phytates and thus indirectly promote plant growth [145]. Furthermore, probiotic can reduce the activity of pathogenic microorganisms through microbial antagonisms and by activating the plant to better defend itself, a phenomenon termed "induced systemic resistance" [146, 147]. Microbial antagonism includes the inhibition of microbial growth, competition for colonization sites and nutrients, competition for minerals, and degradation of pathogenicity factors [38, 43]. In Japanese composting, at least three groups of compositing bacteria were used individually, or in combination. The following species were used: *Bacillus* bacteria groups, Lactic acid bacteria groups and *Actinomycetous* groups. These bacteria species can protect plant products from cropping hazards. They do this by expelling against various bad worms and insects, such as nematodes with potatoes and some types of insects with soybeans and maize. They are also effective in controlling fungi such as powdery mildew, downy mildew, *phythium* (damping off with many plants), *plasmodipophora brosscae* (club-root with the cabbage Jamily); *Crucijert1e* (plants. and fusarium of wilt with tomato and banana) [148].

5. Probiotics application challenges

From a technological standpoint, Champagne has listed many challenges in the development of a probiotic food product including: strain selection, inoculation, growth and survival during processing, viability and functionality during storage, assessment the viable counts of the probiotic strains particularly when multiple probiotic strains are added and when there are also starter cultures added, and the effects on sensory properties [94]. Champagne has focused in his chapter on three of these challenges: inoculation, processing and storage issues. Other challenges such as: maintaining of probiotics, diversity and origin of probiotics, probiotic survival and being active, dealing with endogenous microbiota, and proving health benefits have also been discussed [149]. This section will focus on the viability and sensory acceptance as we have found these are the most important challenges to ensure transferring the health benefits and the commercial success.

5.1. Viability and survival

Probiotics have been proved to provide many health benefits. However, the claimed health benefits can't be achieved without high number of viable cells. Many probiotic bacteria have shown to die in the food products after exposure to low pH after fermentation, oxygen during refrigeration distribution and storage of products, and/or acid in the human stomach [150, 151]. Probiotic products need to be supplemented with additional ingredients to support the viability throughout processing, storage, distribution, and gastrointestinal tract to reach the colon. Several reports have shown that survival and viability of probiotic bacteria is often low in yogurt. The efficiency of added probiotic bacteria depends on dose level and their viability must be maintained throughout storage, products shelf-life and they must survive the gut environment [151]. Several studies have focused on the effect of adding certain compounds to enhance the probiotic viability. Many evidences have shown that inulin, oligosaccharides, and fructooligosaccharides (FOS) have good impacts on the probiotics viability. However, the effect of these compounds are strain specific. Martinez-Villaluenga and others have examined the influence of raffinose on the survival of *Bifidobacteria* and *L. acidophilus* in fermented milk. The results showed that retention of viability of *Bifidobacteria* and *L. acidophilus* greater in fermented milk with raffinose [65]. Supplementing probiotic products with FOS, mannitol, maltodextrin and pectin were found to provide a suitable viability for probiotic bacteria [105]. Inulin and FOS were found to support the growth and viability of *L. acidophilus* but did not significantly affect growth and viability of *Bifidobacterium* and *L. casei* [152]. During food formulation step several things need to be considered such as the composition (nutrients, antimicrobials), structure (oxygen permeability, water activity) and pH of the food matrix, and possible interactions with starter microbes in fermented food matrices. Growth of probiotics in non-fermented foods is not desirable (due to possible off flavor formation), but their growth during the production of fermented foods can lower process costs and increase the adaptation of probiotics leading to enhanced viability. The starter microbes in fermented foods can sometimes inhibit probiotics but they can also enhance their survival by producing beneficial substances or by lowering the oxygen pressure. In beverages the most important factor affecting probiotic viability is probably the pH. Shelf-stable beverages typically have pH values below 4.4 to ensure their microbial stability and this low pH value combined with long storage periods is very demanding for most probiotic strains, especially those representing bifidobacteria. The packaging material should be a good oxygen barrier to promote the survival of especially anaerobic probiotic bacteria (bifidobacteria) [153]. Transportation and storage temperature is an important determinant of the shelf-life; with increasing temperatures viability losses can occur rapidly [154].

The viability and survival of probiotics are strain specific. To maintain the viability of very sensitive strains, encapsulation is often the only option, especially microcapsulation that do not affect the sensory properties of the food produced. Microencapsulation technologies have been developed and successfully applied using various matrices to protect the bacterial cells from the damage caused by the external environment [155]. Overall microencapsulation improved the survival of probiotic bacteria when exposed to acidic conditions, bile salts, and mild heat treatment [156]. The immobilization of probiotics using microencapsulation may improve the survival of these microorganisms in products, both during processing and storage, and during digestion [157, 158].

Some probiotic bacteria, such as the spore-forming bacteria, GanedenBC^{30} provides better viability and stability, making it an ideal choice for product development, compared to other probiotic bacteria strains, such as *L. acidophilus* and bifidobacteria. This spore safeguards the cell's genetic material from the heat and pressure of manufacturing processes, challenges of shelf life and the acid and bile it is exposed to during transit to the digestive system. GanedenBC[30] can withstand manufacturing processes. and survive through high temperature processes such as baking and boiling, low temperature processes such as freezing and refrigeration and high pressure applications like extrusion and roll forming. GanedenBC[30] requires no refrigeration and can be formulated into products to have up to a two-year. Once it is safely inside the small intestine, the viable spore is then able to germinate and produce new vegetative cells or good bacteria [159].

5.2. Sensory acceptance

Probiotic foods must show, at least, the same performance in any sensory test as conventional foods. In most probiotic foods sensory tests are aiming to determine acceptance of the products, without, obtaining details concerning the addition of the probiotics to the food and their interaction with the consumer. Therefore, it is important to development sensory tests for probiotic foods that can be accompanied by specific sensory analyses. Sensory testing must cover all characteristics with regard to change over time during storage. Some studies have reported the possibility of obtaining similar, or even better, performance with probiotic products as compared to conventional products such as: functional yogurt supplemented with *L. reuteri* RC-14 and *L. rhamnosus* GR-1 [160], chocolate mousse with added inulin and *L. paracasei* [86] , curdled milk with inulin, and *L. acidophilus* [152], and milk fermented with *B. animalis* and *L. acidophilus* La-5, and supplemented with inulin [161].

Sensory methodology will allow obtaining important data for developing the probiotic foods. In most cases the developed products need to match similar commercial products in parallel. In general, metabolism of the probiotic culture can result in the production of components that may contribute negatively to the aroma and taste of the food product, probiotic off-flavor. For example, acetic acid produced by *Bifidobacterium* spp. can result in a vinegary flavor in the product, prejudicing the performance in sensory assessments.

Masking is one technique that has been used to reduce the off flavors in foods and it has been performed successfully through the addition of new substances or flavors to reduce the negative sensory attributes contributed by probiotic cultures. The addition of tropical fruit juices, mainly pineapple, but also mango or passion fruit, might positively contribute to the aroma and flavor of the final product and might avoid the identification of probiotic off-flavors by consumers [162]. The influence of exposure has been identified in many consumer studies [91, 163] that the frequency of exposure to a food stimulus is increased, food stimuli have been shown to be better liked. Therefore, repeated exposure and increased familiarity to sensory off-flavors, may influence consumer attitudes in a positive way, therefore increasing willingness to consume probiotic juices. Nonsensory techniques have proven useful in enhancing the sensory quality of products, such as providing consumers with health benefit information associated with probiotic cultures. Health information has been shown to be a vital tool in the consumer acceptance of a variety of probiotic food products [164-166]. Finally, microcapsules of probiotics may help prevent the off flavor of cultures [167].

6. The future of probiotics

Dairy based products containing live bacteria are the main vehicles of probiotics to human. Non-dairy beverages would be the next food category where the healthy bacteria will make their mark. Microencapsulation technologies have provided the necessary protection for probiotics and moved them outside the pharmaceutical and supplemental use to become food ingredients.

6.1. Nanotechnology, encapsulation, and probiotics

The word "nano" comes from the Greek for "dwarf ". A nanometer is a thousandth of a thousandth of a thousandth of a meter (10^{-9} m). Nanoparticles are usually sized below 100 nanometers which will enable novel applications and benefits. Nanotechnology of probiotics is an area of emerging interest and opens up whole new possibilities for the probiotics applications. Their applications to the agriculture and food sector are relatively recent compared with their use in drug delivery and pharmaceuticals. The basic of probiotic nanotechnology applications is currently in the development of nano-encapsulated probiotics. The nanostructured food ingredients are being developed with the claims that they offer improved taste, texture and consistency. Applications of nanotechnology in organic food production require precaution, as little is known about their impact on environment and human health. Some recent food applications of nanotechnology, safety and risk problems of nanomaterials, routes for nanoparticles entering the body, existing regulations of nanotechnology in several countries, and a certification system of nanoproducts were reported [168, 169]. Currently, no regulations exist that specifically control or limit the production of nanosized particles and this is mainly owing to a lack of knowledge about the risks [169]. Nanoencapsulation is defined as a technology to pack substances in miniature using techniques such as nanocomposite, nanoemulsification, and nanoestructuration and provides final product functionality and control the release of the core [170]. Encapsulation of food ingredients may extend the shelf life of the product. Nanoencapsulation of probiotic is desirable technique that could deliver the probiotic bacteria to certain parts of the gastrointestinal tract where they interact with specific receptors [170]. These nanoencapsulated probiotic bacterial may also act as *de novo* vaccines, with the capability of modulating immune responses [171].

Microencapsulation with alginate can be applied to many different probiotic strains and results show better survival than free cells at low pH of 2.0, high bile salt concentrations, and moderate heat treatment of up to 65 ◦C [156]. Microencapsulation may prove to be an important method of improving the viability of probiotic bacteria in acidic food products and help deliver viable bacteria to the host's gastrointestinal tract. Furthermore, microencapsulation appeared to be effective in protecting cells from mild heat treatment and thus could stimulate research in functional food products that receive a mild heat treatment [156]. The microencapsulation allows the probiotic bacteria to be separated from its environment by a protective coating. Several studies have reported the technique of the microencapsulation by using gelatin, or vegetable gum to provide protection to acid-sensitive *Bifidobacterium* and *Lactobacillus* [172-176].

6.2. Biotechnology and probiotics

With the revolution in sequencing and bioinformatic technologies well under way it is timely and realistic to launch genome sequencing projects for representative probiotic microorganisms. The rapidly increasing number of published lactic acid bacterial genome sequences will enable utilizing this sequence information in the studies related to probiotic technology. If genome sequence information is available for the probiotic species of interest, this can be utilized, e.g. to study the gene expression (transcription) profile of the strain during fermenter growth. This will enable better control and optimization of the growth than is currently possible. Transcription profiling during various production steps will allow following important genes for probiotic survival during processing (e.g., stress and acid tolerance genes) and identifying novel genes important for the technological functionality of probiotics [177].

Increasing knowledge of genes important for the technological functionality and rapid development of the toolboxes for the genetic manipulation of *Lactobacillus* and *Bifidobacterium* species will in the future enable tailoring the technological properties of probiotic strains. However, before wide application of tailored strains in probiotic food products, safety issues are of utmost importance and have to be seriously considered for each modified strain [178].

7. Regulations and guidelines for probiotics

Depending on intended use of a probiotic (drug *vs.* dietary supplement), regulatory requirements differ greatly. If a probiotic is intended for use as a drug, then it must undergo the regulatory process as a drug, which is similar to that of any new therapeutic agent. An Investigational New Drug application must be submitted and authorized by the Food and Drug Administration before an investigational or biological product can be administered to humans. The probiotic drug must be proven safe and effective for its intended use before marketing [14]. In the United States, probiotic products are marketed to a generally healthy population as foods or dietary supplements. For dietary supplements, premarketing demonstration of safety and efficacy and approval by the Food and Drug Administration are not required; only premarket notification is required. The law allows that in addition to nutrient content claims, manufacturers of dietary supplements may make structure/function or health claims for their products. The "health claims" must be defensible when placed under the scrutiny by the controlling authorities. Efforts are being made to establish meaningful standards or guideline for probiotic products worldwide (Table 5). The Joint Food and Agriculture Organization of the United Nations/World Health Organization Expert Consultation on Evaluation of Health and Nutritional Properties of Probiotics developed guidelines could be used as the global standards for evaluating probiotics in food that could lead to the substantiation of health claims.

Organization	Region of impact	Action
Food Agriculture Organization (FAO)/ World Health Organization (WHO)	Worldwide	Developed guidelines for the evaluation of probiotics in foods.
International Dairy Federation	Worldwide	Has begun working on methods to determine certain functional and safety properties outlined in the FAO guidelines for the evaluation of probiotics in food.
European Food and Feed Culture Association	Europe	Developed guidelines for use of probiotics in foods.
Codex Standard for Fermented Milks (Codex Stan 243-2003)	Worldwide	Among other composition stipulations, this standard specifies minimum numbers of characterizing and additional labeled microbes in yoghurt, acidophilus milk, kefir, kumys and other fermented milks.
National Yogurt Association	USA	Petition under consideration by the FDA which would change the standard of identity of yoghurt, including the requirement of minimum levels of live cultures in yoghurt, but not specifically levels for any additional probiotic cultures.
International Scientific Association for Probiotics and Prebiotics	Worldwide	Industry Advisory Committee and Board of Directors to consider method validation and establishment of laboratory sites to assess microbiological content of probiotic products.

Table 5. Organizations involved in attempting to establish standards for probiotics in commercial products. Adapted from [179]

8. Conclusion

The uses of probiotics and their applications have shown tremendous increase in the last two decades. Probiotics can turn many health benefits to the human, animals, and plants. Applications of probiotics hold many challenges. In addition to the viability and sensory acceptance, it must be kept in mind that strain selection, processing, and inoculation of starter cultures must be considered. Probiotics industry also faces challenges when claiming the health benefits. It cannot be assumed that simply adding a given number of probiotic bacteria to a food product will transfer health to the subject. Indeed, it has been shown that viability of probiotics throughout the storage period in addition to the recovery levels in the gastrointestinal tract are important factors [3, 48, 83]. For this purpose, new studies must be carried out to: test ingredients, explore more options of media that have not yet been industrially utilized, reengineer products and processes, and show that lactose-intolerant and vegetarian consumers demand new nourishing and palatable probiotic products.

Author details

Danfeng Song*, Salam Ibrahim and Saeed Hayek
Department of Family and Consumer Science,
North Carolina Agricultural and Technical State University, Greensboro, NC, USA

9. References

[1] Arvanitoyannis IS, Van Houwelingen-Koukaliaroglou M (2005) Functional Foods: a Survey of Health Claims, Pros and Cons, and Current Legislation. Crit. rev. food sci. nutr. 45:385-404.
[2] Joint Food and Agriculture Organization of the United Nations/ World Health Organization Working Group report on drafting guidelines for the evaluation of probiotics in food, London, Ontario, Canada, April 30 and May, 2002 [cited 2010 Aug 25]. ftp://ftp.fao.org/es/esn/food/wgreport2.pdf
[3] Kröckel L (2006) Use of Probiotic Bacteria in Meat Products. Fleischwirtschaft. 86:109-113.
[4] FAO/WHO. (2001) Health and Nutritional Properties of Probiotics in Food including Powder Milk with Live Lactic Acid Bacteria. Cordoba, Argentina: Food and Agriculture Organization of the United Nations and World Health Organization Expert Consultation Report.
[5] Prado FC, Parada JL, Pandey A, Soccol CR (2008) Trends in Non-dairy Probiotic Beverages. Food res. int. 41:111-123.
[6] Leroy F, Falony G, Vuyst L (2008) Latest Developments in Probiotics. In: Toldra F, editor. Meat Biotechnology. Brussels, Belgium: Springer. pp. 217-229.
[7] Von Wright A, Axelsson L (2000) Lactic Acid Bacteria: An Introduction. In: Lahtinne S, Salminen, S, Von Wright A, Ouwehand A, editor. Lactic Acid Bacteria: Microbiological and Functional Aspects. London: CRC Press. pp. 1-16.
[8] Salminen S, von Wright A, Morelli L, Marteau P, Brassart D, de Vos WM, et al (1998) Demonstration of Safety of Probiotics—a Review. Int. j. food microbiol. 44:93-106.
[9] Husebye E, Hellström PM, Sundler F, Chen J, Midtvedt T (2001) Influence of Microbial Species on Small Intestinal Myoelectric Activity and Transit in Germ-free Rats. Am. j. physiol-gast. l. 280:G368-G380.
[10] Banasaz M, Norin E, Holma R, Midtvedt T (2002) Increased Enterocyte Production in Gnotobiotic Rats Mono-associated with *Lactobacillus rhamnosus* GG. Appl. environ. microb. 68:3031-3034.
[11] Halvorsen R, Berstad A, Lassen J, Midtvedt T, Narvhus J (2000) The Use of Probiotics for Patients in Hospitals: A Benefit and Risk Assessment. Norwegian Scientific Committee for Food Safety. 07/112-FINAL: 1-29. ISBN: 978-82-8082-291-8.

* Corresponding Author

[12] Desmazeaud M (1983) Nutrition of Lactic Acid Bacteria: State of the Art. Le lait. 63:267-316.

[13] Marshall V, Law B (1984) The Physiology and Growth of Dairy Lactic-Acid Bacteria. In: Davies FL, Law BA, editors. Advances in the Microbiology and Biochemistry of Cheese and Fermented Milk. London: Elsevier Applied Science Publishers Ltd. pp. 67-98.

[14] Barrangou R, Lahtinen SJ, Ibrahim F, Ouwehand AC (2011) Genus *Lactobacillus*. In: Lactic Acid Bacteria: Microbiological and Functional Aspects. London: CRC Press. pp. 77-91.

[15] Holzapfel WH, Haberer P, Geisen R, Björkroth J, Schillinger U (2001) Taxonomy and Important Features of Probiotic Microorganisms in Food and Nutrition. Am. j. clin. nutr. 73:365S-373S.

[16] Gorbach S (2002) Probiotics in the Third Millennium. Digest liver dis. 34:S2-S7.

[17] Guarner F, Khan AG, Garisch J, Eliakim R, Gangl A, Thomson A, et al (2009) World Gastroenterology Organisation Practice Guideline: Probiotics and prebiotics. Arab j. gastroenterol. 10:33-42.

[18] Vasiljevic T, Shah N (2008) Probiotics—from Metchnikoff to Bioactives. Int. dairy j. 18:714-728.

[19] Shah N (2000) Probiotic Bacteria: Selective Enumeration and Survival in Dairy Foods. J. dairy sci. 83:894-907.

[20] de Vrese M, Schrezenmeir J (2008) Probiotics, Prebiotics, and Synbiotics. Adv. biochem. engin/biotechnol. 111:1-66.

[21] Marteau P, Seksik P, Jian R (2002) Probiotics and Intestinal Health Effects: a Clinical Perspective. Brit. j. nutr. 88:51-58.

[22] de Vrese M, Stegelmann A, Richter B, Fenselau S, Laue C, Schrezenmeir J (2001) Probiotics—Compensation for Lactase Insufficiency. Am. j. clin. nutr. 73:421S-429S.

[23] Reid G, Howard J, Gan B (2001) Can Bacterial Interference Prevent Infection? Trends microbiol. 9:424-428.

[24] Gueimonde M, Kalliomäki M, Isolauri E, Salminen S (2006) Probiotic Intervention in Neonates--Will Permanent Colonization Ensue? J. pediatr. gastroenterol. nutr. 42(5):604-606.

[25] Isolauri E, Sütas Y, Kankaanpää P, Arvilommi H, Salminen S (2001) Probiotics: Effects on Immunity. Am. j. clin. nutr. 73:444S-450S.

[26] Ibrahim F, Ruvio S, Granlund L, Salminen S, Viitanen M, Ouwehand AC (2010) Probiotics and Immunosenescence: Cheese as a Carrier. FEMS immunol. med. microbiol. 59(1):53-59.

[27] Twetman S, Stecksen-Blicks C (2008) Probiotics and Oral Health Effects in Children. Int. j. paediatr. dent. 18:3-10.

[28] Krehbiel CR, Rust SR, Zhang G, & Gilliland SE (2003) Bacterial Direct-fed Microbials in Ruminant Diets: Performance Response and Mode of Action. J. anim. sci. 81(14). Electronic Supplement (2), E120.

[29] Kalavathy R, Abdullah N, Jalaludin S, & Ho YW (2003) Effects of *Lactobacillus* Cultures on Growth Performance, Abdominal Fat Deposition, Serum Lipids and Weight of Organs of Broiler Chickens. Br. poult. sci. 44(1), 139-144.

[30] Patterson J, Burkholder K (2003) Application of Prebiotics and Probiotics in Poultry Production. Poult. sci. 82:627-631.

[31] Vine N, Leukes W, Kaiser H, Daya S, Baxter J, Hecht T (2004) Competition for Attachment of Aquaculture Candidate Probiotic and Pathogenic Bacteria on Fish Intestinal Mucus. J. fish dis. 27:319-326.

[32] Jin L, Marquardt R, Zhao X (2000) A strain of *Enterococcus faecium* (18C23) Inhibits Adhesion of Enterotoxigenic *Escherichia coli* K88 to Porcine Small Intestine Mucus. Appl. environ. microb. 66:4200-4204.

[33] Wang YB (2007) Effect of Probiotics on Growth Performance and Digestive Enzyme Activity of the Shrimp *Penaeus vannamei.* Aquacult. 269:259-264.

[34] Hooper LV, Midtvedt T, Gordon J (2002) How Host-microbial Interactions Shape the Nutrient Environment of the Mammalian Intestine. Annu. rev. nutr. 22:283-307.

[35] Musa H, Wu S, Zhu C, Seri H, Zhu G (2009) The Potential Benefits of Probiotics in Animal Production and Health. J. anim. vet. adv. 8:313-321.

[36] Balcázar JL, Rojas-Luna T, Cunningham DP (2007) Effect of the Addition of Four Potential Probiotic Strains on the Survival of Pacific White Shrimp (*Litopenaeus vannamei*) Following Immersion Challenge with *Vibrio parahaemolyticus*. J. invertebr. pathol. 96:147-150.

[37] Balcázar JL, De Blas I, Ruiz-Zarzuela I, Vendrell D, Gironés O, Muzquiz JL (2007) Enhancement of the Immune Response and Protection Induced by Probiotic Lactic Acid Bacteria against Furunculosis in Rainbow Trout (*Oncorhynchus mykiss*). FEMS immunol. med. microbiol. 51:185-193.

[38] Berg G (2009) Plant–microbe Interactions Promoting Plant Growth and Health: Perspectives for Controlled Use of Microorganisms in Agriculture. Appl. microbiol. biot. 84:11-18.

[39] Bloemberg GV, Lugtenberg BJJ (2001) Molecular Basis of Plant Growth Promotion and Biocontrol by *rhizobacteria*. Curr. opin. plant biol. 4:343-350.

[40] Nelson LM (2004) Plant Growth Promoting Rhizobacteria (PGPR): Prospects for New Inoculants. Crop manage. doi.:10.1094/CM-2004-0301-05-RV.

[41] Saleem M, Arshad M, Hussain S, Bhatti AS (2007) Perspective of Plant Growth Promoting Rhizobacteria (PGPR) Containing ACC Deaminase in Stress Agriculture. J. ind. microbial. biot. 34:635-648.

[42] Sheng XF, Xia JJ, Jiang CY, He LY, Qian M (2008) Characterization of Heavy Metal-resistant Endophytic Bacteria from Rape (*Brassica napus*) Roots and Their Potential in Promoting the Growth and Lead Accumulation of Rape. Environ pollut. 156(3):1164-1170.

[43] Compant S, Clément C, Sessitsch A (2010) Plant Growth-promoting Bacteria in the Rhizo-and Endosphere of Plants: Their Role, Colonization, Mechanisms Involved and Prospects for Utilization. Soil biol. biochem. 42:669-678.

[44] Perrig D, Boiero M, Masciarelli O, Penna C, Ruiz O, Cassán F, et al (2007) Plant-growth-promoting Compounds Produced by Two Agronomically Important Strains of *Azospirillum brasilense,* and Implications for Inoculant Formulation. Appl. microbiol. biot. 75:1143-1150.

[45] Gardiner G, Stanton C, Lynch P, Collins J, Fitzgerald G, Ross R (1999) Evaluation of Cheddar Cheese as a Food Carrier for Delivery of a Probiotic Strain to the Gastrointestinal Tract. J. dairy sci. 82:1379-1387.

[46] Ross R, Fitzgerald G, Collins K, Stanton C (2002) Cheese Delivering Biocultures: Probiotic Cheese. Aust. j. dairy technol. 57:71-78.

[47] Saarela M, Virkajärvi I, Alakomi HL, Sigvart-Mattila P, Mättö J (2006) Stability and Functionality of freeze-dried Probiotic Bifidobacterium Cells during Storage in Juice and Milk. Int. dairy i. 16:1477-1482.

[48] Phillips M, Kailasapathy K, Tran L (2006) Viability of Commercial Probiotic Cultures (*L. acidophilus, Bifidobacterium* sp., *L. casei, L. paracasei,* and *L. rhamnosus*) in Cheddar Cheese. Int i. food microbiol. 108:276-280.

[49] Akın M, Akın M, Kırmacı Z (2007) Effects of Inulin and Sugar Levels on the Viability of Yogurt and Probiotic Bacteria and the Physical and Sensory Characteristics in Probiotic Ice-cream. Food chem. 104:93-99.

[50] Akin MS (2005) Effects of Inulin and Different Sugar Levels on Viability of Probiotic Bacteria and the Physical and Sensory Characteristics of Probiotic Fermented Ice-cream. Milchwissenschaft. 60:297-301.

[51] Bais HP, Weir TL, Perry LG, Gilroy S, Vivanco JM (2006) The role of Root Exudates in Rhizosphere Interactions with Plants and Other Organisms. Annu. rev. plant biol. 57:233-266.

[52] Parvez S, Malik K, Ah Kang S, Kim HY (2006) Probiotics and Their Fermented Food Products are Beneficial for Health. J. appl. microbiol. 100:1171-1185.

[53] Sanders ME, Klaenhammer T (2001) Invited Review: The Scientific Basis of *Lactobacillus acidophilus* NCFM Functionality as a Probiotic. J. dairy sci. 84:319-331.

[54] Sanders ME, Huis in't Veld J (1999) Bringing a Probiotic-containing Functional Food to the Market: Microbiological, Product, Regulatory and Labeling Issues. Anton leeuw. 76:293-315.

[55] Özer BH, Kirmaci HA (2010) Functional Milks and Dairy Beverages. Int. j. dairy technol. 63:1-15.

[56] Succi M, Tremonte P, Reale A, Sorrentino E, Grazia L, Pacifico S, et al (2005) Bile Salt and Acid Tolerance of *Lactobacillus rhamnosus* Strains Isolated from Parmigiano Reggiano Cheese. Fems. microbiol. lett. 244:129-137.

[57] Saarela M, Paquin P (2009) Probiotics as Ingredients in Functional Beverages. In: Paquin P, editer. Functional and Speciality Beverage Technology. New York: CRC Press. pp. 55-70.

[58] Roy D (2005) Technological Aspects Related To the Use of Bifidobacteria in Dairy Products. Le lait. 85:39-56.

[59] Østlie HM, Helland MH, Narvhus JA (2003) Growth and Metabolism of Selected Strains of Probiotic Bacteria in Milk. Int. i. food microbiol. 87:17-27.

[60] Sendra E, Fayos P, Lario Y, Fernández-López J, Sayas-Barberá E, Pérez-Alvarez JA (2008) Incorporation of Citrus Fibers in Fermented Milk Containing Probiotic Bacteria. Food microbiol. 25:13-21.

[61] De Boever P, Wouters R, Verstraete W (2001) Combined Use of *Lactobacillus reuteri* and Soygerm Powder as Food Supplement. Lett. appl. microbiol. 33:420-424.

[62] Shimakawa Y, Matsubara S, Yuki N, Ikeda M, Ishikawa F (2003) Evaluation of *Bifidobacterium breve* Strain Yakult-Fermented Soymilk as a Probiotic Food. Int. i. food microbial. 81:131-136.

[63] Stephenie W, Kabeir B, Shuhaimi M, Rosfarizan M, Yazid A (2007) Growth Optimization of a Probiotic Candidate, *Bifidobacterium pseudocatenulatum* G4, in Milk Medium Using Response Surface Methodology. Biotechnol. bioproc. eng. 12:106-113.

[64] Janer C, Pelaez C, Requena T (2004) Caseinomacropeptide and Whey Protein Concentrate Enhance *Bifidobacterium lactis* Growth in Milk. Food chem. 86:263-267.

[65] Martinez-Villaluenga C, Frías J, Gómez R, Vidal-Valverde C (2006) Influence of Addition of Raffinose Family Oligosaccharides on Probiotic Survival in Fermented Milk during Refrigerated Storage. Int. dairy i. 16:768-774.

[66] Donkor O, Henriksson A, Vasiljevic T, Shah N (2006) Effect of Acidification on the Activity of Probiotics in Yoghurt during Cold Storage. Int. dairy j. 16:1181-1189.

[67] Talwalkar A, Kailasapathy K (2004) A Review of Oxygen Toxicity in Probiotic Yogurts: Influence on the Survival of Probiotic Bacteria and Protective Techniques. Compr. rev. food sci. f. 3:117-124.

[68] Donkor O, Nilmini S, Stolic P, Vasiljevic T, Shah N (2007) Survival and Activity of Selected Probiotic Organisms in Set-type Yoghurt during Cold Storage. Int. dairy j. 17:657-665.

[69] Aryana KJ, McGrew P (2007) Quality Attributes of Yogurt with *Lactobacillus casei* and Various Prebiotics. LWT-Food sci technol. 40:1808-1814.

[70] Dave RI, and Shah NP (1997) Effect of Cysteine on the Viability of Yoghurt and Probiotic Bacteria in Yoghurts Made with Commercial Starter Cultures. Int. dairy j. 7(8-9): 537-545.

[71] Sultana K, Godward G, Reynolds N, Animugaswainy R, Peiris P, and Kailasapathy K (2000) Encapsulation of Probiotic Bacteria with Alginate-starch and Evaluation of Survival in Simulated Gastrointestinal Conditions and In Yoghurt. Int. j. food microbiol. 62(1-2): 47-55.

[72] [72]Kailasapathy K, and Sureeta BS (2004) Effect of Storage on Shelf Life and Viability of Freeze-dried and Microencapsulated *Lactobacillus acidophilus* and *Bifidobacterium infantis* cultures. Aust. j. dairy technol. 59(3): 204-208.

[73] Picot A, and Lacroix C (2004) Encapsulation of Bifidobacteria in Whey Proteinbased Microcapsules and Survival in Simulated Gastrointestinal Conditions and in Yoghurt. Int. dairy j. 14 (6):505-515.

[74] Iyer C, and Kailasapathy K (2005) Effect of Co-encapsulation of Probiotics with Prebiotics on Increasing the Viability of Encapsulated Bacteria under in vitro Acidic and Bile Salt Conditions and in Yogurt. J. food sci. 70 (1): 18-23.

[75] Capela P, Hay TKC, and Shah NP (2006) Effect of Cryoprotectants, Prebiotics and Microencapsulation on Survival of Probiotic Organisms in Yogurt and Freeze-dried Yogurt. Food res. int. 39 (2): 203-211.

[76] Kailasapathy K (2006) Survival of Free and Encapsulated Probiotic Bacteria and Their Effect on the Sensory Properties of Yogurt. LWT- Food sci technol. 39(10):1221-1227.

[77] Oliveira RPS, Florence ACR, Silva RC, Perego P, Converti A, Gioeilli LA, Oliveria MN (2009) Effect of Different Prebiotics on the Fermentation Kinetics, Probiotic Survival and Fatty Acids Profiles in Nonfat Symbiotic Fermented Milk. Int. i. food microbio.128 (3): 467-472.

[78] Paseephol T, and Sherkat F (2009) Probiotic Stability of Yoghurts Containing *Jerusalem artichoke* Insulin during Refrigerated Storage. J. funct. foods. 1(3): 311-318.

[79] Sandoval-Castilla O, Lobato-Calleros C, Garcia-Galllido HS, Alvarez-Rainirez J, and Venion-Carter EJ (2010) Textural Properties of Alginate-pectin Beads and Survivability of Entrapped *Lb. casei* in Simulated Gastrointestinal Conditions and in Yogurt. Food res. int. 43(1): 111-117.

[80] Ong L, Henriksson A, Shah NP (2007) Chemical Analysis and Sensory Evaluation of Cheddar Cheese Produced with *Lactobacillus acidophilus, Lb. casei, Lb. paracasei* or Bifidobacterium sp. Int dairy j. 17:937-945.

[81] Heller KJ, Bockelmann W, Schrezenmeir J, DeVRESE M (2003) Cheese and Its Potential as a Probiotic Food. In: Farnworth E, editore. Handbook of Fermented Functional Foods. Boca Raton, CRC Press. pp. 203-225.

[82] Mäkeläinen H, Forssten S, Olli K, Granlund L, Rautonen N, Ouwehand A (2009) Probiotic Lactobacilli in a Semi-soft Cheese Survive in the Simulated Human Gastrointestinal Tract. Int. dairy j. 19:675-683.

[83] Vinderola C, Prosello W, Ghiberto D, Reinheimer J (2000) Viability of Probiotic (*Bifidobacterium, Lactobacillus acidophilus* and *Lactobacillus casei*) and Nonprobiotic Microflora in Argentinian Fresco Cheese. J. dairy sci. 83:1905-1911.

[84] Mäkeläinen H, Ibrahim F, Forssten S, Jorgensen P, Ouwehand AC (2010) Probiotic Cheese Devlopment and Functionality. Nutra. foods. 9(3):15-19.

[85] Đurić MS, Iličić MD, Milanović SD, Carić MĐ, Tekić MN (2007) Nutritive Characteristics of Probiotic Quark as Influenced by Type of Starter. Acta periodica technologica. 38:11-19.

[86] Aragon-Alegro LC, Alarcon Alegro JH, Roberta Cardarelli H, Chih Chiu M, Isay Saad SM (2007) Potentially Probiotic and Synbiotic Chocolate Mousse. LWT-Food Sci. technol. 40:669-675.

[87] Wilson E, Seo C, Shahbazi A, Ibrahim S (2004) Survival and Growth of Probiotic Cultures in Sour Cream Products. IFT Annual Meeting. 17A.

[88] Cruz AG, Antunes AEC, Sousa ALOP, Faria JAF, Saad SMI (2009) Ice-cream As a Probiotic Food Carrier. Food res. int. 42:1233-1239.

[89] Charalampopoulos D, Wang R, Pandiella S, Webb C (2002) Application of Cereals and Cereal Components in Functional Foods: a Review. Int. j. food microbiol. 79:131-141.

[90] Stanton C, Desmond C, Coakley M, Collins JK, Fitzgerald G, Ross RP (2003) Challenges Facing Development of Probiotic-Containing Functional Foods. In: Farnworth E, editor. Handbook of Fermented Functional Foods. Boca Raton: CRC Press. 27-58.

[91] Granato D, Branco GF, Nazzaro F, Cruz AG, Faria JAF (2010) Functional Foods and Nondairy Probiotic Food Development: Trends, Concepts, and Products. Compr. rev. food sci. f. 9:292-302.

[92] [92] Rozada-Sánchez R, Sattur AP, Thomas K, Pandiella SS (2008) Evaluation of *Bifidobacterium* spp. for the Production of a Potentially Probiotic Malt-based Beverage. Process biochem. 43:848-854.

[93] Angelov A, Gotcheva V, Kuncheva R, Hristozova T (2006) Development of a New Oat-based Probiotic Drink. Int. j. food microbiol. 112:75-80.

[94] Champagne CP (2009) 19 Some Technological Challenges in the Addition of Probiotic Bacteria to Foods. In: Charalampopoulos D, Rastall RA, editors. Prepiotics and Probiotics Science and Technology. New York: Springer. pp. 761-804.

[95] Savard T, Gardner N, Champagne C (2003) Growth of Lactobacillus and Bifidobacterium Cultures in a Vegetable Juice Medium, and Their Stability during Storage in a Fermented Vegetable Juice. Sci. aliment. 23(2): 273-283.

[96] Kun S, Rezessy-Szabó JM, Nguyen QD, Hoschke Á (2008) Changes of Microbial Population and Some Components in Carrot Juice during Fermentation with Selected Bifidobacterium Strains. Process biochem. 43: 816-821.

[97] Yoon KY, Woodams EE, Hang YD (2006) Production of Probiotic Cabbage Juice by Lactic Acid Bacteria. Bioresource technol. 97:1427-1430.

[98] Lee S, Ji G, Park Y (1999) The Viability of Bifidobacteria Introduced into Kimchi. Lett. appl. microbiol. 28: 153-156.

[99] Yoon KY, Woodams EE, Hang YD (2005) Fermentation of Beet Juice by Beneficial Lactic Acid Bacteria. LWT-Food sci. technol. 38: 73-75.

[100] Chou CC, Hou JW (2000) Growth of Bifidobacteria in Soymilk and Their Survival in the Fermented Soymilk Drink during Storage. Int. j. food microbiol. 56:113-121.

[101] Wang YC, Yu RC, Chou CC (2002) Growth and Survival of Bifidobacteria and Lactic Acid Bacteria during the Fermentation and Storage of Cultured Soymilk Drinks. Food microbiol. 19: 501-508.

[102] Lin FM, Chiu CH, Pan TM (2004) Fermentation of a Milk–soymilk and *Lycium chinense* Miller Mixture using a New Isolate of *Lactobacillus paracasei* subsp. *paracasei* NTU101 and *Bifidobacterium longum*. J. ind. microbiol biot. 31:559-564.

[103] Tsai J, Lin Y, Pan B, Chen T (2006) Antihypertensive Peptides and γ-aminobutyric Acid from Prozyme 6 Facilitated Lactic Acid Bacteria Fermentation of Soymilk. Process biochem. 41:1282-1288.

[104] Fotiou F, Goulas A, Fountoulakis K, Koutlas E, Hamlatzis P, Papakostopoulos D, et al (1998) Characterization of Bifidobacterium Strains for Use in Soymilk Fermentation. Int. j. food microbiol. 39:213-219.

[105] Yeo SK, Liong MT (2010) Effect of Prebiotics on Viability and Growth Characteristics of Probiotics in Soymilk. J. sci. food agr. 90:267-275.

[106] Verbeke W (2006) Functional Foods: Consumer Willingness to Compromise on Taste for Health? Food qual. prefer. 17:126-131.

[107] Tuorila H, Cardello AV (2002) Consumer Responses to an Off-flavor in Juice in the Presence of Specific Health Claims. Food qual. prefer. 13:561-569.

[108] Luckow T, Delahunty C (2004) Consumer Acceptance of Orange Juice Containing Functional Ingredients. Food res. int. 37:805-814.

[109] Wang CY, Ng CC, Su H, Tzeng WS, Shyu YT (2009) Probiotic Potential of Noni Juice Fermented with Lactic Acid Bacteria and Bifidobacteria. Int. j. food sci. nutr. 60:98-106.

[110] Mousavi Z, Mousavi S, Razavi S, Emam-Djomeh Z, Kiani H (2011) Fermentation of Pomegranate Juice by Probiotic Lactic Acid Bacteria. World j. microb. biot. 27:123-128.

[111] Pereiraa ALF, Maciela TC, Rodriguesa S (2011) Probiotic Cashew Apple Juice. Int. congr. eng. food. 2011: 1-6.

[112] Tsen JH, Lin YP, King VAE. (2003) Banana Puree Fermentation by *Lactobacillus acidophilus* Immobilized in Ca-alginate. J. gen. appl. microbiol.49:357-361.

[113] Kourkoutas Y, Xolias V, Kallis M, Bezirtzoglou E, Kanellaki M. (2005) *Lactobacillus casei* Cell Immobilization on Fruit Pieces for Probiotic Additive, Fermented Milk and Lactic Acid Production. Process biochem 40, 411-416.

[114] Charalampopoulos D, Pandiella S, Webb C (2002) Growth Studies of Potentially Probiotic Lactic Acid Bacteria in Cereal-based Substrates. J appl. microbiol. 92:851-859.

[115] Salovaara H, Gänzle M (2011) Lactic Acid Bacteria in Cereal-based Products. In: Lahtinen S, Salminen S, Ouwehand A, Wright A. Lactic Acid Bacteria: Microbiological and Functional Aspects. London: CRC Press. pp. 227:245.

[116] Kedia G, Wang R, Patel H, Pandiella SS (2007) Use of Mixed Cultures for the Fermentation of Cereal-based Substrates with Potential Probiotic Properties. Process biochem. 42:65-70.

[117] Gotcheva V, Hristozova E, Hrostozova T, Guo M, Roshkova Z, Angelov A (2002) Assessment of Potential Probiotic Properties of Lactic Acid Bacteria and Yeast Strains. Food biotechnol. 16: 211-225.

[118]] Kedia G, Vázquez JA (2008) Pandiella SS. Fermentability of Whole Oat Flour, PeriTec Flour and Bran by *Lactobacillus plantarum*. J. food eng. 89:246-249.

[119] Mårtensson O, Andersson C, Andersson K, Öste R, Holst O (2001) Formulation of an Oat-based Fermented Product and Its Comparison with Yoghurt. J. sci food agr. 81:1314-1321.

[120] Wood PJ (1997) Functional Foods for Health: Opportunities for Novel Cereal Processes and Products. Cereal 8: 233-238.

[121] Helland MH, Wicklund T, Narvhus JA (2004) Growth and Metabolism of Selected Strains of Probiotic Bacteria, in Maize Porridge with Added Malted Barley. Int. j. food microbiol. 91:305-313.

[122] Gerez C, Cuezzo S, Rollán G, Font de Valdez G (2008) *Lactobacillus reuteri* CRL 1100 as Starter Culture for Wheat Dough Fermentation. Food microbiol. 25:253-259.

[123] [123] Hammes W, Hertel C (1998) New Developments in Meat Starter Cultures. Meat sci. 49:S125-S138.

[124] Tyopponen S, Petaja E, Mattila-Sandholm T (2003) Bioprotectives and Probiotics for Dry Sausages. Int. j. food microbiol. 83:233-244.

[125] Axelsson L (2004) Lactic Acid Bacteria: Classification and Physiology. In: Salminen S, Wright A, Ouwehand AC, editors. Lactic Acid Bacteria: Microbiology and Functional Aspects. New York: Marcel Dekker. pp. 1-66.

[126] Leistner L (2000) Basic Aspects of Food Preservation by Hurdle Technology. Int. j. food microbiol. 55:181-186.

[127] Lücke FK (2000) Utilization of Microbes to Process and Preserve Meat. Meat sci. 56:105-115.

[128] Thapa N, Pal J, Tamang JP (2004) Microbial Diversity in Ngari, Hentak and Tungtap, Fermented Fish Products of North-East India. World j. microbiol. biotechnol. 20: 599-607.

[129] Botsoglou NA, Fletouris DJ, ebrary I (2001) Drug Residues in Foods: Pharmacology, Food Safety, and Analysis. New York: Marcel Dekker. 516 p.

[130] Langhout P (2000) New Additives for Broiler Chickens. World poultry. 16(3): 22-27.

[131] Hashemipour H, Khaksar V, Kermanshahi H (2011) Application of Probiotic on Egg Production and Egg Quality of Chukar Partridge. Afr. j. biotechnol. 10(82):19244-19248.

[132] Kawakami S, Yamad T, Nakanishi N, Cai Y (2010) Feedin of Lactic Acid Bacteria and Yeast on Growth and Diarrhea of Hostein Calves. J. Anim. Vet. Adv. 9:1112-1114.

[133] Gaggìa F, Mattarelli P, Biavati B (2010) Probiotics and Prebiotics in Animal Feeding for Safe Food Production. Int. j. Micribiol. 141:S15-S28.

[134] Balcázar JL, Blas I, Ruiz-Zarzuela I, Cunningham D, Vendrell D, Múzquiz JL (2006) The Role of Probiotics in Aquaculture. Vet. microbiol. 114:173-186.

[135] Cassán F, Maiale S, Masciarelli O, Vidal A, Luna V, Ruiz O (2009) Cadaverine Production by *Azospirillum brasilense* and Its Possible Role in Plant Growth Promotion and Osmotic Stress Mitigation. Eur. j. soil biol. 45:12-19.

[136] Long SR (2001) Genes and Signals in the Rhizobium-legume Symbiosis. Plant physiol. 125:69-72.

[137] De Vleesschauwer D, Hofte M (2007) Using *Serratia plymuthica* to Control Fungal Pathogens of Plants. CAB Rev. 2(046): 1-12.

[138] Bai Y, D'Aoust F, Smith DL, Driscoll BT (2002) Isolation of Plant-growth-promoting Bacillus Strains from Soybean Root Nodules. Can. j. microbiol. 48:230-238.

[139] Kloepper JW, Ryu CM, Zhang S (2004) Induced Systemic Resistance and Promotion of Plant Growth by *Bacillus* spp. Phytopathology. 94:1259-1266.

[140] Preston GM (2004) Plant perceptions of plant growth-promoting Pseudomonas. Philos. Trans. R. Soc Lond. B. Biol. Sci. 359:907-918.

[141] Zhao Y, Thilmony R, Bender CL, Schaller A, He SY, Howe GA (2003) Virulence Systems of *Pseudomonas Syringae* pv. Tomato Promote Bacterial Speck Disease in Tomato by Targeting the Jasmonate Signaling Pathway. Plant j. 36:485-499.

[142] Ryan RP, Monchy S, Cardinale M, Taghavi S, Crossman L, Avison MB, et al (2009) The Versatility and Adaptation of Bacteria from the Genus *Stenotrophomonas*. Nat. rev. microbial. 7:514-525.

[143] Schrey SD, Tarkka MT (2008) Friends and Foes: Streptomycetes as Modulators of Plant Disease and Symbiosis. Anton van leeuw. 94:11-19.

[144] Hartmann A, Gantner S, Schuhegger R, Steidle A, Dürr C, Schmid M, Langebartels C, Dazzo FB, Eberl L (2004) N-Acyl Homoserine Lactones of Rhizosphere Bacteria Trigger Systemic Resistance in Tomato Plants. In: Tikhonovich I, Lugtenberg B,

Provorov, editors. Biology of Molecular Plant–microbe Interactions. St. Paul, Minnoesota: IS-MPMI. 4:554-6.
[145] Unno Y, Okubo K, Wasaki J, Shinano T, Osaki M (2005) Plant Growth Promotion Abilities and Microscale Bacterial Dynamics in the Rhizosphere of Lupin Analysed by Phytate Utilization Ability. Environ. microbial. 7:396-404.
[146] Conrath U, Pieterse CM, and Mauch-Mani B (2002) Priming in Plant Pathogen Interactions. Trends plant sci. 7: 210-216.
[147] [147] van Loon LC (2007) Plant Responses to Plant Growth-promoting Rhizobacteria. Eur. j. plant pathol. 119 (3): 243-254.
[148] Matsui S (2009) Probiotics principle that can help organic farming. J. environ. sanit. eng. res .31. The Association of Environmental & Sanitary Engineering Research, Kyoto University.
[149] Antoine JM (2011) Current Challenges for Probiotics in Food In: Lahtinne S, Salminen S, Von Wright A, & Ouwehand A, editor. Lactic Acid Bacteria: Microbiological and Functional Aspects. London: CRC Press p. 213-226.
[150] Shah NP (2007) Functional Cultures and Health Benefits. Int dairy j. 17:1262-1277.
[151] Kailasapathy K, Chin JC (2000) Survival and Therapeutic Potential of Probiotic Organisms with Reference to *Lactobacillus acidophilus* and *Bifidibacerium* spp. . Immunol. cell biol. 78:80-88.
[152] Rodrigues D, Rocha-Santos TAP, Pereira CI, Gomes AM, Malcata FX, Freitas AC (2011) The Potential Effect of FOS and Inulin upon Probiotic Bacterium Performancein Curdled Milk Matrices. LWT - Food sci. technol. 44:100-108.
[153] Saarela M, Mogensen G, Fonden R, Matto J, Mattila-Sandholm T (2000) Probiotic Bacteria: Safety, Functional and Technological Properties. J. biotech. 84:197-215.
[154] [154]Saxelina M, Grenovb B, Svenssonc U, Fondénc R, Renierod R, Mattila-Sandholme T (1999) The Technology of Probiotics. Trends food sci technol. 10 (12): 387-392.
[155] Del Piano M, Morelli L, Strozzi G, Allesina S, Barba M, Deidda F, et al (2006) Probiotics: from Research to Consumer. Digest liver dis. 38:S248-S255.
[156] Ding W, Shah N (2007) Acid, Bile, and Heat Tolerance of Free and Microencapsulated Probiotic Bacteria. J. food sci. 72:M446-M450.
[157] Capela P, Hay T, Shah N (2006) Effect of Cryoprotectants, Prebiotics and Microencapsulation on Survival of Probiotic Organisms in Yoghurt and Freeze-dried Yoghurt. Food res. int. 39:203-211.
[158] Champagne CP, Girard F, Rodrigue N (1993) Production of Concentrated Suspensions of Thermophilic Lactic Acid Bacteria in Calcium-alginate Beads. Int. dairy j. 3:257-275.
[159] http://www.ganedenbc30.com/
[160] Hekmat S, Reid G (2006) Sensory Properties of Probiotic Yogurt is Comparable to Standard Yogurt. Nut. res. 26:163-166.
[161] Oliveira LB, Jurkiewicz CH (2009) Influence of Inulin and Acacia Gum on the Viability of Probiotic Bacteria in Synbiotic Fermented Milk. Braz. j. food technol. 12:138-144.
[162] Luckow T, Sheehan V, Fitzgerald G, Delahunty C (2006) Exposure, Health Information and Flavor-masking Strategies for Improving the Sensory Quality of Probiotic Juice. Appetite 47: 315-325.

[163] Stein LJ, Nagai H, Nakagawa M, Beauchamp GK (2003) Effects of Repeated Exposure and Health-related Information on Hedonic Evaluation and Acceptance of a Bitter Beverage. Appetite 40:119-129.

[164] Kahkonen P, Turoila H, Rita H (1995) How Information Enhances Acceptability of a Low-fat Spread. Food qual. pref. 7:87-94.

[165] Tuorila H, Andersson A, Martikainen A, Salovaara H (1998) Effect of Product Formula, Information, and Consumer Characteristics on the Acceptance of a New Snack Food. Food qual. pref 9:313-320.

[166] Deliza RA, Silva ALS (2003) Consumer Attitude towards Information on Non Conventional Technology. Food sci. technol 14:43-49.

[167] Rokka S, Rantamäki P (2010) Protecting Probiotic Bacteria by Microencapsulation: Challenges for Industrial Applications. Eur. food res. technol. 231:1-12.

[168] Chau CF, Wu SH, Yen GC (2007) The Development of Regulations for Food Nanotechnology. Trends food sci. technol. 18:269-280.

[169] Sozer N, Kokini JL (2009) Nanotechnology and its Applications in the Food Sector. Trends biotechnol. 27:82-89.

[170] Sekhon BS (2010) Food Nanotechnology–an Overview. Nan. sci. appl. 3:1-15.

[171] Sinha VR, Anamika V, Bhinge JR (2008) Nanocochleates: A Novel Drug Delivery Technology. Pharmainfo.net. Available: http://www.pharmainfo.net/reviews/nanocochleates-novel-drug-delivery-technology . Accessed 2012 Mar 12.

[172] O'Riordan K, Andrews D, Buckle K, Conway P (2001) Evaluation of Microencapsulation of a Bifidobacterium strain with Starch as an Approach to Prolonging Viability during Storage. J. appl. microbiol. 91:1059-1066.

[173] Lee J, Cha D, Park H (2004) Survival of Freeze-dried *Lactobacillus bulgaricus* KFRI 673 in Chitosan-coated Calcium Alginate Microparticles. J. agr food chem. 52:7300-7305.

[174] Chen KN, Chen MJ, Lin CW (2006) Optimal Combination of the Encapsulating Materials for Probiotic Microcapsules and its Experimental Verification (R1). J. food eng. 76:313-320.

[175] Chandramouli V, Kailasapathy K, Peiris P, Jones M (2004) An Improved Method of Microencapsulation and its Evaluation to Protect *Lactobacillu* spp. in Simulated Gastric Conditions. J. microbiol meth. 56:27-35.

[176] Heenan C, Adams M, Hosken R, Fleet G (2004) Survival and Sensory Acceptability of Probiotic Microorganisms in a Nonfermented Frozen Vegetarian Dessert. LWT-Food sci. technol. 37:461-466.

[177] Klaenhammer TR, Barrangou R, Buck BL, Azcarate-Peril MA & Altermann E (2005) Genomic Features of Lactic Acid Bacteria Effecting Bioprocessing and Health. FEMS microbial. rev. 29: 393-409.

[178] Ahmed FE (2003) Genetically Modified Probiotics in Foods. Trends biotechnol. 21: 491-497.

[179] Sanders ME, Heimbach JT (2005) Functional Foods in the USA: Emphasis on Probiotic Foods. In: Gibson GR, editor. Food Science and Technology Bulletin - Functional Foods, Vol 1. International Food Information Service (IFIS Publishing).

Chapter 2

Nutritional Programming of Probiotics to Promote Health and Well-Being

Alice Maayan Elad and Uri Lesmes

Additional information is available at the end of the chapter

http://dx.doi.org/10.5772/50051

1. Introduction

The human large intesine is inhabited by a diverse and complex bacterial flora, which includes an outstanding total number of 10^{14} cells, >1000 species and a biomass of more than 1 kg [1, 2]. Thus, the gut microbiota may be conceived as a specialized 'microbial organ' within the gut, affecting human health and disease through its involvement in pathogenesis, nutrition and immunity of the host [1-3]. Recently it has also been recognized that this dynamic yet stable ecosystem plays a role in conditions such as obesity and diabetes as well as in general well-being, from infancy to ageing [1-8]. Consequently, an increasing number of studies which explore the potential of promoting health by nutrition focuses on possible ways to influence and modulate the composition and activity of the gut flora towards a healthier one [4, 9-12].

In this respect, three major dietary approaches have been studied and applied. The first approach of probiotics is to fortify the gut flora through the consumption of exogenous live microorganisms, e.g. *L. acidophilus* in dairy products. The second strategy of prebiotics seeks to selectively stimulate the growth and/or activity of one or a limited number of advantageous indigenous bacteria in the host gut flora [1, 13]. The third approach, known as synbiotics due to its synergistic nature, aims to combine the previous ones by the simultaneous administration of probiotics and prebiotics, which improves the survival and implantation of the live microbes [13].

Over the years, much attention has been drawn to indigestible carbohydrates that evade enzymatic digestion in the upper gastrointestinal tract and become available for fermentation in the colon [13]. These dietary compounds were later termed as prebiotics, a definition of which has been updated into its current form as "a selectively fermented ingredient that allows specific changes, both in the composition and/or activity in the gastrointestinal microflora that confers benefits upon host well-being and health" [14, 15].

Although a more recent development compared to probiotics, prebiotics have been at the heart of various studies and numerous commercial products since they do not share the problem of probiotic survival upon ingestion by the consumer, and they can be added to a broad range of food products (e.g. confectionary and baked foods as well as more traditional fermented milk products and fruit drinks) because the majority of prebiotics are carbohydrates [16].

Amongst the carbohydrates currently marketed as prebiotics, inulin, fructo-oligosaccharides (FOS), galacto-oligosaccharides (GOS) and lactulose are consistently supported by high quality data from *in vitro*, *in vivo* and human trials [10-12, 14, 16-19]. Specifically, human trials have established that dietary consumption of 5-20 g/day of these prebiotics stimulates the growth of *Bifidobacterium* and *Lactobacillus* and promotes the health and well-being of infants, adults, pregnant and lactating women as well as the elderly to varying extents [6, 8, 11, 20, 21].

2. Prebiotics as gut flora management tools

2.1. Established prebiotics

Overall, three major groups of compounds have been consistently established as prebiotics conferring health benefits (as detailed in Table 1): fructans, which include inulin and fructo-oligosaccharides (FOS), galacto-oligosaccharides and lactulose. Under the general term fructans one can classify three established prebiotic carbohydrates: inulin, fructo-oligosaccharides (FOS) and short chain fructo-oligosaccharides (scFOS) [16-18]. The fructans are polymers composed of D-fructose units joined by β-2-1 glycosidic linkages and terminated by an α-1-2-linked D-glucose.

Established prebiotic	Recommended efficaceous intake [g/day]	Key effects in humans	Potential adverse intake [g/day]	Suggested references
Inulin	5-15	• Stimulate bifidobacteria growth • Production of short chain fatty acids • Protection against enteric infections	> 15 (increase in fecal output)	[22]
Fructo-oligosaccharides	10-15		> 15 (increase in fecal output)	[17]
Galacto-oligosaccharides	10-15		> 20 g/human body (diarrhea)	[19]
Lactulose	10		> 20 (laxative)	[23]

Table 1. The main established prebiotics and their beneficial/adverse intakes.

The degree of polymerization (DP), defined by the number of monosaccharide units, is used to distinguish between inulin, FOS and scFOS. Molecules with a DP between 2-60 are referred to as inulin. Inulin is commercially produced from chicory roots, but it is present in varying extent also in onions, garlic, Jerusalem artichoke, tomato and banana [11, 16, 21]. Similarly, oligofructose, commonly referred to as FOS, is prepared from chicory in an enzymatic hydrolysis using inulinase, and defined as oligosaccharide fractions which have a maximal DP of 20 with most common commercial products having an average DP of 9. In contrast, scFOS are synthesized in an enzymatic reaction via transfer of fructosyl units from sucrose molecules to yield mixtures of fructosyl chains with a maximum DP of 5. The mixture produced is usually comprised mainly of 1-kestose (2 units of fructose linked to glucose, GF2), nystose (GF3) and 1-fructosyl nystose (GF4) [16, 17, 24].

Fructans have a long tradition as prebiotics. Since their fructose units are joined by β-linkages, they are resistant to hydrolysis by the human digestive enzymes which mainly cleave α-linkages. As a consequence, when these carbohydrates reach the colon they selectively stimulate the growth of beneficial bacteria such as bifidobacteria, which do contain specific enzymes for their degradation, i.e. β-fructosidases [16, 21, 25]. Therefore, inulin, FOS and scFOS are classified as 'nondigestible' carbohydrates, with a calorie value of 1.5-2.0 kcal/g [24]. FOS fermentation in the colon results in increased levels of short chain fatty acids (SCFA) which lower the pH in the intestinal lumen. This can provide an explanation to the reports that these fructans lead to a decrease in the number of harmful bacteria in the colon (such as *Clostridium*, *Streptococcus faecallis* and *Escherichia coli*) [21, 25].

Galacto-oligosaccharides (GOS) are galactose-containing oligosaccharide mixtures of the form Glu α-1-4[β-Gal-1-6]$_n$ where n can be between two to five. They are produced from lactose syrup using β-galactosidases, which catalyze the hydrolysis of lactose into glucose and galactose, and also the transgalactosylation reactions with lactose as acceptor of galactose units giving rise to a variety of glycosidic linkages and molecular weights [11, 16, 19, 21]. Furthermore, the use of different enzymes in the various production processes of GOS leads to variability in their purity and glycosidic linkages, with β-1-6, β-1-3 and β-1-4 being the dominant [19]. Several *in vitro* and *in vivo* experiments have demonstrated that as in inulin-type fructans, the β-glycosidic linkages in GOS render them resistant to hydrolysis by the human digestive enzymes secreted in the upper gastrointestinal tract [16, 19, 21, 26]. In light of that, manufacturers are obliged by the European regulation to clearly identify GOS-containing food products as dietary fibers, with an estimated low calorie value of 1-2 kcal/g [19].

Most of the health effects related to GOS arise from their selective fermentation by bifidobacteria and lactobacilli. In fact, it has been reported that when added to infant milk formulas, these oligosaccharides replicated the bifidogenic effect of human breast milk, not only in bacterial counts, but also with respect to the metabolic activity of the microflora in the colon [16, 27]. The growth of *Lactobacillus paracasei* and *Bifidobacterium lactis* has been shown to be preferential when grown on tri- and tetrasaccharide fractions of FOS or GOS, which supports the notion that prebiotics selectively promote the proliferation of bacteria possesing an active transport system enabling them to utilize these oligosaccharides [28-30]. In addition, it has been demonstrated that GOS compete for pathogen binding sites that coat the surface of the gastrointestinal epithelial cells [31, 32].

Finally, lactulose (β-1-4-galactosyl-fructose) is a synthetic disaccharide derived from lactose. It is commonly used as a laxative in pharmaceutical products for the treatment of constipation, in doses over 20 g/day [16, 21]. Nevertheless, human trials have shown that at lower doses, lactulose acts as a prebiotic, reaching the colon and increasing bifidobacteria counts [2, 16, 17, 23]. Although this substance is an established prebiotic, it is still heavily confined to applications as a therapeutic agent.

2.2. Novel prebiotics

The search for new and novel prebiotics is constantly driven by the increased interest in management of human health through nutrition, particularly by the modulation of gut flora. In addition, studies of established prebiotics have enabled the better understanding of mechanisms of action and properties, which provided the basis for emerging prebiotics. Among the large array of prebiotic candidates, isomalto-oligosaccharides (IMO), xylo-oligosaccharides (XOS), soy-oligosaccharides (SOS), gluco-oligosaccharides, lactosucrose and resistant starches can be classified as emerging prebiotics [17, 33-40]. Some of these compounds present advantages over established prebiotics; for example, XOS are stable across a wide range of pH, hence they are resistant to degradation in low pH juices, in contrast to inulin [21]. Besides attempts to identify and isolate naturally occurring prebiotics there are also attempts to enhance and extend the functionality of exisiting natural prebiotics through a rational design approach [41].

Promising results have been reported, including the selective growth of bifidobacteria and lactobacilli and/or the formation of beneficial metabolites, such as short chain fatty acids. However, it should be noted that these studies are still limited to *in vitro* models or small scale animal or human trials [16, 17, 40]. For example, thermally produced resistant starch has been demonstrated to possess a bifidogenic and butyrogenic effect in an *in vitro* three stage continuous fermentation system inoculated with human feces [42]. Furthermore, this study suggested that resistant starch crystalline polymorphism, resulting from different thermal treatments, could convey different prebiotic effects on the human colon flora.

3. Efficacy of prebiotics across the life span

3.1. Methods to evaluate prebiotics

Research into the efficacy of prebiotics includes a collection of methods currently in use, from pure cultures to human trials, which can be generally classified into *in vivo* and *in vitro* methods. Overall, the prebiotic effect is mainly evaluated by the presence of beneficial metabolites and measuring the growth of major bacterial groups commonly present in the human gut, in particular a selection for increased numbers of bifidobacteria and lactobacilli in comparison with undesirable bacteria such as certain clostridia and sulfate reducing bacteria [40]. Ultimately, health claims concerning prebiotic effects must rely on comprehensive well-controlled human trials. Thus, *in vivo* studies have evolved over the years to robust experimental designs which combine double blind and placebo-controlled designs with advanced microbial analyses, such as bacterial enumeration using 16s DNA probes in fluorescence *in situ* hybridization (FISH) [16]. In most human studies, the production of short chain fatty acids (SCFA) has been quantified in fecal samples, as a marker of enhanced saccharolytic fermentation in response to prebiotic treatment [13].

In spite of their high significance, *in vivo* human studies are usually limited, mainly due to financial and ethical restrictions. Therefore, animal models have been used as a possible viable alternative to the human GI tract, while allowing the researchers to perform *in vivo* experiments in tightly controlled conditions as well as access intestinal contents, tissues and organs at autopsy. Moreover, many *in vivo* experiments have used germ-free animals dosed with fecal suspensions obtained from human donors, which are considered to be a reliable model for a reconstituted human gut flora. However, data generated from animal models do not necessarily coincide with human or *in vitro* studies, as has been shown for prebiotic resistant starch type III [42].

Consequently, many *in vitro* experimental models have been developed to simulate various aspects of the human GI tract [43-45]. Seeking to closely mimic the conditions of organs along the GI tract, these models include a reactor or a series of reactors under tightly controlled settings, with the large intestine represented by an anaerobic reactor/s inoculated with fecal slurries [40]. Thus, these systems offer researchers a controlled experimental design that is relatively inexpensive, easy to set up, high throughput and raises minimal ethical issues [46].

One of the first *in vitro* GI models described in the literature was termed the simulator of the human intestinal microbial ecosystem (SHIME) [47]. This computer controlled model is composed of a five serially connected vessels simulating the conditions of the stomach, small intestine, ascending, transverse and descending colon [48, 49]. Operators can control various parameters of physiological relevance, including gastric and pancreatic secretions, pH, transit time, feed composition as well as sample different loci along the system on a regular basis [13, 46]. Another comprehensive *in vitro* GI model was developed in the Netherlands [50]. This model is actually comprised of two seperate parts: TIM-1 is a series of four computer controlled chambers simulating the upper GI, i.e. the stomach, duodenum, jejunum and ileum, while TIM-2 models the large intestine. Unlike the SHIME, this model consists of a series of linked glass vessels containing flexible walls, which allow simulation of the peristaltic movements of the GI. The hollow fiber membrane construct of the system enables to simulate absorption of water and nutrients in the lumen as well as their removal from the colon [13, 46]. Similarly, simple glass reactors have been used for in batch and three stage continuous fermentation systems to simulate the lower GI, i.e. the proximal, transverse and distal colon, and have been validated against sudden death victims [51].

Overall, *in vivo* methods and particularly human trials are essential for establishing health claims regarding prebiotic effects on human microflora. However, such methods are hindered by financial, ethical and practical reasons. *In vitro* models fail to fully mimic the GI system, particularly peristaltic movements, mucosal uptake and impact of immune components. Nevertheless, these systems offer relatively low costs, ease of use and have minimal ethical considerations, while providing researchers controllable settings for studying luminal biochemistry and microbiology.

3.2. Prebiotic efficacy in infancy and childhood

At birth, the neonate gut is considered to be sterile with rapid colonization by bacteria believed to occur in three phases: delivery, breastfeeding and weaning to solid foods [52-54]. Immediately after birth, facultative anaerobic bacteria such as *Enterobacteriaceae*, streptococci and staphylococci colonize the gut environment of the newborn, gradually consuming oxygen and producing various metabolites. Consequently, strict anaerobic bacterial population dominated by bifidobacteria, *Clostridium* and *Bacteroides* can be established [55-57]. Within the first year of life, the microflora is highly dynamic, but by the age of two years with the introduction of solid foods, the colonic microbiota is considered complete – it stabilizes and resembles that of the adult [58-60].

The interest in prebiotics as a nutritional strategy to program infant gut microbiota to favor a more advantageous population has been inspired by the beneficial effects attributed to the 200 different human milk oligosaccharides (HMO) [52-54]. Based on the observations of bifidobacteria in the feces of breastfed babies, attempts have been made to reproduce this bifidogenic aspect in infant formulas by adding commercial prebiotics, in particular FOS and GOS [8, 53]. This practical application of prebiotics was evaluated for example in double-blind, randomized and controlled studies in 90 full term infants, which demonstrated that 4 g/L or 8 g/L of FOS, GOS or their combination resulted in a significant decrease in fecal pH and a concomitant increase in bifidobacteria and lactobacilli after 28 days feeding [61, 62]. A GOS and FOS mixture at a ratio of 9:1 (GOS:FOS) has been extensively studied as a prebiotic additive to infant formulas [63], and shown to increase bifidobacteria in infant feces and lower the incidence of pathogens [64-66]. Therefore, the administration of FOS and GOS into commercial infant formulas for their prebiotic effects has spread, and researchers continue exploring additional prebiotics as possible candidates for infant formulas supplementation. Moreover, studies are also looking into the persistence of the prebiotic effects.

3.3. Prebiotic efficacy in adulthood

To date, various studies have determined fructans (inulin and FOS) induce different beneficial effects on the health and general well-being of healthy adult subjects [10, 16-18]. A daily consumption of 5-10 g of fructans has been demonstrated to exert a bifidogenic effect on healthy adults based on dose-response studies, while similar doses of GOS and lactulose have been reported in *in vitro* and human trials as stimulating a bifidogenic effect [16, 67, 68]. As to bowel habit, i.e. the frequency of bowel discharge but not fecal output, constipation and laxative effect, there is some evidence that in constipated subjects, inulin may increase bowel habit [69], whereas lactulose is prescribed at 20 g/day to increase fecal output of chronically constipated patients, however, having a bifidogenic effect on healthy adults at lower doses [68]. Moreover, numerous studies have shown that fecal output remains unchanged at a daily intake of up to 15 g fructans with a slight increase at doses of 15 g/day or higher [11]. Hence, inulin, FOS, GOS and lactulose may be defined as mild laxatives with adverse effects observed only at a consumption of over 20 g/day.

Additionally, established prebiotics have been linked to protection against enteric infections, modification of the host immune response, production of short chain fatty acids, particularly butyrate, increased mineral absorption and even the reduced risk of colon cancer [10, 11, 14, 16, 18]. Prebiotics efficacy has also been studied during pregnancy and lactation. These life periods are sometimes accompanied by irregular gastrointestinal activity, which can be improved by the consumption of dietary fibers and prebiotics [20]. Furthermore, gestational weight gain and postpartum weight retention have been suggested to be affected from prebiotics intake, since they modulate the gut microflora [5]. However, it is also important to note that the prebiotic effect has not been found to extend to neonates and infants, even when solely breast fed.

3.4. Prebiotics efficacy in ageing

At the old age, increased threshold for taste and smell as well as masticatory dysfunction can lead to a nutritionally imbalanced diet. In addition, various physiological functions deteriorate with age and may influence the absorption and/or metabolism of nutrients. Furthermore, the increased intake of drugs results in GI disturbances due to antibiotics undesired effect on indigenous bacteria in the host gut flora. Thus, changes in the GI tract, modification of diet and host immune system inevitably give rise to bacterial population alterations [70, 71]. In spite of the increasing proportion of the elderly in Western countries [72, 73], scarce data exists on the changes that occur in the intestinal microbiota during the ageing process and their possible health outcomes. Overall, an increase in facultative anaerobes and decrease in *Bacteroides* and bifidobacteria (total numbers as well as species diversity) have been reported [74].

Therefore, modulation of the colon microflora by the consumption of prebiotics is increasingly being studied as a potent, cost effective and natural way to improve the health and well-being of elderly people as well as reduce risks for various diseases [70, 71, 74]. FOS and GOS ingestion, as well as synbiotic preparations, were found to significantly increase the number of bifidobacteria at the expense of less beneficial microbiota in ageing individuals [6, 75-78]. In addition, a randomized, double-blind, controlled study with 74 subjects aged 70 and over has indicated that prebiotic addition can improve the low noise inflammatory process frequently observed in this sensitive population [79].

4. Prebiotics as therapeutics

4.1. Prebiotics therapeutic efficacy in human diseases

It is now well documented that the bacteria microflora residing in the human GI has a role not only in promoting health but also in preventing some diseases [3, 80, 81]. Prebiotics have been reported to protect against pathogenic gastrointestinal infections by promoting the growth of probiotics which help displace pathogens from the mucosa, producing antimicrobial agents and competing with pathogens on binding sites and nutrients [3]. In addition to *in vitro* data which supports this disease preventing effect of prebiotics [16, 22], a human study on 140 infants has concluded that consumption of oligofructose and cereal significantly reduced events of fever, frequency of vomiting, regurgitation and abdominal discomfort [82]. Moreover, various studies have shown that prebiotics can beneficially affect patients with antibiotic-associated diarrhea, especially when it arises from *C. difficile* [11].

Prebiotics have also been reported to reduce the risk of colon cancer as a result of gut flora modulation [11, 14, 16-18]. Specifically, they support the metabolism of carcinogenic molecules and the secretion of short chain fatty acids to the lumen by the colon microbiota [83, 84]. Furthermore, human trials have demonstrated that inulin, FOS and scFOS beneficially affect colorectal cell proliferation and genotoxicity [17], hence the potential of prebiotics in prevention and treatment of colon cancer should be further explored.

Inflammatory bowel disease (IBD), which includes ulcerative colitis (UC) and Crohn's disease (CD), has also been researched as a possible target for prebiotics [11, 16, 17]. As mucosal communities significantly change in these diseases, prebiotics may be used in order to manipulate them. For example, patients fed 15 g per day of a prebiotic mixture composed of 7.5 g inulin and 7.5 g FOS for 2 weeks prior to colonoscopy, have had more than a 10-fold increase in bifidobacterial and eubacterial numbers in the mucosa of the proximal and distal colon [85]. Similarly, in a small open-label human trial, 10 patients with active ileum-colonic CD were fed 15 g FOS daily for 3 weeks, after which a significant reduction in the Harvey Bradshaw index of disease activity was observed as well as an increase of fecal bifidobacteria numbers [11].

4.2. Prospective therapeutic targets

4.2.1. Obesity and the metabolic syndrome

Increasing evidence linking gut flora to human health and diseases have inspired further research regarding the possible link between gut flora and obesity, which has led to the notion that prebiotics could be harnessed as potential therapeutic agents or management tools to prevent and treat obesity and the metabolic syndrome [4, 5, 7, 20, 52]. The metabolic syndrome is a cluster of metabolic abnormalities, including abdominal obesity, type 2 diabetes mellitus and cardiovascular diseases [86, 87].

Various studies have shown that prebiotic intervention decreased fat storage in white adipose tissues and in the liver, decreased hepatic insulin resistance as well as systemic inflammation in several nutritional (high-fat diet-fed) and genetic (*ob*/*ob* mice) obese rodents [88-95]. Some beneficial effects of fructans on BMI, fat mass and insulin resistance were also shown in the limited human trials conducted so far [95-98].

However, to date, the mechanisms underlying the complex role of gut microbiota in such conditions are largely unknown [1]. It has been reported that a lower number of bifidobacteria at birth is associated with overweight later in childhood [99], and in adults, the number of bifidobacteria is slightly lower in individuals with obesity than in lean subjects [100]. The number of these bacteria is also decreased in patients with type 2 diabetes mellitus compared with nondiabetic people [101]. Hence, these results seem to suggest that bifidobacteria affects the development of obesity and its related comorbidities [52].

A remarkable increase has been observed in the number of *Bifidobacterium* spp. following inulin-type fructans supplementation to mice with diet-induced or genetically determined obesity [52]. Interestingly, the number of bifidobacteria was inversely correlated with the development of fat mass, glucose intolerance and levels of lipopolysaccharides (LPS) [102]. LPS has been found at a significantly higher level in the serum of obese individuals, which creates a metabolic endotoxemia, leading to obesity, insulin resistance and systemic inflammation [103]. Moreover, it has been reported that the overexpression of numerous host genes that are related to adiposity and inflammation was prevented by prebiotic intake [52].

A pathway involving short chain fatty acids (SCFA) has been proposed to be involved in the interplay between prebiotics, the gut flora and obesity. SCFA act as signaling molecules and are specific ligands for at least two G protein-coupled receptors, GPR41 and GPR43, which have a potential role in fat mass development [13, 104]. In addition, it has been shown that acetate and propionate can modify hepatic lipid metabolism [105]. Interestingly, a recent study has demonstrated that diet-induced obesity and insulin resistance were prevented when mice on a high-fat diet were supplemented with butyrate, which promoted energy expenditure and induced mitochondrial function [106]. Various studies have shown that a diet enriched with prebiotics leads to a greater intestinal SCFA production and thereby migitates body weight gain, fat mass development and the severity of diabetes [89, 90, 100, 107-109]. Numerous peptides secreted by the enteroendocrine cells along the GI are involved in the regulation of energy homeostasis and/or pancreatic function. Three such peptides which can modulate food intake and energy expenditure are glucagon-like peptide-1 (GLP-1), peptide YY (PYY) and ghrelin [110-113]. Thus, it has been suggested that SCFA are related to changes in the gut peptide secretion, namely increased production and secretion of GLP-1 and PYY and the reduction of ghrelin which induce metabolic effects [13, 114, 115]. Piche *et al.* were the first to report that inulin-type fructan feeding of 20 g/day significantly increased plasma GLP-1 in humans [116]. In another study, a 2-week supplementation with inulin-type fructans (16 g/day) to healthy volunteers increased GLP-1, consequently increasing satiety, lowering calorie intake and decreasing postprandial glycemia [90, 108]. Furthermore, prebiotic treatment in obese patients has been found to induce and increase PYY and decrease ghrelin levels [117]. Finally, Tarini and Wolever have demonstrated that a single dose of inulin significantly increased postprandial plasma GLP-1 and decreased plasma ghrelin [118]. This is in contrast to perceived necessity for a prolonged prebiotics administration to modulate gut microbiota and allow effect on gut endocrine function. Thus, it seems that further studies are needed to fully unravel the potential of prebiotics to offer a nutritional means to cope with the worrisome increase in human obesity and the metabolic syndrome.

5. Future challenges

5.1. Harmonization of methods to evaluate efficacy

To date, a plethora of studies have investigated the effects of prebiotics on human health and well-being, leading to the general realization prebiotics could serve as a possible method of therapeutic intervention. However, the majority of these studies cannot be compared due to the variety of methods employed. For example, results obtained by bacterial isolation techniques [39, 119] cannot be compared with data from more advanced methods for the molecular characterization of the microbiota, now considered essential to obtain a comprehensive view of the gut ecosystem. Furthermore, DNA-based techniques including the use of the 16s ribosomal RNA gene are considered as less biased, hence their results are more reliable [54]. In addition, studies focused on the effect of prebiotics on the elderly, even in healthy subjects, lack a clear definition of 'elderly', and various groups are recruited, usually on the basis of 'over 60' [120], 'over 65' [119, 121] or 'over 70' [122]. This makes is difficult to define a 'threshold age' at which the gut environment starts to be influenced from the ageing process [70]. Thus, harmonization of methods to evaluate efficacy is a prequisite step limiting the further application of prebiotics for the prevention and treatment of diseases as well as the development of novel prebiotics.

5.2. The challenge of personalization

A broad range of parameters has been known to affect the bacterial composition of the infant gut, e.g. mode of delivery, type of feeding (exclusive breastfeeding versus formula), antibiotic use and maternal infection [55-57, 123, 124]. Furthermore, various studies have indicated that the low number of certain bacteria at birth such as bifidobacteria is related to overweight later in childhood [99]. Consequently, prebiotics supplementation even in infants may be used as a preventive nutritional programming tool, which will affect the health and well-being also in adulthood. Moreover, it has been increasingly accepted that environmental factors such as nutritional habits and lifestyle may impact the gut microbiota composition, for example striking country-related differences in the effects of age on the microflora have been reported [70]. This wide variety of factors, affecting intestinal microbiota composition from infancy to elderly, drove the need for personalized nutrition, including personalized and tailored prebiotics. In addition, since prebiotics are recently considered even as therapeutic agents for the prevention and treatment of diseases, it may be beneficial to aim personalized prebiotics to people at high-risk to develop these illnesses. Thus, the challenge of personalization includes performing long-term, large-sized, well-controlled and multidisciplinary-collaborated studies which will demonstrate and establish the health promoting effects of prebiotics and enable harnessing them in the clinic or in supermarket shelves.

6. Conclusions

Prebiotics have emerged as cost-effective and efficient nutritional programming tools to beneficially and selectively promote the growth and/or activity of certain bacteria in the indigenous flora of the human GI. So far, prebiotics have been demonstrated to exert various beneficial effects during an individual's lifetime, from infancy to ageing, as well as function as therapeutic agents for the prevention and treatment of different diseases, including obesity and the metabolic syndrome.

In addition to the well-established prebiotics of FOS, GOS and lactulose, novel prebiotics are constantly being developed. State of the art techniques, *in vitro* gastrointestinal models and advanced computerization tools are leading many researchers to adopt more complete and comprehensive approaches, e.g. metabolomics and metagenomics.

Future challenges include the harmonization of methods of evaluating efficacy that will help focus research efforts and enable a more comprehensive understanding of prebiotic mechanisms of action and beneficial effects and how these can be modulated. Another important prospect is personalization, i.e. fitting tailored prebiotics to individual needs in order to nutritionally program and affect their health and well-being from infancy and into old and prosperous age.

Author details

Alice Maayan Elad and Uri Lesmes
Department of Biotechnology and Food Engineering, Technion, Israel Institute of Technology, Haifa, Israel

7. References

[1] Diamant, M., E.E. Blaak, and W.M. de Vos, Do Nutrient–Gut–Microbiota Interactions Play a Role in Human Obesity, Insulin Resistance and Type 2 Diabetes? Obesity Reviews, 2011. 12(4): p. 272-281.

[2] Venema, K., Intestinal Fermentation of Lactose and Prebiotic Lactose Derivatives, including Human Milk Oligosaccharides. International Dairy Journal, 2012. 22(2): p. 123-140.

[3] O'Hara, A.M. and F. Shanahan, The Gut Flora as a Forgotten Organ. EMBO Rep, 2006. 7(7): p. 688-693.

[4] Delzenne, N.M. and P.D. Cani, Nutritional Modulation of Gut Microbiota in the Context of Obesity and Insulin Resistance: Potential Interest of Prebiotics. International Dairy Journal, 2010. 20(4): p. 277-280.

[5] Duncan, S.H., et al., Human Colonic Microbiota Associated with Diet, Obesity and Weight Loss. International Journal of Obesity, 2008. 32(11): p. 1720-1724.

[6] Tuohy, K.M., Inulin-Type Fructans in Healthy Aging. The Journal of Nutrition, 2007. 137(11): p. 2590S-2593S.

[7] Tuohy, K.M., A. Costabile, and F. Fava, The Gut Microbiota in Obesity and Metabolic Disease - A Novel Therapeutic Target. Nutritional Therapy and Metabolism, 2009. 27(3): p. 113-133.

[8] Parracho, H., A.L. McCartney, and G.R. Gibson, Probiotics and Prebiotics in Infant Nutrition. Proceedings of the Nutrition Society, 2007. 66: p. 405-411.

[9] Fooks, L.J., R. Fuller, and G.R. Gibson, Prebiotics, Probiotics and Human Gut Microbiology. International Dairy Journal, 1999. 9(1): p. 53-61.

[10] Gibson, G.R., Prebiotics as Gut Microflora Management Tools. Journal of Clinical Gastroenterology, 2008. 42(6): p. S75-S79.

[11] Macfarlane, S., G.T. Macfarlane, and J.H. Cummings, Review Article: Prebiotics in the Gastrointestinal Tract. Alimentary Pharmacology & Therapeutics, 2006. 24(5): p. 701-714.

[12] Tuohy, K.M., et al., Using Probiotics and Prebiotics to Improve Gut Health. Drug Discovery Today, 2003. 8(15): p. 692-700.

[13] De Preter, V., et al., The Impact of Pre- and/or Probiotics on Human Colonic Metabolism: Does It Affect Human Health? Molecular Nutrition & Food Research, 2011. 55(1): p. 46-57.

[14] Gibson, G.R., From Probiotics to Prebiotics and a Healthy Digestive System. Journal of Food Science, 2004. 69(5): p. M141-M143.

[15] Schrezenmeir, J. and M. de Vrese, Probiotics, Prebiotics, and Synbiotics - — Approaching a Definition. The American Journal of Clinical Nutrition, 2001. 73(2): p. 361S-364S.

[16] Tuohy, K.M., et al., Modulation of the Human Gut Microflora Towards Improved Health Using Prebiotics – Assessment of Efficacy. Current Pharmaceutical Design, 2005. 11: p. 75-90.

[17] Rastall, R.A., Functional Oligosaccharides: Application and Manufacture. Annual Review of Food Science and Technology, 2010. 1(1): p. 305-339.

[18] Sabater-Molina, M., et al., Dietary Fructooligosaccharides and Potential Benefits on Health. Journal of Physiology and Biochemistry, 2009. 65(3): p. 315-328.

[19] Torres, D.P.M., et al., Galacto-Oligosaccharides: Production, Properties, Applications, and Significance as Prebiotics. Comprehensive Reviews in Food Science and Food Safety, 2010. 9(5): p. 438-454.

[20] Champ, M. and C. Hoebler, Functional Food for Pregnant, Lactating Women and in Perinatal Nutrition: A Role for Dietary Fibres? Current Opinion in Clinical Nutrition & Metabolic Care, 2009. 12(6): p. 565-574.

[21] Caselato de Sousa, V.M., E.F. dos Santos, and V.C. Sgarbieri, The Importance of Prebiotics in Functional Foods and Clinical Practice. Food and Nutrition Sciences, 2011. 2: p. 133-144.

[22] Bosscher, D., J. Van Loo, and A. Franck, Inulin and Oligofructose as Prebiotics in the Prevention of Intestinal Infections and Diseases. Nutrition Research Reviews, 2006. 19: p. 216-226.

[23] Panesar, P.S. and S. Kumari, Lactulose: Production, Purification and Potential Applications. Biotechnology Advances, 2011. 29(6): p. 940-948.

[24] Flamm, G., et al., Inulin and Oligofructose as Dietary Fiber: A Review of the Evidence. Critical Reviews in Food Science and Nutrition, 2001. 41(5): p. 353-362.

[25] Van den Ende, W., D. Peshev, and L. De Gara, Disease Prevention by Natural Antioxidants and Prebiotics Acting as ROS Scavengers in the Gastrointestinal Tract. Trends in Food Science & Technology, 2011. 22(12): p. 689-697.
[26] Van Loo, J., et al., Functional Food Properties of Non-Digestible Oligosaccharides: A Consensus Report from the ENDO Project (DGXII AIRII-CT94-1095). British Journal of Nutrition, 1999. 81: p. 121-132.
[27] Knol, J., et al., Colon Microflora in Infants Fed Formula with Galacto- and Fructo-Oligosaccharides: More Like Breast-Fed Infants. Journal of Pediatric Gastroenterology and Nutrition, 2005. 40(1): p. 36-42.
[28] Gopal, P.K., P.A. Sullivan, and J.B. Smart, Utilisation of Galacto-Oligosaccharides as Selective Substrates for Growth by Lactic Acid Bacteria Including Bifidobacterium lactis DR10 and Lactobacillus rhamnosus DR20. International Dairy Journal, 2001. 11(1–2): p. 19-25.
[29] Kaplan, H. and R.W. Hutkins, Fermentation of Fructooligosaccharides by Lactic Acid Bacteria and Bifidobacteria. Applied and Environmental Microbiology, 2000. 66(6): p. 2682–2684.
[30] Kaplan, H. and R.W. Hutkins, Metabolism of Fructooligosaccharides by Lactobacillus paracasei 1195. Applied and Environmental Microbiology, 2003. 69(4): p. 2217-2222.
[31] Searle, L.E.J., et al., Purified Galactooligosaccharide, Derived from a Mixture Produced by the Enzymic Activity of Bifidobacterium bifidum, Reduces Salmonella enterica Serovar Typhimurium Adhesion and Invasion In Vitro and In Vivo. Journal of Medical Microbiology, 2010. 59(12): p. 1428-1439.
[32] Shoaf, K., et al., Prebiotic Galactooligosaccharides Reduce Adherence of Enteropathogenic Escherichia coli to Tissue Culture Cells. Infection and Immunity, 2006. 74(12): p. 6920–6928.
[33] Mountzouris, K.C., et al., Modeling of Oligodextran Production in an Ultrafiltration Stirred-Cell Membrane Reactor. Enzyme and Microbial Technology, 1999. 24(1–2): p. 75-85.
[34] Rycroft, C.E., et al., A Comparative In Vitro Evaluation of the Fermentation Properties of Prebiotic Oligosaccharides. Journal of Applied Microbiology, 2001. 91(5): p. 878-887.
[35] Kaneko, T., et al., Effects of Isomaltooligosaccharides with Different Degrees of Polymerization on Human Fecal Bifidobacteria. Bioscience, Biotechnology, and Biochemistry, 1994. 58(12): p. 2288-2290.
[36] Saito, Y., T. Takano, and I. Rowland, Effects of Soybean Oligosaccharides on the Human Gut Microflora in In Vitro Culture. Microbial Ecology in Health and Disease, 1992. 5(2): p. 105-110.
[37] Sharp, R., S. Fishbain, and G.T. Macfarlane, Effect of Short-Chain Carbohydrates on Human Intestinal Bifidobacteria and Escherichia coli In Vitro. Journal of Medical Microbiology, 2001. 50(2): p. 152-160.
[38] Hopkins, M.J., J.H. Cummings, and G.T. Macfarlane, Inter-Species Differences in Maximum Specific Growth Rates and Cell Yields of Bifidobacteria Cultured on Oligosaccharides and Other Simple Carbohydrate Sources. Journal of Applied Microbiology, 1998. 85(2): p. 381-386.
[39] Hopkins, M.J. and G.T. Macfarlane, Changes in Predominant Bacterial Populations in Human Faeces with Age and with Clostridium difficile Infection. Journal of Medical Microbiology, 2002. 51(5): p. 448-454.

[40] Fuentes-Zaragoza, E., et al., Resistant Starch as Prebiotic: A Review. Starch - Starke, 2011. 63(7): p. 406-415.

[41] Hernandez-Hernandez, O., et al., Characterization of Galactooligosaccharides Derived from Lactulose. Journal of Chromatography A, 2011. 1218(42): p. 7691-7696.

[42] Lesmes, U., et al., Effects of Resistant Starch Type III Polymorphs on Human Colon Microbiota and Short Chain Fatty Acids in Human Gut Models. Journal of Agricultural and Food Chemistry, 2008. 56(13): p. 5415-5421.

[43] Hur, S.J., et al., In Vitro Human Digestion Models for Food Applications. Food Chemistry, 2011. 125(1): p. 1-12.

[44] Kong, F. and R.P. Singh, A Human Gastric Simulator (HGS) to Study Food Digestion in Human Stomach. Journal of Food Science, 2010. 75(9): p. E627-E635.

[45] Macfarlane, G.T. and S. Macfarlane, Models For Intestinal Fermentation: Association between Food Components, Delivery Systems, Bioavailability and Functional Interactions in the Gut. Current Opinion in Biotechnology, 2007. 18(2): p. 156-162.

[46] Benshitrit, R.C., et al., Development of Oral Food-Grade Delivery Systems: Current Knowledge and Future Challenges. Food & Function, 2012. 3(1): p. 10-21.

[47] Molly, K., et al., Validation of the Simulator of the Human Intestinal Microbial Ecosystem (SHIME) Reactor Using Microorganism-associated Activities. Microbial Ecology in Health and Disease, 1994. 7(4): p. 191-200.

[48] De Boever, P., B. Deplancke, and W. Verstraete, Fermentation by Gut Microbiota Cultured in a Simulator of the Human Intestinal Microbial Ecosystem Is Improved by Supplementing a Soygerm Powder. The Journal of Nutrition, 2000. 130(10): p. 2599-2606.

[49] Gmeiner, M., et al., Influence of a Synbiotic Mixture consisting of Lactobacillus acidophilus 74-2 and a Fructooligosaccharide Preparation on the Microbial Ecology Sustained in a Simulation of the Human Intestinal Microbial Ecosystem (SHIME Reactor). Applied Microbiology and Biotechnology, 2000. 53(2): p. 219-223.

[50] Minekus, M., et al., A Multicompartmental Dynamic Computer-Controlled Model Simulating the Stomach and Small Intestine. Atla Alternatives To Laboratory Animals, 1995. 23(2): p. 197-209.

[51] Macfarlane, G.T., S. Macfarlane, and G.R. Gibson, Validation of a Three-Stage Compound Continuous Culture System for Investigating the Effect of Retention Time on the Ecology and Metabolism of Bacteria in the Human Colon. Microbial Ecology, 1998. 35(2): p. 180-187.

[52] Delzenne, N.M., et al., Targeting Gut Microbiota in Obesity: Effects of Prebiotics and Probiotics. Nature Reviews Endocrinology, 2011. 7(11): p. 639-646.

[53] Chichlowski, M., et al., The Influence of Milk Oligosaccharides on Microbiota of Infants: Opportunities for Formulas. Annual Review of Food Science and Technology, 2011. 2(1): p. 331-351.

[54] Marques, T.M., et al., Programming Infant Gut Microbiota: Influence of Dietary and Environmental Factors. Current Opinion in Biotechnology, 2010. 21(2): p. 149-156.

[55] Morelli, L., Postnatal Development of Intestinal Microflora as Influenced by Infant Nutrition. The Journal of Nutrition, 2008. 138(9): p. 1791S-1795S.

[56] Adlerberth, I. and A.E. Wold, Establishment of the Gut Microbiota in Western Infants. Acta Pædiatrica, 2009. 98(2): p. 229-238.

[57] Penders, J., et al., Factors Influencing the Composition of the Intestinal Microbiota in Early Infancy. Pediatrics, 2006. 118(2): p. 511-521.
[58] Palmer, C., et al., Development of the Human Infant Intestinal Microbiota. PLoS Biol, 2007. 5(7): p. e177.
[59] Harmsen, H.J.M., et al., Analysis of Intestinal Flora Development in Breast-Fed and Formula-Fed Infants by Using Molecular Identification and Detection Methods Journal of Pediatric Gastroenterology & Nutrition, 2000. 30(1): p. 61-67.
[60] Rubaltelli, F.F., et al., Intestinal Flora in Breast- and Bottle-Fed Infants. Journal of Pediatric Gastroenterology & Nutrition, 1998. 26(3): p. 186–191.
[61] Moro, G., et al., Dosage-Related Bifidogenic Effects of Galacto- and Fructooligosaccharides in Formula-Fed Term Infants. Journal of Pediatric Gastroenterology and Nutrition, 2002. 34(3): p. 291-295.
[62] Moro, G.E., et al., Effects of a New Mixture of Prebiotics on Faecal Flora and Stools in Term Infants. Acta Pædiatrica, 2003. 92: p. 77-79.
[63] Fanaro, S., et al., Galacto-Oligosaccharides and Long-Chain Fructo-Oligosaccharides as Prebiotics in Infant Formulas: A Review. Acta Paediatrca Supplement, 2005. 94(449): p. 22–26.
[64] Boehm, G., et al., Supplementation of a Bovine Milk Formula with an Oligosaccharide Mixture Increases Counts of Faecal Bifidobacteria in Preterm Infants. Archives of Disease in Childhood - Fetal and Neonatal Edition, 2002. 86(3): p. F178-F181.
[65] Haarman, M. and J. Knol, Quantitative Real-Time PCR Assays To Identify and Quantify Fecal Bifidobacterium Species in Infants Receiving a Prebiotic Infant Formula. Applied and Environmental Microbiology, 2005. 71(5): p. 2318-2324.
[66] Knol, J., et al., Increase of Faecal Bifidobacteria due to Dietary Oligosaccharides Induces a Reduction of Clinically Relevant Pathogen Germs in the Faeces of Formula-Fed Preterm Infants. Acta Paediatrca Supplement, 2005. 94(449): p. 31-33.
[67] Bouhnik, Y., et al., Administration of Transgalacto-Oligosaccharides Increases Fecal Bifidobacteria and Modifies Colonic Fermentation Metabolism in Healthy Humans. The Journal of Nutrition, 1997. 127(3): p. 444-448.
[68] Salminen, S. and E. Salminen, Lactulose, Lactic Acid Bacteria, Intestinal Microecology and Mucosal Protection. Scandinavian Journal of Gastroenterolgy Supplement, 1997. 222: p. 45-48.
[69] Den Hond, E., B. Geypens, and Y. Ghoos, Effect of High Performance Chicory Inulin on Constipation. Nutrition Research, 2000. 20(5): p. 731-736.
[70] Biagi, E., et al., Ageing of the Human Metaorganism: The Microbial Counterpart. AGE, 2012. 34(1): p. 247-267.
[71] Tiihonen, K., A.C. Ouwehand, and N. Rautonen, Human Intestinal Microbiota and Healthy Ageing. Ageing Research Reviews, 2010. 9(2): p. 107-116.
[72] Cohen, J.E., Human Population: The Next Half Century. Science, 2003. 302(5648): p. 1172-1175.
[73] Christensen, K., et al., Ageing Populations: The Challenges Ahead. The Lancet, 2009. 374(9696): p. 1196-1208.
[74] Woodmansey, E.J., Intestinal Bacteria and Ageing. Journal of Applied Microbiology, 2007. 102(5): p. 1178-1186.

[75] Guigoz, Y., et al., Effects of Oligosaccharide on the Faecal Flora and Non-Specific Immune System in Elderly People. Nutrition research, 2002. 22(1): p. 13-25.
[76] Bartosch, S., et al., Microbiological Effects of Consuming a Synbiotic Containing Bifidobacterium bifidum, Bifidobacterium lactis, and Oligofructose in Elderly Persons, Determined by Real-Time Polymerase Chain Reaction and Counting of Viable Bacteria. Clinical Infectious Diseases, 2005. 40(1): p. 28-37.
[77] Bouhnik, Y., et al., Four-Week Short Chain Fructo-Oligosaccharides Ingestion Leads to Increasing Fecal Bifidobacteria and Cholesterol Excretion in Healthy Elderly Volunteers. Nutrition Journal, 2007. 6(1): p. 42.
[78] Vulevic, J., et al., Modulation of the Fecal Microflora Profile and Immune Function by a Novel Trans-Galactooligosaccharide Mixture (B-GOS) in Healthy Elderly Volunteers. The American Journal of Clinical Nutrition, 2008. 88(5): p. 1438-1446.
[79] Schiffrin, E.J., et al., Systemic Inflammatory Markers in Older Persons: The Effect of Oral Nutritional Supplementation with Prebiotics. Journal of Nutrition Health & Aging, 2007. 11(6): p. 475-479.
[80] Fava, F., et al., The Gut Microbiota and Lipid Metabolism: Implications for Human Health and Coronary Heart Disease. Current Medicinal Chemistry, 2006. 13(25): p. 3005-3021.
[81] Guarner, F. and J.-R. Malagelada, Gut Flora in Health and Disease. The Lancet, 2003. 361(9356): p. 512-519.
[82] Saavedra, J.M. and A. Tschernia, Human Studies with Probiotics and Prebiotics: Clinical Implications. British Journal of Nutrition, 2002. 87: p. S241-S246.
[83] Louis, P. and H.J. Flint, Diversity, Metabolism and Microbial Ecology of Butyrate-Producing Bacteria from the Human Large Intestine. FEMS Microbiology Letters, 2009. 294(1): p. 1-8.
[84] McGarr, S.E., J.M. Ridlon, and P.B. Hylemon, Diet, Anaerobic Bacterial Metabolism, and Colon Cancer: A Review of the Literature. Journal of Clinical Gastroenterology, 2005. 39(2): p. 98-109.
[85] Langlands, S.J., et al., Prebiotic Carbohydrates Modify the Mucosa Associated Microflora of the Human Large Bowel. Gut, 2004. 53(11): p. 1610-1616.
[86] Hotamisligil, G.S., Inflammation and Endoplasmic Reticulum Stress in Obesity and Diabetes. International Journal of Obesity, 2008. 32(S7): p. S52-S54.
[87] Shoelson, S.E. and A.B. Goldfine, Getting Away from Glucose: Fanning the Flames of Obesity-Induced Inflammation. Nature Medicine, 2009. 15(4): p. 373-374.
[88] Cani, P.D. and N.M. Delzenne, Interplay Between Obesity and Associated Metabolic Disorders: New Insights into the Gut Microbiota. Current Opinion in Pharmacology, 2009. 9(6): p. 737-743.
[89] Cani, P.D., et al., Improvement of Glucose Tolerance and Hepatic Insulin Sensitivity by Oligofructose Requires a Functional Glucagon-Like Peptide 1 Receptor. Diabetes, 2006. 55(5): p. 1484-1490.
[90] Cani, P.D., et al., Oligofructose Promotes Satiety in Rats Fed a High-Fat Diet: Involvement of Glucagon-like Peptide-1. Obesity Research, 2005. 13(6): p. 1000-1007.
[91] Cani, P.D., et al., Changes in Gut Microbiota Control Inflammation in Obese Mice through a Mechanism involving GLP-2-Driven Improvement of Gut Permeability. Gut, 2009. 58: p. 1091–1128.

[92] Daubioul, C., et al., Dietary Fructans, but Not Cellulose, Decrease Triglyceride Accumulation in the Liver of Obese Zucker fa/fa Rats. The Journal of Nutrition, 2002. 132(5): p. 967-973.

[93] Daubioul, C.A., et al., Dietary Oligofructose Lessens Hepatic Steatosis, but Does Not Prevent Hypertriglyceridemia in Obese Zucker Rats. The Journal of Nutrition, 2000. 130(5): p. 1314-1319.

[94] Delmee, E., et al., Relation between Colonic Proglucagon Expression and Metabolic Response to Oligofructose in High Fat Diet-Fed Mice. Life Sciences, 2006. 79(10): p. 1007-1013.

[95] Roberfroid, M., et al., Prebiotic Effects: Metabolic and Health Benefits. British Journal of Nutrition, 2010. 104: p. S1-S63

[96] Abrams, S.A., et al., Effect of Prebiotic Supplementation and Calcium Intake on Body Mass Index. The Journal of Pediatrics, 2007. 151(3): p. 293-298.

[97] Maurer, A.D., et al., Changes in Satiety Hormones and Expression of Genes Involved in Glucose and Lipid Metabolism in Rats Weaned Onto Diets High in Fibre or Protein Reflect Susceptibility to Increased Fat Mass in Adulthood. The Journal of Physiology, 2009. 587(3): p. 679-691.

[98] Maurer, A., et al., Consumption of Diets High in Prebiotic Fiber or Protein During Growth Influences the Response to a High Fat and Sucrose Diet in Adulthood in Rats. Nutrition & Metabolism, 2010. 7(1): p. 77.

[99] Kalliomaki, M., et al., Early Differences in Fecal Microbiota Composition in Children May Predict Overweight. The American Journal of Clinical Nutrition, 2008. 87(3): p. 534-538.

[100] Cani, P.D., C. Dewever, and N.M. Delzenne, Inulin-Type Fructans Modulate Gastrointestinal Peptides Involved in Appetite Regulation (Glucagon-Like Peptide-1 and Ghrelin) in Rats. British Journal of Nutrition, 2004. 92(3): p. 521-526.

[101] Wu, X., et al., Molecular Characterisation of the Faecal Microbiota in Patients with Type II Diabetes. Current Microbiology, 2010. 61(1): p. 69-78.

[102] Delzenne, N.M. and C.M. Williams, Prebiotics and Lipid Metabolism. Cuurent Opinion in Lipidology, 2002. 13(1): p. 61-67.

[103] Cani, P.D., et al., Metabolic Endotoxemia Initiates Obesity and Insulin Resistance. Diabetes, 2007. 56(7): p. 1761-1772.

[104] Cani, P.D. and N.M. Delzenne, The Gut Microbiome as Therapeutic Target. Pharmacology & Therapeutics, 2011. 130(2): p. 202-212.

[105] Wong, J.M.W., et al., Colonic Health: Fermentation and Short Chain Fatty Acids. Journal of Clinical Gastroenterology, 2006. 40(3): p. 235-243.

[106] Gao, Z., et al., Butyrate Improves Insulin Sensitivity and Increases Energy Expenditure in Mice. Diabetes, 2009. 58(7): p. 1509-1517.

[107] Cani, P.D., et al., Oligofructose Promotes Satiety in Healthy Human: A Pilot Study. European Journal of Clinical Nutrition, 2005. 60(5): p. 567-572.

[108] Cani, P.D., et al., Gut Microbiota Fermentation of Prebiotics Increases Satietogenic and Incretin Gut Peptide Production with Consequences for Appetite Sensation and Glucose Response After a Meal. The American Journal of Clinical Nutrition, 2009. 90(5): p. 1236-1243.

[109] Whelan, K., et al., Appetite During Consumption of Enteral Formula as a Sole Source of Nutrition: The Effect of Supplementing Pea-Fibre and Fructo-Oligosaccharides. British Journal of Nutrition, 2006. 96(2): p. 350-356.

[110] Chaudhri, O.B., B.C.T. Field, and S.R. Bloom, Gastrointestinal Satiety Signals. International Journal of Obesity, 2008. 32(S7): p. S28-S31.

[111] Cowley, M.A., et al., The Distribution and Mechanism of Action of Ghrelin in the CNS Demonstrates a Novel Hypothalamic Circuit Regulating Energy Homeostasis. Neuron, 2003. 37(4): p. 649-661.

[112] Druce, M.R., C.J. Small, and S.R. Bloom, Minireview: Gut Peptides Regulating Satiety. Endocrinology, 2004. 145(6): p. 2660-2665.

[113] Wynne, K., et al., Appetite Control. Journal of Endocrinology, 2005. 184(2): p. 291-318.

[114] Freeland, K.R., C. Wilson, and T.M. Wolever, Adaptation of Colonic Fermentation and Glucagon-Like Peptide-1 Secretion with Increased Wheat Fibre Intake for 1 Year in Hyperinsulinaemic Human Subjects. British Journal of Nutrition, 2010. 103(1): p. 82-90.

[115] Freeland, K.R. and T.M. Wolever, Acute Effects of Intravenous and Rectal Acetate on Glucagon-Like Peptide-1, Peptide YY, Ghrelin, Adiponectin and Tumour Necrosis Factor-Alpha. British Journal of Nutrition, 2010. 103(3): p. 460-466.

[116] Piche, T., et al., Colonic Fermentation Influences Lower Esophageal Sphincter Function in Gastroesophageal Reflux Disease. Gastroenterology, 2003. 124(4): p. 894-902.

[117] Parnell, J.A. and R.A. Reimer, Weight Loss During Oligofructose Supplementation is Associated with Decreased Ghrelin and Increased Peptide YY in Overweight and Obese Adults. The American Journal of Clinical Nutrition, 2009. 89(6): p. 1751-1759.

[118] Tarini, J. and T.M.S. Wolever, The Fermentable Fibre Inulin Increases Postprandial Serum Short-Chain Fatty Acids and Reduces Free-Fatty Acids and Ghrelin in Healthy Subjects. Applied Physiology, Nutrition, and Metabolism, 2010. 35(1): p. 9-16.

[119] Woodmansey, E.J., et al., Comparison of Compositions and Metabolic Activities of Fecal Microbiotas in Young Adults and in Antibiotic-Treated and Non-Antibiotic-Treated Elderly Subjects. Applied and Environmental Microbiology, 2004. 70(10): p. 6113-6122.

[120] Mueller, S., et al., Differences in Fecal Microbiota in Different European Study Populations in Relation to Age, Gender, and Country: A Cross-Sectional Study. Applied and Environmental Microbiology, 2006. 72(2): p. 1027-1033.

[121] Claesson, M.J., et al., Comparative Analysis of Pyrosequencing and a Phylogenetic Microarray for Exploring Microbial Community Structures in the Human Distal Intestine. PLoS ONE, 2009. 4(8): p. e6669.

[122] Mariat, D., et al., The Firmicutes/Bacteroidetes ratio of the human microbiota changes with age. BMC Microbiology, 2009. 9(1): p. 123.

[123] Reinhardt, C., C.S. Reigstad, and F. Backhed, Intestinal Microbiota During Infancy and Its Implications for Obesity. Journal of Pediatric Gastroenterology and Nutrition, 2009. 48(3): p. 249-256

[124] Biasucci, G., et al., Cesarean Delivery May Affect the Early Biodiversity of Intestinal Bacteria. The Journal of Nutrition, 2008. 138(9): p. 1796S-1800S.

Chapter 3

Conjugated Linoleic and Linolenic Acid Production by Bacteria: Development of Functional Foods

Carina Paola Van Nieuwenhove, Victoria Terán and Silvia Nelina González

Additional information is available at the end of the chapter

http://dx.doi.org/10.5772/50321

1. Introduction

Over the years, the biological significance of conjugated fatty acids has been demonstrated. Among them, there are two that are present naturally in milk and dairy products, from ruminant origin which have been intensively studied in recent times: conjugated linoleic acid and conjugated linolenic acid.

Conjugated linoleic acid (CLA) refers to a mixture of positional and geometric isomers of linoleic acid (c9,c12-C18:2, LA) with a conjugated double bond. It is a natural compound mainly found in ruminant products such as meat, milk and other dairy food that represent the main source of CLA for humans. Of the two biologically important isomers, c9,t11 is the most prevalent one comprising around 80 to 90% of total CLA in ruminant products, and t10,c12 is present in lower amounts as 3-5% of total CLA [1].

CLA is formed as an intermediate product of the biohydrogenation (BH) process that occurs in rumen, as multi-step mechanism carried out by different microorganisms on unsaturated fatty acids to produce stearic acid (C18:0). Also , it can be produced by desaturation of trans vaccenic acid (t11-C18:1, TVA), process that occurs in different tissues such as mammary gland,.

Conjugated linolenic acid (CLNA) are representing by different conjugated isomers of the linolenic acid (c9, c12, c15-C18:3, LNA). It is also resulting of the ruminal microbial metabolism on fatty acids present in foods, but they are also found in some plant seed oils, like pomegranate seed oil rich in punicic acid (c9,t11, c13-CLNA) [2-3] and tung seed oil where α-eleostearic acid (c9,t11,t13-CLNA) content is about 70% [2-3].

Both conjugated fatty acids has undoubted effects on health, with important biological functions demonstrated in animal models, making them a target of intensive study.

Over the years, CLA has received great attention due to their beneficial properties on health. There exist near 28 different CLA isomers produced by natural and industrial process during fatty acid hydrogenation [4], but the most important according to their biological effects are c9, t11 and t10,c12 forms. However, CLA isomer in milk fat according to importance are c9,t11 (around 80%) followed by t7,c9, which is quantitatively the second most important reaching level so high as 3 to 16 % of total CLA [5]. Factors affecting CLA content in milk, such as the food of ruminants [6-7], the animal breeding type and the stage of lactation [8] were widely reported.

Many studies demonstrated the action of CLA as anti-carcinogenic [9], anti-diabetic [10] and immune-modulator [11] compound. Although there is no agreement regarding its function on fat metabolism, some authors revealed that its consumption also decreases the fat deposition [12].

In addition, CLA produced trough chemical isomerization of LA is offer as dietary supplement in many countries. However, unexpected isomers are produced by this process. To consider CLA as a nutraceutical or medicinal compound, a selective isomer production must be done.

On the other hand, CLNA showed anti-carcinogenesis effects *in vitro* and *in vivo* models [9, 13-14] and other isomers were reported as hypolipidemic compound in human liver derived HepG2 cells [15]. Moreover, it was demonstrated that CLNA exhibits stronger cytotoxic effect on tumoral cells than CLA isomers [16].

Since the most important sources of both conjugated fatty acids for human consumption are milk and dairy products, and due to the microbial production of these compounds, several attempts are being developed to increase its content in food using natural process for it production. In the field of human and animal health, it is interesting to understand the potential beneficial role of selection of bacteria with the ability to form conjugated fatty acids to be then included in foods. Thus, the processed products could be considered as functional foods and sometimes as probiotics, as we detailed below.

2. Ruminal production of conjugated linoleic and linolenic acid

Fatty acids are present in forages and concentrate feeds, mainly as esterified form, mostly present as phospholipids and glycolipids in forrages and triglycerides in plant seeds, comonly used in concentrates.

The two most abundant fatty acids from animal diet are linoleic and linolenic acids. Both are incorporated through diet and once they reach the rumen, are extensively modified by microbial enzymes, such as lipases. These enzymes produce as results LA and LNA as free form for further reactions of isomerization and hydrogenation.

The biohydrogenation of both fatty acids occur in a similar manner, but differ in the intermediate products, as shown in *Figure 1*.

Hydrogenation of linoleic acid produces as first intermediate c9, t11-CLA isomer, by a process where the double bond at carbon-12 position is transferred to carbon-11, carried out by linoleate isomerase (EC 5.3.1.5, LAI). The second step, is the rapid conversion to t11-C18:1 (trans vaccenic acid, TVA) by a reduction mechanism and further hydrogenated to stearic acid (C18:0) [17].

The other CLA isomer resulting of rumen metabolism is t10, c12, which is produced by different microorganism such as *Butyrivibrio fibrisolvens* [18] and *Megasphaera eldsenii* YJ-4 [19]. But the hidrogenation of this isomer not produced TVA but c6, t10-C18:2 which is further converted to C18:0. According to authors, some bacteria can produce hydroxiacids previously to its conversion to CLA isomers [20-21].

As we mentioned above, the other fatty acid of importance in ruminant feeding is linolenic acid, which is also converted to C18:0 by microbial action. In this pathway, LNA is isomerized at *cis*-12 position forming as first intermediate product c9,t11,c15 isomer, named conjugated linolenic acid (CLNA). This compound is further reduced to t11,c15-C18:2 and after that converted to three different products: t11-C18:1; c15-C18:1 and t15-C18:1. As shown in *Figure 1*, only t11-C18:1 is reduced up to C18:0.

Note that metabolic pathway of both LA and LNA fatty acids produce TVA as result of a reduction process, forming conjugated fatty acids as intermediate products. All conjugated fatty acids are absorbed by intestine cells, reason why they are further present in milk and meat fat [22-23].

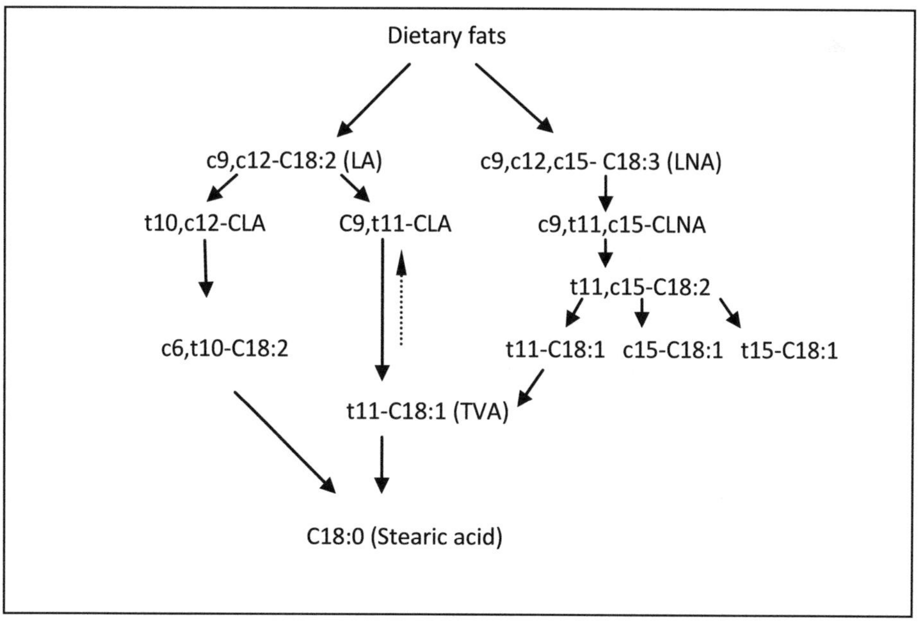

Figure 1. Biohydrogenation process of linoleic and linolenic fatty acid in rumen (adapted from Harfoot and Hazlewood [17]) and endogenous synthesis of CLA in mammary gland (dotted arrow).

Of the rumen microorganism, bacteria are largely responsible for biohydrogenation of unsaturated fatty acids and protozoa seem to be of only minor importance [17].

Kemp and Lander [24] divided bacteria into two groups according to reactions and end products of biohydrogenation. Group A includes those bacteria able to hydrogenate linoleic acid and α-linolenic acid producing t11- C18:1 as an end product. On the other hand, Group B bacteria including those able to use t11-C18:1 as one of the main substrates to produce stearic acid as end product. A listing of the bacteria species of both groups is provided in the review by Harfoot and Hazlewood [17].

Instead of ruminal biohydrogenation, there is another CLA synthesis pathway carried out through the Δ9-desaturase enzyme activity on trans-vaccenic acid (t11-C18:1-TVA) in different tissues, especially in the mammary gland [25]. This endogenous synthesis of CLA is the responsible for most of CLA level found in milk fat, being according to findings around a 64 % [25] to > 80% [26].

But other pathway of CLNA isomers synthesis, in addition to ruminal production, was not yet evidenced. For that reason, its content in ruminant milk apparently comes exclusively from BH of linolenic acid and diet.

As results of microorganism metabolism, many isomers originated in rumen are present in milk. Due to biological properties of both conjugated fatty acids researchers are looking to develop natural foods enriched in CLA and CLNA and thus increase daily intake by humans.

As it was previously mentioned, ruminant milk and meat are the most abundant sources of CLA for humans. Different studies have demonstrated that CLA content of ruminant milk and meat products varies between 4-6 mg/g fat [27-29]. From this value, near the 80 to 90% corresponds to the c9,t11 isomer [30-31]. However, the concentration of CLA can vary widely, where differences are largely related to diet. So, milk fatty acid profile can be modified according to animal feeding.

In the last years, different supplements such as vegetal oils, animal fat, natural pasture and seeds were used to improve fatty acid profile of milk, to attempt higher levels of CLA [32-33] or CLNA [22, 34].

Respect to CLA synthesis in non-ruminant animal, an increase on CLA content in tissues was evidenced in studies using rats [35] and mice [23] after TVA supplementation.

In humans, an endogenous synthesis of CLA was also shown by Adolf *et al.* [36]. But tisular human production in human tissues is so low, that the concentration found in tissues are directly related to food consumption.

Although ruminant foods are the richest source of CLA for humans, it is also found in monogastric animal products, such as swine [37], chicken [30], turkey [30], fish and rabbit [38] meat but in much lower levels. Among south-american camelids, CLA was determined in llama´s (*Lama glama*) milk [39]. Vegetable oils contain little CLA and according to some authors no CLA content were evidenced in vegetal oils. Typical values of CLA in non-ruminant foods are given in *Table 1.*

Product	Total CLA (mg/ 100 g of fat)	Author
Meat		
Turkey	0.25	Chin *et al.* [30]
Fish	0.01-0.09	Fristche and Steinhart [105]
Swine	0.3-0.9	Ross *et al.*[37]
Rabbit	0.11	Fristche and Steinhart [105]
Chicken	0.09-0.2	Chin *et al.* [30]
Eggs yolk	N.D N.D	Raes *et al.* [106] Gultemiriam *et al.* [107]
Milk		
Human	0.1	Park *et al.* [108]
Horse	0.05-0.12	Jahreis *et al.* [8]
Sow	0.19-0.27	Jahreis *et al.* [8]
Llama *(Lama glama)*	0.7	Schoos *et al.* [39]
Vegetable oils	<0.01	Fristche and Steinhart [105]

ND: not determined

Table 1. CLA content in non-ruminant foods

So as CLA, different CLNA isomers occur naturally, some of which could be formed by ruminal biohydrogenation and further incorporated into milk and meat fat.

Only a few studies were done respect to CLNA content in ruminant products and according to data informed, the only isomer present in cow milk is c9,t11,c15 form [40] while in muscle is also present c9,t13,c15 isomer [40].

CLNA content in milk is around of 0.3-0.39 mg/g fat [23, 40]. At the present, the effect of diet on CLNA concentration in milk was only reported by one work, where cows not fed with extruded linseed (control) have no CLNA in milk, but linseed supplementation in diet increased both CLA and CLNA content, reaching the latest fatty acid a value of 0.15% of total fatty acids [22]. In this study, CLNA was also present only as c9, t11, c15 isomer.

CLNA content in non-ruminant products were determined in different seed oils, being the most abundant source of these fatty acid isomers *(Table 2).* Moreover, tung, pomegranate and catalpa oils showed high level of CLNA but in different isomer ratio. On this way, punicic acid (c9, t11, c15-CLNA) is contained about 72% in pomegranate seed oil [3]. In bitter gourd oil and tung seed oil the main isomer present correspond to α-eleostearic acid (c9,t11,t13-CLNA) in about 60% and 70%, respectively [3, 41]. Catalpa seed oil contains CLNA at a level of 31 %, found as catalpic acid (t9, t11, c12-CLNA) isomer.

Product	CLNA (%)	Author
Pomegranate oil	75 86	Suzuki *et al.* [3] Yücel *et al.* [2]
Catalpa oil	27.5 31	Yücel *et al.*[2] Suzuki *et al.* [3]
Bitter gourd oil	60	Yücel *et al.* [2]; Suzuki *et al.* [3]
Tung oil	70	Suzuki *et al.* [3]
Cow milk	0.3-0.39	Loor *et al.* [23] Plourde *et al.* [40]

Table 2. CLNA content in cow milk and seed oils

3. CLA recommended human intake

CLA concentration in dairy products widely varied according to data reported (0.55–9.12 mg/g fat), but even though are lower than required to achieve a biological effect in humans.

Biological properties after CLA administration is depending on isomer and doses administered and the period of study. Those, studies on animal models reported anti-atherosclerosis effect after 0.1-1% of total CLA per day to rabbits [42]. Moreover, anti-carcinogenic effect was determined by authors using levels from 0.5% to 4% into the diet [43-44].

Although the action mechanism is not well understood, CLA was reported as antioxidant compound in animals and *in vitro* models [15].

Just as there are variations in experimental models about effective doses of CLA, depending on animal model and the biological effect evaluated, the recommended dose from human daily intake also widely varied.

In general, by extrapolation of results found in animals, the recommended CLA daily intake is around 0.35 to 1 g/day [15]. Some authors estimated a daily dose of 650 mg [45], but other studies considered that higher doses (3.0 to 4.2 g/day) are adequate to reduce body fat mass [46-47].

However, at the present the real consumption in different countries is lower than recommended dose. Studies on German population estimated a daily CLA intake of 0.35 to 0.43 g for men and women, respectively [38]. In other countries, CLA daily intake was informed so lower as 120 to 140 mg per day [27].

A few epidemiological studies were done in humans, and evidence show that no all isomers are absorbed to a similar extent. According to result is difficult to predict the impact of CLA consumption on humans and the preventive effect of isomers.

Thus, a short-term (4 to 12 weeks) human studies showed that 2.2 g/d, administered as a mixture of c9,t11 and t10,c12 isomers, produces a decrease on inflammatory markers [48]. A higher dose (3 g/d) were used by Moloney *et al.* [49] who found an increase on HDL levels and a decrease on the ratio of LDL cholesterol to HDL cholesterol, but did not show positive effect on insulin levels in diabetics patients.

Smedman *et al.* [50] reported a reduction of body fat in humans after consumption of 4.2 g/d of a mixture of CLA isomer (c9,t11 and t10,c12) during 12 weeks.

Even though there are many positive findings about CLA supplementation by animals, some negative aspects were informed by other authors, such as the induction of fatty liver and spleen and resistance to insulin [51].

Studies concerning to increase CLA content in foods receives great attention since bacterial inclusion improves CLA levels in some fermented dairy products or could generate CLA at intestinal level after a probiotic administration. In this way, studies on bacterial CLA or CLNA production are relevant in this field, and are detailed in this chapter.

4. Bacterial CLA and CLNA production

The ruminal anaerobe *Butyrivibrio fibrisolvens* was the first bacteria were CLA production was evidenced [18]. After years, it was revealed that not only ruminal bacteria were able to form CLA. So, microorganisms isolated from dairy products, human and animal intestine were demonstrated as CLA-producing bacteria, including lactic acid bacteria (LAB) and bifidobacteria. At the present, *Lactobacillus reuteri, Lactobacillus rhamnosus, Lactobacillus plantarum; Lactobacillus brevis, Lactobacillus acidophilus; Lactococcus lactis, Propionibaterium freudenrehichii, Bifidobacterium sp, Streptococcus,* among others, were able to form CLA [52-54].

Some years ago, it was reported the formation of another isomers of conjugated fatty acid from α and γ-linolenic in *Lactobacillus plantarum,* named as CLNA [21]. Even though this conjugated fatty acid production was reported since 2003, it was only recently informed for bifidobacterium strains [55-56].

Conjugation of linoleic and linolenic acid were proposed as a detoxification mechanism to avoid the growth inhibition effect of fatty acid on bacteria [57-58].

CLA/CLNA production varied among strains being influenced by substrate concentration, culture media, temperature and time of fermentation, among other factors. The isomer formed is also strain-dependent, showing some microorganism the production of only one isomer while others produce two or more CLA/CLNA forms.

As an example of the influence of culture condition, Ogawa et al. (2001)[59] informed CLA production by *L. acidophilus* cultured in microaerophilia conditions, but when bacteria were cultured in aerophilia conditions not CLA formation was determined.

Nowadays, different processes are being carried out to increase CLA production by strains. So, Lin et al (2005) [60] immobilized cells of Lactobacilli strains in two matrix (chitosan and poliacrilamide). In this study, *L. delbruekii* ssp. *bulgaricus* and *L. acidophilus* showed higher CLA production than not immobilized cells. Washed cell instead of growth cultures is another way to produce high CLA levels ([20, 59, 61].

Further studies informed that the uses of enzyme extract of *L. acidophilus* at 50ºC and pH 5 produce more than eight CLA isomers, being around 48% as c,t/t,c form [62].

The transformation of linoleic and linolenic acid to the conjugated form is carried out by linoleate isomerase (LAI) enzyme, which is bound to the bacterial membrane [63]. This enzyme will be treated in other section of this chapter.

At the present, the *in vitro* bioproduction of conjugated fatty acids has been shown in lactic acid bacteria (LAB), propionibacteria and bifidobacteria strains.

A different mechanism of CLA production via 10-OH-C18:1 seems to be the most common pathway in human intestine bacteria according to McIntosh *et al.* (2009) [64], who evidenced this metabolic pathway in *Roseburia, Ruminococcus* and other intestinal strains.

5. Lactic acid bacteria (LAB)

CLA production by LAB strains were informed during years. The mechanism, isomer and optimum condition for CLA formation makes these the most variable group on the literature.

Some strains as *L. plantarum* AKU 1009a were informed as CLA-producing bacteria via a two-step reaction: first the hydration of linoleic acid to 10-hydroxy-18:1, followed by dehydration of the resulting hydroxy acid to CLA. In this strain, CLA was formed as c9,t11 (CLA1) and t10,c12 (CLA2) isomers [21].

Xu *et al.* [65] also informed CLA production as c9,t11 and t10,c12 isomer of CLA, at different ratio, in LAB and propionibacteria strains cultured in a fat milk model supplemented with hydrolized soy oil for 24 to 48 h. Among these, *L. acidophilus, L. casei, L. plantarum, E. faecium, L. rhamnosus, Pediococcus (Ped.) acidilactici* and yogurt cultures (mixture of *L. delbruekii* ssp. *bulgaricus* and *Str. salivarius* ssp. *thermophilus,* 1:1 ratio) were reported as CLA-producing bacteria in the mentioned condition. Increasing time from 24 to 48 h did not increase CLA content, except in *Ped. acidilactici* and *L. rhamnosus* strains. The main isomer found was c9,t11 followed by t10,c12 after 24 h of incubation, except in *E. faecium* were t10,c12 were not determined.

The ability to produce CLA in Lactobacilli strains from human origin was also informed by Lee *et al* [66]. In this study, *L. rhamnosus, L. paracasei* and *L. pentosus* also showed different CLA isomer ratio production. So, *L. rhamnosus* and *L. pentosus* were able to transform LA to c9, t11 and t10,c12- CLA, while *L. paracasei* only produce the c9,t11 isomer.

Other study revealed six LAB able to form CLA after 24 h of incubation, varying percentage of LA conversion between 17% and 36%. Here, *L. casei, L. rhamnosus,* and *Strep. thermophilus* showed the highest LA conversion in MRS broth, and increased two- or threefold in milk than MRS broth [53]. *Strep. thermophilus* has importance by it uses as starter culture during fermentation process of dairy products.

Some authors informed a positive correlation between CLA production and tolerance to LA [53, 67] using different substrate concentration. However, the efficiency of CLA production in some LAB and *bifidobacterium* decreases at higher levels of free LA in the medium [53].

Other studies using LAB showed CLA production mainly as c9,t11 form (60-65 %), followed by t10,c12 (30-32%) and other minor isomers like t9,t11 and t10,t12 (2-5%) in *L. acidophilus*, *L. plantarum* and *Lact. lactis* cultured in MRS broth and skim milk during 24 h. [68].

In a recent work, a low CLA production was informed by strains of *L. sakei* and *L. curvatus* (1.6 % and 4.2 %, respectively), commonly present in meat fermentation as starter cultures or natural microorganism [56, 69].

The reaction sequence of isomerization of LA seems to involve different steps according to bacterial strain.

Respect to CLNA production, *L. plantarum* AKU 1009a was able to transform ricinoleic acid to CLA (CLA1 and CLA2) [20]. Further studies demonstrated that this lactobacilli strain has the capacity of use α- and γ-linoleic acids as substrate to generate the corresponding conjugated trienoic acids [21] named CALA and CGLA, respectively. Authors reported a CALA production rate of 40% under two isomer forms: c9, t11, c15-C18:3 (CALA 1, 67% of total CALA) and t9,t11,c15-C18:3 (CALA 2, 33% of total CALA). A higher CGLA production rate was determined in this study (68%) as a mixture of two isomer: c6, c9, t11-C18:3 (CGLA 1, 40 % of total CGLA) and c6, t9, t11-C18:3 (CGLA 2, 60% of total CGLA).

Recently, determination of CLNA production by other LAB strains were informed [69]. Among these, a high production levels were determined in *L. sakei* and *L. curvatus,* reaching a percentage of conversion of 22.4 % and 60.1 %, respectively. Authors evidenced that the isomerization process of LA to CLA and LNA to CLNA is different according to LAB strain, so as isomer resulting after culturing. Some microorganisms were able to form both conjugated fatty acids, but predominantly convert LNA to CLNA, while others not were able to form CLA but effectively converted LNA to CLNA. Results are given in *Table 3*.

Strain	c9,t11	t10,c12	Other isomer	LA conversion (%)	Author
L. curvatus	+	+	-	1.6%	Gorissen *et al.* [69]
L. plantarum	+	+	-	4.6%	Gorissen *et al.* [69]
	+	-	+	N.D	Kishino *et al.* [20]
	+	+	-	N.D	Ogawa *et al.* [109]
	+	+	-	N.D	Rodríguez-alcalá *et al.* [68]
	+	+	-	N.D	Xu *et al.* [65]
L. sakei	+	+	-	4.2	Gorissen *et al.* [69]
L. reuteri	+	+		26	Lee *et al.* [110]
L. rhamnosus	+	+	-		Lee *et al.* [66]
	+	+	-		Ogawa *et al.* [109]
	+	-	-	34	Van Nieuwenhove *et al.* [53]
L. paracasei	+	-	-	N.D	Lee *et al.* [66]
L. pentosus	+	+	-	N.D	Lee *et al.* [66]
	+	+	-		Ogawa *et al.* [109]

Strep. thermophilus	+	-	-	33	Van Nieuwenhove *et al.* [53]
L. brevis	+	+	-	N.D	Ogawa *et al.* [109]
L.curvatus	+	+	-	1.6	Gorissen *et al.* [69]
L. acidophilus	+	-	-	20	Van Nieuwenhove *et al.* [53]
	+	+	-	N.D	Ogawa *et al.* [109]
	+	+	-	N.D	Xu *et al.* [65]
L. reuteri	N.I	N.I	N.I	26	Lee *et al.* [110]
Lact. lactis	+	+	-	N.D	Rodríguez- Alcalá *et al.* [68]

+: positive production. -: no production. N.D: not determined. N.I: not informed

Table 3. CLA production by LAB strains cultured in presence of free LA

6. Propionibacteria

Propionibacteria are commonly present in milk and dairy products and some species play an important role in the creation of cheeses, such as emmental cheese. Propionibacteria represents another important group of bacteria where the capacity of LA isomerization *in vitro* was demonstrated, being relevant since it could be included in fermented products as cheeses. So, *P. freudenrehichii* was able to produce CLA mainly as c9,t11 form according to different studies [58, 65, 70-71] although other author reported eight different isomers of CLA produced by enzyme extract in this bacteria [72].

CLA production in a fat milk model supplemented with hydrolyzed soy oil for 24 to 48 h was demonstrated in two *P. freudenreichii ssp shermanii* and *P. freudenreichii* ssp *freudenreichii* [65]. Higher levels of CLA were determined in skim milk than MRS broth.

The ability of *P. acnes*, isolated from sheep, to form CLA only as t10, c12 form was also evidenced [73].

The results clearly demonstrate that propioniacteria strains show a great variability on CLA production, according to many factors as origin, species, substrate and culture conditions.

To the best of our knowledge, CLNA production by propionibacteria strains was recently evidenced by Henessy *et al.* [74]. In this work, bacteria were culture in presence of different fatty acid used as substrate to evaluate it further conversion into the conjugated form. Thus, LA, α and γ-LNA, stearidonic (c6, c9, c15-C18:4) and other polyunsaturated fatty acids were individually incorporated to culture medium. Strains of *P. freudenreiichii* ssp *shermanii* and *P. freudenreihichii* ssp *freudenreichii* were able to conjugate different PUFA, showing different percentage of conversion of each particular fatty acid. Thus, *P. freudenreiichii* ssp *shermanii* 9093 reached a rate conversion of 50.5; 53.5 and 3.09 for LA, α-LNA and stearidonic acid, respectively. On the other hand, *P. freudenreihichii* ssp *freudenreichii* Propioni-6 reached a conversion rate of 44.65; 8.94; and 3.58 for the same fatty acids. The isomerization process on γ-LNA was not evidenced for these bacteria. The increase of substrate concentration caused a decrease on the percentage of bioconversion (as is shown in *Table 4*).

Strain	c9,t11,c15	t9,t11,c15	CLA (%)	CLNA (%)	Author
L. curvatus LMG 13553	+	+	1.6%	22.4	Gorissen *et al.* [69]
L. plantarum ATCC 8014	+	+	4.6%	26.8	Gorissen *et al.* [69]
L. sakei LMG 13558	+	+	4.2	60.1	Gorissen *et al.* [69]
CG1	+	+	ND	28.4	
B. bifidum LMG 10645	+	+	40.7	78.4	Gorissen *et al.* [55]
B. breve LMG 11040	+	+	44	65.5	Gorissen *et al.* [55]
B. breve LMG 11084	+	+	53.5	72.0	Gorissen *et al.* [55]
B. breve LMG 11613	+	+	19.5	55.6	Gorissen *et al.* [55]
B. breve LMG 13194	+	+	24.2	63.3	Gorissen *et al.* [55]
B. pseudolongum ssp pseudolongum LMG 11595	+	+	42.2	62.7	Gorissen *et al.* [55]
B. breve NCIMB 8807*	+	+	66	68	Hennessy *et al.* [74]
B. breve DPC6330*	+	+	67	83	Hennessy *et al.* [74]
B. longum DPC6315*	-	-	12	0.0	Hennessy *et al.* [74]
P. freudenreichii ssp. *freudenreichii* Propioni-6 **	+	+	44.6	8.9	Hennessy *et al.* [74]
P. freudenreichii ssp. *shermanii* 9093**	+	+	50.5	53.5	Hennessy *et al.* [74]

*: production of conjugated isomers of γ-LNA and stearidonic acid were also reported. **: production of conjugated stearidonic acid was also informed. ND: not determined

Table 4. CLNA isomers production by bacteria cultured in presence of α-LNA

7. Bifidobacterium strains

Bifidobacteria are found as normal inhabitants of the human gut and is among the first colonizers of the sterile gastrointestinal tracks of newborns [75]. Due to their health's benefits on humans, it uses as probiotic strains is indubitable [76]. As results after years of investigations, many functional foods have been developed with the addition of bifidobacteria to the food matrix [77-79].

For this reason, it is not surprising that many studies on the ability of these bacteria to produce CLA have been carried out for a long time.

Bifidobacteria species able to produce CLA was reported at first time by Coakley et al. [57], who informed a considerable interspecies variation. So, *Bifidobacterium breve* and *B. dentium* were the most efficient CLA producers among the range of evaluated strains. The highest percentage of LA conversion was determined for *B. breve,* reaching a value of 65% (c9, t11-CLA). In this study, strains also varied considerably with respect to their tolerance to linoleic acid concentration in the medium.

Other authors showed that strains of *Bifidobacterium breve* and *B. pseudocatenulatum* isolated from human feces, were able to form CLA in a rate conversion of 69% and 78%, respectively [80].

Moreover, CLA production in *B. bifidum* cultured in skim milk, using as substrate hydrolyzed soy oil was reported by Xu *et al.* [65], where authors detected CLA production after 24-48 h only as c9,t11 isomer, and traces of the t10,c12 form.

In a recently study the ability to form CLA in two strains of *B. animalis* were reported [68]. Authors found CLA production from free LA and safflower oil added to MRS broth and skim milk. Strains were able to transform LA to CLA after 24-48 h of incubation. In order to abundance, the most important isomer produced was c9, t11 isomer, followed by t10, c12.

Bifidobacterium breve LMC520 can actively convert linoleic acid to c9,t11-CLA, which is the major isomer derived from microbial conversion according to results from Park *et al.* [81].

The study with the highest number of bifidobacteria were carried out by Gorissen *et al.* [55], which performed a screening of 36 different Bifidobacteria strains to investigate their ability to produce CLA and/or CLNA. As substrate they used free LA and α- LNA, revealing that only six strains were able to convert it to different conjugated fatty acid isomers. Strains were identified as a *Bifidobacterium bifidum, Bifidobacterium pseudolongum* and four *B. breve* strains, named *B. breve* LMG 11084, *B. breve* LMG 11613, *B. breve* LMG 13194, *B. bifidum* LMG 10645 and *B. pseudolongum* subsp. *pseudolongum* LMG 11595. Moreover, all strains have been shown to be more efficient in converting LNA to CLNA than LA to CLA, in percentages from 55.6% to 78.4% and 19.5% to 53.5%, respectively. In addition, the CLNA isomers that were mainly found were in order c9, t11, c15-CLNA followed by t9, t11, c15-CLNA isomer.

Hennessy *et al.* [74] also informed about isomerization process of different fatty acids by bifidobacteria strains. Moreover, different PUFA such as stearidonic, araquidonic and docosapentanoic and docosahexanoic acid were supplemented to the culture. A general patron of isomerization was determined on *B. breve* and *B. longum* strains, being able to transform LA, α and γ-LNA and stearidonic acid to it conjugated form. As was observed in propionibacteria, the percentage of conversion varied among strains, showing around 12 to 67% of LA conversion, mainly into c9, t11 and t10,c12 isomer. α- LNA was converted among 0 to 83% among strains, and lower rate conversion was determined for γ-LNA (0.5- 37%). The conjugation of stearidonic acid varied from 3.8 to 27%. *B. breve* DPC6330 was the most effective conjugated fatty acid producer, showing a bioconversion rate of 70% for LA, 90% of α-LNA, 17% for γ-LNA and 28% for stearidonic acid.

As well as different ability to isomerize fatty acids was determined in LAB and propionibacteria, bifidobacteria also exhibit a wide range of bioconversion rate. Many factors affect the mechanism of the fatty acids isomerization, such as culture conditions and substrate concentration. The production of different isomers ratio was reported for all evaluated strains.

To the best of our knowledge, this is the only work reporting the conjugation of stearidonic acid by bacteria. Results are given in *Table 4*.

8. Alternative substrate to CLA production

Although free fatty acids is the most commonly substrate employed by authors to analyze CLA or CLNA production by strains, alternative substrates are being evaluated. Many studies using vegetable oils (hydrolyzed or not hydrolyzed) and mono or dilinoleins as exogenous source of fatty acids were determined to be further incorporated to food matrix. Therefore, bacteria must have the ability to hydrolyze the triglycerides and liberate linoleic acid or linolenic acid for further conversion. Only hydrolyzed oils can offer the fatty acid as free form.

While vegetable oils are the richest source of linoleic and linolenic acid, data about the utilization of monolinolein by *B. breve* were informed. This strain, from human origin, was able to generate CLA at higher bioconversion rate than free LA or dilinolein was added to the medium [82].

CLA production in milk system models was described by many authors using vegetal oils as substrate for further isomerization. At the present, soy, sunflower, canola, castor and safflower oils were used as source of linoleic acid [20, 61, 68, 83].

Kishino *et al.*[20] determined CLA production in *L. plantarum* using castor oil and ricinoleic acid as substrate, showing the same end product than using free LA. Moreover, the production of the previously reported hydroxyacids as intermediate compounds, were also evidenced in the assay.

CLA formation by LAB and *Bifidobacterium* strain using safflower oil as LA source added to skim milk at 1 mg/ml was reported by other authors [68], where they informed that some bacteria produced higher CLA using safflower oil than free linoleic acid in skim milk broth after 24 h of incubation. Among these group of microorganism was *B. animalis, L. acidophilus* and *Lact. lactis.*

Among bacteria isolated from rumen, *L. brevis* was reported as CLA-producing strain in presence of sunflower oil [84].

However, there was informed no CLA production after the addition of soy oil to skim milk in *P. freudenreihchii; L. casei; L. acidophillus, L. plantarum; P. acidilactici, B. bifidum, L. rhamnosus* and *E. faecium* [65]. But once hydrolyzed soy oil was supplemented to the medium as substrate, CLA production from 0.6 to 2.2 mg/g fat was determined in all selected strains.

According to results, the utilization of vegetable oils by bacteria as source of fatty acid for it further isomerization is also depending on metabolism of strains.

9. LAI enzyme

As we previously mentioned, linoleate isomerase is an enzyme present in some bacteria, which is bound to the membrane. In the most of bacteria, CLA production is primarily located in the extracellular phase [71] but it can be also found in the cellular membrane as an structural lipid [80]. Moreover, both LA and CLA incorporated to the membrane represent less than 1.7% of the total amount of CLA formed [80].

CLNA isomers are also found primarily in the cell-free supernatants compared to the cell pellet [55]. Authors reported that around 7% of LNA and 5% of CLNA corresponding to the cellular pellet in bacteria cultured in presence of LNA.

For this reason, the analysis methods of fatty acids of cultures did not involve the remotion of bacterial cells. Moreover, total fatty acids content is necessary to determine the complete bioavailability of those compounds once bacteria are included in a food matrix or is considered as probiotic strain.

At the present, LAI has been isolated from bacteria such as *L. delbrueckii* subsp. *bulgaricus, But. fibrisolvens, L. acidophilus* and *P. freudenreichii* subsp. *shermanii* showing some differences.

LAI from *But. fibrisolvens* A-38 was isolated by Park *et al [85]*, determining the molecular weight and partial amino acid sequence of the enzyme. According to findings, this LAI consist of a single polypeptide with a molecular weight of 19 kDa.

Other authors isolated and characterized the LAI from *L. reuteri* MRS8 [86], showing a molecular weight of more than 100 kDa. In this study, the optimal activity of the enzyme was in the pH range of 4.7 to 5.4.

A genotypic identification of LAI gene from ten strains able to form CLA and/or CLNA was recently performed [69]. This work presented the homologies of LAI sequence in a dendrogram comparing to other LAI sequence from known LAB.

Moreover, the molecular weight forms LAI from *L. reuteri, P. acnes* and *C. sporogenes* were 68 kDa, 45 kDa and 55 kDa, respectively [87-88].

10. Functional foods and probiotics

The development of healthier food is looking for taking in account their benefits for humans. Among these, dairy products represent a good alternative to manufacture functional and/or probiotic foods. Functional food includes processed food or foods fortified with health-promoting additives. By other hand, probiotics are live microorganisms which when administered in adequate amounts confer a health benefit to the host. Several bacteria are informed as probiotic strains during years, where several positive effects on health have been supported [89].

At the present, conjugated fatty acids have attracted considerable attention because of their potentially beneficial biologic effects. Important properties were attributed to CLA and CLNA, and scientific evidence has been demonstrated both in humans and animal models, including anti-tumor, anti-obese, anti-atherogenic and anti-diabetic activities.

Microbiota present in intestine plays an important physiological rol to the host, modulating some metabolic functions, conferring resistance to microorganism infection and increasing immune response, among other functions.

The bioconversion of LA to CLA and LNA to CLNA by bacteria at intestinal level, result a novel and interesting topic to be developed with the objective to obtain probiotic foods with microorganism able to produce it or functional foods with high levels of CLA and /or CLNA.

The uses of CLA or CLNA-producing bacteria as probiotics have received great attention for nutrition, since many studies evidenced their benefits for the promotion of human health.

It has been demonstrated that isomer of CLA has different function and according to reports t10, c12 is more potent than c9,t11 CLA to prevent cancer cell proliferation [90]. This isomer is also associated to a decrease on body fat in animals [91-92] and humans [93-94].

Previous studies informed that CLA content in cheeses varied according to strain used as starter or adjunct culture [95] and to the ripening time [96]. Therefore, the inclusion of bacteria able to form it during the fermentation process has been received great attention by researchers.

At the present, different functional foods (yogurt, cheese, fermented milk) were manufactured with CLA-producing bacteria, obtaining a final product with a high CLA content. cheeses manufactured with CLA-producing bacteria were developed using sunflower oil as exogenous source of LA, reporting a modification of fatty acids profile in mice tissues after it administration [83]. Mice fed functional cheeses showed a protective effect on viability of intestinal cells after a treatment of 1,2-dimethylhydrazine drug, used as oxidant compound.

Nowadays, CLA production by probiotic bacteria has received special interest in the research field, being well established that bacteria isolated from intestine or fecal samples can form it. However, *in vitro* production was intensely informed, while a few studies have established an *in vivo* CLA production after ingestion of bacteria. Authors revealed that according to administered strain, a high t10, c12 isomer [66, 98] or c9,t11 isomer [99] content in animal tissues occurs.

Linoleic acid excretion in humans is estimate at 340 mg/day [100], being this fatty acid available to further isomerization process by intestinal microbiota. Nevertheless, this local CLA production was only reported after probiotic treatment, but if CLA amount produced is enough to exert a preventive effect require better understanding.

Strains daily administered as probiotic, in a short-term study, produced an increase on CLA systemic content [66]. Authors showed that consumption of *L. rhamnosus* PL60 (10^7-10^9 CFU/day) during 8 weeks increased t10, c12 isomer content in plasma and tissues of diet-induced obese mice. Animals receiving PL60 showed a significant reduction of fat adipose tissue (epididymal and perineal). No liver steatosis were observed in this study, being the most adverse effect informed to t10, c12-CLA. The increasing amount of CLA in tissues after oral treatment with *L. rhamnosus* was explained as an intestinal production once bacterium has been colonized the intestine. Lower leptin levels in PL60 group were also informed. Obese mice selection as animal model was supported by t10, c12-CLA as the main isomer formed by this probiotic strain.

Another work supporting the generation of CLA at intestine using animal models have also reported by same researchers and in this study, they use as probiotic strain *L. plantarum* PL62 in obese mice, at daily dose of 10^7-10^9 CFU/mice. The presence of PL62 was determined in fecal samples after the first week of its intake, and after 5 weeks of feeding a weight reduction in mice receiving PL62 was determined. Similar results were observed after two experimental doses. Respect to CLA, as in the previous study, the main isomer formed by bacterium was t10,c12-CLA [98].

So, both human bacteria *L. rhamnosus* PL60 and *L. plantarum* PL62 were demonstrated to be able to form *in vivo* CLA [66, 98].

But. fibrisolvens from goat rumen was able to rapidly convert LA to CLA and LNA to CLNA, showing similar rate conversion for both fatty acids [101]. In this work, selective strain was administered to mice using a daily dose of 10^{11} CFU/mouse, during 4 weeks. After the trial period, a higher CLA amount in feces was determined. CLA content in tissues was also increased after probiotic treatment. Although a high dose of bacteria was employed, no adverse effect was determined. The aim of this study was to develop a probiotic for animals to generate a continuous CLA production and absorption.

The administration of a mix of bacteria able to form CLA as c9, t11 and t10, c12, called VSL3 was used as probiotic for mice administration [102]. The combination of all strains (*L. casei, L. plantarum, L. acidophilus, L. delbrueckii subsp bulgaricus, B. infantis, B. breve, B. longum, Strep. salivarius subsp. thermophilus*) did not increase CLA production compared with individual strains. Probiotic was prepared as lyophilized form and mice were fed 30 μL of probiotic (0.03 g VSL3 in 10 ml water) for 3 days. Feces were collected at day 0 and 3, and were incubated with LA. Results shown that murine feces with LA after administering VSL3 yielded 100-fold more CLA than feces collected prior to VSL3 feeding. This work also reported that the incorporation of probiotic into conditioned medium produced a reduction of viability and induced apoptosis of HT- 29 and Caco-2 cells.

Another important work using bacteria able to produce CLA as probiotics for animal models was showed by Wall et al. [99]. The administration of *B. breve* NCIMB 702258 to mice and pigs, combined with dietary linoleic acid, showed changes on fatty acid composition of liver and adipose tissues. Higher levels of c9, t11 in liver tissues were determined for both animals receiving *B. breve*, and were also associated with reductions of the pro-inflammatory cytokines level.

Recently, an study to investigate if recombinant lactobacillus expressing LAI (from *P. acnes*, producing t10, c12 isomer) administered to mice produce changes on fatty acids profile was carried out [103]. Authors found that after a daily administration of *L. paracasei* NFBC 338 (10^9 UFC/mouse) during 8 weeks, and 4-fold increase of t10, c12 content in adipose tissue was produced comparing with control mice group. Moreover, in liver a 2.5-fold higher level of the same isomer was reported in treatment group. To the best of our knowledge, this is the only work about using genetically modified strains with the ability to produce t10, c12-CLA, administered as probiotic in mice.

There is few data respect to probiotic administration and *in vivo* CLA production in humans. Lee and Lee [104] reported the effect of PL60 consumption by humans. Here, volunteers consumed PL60 as freeze-dried at a dose of 1g/day (10^{12} CFU/g) during 3 weeks. After one week of uptake, PL60 was recovered from feces, as was previously determined in mice. Respect to CLA content in tissues, both c9,t11 and t10,c12 isomers were higher respect to day 0 of treatment (baseline). Leptin levels were also lower at the end of the study.

11. Conclusion

Although one of the most effective method to increase CLA uptake by humans consist of increase CLA levels in milk and dairy products by modification of animal diet or the inclusion of bacteria able to form it during manufacture process, in the last years the *in vivo* CLA production appears as an alternative way to make it.

Since CLA was recognized as an important biolipid with health benefic properties, there was an increasing interest on this field. However, there is another conjugated fatty acid recently included in studies: conjugated linolenic acid (CLNA). This fatty acid is also generating great attention since anti-atherogenic properties were attributed to them. Some bacteria could produce CLNA using as substrate linolenic acid. CLNA isomers in foods and its biological effects in animal models were lesser understanding than CLA, being the mechanism of it production by bacteria recently investigated. So, in the literature there is not yet recommended dose for this compound for humans.

Development of functional foods enriched on conjugated fatty acids is being extensively studied by researchers, since benefits of health properties were related to humans. The physiological role of conjugated fatty acids like CLA or CLNA is well documented on the literature.

The ability of some species of lactic acid bacteria, propionibacteria and bifidobacteria to *in vitro* conjugate the LA and/or LNA has been established over the years. Manufacture of functional food enriched in conjugated fatty acids by using it as starter or adjunct culture is a promising topic to be developed.

The variation on CLA and CLNA production among bacteria depends on many factors such as intrinsic characteristic of each particular strain, conditions of experimental design and methodology for isomer determination, among others. For this reason, studies must be carefully done before the inclusion of strain during food manufacture.

Few authors have demonstrated the action of bacteria intake on *in vivo* CLA production using experimental animal models and human, but results are promising in this field.

Instead of some technological developments have been performed, many points remain undiscovered at this issue. Some aspects of technological processed foods must be considered, such as CLA-enriched products are also high in fat, being difficult to recommend a single daily dose of CLA after food intake. As we earlier mentioned, not all isomers are incorporated at the same way into tissues fat.

Respect to microorganisms able to form conjugated fatty acids, it is not unreasonable to assume a production of these bioactive compounds at intestinal level, since fatty acid substrate are present in human diet.

Further studies are necessary to understand the kinetic mechanism of it particular production. Questions such as if LAI enzyme is the responsible for both LA and LNA conversion need to be clarified so as the factors determining the isomer production by each strain.

Indeed, taking in account the lack of information respect to some epidemiological and technological aspects of conjugated fatty acids, further studies are required to fully understand the utility of CLA and CLNA in disease prevention. The development of products as probiotic or functional foods to ensure the bioavailability of both compounds for humans is a valuable strategy to be considered.

List of abbreviations

But.: *Butyrivibrio*
B.: *Bifidobacterium*
C.: *Clostridium*
c: cis
CALA: conjugated of alpha linolenic acid
CFU: colony forming unit
CGLA: conjugated of gamma linolenic acid
CLA: Conjugated linoleic acid
CLNA: Conjugated linolenic acid
E.: Enterococcus
L.: *Lactobacillus*
LA: linoleic acid (c9,c12-C18:2)
LAB: lactic acid bacteria
Lac.: *Lactococcus*
LAI: linoleate isomerase
LNA: linolenic acid (c9,c12,c15-C18:3)
P.: Propionibacteria
Ped.: Pediococcus
Strep.: *Streptococcus*
t: trans
TVA: trans vaccenic acid (t11-C18:1)

Author details

Carina Paola Van Nieuwenhove
CERELA-CONICET, S. M. de Tucumán, Argentina
Facultad de Ciencias Naturales e IML- Universidad Nacional de Tucumán, S. M. de Tucumán, Argentina

Victoria Terán
CERELA-CONICET, S. M. de Tucumán, Argentina

Silvia Nelina González
CERELA-CONICET, S. M. de Tucumán, Argentina
Facultad de Bioquímica, Química y Farmacia- Universidad Nacional de Tucumán, S. M. de Tucumán, Argentina

Acknowledgments

This work was supported by grants of Consejo Nacional de Investigaciones Científicas y Técnicas and Consejo de Investigaciones de la Universidad Nacional de Tucumán (PIP 0343 and CIUNT 26/D-429).

12. References

[1] Parodi PW. Conjugated linolenic acid in food. Advances in conjugated linolenic acid research. AOCS Press, Champaign, IL. 2003;2:101-21.
[2] Yücel O. Determination of conjugated linoleic acid content of selected oil seeds grown in Turkey. Journal of the American Oil Chemists Society. 2005;82(12):893-7.
[3] Suzuki R, Noguchi R, Ota T, Abe M, Miyashita K, Kawada T. Cytotoxic effect of conjugated trienoic fatty acids on mouse tumor and human monocytic leukemia cells. Lipids. 2001 May;36(5):477-82.
[4] Bhattacharya A, Banu J, Rahman M, Causey J, Fernandes G. Biological effects of conjugated linoleic acids in health and disease. The Journal of Nutritional Biochemistry. 2006 Dec;17(12):789-810.
[5] Yurawecz MP, Roach JA, Sehat N, Mossoba MM, Kramer JK, Fritsche J, et al. A new conjugated linoleic acid isomer, 7 trans, 9 cis-octadecadienoic acid, in cow milk, cheese, beef and human milk and adipose tissue. Lipids. 1998 Aug;33(8):803-9.
[6] Dhiman TR, Anand GR, Satter LD, Pariza MW. Conjugated linoleic acid content of milk from cows fed different diets. Journal of Dairy Science. 1999 Oct;82(10):2146-56.
[7] Sarrazin P, Mustafa AF, Chouinard PY, Raghavan G, Sotocinal S. Performance of dairy cows fed roasted sunflower seed. Journal of the Science of Food and Agriculture. 2004(84):1179-85.
[8] Jahreis G, Fristche J, Möckel P, Schöne F, Möller U, Steinhart H. The potential anticarcinogenic conjugated linoleic acid, cis-9,trans-11 C18:2, in milk of different species: Cow, goat, ewe, spw, mare, woman. Nutrition Research. October 1999;19(10):1541-9.
[9] Ip C, Chin SF, Scimeca JA, Pariza MW. Mammary cancer prevention by conjugated dienoic derivative of linoleic acid. Cancer Research. 1991 Nov 15;51(22):6118-24.
[10] Houseknecht KL, Vanden Heuvel JP, Moya-Camarena SY, Portocarrero CP, Peck LW, Nickel KP, et al. Dietary conjugated linoleic acid normalizes impaired glucose tolerance in the Zucker diabetic fatty fa/fa rat. Biochemical and Biophysical Research Communications. 1998 Mar 27;244(3):678-82.

[11] Hayek MG, Han SN, Wu D, Watkins BA, Meydani M, Dorsey JL, et al. Dietary conjugated linoleic acid influences the immune response of young and old C57BL/6NCrlBR mice. Journal of Nutrition. 1999 Jan;129(1):32-8.

[12] Chouinard PY, Corneau L, Barbano DM, Metzger LE, Bauman DE. Conjugated linoleic acids alter milk fatty acid composition and inhibit milk fat secretion in dairy cows. Journal of Nutrition. 1999 Aug;129(8):1579-84.

[13] Kelly GS. Conjugated linoleic acid: a review. Alternative Medicine Reviews. 2001 Aug;6(4):367-82.

[14] Ip MM, Masso-Welch PA, Shoemaker SF, Shea-Eaton WK, Ip C. Conjugated linoleic acid inhibits proliferation and induces apoptosis of normal rat mammary epithelial cells in primary culture. Experimental Cell Research. 1999 Jul 10;250(1):22-34.

[15] MacDonald HB. Conjugated linoleic acid and disease prevention: a review of current knowledge. Journal of American College of Nutrition. 2000 Apr;19 (2 Suppl):111S-8S.

[16] Igarashi M, Miyazawa T. Newly recognized cytotoxic effect of conjugated trienoic fatty acids on cultured human tumor cells. Cancer Letters. 2000 Feb 1;148(2):173-9.

[17] Harfoot CG, Hazlewood GP. Lipid metabolism in the rumen. In the rumen microbial ecosystem Elsevier Science Publishers, London, UK. 1988:285-322.

[18] Kepler CR, Hirons KP, McNeill JJ, Tove SB. Intermediates and products of the biohydrogenation of linoleic acid by Butyrinvibrio fibrisolvens. The Journal of Biological Chemistry. 1966 Mar 25;241(6):1350-4.

[19] Kim YJ, Liu RH, Bond DR, Russell JB. Effect of linoleic acid concentration on conjugated linoleic acid production by Butyrivibrio fibrisolvens A38. Applied and Environmental Microbiology. 2000 Dec;66(12):5226-30.

[20] Kishino S, Ogawa J, Ando A, Omura Y, Shimizu S. Ricinoleic acid and castor oil as substrates for conjugated linoleic acid production by washed cells of Lactobacillus plantarum. Bioscience, Biotechnology and Biochemistry. 2002 Oct;66(10):2283-6.

[21] Kishino S, Ogawa J, Ando A, Shimizu S. Conjugated α-linolenic acidproduction from α-linolenic acid by Lactbacillus plantarum AKU1009a. European Journal of Lipid Science and Technology. 2003(105):572–7.

[22] Akraim F, Nicot MC, Juaneda P, Enjalbert F. Conjugated linolenic acid (CLnA), conjugated linoleic acid (CLA) and other biohydrogenation intermediates in plasma and milk fat of cows fed raw or extruded linseed. Animal. 2007 Jul;1(6):835-43.

[23] Loor JJ, Lin X, Herbein JH. Dietary trans-vaccenic acid (trans11-18:1) increases concentration of cis9,transll-conjugated linoleic acid (rumenic acid) in tissues of lactating mice and suckling pups. Reproduction Nutrition Development. 2002 Mar-Apr;42(2):85-99.

[24] Kemp P, Lander DJ. Hydrogenation in vitro of alpha linolenic acid to stearic acid by mixed cultures of pure strains of rumen bacteria. Journal of General Microbiology. 1984(130):527-33.

[25] Griinari JM, Corl BA, Lacy SH, Chouinard PY, Nurmela KV, Bauman DE. Conjugated linoleic acid is synthesized endogenously in lactating dairy cows by Delta(9)-desaturase. Journal of Nutrition. 2000 Sep;130(9):2285-91.

[26] Lock AL, Garnsworthy PC. Independent effects of dietary linoleic and linolenic fatty acids on the conjugated linoleic acid content of cows milk. Animal Science. 2002(74):163-76.

[27] Dhiman TR, Nam SH, Ure AL. Factors affecting conjugated linoleic acid content in milk and meat. Critical Reviews in Food Science and Nutrition. 2005;45(6):463-82.

[28] Jiang J, Bjorck L, Fondén R. Conjugated linoleic acid in Swedish dairy products with special reference to the manufacture of hard cheeses. International Dairy Journal. 1997;(241):863-7.

[29] Schmid A, Collomb M, Sieber R, Bee G. Conjugated linoleic acid in meat and meat products: A review. Meat Science. 2006 May;73(1):29-41.

[30] Chin SF, Liu W, Storkson JM, Ha YL, Pariza MW. Dietary sources of conjugated dienoic isomers of linoleic acid, a newly recognized class of anticarcinogens. Journal of Food Composition and Analysis. 1992;5(3):185-97.

[31] Parodi PW. Conjugated octadecadienoic acids of milk fat. Journal of Dairy Science. 1977(60):1550-3.

[32] Luna P, Fontecha J, Juarez M, Angel de la Fuente M. Changes in the milk and cheese fat composition of ewes fed commercial supplements containing linseed with special reference to the CLA content and isomer composition. Lipids. 2005 May;40(5):445-54.

[33] Khanal RC, Olson KC. Factors affecting conjugated linoleic acid(CLA) content in milk, meat and egg: A review. Pakistan Journal of Nutrition. 2004(3):82-98.

[34] Destaillats F, Trottier JP, Galvez JM, Angers P. Analysis of alpha-linolenic acid biohydrogenation intermediates in milk fat with emphasis on conjugated linolenic acids. Journal of Dairy Science. 2005 Sep;88(9):3231-9.

[35] Corl BA, Barbano DM, Bauman DE, Ip C. cis-9, trans-11 CLA derived endogenously from trans-11 18:1 reduces cancer risk in rats. Journal of Nutrition. 2003 Sep;133(9):2893-900.

[36] Adlof RO, Duval S, Emken EA. Biosynthesis of conjugated linoleic acid in humans. Lipids. 2000 Feb;35(2):131-5.

[37] Ross GR, Van Nieuwenhove CP, Gonzalez SN. Fatty Acid Profile of Pig Meat after Probiotic Administration. Journal of Agriculture and Food Chemistry. 2012 Apr 16.

[38] Fritsche J, Steinhart H. Amounts of conjugated linoleic acid (CLA) in German foods and evaluation of daily intake. European Food Research and Technology. 1998;206(2):77-82.

[39] Schoos V, Medina M, Saad S, Van Nieuwenhove CP. Chemical and microbiological characteristics of Llama's (Lama glama) milk from Argentina. Milchwissenschaft. 2008;63(4):398-401.

[40] Plourde M, Destaillats F, Chouinard PY, Angers P. Conjugated alpha-linolenic acid isomers in bovine milk and muscle. Journal of Dairy Science. 2007 Nov;90(11):5269-75.

[41] Kohno H, Suzuki R, Noguchi R, Hosokawa M, Miyashita K, Tanaka T. Dietary conjugated linolenic acid inhibits azoxymethane-induced colonic aberrant crypt foci in rats. Japanese Journal of Cancer Research. 2002 Feb;93(2):133-42.

[42] Lee KN, Kritchevsky D, Pariza MW. Conjugated linoleic acid and atherosclerosis in rabbits. Atherosclerosis. 1994 Jul;108(1):19-25.

[43] Ip C, Carter CA, Ip MM. Requirements of essencial fatty acid for mammary tumorigenesis in the rat. Cancer Research. 1985(45):1997-2001.

[44] Park HS, Ryu JH, Ha YL, Park JH. Dietary conjugated linoleic acid (CLA) induces apoptosis of colonic mucosa in 1,2-dimethylhydrazine-treated rats: a possible mechanism of the anticarcinogenic effect by CLA. British of Journal Nutrition. 2001 Nov;86(5):549-55.

[45] Van Wijlen RPJ, Colombani PC. Grass-based ruminant production methods and human bioconversion of vaccenic acid with estimations of maximal dietary intake of conjugated linoleic acids. International Dairy Journal. 2010(20):433-48.

[46] Whigham LD, Watras AC, Schoeller DA. Efficacy of conjugated linoleic acid for reducing fat mass: a meta-analysis in humans. Am J Clin Nutr. 2007 May;85(5):1203-11.

[47] Ha YL, Grimm NK, Pariza MW. Newly recognized anticarcinogenic fatty acids: Identification and quantification in natural and processed cheeses. Journal of Agricultural and Food Chemistry. 1989;37(1):75-81.

[48] Mullen A, Moloney F, Nugent AP, Doyle L, Cashman KD, Roche HM. Conjugated linoleic acid supplementation reduces peripheral blood mononuclear cell interleukin-2 production in healthy middle-aged males. The Journal of Nutritional Biochemistry. 2007 Oct;18(10):658-66.

[49] Moloney F, Yeow TP, Mullen A, Nolan JJ, Roche HM. Conjugated linoleic acid supplementation, insulin sensitivity, and lipoprotein metabolism in patients with type 2 diabetes mellitus. The American Journal of Clinical Nutrition. 2004 Oct;80(4):887-95.

[50] Smedman A, Vessby B. Conjugated linoleic acid supplementation in humans--metabolic effects. Lipids. 2001 Aug;36(8):773-81.

[51] Wang Y, Jones PJ. Dietary conjugated linoleic acid and body composition. Am J Clin Nutr. 2004 Jun;79(6 Suppl):1153S-8S.

[52] Kim YJ, Liu RH. Increase of conjugated linoleic acid content in milk by fermentation with lactic acid bacteria. Journal of Food Science. 2002;67(5):1731-7.

[53] Van Nieuwenhove CP, Oliszewski R, Gonzalez SN, Perez Chaia AB. Conjugated linoleic acid conversion by dairy bacteria cultured in MRS broth and buffalo milk. Lett Appl Microbiol. 2007 May;44(5):467-74.

[54] Chung SH, Kim IH, Park HG, Kang HS, Yoon CS, Jeong HY, et al. Synthesis of conjugated linoleic acid by human-derived Bifidobacterium breve LMC 017: utilization as a functional starter culture for milk fermentation. Journal of Agriculture and Food Chemistry. 2008 May 14;56(9):3311-6.

[55] Gorissen L, Raes K, Weckx S, Dannenberger D, Leroy F, De Vuyst L, et al. Production of conjugated linoleic acid and conjugated linolenic acid isomers by Bifidobacterium species. Applied Microbiology and Biotechnology. 2010 Aug;87(6):2257-66.

[56] Gorissen L, Leroy F, Raes K, De Vuyst L, De Smet S. Conjugated linoleic acid and conjugated linolenic acid production by bifidobacteria. Communications in Agriculture and Applied Biological Sciences. 2011;76(1):7-10.

[57] Coakley M, Ross RP, Nordgren M, Fitzgerald G, Devery R, Stanton C. Conjugated linoleic acid biosynthesis by human-derived Bifidobacterium species. Journal of Applied Microbiology. 2003;94(1):138-45.

[58] Jiang J, Bjorck L, Fonden R. Production of conjugated linoleic acid by dairy starter cultures. Journal of Applied Microbiology. 1998 Jul;85(1):95-102.

[59] Ogawa J, Matsumura K, Kishino S, Omura Y, Shimizu S. Conjugated linoleic acid accumulation via 10-hydroxy-12-octadecaenoic acid during microaerobic transformation of linoleic acid by Lactobacillus acidophilus. Applied and Environmental Microbiology. 2001 Mar;67(3):1246-52.

[60] Lin TY, Hung TH, Cheng TSJ. Conjugated linoleic acid production by immobilized cells of Lactobacillus acidophilus. Food Chemistry. 2005;(92):23-8.

[61] Kishino S, Ogawa J, Omura Y, Matsumura K, Shimizu S. Conjugated linoleic acid production from linoleic acid by lactic acid bacteria. JAOCS, Journal of the American Oil Chemists' Society. 2002;79(2):159-63.

[62] Lin TY, Lin CW, Wang YJ. Production of conjugated linoleic acid by enzyme extract of Lactobacillus acidophilus CCRC 14079. Food Chemistry. 2003;83:27-31.

[63] Bauman DE, Baumgard LH, Corl BA, Griinari JM. Biosynthesis of conjugated linoleic acid in ruminants. Proceedings of the American Scociety of Animal Science. 1999.

[64] McIntosh FM, Shingfield KJ, Devillard E, Russell WR, Wallace RJ. Mechanism of conjugated linoleic acid and vaccenic acid formation in human faecal suspensions and pure cultures of intestinal bacteria. Microbiology. 2009;155(1):285-94.

[65] Xu S, Boylston T, Glatz B. Effect of lipid source on probiotic bacteria and conjugated linoleic acid formation in milk model systems. Journal of the American Oil Chemists Society. 2004(81):589-95.

[66] Lee HY, Park JH, Seok SH, Baek MW, Kim DJ, Lee KE, et al. Human originated bacteria, Lactobacillus rhamnosus PL60, produce conjugated linoleic acid and show anti-obesity effects in diet-induced obese mice. Biochimica et Biophysica Acta. 2006 Jul;1761(7):736-44.

[67] Xu H, Lee HY, Hwang B, Nam JH, Kang HY, Ahn J. Kinetics of microbial hydrogenation of free linoleic acid to conjugated linoleic acids. J Appl Microbiol. 2008 Dec;105(6):2239-47.

[68] Rodríguez-Alcalá LM, Braga T, Xavier Malcata F, Gomes A, Fontecha J. Quantitative and qualitative determination of CLA produced by Bifidobacterium and lactic acid bacteria by combining spectrophotometric and Ag +-HPLC techniques. Food Chemistry. 2011;125(4):1373-8.

[69] Gorissen L, Weckx S, Vlaeminck B, Raes K, De Vuyst L, De Smet S, et al. Linoleate isomerase activity occurs in lactic acid bacteria strains and is affected by pH and temperature. Journal of Applied Microbiology. 2011 Sep;111(3):593-606.

[70] Rainio A, Vahvaselka M, Laakso S. Cell-adhered conjugated linoleic acid regulates isomerization of linoleic acid by resting cells of Propionibacterium freudenreichii. Applied Microbiology and Biotechnology. 2002 Dec;60(4):481-4.

[71] Rainio A, Vahvaselkä M, Suomalainen T, Laakso S. Production of conjugated linoleic acid by Propionibacterium freudenreichii ssp. shermanii. Lait. 2002;82(1):91-101.

[72] Lin TY, Lin CW, Wang YJ. Linoleic Acid Isomerase Activity in Enzyme Extracts from Lactobacillus acidophilus and Propionibacterium freudenreichii subsp. shermanii. Journal of Food Science. 2002;(67):1502–5.

[73] Wallace RJ, Chaudhary LC, McKain N, McEwen BS, Richardson AJ, Vercoe PE, et al. Clostridium proteoclasticum: A ruminal bacterium taht forms stearic acid from linoleic acid. FEMS Microbiology Letters. 2006;265:195-201.

[74] Hennessy AA, Barrett E, Paul Ross R, Fitzgerald GF, Devery R, Stanton C. The production of conjugated alpha-linolenic, gamma-linolenic and stearidonic acids by strains of bifidobacteria and propionibacteria. Lipids. 2012 Mar;47(3):313-27.

[75] Favier CF, Vaughan EE, De Vos WM, Akkermans AD. Molecular monitoring of succession of bacterial communities in human neonates. Appl Environ Microbiol. 2002 Jan;68(1):219-26.

[76] Picard C, Fioramonti J, Francois A, Robinson T, Neant F, Matuchansky C. Review article: bifidobacteria as probiotic agents -- physiological effects and clinical benefits. Alimentary Pharmacology and Therapeutics. 2005 Sep 15;22(6):495-512.

[77] Saarela M, Virkajärvi I, Alakomi H-L, Sigvart-Mattila P, Mättö J. Stability and functionality of freeze-dried probiotic Bifidobacterium cells during storage in juice and milk. International Dairy Journal. [doi: 10.1016/j.idairyj.2005.12.007. 2006;16(12):1477-82.

[78] Vinderola G, de los Reyes-Gavilán C, Reinheimer J. Probiotics and prebiotics in fermented dairy products. In Contemporary Food Engineering. 2009:601-34

[79] Vinderola G, Binetti A, Burns P, Reinheimer J. Cell viability and functionality of probiotic bacteria in dairy products. Frontiers in Microbiology. 2011;2:70.

[80] Oh DK, Hong GH, Lee Y, Min S, Sin HS, Cho SK. Production of conjugated linoleic acid by isolated Bifidobacterium strains. World J Microbiol Biotechnol. 2003(19):907-12.

[81] Park HG, Cho SD, Kim JH, Lee H, Chung SH, Kim SB, et al. Characterization of conjugated linoleic acid production by Bifidobacterium breve LMC 520. Journal of Agriculture and Food Chemistry. 2009 Aug 26;57(16):7571-5.

[82] Choi NJ, Park HG, Kim YJ, Kim IH, Kang HS, Yoon CS, et al. Utilization of monolinolein as a substrate for conjugated linoleic acid production by Bifidobacterium breve LMC 520 of human neonatal origin. Journal of Agriculture and Food Chemistry. 2008 Nov 26;56(22):10908-12.

[83] Van Nieuwenhove CP, Gauffin Cano P, Pérez-Chaia AB, González SN. Effect of functional buffalo cheese on fatty acid profile and oxidative status of liver and intestine of mice. Journal of Medicinal Food. 2011 Apr;14(4):420-7.

[84] Puniya AK, Chaitanya S, Tyagi AK, De S, Singh K. Conjugated linoleic acid producing potential of lactobacilli isolated from the rumen of cattle. Journal of Industrial Microbiology and Biotechnology. 2008 Nov;35(11):1223-8.

[85] Park SJ, Park KA, Park CW, Park WS, Kim JO, Ha YL. Purification and aminoacids sequence of the linoleate isomerase from Butyrivbrio fibrisolvens A-38 Journal of Food Science and Nutrition. 1996;1:244-51.

[86] Irmak S, Dunford NT, Gilliland SE, Banskalieva V, Eisenmenger M. Biocatalysis of linoleic acid to conjugated linoleic acid. Lipids. 2006 Aug;41(8):771-6.

[87] Deng MD, Grund AD, Schneider KJ, Langley KM, Wassink SL, Peng SS, et al. Linoleic acid isomerase from Propionibacterium acnes: purification, characterization, molecular cloning, and heterologous expression. Applied Biochemistry and Biotechnology. 2007 Dec;143(3):199-211.

[88] Rosson RA, Deng MD, Grund AD, Peng SS. PCT WO0100846. 2001.

[89] Ross GR, Gauffin Cano P, Gusils León CH, Medina RB, González SN, Van Nieuwenhove CP. Lactic acid bacteria activities to promote health benefits. Multidisciplinary approaches on food science and nutrition for the 21st century. Research signpost ed. 2011:155-74.

[90] Cho HJ, Kim WK, Jung JI, Kim EJ, Lim SS, Kwon DY, et al. Trans-10,cis-12, not cis-9,trans-11, conjugated linoleic acid decreases ErbB3 expression in HT-29 human colon cancer cells. World Journal of Gastroenterology. 2005 Sep 7;11(33):5142-50.

[91] Park Y, Albright KJ, Liu W, Storkson JM, Cook ME, Pariza MW. Effect of conjugated linoleic acid on body composition in mice. Lipids. 1997 Aug;32(8):853-8.

[92] Yamasaki M, Ikeda A, Oji M, Tanaka Y, Hirao A, Kasai M, et al. Modulation of body fat and serum leptin levels by dietary conjugated linoleic acid in Sprague-Dawley rats fed various fat-level diets. Nutrition. 2003 Jan;19(1):30-5.

[93] Mougios V, Matsakas A, Petridou A, Ring S, Sagredos A, Melissopoulou A, et al. Effect of supplementation with conjugated linoleic acid on human serum lipids and body fat. The Journal of Nutritional Biochemistry. 2001 Oct;12(10):585-94.

[94] Thom E, Wadstein J, Gudmundsen O. Conjugated linoleic acid reduces body fat in healthy exercising humans. Journal of International Medical Research. 2001 Sep-Oct;29(5):392-6.

[95] Lin TY, Lin CW, Lee CH. Conjugated Linoleic Acid Concentration as Affected by Lactic Cultures and Added Linoleic Acid. Food Chemistry. 1999;(67):1-5.

[96] Shantha NC, Ram LN, O Leary J, Hicks CL, Decker EA. Conjugated linoleic acid concentrations in dairy products as affected by processing and storage. Journal of Food Science. 1995(60):695-7.

[97] Colakoglu H, Gursoy O. Effect of lactic adjunct cultures on conjugated linoleic acid (CLA) concentration of yogurt drink. Journal of Food, Agriculture and Environment. 2011;9(1):60-4.

[98] Lee K, Paek K, Lee HY, Park JH, Lee Y. Antiobesity effect of trans-10,cis-12-conjugated linoleic acid-producing Lactobacillus plantarum PL62 on diet-induced obese mice. Journal of Applied Microbiology. 2007 Oct;103(4):1140-6.

[99] Wall R, Ross RP, Shanahan F, O'Mahony L, O'Mahony C, Coakley M, et al. Metabolic activity of the enteric microbiota influences the fatty acid composition of murine and porcine liver and adipose tissues. American Journal of Clinical Nutrition. 2009 May;89(5):1393-401.

[100] Edionwe AO, Kies C. Comparison of palm and mixtures of refined palm and soybean oils on serum lipids and fecal fat and fatty acid excretions of adult humans. Plant Foods for Human Nutrition. 2001;56(2):157-65.

[101] Fukuda S, Suzuki Y, Murai M, Asanuma N, Hino T. Isolation of a novel strain of Butyrivibrio fibrisolvens that isomerizes linoleic acid to conjugated linoleic acid without hydrogenation, and its utilization as a probiotic for animals. Journal of Applied Microbiology. 2006 Apr;100(4):787-94.

[102] Ewaschuk JB, Walker JW, Diaz H, Madsen KL. Bioproduction of conjugated linoleic acid by probiotic bacteria occurs in vitro and in vivo in mice. Journal of Nutrition. 2006 Jun;136(6):1483-7.

[103] Rosberg-Cody E, Stanton C, O'Mahony L, Wall R, Shanahan F, Quigley EM, et al. Recombinant lactobacilli expressing linoleic acid isomerase can modulate the fatty acid composition of host adipose tissue in mice. Microbiology. 2011 Feb;157(Pt 2):609-15.

[104] Lee K, Lee Y. Production of c9, t11- and t10, c12- conjugated linoleic acids in humans by Lactobacillus rhamnosus PL60. Journal of Microbiology and Biotechnology. 2009 Oct;19(12):1617-9.

[105] Fritsche J, Steinhart H. Analysis, occurence and physiological properties of trans fatty acids(TFA) with particular emphasis on conjugated linoleic acid isomers (CLA). A review. Fet/Lipid. 1998(100):190-210.

[106] Raes K, Balcaen A, Claeys E, De Smet S, Demeyer D. Effect of duration of feeding diets rich in n-3 PUFA to Belgian Blue double-muscled young bulls, on the incorporation of long-chain n-3 and n-6 PUFA in the phospholipids and triglycerides of the longissimus thoracis. Proceedings of the 48th ICoMST. Rome, Italy. 2002;2:724-5.

[107] Gultemiriam L, Van Nieuwenhove CP, Pérez Chaia A, Apella MC. Physical and chemical characterization of eggs from Araucana hens of free range fed in Argentina. Journal of the Argentine Chemical Society. 2009;97(2):19-30.

[108] Park Y, Storkson JM, Albright KJ, Liu W, Pariza MW. Evidence that the trans-10,cis-12 isomer of conjugated linoleic acid induces body composition changes in mice. Lipids. 1999 Mar;34(3):235-41.

[109] Ogawa J, Kishino S, Ando A, Sugimoto S, Mihara K, Shimizu S. Production of conjugated fatty acids by lactic acid bacteria. Journal of Bioscience and Bioengineering. 2005 Oct;100(4):355-64.

[110] Lee SO, Kim CS, Cho SK, Choi HJ, Ji GE, Oh DK. Bioconversion of linoleic acid into conjugated linoleic acid during fermentation and by washed cells of Lactobacillus reuteri. Biotechnology Letters. 2003 Jun;25(12):935-8.

Chapter 4

Development of New Products: Probiotics and Probiotic Foods

Z. Denkova and A. Krastanov

Additional information is available at the end of the chapter

http://dx.doi.org/10.5772/47827

1. Introduction

Probiotics are live microorganisms that confer a beneficial effect on the host when administered in proper amounts [1, 2]. Their beneficial effects on gastrointestinal infections, the reduction of serum cholesterol, the protection of the immune system, anti-cancer properties, antimutagenic action, anti-diarrheal properties, the improvement in inflammatory bowel disease and suppression of *Helicobacter pylori* infection, Crohn's disease, restoration of the microflora in the stomach and the intestines after antibiotic treatment, etc. are proven by addition of selected strains to food products [3, 4, 5, 6].

Lactobacilli and bifidobacteria are normal components of the healthy human intestinal microflora. They are included in the composition of probiotics and probiotic foods because of their proven health effects on the body [7, 8, 9]. They are the main organisms that maintain the balance of the gastrointestinal microflora [10].

Not all strains of lactobacilli and bifidobacteria can be used as components of probiotics and probiotic foods, but only those that are of human origin, non-pathogenic, resistant to gastric acid, bile and to the antibiotics, administered in medical practice; they should also have the potential to adhere to the gut epithelial tissue and produce antimicrobial substances; they should allow the conduction of technological processes, in which high concentrations of viable cells are obtained as well as to allow industrial cultivation, encapsulation and freeze-drying and they should remain active during storage [11, 12]. This requires the mandatory selection of strains of the genera *Lactobacillus* and *Bifidobacterium* with probiotic properties. Moreover, the concentration of viable cells of microorganisms in the composition of probiotics should exceed 1 million per gram [13] in order for the preparation to exhibit a therapeutic and prophylactic effect.

Along with probiotics probiotic bacteria are most frequently included in the composition of dairy products - yogurt, cheese, etc. [14, 15]. A dairy product that delivers viable cells of

L.acidophilus, L.bulgaricus, Bifidobacterium sp. is bio-yoghurt. Adequate numbers of viable cells, namely the"therapeutic minimum" need to be consumed regularly for transfer of the "probiotic" effect to consumers. This requires, according to Rybka & Kailasapathy, 1995 [16] the consumption of 100 g per day bio-yoghurt containing more than 10^6cfu/cm^3 viable cells.

The species *L.bulgaricus* is a heterogeneous group of bacteria, including strains with probiotic properties [17]. The inclusion of such cultures in yogurt would transform this lactic acid product into a probiotic product.

Probiotic bacteria are included as components of the starter cultures for non-dairy foods [18]. For each type of non-dairy product strains that can grow in the food environment and contribute to the formation of the sensory profile are selected. So in starter cultures for raw-dried meat products probiotic bacteria that are able to grow in the meat environment are included; in soy fermented foods as components of the starter cultures lactobacilli and bifidobacteria strains which can grow and multiply in soy milk are applied; in fruit and vegetables and fruit and vegetable juices microorganisms with probiotic properties suitable for this type of food are used [19].

Some strains of lactobacilli with probiotic potential are used as components of sourdough in bread-making to extend the shelf life and to improve the quality and some technological properties of the final product [20, 21, 22, 23].

In this chapter, the new steps in obtaining probiotics and probiotic foods are discussed. The requirements for the strains of microorganisms which are implemented as components of the probiotics and probiotic foods are listed.

The chapter includes some data from the research of our research team in the field of selection of bifidobacteria and lactobacilli with probiotic properties, developing the technology for obtaining the probiotics "Enterosan", probiotic milk and beverages, probiotic starter cultures for meat foods and non-traditional fermented probiotic foods.

2. Microorganisms with probiotic properties

Enormous amount of microbial biomass inhabits the stomach and the intestines and accompanies individuals throughout their lives. Organisms that are a part of the gastrointestinal microflora, include saprophytic, pathogenic and conditionally pathogenic microorganisms, enterobacteria, lactobacilli, lactic acid cocci, bifidobacteria. They occupy a niche in the digestive tract and enter into complex relationships both among themselves and with the host - man or animal. Depending on the composition of food intake the diversity of species and the ratio between them varies significantly. Upon intake of plant foods fermenting species predominate, while in meat meal representatives of the putrefactive microorganisms take the upper hand. Microbes transform nutrients in food in different ways and excrete metabolites with diverse chemical nature. Through them the gastrointestinal microflora influences the condition and the health of the body. A part of the microflora that includes lactobacilli and bifidobacteria utilizes the substrates and forms metabolites as a result of its vital activity through which it oppresses and expels pathogenic and toxigenic bacteria from the biological niche. The degradation of nutrients from the decay performed by pathogenic and toxigenic microorganisms, which include the pathogenic genera *Clostridium* and *Bacteroides* leads to the formation of toxins and products of decay that inhibit the functioning of the organisms and cause diseases. The balance between these two groups of microorganisms determines to a considerable extent the health of the individuals. Many factors affect this balance - the quality of food, water and air, the neuro-psychological status and stress, the social and personal hygiene, the health and the use of drugs, antibiotics, etc. The age of the individuals also influences the diversity of the microflora in the stomach and intestines.

Maintaining the right balance between the species in the gastrointestinal tract is achieved through the adoption of beneficial flora (lactobacilli and bifidobacteria) in the form of concentrates of viable cells, known as probiotics, or in the composition of foods that can be enriched with them.

Probiotics are biologically active preparations containing high concentrations of beneficial natural microorganisms that allow maintaining a predominantly beneficial microflora in the gastro-intestinal tract, ensuring good health and quality of life. In the last decades, science and health care are paying serious attention to probiotics as preventive and therapeutic tools against many diseases. The first beneficial effect of their adoption is the normalization of the gastrointestinal microflora and the occurrence of recovery processes in the digestive tract. This helps to improve the health status of other organs and systems. The practical application of probiotics clearly speaks in favor of this claim. Probiotic microorganisms should be regarded as an indispensable ingredient of food. Absence, lack or destruction of part or all of the useful microflora poses serious hazards to human health. Therefore, one can neither exist without the normal probiotic microorganisms nor can replace them with something else. Neglecting this requirement is associated with serious consequences for the health and life of humans and animals. Quite often probiotics are the only key to the treatment of some diseases of gastroenterological, functional and deficiency nature.

By applying advanced technologies for fermentation, encapsulation and freeze-drying probiotic preparations (Multibionta, Enterogermina, Reuterina, Enterosan, Florastor) with proven prophilactic and healing action in children and adults against colitis, including ulcerative colitis, gastritis, enteritis, ulcerative disease, intestinal infections, disbacteriosis and some cases of dyspepsia, have been created.

Not all species and strains of lactobacilli and bifidobacteria could act as regulators of the gastrointestinal microflora, but only those who are able to survive and grow under the different conditions of the digestive tract. This requires the selection of strains of lactobacilli and bifidobacteria with probiotic properties, which are reflected in their ability:

1. To be part of the natural microflora in humans and animals.
2. To have the ability to adhere to epithelial cells or cell lines, or at least to be able to colonize the ileum temporarily [24, 25].

Adhesion can be nonspecific - related to the physicochemical factors, and specific - based on specific molecules on the surface of the probiotic cells that adhere to receptor molecules on the surface of the epithelial cells. The strains used in the production of fermented milk products are not with the best adhesion properties, while probiotic bacteria show strong adhesion that is species specific. As far as their ability to adhere is concerned lactic acid bacteria (including lactic acid bacteria used in the manufacture of milk products) show moderate to good adhesion properties when it comes to adhesion on human cell lines [26, 27, 28].

The adhesion of probiotic strains to the surface of the intestine and the subsequent colonization of the gastrointestinal tract of humans creates conditions for better retention in the intestinal tract and implementation of metabolic processes with a strong immunomodulatory effect. Adhesion provides interaction with the mucosa, supporting the contact with the intestine-associated lymphoid tissue, which in turn provides stabilization of the intestinal mucosa that performs a barrier function. The intestine-associated lymphoid tissue can interact with the cells of the probiotic strains and their components and thus has a positive effect on the immune system of the host [29].

In many species of lactic acid bacteria, including those of the genus *Lactobacillus*, the presence of surface-layer proteins [30, 31, 32] has been found. The gene for the S-layer protein has been sequenced and cloned in *Lactobacillus brevis* [33], *Lactobacillus acidophilus* [34], *Lactobacillus helveticus* [35] and *Lactobacillus crispatus* [36].

The thickness of the surface-layer (S-layer) in bacteria is typically 5 to 25 nm and it is composed of subunits arranged in a grid (lattice) with irregular, square or hexagonal symmetry [37]. In the amino acid analysis of the S-layer proteins it was found that they are rich in acidic and hydrophobic amino acids and very poor in sulfur-containing amino acids [38]. In determining the secondary structure it was found that in most S-layer proteins 40% of the amino acids form a β-sheet structure and 10-30% - α-helix [38]. Common feature of all surface-layer proteins characterized so far is their ability to crystallize spontaneously into a two-dimensional layer on the outer side of the bacterial cell wall.

In representatives of the genus *Lactobacillus* some surface located enzymes are established along with the S-layer proteins on the cell surface. The molecular weight of the S-layer proteins in lactobacilli ranges from 40 kDa to 60 kDa and they are one of the smallest known S-layer proteins [31, 36, 39, 40]. Compared with many other S-layer proteins, which are of acidic nature, those in lactobacilli are characterized by high values of their isoelectrical points [41]. The S-layer proteins in some lactobacilli give the cell surface hydrophobicity [40, 42]. Moreover, the hydrophobicity of the cell surface of the strain *Lactobacillus acidophilus* ATCC 4356 can be varied in accordance with the change of the ionic strength of the medium [43]. In a sequencing study of the S-layer proteins in *Lactobacillus acidophilus, Lactobacillus crispatus* and *Lactobacillus helveticus* a high degree of homology in one third of their C-terminus is demonstrated [36].

The functions of the S-layer proteins in lactobacilli are insufficiently studied. The S-layer proteins act as adhesins in many bacteria such as lactobacilli and some representatives of the genus *Bacillus*, so they determine their adhesion to epithelial cells or extracellular matrix proteins [44, 45, 46, 47].

3. To survive in the conditions of the stomach and intestines, i.e. to survive in the conditions of acidic pH in the stomach and to withstand the effects of bile [48, 49, 50].

The survival of bacteria in gastric juice depends on their ability to tolerate the low pH values of the medium. Transition time in these conditions depends on the condition of the individual and the type of the food and it ranges from 1 to 3-4 hours. The lactic acid bacteria *L.sakei, L.plantarum, L.pentosus, P.acidilactici* and *Pediococcus pentosaceus* can survive in acidic conditions [51, 52]. Therefore Klingberg et al., 2006 [51] and Pennacchia et al., 2004 [52] suggest the examination of the survival of the strains for probiotic purposes in cultural medium at pH 2.5, acidified with hydrochloric acid for 4 h. Using this criterion *Lactobacillus* strains resistant to low values of pH (pH 2) and the presence of pepsin [17] are selected.

Bacteria that survive in the conditions of the stomach then enter the duodenum, where bile salts are poured and their concentration is 0.3%. Microorganisms reduce the emulsifiable effect of bile salts by hydrolyzing them, thus reducing their solubility. Some intestinal lactobacilli, such as *L.acidophilus, L.casei* and *L.plantarum* have the ability to hydrolyze bile salts [53]. Moreover, some strains of lactic acid bacteria isolated from sausages such as *L.sakei, L.plantarum, L.pentosus* and *P.acidilactici* are resistant to 0.3% bile salts [51, 52].

The survival of probiotic bacteria in the gastrointestinal tract, their translocational and colonizational properties and the destruction of their active components are essential for the realization of their preventive role.

Different probiotic strains react differently in different parts of the gastrointestinal tract - some strains are killed very quickly in the stomach, while others pass through the entire gastrointestinal tract, retaining high concentrations of viable cells [29, 54, 55, 56, 57, 58, 59, 60].

The natural gastro-intestinal microflora, especially lactobacilli, should have the ability to hydrolyze conjugated bile acids that are present in large quantities in the intestines. Conjugated bile acids provide the emulsification, digestion and absorption of lipids more efficiently than bile acids in non-conjugated form. The hydrolysis of bile acids may be associated with the accumulation of energy in anaerobic conditions and / or the detoxication of bile acids that inhibit bacterial growth.

4. To have the ability to reproduce in the gastrointestinal tract. By primarily utilizing the substrate to oppress and expel from the biological niche the pathogenic and toxigenic microorganisms.

5. To possess antimicrobial activity against conditionally pathogenic, carcinogenic and pathogenic microorganisms, which is associated with inactivation of their enzyme systems, overcoming their adhesion, growth suppression and forcing them out of their biological niche, as a result of which gastrointestinal microflora is normalized.
6. To produce antimicrobial substances.

Probiotic strains should be able to carry out fermentation with lactic acid and bacteriocin production by utilizing the carbohydrates, thus changing the pH of the medium and suppressing the development of pathogenic and toxigenic microorganisms or acting directly on the microbial cells by producing antibacterial substances with peptide nature (bacteriocins) [61, 62].

7. To modulate the immune response.
8. To be safe for clinical and food applications.

Lactic acid bacteria applied in clinical and functional foods must be safe, especially if intended for humans.

9. To allow industrial cultivation, resulting in obtaining concentrates with high concentrations of viable cells that can be included in gel matrices (encapsulation), thus retaining their activity in the process of freeze-drying as well as in the composition of the finished products.

Donald and Brow, 1993 [13] and Wolfson, 1999 [63] conclude that in order to prevail in the balance of gastro-intestinal microbial association the number of live beneficial probiotic bacteria shoud exceed 10^9 per gram product. Achieving this value requires a better understanding of the factors of cultivation, concentration, drying and storage.

3. Lactobacilli and bifidobacteria with probiotic properties – Foundation for the probiotics "Enterosan"

Lactobacilli, bifidobacteria and lactic acid cocci are isolated from different sources (from the intestinal tract of infants naturally fermented raw-meat dried products, naturally fermented sourdough, fermented vegetables, etc.) by contemporary breeding and genetic methods, they are identified using the methods of conventional taxonomy (morphological, physiological, biochemical, cultural) and molecular genetic methods (ARDRA, pulse gel electrophoresis, RAPD).

As a result of extensive breeding work on a wide range of strains of lactobacilli and bifidobacteria, strains suitable for incorporation in starter cultures for fermented milk products, probiotics and probiotic foods and beverages that have the ability to reproduce in the model conditions of digestion, to synthesize lactic and other organic acids, bacteriocins, by inhibiting the growth of pathogens that cause toxicity, toxicoinfections and fungal infections are selected. They allow the accumulation of high concentrations of viable cells in the process of fermentation, immobilisation, freeze-drying that retain their viability in storage conditions (Table 1) [64].

Genus *Lactobacillus*	genus *Bifidobacterium*	Lactic acid cocci
L.bulgaricus BG *L.bulgaricus* GB *L.bulgaricus* BB *L.helveticus* H *L.plantarum* 226-15 *L.plantarum* Sw *L.casei* C *L.acidophilus* 2 *L.acidophilus* A	*B.bifidum* 1H *B.bifidum* L1 *B.infantis* *B.breve* *B.longum*	*Pediococcus pentosaceus* *Lactococcus lactis* L4 *Streptococcus thermophilus* T3

Table 1. Strains of lactobacilli, lactic acid cocci and bifidobacteria with probiotic properties

The human organism is a complex biological system, which requires nutrients, air, water and energy for performing the thousands of biochemical reactions, which provide its normal functioning. The food in the stomach is subjected to transformation under the action of enzyme systems and with the direct participation of microorganisms. A part of them, which are related to the genera *Lactobacillus* and *Bifidobacterium*, form the group of the beneficial microorganisms. They digest substrates and through the metabolites, produced as a result of their vital activity, they inhibit and expel from the biological niche the pathogenic, toxigenic and putrefactive microorganisms.

The assimilation of nutrients by the toxigenic and putrefactive microorganisms, which form the group of the undesired microflora, leads to the synthesis of putrefactive and toxic metabolites, which impede the functions of separate systems and the oragnism as a whole.

Pathogenic microorganisms enter the digestive tract of humans and animals and cause digestive disorders and inflammation of the intestinal mucosa, when present in high concentrations (above 10^5cfu/g).

Some of the metabolites produced by lactic acid bacteria and bifidobacteria are lactic, acetic, citric and other organic acids, through which they acidify the medium and inhibit the growth of pathogens. Another group of substances with antimicrobial action are bacteriocins, which have protein nature.

The interactions between the selected group of lactobacilli [17] and bifidobacteria and the pathogens, representatives of *Enterobacteriaceae*, causing toxicoinfections and toxicoses, as well as fungal pathogens and the cancerogenic *Helicobacter pylori* are of great interest.

Pathogenic microorganisms of human origin - *Salmonella sp., Candida albicans, Proteus vulgaris, Enterococcus faecalis, Staphylococcus aureus* subsp. *aureus, Pseudomonas aeruginosa, Klebsiella pneumoniae* subsp. *pneumoniae, Escherichia coli* with viable cell counts of the suspensions above 10^{10}cfu/cm^3 are used as test-microorganisms. The investigations are conducted using the agar diffusion method. The results from these experiments are presented in Table 2.

Bifidobacteria have inhibitory activities close to that of lactobacilli (Table 2). When cultivated together, *B. breve, B. infantis, B. longum* and *B. bifidum* L1 exhibit greater antimicrobial effect in comparison with each one of the strains separately. The titratable acidity of the liquid supernatant is comparatively higher as well. They demonstrate certain synergism, which has a positive effect on human and animal organisms. Having in mind their distribution in the gastro-intestinal tract, they are the main regulators of the microflora in the colon.

Bifidobacteria belong to the symbionts particularly important to the human and animal organism. They are some of the first inhabitants of the digestive tract of the new-born mammals. Their importance is strengthened by their regulatory role in the colon.

Bifidobacteria have active metabolism, producing other organic acids (acetic, citric, tartaric) beside lactic acid. They exhibit antimicrobial activity against pathogenic and toxigenic microorganisms. Their significant synergism with lactobacilli and the rest of their probiotic properties, as well as their important place of habitat, define the important health-promoting role of bifidobacteria.

Strain / Test-microorganism	*Bif.breve*	*Bif.longum*	*Bif.infantis*	*Bif.bifidum* L1	Bifidobacterium symbiotic culture
Salmonella sp., 1,2.10^{12} cfu/cm^{3}*	14	9	7	10	15
C. albicans, 5.10^{8} cfu/cm^{3}	10	9 – 10	10	10	10
P.vulgaris, 5.10^{11} cfu/cm^{3}	11	9	7	8	13
E. faecalis, 2,2.10^{11} cfu/cm^{3}	13	11	10	9	12 – 13
S. aureus, 1,0.10^{12} cfu/cm^{3}	8	10	10	-	8
P.aeruginosa, 7.10^{10} cfu/cm^{3}	11	7	8	8	10
K. pneumoniae, 1,0.10^{11} cfu/cm^{3}	20	18	19	21	20 – 21
E.coli 1,5.10^{10} cfu/cm^{3}	12	10	9	11	10 – 11

* concentration of the cells of the test-microorganism in the agar medium.

Table 2. Antimicrobial properties of bifidobacteria

The antimicrobial activity of *L. acidophilus* 2, *L. bulgaricus* NBIMCC 3607, the symbiotic culture of *B. bifidum* L1, *B. longum, B. breve* on the growth of 11 strains of *H. pylori* of human origin is determined.

The symbiotic culture of bifidobacteria demonstrates the highest inhibitory effect on *H. Pylori* – the zones of inhibition are >10 mm for 50% of the strains (Table 3.)

L. acidophilus 2 and *L. bulgaricus* GB suppress the growth of half of the investigated strains of *Helicobacter pylori* (Table 3). It must be noted, that the model investigations on the influence of the tested cultures on the cells of *H. pylori* are conducted with liquid concentrates of *L. acidophilus* 2, *L. bulgaricus* GB, symbiotic culture of *B. bifidum* 1, *B. longum, B. breve* with viable cell counts above 10^{10} cfu/cm^3 and pH of the fermentation medium 6,3. This means that the action of part of the metabolites with antimicrobial activity is eliminated.

H. pylori Mc Farland	Inhibition zone of H. pylori, mm			pH
	L. acidophilus 2	*L. bulgaricus* GB	Symbiotic culture of bifidobacteria	
MF=1	7(10)	7(17)	7(12)	6,3
MF=0.5	7	7	7	6,3
MF =0.5	7	7	7	6,3
MF =0.5	7,5	7	7	6,3
MF=0,5	7	7	15,5	6,3
MF =0.5	9,3	10	13	6,3
MF=0.5	26	20	12	6,3
MF=0.5	7	7	20	6,3
MF=2	9	11	10,2	6,3
MF=0.5	11	9,7	14	6,3
MF=0.5	7	7	8,5	6,3

Table 3. Antimicrobial activity of *L. acidophilus* 2, *L. bulgaricus* GB, the symbiotic culture of *Bif. bifidum* 1, *Bif. longum, Bif. breve* against *H.pylori*

4. Antibiotic resistance of bifidobacteria

Antibiotics are substances with antimicrobial action, which influence both Gram-positive and Gram-negative bacteria. They inhibit the growth of or destroy microbial cells. In order to fulfill these functions, the antimicrobial substances must penetrate the cell, conjugate with a certain cell structure, which participates in a vital processes (DNA replication and cell division) or suppress them completely.

The effect of 22 antibiotics - β-lactam (penicillin, ampicillin, cefamndole, ciprofloxacin, amoxicillin, oxacillin, piperacillin, azlocillin), aminoglicoside (streptomycin, gentamicin, kanamycin, lincomycin, clindamycin, amikacin, vancomycin, tobramycin), macrolide (rifampin, erythromycin), tetracycline (tetracycline, doxicycline), aromatic (chloramphenicol) and nalidixic acid, on the growth of the selected lactobacilli - is studied. The 22 antibiotics belong to 3 groups with different mechanism of action – inhibition of the synthesis of the cell walls (penicillin, ampicillin, cefamndole, amoxicillin, oxacillin, piperacillin, azlocillin, vancomycin), inhibition of the protein synthesis (streptomycin, gentamicin, kanamycin, lincomycin, clindamycin, amykacin, tobramycin, rifampin, erythromycin, tetracycline, doxycycline, chloramphenicol), inhibition of the synthesis of DNA and/or cell division (ciprofloxacin and nalidixic acid). The investigated concentrations are equivalent to the actual concentration in *in vivo* antibiotic therapy.

All four strains of bifidobacteria (*Bif.bifidum* L1, *Bif.breve, Bif.infantis, Bif.longum*) are resistant to the action of most of the studied antibiotics with *Bif.bifidum* expressing the best results, followed by *Bif.breve, Bif.infantis* and *Bif.longum*. They show some sensitivity towards the action of aminoglicoside antibiotics. *Bif.bifidum* L1 demonstrates dense growth when tested against 18 out of the 22 antibiotics, weak growth when examined against 3 of the 22 antibiotics and it has single colonies in the clearance zone when tested against vancomycin. *Bif.breve* shows the following results: dense growth - 9 out of 22 antibiotics, weak growth – 11 out of 22 antibiotics, no growth – 2 out of 22 antibiotics. *Bif.infantis* exibits dense growth when tested against 5 out of 22 antibiotics, weak growth – 12 out of 22 antibiotics, single colonies in the clearance zone – 3 out of 22 antibiotics, no growth – 2 out of 22 antibiotics. *Bif.longum* is characterized with dense growth when examined against 15 out of 22 antibiotics, weak growth – 6 out of 22 antibiotics, no growth – 1 out of 22 antibiotics. These results reveal the possibility for the inclusion of the strains in the complex therapy against different diseases.

The resistance of the cells of the different *Lactobacillus* [17] and *Bifidobacterium* strains to 22 of the most frequently applied in medical treatment antibiotics reveals the possibility for their application in the cases of disbacteriosis. Moreover, it is better to use strains with natural polyvalent resistance as components of probiotics for the treatment of disbacteriosis.

5. Survival of bifidobacteria in the model conditions of the digestive tract

Bifidobacteria survive in the model conditions of the digestive tract – at low pH values in the presence of enzymes (pH=2 + pepsin) and at neutral pH values in the presence of enzymes (pH=7 + pepsin) (Fig. 1). The cells of the four strains are more sensitive to pH=2 + pepsin than to pH=7 + pepsin. At pH=2 + pepsin a reduction in the number of viable cells is observed; the reduction is by over 2 to approximately 5 log cfu/g at the 24th hour from the beginning of the cultivation in comparison to the baseline concentration of viable cells in the population; *Bif.infantis* and *Bif.longum* are more sensitive to pH=2 + pepsin than *Bif.breve* and *Bif.bifidum* L1. At pH=7 + pepsin the reduction in the number of viable cells is by over 1 to approximately 3 log cfu/g at the 24th hour from the beginning of the cultivation in comparison to the baseline concentration of viable cells in the population; *Bif.infantis* and *Bif.longum* are more resistant to pH=7 + pepsin than *Bif.breve* and *Bif.bifidum* L1.

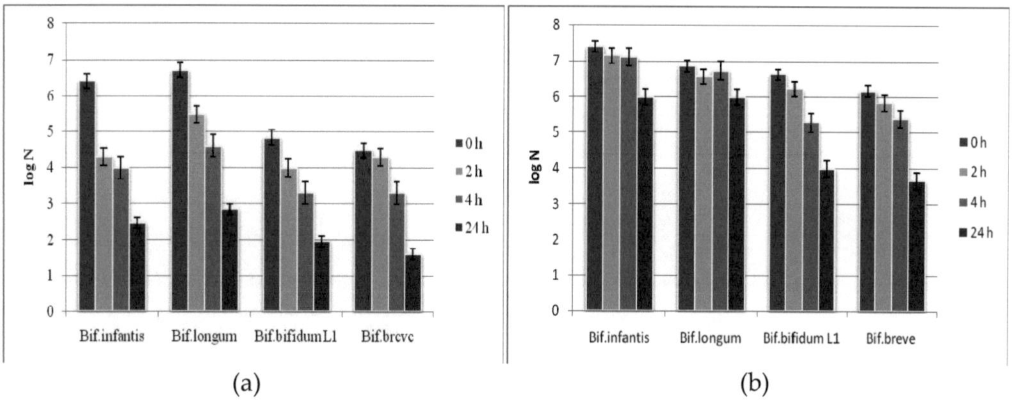

Figure 1. Reduction of the viable cells of bifidobacteria at pH=2 + pepsin (a) and at pH=7 + pepsin (b)

All bifidobacteria strains tested for their resistance to different concentrations of bile salts maintain high levels of viable cells (Fig. 2). An increase in the titre of viable cells at 0,15% bile salts is observed in *Bif.infantis* (Fig. 2a), *Bif.bifidum* L1 (Fig. 2b) and *Bif.longum (Fig. 2c)*, while in *Bif.breve* (Fig. 2d) the number of viable cells at 0,15% bile salts decreases from the very beginning of the experiment. At 0,3% bile salts the number of viable cells of *Bif.bifidum* L1 (Fig. 2b) and *Bif.longum* (Fig. 2c) increases during the first 8 hours, but at the 24th hour the cell count is lower than the value at the 8th hour in both the two strains. In *Bif.infantis* (Fig. 2a) and *Bif.breve* (Fig. 2d), the concentration of viable cells starts decreasing from the beginning of the test.

On the basis of these investigations four groups of probiotics „Enterosan" are developed: probiotics for the gastro-intestinal tract, probiotics for promotion of the functions of some endocrine glands, probiotics for functional usage and probiotics for deficiency diseases [65]. They have high concentration of viable cells of probiotic bacteria (over 10^9cfu/g).

The probiotics "Enterosan" have been tested by leading experts in clinics in our country and abroad and are proven to be beneficial to the human organism - for gastrointestinal infections, rotavirus infections, disbacteriosis due to antibiotics, in chemotherapy, in osteoporosis, arthritis, multiple sclerosis, allergies, anemia, high blood pressure, etc.

The road to developing a probiotic preparation is quite long. It begins with the selection of strains of microorganisms with probiotic properties, the development of probiotic formulations and the implementation of industrial process.

There are several probiotic products on the market but the documentation is often based upon case reports, animal studies or uncontrolled small clinical trials, and only few products declare the content of microorganisms [66].

In the conducted studies on the probiotic properties of different species and strains differences not only between different types of probiotic bacteria, but also between strains within a species are established; differences that should be taken into account in the selection of strains with probiotic properties for industrial use.

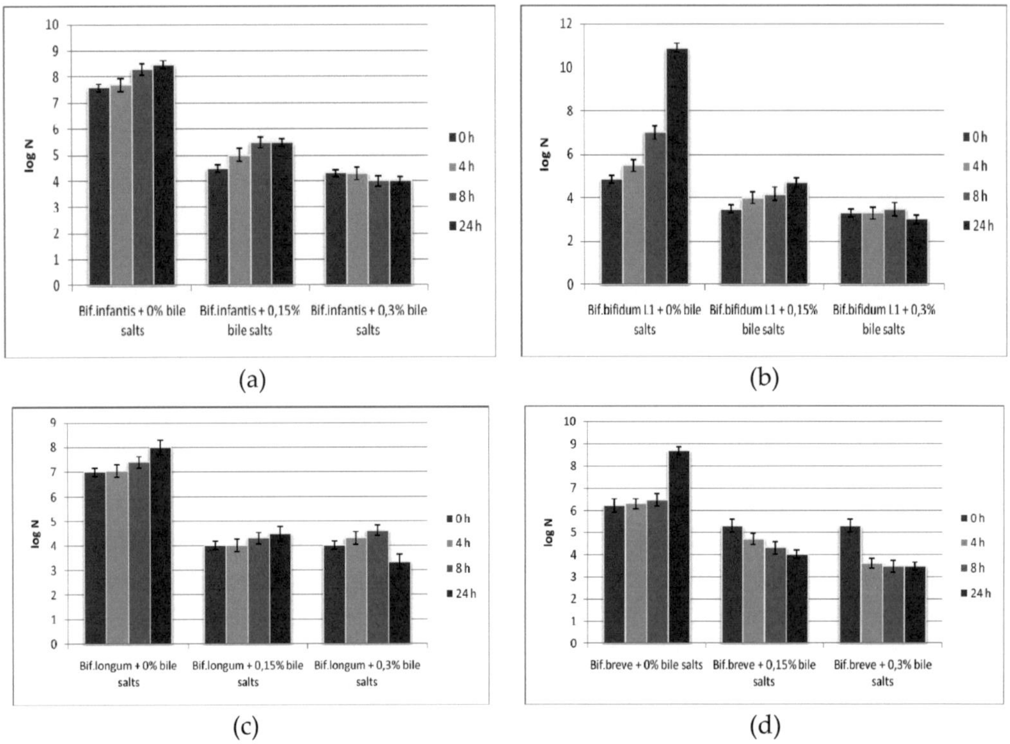

Figure 2. Change in the concentration of viable cells of *Bif.infantis* (a), *Bif.longum* (b), *Bif.bifidum* L1 (c), *Bif.breve* (d) at different concentrations of bile salts

6. Probiotic foods

6.1. Yoghurt with high concentration of viable cells of the probiotic strain *Lactobacillus delbrueckii* subsp.*bulgaricus* NBIMCC 3607

Lactic acid foods occupy a major place in the diet of our contemporaries. About 80% of the population use yoghurt for direct consumption or as a food supplement daily. A characteristic feature of this product is the addition of starters of pure cultures of *Streptococcus thermophilus* and *Lactobacillus delbrueckii* ssp.*bulgaricus* for conducting lactic acid fermentation. By using an appropriate technological process a product with characteristic taste and aroma, physicochemical and biological properties is obtained from milk as a raw material. These traditional lactic acid bacteria have a positive effect on the body, which is a result of the formed metabolites, which inhibit the putrefactive and pathogenic flora or of the improvement of the utilization of lactose [67].

Many functional foods include lactobacilli in their composition (Table 4). Lactobacilli are particularly important in the manufacture of probiotic foods [68]. Several species of the genus *Lactobacillus* are used as starters in the manufacture of yoghurt, cheese and other fermented liquid products [69, 70]. It should be noted that the properties of the strain itself are particularly important in the selection of probiotic cultures. Not all strains can be cultivated on an industrial scale because of the low reproductive capacity in the medium or because of their low survival rate in the processes of freezing and freeze-drying [71]. That is why the cultures used in the production of fermented foods must meet certain requirements (Table 5).

GENUS	SPECIES
Lactobacillus	*L. acidophilus; L. delbrueckii subsp. bulgaricus; L. casei; L. crispatus; L. johnsonii; L. lactis; L. paracasei; L. fermentum; L. plantarum;L. rhamnosus; L. reuteri; L. salivarius.*
Bifidobacterium	*B. adolescentis; B. bifidum; B. breve; B. essensis; B. infantis; B. lactis; B. longum*
Enterococcus	*E. faecalis; E. Faecium*
Pediococcus	*P. acidilactici*
Propionibacterium	*P. freudenreichii*
Saccharomyces	*S. boulardii*
Streptococcus	*S. thermophilus*

Table 4. Probiotic strains used in the production of fermented milk [72, 73, 74]

The selection of probiotic strains is based on microbiological criteria for food safety of the final product. This is achieved by applying non-pathogenic strains with clear health effects and proper hygiene [75].

The high concentration of viable cells and the good survival when passing through the stomach allow lactobacilli and bifidobacteria in fermented milk products to fulfill their biological role in the intestine.

Several properties of bacteria such as oxygen sensitivity, storage stability, resistance to the proteases of the digestive system, sensitivity to aldehyde or phenolic compounds produced by the metabolism of amino acids, antioxidant activity, adhesion to the intestinal mucosa are examined in *in vitro* testings [88,89]. Strains exhibit the specific properties of lactic acid bacteria in a different degree. The combination of strains with different properties allows the increase in the biological activity of fermented foods. This in turn is related to their ability to develop as symbiotic cultures.

Fermented milk products with probiotic properties are designed on the basis of the experience in the field of development of probiotics. Given that yogurt is the most popular food after bread a technology that includes the use of a starter culture with the probiotic strain *Lactobacillus delbrueckii* subsp.*bulgaricus* NBIMCC 3607, which has high reproductive capacity and meets all the requirements for probiotic cultures, has been developed. The technology is piloted for a period of over 1 year in industry. Table 6 presents the change of the acidity and the concentration of viable cells in the finished product during storage.

Industrial field	Criteria	Product	References
Suppliers of probiotic cultures	Cheap cultivation	Cultures for all groups of products	Charteris et al., 1998 [76]
	Easy concentration for obtaining high cellular density	Cultures for all groups of products	Charteris et al., 1998 [76]
Production of probiotic foods	Possibility for industrial production	Products, produced in high quantities (cheese)	Gomes and Malcata, 1999 [77]
	Compatibility with other lactic acid bacteria	All fermented products	Samona and Robinson, 1994 [78] Nighswonger et al., 1996 [79]
	Stability during storage at acidic conditions	Acidophilous milk, yoghurt, cheese	Micanel et al. 1997 [80] Gobbetti et al., 1998 [81]
	Stability during storage in non-fermented milk	Sweetened acidophilous milk	Brashears and Gilliland, 1995 [82]
	Resistance to bacteriophages	All fermented products	Richardson, 1996 [83]
	Survival in the conditions during the maturation and freezing of the ice cream	Ice-cream	Christiansen et al., 1996 [84]
	Tolerance to preservatives	Non-sterilized products	Charteris et al., 1998 [76]
	Stability during storage at temperatures under -20°C	Ice-cream, frozen products	Modler et al., 1990 [85] Christiansen et al., 1996 [84]
	Tolerance towards oxygen during growth	All fermented products	Gomes and Malcata, 1999 [77]
	Low activity at temperatures under 15°C	Cultures for all groups of products	Gomes and Malcata, 1999 [77]
	Utilization of pentanal and n-hexanal	Soy products	Scalabrini et al., 1998 [86] Murti et al., 1993 [87]
	Fermentation of raffinose and stachyose	Soy products	Scalabrini et al., 1998 [86]

Table 5. Some criteria applied in the selection of probiotic strains for fermented foods

Day	Titrable acidity, °T	Concentration of viable cells, cfu/cm^3		Proportion *Str.thermophilus* : *L.bulgaricus*	Extraneous microflora
		Str.thermophilus	*L.bulgaricus*		
1	104	$5x10^{11}$	$5x10^{11}$	1:1	Not found
15	106	$6,5x10^{11}$	$6,45x10^{11}$	1:1	Not found
30	108	$6x10^{11}$	$6x10^{11}$	1:1	Not found

Table 6. Physicochemical and microbiological indicators of yogurt produced using the new technology

The data show that the yoghurt produced according to this technology lasts for one month, during which the acidity is maintained within the standard requirements and the concentration of viable cells of *L.bulgaricus* NBIMCC 3607 in 1 gram of the product exceeds 1 billion by the end of the prolonged storage. Furthermore, the ratio of streptococci to lactobacilli is within the range of 1:1. A similar result can be achieved in any of the currently used technologies.

High concentrations of lactobacilli in yogurt increase its healing and preventive properties.Thus, the most popular product becomes probiotic.

6.2. Bio-yoghurt

Most of the strains of *Streptococcus thermophilus* and *Lactobacillus delbrueckii* subsp.*bulgaricus* do not retain in the intestinal tract, which limits the application of yogurt during antibiotic therapy and for other medical purposes. Therefore, probiotic bacteria are included in the composition of starter cultures for lactic acid products in addition to the traditional microorganisms *L.bulgaricus* and *Str.thermophilus,* which turns them into products with medicinal properties, known as bio-yoghurt (yogurt, dry mixes, ice cream, soft and hard cheeses, products for infant feeding).

The microflora of bio-yoghurt includes mainly *L.acidophilus, L.paracasei* ssp.*paracasei, L.paracasei* biovar *shirota, L.rhamnosus, L.reuteri, L.gasseri, Bifidobacterium infantis, Bif.breve, Bif.longum, Bif. bifidum, Bif.adolescentis* and *Bif.lactis* [90]. In addition to these species, some products contain *Bif.animalis,* which multiplies faster than other bifidobacteria, but unlike them it is not isolated from the intestinal tract of humans, although some *in vitro* studies show that some strains of *Bif.animalis* have the ability to attach to epithelial cells.

Many researchers believe that only species and strains isolated from the gastrointestinal tract of humans, provide probiotic effects on the human body. The digestive system of the fetus in the womb is sterile. It is inhabited within the first 2-3 days after birth. So right after birth the digestive system is inhabited by species and strains that form its gastro-intestinal microflora, as a result of natural selection, and they are better adapted to the conditions of the gastro-intestinal tract. Through this type of functional foods probiotic bacteria enter the body in the form of fermented milk.

Probiotic lactobacilli attach to special receptors on the epithelial wall and fill the vacant spots in the intestine. They utilize nutrients and produce lactic acid and substances with antimicrobial activity [90]. Their prophylactic role consists in changing the conditions, making them unsuitable for the development of bacteria that cause infections such as *Salmonella* sp. [90]. It has been shown that lactobacilli increase the levels of immunoglobulin Ig A and Ig G [91], thus protecting the immune system, lower cholesterol levels [59, 92], etc.

Bifidobacteria are located on the surface of the colon. In this part of the gastrointestinal tract different types of bifidobacteria utilize nutrients and produce lactic and acetic acids and antimicrobial substances (bacteriocins). The large amount of viable cells of bifidobacteria stimulate the walls of the colon to excrete the polysaccharide mucin that facilitates the passage of faeces through the colon, thereby preventing the colonization of cells of *E.coli, Candida* sp. thus protecting the body.

In recent years some yoghurt products have been reformulated to include live cells of strains of *L.acidophilus* and species of *Bifidobacterium* (known as AB-cultures) in addition to the conventional yoghurt organisms, *Str.thermophilus* and *L.bulgaricus*. Therefore bio-yoghurt is yoghurt that contains live probiotic microorganisms, the presence of which may give rise to claimed beneficial health effects [93]. In order to exert its probiotic effect, the number of viable cells of probiotic bacteria in bio-yoghurt should exceed 1 million [94] (10^8-10^9cfu/g) [10]. According to a Japanese standard the number of bifidobacteria in fresh milk must be at least 10^7 viable cells/ml. As far as the National Yoghurt Association (NYA) in the U.S. is concerned in the production of bio-yoghurt the concentration of lactic acid bacteria in the finished products must be 10^8 viable cells of lactic acid bacteria / g. Moreover, the culture must have rapid growth during fermentation as well as acid tolerance in order to maintain high microbial content during storage.

Technologies for obtaining probiotic yogurt from whole milk and lactic acid beverage with bifidobacteria from skimmed cow's milk with the participation of *Streptococcus thermophilus, Lactobacillus bulgaricus* and strains of the genus *Bifidobacterium* have been developed. The microbiological indicators of this probiotic milk are presented in Table 7.

With the inclusion of bifidobacteria in the starter culture for yoghurt a product with high concentration of active cells (more than 10^8cfu/g) with durability of 30 days is obtained (Table 7).

Day	Titrable acidity, °T	Concentration of viable cells, cfu/cm^3			Proportion *Str.thermophilus* : *L.bulgaricus*	Extraneous microflora
		Str.thermophilus	*L.bulgaricus*	Bifidobacteria		
1	114	7x10^{10}	3,5x10^{10}	7x10^9	1:2	Not found
15	120	7,7x10^{10}	3,75x10^{10}	2,9x10^9	1:2	Not found
30	126	6x10^9	3x10^9	5x10^8	1:2	Not found

Table 7. Physicochemical and microbiological indicators of the probiotic yogurt with bifidobacteria during storage at 4 ± 2°C

A technology for obtaining fermented milk beverage with bifidobacteria has been developed and implemented. The concentration of viable cells in the product is over 10^9cfu/cm^3, which is consistent with the requirements for the concentration of viable probiotic cells in bio-yogurt, required to perform health beneficial effects. The beverage retains the concentration of bifidobacteria cells for 40 days when stored at 4 ± 2ºC (Table 8).

Day	Concentration of viable cells, cfu/cm^3		Titrable acidity, °T	Extraneous microflora
	L. delbrueckii subsp. *bulgaricus*	*Bifidobacterium* sp.		
1	8,0x10^{12}	7,0x10^{10}	102	Not found
10	7,7x10^{12}	2,9x10^{10}	104	Not found
20	6,0x10^{12}	5,0x10^{9}	108	Not found
30	1,0x10^{11}	3,0x10^{9}	110	Not found
40	1,2x10^{11}	2,0x10^{9}	120	Not found
90	1,7x10^{8}	7,0x10^{8}	125	Not found

Table 8. Physicochemical and microbiological characterization of the probiotic milk during storage at 4 ± 2°C

A technology for obtaining other probiotic foods - acidophilous milk and milk, containing *Lactobacillus acidophilus* and bifidobacteria - has been developed as well, which expands the range of dairy foods with preventive role for humans, which in turn is the key to protecting public health.

Lactic acid bacteria are applied in the production of different types of cheeses. Other microorganisms that form the specific properties of cheeses are involved as well. Using molds to obtain cheeses not only radically alters the organoleptic characteristics of cheeses, but also requires changes in the production technology. Depending on the types of microorganisms in the composition of starter cultures, cheeses with starter cultures of mesophilic lactic acid bacteria, starter cultures of mesophilic and thermophilic lactic acid bacteria and propionic acid bacteria, with the participation of molds, bifidobacteria and/or *Lactobacillus acidophilus* - dietetic (functional) cheeses [90] are obtained.

In the production of certain hard cheeses with high temperature of the secondary heating propionic acid bacteria participate in the formation of the specific taste, flavor and texture of the product along with lactic acid bacteria. Propionic acid bacteria absorb part of the lactate, forming propionic and acetic acid and carbon dioxide. Therefore, as a component of these starter cultures the propionic acid bacterial species *Propionibacterium frendenreichii* subsp. *frendenreichii, Propionibacterium frendenreichii* subsp. *shermanii* and *Propionibacterium frendenreichii globosum* are included.

The research on the development of starter cultures for yoghurt conducted by our research team shows the importance of achieving symbiosis between the strains in the composition of the starter culture for the quality of the finished product. Few strains of *L. bulgaricus* can be used to obtain a symbiotic culture. The symbiosis between *L. bulgaricus* and *Str. thermophilus* determines the taste-aroma complex of the finished products to a great extent.

A starter culture for hard cheese with the inclusion of the strain *Propionibacterium frendenreichii* subsp. *frendenreichii* NBIMCC 328 with high antioxidant activity (catalase, peroxidase and superoxide reductase), determined by the ORAC method (Oxygen Radical Absorbance Capacity), antimicrobial ability, moderate lipolytic and proteolytic activity is created. The ability of the microorganisms to neutralize free radicals is important for milk production and health, since they enter the gastrointestinal tract with food and their growth continues after intake. Thus another source of antioxidants (bacteria capable of synthesizing antioxidants during growth) is ensured.

Probiotic bacteria are included in a starter culture for hard cheese with high temperature of second heating (50-52^{0}C), providing protection of the product in the process of maturation and storage.

At the end of the ripening process high content of beneficial microorganisms - lactic acid and propionic acid bacteria with concentration of 10^8cfu/g - remains. There are no representatives of the pathogenic microflora. Extraneous microflora is less than 100 cfu/g. Moreover, in the final hard cheese the concentration of viable cells is more than 10^8cfu/g. This opens up new paths for the usage of microorganisms with probiotic potential. The content of short-chain acids in the hard cheese with high temperature of secondary heating is determined by HPLC. The final product contains significant amount of propionate (14,9 mg/kg) and acetate (2 mg/kg).

Goat's milk improves blood composition and exhibits bactericidal properties, strengthens the immunity, accelerates the healing of bone traumas due to its significant levels of calcium, activates the work of the digestive glands and has anti-allergic properties. It also has a positive impact on diseases of the skin, joints, etc. It protects against tooth decay and helps build a healthy enamel. Gastric diseases are rapidly improved with goat's milk. In the cases of arthritis, rheumatism and all conditions in which acidic metabolic products occur predominantly such as diabetes, heart, lung, kidney, liver, etc. the health of the individual improves significantly after the inclusion of goat's milk in the diet. Goat's milk combined with soaked and peeled dates turns out to be useful combination in the case of gastric ulcer and in combination with dried figs in the case of arthritis.

Goat's milk is digested in the stomach 20 min after intake, unlike cow's milk, which requires 2 hours. Great part of the population eats goat's milk.

Yoghurts and yoghurt beverages from goat's milk with lactobacilli and bifidobacteria with probiotic properties are obtained as a result of the work of our research team (Table 9 and Table 10). They are characterized with high concentration of viable cells (above 10^8cfu/cm^3). Probiotic bacteria influence not only the functionality but also the flavor of these products.

Storage time / Bio-yoghurt	10 days			20 days			30 days		
	N, [cfu/cm^3]		TA, [ºT]	N, [cfu/cm^3]		TA, [ºT]	N, [cfu/cm^3]		TA, [ºT]
	LAB	bifidobacteria		LAB	bifidobacteria		LAB	bifidobacteria	
Bif. bifidum L1	5x10^{11}	3,2x10^{11}	97	4x10^{11}	8x10^{11}	110	6x10^{10}	7x10^{8}	108
MZ$_2$ control	2,7x10^{12}	-	68	3x10^{11}	-	112	3x10^{10}	-	126
Lactobacillus acidophilus 2	1,3x10^{12}	-	94	1,1x10^{12}	-	105	6x1010	-	119
Lactobacillus acidophilus 2 + *Bif. bifidum*	9x10^{12}	5,3x10^{11}	82	3,5x10^{12}	4,2x10^{11}	112,5	6x10^{11}	4x10^{8}	120

MZ$_2$ – starter culture for yoghurt containing a probiotic strain of *Lactobacillus delbrueckii* ssp.*bulgaricus* and *Streptococcus thermophilus*

Table 9. Concentration of viable cells (N) and titrable acidity (TA) of goat yoghurt, produced with a probiotic starter cultures during storage

Storage time / Beverage	1day			15 days			30 days		
	N, [cfu/cm^3]		TA, [ºT]	N, [cfu/cm^3]		TA, [ºT]	N, [cfu/cm^3]		TA, [ºT]
	LAB	bifidobacteria		LAB	bifidobacteria		LAB	bifidobacteria	
B. bifidum L1	3,5x10^{12}	1x10^{10}	90	3,75x10^{11}	1,9x10^{10}	98	4x10^{11}	1x10^{9}	102
MZ$_2$ control	7x10^{11}	-	62	6,58x10^{11}	-	90	5,8x10^{10}	-	100
Lactobacillus acidophilus 2	5x10^{11}	-	58	5x10^{11}	-	96	3,8x10^{10}	-	110

MZ$_2$ – starter culture for yoghurt containing a probiotic strain of *Lactobacillus delbrueckii* ssp.*bulgaricus* and *Streptococcus thermophilus*

Table 10. Change in concentration of viable cells (N) of probiotic lactobacilli and bifidobacteria and titrable acidity (TA) of goat yoghurt beverages during storage at temperature 4±2°C

Probiotic goat yoghurt and yoghurt beverages have high concentrations of viable cells of lactobacilli and/or bifidobacteria and can be applied as probiotic foods for 30 days.

6.3. Probiotic bacteria in the composition of the starter cultures for fermented sausages without heating

Biological preservation of ground meat is an important and topical issue for the meat industry. Its solution is associated with the search for suitable strains of microorganisms that provide protective properties and pleasant taste and flavor of the finished products. By applying this method of preservation a number of advantages can be achieved, the most important of which are extending storage, usage of softer modes of cold storage, etc. To achieve targeted fermentation and quality maturation in the production of cured meat products starter cultures of lactic acid bacteria are imported. A new trend in the production of dried meat products is the inclusion of probiotic strains in the composition of starter cultures. They provide proper conduction of the fermentation process in meat foods and significant amounts of microflora beneficial to human health.

Meat products, which are not treated thermally are suitable carriers of probiotic bacteria [95, 96]. Strains of lactic acid bacteria with probiotic properties as starter cultures for fermented sausages are given in Table 11. These species are isolated from the gastrointestinal tract. Human digestive tract is a natural biological environment for *Lactobacillus acidophillus, Lactobacillus casei* and *Bifidobacterium* sp. These microorganisms are found in various fermented foods [90, 97, 98, 99]. According to Anderssen, 1998 [97], however, lactobacilli isolated from the intestines do not grow and contribute to the implementation of fermentation of the meat substrate.

In the preparation of starter cultures for the meat industry various microbial species are included (Table 9).

Microorganisms	Species
Bacteria	
Lactic acid bacteria	*Lactobacillus acidophilus, L.alimentarius, L.casei, L.curvatus, L.plantarum, L.pentosus, L.sakei, Lactococcus lactis, Pediococcus acidilactici, P.pentosaceus*
staphylococci	*Staphylococcus xylosus, S.carnosus subsp. carnosus, S.carnosus subsp. utilis, S.equorum; Halomonadaceae, Halomonas elongata*
enterobacteria	*Aeromonas sp.*
Bifidobacterium sp.	
Actinomycetales	*Kocuria varians* *Streptomyces griseus*
Yeasts	*Debaryomyces hansenii, Candida famata*
Molds	*Penicillium nalgioevnse, Penicillium chrysogenum, Penicillium camemberti*

Table 11. Microbial species involved as components of starter cultures

A study conducted by Hammes et al., 1997 [100] clearly shows the beneficial effects of fermented meat products in the fermentation of which strains of the genus *Lactobacillus* with probiotic properties are used [100].

In studies conducted on representatives of the genus *Lactobacillus* it has been found that *Lactobacillus gasseri* JCM 1131T is suitable for meat fermentation. Moreover, *Lactobacillus gasseri* JCM 1131T and *Lactobacillus acidophilus* are the predominant species in the digestive tract of humans and *Lactobacillus gasseri* JCM 1131T has the ability to adhere to the gastrointestinal mucosa. Further research with this strain in meat environment shows some positive effects, but the culture is sensitive to the addition of NaCl and $NaNO_2$ and can only be used in meat products with low salt concentration without the addition of nitrites [101].

Lactobacillus sakei is widely used in meat industry as a species with probiotic properties, high antimicrobial activity against *Escherichia coli, Staphylococcus aureus* and *Listeria monocytogenes*, and ability to retain the sensory profile of meat products [102, 103, 104].

The strains *Lactobacillus plantarum* NBIMCC 2415 [18] and *Pediococcus pentosaceus* NBIMCC 1441 are selected. They grow well in meat environment at high concentrations of sodium chloride and low temperatures, since under such conditions the processes of salting, ripening and drying of these products are performed. They also have well expressed fermentative activity without gas release, reisitance to low pH, moderate proteolytic and lipolytic activity as well as antioxidant activity, which is associated with the formation of free amino acids, volatile fatty acids, carbonyl compounds and other substances that determine the taste and flavor of meat products, have good antimicrobial activity. The two strains are incorporated as starter cultures in a batch of sausage. *Lactobacillus plantarum* NIBMCC 2415 is imported as a monoculture with concentration of 10^8 cfu/g for the implementation of targeted lactic acid fermentation (batch I) and in a combination with *Pediococcus pentosaceus* NBIMCC 1441 (batch II). The microbiological parameters of the products are tested during the process of fermentation and drying. Experimental data are presented in Table 12 and Table 13.

Day	pH	Total Microbial Count, cfu/g	*S.aureus*, cfu/g	*Salmonella* sp., cfu/g	*E.coli*, cfu/g	*Enterococcus* sp., cfu/g	LAB, cfu/g
8	5,00	$1,1x10^3$	-	-	$3x10^3$	$1,1x10^3$	$2x10^9$
18	4,63	Under 10	-	-	-	$2,3x10^3$	$6x10^{11}$
28	5,1	Under 10	-	-	-	Under 10	$7,8x10^{10}$
40	5,5	Under 10	-	-	-	Under 10	$8x10^8$

Table 12. Microbiological parameters of the first batch of sausage in the process of fermentation and drying

Day	pH	Total Microbial Count, cfu/g	*S.aureus*, cfu/g	*Salmonella* sp., cfu/g	*E.coli*, cfu/g	*Enterococcus* sp., cfu/g	LAB, cfu/g
4	5,2	Under 10	-	-	-	Under 10^4	$8,4x10^9$
14	5,5	Under 10	-	-	-	Under 10^2	$5,4x10^{10}$
48	5,5	Under 10	-	-	-	Under 10	$3,5x10^{10}$

Table 13. Microbiological parameters of the second batch of sausage in the process of fermentation and drying

The extraneous microflora is suppressed and the total number of microorganisms is reduced as well as the number of coliforms and enterococci, which ensures safety of the product on one hand and maintaining its quality during fermentation on the other. The product also contains a high concentration of viable cells ($8x10^8$ - $3,5x10^{10}$cfu/g) of the probiotic strain *Lactobacillus plantarum* NIBMCC 2415, which turns the product into a probiotic and healthy one, and these indicators increase its durability and storage time [18].

6.4. Probiotic bacteria in the composition of bread sourdough

Bread is one of the main products in the diet of contemporary people. The quality of bread depends upon several factors. Intrinsic parameters of the flour, such as carbohydrate [105, 106], gluten [107], mineral element [108], lipid content [109, 110] and endogenous enzyme activity [111], and on the other hand extrinsic parameters referring to the breadmaking procedure, such as temperature, stages and extent of fermentation [112], water activity [113, 114], redox potential and additives [115, 116, 117], and incorporation of nutritional or rheological improvers, such as dairy ingredients [118], affect the quality of the final product. The effect of these factors can be either direct or indirect, by affecting the microflora, either this is supplied as a commercial starter or in traditional sourdough processes. These factors influence the microflora submitted in the form of a starter culture or traditional processes involving sour dough [119].

Bread is considered to be perishable food, microbial spoilage is observed quite often.The growth of molds causing huge economic losses and reduction of the safety of the bread due to the production of mycotoxins. Fungal spoilage of wheat bread is mainly due to *Penicillium* sp., which cause around 90% of wheat bread spoilage [120]. Other common bread spoilage molds belong to the genera *Aspergillus, Monilia, Mucor, Endomyces, Cladosporium, Fusarium* or Rhizopus [121]. At present a number of alternatives are applied to prevent or minimize microbial spoilage of bread, e.g. modified atmosphere packaging, irradiation, pasteurization of packaged bread and/or addition of propionic acid and its salts [121, 122].

Propionic acid has previously been shown to inhibit moulds and *Bacillus* spores, but not yeasts to a large extent, and has therefore been the traditional chemical of choice for bread preservation [123]. Legislation implemented under the European Parliament and Council Directive No. 95/2/EC requires that propionic acid may only be added to bread in a concentration not exceeding 3000 ppm [124]. However, recent studies have shown that under these conditions propionic acid is not effective against common bread spoilage organisms [125]. Additionally, a reduction of preservatives to sub-inhibitory levels might stimulate the growth of spoilage molds [126] and/or mycotoxin production [127, 128, 129]. Recent trends in the bakery industry have included the desire for high-quality foods, which are minimally processed and do not contain chemical preservatives, thus increasing the interest toward natural preservation systems [130].

Among the natural means for preservation of bread is the use of strains of lactic acid bacteria, which are imported in the form of sourdough [131, 132], providing fast and reliable stability of the dominant microflora in the production cycle. As components of the starter cultures selected strains homo- and heterofermentative lactic acid bacteria are applied. The latter utilize substrates with the formation of lactic and acetic acid, resulting in acidification of the medium (pH, total titratable acidity (TTK)) [105, 131, 133]. Acetate production by heterofermentative metabolism is of major importance for the development of flavour. The molar ratio between lactic to acetic acid in bread (fermentation quotient, FQ) is considered optimum in the range between 2.0 and 2.7 [131]. Production of suitable end-products during dough fermentation depends on the availability of soluble carbohydrates, which are attacked by the enzymes of the flour and the microbial enzyme systems [105, 134, 135, 136]. Metabolism of carbohydrates is species specific, even strain specific. It depends on the type of sugars, the co-presence of yeasts and the processing conditions [137].

Besides weak organic acids, i.e. lactic and acetic acid [138, 139, 140], LAB produce a wide range of low molecular weight substances [141], peptides [142] and proteins [143] with antifungal activity.

Sourdough is applied in the production of classic bread, sour bread, snacks, pizza and sweet baked goods. Sourdough fermentation increases the performance of the dough, improves the volume, texture, taste and nutritional value of the final product, slows down the loss of freshness and flavor and protects bread from mold and bacterial spoilage. These beneficial effects result from the appropriate balance between the metabolism of yeast strains and strains of hetero- and homofermentative lactic acid bacteria, which are the predominant microorganisms in natural sourdough. The metabolism of lactic acid bacteria is responsible for the production of organic acids and contributes, together with yeasts, to the production of aromatic components [144, 145, 146].

The activity of the lactobacilli in the composition of sourdough affects the protein fraction of flour during fermentation. This protein is particularly important for the quality of the bread, as the protein network of the bread determines its rheology, gas retention and thus the volume and texture of the bread. The substrates for the microbial conversion of amino acids in taste precursors and antifungal metabolites [147] are provided by proteolytic reactions. The levels of some peptides are reduced, which is helpful in the cases of inability to absorb cereal products by some people [148].

Bread with best quality is obtained by the simultaneous use of homo- and heterofermentative lactic acid bacteria in a certain ratio. Pure cultures of yeasts and lactic acid bacteria, imported in sufficient quantities provide fast and reliable stabilization of the dominant microbiota, normal fermentation process and actively participate in the quality of the finished bread. To observe this effect proper selection of species of lactic acid bacteria and process design, control over the purity and the activity of the cultures are required.

The strains *Lactobacillus casei* C, *Lactobacillus brevis* I, *Lactobacillus plantarum* NBIMCC 2415 and *Lactobacillus fermentum* J are isolated from naturally fermented sourdough, which defines their ability to grow in the mixture of flour and water, reaching high levels of viable cells and accumulating acid. Therefore, the growth of each of the four strains in the mixture of flour and water is examined. The change in the concentration of viable cells and the titratable acidity for 96 hours of cultivation at 30ºC is traced. The proportions for the repeated kneading every 24 hours are: first day - 44% flour: 56% tap water and 5% 48-hour culture of the strain, second to fifth day: 25% starter from the previous day: 75% new mix flour / water with ratio 44% / 56%. All four strains of lactobacilli grow well in the mixture of flour and water, reaching 10^9-10^{15}cfu/g within 96 hours and the TTA of the sourdoughs increases to around 10ºN (Table 14).

Strain \ Time, h	0 h		24 h		48 h		72 h		96 h	
	N, [cfu/g]	TTA, [^{0}N]	N, [cfu/g]	TTA, [^{0}N]	N, [cfu/g]	TTA, [^{0}N]	N, [cfu/g]	TTA, [^{0}N]	N, [cfu/g]	TTA, [^{0}N]
L.casei C	2x10^8	1,8	3x10^{11}	10,4	3,8x10^{11}	12	3,8x10^{12}	11,2	8,1x10^{14}	10,8
L.brevis I	7,6x10^8	1,9	3x10^{10}	13,2	4x10^{10}	13	4,2x10^{10}	8,6	5,6x10^{10}	8
Lactobacillus plantarum NBIMCC 2415	2x10^9	1,7	1,4x10^{11}	11,8	9x10^{12}	11	1,8x10^{13}	9	1,4x10^{15}	8,8
L.fermentum J	1,3x10^8	2	3x10^9	9,6	5,2x10^9	10,2	5,6x10^9	8,8	7x10^9	8

Table 14. Change in the concentration of viable cells (N) of lactobacilli and the total titrable acidity (TTA) of the medium in repeated kneading in flour/water mixture every 24 hours for 96 hours

Based on the results for the four strains of lactobacilli a starter culture for wheat bread is created by mixing them in a certain ratio. The ratio of is 2:1:1:1 = *Lactobacillus plantarum* NBIMCC 2415: *Lactobacillus casei* C: *Lactobacillus brevis* I: *Lactobacillus fermentum* J.

The accumulation of biomass and the change in TTA of the sourdoughs during the repeated kneading every 24 hours is determined. The following experiment scheme is applied: first day - 44% flour: 56% tap water and 10% of the combination; second to fifth day: 25% from the starter culture from the previous day : 75% new mix flour / water with ratio 44% / 56%. On the third day of repeated kneading yeasts are added to the sourdough (1g).

The results of the study on the starter culture for wheat bread are given in Table 15. The four strains develop with the accumulation of high concentrations of viable cells (over 10^{10}cfu / g) of lactobacilli and TTA increases to 17,3^{0}N.

In the sourdough molds have not been established. In addition to that, the metabolites formed by the lactic acid bacteria in the composition of the starter culture inhibit „wild" yeasts that get into sourdough through flours (Table 15). This ability is particularly important in sourdough fermentation of bread in repeated kneading for a long period of time - 6-9 months.

Starter culture \ Time, h	0 h				48 h				96 h			
	N [cfu/g]			TTA [^{o}N]	N [cfu/g]			TTA [^{o}N]	N [cfu/g]			TTA [^{o}N]
	LAB	M	Y		LAB	M	Y		LAB	M	Y	
Wheat sterter culture	3x10^9	3x10^1	Under 10	2,5	6,2x10^{11}	nf	Under 10	8,4	2x10^{10}	nf	Under 10	17,3

Table 15. Concentration of viable cells (N) of lactobacilli and of the Total Titrable Acidity (TTA) in the wheat starter culture and change in the microflora for 96 hours. LAB – lactic acid bacteria, M – molds, Y - yeasts, nf - not found

Along with determining the concentration of viable cells an organoleptic analysis of the starter culture is performed as well. The results show that for 48 to 72 hours of cultivation the starter culture reaches normal consistency of the sourdough and pleasant lactic acid flavor.

The starter culture is probated in industrial production - for the baking of bread 96-hour sourdough with different percentage is used; the percentage is determined by the weight of the used flour - 5%, 7% and 10%, according to the following scheme: 2 kg of flour, 1.5% NaCl, 2% yeasts, the respective percentage from the starter culture and tap water (the amount of water is determined by water absorption of the type of flour). Enhancers are added as well - 2 g/kg flour.

All the indicators of the sourdough and the bread are traced, so that the levels of incorporation of the sourdough would not adversely affect the rheological characteristics of the dough and the technologies adopted by manufacturers for the production of bread. The results of these experimental studies are presented in Table 16.

Wheat bread with the starter culture is baked as well as a control bread (without a starter culture). The data from the evaluation of the final bread with different percentages of the starter culture, including its strength and elasticity, the pieces of bread before and after baking, taste, flavor, etc. are shown in Table 16 and show acceleration of the fermentation process. The bread obtained with the starter culture is healthier, has more elasticity, the loaves of the bread are higher. The final wheat bread has softer and lighter crumb, with pleasant aroma and characteristic lactic acid odour.

	1	2	3	4
Sample	Control (without starter culture)	Starter culture 5%	Starter culture 7%	Starter culture 10%
TTA of the starter culture	-	15,6	15,6	15,6
Dough		Elastic	Elastic	Elastic
Rise of the dough [min]	52	50	48	52
Amount of water [%]	53	51	50	48
Temperature of the dough [ºC]	29.1	28.4	29.5	29.4
Pieces before baking		Higher that the control		
Rise of the dough [cm]	9.0	9.0	9.2	9.2
Baking (upper crust)	Normal	Normal	Normal	Normal
Aroma of the bread	Typical wheat bread aroma	Soft lactic acid aroma	Pleasant, characteristic lactic acid aroma	Strong an sharp characteristic lactic acid aroma
TTA of the bread	1.2	1.5	1.6	1.7

Table 16. Indicators characterizing the rheology of the sourdough, the flavor and aroma of the bread, prepared with 96-hour starter cultures

The created starter culture for sourdough for wheat bread improves its technological and organoleptic characteristics. Along with this it has been found to inhibit "wild" yeasts and mold spores in flour.

6.5. Soy probiotic foods

Soy foods are essential in the diet of the people in the Far East. They are rich in protein, supplying the body with all the essential amino acids for building and maintaining the tissues [149]. They are a source of flavones and isoflavones that exhibit antioxidant activity and can reduce the damage caused by free radicals [150]. Soybeans have stachyose and raffinose, oligosaccharides that are bifidogenic factors. The body is supplied with vitamins from groups B and D, mineral elements - calcium, magnesium, iron, etc. by traditional soy foods. Anti-cancer agents - protease inhibitors, saponins, phytosterols, phenolic acids, phytic acid and isoflavones, most of which are important flavones and isoflavones, which are polyphenolic compounds and relate to the group of plant estrogens, phytoestrogens, are also present in soy foods. The general term phytoestrogens refers to substances which have the effect of female hormones, but are not steroids. It is believed that soy foods play an important role in preventing chronic diseases such as menopausal disorders, cancer, osteoporosis, atherosclerosis.

Soy milk is obtained from dried, ripened, whole soybeans. They are soaked in fresh water for 16-18 hours at room temperature. The beans are washed, drained and ground. Hot potable water is added in a blender of Osterizer. The final suspension is filtered, autoclaved at 121ºC, stored overnight at 5ºC and it is processed to obtain soy milk products.

The dense residual mass is also rich in plant protein, vitamins C and E, calcium, manganese and iron and is a soy enrichment agent.

Soy milk contains no lactose. It replaces cow's milk for all people who suffer from allergies, lactase deficiency and milk protein intolerance. It can be used to carry out lactic acid fermentation with suitable strains of lactic acid bacteria (*Lactobacillus acidophilus, Lactobacillus delbrueckii* ssp. *bulgaricus, Lactobacillus casei, Leuconostoc mesenteroides, Lactococcus lactis* ssp. *lactis, Bifidobacterium longum, Bifidobacterium bifidum*) to obtain various fermented soy foods. It is a suitable environment for the development of new probiotic supplements. Having in mind the fact that it contains oligosaccharides, the obtained concentrates are synbiotics.

Soy milk yoghurt has been studied extensively [151, 152, 153]. Fermented soy milk products may provide economic and nutritional benefits, because they can be preparated at higher protein levels at comparable or lower cost than regular fermented milk products [154]. Soy proteins have favorable amino acid balance, meeting the essential amino acid, require ments, except for methionine [155]. The researches of a number of authors [156, 157, 158] show a lot of advantages of the soy milk products in the nutrition of children and adults, suffering from allergies, diabetes, cancerous, heart and renal diseases. Soy milk products and soy milk yoghurt successfully replace fermented milk products from cow's milk [157, 158].

By selection of strains of lactobacilli (*Lactobacillus acidophilus* A) and bifidobacteria (*Bifidobacterium bifidum* L1) alone and in a combination with streptococci (*Streptococcus thermophilus* T3) soy probiotic milk and beverages, characterized by high concentration of active cells of lactobacilli and bifidobacteria (10^{11} - 10^{14}cfu/g) and moderate titratable acidity, which allows 20 days of storage under refrigerated conditions, are obtained.

It has been shown that the antioxidant activity of fermented soy foods is significantly higher in comparison with unfermented soy foods.

Wang et al., 2006 [149] explores the influence of spray-drying and freeze-drying on fermented soy milk with *L.acidophilus* and *Str.thermophilus* and bifidobacteria - *Bif. longum* and *Bif. infantis*. The authors demonstrate increased antioxidant activity in fermented soy milk and the increase is species specific. Freeze-drying of soy milk leads to lower reduction of the antioxidant activity. This opens up new opportunities to use soy milk for obtaining probiotic supplements and probiotic soy milks and beverages.

Soy cheese can be obtained from soy milk coagulated as a result of the action of lactic acid bacteria. Soy cheese is the result of fermentation with starter cultures for soy cheese and the probiotic strain *L.rhamnosus*.

Probiotic lactobacilli and bifidobacteria may be included in other non-fermented soy foods - soy mayonnaise, soy delicacies, etc.in concentration 10^{6}-10^{7}cfu/g, which provides greater durability of soy foods.

Heenan et al., 2004 [159] includes *L.acidophilus, L.rhamnosus, L.paracasei* subsp.*paracasei, Sacch.boulardii* and *Bif.lactis* in concentrations 10^{6}cfu/g in frozen non-fermented vegetable soy desserts made from soy beverage, sugar, butter, salt and stabilizers.

Thus, the durability of soy foods increases as well as their biological effect on the body since they deliver beneficial microflora as well. That is how the preparation of healthy foods without the application of chemical preservatives is achieved. The role of the chemical preservatives is conducted by the imported probiotic cultures.

6.6. Probiotic bacteria in the fermentation of fruit, vegetables, fruit and vegetable juices

Almost all fruits and vegetables can undergo natural fermentation as they are inhabited by many types of lactic acid bacteria. The latter vary as a function of the microflora of the raw material, the temperature and the storage conditions [160]. Currently fermented cabbage, olives, cucumbers, carrots, lettuce, peas, corn, tomatoes, onions, pickles, radishes, Brussels sprouts, etc. are being produced mainly by natural fermentation. They allow fermentation with starter cultures as well. Lactic acid bacteria including the probiotic strains that are included as components of the starter cultures for fermented fruits and vegetables have the ability to grow in the fruit matrix and the cell vitality depends on the strain, the type of the substrate, the final acidity of the product [73], their resistance to high concentrations of salt in the medium, their ability to grow at temperatures around 18ºC, to reproduce rapidly and to accumulate acids, which acidify the environment and inhibit the growth of extraneous microflora. Most of them belong to the genera *Leuconostoc* (*Leuconostoc mesenteroides*), *Lactobacillus* (*Lactobacillus brevis, lactobacillus plantarum, Lactobacillus casei*) and *Pediococcus* (*Pediococcus pentosaceus*) [161, 162] and can be used as monocultures and as combinations. During its growth in vegetable juice *Leuconostoc* helps the growth of other lactobacilli and bifidobacteria by synthesis of dextranase [163].

Different strains are characterized with different sensitivity to the pH of the juice, to the acidification as a result of the fermentation, to the metabolic products, to the environmental conditions such as temperature, etc. [164, 165]. It has been shown that the optimum temperature for the development of probiotic strains is 35-40ºC and pH varies between 4,0 and 3,6 [6]. To protect the cells from the effects of the environmental factors agar, alginate, chitosan are used [165, 166, 167]. A probiotic banana product fermented with *Lactobacillus acidophilus*, included in alginate gel structures, is obtained. The inclusion of bacteria in alginate gel and carrageenan matrices protects the cells from the damages resulting from freezing and freeze-drying [168]. Encapsulation is applied in the production of probiotics as well [169].

Many fruits and vegetables allow processing to turn into media rich in nutrients, mineral elements, vitamins and antioxidants suitable for the growth of probiotic bacteria [170]. The probiotic strain *Lactobacillus plantarum* NBIMCC 2415 grows well in such medium (tomato juice) [18]. Tomato juice is a suitable medium for the growth of *Lactobacillus acidophilus, Lactobacillus casei, Lactobacillus delbrueckii* [170], which for 48 hours of growth at 30ºC reach concentration of 10^8cfu/ml. This probiotic beverage is kept at refrigerated temperature and maintains the amount of viable cells for 4 weeks. The same author obtained probiotic cabbage juice with the same strains of lactobacilli [171].

According to Rakin et al., 2007 [172] yeast autolysate can be added to vegetable juices before lactic acid fermentation. Its addition stimulates the growth of *Lactobacillus plantarum* and *Lactobacillus delbrueckii*.

Lactobacillus acidophilus and *Lactobacillus plantarum* can grow in red beet juice, reaching up to 10^9 cfu/ml viable cells and reducing the pH from 6.3 to 4.5.

Of course during the growth of probiotic bacteria in fruit and vegetable juices it is possible to obtain a product with specific flavor and aroma. In such cases the addition of fruit juices, which remove the off flavor, is needed.

All this suggests that probiotic bacteria represent a potential for obtaining fruit and vegetable functional foods because of their ability to grow in them and their resistance to acidic environments.

7. Conclusion

Beneficial microorganisms (lactobacilli and bifidobacteria) interact with other members of the intestinal microflora. The ability of the selected strains of lactobacilli and bifidobacteria to inhibit the growth of most representatives of *Enterobacteriaceae* which cause toxemia and toxicoinfections and some molds is a criterion that the microbial strains in the composition of probiotics and probiotic foods must meet. This is particularly important for the industry because of the sustainability of their growth to the majority of antibiotics used in modern health care - while pathogenic microorganisms can develop polyvalent resistance towards antibiotics, they can not do so against probiotic bacteria. The antimicrobial effect of the beneficial microflora is due to the synthesis of lactic, acetic and other organic acids and bacteriocins (proteins associated with microbial cells).

The intact intestinal epithelium with normal intestinal microflora serves as a barrier to the migration of pathogens, antigens and other harmful substances from the intestinal contents. Thus the host is protected and normal functioning of the intestines is provided. The impaired balance of the gastrointestinal microflora leads to diarrhea, intestinal inflammation, problems with the permeability or activation of carcinogens from the intestinal contents.

The future will undoubtedly show the many benefits of the combination of compatible symbiotic bacterial strains and prebiotics in functional foods.

So far probiotics are an effective alternative to antibiotics and chemotherapy, but in the coming years they are expected to demonstrate their suitability as therapeutic and prophylactic agents for many diseases associated with disorders of the digestive system.

As far as the products themselves are concerned future studies should be directed towards the selection of strains of lactobacilli and bifidobacteria with high probiotic effect and the development of technologies for the production of improved probiotics and probiotic foods.

Author details

A. Krastanov
University of Food Technologies, Department "Biotechnology", Plovdiv, Bulgaria

Z. Denkova
University of Food Technologies, Department "Organic Chemistry and Microbiology", Plovdiv, Bulgaria

Acknowledgement

We would like to thank prof. Ivan Murgov for his help in creating the probiotics „Enterosan" and some of the probiotic products.

8. References

[1] Kalliomaki M, Salminen S, Arvilommi H, Kero P, Koskinen P, Isolauri E (2001) Probiotics in primary prevention of atopic disease: a randomised placebocontrolled trial. Lancet 357: 1076–1079.

[2] Brown A C, Valiere A (2004) Probiotics and medical nutrition therapy. Nutr. Clin. Care 7: 56–68.

[3] Agerholm-Larsen L, Raben A, Haulrik N, Hansen A S, Manders M, Astrup A (2000) Effect of 8 week intake of probiotic milk products on risk factors for cardiovascular diseases. Eur. J. Clin. Nutr. 54: 288–297.

[4] Nomoto K (2005) Review prevention of infections by probiotics. J. Biosci. Bioeng.100: 583–592.

[5] Imasse K, Tanaka A, Tokunaga K, Sugano H, Ishida H, Takahashi S (2007) *Lactobacillus reuteri* tablets suppress *Helicobacter pylori* infectionda doubleblind randomised placebo-controlled cross-over clinical study Kansenshogaku zasshi. J. Jpn. Assoc. Infect. Dis. 81: 387–393.

[6] Shah N P (2007) Functional cultures and health benefits. Int. Dairy J. 17: 1262–1277.

[7] Hirayama K, Rafter J (2000) The role of probiotic bacteria in cancer prevention. Microbes Infect. 2: 681–686.

[8] Isolauri E (2001) Probiotics in human disease. American Journal of Clinical Nutrition, 73(6): 1142S–1146.

[9] Marteau P R, de Vrese M, Cellier C J, Schrezenmeir J (2001) Protection from gastrointestinal diseases with the use of probiotics. American Journal of Clinical Nutrition 73(Suppl. 2): 430S–436S.

[10] Rybka S, Kailasapathy K (1995) The survival of culture bacteria in fresh and freeze-dried AB yoghurts. The Australian Journal of Dairy Technology 50(2): 51–57.

[11] Mitsuoka T (1999) The human gastrointestinal tract. In: Wood BJB, editor. The lactic acid bacteria. vol.1, Gaithersburg, MD, USA: Aspen Publishers Inc.: 69-114 p.

[12] Kirtzalidou E, Pramateftaki P, Kotsou M, Kyriacou A (2011) Screening for lactobacilli with probiotic properties in the infant gut microflora. Anaerobe 17: 440 - 443.

[13] Donald J, Brown D (1993) Probiotics and the intestinal ecosystem. Let's live, November, 45 – 47.

[14] Garcıa-Fontan M C, Martınez S, Franco I, Carballo J (2006) Microbiological and chemical changes during the manufacture of Kefir made from cows' milk, using a commercial starter culture. Int. Dairy J. 16: 762–767.

[15] Penna A L B, Rao-Gurram S, Barbosa-Canovas GV (2007) Effect of milk treatment on acidification, physicochemical characteristics, and probiotic cell counts in low fat yogurt. Milchwissenschaft 62: 48–52.

[16] Rybka S, Kailasapathy K (1995) The survival of culture bacteria in fresh and freeze-dried AB yoghurts. The Australian Journal of Dairy Technology 50(2): 51–57.

[17] Denkova Z, Murgov I, Slavchev A (2006) Selection of lactobacilli with probiotic properties. Bulg. J. of agricultural sci. 12: 689-706.

[18] Nedelcheva P, Denkova Z, Denev P, Slavchev A, Krastanov A (2010) Probiotic strain *Lactobacillus plantarum* NBIMCC 2415 with antioxidant activity as a starter culture in the production of dried fermented meat products, Biotechnol. & Biotechnol. Eq. 24, 1: 1624-1630.

[19] Rivera – Espinosa Y, Gallardo-Navarro Y (2010) Non- dairy products. Food Microbiology 27: 1-11.

[20] Ganzle M G, Ehmann M, Hammes W (1998) Modelling in growth of *Lactobacillus sanfranciscensis* and *Candida milleriin* response to process parameters of sourdough fermentation. Applied and Enviromental microbiology 64 (7): 2616-2623.

[21] Katina K, Arendt E, Luikkonen K H, Autio K, Flander L, Poutanen K (2005) Potential sourdough for healthier cereal products. Trends in Food Science and Technology 16 (1-3): 104-112.

[22] Plessas S, Fisher A, Koureta K, Psarianos K, Nigan P, Koutinas A (2008) Application of *Kluveromyces marxianus, Lactobacillus delbrueckii* ssp. *bulgaricus* and *Lactobacillus helveticus* for sourdough bread making, Food chemistry 106: 985-990.

[23] Plessas S, Alexopoulos A, Mantzourani I, Koutinas A, Voidarou C, Stavropoulou E (2011) Application of novel starter cultures for sourdough bread production. Anaerobe 17: 486-489.

[24] Goldin B, Golbach S (1992) Probiotics for human. In R.Fuller (Ed.), Probiotics: the scientific basis. London:Chapman and Hall, 355-376 p.

[25] Ouwehand A, Salminen S, Isolauri E (2002) Probiotics: an overview of beneficial effects. Antonie Van Leeuwenhoek 82: 279-289.

[26] Lehto E M, Salminen S (1997) Adhesion of two *Lactobacillus* strains, one *Lactococcus* and one *Propionobacterium* strain to cultured human intestinal Caco-2 cell line. Biosci. Microflora 16: 13 – 17.

[27] Ouwehand A C, Tolko S, Kulmala J, Salminen S, Salminen E (2000) Adhesion of inactivated probiotic strains to intestinal mucus. Letters in Appl. Microbiology 31: 82-86.

[28] Ouwehand A, Tuomola E, Tolko S, Salminen S (2001) Assessment of adhesion properties of novel probiotic strains to humen intestinal mucus. International J.of Food Microbiology 64: 119 – 126.

[29] Saarela M, Mogensen G (2000) Probiotic bacteria: Safety, functional and technological properties. Journal of biotechnology 84: 197 – 215.

[30] Masuda K (1992) Heterogeneity of S-layer proteins of *Lactobacillus acidophilus* strains. Microbiol. Immunol. 36:297-301.

[31] Masuda K, Kawata T (1983) Distribution and chemical characterization of regular arrays in the cell walls of strains of the genus *Lactobacillus, FEMS Microbiology Letters* 20: 145–150.

[32] Yasui T, Yoda K, Kamiya T (1995) Analysis of S-layer proteins of *Lactobacillus brevis.* FEMS Microbiol. Lett. 133:181-186.

[33] Vidgrén, G, Palva I, Pakkanen R, Lounatmaa K, Palva A (1992) S-layer protein gene of *Lactobacillus brevis*: Cloning by polymerase chain reaction and determination of the nucleotide sequence. J. Bacteriol. 174:7419-7427.

[34] Boot H, Kolen C, van Noort J, Pouwels P (1993) S-layer protein of *Lactobacillus acidophilus* ATCC 4356: purification, expression in *Escherichia coli* and nucleotide sequence of the corresponding gene. J. Bacteriol.175:6089-6096.

[35] Callegari M, Riboli B, Sanders J, Cocconcelli P, Kok J, Venema G, Morelli L (1998) The S-layer gene of *Lactobacillus helveticus* CNRZ 892: cloning, sequence and heterologous expression. Microbiology 144:719-726.

[36] Sillanpää J, Martínez B, Antikainen J, Toba T, Kalkkinen N, Tankka S, Lounatmaa K, Keränen J, Höök M, Westerlund-Wikström B, Pouwels P, Korhonen T (2000) Characterization of the collagen-binding S-layer protein CbsA of *Lactobacillus crispatus*. J. Bacteriol. 182:6440-6450.

[37] Sleytr U, Beveridge T (1999) Bacterial S-layers. Trends Microbiol. 7:253-260.

[38] Sleytr U, Sára M, Pum D, Schuster B (2001) Characterization and use of crystalline bacterial cell surfacelayers. Prog. Surf. Sci. 68:231-278.

[39] Pelletier C, Bouley C, Cayuela C, Bouttier S, Bourlioux P, Bellon-Fontaine M (1997) Cell surface characteristics of *Lactobacillus casei* subsp.*casei, Lactobacillus paracasei* subsp. *paracasei*, and *Lactobacillus rhamnosus* strains. Applied and Environmental Microbiology 63: 1725–1731.

[40] Van der Mei H, van de Belt-Gritter B, Pouwels P, Martinez B, Busscher H (2003) Cell surface hydrophobicity is conveyed by S-layer proteins- a study in recombinant lactobacilli. Colloids Surf. B Biointerfaces. 28:127-134.

[41] Boot H, Pouwels P (1996) Expression, secretion and antigenic variation of bacterial S-layer proteins. Mol. Microbiol. 21:1117-1123.

[42] Ventura M, Jankovic I, Carey Walker D, David Pridmore R, Zink R (2002) Identification and Characterization of Novel Surface Proteins in *Lactobacillus johnsonii* and *Lactobacillus gasseri.* Applied and Environmental microbiology, Vol. 68, No. 12: 6172–6181.

[43] Vadillo-Rodríguez V, Busscher H, Norde W, de Vries J, Van der Mei H (2004) Dynamic cell surface hydrophobicity of *Lactobacillus* strains with and without surface layer proteins. J. Bacteriol. 186:6647-6650.

[44] Hynönen U, Westerlund-Wikström B, Palva A, Korhonen T (2002) Fibronectin-binding function in the SlpA surface protein of *Lactobacillus brevis*. J. Bacteriol. 184:3360-3367.

[45] Kotiranta A, Haapasalo M, Kari K, Kerosuo E, Olsen I, Sorsa T, Meurman J, Lounatmaa K (1998) Surface structure, hydrophobicity, phagocytosis, and adherence to matrix proteins of *Bacillus cereus* cells with and without the crystalline surface layer protein. Infect. Immun. 66:4895-4902.

[46] Schneitz C, Nuotio L, Lounatmaa K (1993) Adhesion of *Lactobacillus acidophilus* to avian intestinal epithelial cells mediated by the crystalline bacterial cell surface layer (S-layer). J. Appl. Bacteriol. 74:290-294.

[47] Toba T, Virkola R, Westerlund R, Bjorkman Y, Sillampaa J, Vartio T (1995) A collagen binding S-layer protein in *Lactobacillus crispatus, Applied and Environmental Microbiology* 61: 2467–2471.

[48] Kashtan H (1990) Manipulation of faecal pH by dietary means. Prev. Med. 19 (6): 607 – 613.

[49] Segal I (1995) Faecal short chain fatty acids in South African urban Africans and whites. Dis. Colon Rectum, 38 (7): 732 – 734.

[50] Erkkila S, Petaja E (2000) Screening of commercial meat starter cultures at low pH and in the presence of bile salts for potential probiotic use. Meat Sci. 55: 297–300.

[51] Klingberg T, Axelsson L, Naterstad K, Elsser D, Budde B (2006) Identification of potential probiotic starter cultures for Scandinavian-type fermented sausages, Int. J. of Food Microbiology 105: 419-431.

[52] Pennacchia C, Ercolini D, Blaiotta G, Pepe O, Mauriello G, Villani F (2004) Selection of *Lactobacillus* strains from fermented sausages for their potential use as probiotics. Meat Science 67: 309-317.

[53] Gilliland S, Speck M (1977) Deconjugation of bile acids by intestinal lactobacilli, Applied and Environmental Microbiology 33: 15-18.

[54] Pochart P, Mavtean P, Bouhnik Y, Goderel I, Bourlioux P, Rambrand J (1992) Survival of bififdobacteria ingested via fermented milk during their passage through the human small intestine: an in vivo study using intestinal perfusion. Am. J. Clin. Nutr. 55: 78-80.

[55] Nielsen E, Schlundt J, Gunvig A, Jacobsen B (1994) Epithelial mucus and lumen subpopulations of *Escherichia coli* in the large intestine of conventional and gnotobiotic rats. Microbial Ecol. Health Dis. 7: 263-273.

[56] Alander M, Korpela R, Saxelin M, Vilpponen-Salmela T, Mattila-Sandholen T, von Wright A (1997) Recovery of *Lactobacillus rhamnosus* GG from human colonic biopsies. Lett. Appl. Microbiol. 24: 361-364.

[57] Donohue D, Salminen S (1996) Safety assessment of probiotic bacteria. Asia Pac. J. Clin. Nutr. 5: 25-28.

[58] Donohue D, Salminen S, Marteau P (1998) Safety of probiotic bacteria. In: Salminen, S., von Wright, A(Eds.), Lactic acid bacteria. Marcel Dekker, New York, 369-384 p.

[59] Adams M R (1999) Safety of industrial lactic acid bacteria. J. Biotechnol., 68: 171-178.

[60] Nikolova D (2010) Probiotic and biotechnological characteristics of strains of the genus *Lactobacillus*, PhD thesis.

[61] Barefoot S, Klaenhammer T (1983) Detection and activity of lactacin B, a bacteriocin produced by *L.acidophilus*. Appl. Environ. Microbiol.: 1808 – 1811.

[62] Eswranandam S, Hettlarachy N, Johnson M (2004) Antimicrobial activity of citric, lactic, malic or martaric acids and nisin-incorporated soy protein film against *L. monocytogenes, E.coli* 0157:H7, and *Salmonella gaminara*. J. of Food Science, 69 (3), FMS 79.

[63] Wolfson N (1999) A probiotics primer. Nutrition Science News, 4(6): 276-280.

[64] Denkova Z, Krastanov A, Murgov I (2004) Immobilized lactic acid bacteria for application as dairy starters and probiotic preparations, J. Gen.Appl.Microbiol.50, № 2: 107-114.

[65] Denkova Z, Murgov I (2005) Probiotics "Enterosan"for protection of the health of today's man. J. Biotechnology & Biotechnical Equipment, vol.1 (19): 188-192.

[66] Ljiung A, Wasdstrom T (2005) Lactic acid bacteria as probiotics. Curr. Issues Intest. Microbiol., 7: 73-90.

[67] Hoier E (1992) Use of probiotic starter cultures in dairy products. Food Australia 44(9): 418–420.

[68] Klaenhammer T (1998) Functional activities of *Lactobacillus* probiotics: Genetic mandate. Int. Dairy J. 8: 497–506.

[69] Champagne C, Gardner N (2005) Challenges in the addition of probiotic cultures to foods. Critical Rev. Food Sci. Nutr. 45: 61-84.

[70] Drake M, Small C, Spence K, Swanson B (1996) Rapid detection and identification of *Lactobacillus* spp. in dairy products by using the polymerase chain reaction. J. Food Protection, 59: 1031–1036.

[71] Saxelin M, Salminen S (1996b) The safety of commercial products with viable *Lactobacillus* strains. Infectious Diseases Clinical Practice 5: 331 – 335.

[72] Tamime A, Robinson R (1999) Yogurt science and technology. 2nd edition. Woodhead Publishing, Cambridge and CRC Press, Boca Raton, 619 p.

[73] Shah N (2001) Functional foods from probiotics and prebiotics. Food Technol., 55(11): 46–53.

[74] Mercenier A, Pavan S, Pot B (2002) Probiotics as biotherapeutic agents: Present knowledge and future prospects. Current Pharmaceutical Design 8: 99–110.

[75] Gibson G, Fuller R (2000) Aspects of in vitro and in vivo research approaches directed toward identifying probiotics and prebiotics for human use. J. Nutr. 130: 391–395.

[76] CharterisW, Kelly P, Morelli L, Collins J (1998) Ingredient selection criteria for probiotic microorganisms in functional dairy foods. Int. J. Dairy Technol., 51(4): 123–136.

[77] Gomes A, Malcata F (1999) *Bifidobacterium* spp. and *Lactobacillus acidophilus*: Biological, biochemical, technological and therapeutical properties relevant for use as probiotics. Trends Food Sci. Technol. 10: 139–157.

[78] Samona A, Robinson R (1994) Effect of yogurt cultures on the survival of bifidobacteria in fermented milks. J. Soc.Dairy Technol. 47: 58–60.

[79] Nighswonger B, Brashears M, Gilliland S (1996) Viability of *Lactobacillus acidophilus* and *Lactobacillus casei* in fermented milk products during refrigerated storage. J. Dairy Sci. 79: 212–219.

[80] Micanel N, Haynes I, Playne M (1997) Viability of probiotic cultures in commercial Australian yogurts. Australian J. Dairy Technol. 52: 24–27.

[81] Gobbetti M, Corsetti A, Smacchi E, Zocchetti A, Angelis M (1998) Production of Crescenza cheese by incorporation of bifidobacteria. J. Dairy Sci. 81: 37–47.

[82] Brashears M, Gilliland S (1995) Survival during frozen and subsequent refrigerated storage of *Lactobacillus acidophilus* cells as influenced by the growth phase. J. Dairy Sci. 78: 2326–2335.

[83] Richardson D (1996) Probiotics and product innovation. Nutr. Food Sci. 4: 27–33

[84] Christiansen P, Edelsten D, Kristiansen J, Nielsen E (1996) Some properties of ice cream containing *Bifidobacterium bifidum* and *Lactobacillus acidophilus*. Milchwissenschaft 51: 502–504.

[85] Modler H, McKellar R, Goff H, Mackie D (1990) Using ice cream as a mechanism to incorporate bifidobacteria and fructooligosaccharides into the human diet. Cultured Dairy Products J., 25(3): 4–6, 8–9.

[86] Scalabrini P, Rossi M, Spettoli P, Matteuzzi D (1998) Characterization of Bifidobacterium strains for use in soymilk fermentation. Int. J. Food Microbiol. 39(3): 213–219.
[87] Murti T, Lamberet G, Bouillanne C, Desmazeaud M, Landon M (1993) Croissance des lactobacilles dans l'extrait de soja. Effets sur la viscosit´e, les compos´es volatils et la prot´eolyse. Sciences des Aliments 13(3): 491–500.
[88] Lin M, Yen C (1999) Antioxidative ability of lactic acid bacteria. J. Agricult. Food Chem. 47: 1460–1466.
[89] Godward G, Sultana K, Kailasapathy K, Peiris P, Arumugaswamy R, Reynolds N (2000) The importance of strain selection on the viability and survival of probiotic bacteria in dairy foods. Milchwissenschaft 55:441–445.
[90] Tamime A, Robinson R (2002) Yoghurt: Science and Technology (Eds.), CRC Press, New York, USA, 469-521 p.
[91] Salminen S, Ouwehand A, Benno Y, Lee Y (1999) Probiotics: How should they be defined? Trends Food Sci. Technol. 10: 107-110.
[92] Alerholm-Larsen L, Bell M, Astrup A (2000) The effect of probiotic milk product on plasma cholesterol: a meta –analysis of short-term intervention studies, Eur. Clin. Nutr. 54: 288-297.
[93] Lourens-Hattingh A, Viljoen B (2001) Yoghurt as probiotic carrier food, International Dairy Journal 11: 1-17.
[94] Speck M (1978) Enumeration of viable *Lactobacillus acidophilus* organisms in dairy products. Journal of Food Protection 41(2): 135–137.
[95] Hugas M, Monfort M (1997) Bacterial starter cultures for meat fermentation. Food Chemisrty 59:547-554.
[96] Incze K (1998) Dry fermented sausages, Meat Science 49: S169-S177.
[97] Anderssen L (1998) Fermented dry sausages with the admixture of probiotic cultures. 44th International Congress of Meat Science and Technology: 826-827.
[98] Arihara K, Luchansky J (1994) Dairy lactobacilli. Food Bacteriology: Microorganisms, Y.H. Hui and G.G. Khachatourians(Ed). VHC Publishers, New York 609-643 p.
[99] Goldin B, Golbach S (1992) Probiotics for human. In: Probiotics, the Saentific basis.(Ed. R. Fuller). Chapman and Hall, London 355-376 p.
[100] Hammes W (1997) Microecol.Therapy 26: 97.
[101] Arihara (1998) *Lactobacillus acidophilus* group Lactic Acid Bacteria Applied to Meat Fermentation. Journal of Food Scienc 63: 544-547.
[102] Bredholt S (2001) Industrial application of an antilisterial strain *Lactobacillus sakei* as protective culture and its effect on the sensory acceptability of cooked, sliced, vacuum-packaged meats. Int. J. Food Microbiol. 66: 191-196.
[103] Niessen H, Dainty R (1995) Comparison of the use of rRNA probes and conventional methods in identifying strains of *Lactobacillus sakei* and *Lactobacillus curvatus* isolated rom meat, Int. J. of Food Microbiology 25: 311-315.
[104] Ray B (2004) Fundamental Food Microbiology, GRC Press, Boca Raton 55–57 p.

[105] Collar C (1996) Biochemical and technological assessment of the metabolism of pure and mixed cultures of yeast and lactic acid bacteria in breadmaking applications. Food Sci Technol Int 2: 349–367.

[106] Henry R, Saini H (1989) Characterization of cereal sugars and oligosaccharides. Cereal Chem 66: 362–365.

[107] Callejo M, Gil M, Rodriguez G, Ruiz M (1999) Effect of gluten addition and storage ime on white pan bread quality: instrumental evaluation. Z Lebensm Unters Forsch A 208:27–32.

[108] Emodi A, Scialpi L (1980) Quality of bread fortified with ten micronutrients.Cereal Chem 57:1–3.

[109] Brooker B (1996) The role of fat in the stabilization of gas cells in bread dough. J Cereal Sci 24:187–198.

[110] Collar C, Armero E, Martinez J (1998) Lipid binding of formula bread doughs. Relationships with dough and bread technological performance. Z Lebensm Unters Forsch A 207: 110–121.

[111] Martinez-Anaya M (1996) Enzymes and bread flavor. J Agric Food Chem 44: 2469–2480.

[112] Spicher G, Bruemmer J (1995) Baked goods. In: Reed G, Nagodawithana TW, editors. Biotechnology. 2nd ed. VCH 243–319 p.

[113] Berland S, Launay B (1995) Rheological prperties of wheat flour doughs in steady and dynamic shear: effect of water content and some additives. Cereal Chem 72: 48–52.

[114] Czuchajowska Z, Pomeranz Y, Jeffers H (1989) Water activity and moisture content of dough and bread. Cereal Chem 66:128–132.

[115] Martinez-Anaya M, Jimenez T (1998) Physical properties of enzymesupplemented doughs and relationship with bread quality parameters. Z Lebensm Unters Forsch A 206: 134–142.

[116] Ravi R, Sai Manohar R, Haridas Rao P (2000) Influence of additives on the rheological characteristics and baking quality of wheat flours. Eur Food Res Technol 210: 202–208.

[117] Wikstrom K, Eliasson A (1998) Effects of enzymes and oxidizing agents on shear stress relaxation of wheat flour dough: additions of protease, glucose oxidase, ascorbic acid, and potassium bromate. Cereal Chem 75: 331–337.

[118] Kenny S, Wehrle K, Stanton C, Arendt E (2000) Incorporation of dairy ingredients into wheat bread: effects on dough rheology and bread quality. Eur Food Res Technol 210: 391–396.

[119] Paramithiotis S, ChouliarasY, Tsakalidou E, Kalantzopoulos G (2005) Application of selected starter cultures for the production of wheat sourdough bread using a raditional three-stage procedure, Process Biochemistry 40: 2813–2819.

[120] Legan J, Voysey P (1991) Yeast spoilage of bakery products and ingredients. Journal of Applied Bacteriology 70: 361–371.

[121] Legan J (1993) Mould spoilage of bread: the problem and some solutions. International Biodeterioration and Biodegradation 32: 33–53.

[122] Pateras I (1998) Bread spoilage and staling. In: Cauvain, S.P., Young, L.S. (Eds.), Technology of Breadmaking. Blackie Academic and Professional, London, 240–261 pp.

[123] Ponte J, Tsen C (1987) Bakery products, In: Beuchat, L. (Ed.), Food and Beverage Mycology, 2nd ed. AVI, New York, N.Y., 233–268 pp.

[124] European Union (1995) European Parliament and Council Directive No. 95/2/EC of 20February 1995 on food additives other than colours and sweeteners, p. 53. http://europa.eu.int/eur-lex/en/ consleg/pdf/1995/en_1995L0002_do_001.pdf..

[125] Lavermicocca P, Valerio F, Evidente A, Lazzaroni S, Corsetti A, Gobbetti M (2000) Purification and characterization of novel antifungal compounds from the sourdough *Lactobacillus plantarum* strain 21B. Applied and Environmental Microbiology 66: 4084–4090.

[126] Marin S, Sanchis V, Sanz D, Castel I, Ramos A, Canela R, Magan N (1999) Control of growth and fumonisin B1 production by *Fusarium verticillioides* and *Fusarium proliferatum* isolates in moist maize with propionate preservatives. Food Additives and Contaminants 16: 555–563.

[127] Yousef A, Marth E (1981) Growth and synthesis of aflatoxin by *Aspergillus parasiticus* in he presence of sorbic acid. Journal of Food Protection 44: 736–741.

[128] Gareis M, Bauer J, von Montgelas A, Gedek B (1984) Stimulation of aflatoxin-B1 and T-2-toxin production by sorbic acid. Applied and Environmental Microbiology 47: 416–418.

[129] Bullerman L (1985) Effects of potassium sorbate on growth and ochratoxin production by *Aspergillus ochraceaus* and *Penicillium* species. Journal of Food Protection 48: 162–165.

[130] Ryan L, Dal Bello F, Arendt E (2008) The use of sourdough fermented by antifungal LAB to reduce the amount of calcium propionate in bread, International Journal of Food Microbiology 125: 274–278

[131] Hammes W, GaËnzle M (1998) Sourdough breads and related products. In B. J. B. Woods (Ed.), Microbiology of fermented foods, (2nd ed.) London: Blackie Academic/Professional 199–216 pp.

[132] Messens W, De Vuyst L (2002) Inhibitory substances produced by Lactobacilli isolated rom sourdoughs—a review. International Journal of Food Microbiology 72: 31–43.

[133] Corsetti A, Gobbetti M, Balestrieri F, Paoletti F, Russi L, Rossi J (1998) Sourdough lactic acid bacteria effects on bread firmness and staling. Journal of Food Science 63: 347–351.

[134] Martinez-Anaya M (1996b) Carbohydrates and nitrogen related components in wheat sourdough processes. Advances in Food Sciences 18: 185–200.

[135] Gobbetti M (1998) The sourdough microflora: interactions of lactic acid bacteria and yeasts. Trends in Food Science and Technology 9: 267–274.

[136] Robert H, Gabriel V, Lefebvre D, Rabier P, Vayssier Y, Fontagner-Faucher C (2006) Study of the behaviour of *Lactobacillus plantarum* and *Leuconostoc* starters during a complete wheat sourdough breadmaking process, LWT 39: 256–265.

[137] Gobbetti M, Corsetti A, Rossi J (1994) The sourdough microflora, evolution of soluble carbohydrates during the sourdough fermentation. Microbiologie Aliments Nutrition 12: 9–15.

[138] Rocken W (1996) Applied aspects of sourdough fermentation. Advances in Food Science 18: 212–216.

[139] Rocken W, Voysey P (1995) Sourdough fermentation in bread making. Journal of Applied Bacteriology 79: 38S–48S.

[140] Stiles M (1996) Biopreservation by lactic acid bacteria. Antonie van Leeuwenhoek 70: 331–345.

[141] Niku-Paavola M, Laitila A, Mattila-Sandholm T, Haikara A (1999) New types of antimicrobial compounds produced by *Lactobacillus plantarum*. Journal of Applied Microbiology 86: 29–35.

[142] Okkers D, Dicks L, Silvester M, Joubert J, Odendaal H (1999) Characterisation of pentocin TV35b, a bacteriocin-like peptide isolated from *Lactobacillus pentosus* with a ungistatic effect on *Candida albicans*. Journal of Applied Microbiology 87: 726–734.

[143] Magnusson J, Schnurer J (2001) *Lactobacillus coryniformis* subsp. *coryniformis* strain Si3 produces a broad-spectrum proteinaceous antifungal compound. Applied and Environmental Microbiology 67: 1–5.

[144] Damiani P, Gobbetti M, Cossignani L, Corsetti, Simonetti M, Rossi J (1996) The sourdough microflora. Characterization of hetero- and homofermentative lactic acid bacteria yeasts and their interactions on the basis of the volatile compounds produced. Lebensmittel-Wissensschaft und Technologie 29: 63–70.

[145] Martinez-Anaya M (1996a) Enzymes and bread flavor. Journal of Agricultural and Food Chemistry 44: 2469–2480.

[146] Meignen B, Onno B, Gelinas P, Infantes M, Guilois S, Cahagnier B (2001) Optimization of sourdough fermentation with *Lactobacillus brevis* and baker's yeast. Food Microbiology 18: 239–245.

[147] Thiele C, Ganzle M, Vogel R (2002) Contribution of sourdough lactobacilli, yeast, and cereal enzymes to the generation of amino acids in dough relevant for bread flavour. Cereal Chemistry 79: 45–51.

[148] Vermeulen N, Kretzer J, Machalitza H, Vogel R, Ganzle M (2006) Influence of redox-reactions catalysed by homo- and hetero-fermentative lactobacilli on gluten in wheat sourdoughs, Journal of Cereal Science 43: 137–143.

[149] Wang Y, Yu R, Yang H, Chou C (2006) Antioxidatives activities of soymilk fermented with lactic acid bacteria and bifidobacteria. Food Microbiol. 23: 128–135.

[150] Lin M, Yen C (1999) Antioxidative ability of lactic acid bacteria. J. Agric. Food Chem. 47: 1460–1466.

[151] Shirai K, Pedrasa G, Duran M, Marshall V, Moiseev S, Garibay M (1992a) Production of a yogurt-like product from plant foodstuffs and whey: Substrate preparation and ermentation. Sci.Food Agric. 59:199-204.

[152] Shirai K, Duran M, Marshall V, Moiseev S, Garibay M (1992b) Production of a yogurt-ike product from plant foodstuffs and whey: Sensory evaluation and physical attributes. Sci.Food Agric. 59: 205-210.

[153] Yadav V, Jha Y, Garg S, Mital B (1994) Effect of soy milk supplementation and additives on sensory characteristic and biochemical changes of yogut during storage. The Australian J. Dairy Technol. 49: 34-38.

[154] Karleskind K, Laye I, Halpin E and Morr C (1991) Improving acid production in soy based yoghurt by adding cheese whey proteins and mineral salts. J. Food Sci. 56: 999-1001.

[155] Snyder H, Kwon T (1987) Soybeen utilization. AVI Publication, New York.

[156] Chang C, Stone M (1990) Effect of total soymilk solids on acid production by selected Lactobacilli. J. of Food Science 55, 6: 1643-1646.

[157] Kim K, Ko Y (1987) Study of growth and acid production by lactic acid bacteria in soy milk. Korean J. Food Sci. Technol. 19,2: 151-155.

[158] Nsofor L, Nsofor O, Udegbe C, Nwoke E (1996) Evalution of pure bacterial culture rom fermented cassava as soy-yoghut starter: a research note. J. Food Research nternational 29, 5-6: 549-553.

[159] Heenan C, Adams M, Hosken R, Fleet G (2004) Survival and sensory acceptability of probiotic microorganisms in a nonfermented frozen vegetarian dessert. Lebensm. Wiss. U. Technol. 37: 461–466.

[160] Bisakowski B, Atwal A, Gardner N, Champagne C (2007) Effect of lactic acid ermentation of onions (Allium cepa) on the composition of flavonol glucosides. Int. J. Food Sci. Technol. 42: 783–789.

[161] Yan P, Xue W, Tan S, Zhang H, Chang X, (2008) Effect of inoculating lactic acid bacteria starter cultures on the nitrite concentration of fermenting Chinese paocai. Food Control 19: 50–55.

[162] Randazzo C, Restuccia C, Romano A, Caggia C (2004) *Lactobacillus casei*, dominant species in naturally fermented Sicilian green olives. Int. J. Food Microbiol. 90: 9–14.

[163] Sanz M, Cote G, Gibson G, Rastall R (2006) Influence of glycosidic linkages and molecular weight on the fermentation of maltose-based oligosaccharides by human gut bacteria. J. Agric. Food Chem. 54: 9779–9784.

[164] Vinderola C, Reinheimer J (2003) Lactic acid bacteria: a comparative "in vitro" study of probiotic characteristics and biological barrier resistance. Food Res. Int. 36: 895–904.

[165] Guerin D, Vuillemard J, Subirade M (2003) Protection of Bifidobacteria encapsulated in polysaccharide–protein gel beads against gastric juice and bile. J. Food Protect. 66: 2076–2084.

[166] Kailasapathy K (2002) Microencapsulation of probiotic bacteria: technology and potential application. Curr. Issues Intest. Microbiol. 3: 39–49.

[167] Kourkoutas Y, Xolias V, Kallis M, Bezirtzoglou E, Kanellaki M (2005) *Lactobacillus casei* cell immobilization on fruit pieces for probiotic additive, fermented milk and lactic acid production. Process. Biochem. 40: 411–416.

[168] Tsen J, Lin Y, Huang H, King V (2007) Accelerated storage testing of freeze-dried mmobilized *Lactobacillus acidophilus*-fermented banana media. J. Food Process. Preserv. 31: 688–701.

[169] Favaro-Trindade C, Heinemann R, Pedroso D (2011) Developments in probiotic encapsulation, CAB Reviews: Perspectives in Agriculture, Veterinary Science, Nutrition and Natural Resources 6, No 004.

[170] Yoon K, Woodams E, Hang Y (2004) Probiotication of tomato juice by lactic acid bacteria. J. Microbiol. 42: 315–318.

[171] Yoon K, Woodams E, Hang Y (2006) Production of probiotic cabbage juice by lactic acid bacteria. Bioresource Technol. 97: 1427–1430.

[172] Rakin M, Vukasinovic M, Siler-Marinkovic S, Maksimovic M (2007) Contribution of lactic acid fermentation to improved nutritive quality vegetable juices enriched with brewer's yeast autolysate. Food Chem. 100: 599–602.

Chapter 5

Dairy Probiotic Foods and Coronary Heart Disease: A Review on Mechanism of Action

Fariborz Akbarzadeh and Aziz Homayouni

Additional information is available at the end of the chapter

http://dx.doi.org/10.5772/50432

1. Introduction

Coronary heart disease (CHD) is one of the major causes of death in adults in the developed and developing countries which is referred to the condition in which the main coronary arteries supplying the heart are no longer able to supply sufficient blood and oxygen to the heart muscle (myocardium). The main cause of the reduced flow is an accumulation of plaques, mainly in the intima of arteries, a disease known as atherosclerosis (Akbarzadeh and Toufan, 2008). A number of risk factors known to affect an individual to CHD have been categorized such as hyperlipidaemia (high levels of lipids in the blood), hypertension (high blood pressure), obesity, cigarette smoking and lack of exercise. Probiotics as a live microbial food supplement beneficially affects the host by improving its intestinal microbial balance and is generally consumed as fermented milk products containing lactic acid bacteria such as bifidobacteria and/or lactobacilli. The supposed health benefits of probiotics include improved resistance to gastrointestinal infections, reduction in total cholesterol and TAG levels and stimulation of the immune system. A number of mechanisms have been proposed to explain their putative lipid-lowering capacity and these include a 'milk factor', which has been thought to inhibit HMG-CoA reductase and the assimilation of cholesterol by certain bacteria. The mechanism of action of probiotics on cholesterol reduction include physiological actions of the end products of fermentation SCFAs, cholesterol assimilation, deconjugation of bile acids and cholesterol binding to bacterial cell walls. It has been well documented that microbial bile acid metabolism is a peculiar probiotic effect involved in the therapeutic role of some bacteria. The deconjugation reaction is catalyzed by conjugated bile acid hydrolase enzyme, which is produced exclusively by bacteria. The mechanism of cholesterol binding to bacterial cell walls has also been suggested as a possible explanation for hypocholesterolaemic effects of probiotics. Probiotics have received attention for their beneficial effects on the gut microflora and links to their systemic effects on the lowering of lipids known to be risk factors for CHD, notably cholesterol and TAG. The incorporation of probiotics into dairy products such as fermented milk products controlled nutrition studies need to be carried out to determine the beneficial effects of prebiotics, probiotics and synbiotics before substantial health claims can be made (Ranjbar et al., 2007a).

2. Probiotics

Probiotics are distinct as live microorganisms which, when administered in sufficient amounts present a health benefit on the host (FAO/WHO, 2002; Homayouni, 2008a; Homayouni, 2009). In recent years probiotic bacteria have increasingly been incorporated into dairy foods as dietary adjuncts. *Lactobacillus* and *Bifidobacterium* are the most common species of probiotic bacteria that were used in the production of fermented and non-fermented dairy products. Consumption of probiotic bacteria via dairy food products is an ideal way to re-establish the intestinal microflora balance (Homayouni, 2008a).

Probiotics have been shown to be effective against a number of disorders. Some mostly documented effects are relieving diarrhea, improving lactose intolerance and its immunomodulatory, anticarcinogenic, antidiabetic, hypocholesterolemic and hypotensive properties (Shah, 2007; Mai, and Draganov, 2009; Lye, et al., 2009). Probiotic bacteria, by competing with enteric pathogens for available nutrients and binding sites, reducing the pH of the gut, producing a variety of chemicals which inactivate viruses, enhancing specific and non-specific immune responses and increasing mucin production, can reduce incidence, severity and duration of diarrhea (Homayouni, et al., 2007; Allen, et al., 2010; Ejtahed, and Homayouni Rad, 2010). Alleviation of lactose intolerance symptoms by probiotic bacteria is attributed to their intracellular β-galactosidase content (Mustapha, et al., 1997). Studies have revealed that probiotic bacteria can induce many immunological changes and affect both Th1 and Th2 cytokine production and that these effects are strongly strain-specific (Lebeer, et al., 2010). Some major routes through which probiotic bacteria have been assumed to prevent cancer are: binding to mutagenic compounds thus decreasing their absorption, suppression of the growth of bacteria which convert procarcinogens to carcinogens, decreasing the activity of enzymes predictive of neoplasm including β-glucuronidase, nitroreductase and choloylglycine hydrolase as well as enhancing immune responses (Roos, and Katan, 2000). Inflammation plays a major role in both initiation and progression of diabetes (Duncan, et al., 2003; Pickup, and Frcpath, 2004). By reducing inflammatory responses, probiotics have been shown to correct insulin sensitivity and reduce development of diabetes mellitus. This anti-inflammatory effect has been proposed to be rooted in immunomodulatory properties of probiotic bacteria (Lye, et al., 2009). By reducing cholesterol absorption in the gut, incorporation of cholesterol into cell membranes, enzymatically deconjugation of bile salts and conversion of cholesterol to coprostanol, probiotics can reduce blood cholesterol (Lye, et al., 2009; Ooi, and Liong, 2010). Release of angiotensin converting enzyme (ACE) inhibitory peptides from the parent protein through proteolytic action explains how probiotics can exert antihypertensive effects (Lye, et al., 2009).

3. Dairy probiotic foods

Dairy probiotic foods are scientifically documented as having physiological benefits beyond those of basic nutritional values. Dairy products such as ice cream, cheese, yogurt, acidophilus-bifidus-milk, ayran, kefir, kumis, doogh containing probiotics and milk having omega-3, phytosterols, isoflavins, CLA, minerals, and vitamins have an outstanding position in the development of functional foods (Homayouni, et al., 2008b; Homayouni, et al., 2008c). Dairy beverages (both fermented and non-fermented) have long been considered as important vehicles for the delivery of probiotics. In fermentation process, lactic acid, acetic acid and citric acid are naturally produced which are commonly used organic acids to enhance organoleptic qualities as well as safety of many food products. Lactic acid bacteria are found to be more tolerant to acidity and organic acids than most of the pathogens and spoilage microorganisms (Homayouni, et al., 2008d).

4. Coronary heart disease (CHD)

Coronary heart disease (CHD) is one of the major causes of death in adults in the developed and developing countries which is referred to the condition in which the main coronary arteries supplying the heart are no longer able to supply sufficient blood and oxygen to the heart muscle (myocardium). The main cause of the reduced flow is an accumulation of plaques, mainly in the intima of arteries, a disease known as atherosclerosis (Akbarzadeh etal., 2003; Ranjbar et al., 2007b; Akbarzadeh etal., 2010; Ghaffari etal., 2010).

5. Main risk factors of coronary heart disease

CHD has assumed almost epidemic proportions in wealthy societies, whereas rheumatic heart disease is common in developing countries (Akbarzadeh etal., 2003; Akbarzadeh etal., 2008). Known risk factors of CHD can be classified into those that cannot be modified (being male increasing age, genetic traits including lipid metabolism abnormalities, body build, ethnic origin), those that can be changed (cigarette smoking, hyperlipidaemia, low levels of high density lipoprotein, obesity, hypertension, low physical activity, increased thrombosis, stress, alcohol consumption), those associated with disease states (diabetes and glucose intolerance) and those related to geographic distribution (climate and season, cold weather, soft drinking water) (Lovegrove and Jackson, 2003; Akbarzadeh etal., 2009a). It has been demonstrated that there is a strong and consistent relationship between total plasma cholesterol and CHD risk (Martin et al., 1986). Accumulation of LDL in the plasma leads to a deposition of cholesterol in the arterial wall, a process that involves oxidative modification of the LDL particles. The oxidized LDL is taken up by macrophages, which finally become foam cells and forms the basis of the early atherosclerotic plaque. It has been estimated that every 1% increase in LDL cholesterol level leads to a 2-3% increase in CHD risk (Gensini et al., 1998; Akbarzadeh etal., 2009b). HDL cholesterol levels are higher in women than in men. Factors that may lead to reduced HDL cholesterol levels include smoking, low physical activity and diabetes mellitus; whereas those that increase levels include moderate alcohol consumption (Assmann et al., 1998; Akbarzadeh etal., 2009c).

6. Probiotics and CHD: Mechanism of action

Diet is considered to control the risk of CHD through its effects on certain risk factors including blood lipids, blood pressure and probably also through thrombogenic mechanisms. New evidences suggest a protective role for dietary antioxidants such as vitamins E and C and carotenes, possibly through a mechanism that prevents the oxidation of LDL cholesterol particles (Lovegrove and Jackson, 2003). The diet is one of the adjustable risk factors associated with CHD risk which is recommends to reduce total fat (especially saturated fat), increasing Non-starch polysaccharides (NSP) intake and consumption of fruit and vegetables is advice that is expected to be associated with overall benefits on health.

As a result of low consumer compliance of low-fat diets, attempts have been made to identify other dietary components that can reduce blood cholesterol levels. These have included investigations into the possible hypocholesterolaemic properties of milk products, especially in a fermented form. 18% fall in plasma cholesterol after feeding 4-5 liters of fermented milk per day for three weeks (Mann, and Spoerry, 1974).

The mechanisms of action of probiotics on cholesterol reduction are physiological actions of the end products of fermentation SCFAs, cholesterol assimilation, deconjugation of bile acids and cholesterol binding to bacterial cell walls. The SCFAs that are produced by the bacterial anaerobic breakdown of carbohydrate are acetic, propionic and butyric. It has been well documented that microbial bile acid metabolism is a irregular probiotic effect involved in the therapeutic role of some bacteria. The deconjugation reaction is catalyzed by conjugated bile acid hydrolase enzyme, which is produced exclusively by bacteria. Deconjugation ability is widely found in many intestinal bacteria including genera Enterococcus, Peptostreptococcus, Bifidobacterium, Fusobacterium, Clostridium, Bacteroides and Lactobacillus (Hylemond, 1985). This reaction releases the amino acid moiety and the deconjugated bile acid, thereby reducing cholesterol reabsorption, by increasing faecal excretion of the deconjugated bile acids. Many in vitro studies have investigated the ability of various bacteria to deconjugate a variety of different bile acids. Grill et al. (1995) reported Bifidobacterium longum as the most efficient bacterium when tested against six different bile salts. Another study reported that Lactobacillus species had varying abilities to deconjugate glycocholate and taurocholate (Gilliland et al., 1985). Studies performed on in vitro responses are useful but in vivo studies in animals and humans are required to determine the full contribution of bile acid deconjugation to cholesterol reduction. Intervention studies on animals and ileostomy patients have shown that oral administration of certain bacterial species led to an increased excretion of free and secondary bile salts (De Smet, et al., 1998; Marteau, et al., 1995).

There is also some in vitro evidence to support the hypothesis that certain bacteria can assimilate (take up) cholesterol. It was reported that L. acidophilus and B. bifidum had the ability to assimilate cholesterol in in vitro studies, but only in the presence of bile and under anaerobic conditions (Gilliland, et al., 1985; Rasic, et al., 1992). However, despite these reports there is uncertainty whether the bacteria are assimilating cholesterol or whether the cholesterol is co-precipitating with the bile salts. Studies have been performed to address this question. Klaver and Meer (1993) concluded that the removal of cholesterol from the growth medium in which L. acidophilus and a Bifidobacterium sp. were growing was not due to assimilation, but due to bacterial bile salt deconjugase activity. The same question was addressed by Tahri et al., (1995) with conflicting results, and they concluded that part of the removed cholesterol was found in the cell extracts and that cholesterol assimilation and bile acid deconjugase activity could occur simultaneously.

The mechanism of cholesterol binding to bacterial cell walls has also been suggested as a possible explanation for hypocholesterolaemic effects of probiotics. Hosona and Tono-oka (1995) reported Lactococcus lactis subsp. biovar had the highest binding capacity for cholesterol of bacteria tested in the study. It was speculated that the binding differences were due to chemical and structural properties of the cell walls, and that even killed cells may have the ability to bind cholesterol in the intestine. The mechanism of action of probiotics on cholesterol reduction could be one or all of the above mechanisms with the ability of different bacterial species to have varying effects on cholesterol lowering. However, more research is required to elucidate fully the effect and mechanism of probiotics and their possible hypocholesterolaemic action.

It has been demonstrated that microbial bile acid metabolism is a main effect in the therapeutic role of probiotic bacteria. The deconjugation reaction is catalysed by conjugated bile acid hydrolase enzyme, which is produced by Bifidobacterium and Lactobacillus. This reaction releases the amino acid and deconjugated bile acid, which is reducing cholesterol reabsorption, by increasing faecal elimination of the deconjugated bile acids.

7. Conclusions and future trends

Risk factors known to affect an individual to CHD have been categorized such as hyperlipidaemia, hypertension, obesity, cigarette smoking and lack of exercise. Probiotics may prevent coronary heart disease by cholesterol reduction and microbial bile acid metabolism. The mechanism of action of probiotics on cholesterol reduction include physiological actions of the end products of fermentation SCFAs, cholesterol assimilation, deconjugation of bile acids and cholesterol binding to bacterial cell walls. It has been demonstrated that microbial bile acid metabolism is a peculiar probiotic effect involved in the therapeutic role of some bacteria. Deconjugation reaction is catalyzed by conjugated bile acid hydrolase enzyme, which is produced exclusively by bacteria. The mechanism of cholesterol binding to bacterial cell walls has also been suggested as a possible explanation for hypocholesterolaemic effects of probiotics. Probiotics have beneficial effects on the gut microflora and links to their systemic effects on the lowering of lipids known to be risk factors for CHD, notably cholesterol and TAG. In recent years, several probiotic foods were produced industrially. These foods have received attention for their beneficial effects on the gut microflora and links to their systemic effects on the lowering of lipids known to be risk factors for CHD. For progress to be made, the consumers need to be educated about the various health benefits and how they will be able to use these products in their own diet without adverse consequences. Also to make these foods attractive to the consumer, the products need to be priced in such a way that they are accessible to the general public.

Author details

Fariborz Akbarzadeh*
Faculty of Medicine, Tabriz University of Medical Sciences, Tabriz, I.R. Iran
Cardiovascular Research Center, Tabriz University of Medical Sciences, Tabriz, Iran

Aziz Homayouni
Department of Food Science and Technology, Faculty of Health and Nutrition, Tabriz University of Medical Sciences, Tabriz, I.R. Iran

8. References

Akbarzadeh, F. and Toufan, M. (2008). Atrioventricular Delays, Cardiac Output and Diastolic function in patient with implanted dual chamber pacing and sensing pacemakers, Pakistan Journal of Biological Sciences, 11 (20): 2407-2412.

Akbarzadeh, F., Hejazi, M. E., Koshavar, H. and Pezeshkian, M. (2003). Prevalence of cardiovascular diseases and cardiac risk factors in north western Tabriz, Medical Journal of Tabriz University of Medical Sciences, 11-15.

Akbarzadeh, F., Kazemi, B. and Pourafkari, L. (2009a). Supraventricular Arrhythmia Induction by an implantable cardioverter defibrillator in a patient with hypertrophic cardiomyopathy, Journal compilation, 1-5.

Akbarzadeh, F., Kazemi-arbat, B., Golmohammadi, A. and Pourafkari, L. (2009b). Biatrial Pacing vs. Intravenous amiodarone in prevention of atrial fibrilation after coronory artery bypass surgery, Pakistan Journal of Biological Sciences, 12 (19): 1325-1329.

Akbarzadeh, F., Pourafkari, L., Mohammad Hashemi Jazi, S., Hesami, L. and Habibi, H. (2010). Prevalence and severity of cad among hypertensive and normotensive patients undergoing elestive coronary angiographi in Tabrize madani heart center, ARYA Atherosclerosis Journal, 5: 1-5.

Akbarzadeh, F., Ranjbar kouchaksaraei, F., Bagheri, Z. and Ghezel, M. (2009c). Effect of Preoperative information and reassurance in decreasing anxiety of patients who are candidate for coronary artery bypass graft surgery, J. Cudirwsc. Thoruc. Rs. 25-28.

Akbarzadeh, F., Toufan, M. and Afsarpour, N. (2008). AV Interval and cardiac output in patient with implanted DDD pacemaker, Research Journal of Biological Sciences, 3 (12): 1381-1386.

Allen, S. J., Martinez, E. G., Gregorio, G. V. and Dans, L. F. (2010). Probiotics for treating acute infectious diarrhoea. Chocrane Collaboration.

Assmann, G., Cullen, P. and Schulte, H. (1998). The Münster Heart Study (PROCAM): Results of follow-up at 8 years, Eur. Heart J. 19: 2-11.

* Corresponding Author

De Smet, I., De Boever, P. and Verstraete, W. (1998). Cholesterol lowering in pigs through enhanced bacterial bile salt hydrolase activity, BJN, 79: 185-194.

Duncan, B. B, Schmidt, M. I., Pankow, J. S. and Ballantyne, C. M. (2003). Low-Grade Systemic Inflammation and the Development of Type 2 Diabetes. Diabetes; 52: 1799-1805.

Ejtahed, H. S. and Homayouni Rad, A. (2010). Effects of Probiotics on the Prevention and Treatment of Gastrointestinal Disorders. Microbial biotechnological journal of Islamic Azad University, 2(4): 53-60 [Persian].

Gensini, G. F., Comeglio, M. and Colella, A. (1998). Classical risk factors and emerging elements in the risk profile for coronary artery disease, Eur. Heart J. 19: 52-61.

Ghaffari, S., Akbarzadeh, F. and Pourafkari, L. (2010). Aneurysmal coronary arteriovenous fistula closing with covered stent deployment: A case report and review of literature, Cardiology Journal, 17: 1-4.

Gilliland, S. E., Nelson, C. R. and Maxwell, C. (1985). Assimilation of cholesterol by Lactobacillus acidophilus, Appl. Environ. Microbiol. 49(2): 377-381.

Grill, J. P., Manginot-durr, C., Schneider, F. and Ballongue, J. (1995). Bifidobacteria and probiotic effects: action of Bifidobacterium species on conjugated bile salts, Curr. Microbiol. 31: 23-27.

Homayouni, A. (2008a). Therapeutical effects of functional probiotic, prebiotic and symbiotic foods. (1st ed.). Tabriz University of Medical Sciences. Tabriz. Iran.

Homayouni, A. (2009). Letter to the editor. Food Chemistry, 114: 1073-1073.

Homayouni, A., Azizi, A., Ehsani, M. R., Razavi, S. H. and Yarmand, M. S. (2008b). Effect of microencapsulation and resistant starch on the probiotic survival and sensory properties of synbiotic ice cream. Food Chemistry, 111: 50-55.

Homayouni, A., Ehsani, M. R., Azizi, A, Yarmand, M. S. and Razavi, S. H. (2007). Effect of lecithin and calcium chloride solution on the microencapsulation process yield of calcium alginate beads. Iranian Polymer Journal, 16(9): 597-606.

Homayouni, A., Ehsani, M. R., Azizi, A., Razavi, S. H. and Yarmand, M. S. (2008c). Spectrophotometrically evaluation of probiotic growth in liquid media. Asian Journal of Chemistry, 20(3): 2414-2420.

Homayouni, A., Ehsani, M. R., Azizi, A., Razavi, S. H. and Yarmand, M. S. (2008d). Growth and survival of some probiotic strains in simulated ice cream conditions. Journal of Applied Sciences, 8(2): 379-382.

Hosono, A. and Tono-Oka T. (1995). Binding of cholesterol with lactic acid bacterial cells, Milchwissenschaft, 50(20): 556-560.

Hylemond, P. B. (1985). Metabolism of bile acids in intestinal microflora. In H. Danielson and J. Sjovall (eds), pp. 331-343, Sterols and Bile Acids, New York, Elsevier Science.

Klaver, F. A. M. and Van Der Meer, R. (1993). The assumed assimilation of cholesterol by lactobacilli and Bifidobacterium bifidum is due to their bile salt-deconjugating activity, Appl. Environ. Microbiol. 59(4): 1120-1124.

Lebeer, S., Vanderleyden, J. and Keersmaecker, S. C. J. (2010). Host interactions of probiotic bacterial surface molecules: comparison with commensals and pathogens. Nature Reviews Microbiology, 8: 171-184.

Lovegrove, J. and Jackson, K. (2003). Coronary heart disease in: Functional dairy products, (Eds: Tiina Mattila-Sandholm and Maria Saarela). Woodhead Publishing Ltd and CRC Press LLC. England, pp: 54-93.

Lye, H. S., Kuan, C. Y., Ewe, J. A., Fung, W. Y. and Liong, M. T. (2009). The improvement of hypertension by probiotics: effects on chelesterol, diabetes, renin and phytoesterogens. International Journal of Dairy Sciences, 10: 3755-3775.

Mai, V. and Draganov, P. V. (2009). Recent advances and remaining gaps in our knowledge of associations between gut microbiota and human health. World J Gastroenterol, 15(1): 81-85.

Mann, G. V. and Spoerry A. (1974). Studies of a surfactant and cholesteremia in the Maasai, Am J Clin Nutr 27: 464-469.

Marteau, P., Gerhardt, M. F., Myara, A., Bouvier, E., Trivin, F. and Rambaud, J. C. (1995). Metabolism of bile salts by alimentary bacteria during transit in human small intestine, Microbiol. Ecol. in Health and Disease, 8: 151-157.

Martin, M. J., Browener, W. S., Wentworth, J., Hulley, S. B. and Kuler, L. H. (1986). Serum cholesterol, blood pressure and mortality: implications from a cohort of 361662 men, Lancet, 2: 933-936.

Mustapha, A., Jiang, T. and Savaiano, D. A. (1997). Improvement of lactose digestion by humans following ingestion of unfermented acidophilus milk: influence of bile sensitivity, lactose transport and acid tolerance of lactobacilus acidophilus. Journal of Dairy Sciences, 80: 1537-1545.

Ooi, L. G. and Liong, M. T. (2010). Cholesterol-lowering effects of probiotics and prebiotics: a review of in vivo and in vitro findings. International Journal of molecular Sciences, 11: 2499-2522.

Pickup, J. C. and Frcpath, D. (2004). Inflammation and Activated Innate Immunity in the Pathogenesis of Type 2 Diabetes. Diabetes Care, 27: 813-823.

Ranjbar, F., Akbarzade, F. and Hashemi, M. (2007a). Quality of life of patients with impianted cardiac pacemakers in north west of Iran, ARYA Atherosclerosis Journal, 2 (4):197-203.

Ranjbar, F., Akbarzadeh, F., Kazemi, B. and Safaeiyan, A. (2007b). Relaxation therapy in the background of standard antihypertensive drug treatment is effective in management of moderate to severe essential hypertension, Relaxation therapy in management of hypertension, 32: 120-124.

Rasic, J. L., Vujicic, I. F., Skrinjar, M. and Vulic, M. (1992). Assimilation of cholesterol by some cultures of lactic acid bacteria and bifidobacteria, Biotech. Lett. 14 (1): 39-44.

Roos, N. M. and Katan, M. B. (2000). Effects of probiotic bacteria on diarrhea, lipid metabolism and carcinogenesis: a review of papers published between 1988 and 1998. American journal of clinical nutrition, 71: 405-411.

Shah, N. P. (2007). Functional cultures and health benefits. International dairy journal, 17: 1262-1277.

Tahri, K., Crocciani, J., Ballongue, J. and Schneider, F. (1995). Effects of three strains of bifidobacteria on cholesterol, Lett. Appl. Microbiol. 21: 149-151.

Chapter 6

Probiotics in Dairy Fermented Products

Emiliane Andrade Araújo, Ana Clarissa dos Santos Pires,
Maximiliano Soares Pinto, Gwénaël Jan and Antônio Fernandes de Carvalho

Additional information is available at the end of the chapter

http://dx.doi.org/10.5772/51939

1. Introduction

Since ancient times, food has been considered essential and indispensable to human life. Numerous studies clearly show that an individual's quality of life is linked to daily diet and lifestyle (Moura, 2005).

Interest in the role of probiotics for human health began as early as 1908 when Metchnikoff associated the intake of fermented milk with prolonged life (Lourens-Hattingh and Vilijoen, 2001b). However, the relationship between intestinal microbiota and good health and nutrition has only recently been investigated. Therefore, it was not until the 1960's that health benefit claims began appearing on foods labels.

In recent years, there has been an increasing interest in probiotic foods, which has stimulated innovation and fueled the development of new products around the world. Probiotic bacteria have increasingly been incorporated into foods in order to improve gut health by maintaining the microbial gastrointestinal balance. The most popular probiotic foods are produced in the dairy industry because fermented dairy products have been shown to be the most efficient delivery vehicle for live probiotics to date.

In this chapter, we will discuss the application of probiotic microorganisms in fermented dairy products, particularly cheeses. In addition, we will also discuss the benefits of probiotic fermented foods on human health.

2. Probiotic concepts

The word "probiotic" comes from Greek and means "for life" (Fuller, 1989). Over the years, the term "probiotic" has been given several definitions. "Probiotic" is used to refer to cultures of live microorganisms which, when administered to humans or animals, improve properties of indigenous microbiota (Margoles and Garcia, 2003). In the food industry, the term is described as "live microbial food ingredients that are beneficial to health" (Clancy, 2003).

It is important to mention that for a microorganism to be considered probiotic, (Figure 1), it must survive passage through the stomach and maintain its viability and metabolic activity in the intestine (Hyun and Shin, 1998). Native inhabitants of the human or animal gastrointestinal tract, such as lactobacilli and bifidobacteria, are considered to be probiotic, but often display low stress tolerance, which reduces their viability in probiotic applications. Microorganisms traditionally grown in fermented foods, such as lactic acid bacteria, propionibacteria and yeasts, are also considered for these applications..

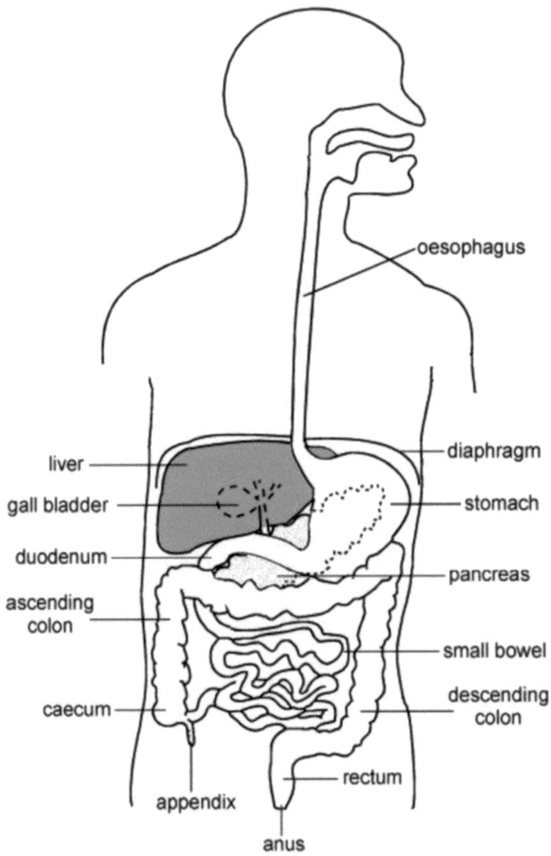

Figure 1. Schematic representation of gastrointestinal tract

It is essential that commercialized probiotic products which make health claims meet the minimum criterion of one million viable probiotic cells per milliliter of product at the expiration date. Accordingly, the minimum dosage of probiotic cells per day for any beneficial effect on the consumer is considered to be 10^8 – 10^9 probiotic CFU ml^{-1} or CFU g^{-1}, which corresponds to an intake of 100 g product containing 10^6 – 10^7 CFU ml^{-1} or CFU g^{-1} per day (Lorens-Hattingh and Viljoen, 2001a).

2.1. Selection of probiotic microorganisms

The human intestinal tract constitutes a complex ecosystem of microorganisms. The bacterial population in the large intestine is very high and can reach maximum counts of 10^{12} CFU g^{-1}. In the small intestine, the bacterial content is considerably lower at only 10^{4}–10^{8} CFU g^{-1}. In the stomach only 10^{1}-10^{2} CFU g^{-1} are found due to the low pH of the environment (Lorens-Hattingh and Viljoen, 2001b).

It is known that microbiota in the human intestine changes during human development. The intestine of newborn babies is fully sterile, however immediately after birth, colonization of many kinds of bacteria begins. On the first and second days after birth, coliforms, enterococci, clostridia and lactobacilli have been shown to be present present in infants' feces. Within three to four days, bifidobacteria begins colonization and becomes predominant around the fifth day. Simultaneously, coliform counts decrease. Breast-fed babies show 1 log-count more of bifidobacteria in feces than bottle-fed babies. Enterobacteriaceae, streptococci, and other putrefactive bacteria counts are higher in bottle-fed babies, suggesting that breast-fed babies are more resistant to gastrointestinal infections than the bottle-fed infants (Lorens-Hattingh and Viljoen, 2001b).

In addition to the microbiota changes that occur during human aging, the microbiota in the gastrointestinal system can also change because of the food and health conditions of an individual. For example, use of antibiotics can damage the equilibrium of intestinal microbiota, reducing counts of bifidobacteria and lactobacilli and increasing clostridia. The ensuing imbalance can cause diarrhea in elderly and immunocompromised people.

To help improve the balance of intestinal microbiota, probiotic microorganisms can be added to the human diet in order to stimulate the growth of preferred microorganisms, crowd out potentially harmful bacteria, and reinforce the body's natural defense mechanisms.

The selection of probiotic microorganisms is based on safety, functional and technological aspects, as reported by (Saarela et al., 2000). These are summarized in Figure 2.

Certain probiotic bacteria have been extensively studied and are already on the market, as shown in Table 1.

Before probiotic strains can be delivered to consumers, they must first be able to be manufactured under industrial conditions. They must then survive and retain their functionality during storage as frozen or freeze-dried cultures, as well as in the food products into which they are finally formulated. Moreover, they must be able to be incorporated into foods without producing off-flavors or textures (Saarela et al., 2000).

Functional food requirements must take into consideration the following aspects in relation to the probiotics: The preparation should remain viable for large-scale production; it should remain stable and viable during storage and use; it should be able to survive in the intestinal ecosystem (Prado et al., 2008).

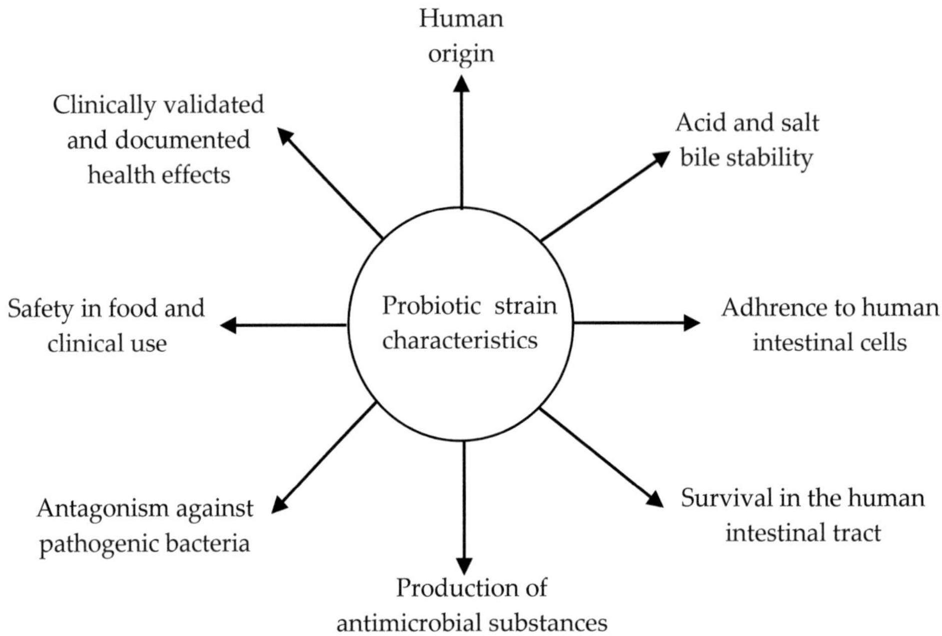

Figure 2. Theoretical basis for selection of probiotic microorganism selection (adapted from Saarela et al., 2000).

Strains	Origin
Lactobacillus casei Shirota	Yakult, Japan
Lactobacillus reuteri MM53	BioGaia, Sweden
Bifidobacterium lactis HN019	Danisco, France
Lactobacillus rhamnosus GG	Valio, Finland
Lactobacillus acidophilus NCFM	Nestle, Switzerland
Lactobacillus casei DN-173 010	Danone, France
Lactobacillus casei CRl-431	Chr. Hansen, USA
Bifidobacterium animalis BB12	Chr. Hansen, Denmark
Bifidobacterium animalis DN173010	Danone, France

Source: Prado et al., 2008

Table 1. Probiotic bacteria marketed worldwide

3. Beneficial effects of probiotics

The role of balanced nutrition for health maintenance has attracted the attention of the scientific community, which in turn has produced numerous studies in order to prove the performance of certain foods in reducing the risk of Some diseases. There has also been considerable growing interest in encouraging research into new natural components (Thamer and Penna, 2006).

In a healthy host, a balance exists among members of the gut microbiota, such that potential pathogenic and non-pathogenic organisms can be found in apparent harmony. In the case of bacterial infection, this balance can become disturbed, leading to often dramatic changes in the composition.

For most bacterial infections, nonspecific antibiotics are used, killing both non-pathogenic members of gut microbiota as well as pathogenic members. This can lead to a substantial delay in the restoration of healthy gut microbiota (Reid et al, 2011). The restoration of the gut microbiota balance is believed to be important because maintaining a healthy and balanced gut microbiota throughout life is thought to help preserve health and favor longevity.

The most comprehensive analysis of human microbiota to date examined 27 distinct sites in the body and revealed the presence of 22 bacterial phyla, with most sequences (92.3%) related to just four phyla: Actinobacteria (36.6%), Firmicutes (34.3%), Proteobacteria (11.9%) and Bacteroidetes (9.5%) (Costelo, 2008).

The metabolic capacity of gut bacteria is extremely diverse. This diversity is influenced by the large number of bacterial genera and species. Lactic acid species are present, as well as peptide-degrading bacteria, amino acids, and other methanogenic bacteria components of the gut microbiota which grow with the intermediate products of fermentation such as hydrogen, lactate, succinate and ethanol (Topping and Clifton, 2001).

In host's diet residue (matter undigested by its digestive system including resistant starch, fibers, proteins and peptides) substrates for primary fermentation can be found. Other important available substrates derive from mucin glycoproteins, exfoliated epithelial cells and pancreatic Secretions (MacFarlane et al., 1992).

Hydrolysis and carbohydrate metabolism in the large intestine is influenced by a variety of physical, chemical, biological and environmental parameters. Probably the nature and quantity of available substrate that has greater meaning, making the diet easier and the main mechanism by which to influence the profile of fermentation. Other factors affecting the colonization and growth of bacteria in the intestine are intestinal pH, which inhibits the production of metabolites (acids and peroxides) and specific inhibitory substances (bacteriocins), bile salts and molecules and cells which constitute the immune system (Rastall et al., 2000) .

Knowledge of intestinal gut microbiota and their interactions led to the development of food strategies aimed at the stimulation and maintenance of normal bacteria present in the gut (Gibson and Fuller, 2000).

According to Wohlgemuth (2010), strategies for studying mechanisms of probiotic action involve in-vitro models, or conventional or gnotobiotic animal models, plus development of a simplified human intestinal gut microbiota. Wohlgemuth's article proposes certain requirements that a model should ideally fulfill:

- Selected bacterial species should represent numerically dominant organisms of the human gut microbiota.
- By and large, the metabolic activity of this community should mimic that of normal human gut microbiota.
- The genome sequence of all members of the microbial community should be known.
- The members of this consortium should form a stable community in rodents. It should be possible to maintain this community under gnotobiotic conditions from generation to generation.
- The composition of the microbial community should be modifiable when required.

It is possible to increase the number of health-promoting microorganisms in gut microbiota through the introduction of probiotics in the diet. The probiotics will selectively modify the composition of the gut microbiota, providing the probiotic microorganisms demonstrate a competitive advantage over other bacteria in the ecosystem (Crittenden, 1999). Probiotic therapeutic properties are listed in Table 2.

Probiotic therapeutic properties
Influence on host gut microbiota and pathogenic bacteria
Improvement of specific enzymatic activities
Production of antibacterial substances
Competitive exclusion of pathogenic bacteria
Induction of defensin production
Improvement of intestinal barrier function
Modulation of host immune functions
Modulation of intestinal carcinogenesis
Modulation of cholesterol uptake

Wohlgemuth et al. (2010); Reddy and Rivenson (1993); Chen et al. (1984); Zhu et al., Cancer letters (2011); Jones et al., Br J Nutr (2012)

Table 2. Therapeutic Properties of Probiotics

There is a growing body of evidence that ingested beneficial bacteria, called probiotics, can beneficially modulate chronic intestinal inflammation, diarrhea, constipation, vaginitis, irritable bowel syndrome, atopic dermatis, food allergies and liver disease (Wallace et al., 2011, Nutrition reviews).

Probably the most promising area is the alleviation of symptoms linked to inflammatory bowel diseases (IBD), a growing health concern. As an example, the probiotic preparation VSL#3 induced remission in children (n=18) with mild to moderate ulcerative colitis (UC) (Huynh et al., 2009, Inflamm. Bowel Dis.) Accordingly, VSL#3 was tested in a 1-year, placebo-controlled, double-blind clinical study on UC children (n=29). Remission was achieved in 36.4% of children receiving IBD therapy and placebo, but in 92.8% of children receiving IBD therapy and VSL#3 (Milele et al., 2009, Am J Gastroeterol.) Similar promising results were obtained with the probiotic *Escherichia coli* Nissle 1917 strain (Kruis et al., 2004, Gut ; Do et al., Ann Pharmacother, 2010). However, a review of available data indicates that more clinical studies are needed to confirm the beneficial effects of these products in UC and in inactive pouch patients (Jonkers et al., 2012, Drugs). This review also states that there is no evidence to support the use of probiotics in Crohn's disease.

Other studies confirm these findings. Miele et al. (2009) reported that all of 29 patients studied responded to inflammatory bowel disease therapy. Remission was achieved in 92.8% of patients treated with mixed probiotics and 36.4% of patients treated with placebo. Overall, 21.4 % patients treated with a mix of probiotics and 73.3 % patients treated with placebo relapsed within 1 year of follow-up.

Urinary tract infections (UTIs) are a common and frequently recurrent infection among women. Depletion of vaginal lactobacilli is associated with UTI risk, which suggests that repletion of the bacteria may be beneficial. Young women with a history of recurrent UTI were randomized to receive either a probiotic or placebo daily. Recurrent UTI occurred in 15% of women receiving probiotic compared with 27% of women receiving placebo (Stapleton et al., 2011).

Probiotics have considerable potential for preventive and therapeutic applications in gastrointestinal disorders. However, it is important to note that many probiotic health claims have not yet been substantiated through experimental evidence. In addition, the efficacy demonstrated for a single given bacterial strain cannot be extrapolated to other probiotic organisms. Moreover, the mechanisms underlying probiotic action have not yet been fully elucidated. A better understanding of these mechanisms will be able to shed light on the disparate clinical data and provide new tools to help the prevention or treatment of health disorders (Wohlgemuth et al., 2010; Yan et al., 2011).

4. Application of probiotic bacteria in dairy foods

There is evidence that food matrices play an important role in the beneficial health effects of probiotics on the host (Espirito Santo et al., 2011).

Fermented foods, particularly dairy foods, are commonly used as probiotic carriers. Fermented beverages provide an important contribution to the human diet in many countries because fermentation is an inexpensive technology which preserves food, improves its nutritional value and enhances its sensory properties (Gadaga et al., 1999). However, the increasing demand for new probiotic products has encouraged the development of other matrices to deliver probiotics, such as ice cream, infant milk power and fruit juice.

Davidson et al. (2000) evaluated the viability of probiotic strains in low-fat ice cream. They used cultures containing *Streptococcus salivarius* ssp. *thermophilus* and *Lactobacillus delbrueckii* ssp. *Bulgaricus, Bifidobacterium longum* and *Lactobacillus acidophilus,* and verified that culture bacteria did not decrease in the yogurt during frozen storage. Also, the presence of probiotic bacteria did not alter the sensory characteristics of the ice cream. The ice cream matrix may offer a good vehicle for probiotic cultures due to its composition, which includes milk proteins, fat and lactose, as well as other compounds. Moreover, its frozen state contributes to its efficiency. However, a probiotic ice cream product should have relatively high pH values - 5.5 to 6.5, in order to favor an increased survival of lactic cultures during storage. The lower acidity also results in increased consumer acceptance, especially among consumers who prefer milder Products. (Cruz et al., 2009b).

Growth of a probiotic yeast, *Saccharomyces boulardii*, in association with the bio-yogurt microflora, which is done by incorporating the yeast into commercial bio-yogurt, has been suggested as a way to stimulate growth of probiotic organisms and to assure their survival during storage. Lorens-Hattingh and Viljoen (2001a) studied the ability of probiotic yeast to grow and survive in dairy products, namely bio-yogurt, UHT yogurt and UHT milk. *S. boulardii* was incorporated into these dairy products and stored at 4 ºC over a 4-week period. It was observed that the probiotic yeast species, *S. boulardii*, had the ability to grow in bio-yogurt and reach maximum counts exceeding 10^7 CFU g^{-1}. The number of yeast populations was substantially higher in the fruit-based yogurt, mainly due to the presence of sucrose and fructose derived from the fruit. Despite the inability of *S. boulardii* to utilize lactose, the yeast species utilized available organic acids, galactose and glucose derived from bacterial metabolism of the milk sugar lactose present in the dairy products.

The viability of strains of *L. acidophilus* and *Bifidobacterium animalis* ssp. *lactis* in stirred yoghurts with fruit preparations of mango, mixed berry, passion fruit and strawberry was evaluated during shelf-life (Godward et al., 2000; Kailasapathy et al., 2008). The authors observed that regardless of concentrations, the addition of any of the fruit preparations had no effect on the counts of the two probiotics tested.

Fermented milks supplemented with lemon and orange fibers increased the counts of *L. acidophilus* and *L. casei* during cold storage compared to the control set. This was not the case for *B. bifidum*, possibly owing to the well-known sensitivity of bifidobacteria species to an acidic environment (Sendra et al., 2008).

5. Probiotic cheeses

Probiotic foods are currently primarily found in fermented milk drinks and yogurt, both of which have limited shelf life compared to cheeses. Incorporation of probiotic cultures in cheeses offers the potential not only to improve health but also product quality. It also opens the way to increasing the range of probiotic products on the market. The manufacture of most cheeses involves combining four ingredients: milk, rennet, microorganisms and salt These are processed using a number of common steps such as gel formation, whey expulsion, acid production and salt addition. Variations in ingredient blends and subsequent processing have led to the evolution of all cheese varieties.

Cheeses are dairy products which have a strong potential for delivering probiotic microorganisms into the human intestine, due to their specific chemical and physical characteristics. Cheeses have higher pH levels, lower titratable acidity, higher buffering capacity, more solid consistency, relatively higher fat content, higher nutrient availability and lower oxygen content than yogurts. These qualities protect probiotic bacteria during storage and passage through the gastrointestinal tract (Karimi et al., 2011; Ong et al., 2006).

As mentioned above, the physicochemical properties of food influence probiotic bacteria survival in the digestive tract, due to the low pH in the stomach, typically between 2.5 and 3.5 (Holzapfel et al., 1998), and the anti-microbial activity of pepsin that serve as effective barriers against the entrance of bacteria into the intestinal tract. Values of pH between 1 and 5 are commonly employed in determining the *in vitro* acid tolerance of *Lactobacillus* and *Bifidobacterium* spp. (Charteris et al., 1998). Bile salt concentrations between 0.15% and 0.3% have been recommended as appropriate for selection of probiotic bacteria for human consumption (Yang and Adams, 2004).

A variety of microorganisms, typically food-grade lactic acid bacteria (LAB), have been evaluated for their probiotic potential and have been applied as adjunct cultures in various food products or therapeutic preparations (Rodgers, 2008). *Lactobacillus* and *Bifidobacterium* species may be found in many foods; some are frequently regarded as probiotics due to their capacity to improve certain biological functions in the host. Complex interactions occur among resident microbiota, epithelial and immune cells and probiotics. These interactions play a major role in the development and maintenance of the beneficial activities for healthy humans (Medici el al., 2004).

According to Karimi et al. (2012), recommendations for the minimum viable counts of each probiotic strain in gram or millilitre of probiotic products vary when it comes to providing health benefits related to probiotic organisms. For example, the minimum viable levels of 10^5 cfu g^{-1} have been recommended (Shah, 1995); while 10^6 cfu g^{-1} (Karimi and Amiri-Rigi, 2010; Talwalkar and Kailasapathy, 2004) and 10^7 cfu g^{-1} (Samona and Robinson, 1994) have been suggested for probiotics in different products. However, populations of 10^6-10^7 CFU/g in the final product have been shown to be more acceptable as efficient levels of probiotic cultures in processed foods (Talwalkar, Miller, Kailasapathy and Nguyen, 2004), with numbers attaining 10^8 - 10^9 CFU when provided by a daily consumption of 100 g or 100 mL of probiotic food, and hence benefiting human health (Jayamanne & Adams, 2006). It is important to emphasize that the incorporation of probiotic cultures into cheeses would produce functional foods only if the cultures remained viable in recommended numbers during maturation and shelf life of the products.

One of the preconditions for a bacterial strain to be called probiotic is the strain's ability to survive in the gastrointestinal environment, although the importance of viability for the beneficial effects of probiotics has not been well defined since inactivated and dead cells can also have immunological and health-promoting effects (Ghadimi et al., 2008; Lopez et al., 2008). Moreover, there are significant technological challenges associated with the introduction and maintenance of high numbers of probiotic microorganisms in foods that depend on the form of the probiotic inoculant, and with the viability and maintenance of probiotic characteristics in the food product up to the time of consumption. Spray drying has been used as a preservation method for microbial cultures. Gardiner et al. (2002) produced spray-dried probiotic milk powder containing the probiotic *Lactobacillus paracasei* NFBC 338. The powder contained 1 x 10^9 CFU.g^{-1} *L. paracasei* which was used as adjunct inoculums during probiotic Cheddar cheese manufacture. After three months of ripening, the count was 7.7 x 10^7 CFU.g^{-1}, without any adverse effects on the cheese. The researchers' data shows that probiotic spray-dried powder may be a useful means for adding probiotic strains to dairy products.

In order to use probiotic bacteria in the manufacture of cheese products, the process may have to be modified and adapted to the requirements of the strains employed. Overall, probiotic strains should be technologically compatible with the food manufacturing process involved. With regard to the development of probiotic cheeses, this means that such strains should be cultivable to high cell density for inoculation into the cheese vat, or that the strains are capable of proliferating during the manufacturing and/or ripening process (Ross et al., 2002). In general, a probiotic cheese should have the same attributes as a conventional cheese: the incorporation of probiotic bacteria should not imply a loss of quality of the product. In this context, the level of proteolysis and lipolysis must be the same or even better than for cheese which does not have functional food appeal (Cruz et al., 2009a).

Proteolysis plays a critical role in determining typical sensory characteristics and represents a significant quality indicator for certain cheeses. Proteolysis is caused by enzymes found in milk (plasmin), rennet (pepsin and chymosin) and microbial enzymes released by starter cultures. The activities of these enzymes hydrolyze the fractions of caseins, which leads to the formation of peptides. These peptides may be further hydrolyzed with proteolytic enzymes originating from microbiota such as starter bacteria, non-starter lactic acid bacteria (NSLAB) and probiotic adjuncts to the cheeses, into smaller peptides and free amino acids, which are important for flavor development in some cheeses (Ong et al., 2007; Cliffe et al., 1993; Lynch et al., 1999).

Three batches of Cheddar cheeses (Batch 1, with only starter lactococci; Batch 2, with lactococci and *Lactobacillus acidophilus* 4962, *Lb. casei* 279, *Bifidobacterium longum* 1941; Batch 3, with lactococci and *Lb. acidophilus* LAFTIs L10, *Lb. paracasei* LAFTI L26, *B. lactis* LAFTI B94) were manufactured in triplicate to study the survival and influence of probiotic bacteria on proteolytic patterns and production of organic acid during a ripening period of 6 months at 4 ºC. All probiotic adjuncts survived the manufacturing process and maintained their viability of 7.5 log10 cfu g^{-1} at the end of the ripening term. The number of lactococci decreased by one to two log cycles, but their counts were not significantly different ($P > 0.05$) in either the control or the probiotic cheeses. No significant differences were observed in composition (fat, protein, moisture, salt content), although acetic acid concentration was higher in the probiotic cheeses. Proteolysis assessment during ripening showed no significant differences ($P > 0.05$) in the level of water-soluble nitrogen (primary proteolysis), but the levels of secondary proteolysis indicated by the concentration of free amino acids were significantly higher (P> 0.05) in probiotic cheeses. These data thus suggested that Cheddar cheese is an effective vehicle for the delivery of probiotic organisms (Ong et al., 2006).

Phillips et al. (2006) have also studied probiotic Cheddar cheese. They manufactured six batches of Cheddar cheese containing different combinations of commercially-available probiotic cultures. Duplicate cheeses contained organisms from each supplier, *Bifidobacterium* spp., *Lactobacillus acidophilus* and either *Lactobacillus casei, Lactobacillus paracasei,* or *Lactobacillus rhamnosus*. Using selective media, the different strains were assessed for viability during Cheddar cheese maturation over 32 weeks. *Bifidobacterium* sp. remained at high numbers with the three strains present in cheese at 4×10^7, 1.4×10^8, and 5×10^8 CFU/g respectively after 32 weeks. Similarly, the *L. casei* (2×10^7 CFU/g), *L. paracasei* (1.6×10^7 CFU/g), and *L. rhamnosus* (9×10^8 CFU/g) strains survived well. However, the *L. acidophilus* strains performed poorly. Both decreased in a similar manner and were recorded at 3.6×10^3 CFU/g and 4.9×10^3 CFU/g after 32 weeks.

Numerous scientific papers have been published on the development of fresh cheeses containing recognized and potentially probiotic cultures. They have described suitable viable counts as well as a positive influence on texture and sensorial properties of the cheeses. Cottage cheese in particular shows an adequate profile for the incorporation of probiotic cells and/or prebiotic substances. In addition, cottage cheese is a healthy alternative to many other cheeses by virtue of its low fat content.

Araújo et al. (2010) developed a symbiotic cottage cheese containing *Lactobacillus delbrueckii* UFV H2b20 and inulin, and evaluated the survival of this bacterium when the cheese was exposed to conditions simulating those found in the gastro-intestinal tract. Throughout the entire storage period of the cheese, the probiotic cell counts were higher than recommended levels for probiotic products. The probiotic bacterium exhibited satisfactory resistance to low pH values and to high concentrations of bile salts. The addition of probiotic cells and inulin generated no alterations in the physicochemical characteristics of cheese. By allowing the viable microorganism has characteristics desirable for incorporation of a probiotic strain. Probiotic cells could be added to the dressing, creamy liquid that surrounds the granules of cheese because after this step there is not exposition at high temperature.

Although cottage cheese is well adapted to the health requirements of modern populations, its consumption has been in decline over the past few years. By developing new production processes, cottage cheese, apart from carrying the nutritional qualities of milk, may also furnish consumers with a source of lactic acid bacteria, probiotic microorganisms and prebiotics. The lactic acid bacteria perform more critical functions in cottage cheese than just producing lactic acid. They also aid the manufacture process and increase the final rheological and sensorial qualities of the cheese. Controlling of the fermentation process with lactic acid bacteria allows for the enhancement of the sensorial quality of the cheese and could hence play a crucial role in increasing consumption of cottage cheese.

Souza, et al. (2008) and Souza and Saad (2009) studied the manufacture of Minas fresh cheese supplemented solely with the probiotic strain of *L. acidophilus* La-5. Cheeses manufactured solely with La-5 presented populations above 1 x 10^6 CFU/g, reaching 1 x 10^7 CFU/g on the 14th day of storage.

The Argentinean fresh cheese is a soft rindless cheese with a ripening period of 12 days at 5 ºC before its commercial distribution. This cheese presents the following physicochemical characteristics: pH 5.29, moisture 58% (w/w), fat 12% (w/w), proteins 23% (w/w), salt 0.9% (w/w), ashes 3.4% (w/w), dry matter 40.8% (w/w) and calcium 0.6% (w/w). This product has proven to be an adequate vehicle for probiotic bacteria during storage and until consumption. It offers offer a certain degree of protection of the viability of bacteria during the *in vitro* simulation of gastric transit (Vinderola et al., 2000).

Kasimoglu et al. (2004) have shown that *L. acidophilus* strain can be used for the manufacture of probiotic Turkish white cheese. The final numbers of *L. acidophilus* were greater than the minimum (10^7 cfu g^{-1}) required to make health benefits claims. Furthermore, *L. acidophilus* can be used to enhance flavor, texture, and a produce a high level of proteolysis. Moreover, probiotic cheese which was vacuum packed following salting was shown to be more acceptable than the corresponding cheese stored in brine following salting. Therefore, vacuum packaging is the preferred means for storing probiotic Turkish white cheeses.

6. Concluding remarks and future trends

In conclusion, probiotic microorganisms, including bacteria and yeasts, are attracting a growing interest due to their promising physiological effects as well as the value they add to probiotic-containing food products. There is a growing body of evidence that probiotics may play a beneficial role in human health (Ouwehand et al., 2002; Collado et al., 2009). Established effects in humans include alleviation of symptoms linked to lactose intolerance or to irritable bowel syndrome. They also include reduced diarrhea associated with antibiotic treatment, rotavirus or traveler's diseases. It should be emphasized that the beneficial properties of probiotic microorganisms are highly dependent on the strains, which means that each strain or product requires demonstration of the specific effects *in vivo*. The possibility of using certain probiotics to modulate the immune system, particularly at the mucosal level (O'Flaherty et al., 2010) is the most promising application. In this respect, promising healing effects were obtained using the probiotic mixture VSL#3 on ulcerative colitis patients (Miele et al., 2009; Huynh et al., 2009; Ng et al., 2010). These clinical studies, which still need to be confirmed by larger studies, strongly suggest that selected strains of probiotics may help in treating the bowel diseases which constitute a growing health concern in developing countries. Clearly, animal studies suggest other promising probiotic effects incuding inflammatory diseases, allergies and associated asthma, and colorectal cancer. These applications open exciting avenues that must be investigated at both molecular and clinical levels.

Understanding the impact of ingested bacteria on health, as well as the impact of gut microbiota perturbation (dysbiosis) on emerging diseases, including immune disorders and cancer remains a great challenge. In developed countries, gut microbiota have evolved with a reduced diversity of bacterial species (Yatsunenko et al., 2012). This is particularly true in Crohn's disease patients (Manichanh et al., 2006), who lack immunomodulatory anti-inflammatory bacteria, including *Faecalibacterium prausnitzii* (Sokol et al., 2008). A similar reduced diversity was also described in the case of colorectal cancer, (Chen et al., 2012) confirming the involvement of dysbiosis in digestive cancers (Azcarate-Peril et al., 2011). The composition of gut microbiota is linked to long term dietary patterns (Wu et al., 2011). This suggests that ingested bacteria can participate in the prevention and/or treatment of emerging diseases. This hypothesis has been reinforced by recent epidemiological studies which show that raw milk prevents the onset of allergy and asthma in children (Loss et al., 2011; Waser et al., 2007; Braun-Fahrlander et al., 2011). The authors suggested a protective immunomodulatory role of raw milk bacteria (Braun-Fahrlander et al., 2011).

Most interestingly, bacterial species used as dairy starters display promising properties in this field. For example, immunomodulatory anti-inflammatory properties were described in certain strains of *Propionibacterium freudenreichii* (Foligné et al., 2010; Deutsch et al., 2012), *Streptococcus thermophilus* (Ogita et al., 2011), *Lactobacillus delbrueckii* subsp. *bulgaricus* and subsp. *lactis* (Santos-Rocha et al., 2012), as well as *Lactobacillus helveticus* (Guglielmetti et al., 2010). Modulation of colon cancer cell growth was also reported in vitro and/or in animal models for *P. freudenreichii* (Cousin et al., 2010; Lan et al., 2008), when the cells were exposed to yogurt containing *S. thermophilus and L. bulgaricus* (Narushima et al., 2010; Perdigon et al., 2002) and *L. helveticus* (de Moreno et al., 2010). Future trends may thus include the development of specific fermented dairy products designed for specific population. These could use bacteria strains and employ both technological capabilities and probiotic potential to affect immune system modulation, gut physiology and cancer cells.

Author details

Emiliane Andrade Araújo
Universidade Federal de Viçosa, Campus Rio Paranaíba, Rio Paranaíba, MG, Brazil

Ana Clarissa dos Santos Pires and Antônio Fernandes de Carvalho*
Departamento de Tecnologia de Alimentos, Universidade Federal de Viçosa, Viçosa, MG, Brazil

Maximiliano Soares Pinto
Instituto de Ciências Agrárias, Universidade Federal de Minas Gerais, Montes Claros, MG, Brazil

Gwénaël Jan
INRA, UMR1253 Science et Technologie du Lait et de l'Œuf, Rennes, France

Acknowledgement

We would like to thank to Mary Margaret Chappell for reading and contributing. The authors are supported by grants from the FAPEMIG, CAPES and CNPq.

* Corresponding Author

7. References

Araujo, E. A., Carvalho, A. F., Leandro, E. S., Furtado, M.M., Moraes, C. A. (2010). Development of a symbiotic cottage cheese added with *Lactobacillus delbrueckii* UFV H2b20 and inulin. *Journal of Functional Foods,* 2, 85-89.

Azcarate-Peril, M. A., Sikes, M., Bruno-Barcena, J. M. (2011). The intestinal microbiota, gastrointestinal environment and colorectal cancer: a putative role for probiotics in prevention of colorectal cancer? *Am. J. Physiol Gastrointest. Liver Physiol,,* 301:G401-G424.

Benton, D., Williams, C., Brown, A. (2007). Impact of consuming a milk drink containing a probiotic on mood and cognition. *European Journal of Clinical Nutrition,* 61, 355–361.

Braun-Fahrlander, C., & Von, M. E. (2011). Can farm milk consumption prevent allergic diseases? *Clin. Exp. Allergy,* 41:29-35.

Charteris, W. P., Kelly, P. M., Morelli, L., Collins, J. K. (1998). Development and application of in vitro methodology to determine the transit tolerance of potentially probiotic *Lactobacillus* and *Bifidobacterium* species in the upper human gastrointestinal tract. *Journal of Applied Microbiology,* 84, 759–768.

Chen, W. J. L., Anderson, J. W., Jennings, D. (1984). Propionate may mediate the hypocholesterolemic effects of certain soluble plant fibers in cholesterol-fed rats. *Proc. Soc. Exp. Biol. Med,*175, 215-218.

Chen, W., Liu, F., Ling, Z., Tong, X., Xiang, C. (2012). Human intestinal lumen and mucosa-associated microbiota in patients with colorectal cancer. *PLoS. ONE.,* 7:e39743.

Clancy, R. (2003) Immunobiotics and the probiotic evolution. *FEMS Immunology and Medical Microbiology,* 38, 9-12.

Cliffe, A. J., Marks, J. D.; Mulholland, F. (1993). Isolation and characterization of non-volatile flavors from cheese: Peptide profile of flavor fractions from Cheddar cheese, determined by reverse-phase high performance liquid chromatography. *International Dairy Journal,* 3, 379–387.

Collado, M. C., Isolauri, E., Salminen, S., Sanz, Y. (2009). The impact of probiotic on gut health. *Curr. Drug Metab,* 10:68-78.

Costello, E. K., Lauber, C. L., Hamady, M., Fierer, N., Gordon, J. I., Knight, R. (2009) Bacterial community variation in human body habitats across space and time. *Science,* 326, 1694-1697.

Cousin, F. J., Jouan-Lanhouet, S., Dimanche-Boitrel, M. T., Corcos, L., Jan, G. (2012). Milk Fermented by *Propionibacterium freudenreichii* Induces Apoptosis of HGT-1 Human Gastric Cancer Cells. *PLoS. ONE.,* 7:e31892.

Crittenden, R. G. Prebiotics. In: Tannock, G. W. (Ed.). (1999). *Probiotics: a critical review.* Norfolk: Horizon Scientific Press, 141-156.

Cruz, A. G., Buriti, F. C. A., Souza, C. H. B., Faria, J. A. F., Saad, S. M. I. (2009a). Probiotic cheese: health benefits, technological and stability aspects. *Trends in Food Science & Technology,* 20, 344-354.

Cruz, A.G., Antunes, A.E.C., Sousa, A.L.O.P., Faria, J.A.F., Saad, S.M.I. (2009b). Ice-cream as a probiotic food Carrier. *Food Research International,* 42, 1233-1239.

Cruz, A.G., Faria, J.A.F., Van Dender, A.G.F. (2007). Packaging system and probiotic dairy foods. *Food Research International*, 40, 951–956.

Davidson, R.H., Duncan, S.E., Hackney, C.R., Eigel, W.N., Boling, J.W. (2000). Probiotic Culture Survival and Implications in Fermented Frozen Yogurt Characteristics. *Journal of Dairy Science*, 83, 666–673.

de Moreno, L. A., & Perdigon, G. (2010). The application of probiotic fermented milks in cancer and intestinal inflammation. *Proc. Nutr. Soc.*, 69:421-428.

Deutsch, S. M., Parayre, S., Bouchoux, A., Guyomarc'h, F., Dewulf, J., Dols-Lafargue, M., Baglinière, F., Cousin, F. J., Falentin, H., Jan, G., Foligné, B. (2012). Contribution of surface beta-glucan polysaccharide to physicochemical and immunomodulatory properties of *Propionibacterium freudenreichii*. *Appl. Environ. Microbiol.*, 78:1765-1775.

Espirito Santo, A.P., Perego, P., Converti, A., Oliveira, M.N. (2011). Influence of food matrices on probiotic viability: A review focusing on the fruity bases. *Trends in Food Science & Technology*, 22, 377-385.

Foligné, B., Deutsch, S. M., Breton, J., Cousin, F. J., Dewulf, J., Samson, M., Pot, B., Jan, G. (2010). Promising immunomodulatory effects of selected strains of dairy propionibacteria as evidenced *in vitro* and *in vivo*. *Appl. Environ. Microbiol.*, 76:8259-8264.

Fuller, R. (1989). Probiotics in man and animals. *Journal of Applied Bacteriology*, 66, 365–378.

Gadaga, T. H., Mutukumira, A. N., Narvhus, J. A., Feresu, S. B. (1999). A review of traditional fermented foods and beverages of Zimbabwe. *International Journal of Food Microbiology*, 53, 1–11.

Gardiner, G. E., Bouchier, P., O'sullivan, E., Kelly, J., Collins, K., Fitzgerald, G., Ross, R.P., Stanton, C. (2002). A spray-dried culture for porbiotic Cheddar chesse manufacture. *International Dairy Journal*, 12, 749-756.

Ghadimi, D., Folster-Holst, R., De Vrese, M., Winkler, P., Heller, K.J., Schrezenmeir, J. (2008). Effects of probiotic bacteria and their genomic DNA on TH1/TH2-cytokine production by peripheral blood mononuclear cells (PBMCs) of healthy and allergic subjects. *Immunobiology*, 213, 677–692.

Gibson, G. R., Fuller, R. (2000) Aspects of in vitro and in vivo research approaches directed toward identifying probiotics and prebiotics for human use. *Journal Nutrition*, 130, 391-395.

Godward, G., Sultana, K., Kailasapathy, K., Peiris, P., Arumugaswamy, R., & Reynolds, N. (2000). The importance of strain selection on the viability and survival of probiotic bacteria in dairy foods. *Milchwissenschaft*, 55, 441-445.

Guglielmetti, S., Taverniti, V., Minuzzo, M., Arioli, S., Zanoni, I., Stuknyte, M., Granucci, F., Karp, M., Mora, D. (2010). A dairy bacterium displays in vitro probiotic properties for the pharyngeal mucosa by antagonizing group A streptococci and modulating the immune response. *Infect. Immun.*, 78:4734-4743.

Holzapfel, W. H., Haberer, P., Snel, J., Schillinger, U., & Huis In't Velt, J. H. J. (1998). Overview of gut flora and probiotics. *International Journal of Food Microbiology*, 41, 85–101.

Huynh, H. Q., Debruyn, J., Guan, L., Diaz, H., Li, M., Girgis, S., Turner, J., Fedorak, R., Madsen, K. (2009). Probiotic preparation VSL#3 induces remission in children with mild to moderate acute ulcerative colitis: a pilot study. *Inflamm. Bowel. Dis.,* 15:760-768.

Hyun, C., Shin, H. (1998). Utilization of bovine plasma obtained from a slaughterhouse for economic production of probiotics. *Journal of Fermentation and Bioenginering,* 86, 34–37.

Jayamanne, V. S., & Adams, M. R. (2006). Determination of survival, identity, and stress resistance of probiotic bifidobacteria in bio-yoghurts. *Letters in Applied Microbiology,* 42(3),189-194.

Kailasapathy, K., Harmstorf, I., & Phillips, M. (2008). Survival of *Lactobacillus acidophilus* and *Bifidobacterium animalis* ssp *lactis* in stirred fruit yogurts. *LWT-Food Science and Technology,* 41, 1317-1322.

Karimi, R., Amiri-Rigi, A., 2010. *Probiotics in Dairy Products.* Marz Danesh Publication,Tehran.

Karimi, R., Mortazavian, A.M., Amiri-Rigi, A. (2012). Selective enumeration of probiotic microorganisms in cheese. *Food Microbiology,* 29,1-9.

Karimi, R., Mortazavian, A.M., Da Cruz, A.G. (2011). Viability of probiotic microorganisms in cheese during production and storage: a review. *Dairy Science and Technology,* 91, 283-308.

Kasimoglu, A., Goncuoglu, M., Akgun, S. (2004). Probiotic White cheese with *Lactobacillus acidophilus. International Dairy Journal,* 14, 1067-1073.

Lan, A., Bruneau, A., Bensaada, M., Philippe, C., Bellaud, P., Rabot, S., Jan, G. (2008). Increased induction of apoptosis by *Propionibacterium freudenreichii* TL133 in colonic mucosal crypts of human microbiota-associated rats treated with 1,2-dimethylhydrazine. *Br. J Nutr.,* 100:1251-1259.

Lopez, M., Li, N., Kataria, J., Russell, M., Neu, J. (2008). Live and ultraviolet-inactivated *Lactobacillus rhamnosus* GG decrease flagellin-induced interleukin-8 production in Caco-2 cells. *Journal of Nutrition,* 138, 2264–2268.

Loss, G., Apprich, S., Waser, M., Kneifel, W., Genuneit, J., Buchele, G., Weber, J., Sozanska, B., Danielewicz, H., Horak, E., Van Neerven, R. J., Heederik, D., Lorenzen, P. C., Von, M. E., Braun-Fahrlander, C. (2011). The protective effect of farm milk consumption on childhood asthma and atopy: the Gabriela study. *J. Allergy Clin. Immunol.,* 128:766-773.

Lourens-Hattingh,A., Viljoen, B.C. (2001a). Growth and survival of a probiotic yeast in dairy products. *Food Research International,* 34, 791-796.

Lourens-Hattingh,A., Viljoen, B.C. (2001b). Yogurt as probiotic carrier food. *International Dairy Journal,* 11, 1-17.

Lynch, C. M., Muir, D. D., Banks, J. M., Mcsweeney, P. L. H.; Fox, P. F. (1999). Influence of adjunct cultures *of Lactobacillus paracasei* ssp. *paracasei* or *Lactobacillus plantarum* on Cheddar cheese ripening. Journal of Dairy Science, 82, 1618–1628.

Macfarlane, G. T. Gibson, G. R. Cummings, J. H. (1992) Comparison of fermentation reactions in different regions of the human colon. *Journal Applied Bacteriology,* 72(3), 57-64.

Manichanh, C., Rigottier-Gois, L., Bonnaud, E., Gloux, K., Pelletier, E., Frangeul, L., Nalin, R., Jarrin, C., Chardon, P., Marteau, P., Roca, J., Dore, J. (2006). Reduced diversity of faecal microbiota in Crohn's disease revealed by a metagenomic approach. *Gut*, 55:205-211.

Margoles, A., Garcia, L. (2003). Characterisation of a bifidobacterium strain wish acquired resistance to cholate: A preliminary study. *International Journal of Food Microbiology*, 80, 191–198.

Medici, M., Vinderola, C. G., Perdigon, G. (2004). Gut mucosal immunomodulation by probiotic fresh chesse. *International Dairy Journal*, 14, 611-618.

Miele, E., Pascarella, F., Giannetti , E., Quaglietta , L Robert N. Baldassano , R. N., Annamaria Staiano , A. (2009). Effect of a Probiotic Preparation (VSL#3) on Induction and Maintenance of Remission in Children With Ulcerative Colitis. *American Journal of Gastroenterology*, 104, 437-443.

Miele, E., Pascarella, F., Giannetti, E., Quaglietta, L., Baldassano, R. N., Staiano, A. (2009). Effect of a probiotic preparation (VSL#3) on induction and maintenance of remission in children with ulcerative colitis. *Am. J. Gastroenterol.*, 104:437-443.

Moura, M.R. L., (2005). Alimentos Funcionais: seus benefícios e a legislação: Avaiable in: http://acd.ufrj.br/consumo/leituras/ld.htm#leituras.

Narushima, S., Sakata, T., Hioki, K., Itoh, T., Nomura, T., Itoh, K. (2010). Inhibitory effect of yogurt on aberrant crypt foci formation in the rat colon and colorectal tumorigenesis in RasH2 mice. *Exp. Anim.*, 59:487-494.

Ng, S. C., Plamondon, S., Kamm, M. A., Hart, A. L., Al-Hassi, H. O., Guenther, T., Stagg, A. J., Knight, S. C. (2010). Immunosuppressive effects via human intestinal dendritic cells of probiotic bacteria and steroids in the treatment of acute ulcerative colitis. *Inflamm. Bowel. Dis.*, 16:1286-1298.

O'flaherty, S., Saulnier, D. M., Pot, B., Versalovic, J. (2010). How can probiotics and prebiotics impact mucosal immunity? *Gut Microbes*, 1:293-300.

Ogita, T., Nakashima, M., Morita, H., Saito, Y., Suzuki, T., Tanabe, S. (2011). *Streptococcus thermophilus* ST28 ameliorates colitis in mice partially by suppression of inflammatory Th17 cells. *J. Biomed. Biotechnol.*, 2011:378417.

Ong, L., Henriksson, A., Shah, N.P. (2006) Development of probiotic Cheddar cheese containing *Lactobacillus acidophilus, Lb. casei, Lb. paracasei* and *Bifidobacterium* spp. and the influence of these bacteria on proteolytic patterns and production of organic acid. *International Dairy Journal* 16, 446–456.

Ong, L., Henrikssonb, A., Shaha, N. P. Chemical analysis and sensory evaluation of Cheddar cheese produced with *Lactobacillus acidophilus, Lb. casei, Lb. paracasei* or *Bifidobacterium* sp. *International Dairy Journal* 17 (2007) 937–945.

Ouwehand, A. C., Salminen, S., Isolauri, E. (2002). Probiotics: an overview of beneficial effects. *Anton. Leeuw. Int. J. G.*, 82:279-289.

Perdigon, G., De Moreno, D. L., Valdez, J., Rachid, M. (2002). Role of yoghurt in the prevention of colon cancer. *Eur. J. Clin. Nutr.*, 56 Suppl 3:S65-S68.

Prado, F.C., Parada, J.L., Pandey, A., Soccol, C.R. (2008). Trends in non-dairy probiotic beverages. *Food Research International*, 41, 111–123.

Rastall, R. A., Fuller, R., Gaskins H. R., Gibson G. R. (2000) Colonic functional foods. In *Functional Foods*, 71–89 [GR Gibson and CM Williams, editors]. Cambridge: Woodhead Publishing Limited.

Reddy, B. S., Rivenson, A. (1983) Inhibitory effect of Bifidobacterium longum on colon, mammary, and liver carcinogenesis induced by 2-amino-3-methyllimidazo [4,5-f]quinoline, a food mutagen. *Cancer Research*, 53, 3914-3918.

Reid G., Jessica A., Younes, J. A., Van Der Mei, H. C., Gloor, G. B., Knight, R., Busscher, H. J. (2011) Gut flora restoration: natural and supplemented recovery of human microbial communities. *Nature Reviews*. 9, 27-38.

Rodgers, S. (2008). Novel applications of live bacteria in food services: probiotics and protective cultures. *Trends Food Sci. Tech.* 19, 188-197.

Ross, R. P., Fitzgerald, G., Collins, K., Stanton, C. (2002). Cheese delivering biocultures: probiotic cheese. *Australian Journal of Dairy Technology*, 57(2), 71-78.

Saarela, M., Mogensen, G., Fonden, R., Matto, J., Mattila-Sandholm, T. (2000). Probiotic bacteria: safety, functional and technological properties. *Journal of Biotechnology*, 84, 197–215.

Samona, A., Robinson, R.K. (1994). Effect of yogurt cultures on the survival of bifidobacteria in fermented milks. *Journal of the Society of Dairy Technology* 47, 58-60.

Santos-Rocha C., Lakhdari, O., Blottiere, H. M., Blugeon, S., Sokol, H., Bermu'dez-Humara'n, L. G., Azevedo, V., Miyoshi, A., Dore, J., Langella, P., Maguin, E., Van De, G. M. (2012). Anti-inflammatory properties of dairy lactobacilli. *Inflamm. Bowel. Dis.*, 18:657-666.

Sendra, E., Fayos, P., Lario, Y., Fernandez-Lopez, J., Sayas-Barbera, E., & Perez-Alvarez, J. (2008). Incorporation of citrus fibers in fermented milk containing probiotic bacteria. *Food Microbiology*, 25, 13-21.

Shah, N.P., Lankaputhra, W.E.V., Britz, M.L., Kyle, W.S.A. (1995). Survival of *Lactobacillus acidophilus* and *Bifidobacterium bifidum* in commercial yoghurt during refrigerated storage. *International Dairy Journal*, 5, 515-521.

Sokol, H., Pigneur, B., Watterlot, L., Lakhdari, O., Bermudez-Humaran, L. G., Gratadoux, J. J., Blugeon, S., Bridonneau, C., Furet, J. P., Corthier, G., Grangette, C., Vasquez, N., Pochart, P., Trugnan, G., Thomas, G., Blottiere, H. M., Dore, J., Marteau, P., Seksik, P., Langella, P. (2008). *Faecalibacterium prausnitzii* is an anti-inflammatory commensal bacterium identified by gut microbiota analysis of Crohn disease patients. *Proc. Natl. Acad. Sci. U. S. A*, 105:16731-16736.

Souza, C. H. B., & Saad, S. M. I. (2009). Viability of Lactobacillus acidophilus La-5 added solely or in co-culture with a yoghurt starter culture and implications on physico-chemical and related properties of Minas fresh cheese during storage. *LWT e Food Science and Technology*, 42(2), 633-640.

Souza, C. H. B., Buriti, F. C. A., Behrens, J. H., & Saad, S. M. I. (2008). Sensory evaluation of probiotic Minas fresh cheese with *Lactobacillus acidophilus* added solely or in co-culture with a thermophilic starter culture. *International Journal of Food Science and Technology*, 43(5), 871-877.

Stapleton, A. E., Au-Yeung, M., Hooton, T. M., Fredricks, D. N., Roberts, P. L., Czaja, C. A., Yarova-Yarovaya, Y., Fiedler, T., Cox, M., Stamm, W. E. (2011). Randomized, Placebo-controlled Phase 2 Trial of a Lactobacillus crispatus Probiotic Given Intravaginally for Prevention of Recurrent Urinary Tract Infection. *Clinical Infectious Diseases*, 52 (10), 1212-1217.

Talwalkar, A., Kailasapathy, K. (2004). Comparison of selective and differential media for the accurate enumeration of strains of *Lactobacillus acidophilus*, *Bifidobacterium* spp. and Lactobacillus casei complex from commercial yoghurts. *International Dairy Journal*, 14, 142-149.

Talwalkar, A., Miller, C. W., Kailasapathy, K., & Nguyen, M. H. (2004). Effect of packaging materials and dissolved oxygen on the survival of probiotic bacteria in yoghurt. *International Journal of Food Science and Technology*, 39(6), 605-611.

Thamer, K. G.; Penna, A. L. B. (2005). Efeito do teor de soro, acucar e de frutooligossacarideos sobre a populacao de bacterias lacticas probioticas em bebidas fermentadas. *Revista Brasileira de Ciências Farmacêuticas*, 41(3), 393-400.

Topping, D. L. Clifton, P. M. (2001) SHorty-chain fatty acids and human colonic function: roles of resistant starch and nonstarch polysaccharides. *Physiological Reviews*, 81(3), 1031-1064.

Vinderola, C. G., Prosello, W., Ghiberto, D., Reinheimer, J. (2000). Viability of probiotic (*Bifidobacterium*, *Lactobacillus acidophilus* and *Lactobacillus casei*) and nonprobiotic microflora in argentinian Fresh cheese. *Journal of Dairy Science*, 83, 1905–1911.

Waser, M., Michels, K. B., Bieli, C., Floistrup, H., Pershagen, G., Von, M. E., Ege, M., Riedler, J., Schram-Bijkerk, D., Brunekreef, B., Van, H. M., Lauener, R., Braun-Fahrlander, C. (2007). Inverse association of farm milk consumption with asthma and allergy in rural and suburban populations across Europe. *Clin. Exp. Allergy*, 37:661-670.

Wohlgemuth, S., Gunnar Loh, G., Blaut, M. (2010). Recent developments and perspectives in the investigation of probiotic effects. *International Journal of Medical Microbiology*, 300, 3–10

Wu, G. D., Chen, J., Hoffmann, C., Bittinger, K., Chen, Y. Y., Keilbaugh, S. A., Bewtra, M., Knights, D., Walters, W. A., Knight, R., Sinha, R., Gilroy, E., Gupta, K., Baldassano, R., Nessel, L., Li, H., Bushman, F. D., Lewis, J. D. (2011). Linking long-term dietary patterns with gut microbial enterotypes. *Science*, 334:105-108.

Yan, F., Cao, H., Cover, T, L., Washington, M. K., Shi, Y, Liu, L., Chaturvedi, R., Peek Jr, R. M., Wilson, K. T., Polk, D. B. (2011). Colon-specific delivery of a probiotic-derived soluble protein ameliorates intestinal inflammation in mice through an EGFR-dependent mechanism. *Journal of Clinical Investigation*, 121(6), 2242-2253.

Yang, H., & Adams, M. C. (2004). In vitro assessment of the upper gastrointestinal tolerance of potential probiotic dairy propionibacteria. *International Journal of Food Microbiology*, 91, 253–260.

Yatsunenko, T., Rey, F. E., Manary, M. J., Trehan, I., Dominguez-Bello, M. G., Contreras, M., Magris, M., Hidalgo, G., Baldassano, R. N., Anokhin, A. P., Heath, A. C., Warner, B., Reeder, J., Kuczynski, J., Caporaso, J. G., Lozupone, C. A., Lauber, C., Clemente, J. C., Knights, D., Knight, R., Gordon, J. I. (2012). Human gut microbiome viewed across age and geography. *Nature*, 486:222-227.

Chapter 7

Probiotics and Lactose Intolerance

Roel J. Vonk, Gerlof A.R. Reckman,
Hermie J.M. Harmsen and Marion G. Priebe

Additional information is available at the end of the chapter

http://dx.doi.org/10.5772/51424

1. Introduction

Lactose is the main sugar in milk and therefore the main energy source for the newborn. Milk contains 4,8% lactose [1]. Lactose is a disaccharide consisting out of glucose and galactose.

In normal physiological conditions lactose is hydrolyzed by lactase also known as lactase-phlorizin hydrolase and under its systemic name lactose- galactosehydrolase (EC 3.2.1.108), which is a brush-border membrane bound enzyme. Glucose and galactose are taken up by the intestinal cells and transported into the bloodstream (Fig. 1). A considerable part of glucose and most galactose is cleared by the liver after the first pass. Lactose which is not hydrolyzed in the small intestine is passing into the colon where it is fermented. Lactose itself and its metabolites are osmotic active products causing an osmotic pressure; excessive amounts present in the colon are related to the development of clinical symptoms as diarrhea.

The apparent lactase enzyme activity is affected by various factors like a. age, b. genetic background, c. integrity of the small intestinal membrane and d. the small-intestinal transit time

a. The activity of the enzyme lactase is age dependent. The activity is high in the first year of age and declines until adulthood is reached. It is not clear what the physiological advantage is of the age dependency of the lactase activity in relation to the disaccharide glucose-galactose.

Several remarkable aspects can be brought up in this respect:

- Galactose has a higher hepatic clearance than glucose, which prevents a significant postprandial increase in blood glucose in the systemic circulation.
- Galactose does not lead to an induction of the pancreatic insulin response.

- Lactase might have a variable and limited capacity which leads to a regulated spill over of lactose into the colon. There lactose might act as a primer (prebiotic) for the colonic microbiota in the first period of life (See also the subchapter of colonic fermentation of lactose).

b. The role of the genetic background of the lactase activity has been described in detail elsewhere [2].
c. The lactase activity is strongly affected by the integrity of the small intestinal membrane. This is the reason why in patients with celiac disease, which have not been treated optimally, symptoms of lactose intolerance may appear [3].
d. Finally, the turnover of lactose by the enzyme is dependent on the small intestinal transit time (apparent enzyme activity).

Fermented milk products can alleviate symptoms by delaying gastric emptying, orocecal transit time, or both. Delay of gastric emptying is due to the higher viscosity of the fermented milk product as compared to milk. Decrease of orocecal transit time is due to the metabolic products of probiotics or a lower osmotic force due to improved lactose digestion. A longer passage time in lactose maldigesters aids in hydrolyzing as much lactose as possible before spill over into the colon occurs. These findings support that pasteurized yogurt already provides alleviation of symptoms and that yogurt containing living probiotics improves this alleviation [4]. The effect of sugars, including lactose, on the small intestinal transit time is not well documented [5]. Changes in intestinal transit time due to the sugar molecules might especially play a role in other pathological conditions like irritable bowel disease.

Lactose intolerance is the pathophysiological situation in which the small intestinal digestion and / or colonic fermentation is altered which leads to clinical symptoms.

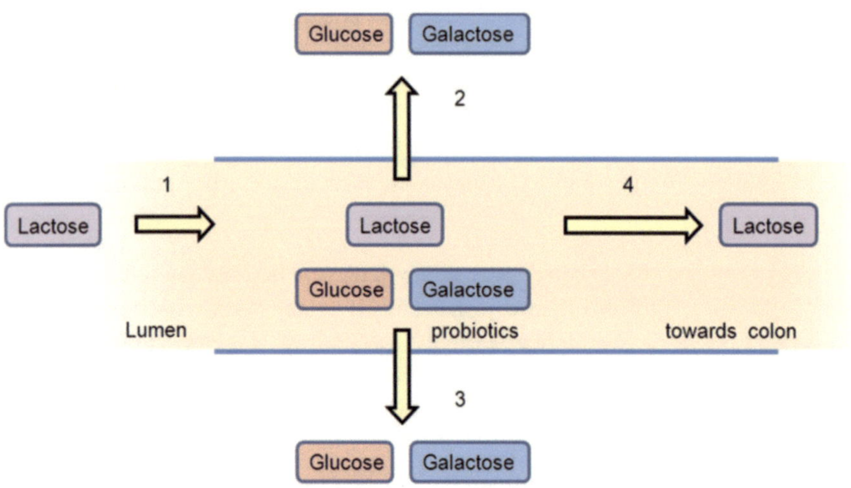

Figure 1. Small intestinal metabolism of lactose. Lactose enters the small intestine (1), lactose is then coverted by lactase from the host (2) or by probiotics (3). Excess amounts of lactose spill over into the colon (4).

2. Colonic fermentation of lactose

Lactose which is spilled over into the colon can be hydrolyzed by the colonic bacterial enzyme β-galactosidase resulting in the formation of glucose and galactose. Glucose and galactose are subsequently converted into lactate as well as into the short chain fatty acids (SCFA) acetate, propionate and butyrate (see Fig. 2). Additionally, microbial biomass will be formed. The original substrate lactose, the intermediate products glucose and galactose and the final products can all contribute to the osmotic load in the colon. This might lead to increased colonic transit time, altered fermentation profiles and ultimately to diarrhea.

The central question is which molecule contributes most to the pathological symptoms, like diarrhea: the original substrate lactose and / or one of the metabolites.

As indicated in Fig. 2 the number of molecules is doubled after the first conversion by β-galactosidase and tripled after the second conversion. A rapid conversion to the final metabolites enhances the osmotic force considerable.

We first analyzed the role of lactose itself assuming that β-galactosidase is the rate limiting step. In a recent paper of us [6], we describe that inducing the colonic β-galactosidase by administration of yogurt and additional probiotics alleviates the clinical symptoms of lactose intolerance in an adult Chinese population. This suggests a specific role of lactose itself in the development of clinical symptoms. Our observation was confirmed by [7], who observed in post-weaning Balb/c mice that symptoms of diarrhea were reduced by inducing the β-galactosidase activity by administration of a recombinant *Lactococcus lactis MG1363/FGZW* strain expressing β-galactosidase.

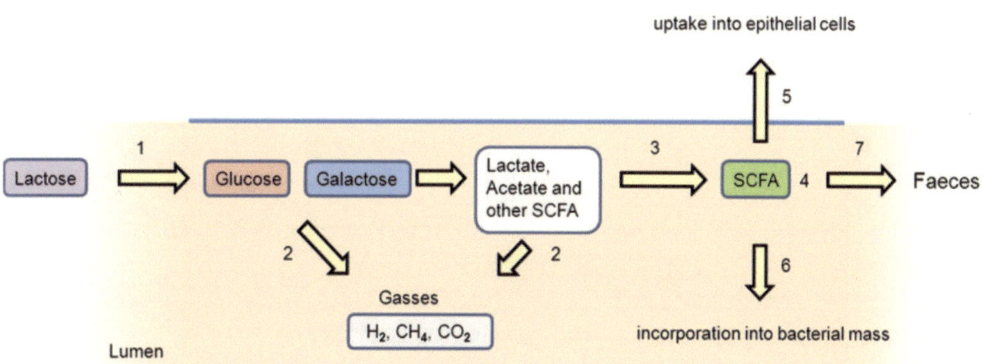

Figure 2. Colonic metabolism of lactose. Lactose enters the colon (1) and is fermented by the microbiota into glucose and galactose. Gasses such as hydrogen, methane and carbondioxide are formed (2). Lactate is also formed and converted into short chain fatty acids (SCFA)(3,4), also in this stage gasses are formed (2). These SCFAs can be taken up by epithelial cells (5) or can be used by the microbiota (6) or excreted in the faeces (7).

In contrast with these observations is the fact that β-galactosidase is an abundant enzyme in the colonic microbiota. It is present in many phylogroups of bacteria which in total might contribute to more than 40% of the total population of the colonic microbiome (Table 1). However, relative abundance and composition of bacteria with β-galactosidase in the distal colon do not seem to be related to lactose intolerance [8]. Another argument to consider is that the conversion of lactose into glucose, galactose and subsequently SCFA / lactate doubles and triples respectively the osmotic pressure. This aspect will be discussed in more detail under the chapter administration of pre- and probiotics.

Domain	Phylum	Order	Family
Bacteria (1718)			
	CFB group bacteria (1006)		
		Bacteroidales (994)	
			Bacteroidaceae (830)
			Porphyromonadaceae (102)
			more... *(62)*
		Flavobacteriales (12)	
	Firmicutes (394)		
		Clostridiales (327)	
			Lachnospiraceae (146)
			Clostridiaceae (86)
			more... *(95)*
		Erysipelotrichales (36)	
		Lactobacillales (29)	
		Selenomonadales (2)	
	Proteobacteria (190)		
		g-proteobacteria (182)	
			Enterobacteria (176)
			more... *(6)*
		more... *(8)*	
	Actinobacteria (95)		
		Bifidobacteriales (74)	
		Coriobacteriales (15)	
		Actinomycetales (6)	
	Spirochetes (32)		
	more... *(1)*		
unclassified (4)			

Table 1. Overview of all bacteria known to produce β-galactosidase.

Considering the physiological aspects of lactose digestion and fermentation it is clear that sufficient small intestinal hydrolysis of lactose related to the dose consumed will prevent symptoms of lactose intolerance. In case of relative insufficient lactase activity in the small intestine, spillover into the colon will occur. Adequate removal of osmotic active molecules, however, can prevent development of clinical symptoms of diarrhea.

3. Clinical symptoms of lactose intolerance

Symptoms of intestinal discomfort, abdominal pain and / or diarrhea can occur in case of lactose intolerance. These complaints are, however, not specific and can also be noticed in several other clinical conditions (for example irritable bowel syndrome, coeliac disease, Crohn's disease). For proper treatment and correct interpretation of interventions accurate diagnosis of the underlying pathophysiology is therefore very important.

4. Diagnostics of lactose intolerance

The most direct diagnosis is the analysis of lactase activity. However, the enzyme activity derived from a small intestinal biopsy does not reflect the overall lactase activity in the small intestine because of the patchy character of the distribution of this enzyme. This can lead to false positive and negative estimation of the overall physiological capacity to hydrolyze lactose.

Screening the genotype of people with lactose intolerant-like symptoms can aid in the correct diagnosis of lactose intolerance. The lactase gene can contain single-nucleotide polymorphisms (SNP) in the promotor region which leads to a high capacity to digest lactose. The most common SNP C/T-13910 is found in many Northwest European people. Several methods have been developed to detect this most common SNP. Järvelä et al. [2] sum up in their review the different methods for detection: minisequencing, enzyme digest, polymerase chain reaction-restriction fragment length polymorphism and pyrosequencing. For detection of all known SNPs, sequencing is the most reliable technique. Because there is a poor correlation between abdominal symptoms and lactase activity, genetics alone is not sufficient for a correct clinical diagnosis of adult lactose intolerance.

For congenital lactase deficiency genetic screening is effective, mutations occur in the lactase gene itself and symptoms start shortly after birth [2]. The prevalence of this syndrome however, is very low.

The analysis of the capacity to digest lactose in vivo by using two stable isotopes might be theoretically the best diagnostical method [9]. This test consists of the administration of ^{13}C-lactose and ^{2}H-glucose and calculation of the ratio of the ^{13}C-glucose/^{2}H-glucose concentrations measured in plasma. This test can be used to analyze the effect of interventions and to demonstrate changes in the capacity to digest lactose. However, as a routinely used diagnostic tool this test is not applicable because of its complex character.

The most commonly used diagnostic method for lactose intolerance is the hydrogen breath test. This test is easy to apply in clinical practice, but as discussed in detail by us [12] others [11] this test leads to false positive and false negative results.

A way to improve the precision of the breath test is to use ^{13}C-lactose as a substrate and measure both H_2 and $^{13}CO_2$ in breath as first described by Hiele et al. [12]. This might be the best applicable test in daily practice.

5. Application of pre- and probiotics to improve the clinical symptoms of lactose intolerance

An effect of an intervention with probiotics can be expected at two levels:

a. hydrolysis of lactose in the milk product and in the small intestine
b. at the level of colonic fermentation

a. The hydrolytic capacity of probiotic strains can be used to reduce the actual amount of lactose in the product, as occurs in yogurt. It can also be used to increase the overall hydrolytic capacity in the small intestine. The probiotic strain can be alive or can be lysed in the intestinal tract for its effect. *Lactobacillus acidophilus* is a bile-salt tolerant bacterium which hardly increases lactose digestion. However, sonication of *Lactobacillus Acidophilus* milk weakens their membranes and improves lactose-intolerance symptoms [4]. *Lactobacillus delbrüeckii* in a milk product can deliver β-galactosidase activity. These microorganisms do not have to be alive as long as their membranes are intact which helps to protect β-galactosidase during gastric passage [4]. Yogurt improves the lactose intolerance due to the presence of a group of lactobacillus bacteria it contains, i.e., *Lactobacillus acidophilus* [13]. Kinova et al. [14] described the beneficial effects of *Lactobacillus* present in fermented milk products. In [15] is described that consumption of yogurt containing *Lactobacillus bulgaricus* and *Streptococcus thermophiles* alleviate the lactose intolerance through their enzyme lactase when the product reaches the intestinal tract. Also Masood et al. [16] describe the beneficial effects of lactic acid bacteria in their review. From these findings it is inferred that lactose intolerance can be reduced by regularly consuming the fermented dairy products due to the production of β-galactosidase enzyme by lactic acid bacteria present in them.

In general, it can be stated that in yogurt several probiotic strains are present which results in a better tolerance of lactose in lactose intolerant persons.

b. Application of probiotics to manipulate the colonic fermentation.

As suggested before [17], one of the problems in studies concerning this topic is that it is difficult to prove that the intervention only has an effect at the level of the colon and not at the level of the small intestine.

As discussed before it is not clear which compound, lactose or one of its fermentation metabolites contributes most to the development of symptoms of lactose intolerance. The hypothesis is that removal of these product(s) can reduce the clinical symptoms.

1. Removal of lactose

Lactose is hydrolyzed by β-galactosidase. We recently published [6] that a mix of probiotics in yogurt together with *Bifidobacterium longum* capsules could increase the β-galactosidase activity in faeces and alleviate the complaints of lactose intolerance.

Together with the observation that the capacity to digest lactose, which was measured by the 13C-lactose/2H-glucose test, was not changed, it could be concluded that this intervention has an effect on colonic metabolism, possibly by enhancing the β-galactosidase activity. A study with mice [7] suggested the same mechanism. However, after analyzing the presence of β-galactosidase in the common bacterial strains in humans it can be concluded that β-galactosidase is abundantly present and it seems that administration of exogenous β-galactosidase from probiotics is not important. Alleviation of complaints and enhanced β-galactosidase concentration in stool therefore might have been a coincidence in our study.

2. Removal of glucose and galactose

Glucose is a preferred substrate for many bacterial strains and it is not likely that enhanced glucose removal by probiotic administration might play a role in alleviation of symptoms. Also galactose is easily consumed by most bacteria. Our in vitro studies [18] also indicated that accumulation of glucose and galactose does not occur during the breakdown of lactose, which confirms that these molecules once formed are subsequently metabolized very fast.

3. Removal of acetate, propionate, butyrate and lactate

As illustrated in Fig. 2 removal of SCFA takes place at the epithelium by uptake in the colonocytes and through the uptake and metabolism by various bacteria ("bacterial mass").

Uptake of SCFA into the epithelial cells is very effective because of co-transport of fluid which reduces the osmotic force [19]. The maximal epithelial uptake rate is not known and it is not known if this varies in persons with hypolactasia with and without symptoms after lactose consumption.

Another major way by which SCFA are removed is via the uptake and metabolism by bacteria. SCFA serve as a carbon and energy source for the anaerobic bacteria and this may increase the "bacterial mass". In the presence of sulphate, lactate may be metabolized by sulphate-reducing bacteria, producing toxic sulphide as byproduct [20]. On the other hand lactate together with acetate can be converted by different groups of bacteria into butyrate; for instance by bacteria such as *Eubacterium hallii* and *Anaerostipes cacca* [21]. Butyrate is thought to be beneficial for colonic health. Also *Bacteroides* several subspecies are capable of metabolizing lactate, but produce propionate. The metabolism of intermediates like lactate and acetate are an important step in the breakdown of sugars by gut bacteria [22]. For gut health it is important that from lactate a balanced mixture of SCFA are formed and for this correct conditions should be present. The hypothesis that for the prevention of diabetes type 1 butyrate production is preferred over propionate production is stated by [23]. They stated this because butyrate production enforces the barrier function of the gut. Therefore, conditions that stimulates these metabolic associations should be enforced. This implies that a mixture of pro- and prebiotics as occurs in yogurt might be an efficient approach, since it favors acetate and lactate formation, and in this way stimulate butyrate formation. If lactate removal via for instance butyrate production, does not occur this may impact functioning of the epithelium. It then can be speculated that an impaired epithelial function will hamper the uptake of lactate, and causes an increased osmotic pressure in the gut.

Several other studies have reported the beneficial effect of a probiotic intervention on symptoms of lactose intolerance but without describing a precise mechanism. In some of these studies the observation that the specific strains under study survive the small intestinal passage is used as an argument that the effect occurs at the colonic level.

The combination of *Lactobacillus casei* Shirota and *Bifidobacterium breve* Yakult has been shown to survive gastrointestinal transit and to improve symptoms of lactose intolerance. This effect persists after the intervention is ceased [24]. Other probiotic strains have shown beneficial effects on lactose digestion and symptoms in lactase deficient persons [12,25,26]. Further investigation with different strains of *bifidobacteria* or *lactobacilli* on symptoms of lactose intolerance showed contradictory results. [27] observed that 7 day supplementation with *Lactobacillus acidophilus* did not change hydrogen production or symptoms. [28] however found a decrease in hydrogen production after 7 days of milk intake supplemented with *Lactobacillus acidophilus*, but not all individuals had relief of their symptoms. *Bifidobacterium breve* for 5 days did not improve lactose intolerance symptoms, but reduction in breath hydrogen was measured [29]. Overall these contradictions have not led to a general acceptance of probiotics as a efficient treatment for lactose intolerance [30-32].

The observation of adaptation seen in lactose intolerant persons consuming regularly small amounts of dairy products might be in accordance with the concept of adaptation of the colonic metabolism by increased lactate metabolizing populations in the gut. This allows efficient metabolism of increased amounts of lactose [33]. The observation that lactulose fermentation is impaired during ingestion of ampicillin (2g / day) gives rise to the idea that antibiotics can disrupt the microbiota in the colon. There is no evidence in the literature that antibiotics have a negative effect on the fermentation of lactose, however it would not be surprising if such a phenomenon was found [34].

6. Conclusion

There is evidence that probiotics can alleviate symptoms of lactose intolerance. This can occur by increased hydrolysis of lactose in the dairy product and in the small intestine. It can also be achieved by manipulation of the colonic metabolism. However, the precise mechanism how colonic metabolism influences lactose intolerance symptoms is not yet known. The reported studies are not consistent in their experimental set-up, results and conclusions.

The diagnosis of lactose maldigestion and the relation to complaints is highly complex. For an effective treatment of lactose intolerance and a correct interpretation of the effects of an intervention, knowledge of the underlying mechanisms of lactose intolerance is essential. Development of new strategies concerning the treatment with probiotics should therefore include an analysis of the relevant intermediate endpoints. In this way applications of probiotics for treatment of lactose intolerance could lead to a promising strategy.

Author details

Roel J. Vonk *, Gerlof A.R. Reckman, Hermie J.M. Harmsen and Marion G. Priebe
Dept. of Cell Biology, Centre for Medical Biomics and Dept. of Medical Microbiology, UMCG, Groningen, The Netherlands

7. References

[1] Chandan R. Dairy-based ingredients. Eagan Press, St. Paul, MN; 1997.

[2] Järvelä I, Torniainen S, Kolho K. Molecular Genetics of Human Lactase Deficiencies. Annals of Medicine 2009;41 568-575.

[3] Koetse HA, Vonk RJ, Gonera-De Jong GBC, Priebe MG, Antoine JM, Stellaard F, Sauer PJJ. Low Lactase Activity in a Small-Bowel Biopsy Specimen: Should Dietary Lactose Intake Be Restricted in Children with Small Intestinal Mucosal Damage? Scandinavian Journal of Gastroenterology 2006;41(1) 37-41.

[4] de Vrese M, Stegelmann A, Richter B, Fenselau S, Laue C, Schrezenmeir J. Probiotics-Compensation for Lactase Insufficiency. The American Journal of Clinical Nutrition 2001;73(suppl) 421S-429S.

[5] He T, Priebe MG, Welling GW, Vonk RJ. Effect of Lactose on Oro-Cecal Transit in Lactose Digesters and Maldigesters. European Journal of Clinical Investigation 2006;36(10) 737-742.

[6] He T, Priebe MG, Zhong Y, Huang C, Harmsen HJM, Raangs GC, Antoine JM, Welling GW, Vonk RJ. Effects of Yogurt and Bifidobacteria Supplementation on the Colonic Microbiota in Lactose-Intolerant Subjects. Journal of Applied Microbiology 2008;104 595-604.

* Corresponding Author

[7] Li J, Zhang W, Wang C, Yu Q, Dai R, Pei X. Lactococcus Lactis Expressing Food-grade β-galatosidase Alleviates Lactose Intolerance Symptoms in Post-weaning Balb/c Mice. Applied Microbiology and Biotechnology 2012;DOI 10.1007/s00253-012-3977-4.

[8] He T, Priebe MG, Vonk RJ, Welling GW. Identification of Bacteria with β-galactosidase Activity in Faeces from Lactas Non-Persistent Subjects. FEMS Microbiology Ecology 2005;54(3) 463-469.

[9] Vonk RJ, Stellaard F, Priebe MG, Koetse HA, Hagedoorn RE, De Bruijn S, Elzinga H, Lenoir-Wijnkoop I, Antoine J-M. The 13C/2H-glucose Test for Determination of Small Intestinal Lactase Activity. European Journal of Clinical Investigation 2001;31(3) 226-233.

[10] He T, Venema K, Priebe MG, Welling GW, Brummer R-JM, Vonk RJ. The Role of Colonic Metabolism in Lactose Intolerance. European Journal of Clinical Investigation 2008;38(8) 541-547.

[11] Ojetti V, LaMura R, Zocco MA, Cesaro P, De Masi E, La Mazza A, Cammarota G, Gasbarrini G, Gasbarrini A. Quick Test: New Test for the Diagnosis of Duodenal Hypolactasia. Digestive Diseases and Sciences. 2008;53(6) 1589-1592.

[12] Hiele M, Ghoos Y, Rutgeerts P, Vantrappen G, Carchon H, Eggermont E. 13CO2 Breath Test Using Naturally 13C-Enriched Lactose for Detection of Lactase Deficiency in Patients with Gastrointestinal Symptoms. Journal of Laboratory and Clinical Medicine 1988;122(2) 193-200.

[13] Fuller R. Probiotics in Human Medicine. Gut 1991;32 439-442.

[14] Kinová Sepová H, Bilková A, Bukovský M. Lactobacilli and their Probiotic Properties. Ceska Slov Farm. 2008;57(2) 95-98.

[15] Schaafsma G. Lactose Intolerance and Consumption of Cultured Dairy Products — a Review. International Dairy Federation Newsletter 1993;2 15–16.

[16] Masood MI, Qadir MI, Shirazi JH, Khan IU. Beneficical Effects on Lactic Acid Bacteria on Human Beings. Critical Reviews in Microbiology 2011;37(1) 91-98.

[17] Priebe MG, Vonk RJ, Sun X, He T, Harmsen HJM, Welling GW. The Physiology of Colonic Metabolism. Possibilities for Interventions with Pre- and Probiotics. European Journal of Nutrition 2002;41(s1) 2-10.

[18] He T, Priebe MG, Harmsen HJM, Stellaard F, Sun X, Welling GW, Vonk RJ. Colonic Fermentation May Play a Role in Lactose Intolerance in Humans. The Journal of Nutrition 2006;136(1) 58-63.

[19] Binder HJ. Role of Colonic Short-Chain Fatty Acid Transport in Diarrhea. Annual Review of Physiology 2010;72 297-313.

[20] Marquet P, Duncan SH, Chassard C, Bernalier-Donadille A, Flint HJ. Lactate Has the Potential to Promote Hydrogen Sulphide Formation in the Human Colon. FEMS Microbiology Letters 2009;299(2) 128-134.

[21] Muñoz-Tamayo R, Laroche B, Walter E, Doré J, Duncan SH, Flint HJ, Leclerc M. Kinetic Modelling of Lactate Utilization and Butyrate Production by Key Human Colonic Bacterial Species. FEMS Mircobiology Ecology 2011;76(3) 615-624.

[22] Flint HJ, Bayer EA, Rincon MT, Lamed R, White BA. Polysaccharide Utilization by Gut Bacteria: Potential for New Insights from Genomic Analysis. Nature Reviews Microbiology. 2008;6(2) 121-31.

[23] Brown CT, Davis-Richardson AG, Giongo A, Gano KA, Crabb DB, Mukherjee N, Casella G, Drew JC, Ilonen J, Knip M, Hyöty H, Veijola R, Simell T, Simell O, Neu J, Wasserfall CH, Schatz D, Atkinson MA, Triplett EW. Gut Microbiome Metagenomics Analysis Suggests a Functional Model for the Development of Autoimmunity for Type 1 Diabetes. PLoS ONE. 2011;6(10) e25792.

[24] Almeida CC, Lorena SLS, Pavan CR, Akasaka HMI, Mesquita MA. Beneficial Effects on Long-Term Consumption of a Probiotic Combination of *Lactobacillus casei* Shirota and *Bifidobacterium breve* Yakult May Persist After Suspension of Therapy in Lactose-Intolerant Patients. Nutrition in Clinical Practice 2012;27(2) 247-251.

[25] Lin M-Y, Dipalma JA, Martini MC, Gross CJ, Harlander SK, Savaiano DA. Comparative Effects of Exogenous Lactase (β-galactosidase) Preparations on in Vivo Lactose Digestion. Digestive Diseases and Sciences 1993;38(11) 2022-2027.

[26] Rabot S, Rafter J, Rijkers GT, Watzl B, Antoine J-M. Guidance for Substantiating the Evidence for Beneficial Effects of Probiotics: Impact of Probiotics on Digestive System Metabolism. Journal of Nutrition 2010;140(3) 677S-689S.

[27] Saltzman JR, Russell RM, Golner B, Barakat S, Dallal GE, Goldin BR. A Randomized Trial of *Lactobacillus acidophilus* BG2FO4 to Treat Lactose Intolerance. The American Journal of Clinical Nutrition. 1999;69(1) 140-146.

[28] Kim HS, Gilliland SE. *Lactobacillus acidophilus* as a Dietary Adjunct for Milk to Aid Lactose Digestion in Humans. Journal of Dairy Science 1983;66 959-966.

[29] Park MJ, Lee JH, Kim KA, Kim JS, Jung HC, Song IS, Kim CY. The Changes in the Breath Hydrogen Concentration After the Ingestion of *Bifidobacterium breve* KY-16 in the Lactose Malabsorbers. The Korean Journal of Gastroenterology. 1999;34(6) 741-748.

[30] Levri KM, Ketvertis K, Deramo M, Merenstein JH, D'amico F. Do Probiotics Reduce Adult Lactose Intolerance? The Journal of Family Practice. 2005;54 613-620.

[31] Shaukat A, Levitt MD, Taylor BC, MacDonald R, Shamliyan TA, Kane RL, Wilt TJ. Systematic Review: Effective Management Strategies for Lactose Intolerance. Annals of Internal Medicine. 2010;152(12) 797-803.

[32] Wilt TJ, Shaukat A, Shamliyan T, Taylor BC, MacDonald R, Tacklind J, Rutks I, Schwarzenberg SJ, Kane RL, Levitt M. Lactose Intolerance and Health. Evidence Report – Technology Assesment 2010;192 1-410.

[33] Rong Q, Cheng Yu H, HuiZhang D, Guo Z, Ling LI, Sheng YE. Milk Consumption and Lactose Intolerance in Adults. Biomedical and Environmental Sciences 2011;24(5) 512-517.

[34] Rao SS, Edwards CA, Austen CJ, Bruce C, Read NW. Impaired Colonic Fermentation of Carbohydrate After Ampicillin. Gastroenterology 1988;94(4) 928-932.

Chapter 8

Cereal-Based Functional Foods

R. Nyanzi and P.J. Jooste

Additional information is available at the end of the chapter

http://dx.doi.org/10.5772/50120

1. Introduction

Functional foods are defined as foods that, in addition to their basic nutrients, contain biologically active components, in adequate amounts, that can have a positive impact on the health of the consumer [1, 2, 3, 4]. Such foods should improve the general and physical conditions of the human organism and/or decrease the risk of occurrence of disease [5]. Functional foods have also been referred to as medicinal foods, nutritional foods, nutraceuticals, prescriptive foods, therapeutic foods, super-foods, designer foods, foodceuticals and medifoods [4]. These foods generally contain health-promoting components beyond traditional nutrients [1]. Various criteria for defining functional foods have been mooted by [6] and a number of published reports have indicated the benefits of functional foods to the consumer [7, 8].

One way of creating a functional food is by inclusion of ingredients such as probiotics and prebiotics to levels that enable the consumer to derive optimal health benefits [2]. Probiotics are defined as live microorganisms which upon ingestion in adequate numbers impart health benefits to the host animal beyond inherent basic nutrition [4, 9,10]. Most of the probiotic species belong to the genera *Lactobacillus* and *Bifidobacterium* [11, 12,13]. Benefits of probiotic intake include prevention and treatment of infantile diarrhoea, travelers' diarrhoea, antibiotic induced diarrhoea, colon cancer, constipation, hypercholesterolaemia, lactose intolerance, vaginitis and intestinal infections [14, 15, 16]. Prebiotics, on the other hand, are non-digestible food ingredients that affect the host by selectively targeting the growth and/or activity of one or a limited number of beneficial bacteria in the colon, and thus have the potential to improve health [2, 7, 17, 18, 19]. Potential benefits of prebiotic intake include reduction of cholesterol absorption, control of constipation, bioavailability of minerals and reduction in blood glucose levels when used to replace sucrose in diabetic diets [8, 15, 20, 21]. The main aim of this chapter is therefore to discuss the possibility of converting cereal-based fermented foods into functional foods similar to the existing commercial dairy products. The fermentation of cereal based foods and the beneficial attributes of such foods will be discussed. The latter attributes include the use of such foods as delivery vehicles for probiotic bacteria to the consumer.

2. Fermentation of cereal based foods

Generally, fermentation is a food preservation method intended to extend shelf-life, improve palatability, digestibility and the nutritive value of food [22, 23, 24]. Lactic acid fermentation comprises of the chemical changes in foods accelerated by enzymes of lactic acid bacteria resulting in a variety of fermented foods [11, 25]. Lactic acid fermentation processes are the oldest and most important economical forms of production and preservation of food for human consumption ([11, 23, 26, 27]. It is, therefore, not surprising that fermented foods and beverages make a big contribution to people's diets in Africa [28]. It is reported that fermented foods globally contribute 20 to 40% of the food supply and usually, a third of the food consumed by man is fermented [29]. This renders fermented foods and beverages a significant component of people's diets globally. It is estimated that the largest spectrum of lactic acid fermented foods occurs in Africa [23, 30]. However, in Africa, fermented foods and beverages are often prepared by employing spontaneous fermentation processes at household level or by small-scale industries using maize, sorghum and millet as the main cereals [11, 31, 32]. In sections 3 and 4 of this chapter, a description will be given of acid-fermented cereal-based foods and beverages and the major bacteria involved in the fermentation of such foods. In section 5 of this chapter, probiotic cereal beverages will be dealt with.

2.1. Some beneficial attributes of African fermented cereal-based foods

Lactobacillus species are the predominant organisms involved in the fermentation of cereal-based foods and beverages in Africa (see section 4.1). These organisms are reported to have bacteriostatic, bactericidal, viricidal, anti-leukaemic and antitumor effects in the consumer [25, 28, 33]. Beneficial starter cultures are not usually used in the fermentation of traditional cereal-based foods and beverages. However, it is reported that fermented foods have a probiotic potential [34] due to the probiotic *Lactobacillus* species that may be contained in them, some of which are of human intestinal origin [11].

The quality of some traditional African fermented products (see section 3.2) can be enhanced using beneficial cultures. '*Dogik*' for example is '*ogi*' enhanced with a lactic acid starter culture reputed to have antimicrobial activities against diarrhoeagenic bacteria [11]. *Lactobacillus paracasei* ssp. *paracasei*, a probiotic *Lactobacillus* species [11] was present together with other LAB in *uji* [35]. Strains of *Lb. acidophilus,* which are probiotic, were also isolated from an African sorghum-based product in which accelerated natural lactic fermentation was observed [36].

Improved production of milk by nursing mothers has been attributed to consumption of fermented *uji*, one of the traditional fermented beverages in Africa. *Kanun-Zaki*, a fermented non-alcoholic cereal-based beverage widely consumed in Northern Nigeria is also popularly believed to enhance lactation in nursing mothers [37]. Restoration of the normal blood level and resultant compensation for blood lost during traditional tribal circumcision operations in parts of Africa is attributed to drinking large quantities of fermented *uji* [38].

It is reported that several B vitamins including niacin (B3), panthothenic acid (B5), folic acid (B9), and also vitamins B1, B2, B6 and B12 are released by LAB in fermented foods. These vitamins are co-factors in some metabolic reactions, for instance, folates prevent neural tube defects in babies and provide protection against cardiovascular disease and some cancers [39].

2.1.1. Shelf-life extension and improved nutritional and sensory properties

Generally, shelf-life, texture, taste, aroma and nutritional value of food products can be improved by fermentation [11, 23, 25, 40, 41]. The metabolic activities of microbial fermenters are responsible for the improvement in taste, aroma, appearance and texture [23, 30]. During fermentation, there is production of lactic, acetic and other acids and this enhances the flavour and lowers the pH of the final product. The acids also prolong food shelf-life by lowering the pH to below 4 and this restricts the growth and survival of spoilage organisms and some pathogenic organisms such as *Shigella, Salmonella* and *E. coli* [11, 25, 28, 33, 42]. Fermented foods, unlike non-fermented foods, have a longer shelf-life, making fermentation a key factor in the preservation of such foods [23, 43]. Because fermentation improves keeping quality and nutritional value, it is a predominant food processing and preservation process [44, 45]. During fermentation, enzymes such as lipases, proteases, amylases and phytases are produced and these in turn hydrolyse lipids, proteins, polysaccharides and phytates respectively [46]. The released nutrients contribute to the enhancement of sensory quality and nutritional value of the product [46, 47].

2.1.2. Inhibition of pathogenic microorganisms in fermented foods.

Spontaneous fermentation may involve species of *Lactobacillus, Lactococcus, Pediococcus* as well as certain yeasts and moulds [48]. Lactic acid bacteria involved in fermentation are able to produce hydrogen peroxide, but lack the true catalase to break down the hydrogen peroxide. The hydrogen peroxide can, therefore, accumulate and be inhibitory to some harmful bacteria and to the LAB themselves [11].

The organic acids released (e.g. lactic, acetic, propionic and butyric acids), as by-products during lactic acid fermentation, lower the pH to levels of 3 to 4 with a titratable acidity of about 0.6% (as lactic acid) [23, 40, 48]. The undissociated forms of the acetic and lactic acids at low pH exhibit inhibitory activities against a wide range of pathogens [23 48]. This improves food safety by restricting the growth and survival, in fermented cereal beverages, of spoilage organisms and some pathogenic organisms such as *Shigella, Salmonella* and *E. coli* [11, 25, 28, 33, 43, 47]. Fermented maize gruel and high-tannin sorghum gruel at pH 3.8 inhibited *E. coli, Campylobacter jejuni, Shigella flexneri, Salmonella typhimurium* and *Staphylococcus aureus* [30]. When starter cultures were used to ferment sour maize bread, it was found out that *Lb. plantarum* lowered the pH to 3.05 [40]. The fermented maize dough also showed growth inhibitory activity against *Salmonella typhi, S. aureus, E. coli,* and the aflatoxigenic *Aspergillus flavus* [40].

Although *Koko* sour water (KSW) fed to Ghanaian children did not seem to halt diarrhoea, improved well-being was claimed after 14 days of consumption of this product [44]. Conflicting results about the efficacy of fermented beverages against pathogens and diarrhoea is attributed to the unpredictable nature of spontaneous fermentation. Spontaneous fermentation results in a variety of species and strains with varying degrees of antibacterial activity and ability to adhere to intestinal membranes [44]. Other studies have however, reported positive outcomes of consuming fermented cereal beverages. It was reported that a fermented cereal gruel in Tanzania reduced diarrhoea by 40% in consuming children compared to those children that did not consume it over a period of 9 months [44, 48]. This was attributed to better beverage microbial safety as well as protection against intestinal enteropathogenic colonization [48]. In a review by [25] information gathered revealed that fermented cereal-based products which contained *Lactobacillus* spp. and lactic acid had viricidal, anti-leukemic, antitumor and antibacterial activities. .

Lactobacillus isolates including *Lb. fermentum,* and *Lb. plantarum,* from maize-based ogi (West Africa) and *Lb. fermentum, Lb. paracasei* and *Lb. rhamnosus* from maize-based *boza* (Eastern Europe) were active against potential pathogens such as *Escherichia coli, Klebsiella pneumoniae, Pseudomonas aeruginosa, Enterococcus faecalis and Bacillus cereus* due to the low pH in these products and the production of bacteriocins by the *Lactobacillus* spp [49].

2.1.3. Production of bacteriocins by lactic acid bacteria

Bacteriocinogenic lactic acid bacteria (LAB) isolated from fermented foods produce proteinaceous, antimicrobial substances (Table 1) called bacteriocins [23, 31, 50, 51]. It was reported that bacteriocinogenic LAB prevent the growth of pathogens such as *Listeria monocytogenes, Bacillus cereus, Staphylococcus aureus* and *Clostridium dificile* [23].

Bacteriocins have the ability to form pores in the membrane of target bacteria, in this way exerting bactericidal and bacteriostatic effects against the growth of pathogens in the intestinal tract [52]. Bacteriocins also reduce or prevent post-production microbial contamination of feed and food fermentation products in the food chain [51]. It was observed that bacteriocins from *Lb. plantarum* and *Lb. casei* isolated from fermented maize products, *kenkey* and *ogi* respectively inhibited and acted against a number of food borne pathogens [51]. However, bacteriocins have a narrow antimicrobial spectrum and of all bacteriocins, nisin produced by *Lactococcus lactis* is the only one generally used as a preservative by food manufacturers [46, 50]. A range of characterized bacteriocins that have potential benefits, have been reported to be produced by the *Lactobacillus* spp. and these are referred to in Table 1. While some LAB may show bacteriocin-linked inhibition of food spoilage and pathogenic bacteria *in vitro* in laboratory media, inhibitory activity in the food matrices may not be equally effective. This may be due to poorer diffusion of the bacteriocin into the cells of pathogenic bacteria in the food matrix or be the result of bacteriocin inactivation by nutrient components in the food [53].

Bacteriocin	Bacterial Species	Active against
Bulgarican	*Lb. delbrueckii* subsp. *bulgaricus*	Broad, including G (-).
N.N	*Lb. fermentum*	Broad G (+) incl *Listeria* spp
Acodophillin	*Lb. acidophilus* DDS 1	Disease-causing M/Os
Lactocidin	*Lb. acidophilus*	Disease-causing M/Os
Acidolin	*Lb. acidophilus*	Disease-causing M/Os
Lactobacillin	*Lb. acidophilus*	Disease-causing M/Os
Lactacin B	*Lb. acidophilus*	LAB
Nisin	*Lactococcus lactis*	Broad G(+) incl *Listeria* spp
Lactabacillin	*Lb. brevis*	LAB
Brevicin	*Lb. brevis*	LAB
Caseicin 80	*Lb. casei*	*Lb. brevis*
Plantaricin A	*Lb. plantarum*	LAB
Reuterin	*Lb. reuteri*	Broad G (+), G (-) and fungi

Source: [22, 27, 52, 119], G+, Gram positive bacteria; G-, Gram negative bacteria; MOs, microorganisms

Table 1. Some of the bacteriocins produced by lactic acid bacteria (LAB)

2.1.4. The effect of fermentation on toxic, antinutritional and indigestible compounds in cereal foods

During fermentation, microbial activity may lead to the elimination of toxic compounds from food products [28, 31]. For example it was reported that fermentation with *Lb. plantarum* starter cultures significantly reduced the cyanogenic glucoside content of cassava [23]. High cyanide content in a diet can cause acute poisoning, tropical ataxic neuropathy and konzo (a paralytic disease). It may also exacerbate iodine deficiency resulting in goitre and cretinism [54]. During 'gari' and 'lafun' production from cassava, the cyanogenic glucoside, linamarin, is hydrolysed by the linamarinase enzyme to glucose and cyanohydrin. The latter product is then broken down to acetone and hydrocyanic acid by hydroxynitrile lyase at pH 5-6 and the free cyanide is released faster by gentle heating [25, 55]. If the cyanogenic glucoside linamarin were to be hydrolysed in the gastro-intestinal tract (GIT), the released cyanide anion would be absorbed and halt the functioning of cytochrome oxidase enzymes in the body [23, 29].

Legumes and cereals contain indigestible oligosaccharides such as stachyose, verbascose, and raffinose which cause flatulence, diarrhoea and digestion problems [23]. The α-D-galactosidic bonds in the above-mentioned sugars are relatively heat-resistant, but they can be degraded by the galactosidase enzymes of some LAB including strains of *Lb. fermentum, Lb. plantarum, Lb. salivarius, Lb. brevis, Lb. buchneri* and *Lb. cellobiosus* [23]. During fermentation, the microorganisms disintegrate these flatulence-causing and indigestible oligosaccharides into utilisable di- and mono-saccharides [25, 29, 53].

Phytic acid, tannins and phenolic acids are polyphenols that are considered to be antinutritional factors (ANFs) and are found in cereals and legumes and the foods prepared therefrom [56]. The ANFs contribute to malnutrition and reduced growth rate due to the promotion of poor protein digestibility and by limiting mineral bioavailability [23, 46, 56, 57]. Phytic acid in cereals and legumes, for example, (Table 2) affects the nutritional quality due to chelation of phosphorus and other minerals such as Ca, Mg, Fe, Zn, and Mo [41, 56, 58, 59]. The resultant low mineral bioavailability can result in mineral deficiency [47, 59]. Deficiency in a mineral such as iron can result in anaemia, a decrease in immunity against disease and impaired mental development. Poor calcium bioavailability on the other hand prevents optimal bone development and can cause osteoporosis in adults. Insufficient zinc brings about recurring diarrhoea and retarded growth [59].

Product	Range (%)
Sorghum	0.57-0.96
Maize	0.44-1.2
Millet	0.85-1.1
Cowpeas	0.89-1.5

Adapted from reference [30]

Table 2. Approximate phytate content of sorghum, maize, millet and cowpeas

Other negative effects of the presence of phytate in the diet, include the reduction of the activity of digestive enzymes such as trypsin, alpha-amylase and beta-galactosidase in the GIT. This is due to the formation of complexes of phytate with the enzymes and other nutrients that negatively affect digestive processes [57, 58]. Similarly tannins and polyphenols are enzyme inhibitors of plant origin that form complexes with proteins, resulting in deactivation of digestive enzymes, reduction in protein solubility and digestibility and reduction of absorbable ions [57, 60, 61]. The enzymes inhibited by tannins and/or polyphenols include pepsin, trypsin, chymotrypsin, lipases, glucosidase and amylase [57, 62]. Inhibition of the amylase enzymes results in low starch breakdown and hence, less sugar release in the GIT [117]. In fermented products this amylase inhibition by tannins impairs microbial proliferation [83]. This in turn decelerates pH decrease and acidity production in the medium [83].

Fermentation, by certain LAB and yeasts, removes or reduces the levels of antinutritional factors such as phytic acid, tannins and polyphenols present in some cereals meant for weaning purposes [23, 31, 41, 47, 53, 56, 59, 63]. During fermentation, optimal pH conditions prevail for enzymatic degradation of the antinutritional factors. This results in better bioavailability of minerals such as iron, zinc and calcium [11, 23]. Strains of *Lb. plantarum* degraded phytic acid in the cereals after incubation at 37 °C for 120 hours [23]. This degradation can be ascribed to the hydrolysis of the phosphate group by phytases from the raw cereal substrate and produced by the fermenting microorganisms [46, 47, 57]. Fermentation alone reduced the phytate content by 39%. The combined effect of fermentation plus the addition of exogenous phytase, resulted in a reduction of 88% of the phytates in tannin sorghum gruel [47].

Fermentation reduced phenolic compounds and tannins in finger millet by 20% and 52% respectively [60]. Fermentation coupled with methods such as decortication, soaking and germination reduced the tannins in sorghum, other cereals and in beverages made from these cereals [57, 60, 61, 62, 83]. Fermentation of porridges from whole and decorticated tannin sorghum led to significant reduction of total phenols [61].

The use of *Rhizopus oligosporus* to ferment cooked soybean in *tempe* production reduced residual trypsin inhibitor activity (TIA) by 91% in addition to the 86.4% reduction attributed to steaming [57]. The reduction of the TIA was ascribed to hydrolysis of the trypsin inhibitor by the fungi fermenting the *tempe* [57]. In another study [63], *Lb. brevis, Lb. fermentum, Streptococcus thermophilus* and *Pediococcus pentosaceus* were observed to have improved the nutritional quality of fermented sorghum products. Table 3 shows that some strains of LAB significantly degraded trypsin inhibitors. This illustrates the possibility that using carefully selected probiotic bacteria to ferment cereal foods may reduce the antinutritional factors in such products.

Fermentation can also decrease the activity of the proteinase and amylase inhibitors in cereals resulting in an increase in the availability of starch and essential amino acids such as lysine, leucine, isoleucine and methionine [23, 46, 53]. The protein quality and nutritive value of fermented products such as *kenkey; iru;* and *ugba* [25] and *ogi* [64] was improved during fermentation due to either microbial protein synthesis or loss of non-protein material. In support of the above, [39] reported that fermenting with *Lb. plantarum* OG 261-5 significantly improved the levels of tryptophan, lysine and tyrosine even though other amino acids such as isoleucine, leucine, valine and phenylalanine decreased.

LAB isolate	Reduction of TI (mg)	Percent reduction
Lb. plantarum 91	2.41	48.0
Lb. fermentum 103	1.22	24.4
Pediococcus sp. 90	0.89	17.8
Pediococcus sp. 19	1.08	21.6
Leuconostoc sp. 106	2.68	53.6
Lactobacillus sp. 41	0.65	13.0
Lactobacillus sp. 17	1.86	37.2
Lactobacillus sp. 62	1.34	26.8

Adapted from references [23, 30]; **Aflata* is a gelatinized maize paste intermediate in kenkey production.

Table 3. Degradation of trypsin inhibitor (TI) by lactic acid bacteria isolated from **aflata* in Ghana

Fermentation in many instances results in an increased vitamin content of the final product [23]. Lactobacilli involved in fermentation may require vitamins for growth, but several of them are capable of bio-synthesizing B-vitamins in excess. It is reported that several B vitamins including niacin (B3), panthothenic acid (B5), folic acid (B9), and also vitamins B1, B2, B6 and B12 are released by LAB in fermented foods [39]. Cereal-based products such as *ogi; mageu;* and *kenkey* have thus been reported to have an improved B-vitamin content [25, 29]. Fermentation therefore improves the nutritive value of cereal foods.

2.1.5. Reduction, binding or detoxification of mycotoxins in fermented foods

Maize (*Zea mays*), sorghum (*Sorghum vulgare*), pearl millet (*Pennisetum glaucum*) and finger millet (*Eleusine coracana*) constitute the most important cereals for the preparation of fermented foods in the developing world [41, 65, 66, 67]. These cereal grains are however, exposed to pre- and post-harvest mycotoxin contamination which end up in the fermented foods [23, 54. 67]. Among the cereals, maize is the most prone to mycotoxin contamination [66].

Mycotoxins are secondary metabolites released into cereal grains and legume seeds by species of the genera *Aspergillus, Fusarium* and *Penicillium* [54, 66]. Aflatoxins and fumonisins are the mycotoxins, in cereals, of major health and economic concern in the developing world [23, 24, 48, 54, 66, 68, 69]. Table 4 shows the deaths linked to mycotoxins in foods. Aflatoxin B1 (AFB1) is toxic, carcinogenic, mutagenic and teratogenic [45, 69]. Fumonisins have been linked to oesophageal cancer in South Africa and liver cancer in China [66, 68]. Kwashiorkor in children is aggravated by long term exposure to aflatoxin [66]. The development and propagation of cereal-based probiotic and/or synbiotic (prebiotics and probiotics combined) beverages may consequently, to some extent, be hampered by mycotoxin-contamination of the cereals used in making such beverages.

Bacterial and fungal (biological) decontamination is one of the mycotoxin-reducing strategies that have been and are being investigated [24]. *Flavobacterium aurantiacum (Nocardia corynebacterioides), Corynebacterium rubrum, Saccharomyces cerevisiae, Candida lipolitica, Candida krusei, Aspergillus niger, Mucor* spp., *Rhizopus* spp., *Nurospora* spp., *Amillariella tabescens,* and *Trichoderma viride* are bacterial and fungal species reported to have the capability to degrade mycotoxins enzymatically ([23, 24, 45, 69]. Extracellular extracts of *Rhodococcus erythropolis* reduced Aflatoxin B1 (AFB1) by 66.8% after 72 hours of incubation [69]. Fermentation by *R. oryzae* and *R. oligosporus* was reported to reduce aflatoxins to aflatoxicol A which, under conditions created by organic acids, gets permanently converted to aflatoxicol B [54]. It was claimed that aflatoxin B1 is 18 times more toxic than aflatoxicol B and it is also possible that the former, during lactic acid fermentation to pH < 4.0, gets transformed into a less toxic isomer, aflatoxin B2 [54].

A heat-treated *Saccharomyces* yeast species was said to absorb more than 90 % (w/w) of ochratoxin A in grape juice while live cells could only bind 35 % (w/w) [24, 45]. Other workers have indicated that binding of Aflatoxin B1 was better at low pH and when cells were subjected to acid or heat treatment [24]. The implication is that food beverage preparation, which involves cooking after fermentation, together with the highly acidic conditions of the fermented food beverage, may physically alter the microbial cell structure thereby increasing the binding sites for AFB1 [45]. This provides a way of reducing aflatoxins in African fermented foods and beverages. However, some of the microorganisms indicated in the above paragraphs may not necessarily be GRAS (generally recognised as safe) in the human GIT.

Country	Year	Food source	Mycotoxin content	Percentage of samples contaminated	Mycotoxin	Deaths	Case patients
India	1974	maize	NA	NA	Aflatoxin B1	106	397
Kenya	1981	maize	NA	NA	Aflatoxin B1	NA	20
Kenya	2004	maize	~4400ppb	NA	Aflatoxin B1	215	317
Nigeria	2005	maize	NA	NA	Aflatoxin B1	100	NA
Kenya	2005	maize	NA	NA	Aflatoxin B1	30	8
Kenya	2006	maize	NA	NA	Aflatoxin B1	9	NA
Kenya	NA	3 maize brands	0.4-2.0 µg/Kg	NA	Aflatoxins	NA	NA
South Africa	NA	Peanut butter	< 300 ppb	NA	Aflatoxin B1	NA	NA
Togo, Benin	NA	Household maize	NA	30%	Aflatoxin B1	NA	NA
Nigeria	NA	Maize samples	NA	33%	Aflatoxin B1	NA	NA
Benin	NA	Agro-zone sample	> 5 µg/Kg	9.9 - 32.2%	Aflatoxins	NA	NA
Ghana	NA	Maize silos	20-335 µg/Kg	NA	Aflatoxins	NA	NA
Togo, Benin	NA	Maize samples	> 100 ppb	50%	Aflatoxins	NA	NA

Source: reference [66]

Table 4. Deaths and ill health linked to mycotoxin contamination of samples in African countries

Aflatoxin B1 could not be detected in fermented maize porridge (*amahewu*) that had been made from maize meal samples containing 0.55 and 0.84 µg/g aflatoxin B1. In the same study, the levels of fumonisin B1, in contaminated maize meal samples containing 12.1, 24.6, 4.1, 20.6, 47.2 µg/g of this mycotoxin, were drastically reduced in fermented maize porridge to levels of 1.4, 1.4, 0, 6.9, 6.3 µg/g respectively [46]. This exemplifies the detoxification potential for cereal beverages by lactic acid fermentation. The mechanism of mycotoxin removal from fermented food matrices is not clear.

Without forgetting the above paragraph relating to the effect of probiotic fermentation on mycotoxin levels, some reports on fermentation-linked reduction of aflatoxins in cereal food matrices are controversial. There are reports indicating no significant aflatoxin reduction during fermentation [54]. It was observed that fermentation only enabled a reduction of 18% and 13% of aflatoxin and fumonisin respectively in *ogi* [68]. It was reported that under acidic conditions, aflatoxins persist due to aflatoxin precursors and on the other hand, aflatoxin only undergoes reformation but not reduction under acidic conditions created by organic acid metabolites of LAB [68]. There are also fears that fumonisin binds with starch to form an undetectable complex and besides this, they may react with reducing sugar (D-glucose) to form sugar adducts or are hydrolysed to aminopolyols AP1 and AP2 [68].

The foregoing findings indicate that mycotoxin-reduction in fermented cereal food matrices has not yet been properly elucidated. It is therefore necessary to screen probiotic microbial isolates to find those strains that have a definite potential to degrade aflatoxins during fermentation in food matrices. Such mycotoxin-degrading species need to be fully compatible with the human GIT ecosystem. Some workers recommended the use of probiotic microorganisms with high aflatoxin B1 binding capability in fermented foods [24].

However, binding is not degradation and the binding probiotic cells are consumed along with the food matrix. The fate of bound toxins in fermented food matrices needs to be investigated. Probiotics and/or LAB suitably screened for their biological mycotoxin degradation, among other technological and health benefits could be better applied in human food fermentation, even though, prevention of mycotoxin contamination is the better option. Besides fermentation and contamination-preventive measures, it was noted that processing operations including sorting, winnowing, washing, crushing and dehulling [68] significantly reduced mycotoxin levels in several cereal foods.

3. Cereal-based beverages with a probiotic potential

3.1. Selected non-African cereal foods

Most of the commercial products containing probotics and prebiotics available today are dairy-based [70]. Several workers have, however, endeavoured to develop non-dairy, cereal-based probiotic and/or synbiotic products [4, 57, 70-76]. The following non-African fermented cereal beverages have a probiotic potential or in other words, the potential to be transformed into functional beverages.

3.1.1. Boza

Boza is consumed in countries of the Balkan region including Bulgaria, Romania, Albania and Turkey [4, 77]. Reports indicated that *boza* in Turkey contained 0.03-0.39% (w/v) alcohol but the country's national regulations allow beverages with an alcohol content of not more than 5.0 g/L to be considered non-alcoholic [78].

Boza is a highly viscous traditional fermented product, made from millet, maize, wheat, rye, or rice and other cereals mixed with sugar [79, 78, 80]. In the preparation of *boza,* the milled cereals are mixed in water and then cooked in an open or steam-jacketed boiler. The gruel is cooled and strained to remove the bran and hull. Sugar is added and then fermented at 30 °C for 24 hours by back-slopping or use of sourdough and/or by adding yoghurt starter cultures [78]. Fermented *boza* is then cooled to refrigeration temperatures and distributed into 1L plastic bottles to be consumed within 3-5 days [78]. *Boza* is popularly accepted in the countries referred to above due to its pleasant taste, flavour and nutritional value [4].

Spontaneous fermentation involves LAB and yeasts [80]. Lactic acid bacterial species isolated from *boza* included *Leuconostoc paramesenteroides, L. mesenteroides* subsp. *mesenteroides, L. mesenteroides* subsp. *dextranicum, L. oenus, L. raffinolactis Lb. coryniformis, L. confusus, L. sanfrancisco, Lb. fermentum Lb. plantarum, Lb. acidophilus, Lb. coprophilus* and *Lb. brevis* [4, 79, 80]. The yeast isolates included *Saccharomyces cerevisiae, Candida tropicalis, Candida glabrata, Geotrichum penicillatum* and *G. candidum* [4, 80]. The microflora in *boza* [4, 80] can vary depending on the region and/or country as well as the combination of cereals used and other factors. Only three species were however recommended for inclusion in a mixed starter culture for *boza* production namely: *S. cerevisiae, L. mesenteroides* subsp. *mesenteroides* and *L. confusus* [80].

3.1.2. *Kvass*

Kvass is a non-alcoholic fermented cereal-based beverage made from rye and barley malt, rye flour, stale rye bread, and sucrose and is most often consumed in Eastern Europe [81]. *Kvass* is manufactured using two techniques. One technique involves the use of stale dough bread in which the sugars for the yeast fermentation are obtained from the bread-making process, while the second technique involves the use of malt enzymes to hydrolyse the gelatinized starch [81]. Before fermentation is initiated by the addition of baker's yeast or back-slopping, sucrose is added to the *kvass* wort [81]. The fermentation process is terminated by cooling the *kvass* to 4 °C and the product contains proteins, amino acids, vitamins and organic acids either from the raw materials or from the activity of the fermenting microorganisms [81].

The *kvass* alcohol content is less than 1% while the carbohydrate components predominantly include maltose, maltotriose, glucose and fructose [81]. Maltose and maltotriose components are categorized as isomalto-oligosaccharides that are not completely broken down by digestive enzymes in the GIT [81]. Isomalto-oligosaccharides can hence serve as bifidogenic (prebiotic) factors for the proliferation of probiotic bifidobacteria in the intestines [81].

The predominant microorganisms in *kvass* fermentation were found to be *Lb. casei, L. mesenteroides* and *S. cerevisiae*. *Kvass* is not heat-treated after fermentation and as a result high counts of viable cells can be found in the beverage [81]. The isolation of *Lb. casei* from *kvass* (in which it was highly viable), is indicative of the potential of cereal-based beverages such as this to be used as alternatives to milk products in the delivery of probiotics and other functional ingredients to the consumer in the developing world [81].

3.1.3. *Pozol*

Pozol is a traditional fermented maize dough consumed in South-eastern Mexico [4]. *Pozol* is made mainly by Indian and Mestizo populations of Mexico [82]. During *pozol* preparation, maize grains are cooked in lime water to obtain nixtamal (nixtamalization is a process in which maize (corn), or other grains are treated by soaking and cooking in limewater). This results, *inter alia,* in the grain being more easily ground and the nutritional value being improved). The nixtamalized product is then cleaned by washing in water to separate the husks. The grains are ground, moulded into balls, then wrapped in banana leaves and spontaneously fermented at room temperature for about 7 days [82]. The pH of *pozol* is usually in the range of 3.7-4.7 after 48 hours of fermentation [82]. *Pozol* balls at different stages of fermentation can be mixed with water to make a gruel of desired viscosity and then consumed as a beverage by adults, children and infants [82]. Although African fermented maize gruels are not nixtamalized, *pozol* is similar to African traditional products such as *mageu/mahewu, ogi, kenkey* and *koko* that will be discussed in the next section of this chapter.

Escherichia coli was isolated from *pozol* after 48 hours of fermentation [82]. This was linked to the high pH in the initial stages of fermentation and the possibility of the presence of high pH-localities in the dough after 48 hours even though the measured pH was 3.4-4.7 [82]. It is also possible that acid fermented doughs can harbor some pathogenic bacterial strains resistant to high acidity and/or strains adapted to low pH [82].

3.2. African traditional fermented foods

In Table 5 a number of African traditional lactic acid-fermented cereal-based foods and beverages and the major lactobacilli involved in fermentation are listed. Cereals including maize, sorghum and millet have been used individually or in combination in the preparation of a variety of fermented beverages in Africa [83].

3.2.1. Ben-saalga

Ben-saalga is a pearl millet (*P. glaucum*)-based fermented beverage mainly consumed in Burkina Faso [41, 43, 84]. It is popularly consumed by the young, elderly, the sick and the general populace [41, 84]. The traditional way of producing *ben saalga* involves washing the pearl millet, soaking, wet-milling, kneading and sieving moistened flour, and fermenting the settled, but diluted slurry prior to cooking. This then becomes the *ben-saalga* beverage [41, 43, 84]. The pH decreases from 6 to to a pH of 3.6 – 4.0 during a 24-hour fermentation period [84, 85]. In terms of the LAB responsible for the fermentation, spontaneously fermented *ben saalga* is dominated by *Lb. fermentum, Lb. plantarum* and *Pediococcus pentosaceus* [41]. Ethanol, lactic acid and acetic acid were the main products of fermentation in *ben saalga* [84].

Ben saalga has a solids content of 8-10 g/ 100 mL and like other cereal beverages discussed in this chapter, it has a poor energy density and nutrient content [41]. However, the preparation of *ben-saalga* results in a reduction of millet's antinutritional factors, such as phytic acid, by about 50% [41]. Thirty three of the 99 bacterial isolates from *ben-saalga* showed antimicrobial activity against at least one of the indicator pathogens used in the study [43]. Seven of the isolates, identified as *Lb. plantarum,* were bacteriocinogenic against indicator pathogens which included *Escherichia coli* U-9, *Listeria monocytogenes* CECT 4032, *L. innocua, Salmonella typhimurium, S. aureus* CECT 192 and *B. cereus* LWL1 [43]. These findings indicate the probiotic and/or the prophylactic and the therapeutic potential of intake of this fermented cereal beverage. These characteristics may even be improved by using selected starter cultures that can benefit the health of the consumer and enhance the preservation and safety of the food.

3.2.2. Dégué

Dégué is a millet-based fermented food consumed in Burkina Faso [86]. Preparation of *dégué* involves dehulling and grinding of the millet grains, modeling into balls with water and steam cooking to produce gelatinized balls. The balls are then stored to allow a further 24-hour spontaneous fermentation [86]. The pH of *dégué* is usually in the range of 4.57-4.72 and the following microorganisms have been found in the product: *Lb. fermentum, Lb. brevis, Lb. gasseri, Lb. casei, E. coli* and *Enterococcus* sp. [86].

Fermented food product name	Raw materials	Lactobacilli involved	Nature of use	Country or region	References
Ogi, Ogi-baba	Maize, millet or sorghum	*Lb. plantarum*	Paste as staple, breakfast or weaning food	Nigeria, W. Africa	[11, 26, 99]
Uji	Maize, millet or sorghum	*Lb. plantarum*	Porridge	Uganda, Kenya, Tanzania	[11]
Koko	Maize	*Lb. plantarum, Lb. brevis*		Ghana	[11]
Kenkey	Maize	*Lb. fermentum Lb. reuteri*	Mush steamed, eaten with vegetables	Ghana	[11]
Kwunu-Zaki	Millet, sorghum or maize	LAB*	Paste used as breakfast cereal	Northern Nigeria	[37]
Mahewu	Maize, sorghum, millet	*Lb.delbrueckii, Lb. bulgaricus Strep. lactis*	Gritty gruels, Solid staple	S. Africa	[28, 99]
Mawe	Maize	LAB*	Basis of preparation of many dishes	S. Africa, Togo	[11]
Mangisi	Millet	Unknown	Sweet-sour non-alcoholic drink	Zimbabwe	[11]
Munkoyo	Sorghum,millet or maize plus munkoyo roots	Unknown	Liquid drink	Zambia, Africa	[11]
Mutwiwa	Maize	LAB*	Porridge	Zimbabwe	[11]
Tobwa	Maize	LAB*	Non-alcoholic drink	Zimbabwe	[11]
Togwa	Sorghum, millet, maize		Acid fermented gruel for refreshment and weaning	Tanzania	[34]
Liha	Maize	Unknown	Sweet-sour non-alcoholic drink	Ghana, Togo, Benin, Nigeria	[118]

Table 5. African acid-fermented non-alcoholic cereal-based foods and beverages and the lactic acid bacteria involved in the fermentation (LAB*, lactic acid bacteria)

3.2.3. *Kanun-Zaki*

Kanun-zaki is a non-alcoholic fermented cereal-based beverage consumed in Northern Nigeria [11, 37]. *Kanun-zaki* can be prepared from pearl millet, sorghum or maize ([37]:49). This product is popularly served as a breakfast dish [25]. In the preparation of *Kanun-zaki,* the kernels are washed and dried in the sun, then coarsely ground in a mortar and pestle. The flour is then is mixed with hot water to form a paste which is spontaneously fermented for 1-3 days resulting in a sour beverage [25]. It was reported that this beverage is nutritionally, medically and economically important in the regions where it is widely consumed [39].

3.2.4. *Kenkey*

Kenkey is a fermented maize dough product eaten by the people of Ghana, primarily the Gas, Fantis and Ewes [38, 41]. The preparation of the two main types of *kenkey* (Ga-*kenkey* and Fanti-*kenkey*) was described in reference [41].The Fanti people's name for *kenkey* is *dokon* interpreted to mean 'mouth-watering' because of its pleasant odour and flavour [38]. Similar products to *kenkey* made from sour maize dough include *akasa, koko, banku, abele, akple,* and *kpekpe* though these are not as popular as *kenkey* [38]. *Kenkey* fermentation is spontaneous and is dominated by lactic acid bacteria, particularly *Lb. fermentum* and *Lb. reuteri,* and yeasts that include *C. krusei* (*Issatchenkia orientalis*) as the dominant yeast species, while *S. cerevisiae* also contributes to the flavour [11, 41]. Apart from improvement in the protein content from 1.3 to 3.3 g per 16 g nitrogen in ready-to-eat *kenkey,* the *kenkey* flavour is attributed to the formation of flavour compounds, during fermentation, such as 2,3-butanediol, butanoic acid, lactic acid, 3-methylbutanoic acid, octanoic acid, 2-phenylethanol, and propanoic acid [41].

3.2.5. *Koko*

Koko is a millet-based spontaneously fermented beverage mainly consumed in Northern Ghana [44]. The predominant microbial species during fermentation are *Lb. fermentum* and *Weissella confusa* [44]. It was reported that isolates from *koko* showed good antimicrobial activity, tolerance to 0.3% oxgall bile and acid resistance at pH 2.5, which are characteristics of good probiotic strains [44].

3.2.6. *Mageu (mahewu)*

Mageu is a non-alcoholic largely maize-based beverage popular among the indigenous people of Southern Africa, but is also consumed in some Arabian Gulf countries [4, 74, 83]. It is consumed at schools and mines and on farms. It is a refreshing drink and a traditional weaning beverage for infants. *Mageu* is prepared by using 8% to 10% (w/v) maize flour as the major solid substrate in aqueous suspension. Wheat flour or maize bran is added to initiate the lactic acid fermentation [32]. Some ethnic groups also use sorghum and millet flours instead of maize flour and *mageu* is known by different names (Table 6) among the ethnic groups in Southern Africa. Acceptable *mageu* contains 0.4 – 0.5% lactic acid corresponding to an average pH of 3.5 [87, 88, 89].

Several studies have been conducted on *mageu*. One of these included an investigation of the survival of bacterial enteric pathogens in fermented *mageu*, from which it was concluded that fermented *mageu* had bacteriostatic and bactericidal properties [33]. Another study targeted the growth and survival of *Bacillus cereus* in fermented *mageu* in which growth inhibition of the organism was observed [32]. Studies on the development of a starter culture for *mageu* [88, 90, 91] led to the production of mahewu on a commercial scale [92].

Ethnic group	Local name of product	Reference
Zulu	Amahewu	[91]
Swazi	Emahewu	[89]
Xhosa	Emarewu	[91]
Venda	Mabundu	[70]
Pedi	Mapotho	[70]
Sotho	Machleu	[89]

Table 6. Local names for sour maize porridge (*mageu*) in Southern Africa

3.2.7. *Mawe*

This is fermented maize dough consumed in the form of a variety of dishes in Togo, Benin and Nigeria [68]. Making the *mawe* (maize dough) involves washing, wet extraction of the endosperm and kneading to a dough which is then spontaneously fermented for about 3 days [41]. In Bennin, *mawe* dough is used for the preparation of cooked beverages (*koko*), stiff gels (*akassa, agid and, eko*) and steam cooked bread (*ablo*) [41]. The predominant LAB in the fermented *mawe* dough included *Lb. fermentum, Lb. cellobiosus, Lb. brevis, Lb. curvatus, Lb. buchneri* and *Weissella confusa.* Other microorganisms in the dough included pediococci and yeasts such as *Candida krusei, C. kefyr, C. glabrata* and *Saccharomyces cerevisiae* [41]. It was reported that in a study of *mawe* production using starter cultures, *C. krusei,* stimulated the growth of *Lb. fermentum* and *Lb. brevis* [41]. Fermentation of this product offers a number of benefits that include flavour enhancement, nutrient bioavailability (including that of some proteins, minerals and B vitamins) as well as protection against some pathogens due to reduction of the pH to 3.5-4.0 [41]. Maize products are however, deficient in some amino acids such as lysine, tryptophan and methionine, which are found more abundantly in legumes such as cowpeas and sybeans. Co-fermentation with legumes can therefore be expected to improve the quality of the protein and protein levels significantly.

3.2.8. *Munkoyo*

Munkoyo is a traditional fermented maize-based beverage popularly consumed in Zambia and the Democratic Republic of Congo's Katanga province in the south [93, 94]. In Zambia, tree species of *Eminia, Vigna* and *Rhynchosa,* generally referred to as *munkoyo,* are extracted and the extract, high in α- and β-amylases, is used for the liquefaction of maize porridge gel

[93, 94]. The thinned porridge is then spontaneously Fermented, mainly by LAB, for 24-48 hours at room temperature. The sweet-sour *Munkoyo* flavoured drink has a mean pH of 3.5 due to organic acids produced during fermentation, but alcohol (14-26 g/kg) is also detectable. The beverage is consumed by people of all ages [93].

Introduction of *Rhynchosia heterophylla* root extract, *Lb. confusus* LZ1 and *Sacchromyces cerevisiae* YZ20 to the fermentation mix, resulted in a *munkoyo* beverage of pH 3.3, 60 mmol/l lactic acid and an ethanol content of 320-410 mmol/l [93]. The workers observed that a ratio of not more than 1:1000 (yeast: LAB starter culture) fermented for not more than 24 hours resulted in an acceptable *munkoyo* beverage [93]. *Munkoyo* was found to have antibacterial activities. Total coliforms in the *munkoyo* mash initially were 10 cfu/mL but were absent when tested after 15 hours of fermentation due to acidification of the product [94]. The microorganisms in *munkoyo* were not recognised probiotics and it was therefore recommended that the incorporation of probiotic starter cultures producing D (+) lactate be investigated to improve the nutritional, sensory and health benefits of *munkoyo* [94].

3.2.9. Obushera (bushera)

Obushera fermented spontaneously from malted sorghum or millet flour is consumed by young people and adults in Western Uganda [95]. *Obushera* is prepared using sorghum or millet flour. The flour is mixed with water and cooked into a thin porridge and then mixed with a portion of previously fermented porridge. The added fermented portion acts as a 'starter culture' for fermentation to commence and the result is the '*obushera*' beverage consumed by people of any age [48]. Obushera, produced on a small commercial scale, can be used as a thirst quencher, social drink, energy drink and weaning food [95]. The household *bushera,* with a pH in the range 3.7-4.5, had LAB counts varying from 7.1 to 9.4 $\log_{10}$ cfu/mL and coliform counts that were in the range of <1 to 5.2 $\log_{10}$ cfu/mL [96]. The LAB species from household *bushera* included *Lb. plantarum, Lb. paracasei* subsp. *paracasei, Lb. fermentum, Lb. brevis, Lb. delbrueckii* subsp. *delbrueckii* and *Streptococcus thermophilus.* The isolates from laboratory fermented *bushera* belonged to the genera *Lactococcus, Leuconostoc, Lactobacillus, Weissella* and *Enterococcus* [96]. This is indicative of the probiotic potential of *obushera.*

3.2.10. Ogi

Ogi is another traditional African acid-fermented cereal gruel prepared from maize, although sorghum and millet flours are also used [11, 25]. During fermentation, *Lb. plantarum* is the predominant microorganism although bacteria such as *Corynebacterium* spp hydrolyse the corn-starch following which yeast genera such as *Saccharomyces* and *Candida* contribute to the flavour [11, 27]. *Ogi* is traditionally produced by washing the grains, steeping for 12 to 72 hours, wet-milling, wet-sieving and sedimenting the filtrate for 1-3 days to obtain sour *ogi* [64, 97]. The pH of *ogi* is 3.0 – 4.0 after fermentation depending on the time of fermentation and the presence of LAB [64, 68]. *Ogi* has a sour flavour and a characteristic aroma [25, 38, 98]. In Nigeria the name of '*ogi*' depends on the locality and the type of cereal. *Ogi* is the generic name in the Western states of Nigeria where it is usually processed from white maize. *Ogi* from sorghum is known as '*ogi*-baba' [99] while 'ogi-gero' is prepared from millet. In Northern Nigeria, *ogi* is known as 'akamu' or 'eko gbona', while in the Republics of Togo, Benin and Ghana, *ogi* from maize is known as '*koko*' [38, 98]. *Ogi* is the major traditional weaning food commonly served to babies in West Africa. It is also eaten as a breakfast meal and it is a food of choice for the sick [25, 31, 64].

It was observed that use of *Lb. brevis* alone to ferment sterile maize slurry for *ogi* production rapidly reduced the pH to 3.0 in 48 hours compared to the sterile slurry fermented by *S. cerevisiae* [64]. In this study, it was illustrated that it is possible to use starter cultures, such as *Lb. brevis*, to produce *ogi* without compromising its acceptability [64]. The use of starter cultures results in rapid drop in the pH of the food matrix [40]. Rapid pH decline may imply significant increase in the *Lactobacillus* population and increased concentration of organic acids can be indicative of the anti-pathogenic and/or prophylactic and therapeutic potential of *ogi* or other fermented cereal beverages.

3.2.11. Poto poto

This is a traditional fermented maize dough used in homes by the people of the Congo for weaning and for other purposes [86, 100]. *Poto poto* is prepared by soaking maize kernels for about 55 hours followed by milling and sedimentation of the paste in water [86]. The paste is fermented for about 11 hours and then cooked to produce maize gruel [86, 100]]. The fermented paste can be made into *poto poto* balls for selling to make *poto poto* gruel through addition of water and sugar [86, 100]. The pH of *poto poto* samples was found to be in the range 3.48-3.66 [86].

When DNA bands from TTGE gels of *poto poto extracts* were sequenced, the following microorganisms were observed to be present in the fermented product namely: *Lb. plantarum* (predominant), *Lb. gasseri, Enterococcus* sp., *E. coli, Lb. acidophilus, Lb. delbrueckii, Lb. reuteri* and *Lb. casei* [86]. It was established that *Lb. plantarum* and *Lb. fermentum* isolated from *poto poto* produced bacteriocins that were variably inhibitive against strains of *E. coli, Salmonella typhi, Enterobacter aerogenes, Bacillus cereus, Staphylococcus aureus, Listeria monocytogenes* and *Enterococcus faecalis* [100]. The *E. coli, B. cereus* and other food pathogens reported to be in *poto poto* can consequently be inactivated by the bacteriocin-producing LAB from the same food source and make it safer for human consumption [86, 100].

3.2.12. Thobwa

This is a non-alcoholic thin porridge drink prepared from sorghum in Malawi and is popularly consumed by people of all demographics in the country. It is important to note however, that there is an alcoholic version of the *thobwa* in Malawi [67]. *Thobwa* may be similar to *togwa* reportedly made from maize or cassava flour and finger millet malt and consumed in Southern Tanzania [4].

3.2.13. *Ting*

Ting is a fermented traditional sorghum food of Botswana and South Africa [101, 102]. *Ting* is prepared by combining sorghum flour (40-45%, w/v) with warm water and the slurry formed is kept in a warm place (~30-37 °C) for spontaneous fermentation to take place over a period of 2-3 days [102]. *Bogobe* and *motogo* (stiff and soft porridge respectively) are the two types of porridge that can be prepared and/or cooked from *ting* previously soured to pH 3.5-4.0 mainly by LAB and yeasts [102]. *Motogo* (soft) is usually consumed for breakfast and administered to weaning infants while *bogobe* (stiff) is consumed at lunchtime and supper by adults [101, 102]. In recent studies, the dominant microbiota during *ting* fermentation consisted of *Lb. reuteri, Lb. fermentum, Lb. harbinensis, Lb. plantarum, Lb. parabuchneri, Lb. casei* and *Lb. coryniformis, Lb. rhamnosus, Lb. curvatus* and *Weissella cibaria* [101, 102]. The presence of these microorganisms and the low pH (3.5-4.0) inhibits proliferation of a number of pathogens, in this manner maintaining the safety of the food. Fermentation of sorghum for *ting* production improves nutrient levels and reduces antinutritional factors thus increasing the bioavailability of macro-and micronutrients as well as enhancing the sensory attributes [101].

3.2.14. *Uji*

Uji is a non-alcoholic beverage consumed widely in East Africa (Uganda, Kenya and Tanzania). It is usually prepared from maize [41, 103] although sorghum and/or millet could be mixed with the maize flour [35, 41]. There are two types of *uji*, fermented and unfermented. The unfermented *uji* is prepared by boiling water and adding the flour while stirring to obtain the desired drinkable viscosity [41]. Fermented *Uji* can be obtained by fermenting before or after cooking the porridge [38, 41].

Finely ground cereal is slurried with water at a concentration of about 30% w/v. The slurry is spontaneously fermented for two to five days at room temperature (25 °C). During fermentation of *uji*, *Lb. plantarum* has been found to be the dominant *Lactobacillus* species [35] while *Lb. fermentum, Lb. cellobiosus* and *Lb. buchneri, Pediococcus acidilactici* and *P. pentosaceus* are also reported to be part of the fermenting microorganisms in *uji* [41]. The pH of *uji* decreases to 3.5 to 4.0 whereas total acidity (as lactic acid) reaches 0.3 to 0.6% in 32 to 40 hours [38]. After fermentation, *uji* is diluted to about 8 to 10% solids and brought to boil. It is further diluted to 4-5% solids and then sweetened by the addition of 6% sucrose and consumed while still warm [38]. Like other maize beverages, *uji* is of low energy density and is deficient in essential amino acids. Fortification with legumes can improve the protein quality and content while the involvement of α-amylase-rich malt flour and/or fermenting with starch-hydrolyzing starter cultures can increase the rate of fermentation [41]. Fermented and non-fermented *uji* is mainly consumed by rural and urban housewives. Non-fermented cooked *uji* is also consumed in boarding schools, hospitals and hostels. As is the case with *mageu* in South Africa [89], *uji* is also known by different names in different localities in Kenya (see Table 7).

Ethnic group	Local name of product
Embu	Ucuru
Kamba	Uccu
Luo	Nyuka
Luhya	Obusera
Swahili	Uji[a]

Source: reference [38], [a] the common name of sour porridge in East Africa

Table 7. Local names for sour porridge in Kenya

4. Microorganisms involved in cereal-based food fermentations

4.1. Lactic acid bacteria (LAB) involved in African food fermentations

Microorganisms of major importance in lactic acid fermentations belong to the genera *Lactobacillus, Lactococcus, Leuconostoc* and *Pediococcus* [30, 31]. Others include *Streptococcus, Aerococcus, Carnobacterium, Enterococcus, Tetragenococcus, Weisella* and *Vagococcus* [42]. These genera are lactic acid bacteria (LAB) that are widely used in the production of fermented food [39, 52]. The LAB are described as Gram positive, catalase-negative non-sporing rods and cocci, which are usually non-motile [31]. The LAB starter cultures are significant in the production of desired preservative organic acids in the food product during food fermentation [52]. Starter cultures are, however, not usually employed in food fermentations in Africa. Table 8 below shows the lactic acid bacterial species that are dominant in the spontaneous fermentations of several African traditional foods.

Product name	Dominant bacteria	Reference
Fufu	*Lb. plantarum*	[26]
Gari	*Lb. plantarum*	[27]
Mageu	*Lactococcus lactis*	[99]
Mawe	*Lb. fermentum,* *Pediococcus pentosaceus,* *Lactococcus lactis*	[31]
Ogi	*Lb. plantarum*	[26]
Ogi-baba	*Lb. plantarum,* *Lactococcus lactis*	[99]
Togwa	*Lb. plantarum*	[34]
Uji	*Lb. plantarum*	[35]

Table 8. Lactic acid bacteria (LAB) dominant in the spontaneous lactic acid fermentation of African traditional foods

Strains of *Lb. plantarum, Lb. fermentum, Lb. brevis, Pediococcus pentosaceus* and *P. acidilactici* are reported to be among the most predominant species in most African cereal-based fermented beverages [23, 39]. The strains of some of these species have several reported probiotic properties and/or characteristics. Species such as *Lb. plantarum* and *Lb. fermentum* are characterized by being less fastidious, relatively acid resistant, bile tolerant and can thrive on the substances provided in the cereal matrices [39]. It was reported that *Lb. plantarum* showed rapid acidification and produced inhibitory compounds that were active against *Penicillium* and *Aspergillus* strains [40].

Although most of the lactobacilli are generally poor starch fermenters [104], *Lb. plantarum* and *Lb. fermentum* are reported to be the most dominant bacterial species in acid-fermented cereal-based foods. This can be attributed to the degree of acid tolerance and superiority of these species in the utilization of starchy substrates [34, 39]. *Lactobacillus plantarum* isolates from starchy foods such as 'togwa' [34], 'ogi' [104] and cassava [34, 104] have been shown to have good starch-fermenting abilities. The fact that several cereal-based beverages are high in starch, has resulted in several α-amylase-containing lactic acid bacteria, termed amylolytic LAB, becoming sought-after in Africa and elsewhere globally. It has been reported that several strains of *Lb. plantarum, Lb. fermentum,* and *Lb. manihotivorans* with amylolytic capabilities have been isolated from maize-, cassava-, sorghum- and millet-based fermentations [39, 42]. Such strains can ferment starch from a variety of different sources.

4.2. Other microorganisms and combinations of microbial species involved in cereal based food fermentations

Besides LAB, *Saccharomyces cerevisiae* is notable as a predominant yeast species involved in food fermentation in Africa [45]. However, it is important to note that there are several factors determining the predominant microbial species and these include the type of cereal, the geographical location or region, conditions in the fermentation medium, moisture content and the season of the year. Yeast species isolated from an ogi maize fermention mix included *Geotrichum fermentans, G. candidum, Rhodotorula graminis, Saccharomyces cerevisiae, Candida krusei,* and *C. tropicalis* [97]. Further investigations revealed that *Candida krusei* was better than *S. cerevisiae,* but both species improved the growth of *Lb. plantarum* in maize slurry when each of the yeast species were in combination with the lactobacilli [97]. This was attributed to the capability of the two yeast strains to produce amylolytic enzymes which enabled starch breakdown into simpler sugars for the lactobacilli to metabolise into organic acids [97]. For the same reason, during the mixed culture fermentation of *mawe, Candida krusei* improved the growth of *Lb. fermentum* and *Lb. brevis* [23, 41]. During yeast and *Lactobacillus* mixed culture fermentation, the yeasts were also able to provide vitamins and other nutrients for the metabolic activities of the lactobacilli [40].

Certain yeasts were important in producing enzymes such as lipase, esterase and phytase [97]. The lipolytic activity resulted in fatty acids which are precursors of flavour while esterase activity determined aroma and flavour. On the other hand, phytase, produced by these organisms, lowers phytic acid which can form complexes with minerals that in turn can negatively affect protein digestibility [97]. A mixture of *Lb. fermentum* and *Saccharomyces cerevisiae* as starters in the fermentation of *kenkey* and *koko* achieved more rapid pH reduction in 24 hours than spontaneously fermented preparations in 48 hours [39].

4.3. Safety concerns around the use of bacterial strains that could be used as probiotics

The cereal fermented foods and the predominant LAB are generally regarded as safe (GRAS, [23]. Some of the LAB in the fermented food beverages are of human origin and have been used for centuries knowingly or unknowingly [30]. The dominant microorganisms involved in the fermentation of cereal-based beverages have no reported health risk to human life [23]. It was however, noted that some strains of *Enterococcus faecium, E. faecalis,* and *Lb. rhamnosus* were in isolated, highly questionable, cases linked to endocarditis [30]. *Escherichia coli Nissle, Saccharomyces boulardii, Streptococcus thermophilus, Enterococcus francium, Propionibacterium, Pediococcus* and *Leuconostoc* have also been categorized as probiotic species or genera [10].

Most of the bacteria used as probiotics, such as *Lactobacillus* and *Bifidobacterium,* are of human or animal origin and are generally recognized as safe [105]. Apart from *Lactobacillus* and *Bifidobacterium,* other genera such as *Enterococcus* have safety concerns as some of the species are pathogenic [10]. It was reported that even though some enterococci are of technological importance in cheese making, some clinical isolates are regarded as opportunistic pathogens [105]. On that basis LAB, but not enterococci, are generally regarded as safe (GRAS, [105] and can be used in the preparation of cereal-based probiotic beverages.

4.4. Concerns relating to the isomeric type of lactic acid produced by lactic acid bacteria

The organic acids contribute to preservation and food safety, however, it is important to note the concerns relating to L (+) and D (-) lactic acid isomers. The LAB predominantly found in spontaneously fermented African cereal beverages produce lactic acid as one of the major organic acids. Lactic acid contributes to preservation, taste and safety of the fermented foods and beverages [46]. However, lactic acid can occur in two isomers namely L (+) and D (-) isomers and it is only the former isomer that can be degraded in the human system due to the presence of L-lactate dehydrogenase in the gastro-intestinal canal [27, 42, 94]. The genera *Streptococcus, Enterococcus, Lactococcus* and *Carnobacterium* mainly produce the L(+) isomer while *Leuconostoc* spp. and all subspecies of *Lb. delbrueckii* produce the D (-) isomer [23]. The *Weissella* species, *Lb. sakei* and heterofermentative lactobacilli produce a racemate (DL) of isomers [23]. Reports indicate that industrial production of mahewu, a fermented maize beverage, using *Lb. delbrueckii,* creates a challenge of D (-) lactate production [94]. The D (-) lactate producing *Lb. delbrueckii* (ID12441) was also the major fermenting organism isolated from *munkoyo* (see section 3.2.8) [94]. This is a concern since the organisms involved in spontaneous fermentation and the major lactic acid isomer produced in cereal beverages for weaning infants and children may not be known. Lactobacilli and pediococci produce lactic acid isomers that are species specific [23, 30]. In beverages used for weaning purposes, it needs to be established whether LAB strains produce the D (-) or the L (+) lactic acid isomer [53]. An acid-base imbalance can be induced in children consuming excessive amounts of beverages containing D (-) lactic acid and therefore L (+) lactic acid is the most recommended isomer for man [94]. It is therefore necessary to screen any probiotic cultures used in foods due to the disadvantages (possible acidosis) of offering children foods containing D (-) lactic acid [53].

5. Probiotic cereal-based beverages

5.1. Introduction

It is estimated that over 60 million people use sorghum and millet as part of their staple food in Africa in the fermented or unfermented form [63]. This is in addition to maize which is a staple cereal for the majority of the people in Africa and elsewhere in the world. This extensive consumption of cereals is partially the basis for the mounting research into the development of non-dairy cereal-based probiotic beverages. Consumers are becoming more aware of the need to eat food for health reasons. This implies that apart from good taste and nutrients provided, food needs to impart additional health benefits to the consumer. Such benefits can be realized by processing the food in such a way that its functionality is improved, for example by incorporating ingredients such as prebiotics and probiotics.

Probiotic bacteria have several reported potential health benefits [70]. Besides probiotics, prebiotic oligosaccharides also impart reported health benefits to the consumer [70]. However, in terms of foods that are used to deliver probiotic bacteria to the consumer, milk and milk products are almost exclusively used for this purpose [4, 10]. Such dairy products however have limitations that include cost (especially in the developing world), allergens, cultural food taboos against milk consumption, requirement of cold-chain facilities, the need to use beverages that form part of the people's daily diets as well as the need to maintain viability of the probiotic bacterial population in excess of the physiologically required therapeutic minimum of 10^6 -10^7 cfu/mL viable cells in the product when consumed [106].

Probiotic microorganisms need to be consumed regularly and adequately (10^6 cfu/mL per serving) to maintain the intestinal population and to ensure that health benefits will be derived by the consumer [105]. The increasing need to eat food for health reasons, the demand for vegetarian probiotic foods, the growing lactose intolerance in the world population, and the arguable concern about the cholesterol content of fermented dairy products, are other factors that increase the need for the development of non-dairy cereal-based foods [4, 10, 105]. The following paragraphs illustrate the investigations that have been directed towards cereal- and/or legume-based probiotic beverage development.

5.2. Oats-based probiotic beverages

5.2.1. *Proviva*

Proviva is known to be the first commercial oats-based probiotic food beverage [4]. *Proviva* is produced by Skane Dairy and it has been a commercial product in Sweden since 1994. *Proviva* has malted barley added as liquefying agent and the active probiotic component is

Lactobacillus plantarum 299v. The final product which is a mixture of fruit juice and 5% oat meal has a probiotic bacterial population count in the region of 5 x 10^{10} cfu/L [4, 76].

5.2.2. *Yosa*

Yosa is a probiotic oat snack food marketed in Finland and other Scandinavian countries. *Yosa*, which has a flavour and texture comparable to that of dairy yoghurt, is made by cooking the oat bran pudding in water and fermenting with lactic acid bacteria and bifidobacteria. The probiotic species are reported to be *Lb. acidophilus* LA5 and *Bf. lactis* Bb12 [11, 76]. Apart from probiotic bacteria, *yosa* also contains oat fibre, a source of β-glucan that has the potential to lower blood cholesterol and so reduce the chances of heart disease [11, 49].

5.2.3. *Other experimental probiotic oats products*

Several workers have endeavoured to develop non-dairy cereal-based probiotic food products. An oats-based synbiotic functional drink made by fermenting an oats substrate with *Lactobacillus plantarum* B28 was developed [4]. At the end of 21 days of refrigerated storage the bacterial cell counts were still at a level of 7.5 x 10^{10} cfu/ml. The drink was referred to as synbiotic due to the presence of β-glucan, a functional component in cereals and usually highest in oats and barley in addition to the probiotic organism [4, 105]. Oats therefore appears to be a suitable substrate for the growth of probiotic bacteria [71].

It is important, however, to take the probiotic species into consideration when developing cereal based probiotic beverages. The probiotic bacterial population levels were studied in an envisaged synbiotic oats beverage consisting of 5% oats, 2% inulin, 0.5% whey protein concentrate and 4% sugar [107]. After a storage period of 10 weeks at 4 °C the population levels for two probiotic species (*Lb. plantarum* B-28 and *Lb. paracasei* ssp. *casei* B-29) were 1.77 x 10^6 – 1.29 x 10^7 cfu/mL and 7.39 x 10^7 – 4.49 x 10^8 cfu/mL respectively. However when *Lb. acidophilus* ATCC 521 was inoculated into the same oats beverage, the initial population level of 6.77 x 10^7 cfu/mL declined to 1.55 x 10^5 cfu/mL by the 4th week of storage at 4 °C. This decline gradually continued during a subsequent storage period [107]. This tendency was confirmed by other workers [71] who also found that, *Lb. acidophilus* showed slower rates of pH reduction and lower viable counts in oats due to its higher requirement for nutrients in comparison with *Lb. plantarum* and *Lb. reuteri*. To be referred to as a probiotic beverage at the time of consumption such beverages should have a population level of at least 10^6 cfu/mL viable cells [107]. These findings illustrated that the survival of probiotics in cereal beverages is species and strain specific and this should be kept in mind in developing such products.

5.3. Probiotic beverages incorporating malted cereals and hidrolysates

The potential of four bifidobacterial species of human origin to ferment a barley malt hidrolysate similar to that obtained in the brewery was investigated [76]. These species included *Bf. adolescentis* NCIMB 702204, *Bf. infantis* NCIMB 702205, *Bf. breve* NCIMB 702257 and *Bf. longum* NCIMB 702259. The workers found that the addition of yeast extract to the malt hidrolysate as a growth promoter was necessary for the population levels to increase by 1.5 - 2.0 log_{10} cycles to 8.73 – 9.00 log_{10} cfu/ml after 24 hours of fermentation at 37 °C. Their work illustrated the potential of using bifidobacteria to develop a probiotic malt-based beverage by way of looking at the population levels attained in the study [76]. The study did not include product characterisation to establish its sensory attributes neither was the acceptance of the product tested among the target consumers. In addition to this, shelf-life studies in terms of viable bacterial cells were not conducted. On the other hand the barley-malt hidrolysate used as the substrate may not be commercially feasible for use in the developing world and if it were, its protein deficiencies would have malnutrition implications for the African consumer [76].

In another study relating to barley malt, the potential of using *Lactobacillus reuteri* (probiotic) and yeast to develop a cereal-based probiotic drink by fermenting a 5% (w/v) malt suspension was investigated [75]. The workers observed that using a mixed culture of *Lb. reuteri* and yeast resulted in a better decrease in pH, increased lactic acid production and increased ethanol production compared to that observed with pure cultures.The protective effect of extracts of malt, barley and wheat on the bile tolerance of *Lactobacillus reuteri, Lb. acidophilus* and *Lb. plantarum* has also been investigated [108]. It was illustrated that the cereal extracts, particularly from malt, exerted a protective effect, against bile salts, on the studied lactobacilli. The protection was attributed to the presence, in cereal malt extracts, of non-reducing sucrose and soluble oligosaccharides (non-digestible carbohydrates) that have been reported to improve bile tolerance. The study indicated the potential of malt, barley and wheat extracts to offer protection against bile to the probiotics when ingested together.

The factors that influence the growth of selected potential probiotic lactobacilli (e.g. *Lb. fermentum, Lb. reuteri, Lb. acidophilus* and *Lb. plantarum*) in selected cereal substrates as a way of assessing the potential of producing a probiotic cereal-based beverage was investigated [72]. In their study, a malt medium enabled the tested lactobacilli to attain higher counts (8.10 – 10.11 log_{10} cfu /mL) than in non–malted barley and wheat media (7.20 – 9.43 log10 cfu /mL). The differences in counts were attributed to a higher level of sugars (15 g/L total fermentable sugars) and an increased free amino nitrogen concentration (80 mg/L) in malt medium than in the non-malted barley or wheat media (3 – 4 g/L total fermentable sugars and free amino nitrogen concentration of 15.3 – 26.6 mg/L). The sugars were present in the form of maltose, sucrose and also in the form of their monomeric components (glucose and fructose). Growth limitation was a result of either a low pH or a substrate deficiency. In malt medium, where sugars were abundant, the microbial growth was limited by low pH (3.40 – 3.77) while in barley and wheat media, growth was limited by insufficient fermentable sugars and free amino nitrogen. This was based on the observation that growth was halted at a higher pH (3.73 – 4.88) in barley and wheat media than in malt medium [72]. Barley is not abundant in the developing world and therefore a barley-malt probiotic beverage production would not be feasible [72] in this part the world.

5.4. Maize (corn)-based probiotic beverages

5.4.1. *Synbiotic mahewu (mageu)*

Mageu is commercially produced in South Africa which provides it with the potential to deliver probiotic bacteria to the consumers for whom it is part of their daily diets. The commercial *mageu* is prepared using *Lactobacillus delbrueckii* and the product is pasteurized after fermentation and it is therefore not a probiotic product. The possible enhancement of the functional quality of *mageu* was investigated [70]. To this end, six pure probiotic *Lactobacillus* starter cultures and prebiotic oligosaccharides in developing six fermented synbiotic maize-based *mageu*-like beverages were tested. The strains included *Lb. casei* BGP93, *Lb. casei* (Shirota strain), *Lb. rhamnosus* LRB, *Lb. paracasei* BGPI, *Lb. plantarum* BG112, *Lb. acidophilus* PRO and *Lb. delbrueckii* subsp. *lactis* C09 (used to prepare the control). The suitable prebiotic ingredient and the factors affecting the growth of these organisms in the maize gruel, as well as the sustained viability of these organisms in the product during extended refrigerated storage were investigated [70].

The viability of the probiotic strains, in terms of population level, in the fermented synbiotic maize-based beverages at the end of a 90-day storage period at 5 °C exceeded 7.5 $\log_{10}$ cfu/mL [70]. This was well above the recommended therapeutic minimum of 6 $\log_{10}$ cfu/mL at the time of consumption [109, 110]. Intake of a portion of 200 – 300 ml of the experimental synbiotic *mageu* products would potentially enable the consumer to derive 7 to 10.5 g d^{-1} of prebiotic Raftiline® GR (inulin) and $2 \times 10^{10} - 3 \times 10^{11}$ viable probiotic bacterial cells d^{-1}. A trained sensory panel found that the synbiotic maize-based beverages fermented by *Lb. acidophilus* PRO and *Lb. rhamnosus* LRB were the most similar to the control (*Lb. delbrueckii*). This was confirmed by a larger consumer acceptance panel [111]. This illustrated that *mageu* can be converted to an acceptable synbiotic beverage and that it was able to sustain a population of viable probiotic cells, exceeding the therapeutic minimum level, during an extended storage period.

5.4.2. *Mahewu (mageu) with bifidobacteria*

The survival of probiotic *Bifidobacterium lactis* DSM 10140 as harvested and inoculated free cells or as microencapsulated cells in mahewu (*mageu*) was studied [74]. The workers observed that the counts of free cells of *B. lactis* reduced significantly during the 21day storage at 4 °C and 22 °C both in the presence or absence of oxygen. Poor viability of *Bf. lactis* in mahewu was attributed to exposure to the low pH (3.5) of mahewu and the inadequate buffering capacity as a result of a low protein content (5.2 g/L) in a medium containing 78.4 g/L of carbohydrates [74]. The workers then recommended the use of microencapsulation coupled with storage at 4 °C as being optimal for the delivery of *Bf. lactis* to the consumer [74]. However, microencapsulation is not without its technological challenges and added cost. *Bifidobacterium lactis* has also been said to be closely related to *Bf. animalis* which is a probiotic of animal origin [112]. It is therefore important that the potential of using bifidobacteria of human origin as starters in combination with lactobacilli are investigated in providing a probiotic enhanced *mageu* product.

5.4.3. Fermented maize weaning porridge

In a fermented "maize porridge" (18.5% w/w maize meal) mixed with malted barley (1.5% w/w), the growth and metabolism of four strains of probiotic lactobacilli (*Lb. reuteri* SD 2112, *Lb. rhamnosus* GG, *Lb. acidophilus* LA5 and *Lb. acidophilus* 1748) were studied in terms of cell counts, pH and metabolites [73]. Bacterial cell counts attained maximum levels of 7.2-8.2 $\log_{10}$ cfu within 12 hours of fermentation at 37 °C [73]. The lowest pH range attained after 24 hour fermentation period at 37 °C was 3.1-3.7 [73]. The products were of low viscosity that could be attributed to the use of the barley malt expected to be the source of amylase for the enzymatic hydrolysis of maize starch. Whereas the malt may have increased the level of fermentable sugars, it also led to a product of low viscosity (too watery) that may not have consumer appeal in the developing world either as porridge or a beverage. This product was not subjected to sensory evaluation, consumer preference evaluation or shelf-life testing. 'Maize weaning porridge' as it was referred to by the workers would not be nutritionally suitable for this purpose due to the inherent protein deficiency of maize that was the principal ingredient. It should also be noted that barley malt may not be readily available in the developing world.

5.5. Probiotic soy-based probiotic beverages

Soybeans and rice fermentation media are also reported to be suitable substrates for the growth of certain probiotic lactobacilli and bifidobacteria [49]. Soybean usage is however hampered by the presence of raffinose and stachyose, which can cause flatulence [105]. The non-inactivated lipoxygenase enzyme in the soybean is the causative agent of the beany off-flavour (as perceived in Western societies) in soy-containing products [105]. These limiting factors can, however, be significantly reduced by fermenting with technologically suitable LAB. Soy yoghurt and/or "sogurt" developed using soymilk, is characterized by a hard and coarse texture in addition to a beany "off-flavour". Coupled with inadequate acid development, this has resulted in a lower sensory appeal of these products [105]. Reports indicate that inclusion of fructose, calcium, cheese whey proteins, gelatin and lactose as well as probiotic bacteria improved the textural and sensory properties of sogurt [105].

Soymilk is suitable for the growth of lactobacilli and bifidobacteria and a probiotic soymilk and soybean yoghurt with added prebiotic oligofructose and inulin was developed [4]. This was found to be the case with several lactobacilli that included *Lb. casei, Lb. fermentum, Lb. reuteri*, and *Lb. acidophilus* [49]. Probiotic bacteria were also introduced into a non-fermented vegetarian frozen soy dessert. This product was composed of a soymilk beverage, sugar, oil, stabilizer and salt. The probiotic organisms introduced included *Lactobacillus acidophilus, Lb. rhamnosus, Lb. paracasei* ssp. *paracasei, Saccharomyces boulardi* and *Bifidobacterium lactis.* Bacterial population levels after 6 months' storage exceeded 10^7 cfu/g for all species except for *S. boulardi* [49]. The population level of the yeast species was below the therapeutic minimum of 10^6 cfu/g and this was attributed to the absence of 'cell shielding'.

In summary it can be stated that generally speaking, cereals are good growth-substrates of probiotic bacteria [108]. This is illustrated by the Yosa oats-based product, which to date is the only cereal-based commercial product known to contain both LAB and bifidobacteria. Since cereal-nutrient components vary, growth rates of probiotic organisms may also vary. Further research is therefore imperative to investigate the growth factors that may enhance the growth and survival of lactobacilli and bifidobacteria in cereal-based gruels. The indigestible variable fractions of the cereals can be utilised as prebiotics by probiotics in the GIT of the host upon ingestion of the fermented cereal-based beverage and these should also be defined and tested.

5.6. Therapeutic minimum levels of bacterial species in probiotic beverages

The therapeutic minimum population level for bacterial species in probiotic beverages is recommended to be 10^6 cfu ml^{-1}. This is the lowest probiotic bacterial count in a probiotic product that may adequately impart prophylactic and therapeutic benefits to the host. In order to realize therapeutic effects of probiotic bacteria in a product, the bacterial counts should exceed 10^6 cfu ml^{-1} [113]. Such a dose should be consumed regularly to ensure permanent colonisation in the small intestines. These high bacterial cell counts of probiotic bacteria are proposed to allow for the possible reduction in numbers during passage through the stomach and the intestines [114]. The need to have live probiotic cultures in products claimed to be probiotic has resulted in the formation of regulatory bodies and food legislation in some countries.

The Swiss Food Regulation and the International Standard of FIL/IDF require probiotic products to contain at least 10^6 cfu ml^{-1} [115]. The Fermented Milks and Lactic Acid Bacteria Beverages Association of Japan specifies a minimum of 10^7cfu ml^{-1} to be present in fresh probiotic dairy products [114, 115]. Japan has the FOSHU (Foods for Specified Health Use) programme for approving functional foods for marketing. A product with a "FOSHU" tag is defined as a food, which is expected to have certain functional benefits and has been licensed to bear a label to that effect [1]. The USA's National Yoghurt Association (NYA) specifies a population level of 10^8 cfu/g of lactic acid bacteria, at the time of manufacture, before placing a "Live and Active Culture" logo on the containers of the product [14]. However, in the USA, no indication is given as to what the viable count should be at the end of shelf-life. In the South African context, the South African Food and Health Draft Regulation (regulation 63) stipulates that selected probiotic microbes must be present at levels of at least 10^6 cfu ml^{-1} of product in order to exert a beneficial effect [110].

6. Conclusions and recommendations

Cereals and fermented cereal beverages can be advocated for use as delivery vehicles of health-benefiting functional ingredients such as probiotics and prebiotics. However, it is important to note some of the challenges associated with cereal grains and how they may be circumvented in improving probiotic cereal food delivery to masses in Africa and the developing world. It was noted that there is no known distribution channel for starter cultures to small scale or household scale processers of cereal-based fermented beverages in Africa and the developing world [30]. The other bottleneck is the fact that probiotic strains that have been technologically used successfully in dairy products may not exhibit similar acceptable growth and viability in cereal beverages. This accentuates the need for doing further screening [105]. The developed plant-cereal-based synbiotic beverages may also not have the necessary acceptable sensory attributes [3, 105, 116]. In a recent study, the use of a strain of *Lb. paracasei* BGP1 in a maize based fermented synbiotic experimental product resulted in off-flavours detected by a trained sensory panel [70, 111].

The use of probiotic strains in a combination of cereals and legumes in fermented products needs to be based on a number of considerations including technological and functional properties; sensory properties, growth rate; capability to deal with antinutritional factors; reduction of toxic substances in cassava; reduction of mycotoxins in cereals; reduction of flatulence causing compounds in legumes; pathogen inhibitory capabilities; co-existence and growth in mixed cultures [30]. These determinations however are hampered by the lack of facilities, expertise and the cost-benefit ratio that, in most cases, is not favourable to small scale and household scale cereal beverage producers in the developing world [30].

Author details

R. Nyanzi and P.J. Jooste*
Department of Biotechnology and Food Technology,
Tshwane University of Technology, Pretoria, South Africa

7. References

[1] Berner LA., O'Donnell JA. Functional foods and health claims legislation: application to dairy foods. International Dairy Journal 1998;8, 355-362.
[2] Marchand J., Vandenplas Y. Microorganisms administered in the benefit of the host: myths and facts. European Journal of Gastroenterology and Hepatology 2000;12(10), 1077-1088.
[3] Tuorila H., Cardello AV. Consumer responses to an off-flavour in the presence of specific health claims. Food Quality and Preference 2002;13, 561-569.
[4] Prado FC., Parada JL., Pandey A., Soccol CR. Trends in non-dairy probiotic beverages. Food Research International 2008;41, 111-123.
[5] Siró IN., Kápolna E., Kápolna BT., Lugasi A. Functional food. Product development, marketing and consumer acceptance-A review. Appetite, 2008;51(3), 456-467.
[6] Goldberg, I. Introduction. In: Goldberg, I. (Ed.) Functional Foods. Chapman and Hall; 1994. pXV-XVII, 6-13.

* Corresponding Author

[7] Salminen S., Roberfroid M., Ramos P., Fonden R. 1998. Prebiotic substrates and lactic acid bacteria. In: Salminen, S., van Wright A. (Eds.) Lactic acid bacteria: microbiology and functional aspects. Marcel Dekker; 1998. p343 - 358.

[8] Niness KR. Inulin and Oligofructose: what are they? Journal of Nutrition 1999;129, 1402S-1406S.

[9] Sanz Y. Ecological and functional implications of the acid-adaptation ability of *Bifidobacterium*: away of selecting improved probiotic strains. International Dairy Journal 2007;17, 1284-1289.

[10] Ranadheera RDCS., Baines SK., Adams MC. Importance of food in probiotic efficacy. Food Research International 2010;43(1), 1-7.

[11] Blandino A., Al-Aseeri ME., Pandiella SS., Cantero D., Webb, C. Cereal-based fermented foods and beverages. Food Research International 2003;36(6), 527-543.

[12] Manning TS., Gibson GR. Prebiotics. Best practice and research. Clinical Gastroenterology 2004;18(2), 287–298.

[13] Commane D., Hughes R., Shortt C., Rowland J. The potential mechanisms involved in the anticarcinogenic action of probiotics. Mutation Research 2005;591, 276–289.

[14] Kailasapathy K., Rybka S. *L. acidophillus* and *Bifidobacterium spp.*- their therapeutic potential and survival in yoghurt. The Australian Journal of Dairy Technology 1997;52(April), 28-35.

[15] Zeimer CJ., Gibson GR. An overview of probiotics, prebiotics and synbiotics in functional food concept: perspectives and future strategies. International Dairy Journal 1998;8, 473-479.

[16] Franz CMAP., Stiles ME., Schleifer KH., Holzapfel WH. Enterococci in foods - a conundrum for food safety. International Journal of food Microbiology 2003;88, 105–122.

[17] Collins JK., Thornton G., Sullivan GO. Selection of probiotic strains for human Applications. International Dairy Journal 1998;8, 487-490.

[18] Young J. European Market Development in prebiotic- and probiotic- containing foodstuffs. British Journal of Nutrition 1998;80(Supp 2), S231-233.

[19] Bosscher D., Van Loo J., Franck A. Inulin and oligofructose as functional ingredients to improve bone mineralization. International Dairy Journal 2006;16, 1092–1097.

[20] Modler HW., Mckeller RC., Yaguchi M. Bifidobacteria and bifidogenic factors. Canadian Institute of Food Science and Technology Journal 1990;24(1), 29-41.

[21] Gibson RG., Willis CL., Van Loo J. Non-digestible oligosaccharides and bifidobacteria: implications for health. International Sugar Journal 1994;96(1150), 381-387.

[22] Nout MJR., Rombouts FM. Fermentative preservation of plant foods. Journal of Applied Bacteriology 1992;73, 136-147.

[23] Holzapfel WH. Appropriate starter culture technologies for small-scale fermentation in developing countries. International Journal of Food Microbiology 2002;75, 197-212.

[24] Shetty PH., Jespersen L. *Saccharomyces cerevisiae* and lactic acid bacteria as potential mycotoxin decontaminating agents. Trends in Food Science & Technology 2006;17, 48-55.

[25] Iwuoha CI., Eke OS. Nigerian indigenous fermented foods: their traditional process of operation, inherent problems, improvements and current status. Food Research International 1996;29(5/6), 527-540.

[26] Adegoke GO., Babalola AK. Characterisation of microorganisms of importance in the fermentation of fufu and ogi , two Nigeria foods. Journal of Applied Bacteriology 1988;65, 449-453.

[27] Caplice E., Fitzgerald GF. Food fermentations: role of microorganisms in food production and preservation. International Journal of Food Microbiology 1999;50, 131-149.

[28] Sanni AI. The need for process optimisation of African fermented foods and beverages. International Journal of Food Microbiology 1993;18, 85-95.

[29] Campbell-Platt G. Fermented foods - a world perspective. Food Research International 1994;27, 253-257.

[30] Holzapfel WH. Use of starter cultures in fermentation on a household scale. Food Control 1997;8(5/6), 241-258.

[31] Oyewole OB. Lactic fermented foods in Africa and their benefits, Food Control 1997;8(5/6), 289-297.

[32] Byaruhanga YB., Bester BH., Watson TG. Growth and survival of *Bacillus cereus* in *mageu*, a sour maize beverage. World Journal of Microbiology & Biotechnology 1999;15, 329-333.

[33] Simango C., Rukure G. Survival of bacterial enteric pathogens in traditional fermented food. Journal of Applied Bacteriology 1992;73, 37-40.

[34] Mugula JK., Narvhus JA., Sørhaug T. Use of starter cultures of lactic acid bacteria and yeasts in the preparation of *togwa*, a Tanzanian fermented food. International Journal of Food Microbiology 2003;83(3), 307-318.

[35] Onyango C., Bley T. Raddatz H., Henle T. Flavour compounds in backslop fermented uji (an East African sour porridge). European Food Research & Technology 2004;218, 579-583.

[36] Nout MJR. Ecology of accelerated natural lactic fermentation of sorghum-based infant food formulas. International Journal of Food Microbiology 1991;12, 217-224.

[37] Efiuvwevwere BJ., Akona O. The microbiology of 'Kunun-Zaki', a cereal beverage from Northern Nigeria, during the fermentation (production) process. World Journal of Microbiology & Biotechnology 1995;11, 491-493.

[38] Steinkraus KH. Handbook of Indigenous Fermented Foods. Marcel Dekker; 1983. p1-667.

[39] Ezeogu LI. Research in ethnic West African fermented cereal-based foods. In: Taylor JRN., Cracknell RL. (Eds.) The ICC book of ethnic cereal-based foods and beverages across the continents. The University of Pretoria, Pretoria, South Africa; 2009. p 63-92.

[40] Edema MO., Sanni AI. Functional properties of selected starter cultures for sour maize bread. Food Microbiology 2008;25(4), 616-625.

[41] Nout MJR. Rich Nutrition from the poorest - Cereal fermentations in Africa and Asia. Food Microbiology 2009;26, 685-692.

[42] Reddy G., Altaf M., Naveena BJ., Venkateshwar M., Kumar E.V. Amylolytic bacterial acetic acid fermentation - A review. Biotechnology Advances 2008;26(1), 22-34.

[43] Omar NB., Abriouel H., Lucas R., Martinez-Cañamero, M., Guyot, J., Gálvez A. Isolation of bacteriocinogenic *Lactobacillus plantarum* strains from ben saalga, a traditional fermented gruel from Burkina Faso. International Journal of Food Microbiology 2006;112, 44-50.

[44] Lei V., Friis H., Michaelsen KF. Spontaneously fermented millet product as a probiotic treatment for diarrhea in young children: An intervention study in Northern Ghana. International Journal of Food Microbiology 2006;110, 246-253.

[45] Shetty PH., Hald B., Jespersen L. Surface binding of aflatoxin B1 by *Saccharomyces cerevisiae* strains with potential decontaminating abilities in indigenous fermented foods. International Journal of Food Microbiology 2007;113, 14-46.

[46] Chelule PK., Mbongwa HP., Carries S., Gqaleni N. Lactic acid fermentation improves the quality of amahewu, a traditional South African maize-based porridge. Food Chemistry 2010;122(3), 656-661.

[47] Towo E., Matuschek E., Svanberg U. Fermentation and enzyme treatment of tannin sorghum gruels: effects on phenolic compounds, phytate and in vitro accessible iron. Food Chemistry 2006;94(3), 369-376.

[48] Mensah P. Fermentation-the key to food safety assurance in Africa? Food Control 1997;8(5/6), 271-278.

[49] Rivera-Espinoza, Y., Gallardo-Navarro Y. Non-dairy probiotic products. Food Microbiology 2010;27(1), 1-11.

[50] Cook PE. Fermented foods as biotechnological resources. Food Research International 1994;27, 309-316.

[51] Olasupo NA., Olukoya, DK., Odunfa SA. Studies on bacteriocinogenic *Lactobacillus* isolates from selected Nigerian fermented foods. Journal of Basic Microbiology 1995;35(5), 319-324.

[52] Hansen EB. Commercial bacterial starter cultures for fermented foods of the future. International Journal of Food Microbiology 2002;78, 119-131.

[53] Leroy FDR., De Vuyst L. Lactic acid bacteria as functional starter cultures for the food fermentation industry. Trends in Food Science & Technology 2004;15(2), 67-78.

[54] Westby A., Reilly A., Bainbridge Z. Review of the effect of fermentation on naturally occurring toxins. Food Control 1997;8(5/6), 329-339.

[55] Onyekwere OO., Akinrele IA., Koleoso OA., Heys G. Industrialisation of gari fermentation. In: Steinkraus K.H. (Ed.). Industrialisation of Indigenous Fermented Foods. Marcel Dekker; 1989. p363-389.

[56] Claver IP., Zhou H.-M., Zhang H.-H., Zhu K.-X., Li Q., Murekatete N. The Effect of Soaking with Wooden Ash and Malting upon Some Nutritional Properties of Sorghum Flour Used for Impeke, a Traditional Burundian Malt-Based Sorghum Beverage. Agricultural Sciences in China 2011;10(11), 1801-1811.

[57] Reddy NR., Pierson MD. Reduction of antinutritional and toxic components in plant foods by fermentation. Food Research International 1994;27, 284-290.

[58] Ologhobo AD., Fetuga BL. Investigations on the trypsin inhibitor, hemagglutinin, phytic and tannic acid contents of cowpea *Vigna unguiculata*. Food Chemistry 1983;12, 249-254.

[59] Frontela C., García-Alonso FJ., Ros G., Martínez C. Phytic acid and inositol phosphates in raw flours and infant cereals: The effect of processing. Journal of Food Composition and Analysis 2008;21, 343-350.

[60] Dykes L., Rooney LW. Sorghum and millet phenols and antioxidants. Journal of Cereal Science 2006;44, 236 – 251.

[61] Dlamini NR., Taylor JRN., Rooney, LW. The effect of sorghum type and processing on the antioxidant properties of African sorghum-based foods. Food Chemistry 2007;105(4), 1412-1419.

[62] Chethan S., Sreerama YN., Malleshi N.G. Mode of inhibition of finger millet malt amylase by the millet phenolics. Food Chemistry 2008;111, 187-191.

[63] Correia MITD., Liboredo JC., Consoli MLD. The role of probiotics in gastrointestinal surgery. Nutrition 2010; 28(3), 230-234.

[64] Teniola OD., Odunfa SA. The effects of processing methods on the levels of lysine, methionine and the general acceptability of *ogi* processed using starter cultures. International Journal of Food Microbiology 2001;63, 1-9.

[65] Bvochora JM., Zvauya R. Biochemical changes occurring during the application of high gravity fermentation technology to the brewing of Zimbabwean traditional opaque beer. Process Biochemistry 2001;37(4), 365-370.

[66] Wagacha JM., Muthomi JW. Mycotoxin problem in Africa: Current status, implications to food safety and health and possible management strategies. International Journal of Food Microbiology, 2008;124(1), 1-12.

[67] Matumba L., Monjerezi M., Khonga EB., Lakudzala DD. Aflatoxins in sorghum, sorghum malt and traditional opaque beer in southern Malawi. Food Control 2011;22(2), 266-268.

[68] Fandohan P., Zoumenou D., Hounhouigan DJ., Marasas WFO., Wingfield MJ., Hell K. Fate of aflatoxins and fumonisins during the processing of maize into food products in Benin. International Journal of Food Microbiology 2005;98(3), 249-259.

[69] Alberts JF., Engelbrecht Y., Steyn PS., Holzapfel WH., Van Zyl WH. Biological degradation of aflatoxin B1 by *Rhodococcus erythropolis* cultures. International Journal of Food Microbiology 2006;109, 121-126.

[70] Nyanzi R. Enhancing the functional quality of mageu. MTech d*issertation*. Tshwane University of Technology. Pretoria; 2007.

[71] Charalampopoulos D., Wang R., Pandiella SS., Webb C. Application of cereal components in functional foods: a review. International Journal of Food Microbiology 2002;79, 131-141.

[72] Charalampopoulos D., Pandiella SS., Webb C. Growth studies of potentially probiotic lactic acid bacteria in cereal-based substrates. Journal of Applied Microbiology 2002;92, 851-859.

[73] Helland MH., Wicklund T., Narvhus J. Growth and metabolism of selected strains of probiotic bacteria, in maize porridge with added malted barley. International Journal of Food Microbiology 2004;91, 305-313.

[74] Mcmaster LD., Kokott SA., Reid SJ., Abratt VR. Use of traditional African fermented beverages as delivery vehicles for *Bifidobacterium lactis* DSM 10140. International Journal of Food Microbiology 2005;102(2), 231-237.

[75] Kedia G., Wang R., Patel H., Pandiella SS. Use of mixed cultures for the fermentation of cereal-based substrates with potential probiotic properties. Process Biochemistry 2007;42, 65-70.

[76] Rozada-Sánchez R., Sattur AP., Thomas K., Pandiella SS. Evaluation of *Bifidobacterium* spp. For the production of a potentially probiotic malt-based beverage. Process Biochemistry 2008;43, 848-854.

[77] Todorov SD., Dicks LMT. Screening for bacteriocin-producing lactic acid bacteria from boza, a traditional cereal beverage from Bulgaria: Comparison of the bacteriocins. Process Biochemistry 2006;41(1), 11-19.

[78] Yeğin SR., Üren A. Biogenic amine content of boza: A traditional cereal-based, fermented Turkish beverage. Food Chemistry 2008;111(4), 983-987.

[79] Hancioğlu Ö., Karapinar M. 1997. Microflora of Boza, a traditional fermented Turkish beverage. International Journal of Food Microbiology 1997;35, 271-274.

[80] Zorba M., Hancioglu O., Genc M., Karapinar M., Ova G. The use of starter cultures in the fermentation of boza, a traditional Turkish beverage. Process Biochemistry 2003;38(10), 1405-1411.

[81] Dlusskaya E., Jänsch A., Schwab C., Gänzle MG. 2008. Microbial and chemical analysis of kvass fermentation. European Food Research & Technology 2008;227, 261-266.

[82] Sainz T., Wacher C., Espinoza J., Centurión D., Navarro A., Molina, J., Inzunza A., Cravioto A., Eslava C. Survival and characterization of Escherichia coli strains in a typical Mexican acid-fermented food. International Journal of Food Microbiology 2001;71, 169-176.

[83] Bvochora JM., Reed JD., Read JS. & Zvauya R. Effect of fermentation processes on proanthocyanidins in sorghum during preparation of mahewu, a non-alcoholic beverage. Process Biochemistry 1999;35, 21-25.

[84] Tou EH., Mouquet-Rivier C., Rochette I., Traoré AS., Tréche S., Guyot JP. Effect of different process combinations on the fermentation kinetics, microflora and energy density of ben-saalga, a fermented gruel from Burkina Faso. Food Chemistry 2007;100(3), 935-943.

[85] Tou EH., Guyot JP., Mouquet-Rivier C., Rochette I., Counil E., Traoré, AS., Tréche S. Study through surveys and fermentation kinetics of the traditional processing of pearl millet (*Pennisetum glaucum*) into ben-saalga, a fermented gruel from Burkina Faso. International Journal of Food Microbiology 2006;106(1), 52-60.

[86] Abriouel H., Omar NB., López, RL., Martinez-Cañamero M., Keleke S., Gálvez A. Culture-independent analysis of the microbial composition of the African traditional fermented foods *poto poto* and *dégué*. International Journal of Food Microbiology 2006;111, 228-233.

[87] Schweigart F., De Wit JP. 1960. Favourite beverages of the Bantu: preparing and drying of mahewu and its nutritional value. National Nutritional Research Institute, CSIR. Pretoria (May 1960).

[88] Schweigart F. The drying of lactic acid bacteria cultures for mahewu production. Lebensm.Wiss.u. Technology 1971;4(1):20-23.

[89] Holzapfel WH. Industrialization of mageu fermentation in South Africa. In: Steinkraus KH. (Ed.). Industrialization of Indigenous Fermented Foods. Marcel Dekker; 1989. p285-328.

[90] Schweigart F., Fellingham SA. A study of fermentation in the production of mahewu, an indigenous sour maize beverage of Southern Africa. Milchwissenschaft 1963;18, 241-246.

[91] Van Noort G., Spence C. The mahewu industry. South African Food Review 1976;October, 129-133.

[92] Edwards C. Mageu – where to for Africa's energy drink? South African Food Review 2003;30 (3), 25-27.

[93] Zulu RM., Dillon VM., Owens JD. 1997. Munkoyo beverage, a traditional Zambian fermented maize gruel using *Rhynchosia* root as amylase source. International Journal of Food Microbiology 1997;34(3), 249-258.

[94] Foma RK., Destain J., Mobinzo PK., Kayisu K., Thonart P. Study of physicochemical parameters and spontaneous fermentation during traditional production of *munkoyo*, an indigenous beverage produced in democratic Republic of Congo. Food Control 2012;*25*, 334-341.

[95] Mukisa IM., Nsiimire DG., Byaruhanga YB., Muyanja CMBK., Langsrud T., Narvhus JA. Obushera: Descriptive sensory profiling and consumer acceptance. Journal of Sensory Studies 2010;25, 190-214.

[96] Muyanja CMBK., Narvhus JA., Treimo J., Langsrud T. Isolation, characaterisation and identification of lactic acid bacteria from *bushera*: a Ugandan traditional fermented beverage. International Journal of Food Microbiology 2003;80, 201-210.

[97] Omemu AM., Oyewole OB., Bankole MO. Significance of yeasts in the fermentation of maize for ogi production. Food Microbiology 2007;24(6), 571-576.

[98] Onyekwere OO., Akinrele IA., Koleoso OA. Industrialisation of ogi fermentation. In: Steinkraus KH. (Ed.) Industrialisation of Indigenous Fermented Foods. Marcel Dekker; 1989. p329-362.

[99] Odunfa SA., Adeyele S. Microbiological changes during the traditional production of ogi-baba, a West African fermented sorghum gruel. Journal of Cereal Science 1985;3, 173-180.

[100] Omar NB., Abriouel H., Keleke S., Venuezela AS., Martinez-Cañamero M., López RL., Ortega E., Gálvez A. Bacteriocin-producing *Lactobacillus* strains isolated from *poto poto*, a Congolese fermented maize product, and genetic fingerprinting of their plantaricin operons. International Journal of Food Microbiology 2008;127, 18-25.

[101] Madoroba E., Steenkamp ET., Theron J., Scheirlinck I., Cloete TE., Huys G. Diversity and dynamics of bacterial populations during spontaneous sorghum fermentations

used to produce ting, a South African food. Systematic and Applied Microbiology, 2011;34(3), 227-234.

[102] Sekwati-Monang B., Gänzle MG. Microbiological and chemical characterisation of ting, a sorghum-based sourdough product from Botswana. International Journal of Food Microbiology 2011;150, 115-121.

[103] Mbugua SK. Microbial growth during spontaneous Uji fermentation and its influence on the end product. East African Agriculture and Forestry Journal 1985;50(4), 101-110.

[104] Johansson M-L., Sanni A., Lonner C., Molin G. 1995. Phenotypically based taxonomy using API 50 CH of Lactobacilli from Nigerian ogi and the occurrence of starch fermenting strains. International Journal of Food microbiology 1995;25, 159-168.

[105] Peres CTM., Peres CL., Hernández-Mendoza AN., Malcata FX. Review on fermented plant materials as carriers and sources of potentially probiotic lactic acidbacteria – With an emphasis on table olives. Trends in Food Science & Technology 2012; in press.

[106] Lavermicocca P. Highlights on new food research. Digestive and Liver Disease 2006;38(Suppl. 2), S295-S299.

[107] Gokavi S., Zhang L., Zhang L., Huang M.-K, Zhao X., Guo M. Oat-based synbiotic beverage fermented by *Lactobacillus plantarum, Lactobacillus paracasei ssp. casei,* and *Lactobacillus acidophilus*. Journal of Food Science 2005;70(4), 216-223.

[108] Patel HM., Pandiella SS., Wang RH., Webb C. Influence of malt, wheat and barley extracts on the bile tolerance of selected strains of lactobacilli. Food Microbiology 2004;21, 83-89.

[109] Shah NP., Lankaputhra WEV. Improving viability of *Lactobacillus acidophilus* and *Bifidobacterium* spp. in yoghurt. International Dairy Journal 1997;7, 349-356.

[110] Theunissen J., Witthuhn C. Probiotic content: truth or fiction? South African Food Review 2004;31 (5):15-17.

[111] Nyanzi R., Jooste PJ., Abu JO., Beukes EM. Consumer acceptability of a synbiotic version of the maize beverage mageu. Development Southern Africa 2010;27 (3), 447-463.

[112] Gomes AMP., Malcata FX 1999. *Bifidobacterium* spp. And *Lactobacillus acidophilus*: biological, biochemical, technological and therapeutical properties relevant for use as probiotics. Trends in Food Science & Technology, 1999;10, 139-157.

[113] Kailasapathy K., Supriadi D. Effect of whey protein concentrate on the survival of *Lactobacillus acidophilus* in lactose hydrolysed yoghurt during refrigerated storage. Milchwissenschaft 1996;51(10), 565-568.

[114] Shah NP. Probiotic bacteria: selective enumeration and survival in dairy foods. Journal of Dairy Science 2000;83, 894-907.

[115] Shin HS., Lee JH., Pestka JJ., Ustunol, Z. Viability of Bifidobacteria in commercial dairy products during refrigerated storage. Journal of Food Protection 2000;63(3), 327-331.

[116] Luckow T., Delahunty C. Consumer acceptance of orange juice containing functional ingredients. Food Research International 2004;37, 805-814.

[117] Taylor JRN., Schober TJ., Bean SR. Novel food and non-food uses for sorghum and millets. Journal of Cereal Science 2006;44, 252-271.

[118] Tagbor AAK. Shelf-life extension and fortification of *Liha,* a fermented Ghanaian maize malt beverage. MTech d*issertation,* Technikon Pretoria; 2001.
[119] Chaitow L., Trenev N. Probiotics: the revolutionary 'friendly bacteria' way to vital health and well-being. Thorsons Publishers Ltd., England; 1990.

Chapter 9

Functional Dairy Probiotic Food Development: Trends, Concepts, and Products

Aziz Homayouni, Maedeh Alizadeh, Hossein Alikhah and Vahid Zijah

Additional information is available at the end of the chapter

http://dx.doi.org/10.5772/48797

1. Introduction

In recent years, scientific investigators have moved from primary role of food as the source of energy and nutrients to action of biologically active food components on human health. On the other hand, consumer interest about the active role of food in well-being and life prolongation has been increased. In this way, a novel term -functional food- was introduced which refers to preventional and/or curing effects of food beyond its nutritional value. There is a wide rage of functional foods that were developed recently and many of them are being produced in all over the world including probiotic, prebiotic and symbiotic foods as well as foods enriched with antioxidants, isoflavones, phytosterols, anthocyanins and fat-reduced, sugar-reduced or salt-reduced foods. Among these foods, probiotic functional food has exerted positive effects on the overall health. We can divide it in both probiotic dairy foods and probiotic non-dairy foods. The market of probiotic dairy foods is increasing annually. An increased demand for dairy probiotic products comes from health promotion effects of probiotic bacteria which are originally initiated from milk products, bioactive compounds of fermented dairy products and prevention of lactose intolerance. Therefore, development of these products is a key research priority for food design and a challenge for both industry and science sectors.

Literatures about probiotic application in pediatrics have some characteristics including numerous, randomized, controlled clinical trials or meta-analyses but the substantial heterogeneity of these works greatly complicates the interpretation of the results and thus makes it difficult to draw univocal and general conclusions. Despite these complications, it is possible to draw some conclusions about the clinical effectiveness of probiotics by examining the most significant literature on each pathology. In particular, there is strong evidence indicating that probiotics have preventive and therapeutic effect on pathologies such as acute diarrhea, antibiotic-associated diarrhea, NEC, and allergic pathology. It was reported that administration of L.GG to 50 infants, for a period of 6 weeks, did not improve abdominal pain but did reduce the incidence of abdominal tension compared to the placebo (Bausserman and Michail, 2005) But in other works it was clearly demonstrated that L. acidophilus did improve the symptoms in about half of the patients with IBS, that the blend of VLS#3 probiotics decreased abdominal swelling, while the combined use of L. plantarum and B. breve reduced pain intensity (Halpern, et al., 1996; Kim, et al., 2003; Saggioro, 2004). L. acidophilus and B. infantis for 4 weeks were administered alone or in combination with antibiotics ciprofloxacin for the first week to three different groups with IBS: diarrhea, constipation, and alternating diarrhea and constipation. Both therapeutic approaches have improved the quality of life and reduced symptoms in all three groups (Faber, 2000). In conclusion, although the use of some types of probiotics on IBS appears promising, additional studies are needed. Food supplementation with pre- and probiotics may reduce the prevalence for the infant in high-risk families developing an atopic eczema during the first 2 years of life. Those pregnant women should be advised to take probiotics (L. GG) in late pregnancy and the first 6 months postnatally during nursing. If breast-feeding is not possible, pro- or prebiotics can be supplemented to the infant. There are no known adverse reactions and it might prevent atopic eczema, especially in neonates after cesarean delivery. Therapeutic use of probiotics to improve atopic eczema is only supportive in infants 18 months and with IgE sensitization.

Recent experimental studies have shown that certain gut bacteria, in particular species of Lactobacillus and Bifidobacterium, may exert beneficial effects in the oral cavity by inhibiting Streptococci and Candida sp. Probiotic lactic acid bacteria can produce different antimicrobial components such as organic acids, hydrogen peroxide, carbon peroxide, diacetyl, low molecular weight antimicrobial substances, bacteriocins, and adhesion inhibitors, which also affect oral microflora. However, data is still sparse on the probiotic action in the oral cavity. More information is needed on the colonization of probiotics in the mouth and their possible effect on and within oral biofilms. There is every reason to believe that the putative probiotic mechanisms of action are the same in the mouth as they are in other parts of the gastrointestinal tract. Because of the increasing global problem with antimicrobial drug resistance, the concept of probiotic therapy is interesting and pertinent, and merits further research in the fields of oral medicine and dentistry (Meurman, 2005).

The number of microbial cells in the human gut is 10 times more than the number of cells in the adult body (Mountzouris and Gibson, 2003). So, the change of microbial balance in human intestine can impress the host health. The ratio between the beneficial microbes (probiotics) and harmful microbes would have an important effect on host health. One way to keeping up the probiotic cells in the gut, is to entering probiotics into the intestine through the regular consumption of food containing these bacteria. Among the functional foods, the dairy probiotic products, especially ice cream and cheese are good vehicle to transfer probiotics to the human intestinal tract (Homayouni, 2008a; Homayouni et al., 2012). Dairy products have an important role in human health and form the main part of the food pyramid. The therapeutical and health care characteristic of fermented dairy products has been used over long years. Another way to keeping up the probiotic cells in the gut is to entering prebiotics into the intestine through the regular consumption of foods containing these components. It is clear that versus probiotics the amounts of prebiotics do not changes during the passage from upper intestinal tract (Homayouni, 2008a).

The main role of food is providing enough nutrients to meet metabolic requirements in human body, while giving the consumer a satisfaction feeling and well-being (Homayouni, 2008a). Beyond meeting nutrition needs, food may have different physiological functions and may play detrimental or beneficial roles in some diseases (Koletzko et al., 1998). Functional foods were developed in order to promote a well-being state, improving health, and reducing the diseases risk. "Functional food" means; special foods which have preventional and/or curing effects beyond its nutritional (Homayouni, 2008a). There is a wide rage of functional foods that were developed recently and many of them are being produced in all over the world including probiotic, prebiotic and symbiotic foods as well as foods enriched with antioxidants, isoflavons, phytosterols, anthosyanins and fat-reduced, sugar-reduced or salt-reduced foods. Among these foods, probiotic functional foods are the first choice to exert positive effects on the human health. Probiotic functional foods were divided into dairy probiotic foods and non-dairy probiotic foods. Some of dairy probiotic foods including probiotic ice cream, frozen fermented dairy deserts, probiotic cheese, bio-yoghurt, drinking yoghurt, kefir, Freeze-dried yoghurt and spray dried milk powder have been employed as possible delivery vehicles for probiotic bacteria (Haynes and Playne, 2002; Homayouni et al., 2008b; Homayouni et al., 2012; Ejtahed et al., 2011; Ejtahed et al., 2012; Mirzaei et al., 2012 Kailasapathy and Rybka, 1997; Ravula and Shah, 1998; Stanton et al., 2001). Probiotics are distinct as live micro-organisms which, when administered in sufficient amounts present a health benefit on the host (Food and Agriculture Organization of United Nations; World Health Organization - FAO/WHO, 2002; Homayouni, 2009). In recent years probiotic bacteria have increasingly been incorporated into dairy foods as dietary adjuncts. *Lactobacillus* and *Bifidobacterium* are the most common probiotic bacterial cells that were used in the production of fermented and non-fermented dairy products.

Consumption of probiotic bacteria via dairy food products is an ideal way to re-establish the intestinal micro-floral balance. It must conform to certain requirements for a dairy food product to be considered as a valuable alternative for delivery of probiotic bacteria in one hand and for variety of probiotic cultures to use as a dietary adjunct and to exert a positive influence in the other hand. The culture must be native of the human gastrointestinal tract, having the ability to ferment prebiotics, survives passage through the stomach and small bowel in adequate numbers, be capable of colonizing in site of action, and have beneficial effects on human health. In order to survive, the strain must be resistant to acidic conditions (gastric pH 1-4), alkaline conditions (bile salts present in the small bowel), enzymes present in the intestine (lysozyme) and toxic metabolites produced during digestion (Homayouni et al., 2008d). For example in traditional yoghurt production, *Lactobacillus bulgaricus* and *Streptococcus thermophilus* were used as starter culture. These bacteria do not belong to the indigenous intestinal flora, are not bile-acid resistant and do not survive passage through the gut. So, the traditional yoghurt culture is not to be considering as probiotic. In the case of dairy food product to be considered as a valuable alternative for delivery of probiotics, it must to match definite necessities such as neutral pH, high enough total solids level, absence of oxygen and near to ambient temperatures (Homayouni et al., 2008b; Homayouni et al., 2008d; Homayouni et al., 2012). A number of dairy food bio-products have been employed and developed as delivery vehicles of probiotic bacteria. Around 80 bifido containing products are estimated to be on the world markets. Most of these products are from dairy origin including fresh milk, fermented milk, dairy beverages, ice cream, dairy desserts, cheese, cottage cheese and powdered milk (Tamime et al., 1995). Since the more interest in probiotics, different types of functional products were proposed as carrier foods for probiotic micro-organisms by which consumers can take in large amounts of probiotic bacteria for the therapeutic effects. Therefore, development of these products is a key research priority for food design and a challenge for both industry and science sectors. This chapter presents an overview of functional foods development with emphasizing probiotic dairy foods.

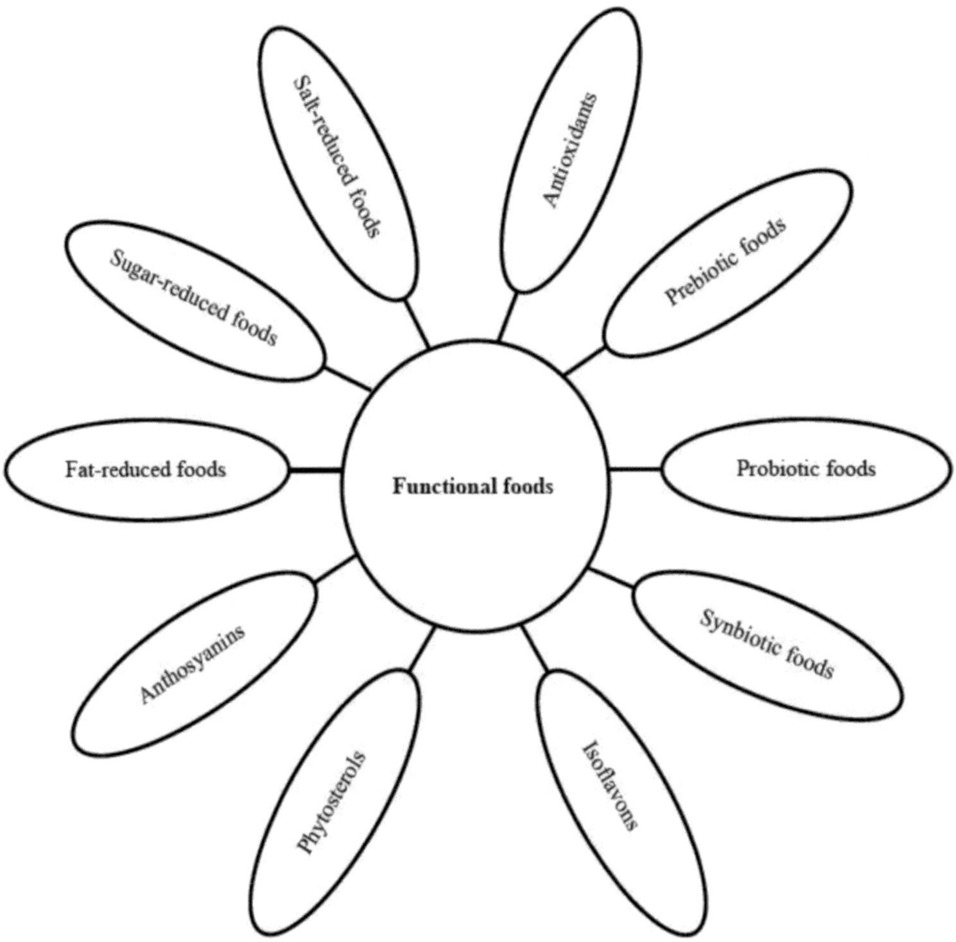

Figure 1. Classification of functional foods

2. Dairy probiotic foods

As mentioned before, dairy functional foods beyond its basic nutritional value has physiological benefits. Milk has an outstanding position in the development of functional foods because it has Omega-3, phytosterols, isoflavins, conjugated linoleic acid, minerals, and vitamins. Dairy products such as ice cream, cheese, yogurt, Acidophilus-Bifidus-milk, Ayran, Kefir, Kumis, Doogh containing probiotics and dairy beverages (both fermented and non-fermented) have long been considered as important vehicles for the delivery of probiotics. In fermentation process, acids such as lactic acid, acetic acid and citric acid are naturally produced. These acids are commonly used as organic acids to enhance organoleptic qualities as well as safety of food products. Lactic acid bacteria are found to be more tolerant to acidity and organic acids than most of the pathogens and spoilage micro-organisms.

2.1. Probiotic ice cream

Probiotic ice cream can be produced by incorporation of probiotic bacteria in both of fermented and unfermented mix (Homayouni et al., 2008b; Homayouni et al., 2012). Ice cream is ideal vehicle for delivery of these micro-organisms in the human diet (Akin et al., 2007; Kailasapathy and Sultana, 2003; Ravula and Shah, 1998; Homayouni et al., 2008d; Homayouni et al., 2012). *Lactobacillus* and *Bifidobacterium* are the most common species of lactic acid bacteria used as probiotics for fermented dairy products. Among the frozen dairy products with live probiotics, probiotic ice cream is also gaining popularity for its neutral pH. The pH of non-fermented ice cream is near to seven which is providing to survive probiotic bacteria (Akin et al., 2007; Christiansen et al., 1996; Homayouni et al., 2008b; Homayouni et al., 2008c; Homayouni et al., 2012). The high total solids level in ice cream including the fat and milk solids provides protection for the probiotic bacteria (Homayouni et al., 2012). Because the efficiency of added probiotic bacteria depends on dose level, type of dairy foods, presence of air and low temperature (Homayouni et al., 2008b), their viability must be maintained throughout the product's shelf-life and they must survive the gut environment (Kailasapathy and Chin, 2000). The therapeutic value of live probiotic bacteria is more than unviable cells; therefore, International Dairy Federation (IDF) recommends that a minimum of 10^7 probiotic bacterial cells should be alive at consumption time per gram/mililiter of product. Studies indicate, however, the bacteria may not survive in high enough numbers when incorporated into frozen dairy products unless a suitable method is used against freeze injury and oxygen toxicity (Dave and Shah, 1998; Kailasapathy and Sultana, 2003; Ravula and Shah, 1998; Homayouni et al., 2008d). The methods of increasing probiotic survival depend on type of food products. Selection of resistant probiotic strains to tolerate production, storage and gastrointestinal tract conditions, is one of the important methods (Homayouni et al., 2008d). Another way is to adjust the conditions of production and storage for more survival rates. The physical protection of probiotics by microencapsulation is a new method for increasing the survival of probiotics (Homayouni et al., 2007; Homayouni et al., 2008b). Encapsulation helps to isolate the bacterial cells from the adverse environment of the product and gastrointestinal tract, thus potentially reducing cell loss. Encapsulation thus may enhance the shelf-life of probiotic cultures in frozen dairy products (Kebary et al., 1998; Shah and Ravula, 2000; Homayouni et al., 2008b). Selecting of suitable probiotic strains depends to ability survive simulated conditions of ice cream (high sucrose concentrations, high oxygen, refrigeration and freezing temperatures), acidic (to simulate gastric) and alkaline conditions (to simulate intestinal). Microencapsulation of probiotics can further protect these bacteria from the mentioned conditions (Homayouni et al., 2008d).

Homayouni et al. (2008d) studied the survival of probiotics in simulated ice cream and gastrointestinal conditions in order to select appropriate probiotic strains for use in probiotic ice cream. The growth and survival rate of *Lactobacillus acidophilus, Lactobacillus casei, Bifidobacterium lactis* and *Bifidobacterium longum* in varying amount of sucrose concentrations (10, 15, 20 and 25%), oxygen scavengering components (0.05% L-cysteine and 0.05% L-ascorbate) and low temperatures (4°C and 20°C) during different periods of time (30, 60 and 90 days) in MRS-broth medium was studied. All of above stress factors have been able to influence the growth and survival of four probiotic strains. Results have demonstrated that it is possible to select the appropriate probiotic strains for use in probiotic ice cream. *Lactobacillus casei* (Lc01) and *Bifidobacterium lactis* (Bb12) had the highest resistance to simulated acidic, alkaline and ice cream conditions in comparison with other probiotic strains, making them suitable probiotic strains for use in probiotic ice cream (Homayouni et al., 2008b; Homayouni et al., 2008d).

2.2. Probiotic cheese

Survival in processing conditions, presence of oxygen, degree of acidity, ability to grow well in milk-based products and to rapidly acidify milk, thus reducing the fermentation time and, consequently, contamination risk during preparation of inoculums are important factors for probiotic bacteria such as *Lactobacillus spp.* and *Bifidobacterium spp* in order to apply these bacteria in probiotic dairy products. Probiotic bacterial cells have to fulfill the basic technological necessities when used in commercial probiotic dairy products. Since probiotic bacteria have to be presented in sufficient numbers in product at consumption time, their survival have to be maintained up to shelf-life date. In addition, no adverse effects on taste and aroma of the product should be exerted by the probiotic organisms. Various types of cheese have a good potential to maintain the probiotic survival. So, it is a good vehicle to transfer probiotics to the human intestinal tract. There are two ways for development of probiotic cheese: in the first step, the manufacture processes of cheese products may have to be modified and adapted to the requirements of probiotics and in second step, appropriate probiotic strains may be applied or new cheese products may have to be developed. Dairy products containing living bacteria have to be cooled during storage. Cooling is necessary to guarantee high survival rates of probiotics and to bring sufficient stability of the product (Roy et al., 1997). In addition, oxygen content and water activity of the probiotic cheese have to be considered in prepackaged cheese (Dave and Shah, 1997a). Interaction of the live probiotic microorganisms with the components of the cheese have to be inhibited by cooling of product. The degree of interaction depends on the kind and amount of carbohydrates available, degree of hydrolysis of milk proteins and thus availability of essential amino acids, and composition and degree of hydrolysis of milk lipids, determining the availability of short chain fatty acids (Fox et al., 1996). However, the proteolytic and lipolytic properties of the probiotic bacterial cells may have important effects on taste and flavor of the probiotic cheese (Kunji et al., 1996). The strength of interactions between probiotics and starter organisms in probiotic cheese depends on when the probiotics are added to the product. If they are added after fermentation, interactions may be kept to a minimum, since addition is possible immediately before or even after cooling below 8°C and metabolic activities of starters and probiotics are considerably reduced at refrigerated temperatures.

Antagonism between bacteria is often based on the production of metabolites that inhibit or inactivate more or less specifically other related starter organisms or even unrelated bacteria. While antagonism caused by bacteriocins, peptides, or proteins exhibiting antibiotic properties has been described as a limiting factor for combinations of starters and probiotics (Joseph et al., 1998), antagonism caused by hydrogen peroxide, benzoic acid, biogenic amines, and lactic acid may have considerable effects on probiotics in probiotic cheese. The physiological state of the probiotics may be of considerable importance for survival during ripening and/or storage if probiotics are added to the probiotic cheese after fermentation (Desmazeaud, 1996; Lankaputhra et al., 1996; Leuschner et al., 1998; Weber, 1996).

In probiotic cheese, probiotic cells must be able to grow and/or multiply in the human intestine and therefore should be able to survive during the passage through the gastrointestinal tract (GIT), which involves exposure to hydrochloric acid in stomach and bile in small intestine (Stanton et al., 2003). In fact, cheese provides a valuable vehicle for probiotic delivery, due to creation of a buffer against the high acidic environment in the gastrointestinal tract, and thus creates a more favorable environment for probiotic survival throughout the gastric transit, ought to higher pH. Moreover, the dense matrix and relatively high total solids as well as fat content of cheese may offer additional protection to probiotic bacteria in stomach (Bergamini et al., 2005; Ross et al., 2002). The presence of the prebiotics inulin and oligofructose can promote growth rates of *bifidobacteria* and *lactobacilli*, besides increased lactate and short chain fatty acids production in petit-suisse cheese (Cardarelli et al., 2007).

2.3. Probiotic yoghurt

Yoghurt has been historically recognized to be 'a healthy food' with therapeutically effects. There has been a considerable increase in the popularity of yoghurt especially probiotic yoghurt in recent years. The conventional yoghurt starter bacteria, *L. bulgaricus* and *Streptococcus thermophilus*, do not have ability to survive passage through intestinal tract and consequently so, they are not considered as probiotics. But the addition of *L. acidophilus* and *B. bifidum* into yoghurt can add extra nutritional and physiological values.

Similar processing to traditional yoghurt is applied for production of bio-yoghurt with incorporation of live probiotic starter cultures. Heat treated homogenized milk with an increased protein content (3.6–3.8%) is inoculated with the conventional starter culture at 45°C or 37°C and incubated for 3.5 and 9 h, respectively. The probiotic culture can be added prior to fermentation simultaneously with the conventional yoghurt cultures or after fermentation to cooled (4°C) product before packaging. Bio-yoghurt, containing *L. acidophilus* and *B. bifidum* is a potential vehicle for delivery of these probiotic cells to consumers. *L. acidophilus* and *B. bifidum* have to retain viability and activity in yoghurt as a probiotic at consumption time. Viability of probiotic bacteria in yoghurt products at refrigeration temperature is reported to be unsatisfactory over a long shelf life (Dave and Shah, 1997a). The survival of probiotic bacteria in fermented dairy products depends on the chemical composition of the fermentation medium (e.g. carbohydrate source), final acidity, milk solids content, availability of nutrients, growth promoters and inhibitors, strains used, interaction between species present, culture conditions, concentration of sugars (osmotic pressure), dissolved oxygen (especially for *Bifidobacterium* spp.), level of inoculation, incubation temperature, fermentation time and storage temperature. The lack of acid tolerance of some probiotic species and strains in fermented products based on milk is an important factor. During fermentation, pH levels decreases when the lactic acid content increases. 'Over-acidification' or 'post acidification' is due to decrease in pH after fermentation and during storage at refrigerated temperature. Excessive acidification is mainly due to the uncontrollable growth of strains of *L. bulgaricus* at low pH values and refrigerated temperatures. The 'overacidification' can be prevented to a limited extent by applying 'good manufacturing practice' and by using cultures with reduced 'over-acidification' behavior.

Viability of both *Lactobacillus* and *Bifidobacterium* species reduces at low pH levels during refrigerated storage. So, strain selection and survival monitoring are necessary to produce high quality bio-yoghurt. Probiotic yoghurt contains metabolic products secreted by starter microorganisms, which influence the viability of *L. acidophilus* and *B. bifidum*. The inhibition of *bifidobacteria* in probiotic yoghurt is due to antagonism effects among starter bacteria rather than hydrogen peroxide or organic acids (Dave and Shah, 1997a). The ideal procedure for probiotic yoghurt manufacturing is growing the *Bifidobacterium* spp. separately, followed by washing out of free metabolites and the transfer of the cells to the probiotic yoghurt. Oxygen toxicity is a critical problem for *Bifidobacterium* spp. because they are strictly anaerobic. Low initial oxygen content in milk may obtain the low redox potential required in the early phase of incubation to guarantee *Bifidobacteria* growth. Oxygen easily dissolves in milk during yoghurt production and also permeates through packages during storage. It has been suggested to inoculate *S. thermophilus* and *Bifidobacterium* simultaneously during fermentation to avoid the oxygen toxicity problem. *S. thermophilus* has a high oxygen utilization ability, which results in reduction of dissolved oxygen in probiotic yoghurt and an enhancement in viability of *bifidobacteria*. Higher survival rates of lactic acid bacteria were obtained at lower storage temperatures (Foschino et al., 1996). Low storage temperature restricts the growth of *L. bulgaricus* and consequently also over-acidification. *Bifidobacteria* are substantially less tolerant to low storage temperature when compared to *L. acidophilus*.

2.4. Probiotic milk

Lactobacillus acidophilus does not rapidly grow in milk because it is an acid-loving bacterium. Therefore, it is essential to maintain the inoculum active by daily transfers of mother culture in acidophilus milk production. The probiotic milk is to market in liquid form. During fermentation, milk pH often goes beyond the narrow range of optimal pH of *Lactobacillus acidophilus* (5.5-6.0). This eventually leads to decrease these bacterial counts. In traditional acidophilus milk production, the milk is heated at 95°C for 1 h or at 125°C for 15 min (Vedamuthu, 2006). Such a high heat treatment stimulates the growth of *Lactobacillus acidophilus* by providing denatured proteins and released peptides. High-heat-treated milk is cooled to 37°C and kept at this temperature for a period of 3-4 h to allow any spores present to germinate. Then, milk is re-sterilized to destroy almost all vegetative cells. Unless skim milk is used, the heat-treated milk is homogenized and cooled down to inoculation temperature (37°C). *Lactobacillus acidophilus* is added as active bulk culture. The level of inoculation is usually 2-5% and the inoculated milk is left to ferment until pH 5.5-6.0 or ~1.0% lactic acid is obtained, with no alcohol (Surono and Hosono, 2002). The fermentation takes about 18-24 h under inactive conditions. After the fermentation, the number of viable *Lactobacillus acidophilus* colonies is about 2-3×10^9 cfu mL^{-1}, but this number decreases up to consumption time. In extended incubation period reduction in counts of *Lactobacillus acidophilus* may occur. To overcome this problem, replacement of 25% of *Lactobacillus acidophilus* culture by a mixture of *Streptococcus thermophilus* and *Lactobacillus delbrueckii* subsp. *Bulgaricus* can be used. Following fermentation, the warm product is rapidly cooled to <7°C before agitation and pumped to a filler where it is filled into bottles or cartons (Kosikowski and Mistry, 1997; Vedamuthu, 2006). Protein quality and total amino acid content are similar in both fermented and non-fermented milk. Acidophilus milk has higher free amino acids than milk. As the milk lactose is hydrolyzed by β-galactosidase of *Lactobacillus acidophilus,* acidophilus milk is more suitable for individuals suffering from lactose intolerance. It is also possible to enrich acidophilus milk with calcium, iron and vitamins. Undesirable sour milk flavor caused acidophilus milk is gained limited popularity by consumers. So, sweet acidophilus milk has been developed. When *Lactobacillus acidophilus* is incorporated into pasteurized milk at about 5°C and bottled aseptically, these bacteria are able to keep their viability up to 14 days without reducing the pH of milk due to it does not grow at low temperatures (<10°C). Freeze-dried cultures may keep their viability up to 58% after 23 days at 4°C in sweet acidophilus milk. *Lactobacillus acidophilus* remained viable in sweet acidophilus milk over 28 days at 7°C. Addition of 200 g of frozen culture concentrate to 2000 L of pasteurized milk is satisfactory to reach the target level of *Lactobacillus acidophilus* in the probiotic milk (Vedamuthu, 2006).

Technology of bifidus milk and acidophilus-bifidus milk manufacturing is similar to acidophilus milk. Milk is standardized to desire protein and fat levels in both products. Then, for manufacture of bifidus milk, milk is heat-treated at 80-120°C for 5-30 min and rapidly cooled to 37°C. Heat-treated milk is inoculated with frozen culture of *Bifidobacterium bifidum* and *Bifidobacterium longum* at a level of 10% and left to ferment until pH 4.5. After fermentation, the product is cooled to <10°C and packaged. Final product has a slightly acidic flavor and the ratio of lactic acid to acetic acid is 2:3. Milk used for acidophilus-bifidus milk production is usually enriched with protein prior to fat standardization and homogenization. The standardized milk is heat-treated at 75°C for 15 s or 85°C for 30 min. After cooling the milk to 37°C, frozen cultures of *Lactobacillus acidophilus* and *Bifidobacterium bifidum* are inoculated and fermentation is allowed until pH 4.5–4.6 is reached (~16 h). Following fermentation, the fermented milk is cooled to <10°C. The shelf life of the product is about 20 days. Acidophilus-bifidus milk has a characteristic aroma and slightly acidic flavor. High viscosity of product cause to producing it in set form. It is also possible to produce probiotic milks by simply adding mix culture of *Lactobacillus acidophilus* and *Bifidobacterium bifidum* to cold pasteurized milk.

3. Development of functional dairy foods

Innovation is today's business demand and development of a new functional food is an expensive process and is very important for both food companies and consumers. Regulations should encourage food companies to follow functional food development. Development of dairy probiotic products requires detailed knowledge of both products and customers. It needs to manage customer knowledge effectively (Walzem, 2004; Jousse, 2008). Fundamental risks can affect the development of new functional food products and may leads to fail the development process. Development of new functional food products is very challenging and it has to complete the consumer's expectations for palatable and healthy products (Fogliano and Vitaglione, 2005; Granato et al., 2010; Shah, 2007). So, the development and commerce of functional food products is rather complex, expensive, and uncertain. Key points regarding for a successful functional food product development are consumer demands, technological conditions, and legislative regulatory background. However, consumer's knowledge of the health effects of specific ingredients can affect the acceptance of specific functional food. Therefore, functional ingredients that are in consumers mind for a long period of time, such as minerals, fiber, and vitamins, achieve considerably higher rates of consumer acceptance than new products, such as foods enriched with probiotics, prebiotics, flavonoids, carotenoids, and conjugated linilenic acid (CLA). Several ways to make a functional food product is to eliminating an allergenic protein, lactose, phenylanine and etc from the natural food product; by fortification with a micronutrient; by adding antioxidants, probiotics or prebiotics); by replacing a component, or by increasing bioavailability or stability of a component known to produce a functional effect or to reduce the disease-risk potential of the food (Roberfroid, 2000; Siro, et al., 2008; Granato, et al., 2010). Field of functional probiotic foods requires the cooperation of food technologists, nutritionists, medical doctors, and food chemists in order to obtain innovative products. In this way, these foods may be able to adjust physiological parameters related to health status or disease prevention in human. So, the design and development of functional probiotic foods is a scientific work (Hasler, 1998; Walzem, 2004; Fogliano and Vitaglione, 2005) which is an expensive and multistage process that takes into account many factors, such as sensory acceptance, physical and microbial stability, price, and chemical and other intrinsic functional properties to be successful in the marketplace. Moreover, consumer attitude toward the functional probiotic product also needs to be understood and taken into consideration.

4. Consumer attitude toward functional dairy foods

The development of functional probiotic foods is increasing, as their market increases day by day, although the consumer's information about these foods is increasing without relation to gender, age, and educational or economic levels of the consumers. The therapeutical effect of a functional probiotic food may depend on the consumer's characteristics and the type of carrier and enrichment considered. For instance, yoghurt is most preferred by its enrichment with calcium and fiber. Ingredients such as vitamins and minerals applied in fortification of functional foods are widely recognized and accepted by consumers, but new functional ingredients such as probiotics and prebiotics are not common to them. So, there is a need for increasing the consumer knowledge with respect to these new special ingredients (Hillian, 2000; Luckow and Delahunty, 2004; Ares and Gambaro, 2007; Vianna et al., 2008).

The sensory properties of prebiotic functional foods in comparison with conventional products can lead to different acceptance level. Oligofructose provides some suitable sensory properties such as rounder mouth feel, reduced aftertaste, and slight sweetness to the products. These properties are responsible for high score values for taste, creaminess, and overall acceptability of functional food products. The first important marker in choosing a functional food is flavor, and health consideration is in the second order. If the ingredients added give unpleasant flavors to the product, consumers are not interested in consume such functional probiotic food even if this results in health advantages. This means that flavor is correlated to intrinsic sensory properties of the product such as overall acceptability. In general, as functional products consumption increases, the acceptance of such products may increase, even if the sensory profiles are different from conventional products. When functional ingredients such as probiotics are added to dairy foods, consumers must be aware of probiotics health benefits in order to recognize the functional probiotic foods as being more beneficial than the conventional ones. Functional probiotic food industry should communicate with consumer in a clear way and this is one of the most important aspects for success (Tepper and Trail, 1998; Matilla-Sandholm et al., 1999; Roberfroid, 2000; Tuorila and Cardello, 2002; Nicolay, 2003; Vieira, 2003; Homayouni, 2008a).

5. Conclusion

The future success of functional probiotic dairy foods in marketplace depends on consumer acceptance of such products. The consumers must be convinced by its health claims through clear, honest, and definite messages to agree to pay the cost associated with functional probiotic dairy foods. Development of probiotic dairy products is a key research priority for food design and a challenge for both industry and science sectors. Among the functional foods, the dairy probiotic products, especially ice cream and cheese are good vehicle to transfer probiotics to the human intestinal tract. Additional way to keeping up the probiotic cells in the gut is to entering prebiotics into the intestine through the regular consumption of food containing these components. It is clear that versus probiotics the amounts of prebiotics do not changes during the passage from upper intestinal tract.

Author details

Aziz Homayouni
Department of Food Science and Technology, Faculty of Health and Nutrition, Tabriz University of Medical Sciences, Tabriz, I.R. Iran

Maedeh Alizadeh
Pediatric Nursing, Faculty of Nursing and Midwifery of Maragheh, Tabriz University of Medical Sciences, Tabriz, I.R. Iran

Hossein Alikhah
Faculty of Medicine, Tabriz University of Medical Sciences, Tabriz, I.R. Iran

Vahid Zijah*
Department of Dentistry, Behbood Hospital, Tabriz University of Medical Sciences, Tabriz, I.R. Iran

6. References

Akin, M. B., Akin, M. S., and Kirmaci, Z. Effects of inulin and sugar levels on the viability of yogurt and probiotic bacteria and the physical and sensory characteristics in probiotic ice cream, *Food Chemistry* 104 (2007), pp. 93-99.

Ares, G., and G′ambaro, A. Influence of gender, age and motives underlying food choice on perceived healthiness and willingness to try functional foods, *Appetite* 49 (2007), pp. 148-158.

Bausserman, M. and Michail, S. (2005). The use of Lactobacillus GG in irritable bowel syndrome in children: a double-blind randomized control trial. The Journal of Pediatrics, 147(2): 197-200.

Bergamini, C. V., Hynes, E. R., Quiberoni, A., Sua′rez, V. B., and Zalazar, C. A. Probiotic bacteria as adjunct starters: influence of the addition methodology on their survival in a semi-hard Argentinean cheese, *Food Research International* 38(5) (2005), pp. 597-604.

Cardarelli, H. R., Saad, S. M. I., Gibson, G. R., and Vulevic, J. Functional petitsuisse cheese: Measure of the prebiotic effect, *Anaerobe* 13 (2007), pp. 200-207.

Christiansen, P. S., Edelsten, D., Kristiansen, J. R., and Nielsen, E. W. Some properties of ice cream containing Bifidobacterium bifidum and Lactobacillus acidophilus, *Milschwissenschaft* 51 (1996), pp. 502-504.

Cummings, J. H., Macfarlane, G. R., and Englyst, H. N. Prebiotic digestion and fermentation, *American Journal of Clinical Nutrition* 73 (2001), pp. 415-420.

* Corresponding Author

Dave, R., and Shah, N. P. Viability of probiotic bacteria in yoghurt made from commercial starter cultures, *International Dairy Journal* 7 (1997a), pp. 31-41.

Dave, R. I., and Shah, N. P. Effect of cysteine on the viability of yoghurt and probiotic bacteria in yoghurts made with commercial starter cultures, *International Dairy Journal* 7 (1997b), pp. 537-545.

Dave, R. I., and Shah, N. P. Ingredient supplementation effects on viability of probiotic bacteria in yogurt, *Journal of Dairy Science* 81 (1998), pp. 2804-2816.

Desmazeaud, M. (1996). *Growth inhibitors of lactic acid bacteria, in Dairy Starter Cultures, Cogan, T.M. and Accolas*. New York: VCH Publishers, (pp. 131–155).

Ejtahed, H. S., Mohtadi-Nia, J., Homayouni-Rad, A., Niafar, M., Asghari-Jafarabadi, M., Mofid, V. and Akbarian-Moghari, A. (2011). Effect of probiotic yogurt containing Lactobacillus acidophilus and Bifidobacterium lactis on lipid profile in individuals with type 2 diabetes mellitus. Journal of Dairy Science, 94: 3288-3294.

Ejtahed, H. S., Mohtadi-Nia, J., Homayouni-Rad, A., Niafar, M., Asghari-Jafarabadi, M. and Mofid, V. (2012). Probiotic yogurt improves antioxidant status in type 2 diabetic patients. Nutrition, 28: 539- 543.

Faber, S. M. (2000). Comparison of probiotics and antibiotics to probiotics alone in treatment of diarrhea predominant IBS (D-IBS), alternating (A-IBS) and constipation (C-IBS) patients. Gastroenterology, 118: 687-688.

Food and Agriculture Organization of United Nations; World Health Organization. FAO/WHO (2002). *Guidelines for the evaluation of probiotics in food, Food and Agriculture Organization of the United Nations and World Health Organization Expert Consultation Report*. Available from. http://www.who.int/foodsafety/publications/fs_management/probiotics2/en.

Fogliano, V., and Vitaglione, P. Functional foods: planning and development, *Molecular Nutrition and Food Research* 49 (2005), pp. 256-262.

Foschino, R., Fiori, E., and Galli, A. Survival and residual activity of Lactobacillus acidophilus frozen cultures under different conditions, *Journal of Dairy Research* 63 (1996), pp. 295-303.

Fox, P. F., Wallace, J. M., Morgan, S., Lynch, C. M., Niland, E. J., and Tobin, J. Acceleration of cheese ripening, *Antonie van Leeuwenhoek* 70 (1996), pp. 175-201.

Granato, D., Castro, I. A., Ellendersen, L. S. N., and Masson, M. L. Physical stability assessment and sensory optimization of a dairy-free emulsion using response surface methodology, *Journal of Food Science* 73 (2010), pp. 149-155.

Granato, D., Branco, G. F., Cruz, A. G., Faria, J. A. F., and Shah, N. P. Probiotic dairy products as functional foods, *Comperhensive reviews in Food Science and food safety* 9 (2010), pp. 455-470.

Halpern, G. M., Prindville, T. and Blankenburg, M. (1996). Treatment of irritable bowel syndrome with Lacteol forte: A randomized, double-blind, crossover trial. The American Journal of Gastroenterology, 91: 1579-1585.

Hansen, L. T., Wojtas, P. M. A., Jin, Y. L. and Paulson, A. T. Survival of Ca-alginate microencapsulated Bifidobacterium spp. In milk and simulated gastrointestinal conditions, *Food Microbiology* 19 (2002), pp. 35-45.

Hasler, C. M. Functional foods: their role in disease in developing new food products for a changing prevention and health promotion, *Food Technology* 52 (1998), pp. 57-62.

Haynes, I. N., and Playne, M. J. Survival of probiotic cultures in low-fat ice-cream, *Australian Journal of Dairy Technology* 57 (2002), pp. 10-14.

Hillian, M. Functional food: how big is the market? *World Food Ingredients* 12 (2000), pp. 50-53.

Homayouni, A., Ehsani, M. R., Azizi, A., Yarmand, M. S., and Razavi, S. H. Effect of lecithin and calcium chloride solution on the microencapsulation process yield of calcium alginate beads, *Iranian Polymer Journal* 16(9) (2007), pp. 597-606.

Homayouni, A. (2008a). *Therapeutical effects of functional probiotic, prebiotic and symbiotic foods.* (1st ed.). Tabriz: Tabriz University of Medical Sciences.

Homayouni, A., Azizi, A., Ehsani, M. R., Razavi, S. H. and Yarmand, M. S. Effect of microencapsulation and resistant starch on the probiotic survival and sensory properties of synbiotic ice cream, *Food Chemistry* 111 (2008b), pp. 50-55.

Homayouni, A., Ehsani, M. R., Azizi, A., Razavi, S. H., and Yarmand, M. S. Spectrophotometrically evaluation of probiotic growth in liquid media, *Asian Journal of Chemistry* 20(3) (2008c), pp. 2414-2420.

Homayouni, A., Ehsani, M. R., Azizi, A., Razavi, S. H., and Yarmand, M. S. Growth and survival of some probiotic strains in simulated ice cream conditions, *Journal of Applied Sciences* 8(2) (2008d), pp. 379-382.

Homayouni, A. Letter to the editor, *Food Chemistry* 114 (2009), pp. 1073.

Homayouni, A., Azizi, A., Javadi, M., Mahdipour, S. and Ejtahed, H. (2012). Factors influencing probiotic survival in ice cream: A Review. International Journal of Dairy Science, doi: 10.3923/ijds.2012.

Joseph, P. J., Dave, R. I., and Shah, N. P. Antagonism between yogurt bacteria and probiotic bacteria isolated from commercial starter cultures, commercial yogurts, and a probiotic capsule, *Food Australia* 50 (1998), pp. 20-23.

Jousse, F. Modeling to improve the efficiency of product and process development, *Comprihensive Review of Food Science and Food Safety* 7 (2008), pp. 175-181.

Kailasapathy, K., and Rybka, S. L. acidophilus and Bifidobacterium spp. Their therapeutic potential and survival in yoghurt, *Australian Journal of Dairy Technology* 52 (1997), pp. 28-35.

Kailasapathy, K., and Chin, J. Survival and therapeutic potential of probiotic organisms with reference to Lactobacillus acidophilus and Bifidobacterium spp., *Immunology and Cell Biology* 78 (2000), pp. 80-88.

Kailasapathy, K., and Sultana, K. Survival and β-D-galactosidase activity of encapsulated and free Lactobacillus acidophilus and Bifidobacterium lactis in ice cream, *Australian Journal of Dairy Technology* 58 (2003), pp. 223-227.

Kebary, K. M. K., Hussein, S. A., and Badawi, R. M. Improving viability of bifidobacterium and their effect on frozen ice milk, *Egyptian Journal of Dairy Science* 26 (1998), pp. 319-337.

Kim, H. J., Camilleri, M. and McKinzie, S. (2003). A randomized controlled trial of a probiotic, VSL 3, on gut transit and symptoms in diarrhoea-predominant irritable bowel syndrome. Alimentary Pharmacology and Therapeutics, 17: 895-904.

Koletzko, B., Aggett, P.J., and Bindels, J.G. Growth, development and differentiation: a functional food science approach, *Brazilian Journal of Nutrition* 80 (1998), pp. 35-45.

Kosikowski, F. V. and Mistry, V. V. (1997). *Cheese and fermented milk foods, in Origin and Principles: Fermented Milks*. Westport: F.V. Kosikowski, (pp. 57-74).

Kunji, E. R. S., Mierau, I., Hagting, A., Poolman, B., and Konings, W. N. The proteolytic systems of lactic acid bacteria, *Antonie van Leeuwenhoek* 70 (1996), pp. 187-221.

Lankaputhra, W. E. V., Shah, N. P., and Britz, M. L. Survival of bifidobacteria during refrigerated storage in the presence of acid and hydrogen peroxide, *Milchwissenschaft* 51 (1996), pp. 65-70.

Leuschner, R. G., Heidel, M., and Hammes, W. P. Histamine and tyramine degradation by food fermenting microorganisms, *International Journal of Food Microbiology* 39 (1998), pp. 1-10.

Luckow, T., and Delahunty, C. Consumer acceptance of orange juice containing functional ingredients, *Food Research International* 37 (2004), pp. 805-14.

Matilla-Sandholm, T., Blum, S., Collins, J.K., Crittenden, R., DeVos, W., Dunne, C., et al. Probiotics: towards demonstrating efficacy, *Trends in Food Science and Technology* 10 (1999), pp. 393-399.

Meurman, J. H. (2005). Probiotics: do they have a role in oral medicine and dentistry? Europian Journal of Oral Sciences, 113(3): 188-196.

Mirzaei, H., Pourjafar, H. and Homayouni, A. (2012). Effect of calcium alginate and resistant starch microencapsulation on the survival rate of Lactobacillus acidophilus La5 and sensory properties in Iranian white brined cheese. Food Chemistry, 132: 1966-1970.

Mountzouris, K. C. and Gibson, G. R. Colonization of the gastrointestinal tract, *Annales Nestle* 61 (2003), pp. 43-54.

Nicolay, C. Language is a key to marketing digestive health products, *Functional Foods and Nutraceuticals* 6 (2003), pp. 20-22.

Ravula, R. R. and Shah, N. P. Viability of probiotic bacteria in fermented frozen dairy desserts, *Food Australia* 50 (1998), pp. 136-139.

Roberfroid, M. (2000). *Inulin-type fructans*. Boca Raton: CRC Press.

Roberfroid, M.B. Concepts and strategy of functional food science: the European perspective, *American Journal of Clinical Nutrition* 71 (2000), pp. 1660-1664.

Ross, R. P., Fitzgerald, G., Collins, K., and Stanton, C. Cheese delivering biocultures: probiotic cheese, *Australian Journal of Dairy Technology* 57(2) (2002), pp. 71-78.

Roy, D., Mainville, I., and Mondou F. Bifidobacteria and their role in yogurt-related products, *Microecology Therapy* 26 (1997), pp. 167-180.

Saggioro, A. (2004). Probiotics in the treatment of irritable bowel syndrome. Journal of Clinical Gastroenterology, 38: 104-106.

Shah, N. P., and Ravula, R. R. Microencapsulation of probiotic bacteria and their survival in frozen fermented dairy desserts, *Australian Journal of Dairy Technology* 55 (2000), pp. 139-144.

Shah, N. P. Functional cultures and health benefits, *International Dairy Journal* 17 (2007), pp. 1262-1277.

Siro, I., Kapolna, E., Kapolna, B., and Lugasi, A. Functional food: product development, marketing and consumer acceptance-A review, *Appetite* 51 (2008), pp. 456-467.

Stanton, C., Gardiner, G., Meehan, H., Collins, K., Fitzgerald, G., Lynch, P. B., and et al. Market potential for probiotics, *American Journal of Clinical Nutrition* 73 (2001), pp. 4765-4835.

Stanton, C., Desmond, C., Coakley, M., Collins, J. K., Fitzgerald, G., and Ross, R. P. (2003). Challenges facing development of probioticcontaining functional foods. In E. R. Farnworth (Eds.). *Handbook of fermented functional foods* (pp. 27-58). Boca Ranton: CRC Press.

Surono, I.S. and Hosono, A. (2002). Fermented milks: Types and standards of identity. In H. Roginski, J. Fuquay, and P.F. Fox (Eds.). *Encyclopedia of Dairy Microbiology*. (pp. 1018-1023).

Tamime, A. Y., Marshall, V. M. E., and Robinson, R. K. Microbiological and technological aspects of milks fermented by bifidobacteria, *Journal of Dairy Research* 62 (1995), pp. 151-187.

Tepper, B., and Trail, A. Taste or health: a study on consumer acceptance of corn chips, *Food Quality and Preference* 9 (1998), pp. 267-272.

Tuorila, H., and Cardello, A. V. Consumer responses to an off-flavour in juice in the presence of specific health claims, *Food Quality and Preference* 13 (2002), pp. 561-569.

Vedamuthu, E.R. (2006). Other fermented and culture-containing milks. In R. Chandan, C.H. White, A. Kilara, and Y.H. Hui (Eds.), *Manufacturing Yogurt and Fermented Milks* (pp. 295-308). Blackwell Publishing.

Vernazza, C. L., Rabiu, B. A., and Gibson, G. R. (2006). Human colonic microbiology and the role of dietary intervention: Introduction to prebiotics. In G. R. Gibson and R. A. Rastall (Eds.), *Prebiotics: Development and application* (pp. 1-12). England: John Wiley and Sons Ltd.

Vianna, J. V., Cruz, A. G., Zoellner, S. S., Silva, R., and Batista, A. L. D. Probiotic foods: consumer perception and attitudes, *International Journal of Food Science and Technology* 43 (2008), pp. 1577-1580.

Vickers, Z., Mullan, L., and Holton, E. Impact of differences in taste ratings on the consumption of milk in both a laboratory and a foodservice setting, *Journal of Sensory Studies* 14 (1999), pp. 249-262.

Vieira, P. How to create brand awareness for new products, *Functional Foods and Nutraceuticals* 6 (2003), pp. 38-40.

Walzem, R. L. Functional foods, *Trends In Food Science and Technology* 15 (2004), pp. 518.

Weber, H. (1996). Starter cultures in dairy industry. In H. Weber (Eds.), *Mikrobiologie der Lebensmittel: Milch und Milchprodukte (in German)* (pp. 105-152). Hamburg: Behr's Verlag.

Chapter 10

Innovative Dairy Products Development Using Probiotics: Challenges and Limitations

Esteban Boza-Méndez, Rebeca López-Calvo and Marianela Cortés-Muñoz

Additional information is available at the end of the chapter

http://dx.doi.org/10.5772/50104

1. Introduction

Probiotic foods are food products that contain a living probiotic ingredient in an adequate matrix and in sufficient concentration, so that after their ingestion, the postulated effect is obtained, and is beyond that of usual nutrient suppliers (Saxelin et al., 2003).

Probiotic delivery has been consistently associated with foods (especially dairy). However, nowadays there is an increasing trend toward using probiotics in different food systems despite its original sources and even as nutraceuticals, such as in capsules. According to Ranadheera et al. (2010) this changing trend in delivering probiotics may lead to a reduction in functional efficacy due to the exclusion of the potential synergistic effect of the food. Selection of the adequate food system to deliver probiotics is a vital factor that should be considered when developing functional products.

Foods are carriers for the delivery of probiotic microorganisms to the human body. The growth and survival of probiotics during gastric transit is affected by the characteristics of the food carriers, like chemical composition and redox potential. Same probiotic strains could vary in functional and technological properties in the presence of different food ingredients or in different food environments (Ranadheera et al., 2010). Thus, variation between different strains' behavior in different conditions would be expected.

Dairy products have been considered as a good carrier for probiotics since fermented foods and dairy products have particularly a positive image. A major advantage is that consumers are already familiar with them and many believe that dairy products are healthy, natural products. Table 1 shows some of the beneficial physiological properties that have been associated with milk components.

Others advantages of dairy products as vehicles for probiotics are that fermentation acts to retain and optimize microbial viability and productivity, while simultaneously preserving the probiotic properties. Consumers are familiarized with the fact that a fermented dairy product contains living microorganisms, and they are also able to protect probiotics through the gastrointestinal transit. This protection comes as a result from the buffering capacity that increases survival chances. The refrigerated storage recommended for these products helps to stabilize probiotic bacteria (Ross et al., 2002; Stanton et al., 2003).

Ingredient	Source	Claim areas examples
Minerals	Calcium Casein peptides	Optimum body growth and development, dental health, osteoporosis
Fatty acids	Conjugated linoleic acid	Heart disease, cancer prevention, weight control
Prebiotics/carbohydrates	Galactooligosaccharides Lactulose Lactose	Digestion, pathogen prevention, gut flora balance, immunity, lactose intolerance
Probiotics	Lactic acid bacteria Bifidobacteria	Digestion, immunity, vitamin production, heart disease, antitumor activity, remission of inflammatory bowel disease, prevention of allergy, alleviation of diarrhea
Proteins/Peptides	Caseins, whey proteins, immunoglobulins, lactoferrin, glycoproteins, specific peptides	Immunomodulation, body growth, antibacterial activity, dental health, hypertension regulation (angiotensin inhibitors)

Table 1. Selection of ingredients and claims associated with functional dairy foods (adapted from Shortt et al., 2003).

Besides, according to Shortt et al. (2003) significant opportunities exist for dairy products whose functionalities have widespread appeal. This means that a product encapsulating the needs of every member of a family is extremely likely to be a success. The broad potential interest in functional dairy products is an important market advantage. Functional dairy products that affect conditions such as osteoporosis, heart disease and cancer are attractive specifically to adults, while products with claims on tooth health, bone health and immunity appeal to adults and children in a similar way. The possible range of sensory characteristics with dairy ingredients also allows the production of diverse textures and aromas, adding another benefit.

Current knowledge on probiotics support a number of potential health benefits. They help to maintain good balance and composition of intestinal flora increasing the ability to resist pathogens invasion and maintain the host's well being. Reduction of blood pressure, cholesterol and/or triglycerides levels, reduction of lactose intolerance problems, immune system enhancement, anti carcinogenic activity and improve nutrients utilization are well described in literature. The use of probiotics for preventing and treating illnesses related to gastrointestinal, respiratory and urogenital tracts have been studied. They have been widely used in therapeutic applications as constipation, diarrhea control, bowel syndrome, control of inflammatory processes, prevention of eczema, osteoporosis and food allergy (Aureli et al., 2011; Ranadheera et al., 2010; Rastall et al., 2000; Vasiljevic and Shah, 2008).

The most common probiotic strains used in dairy foods belong to *Lactobacillus* (*L. acidophillus, L. johnsonii, L. gasseri, L. crispatus, L. casei/paracasei, L. rhamnosus, L. reuteri, L.plantarum*) and *Bifidobacterium* (*Bifidobacterium lactis, B. bifidum, B. infantis, B. breve, B. animalis, B. adolescentis*) genera (Saxelin, 2008).

In Europe EFSA is responsible for the evaluation procedure that accepts or rejects applications for health and nutrition claims on food and beverages (EU Regulation 1924/2006). In recent years this European authority has rejected probiotic health claims adducing that there is no sufficient scientific evidence for the declared beneficial effects. This situation obliged food companies from probiotic industry to perform new clinical studies trying to generate solid scientific evidence for specific probiotic strains and health benefits for submission to the EFSA approval. Consumers still identify probiotic dairy products as healthy despite of this situation.

According to Shortt et al. (2003), the dairy industry is in an excellent position to develop and exploit the functional food market. These products are significant players in the functional food market; for example, they were estimated to account for approximately 60% of functional food sales in Europe by 2000. In 2008, consumers market for probiotic foods was over 1.4 billion Euros in Western Europe, and their annual sales growth was forecast at 7-8% for a 5 year period (Saxelin, 2008). Developing new technologies and new functional dairy products is nowadays relevant.

This chapter focuses on the development of innovative probiotic dairy products considering limiting factors for the survival of probiotics, techniques for the addition and protection of these microorganisms, the quality modifications of final products, the application of sensory analysis and finally how to determine probiotic populations in dairy products.

2. Limiting factors for the survival of probiotics

The food industry has an important market created by the incorporation of probiotic microorganisms into products. However, the addition of this kind of cultures in a food product could be difficult because of the bacteria conditions required in order to survive or to grow in food. Some authors have suggested that more research regarding the challenges that represent incorporating a probiotic culture is necessary because most of the information available is focused on health benefits of the probiotics (Champagne et al., 2005). Evaluation of technological traits such as growth and survival in milk-based media and during product manufacture and shelf life can be important considerations for the selection of strains for food applications (Stanton et al., 2003).

Successful marketing of probiotic products require a minimal amount of viable probiotic cells guaranteed throughout shelf life. To obtain the beneficial effects associated with this type of food, the bacteria must remain viable and in a proper concentration when the host consumes the product. This fact could determine the shelf life of the developed product, because the survival of the probiotics depends on many factors in the food (Talwaker and Kailasapathy, 2004).

Champagne et al. (2005) list seven factors that culture distributors and food manufacturers need to consider in order to add probiotics successfully into products. These factors include: type and form of the culture, the amount of bacteria required to obtain a beneficial effect, toxicity, production process effect on viability, the determination of probiotic cells used in the product, stability during storage and possible changes in sensory properties of the food.

To use a probiotic strain compatible with food production processes technologies is ideal. This means that the elaboration, distribution and commercialization of the product should not have any effect in the viability of bacteria. For example, in the specific case of dairy products, the probiotic should have the capacity to grow in milk (or dairy) but also have a low metabolic activity at low temperatures, in order to guarantee the proper amount of bacteria in the product with no significant changes in quality during shelf life. However, probiotic bacteria generally do not grow well in milk and are adversely affected by storage conditions in some dairy products (Champagne, 2008).

The compatibility and adaptability between the selected strain(s) and the food used as carrier is fundamental, and may represent a significant technological challenge since many probiotic microorganisms are sensitive to the concentration of oxygen, carbon dioxide and salt, high and freezing temperatures and acidic environments (Corrales et al., 2007; Cruz et al., 2009a; Fortin et al., 2011; Talwaker and Kailasapathy, 2004).

Since many dairy products are fermented, it is common to found levels of acidity that may affect the probiotics viability. Numerous studies have reported large losses in viability during storage of fermented milk, yogurt and alike (dairy products known as acid). It is believed that the pH is actually a critical stress factor in the probiotics viability through storage, although there are variations between species and strains for the survival in acidic environments (Roy, 2005). Donkor et al. (2006) evaluated the effect of the acidity of yogurt on the viability of some *Lactobacilli* and *Bifidobacteria* strains. They concluded that *Lactobacilli* strains showed a good cellular stability maintaining constant concentration throughout the storage period regardless of final pH. On the other hand, the cell counts of *Bifidobacteria* decreased by one log cycle at the end of the storage period, due to the high production of organic acids.

Boza et al. (2010) studied the effect of adding *Lactobacillus paracasei* subsp. *paracasei* to a semi hard cheese. Figure 1 presents the pH variation found in cheese during ripening at controlled conditions of 12°C and 85% RH. An important initial decrease is observed (day 0 to 13), pH values tend then to stabilize during cheese ageing process.

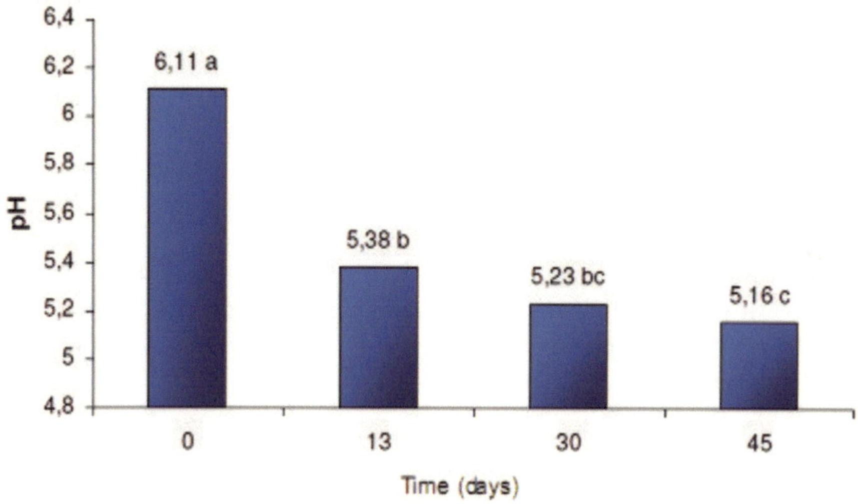

Figure 1. Values of pH for semi hard cheese with *Lactobacillus paracasei* subsp. *paracasei* aged for different periods at 12°C and 85% RH [18]. Different letters in the columns indicate significant differences (*P*<0,05).

Corriols (2004) studied the survival of *Bifidobacterium lactis* in a light sour cream (12% fat, w/w) during 40 days at 5ºC. In this study, product behavior considering pH of a regular sour cream inoculated with a starter culture mix of *Lactococcus lactis* subsp. *cremoris, Lactococcus lactis* subsp. *lactis, Leuconostoc mesenteroides* subsp. *cremoris, Lactococcus lactis* subsp. *diacetylactis* and a probiotic sour cream (starter culture + *Bifidobacterium lactis*) was performed. Table 2 presents pH values for probiotic light sour cream during storage time at 4ºC. Evaluating pH at day 8, 15 and 22 showed that there was no significant difference (*P*>0.05) in these values.

Storage time (days)	pH
0	4.51 a
8	4.37 b
15	4.36 b
22	4.39 b

Table 2. Variation of pH for 12% fat (w/w) sour cream with *B. lactis* during refrigerated storage at 4ºC. Average of 5 measurements of three independent experiments. Values followed by the same letter within a column are not significant at *P*<0.05.

Since there was a slight product post-acidification (see table 2) *B. lactis* survival was possible as acidity could be a cause of probiotics viability loss in fermented products. No significant difference (*P*>0.05) was found in probiotic and regular sour cream pH values. Finally, this study showed that it was possible to preserve a probiotic population around 7 x 10^6 CFU/g after 40 days of storage indicating that this cheese could be considered a functional product along its shelf life. Author reported an increase of 12% on final cost of probiotic light sour cream when compared to regular product.

It is also important to note the relationship between probiotics and other fermenting microorganisms, as there may be synergistic or antagonistic effects between them (Heller, 1998). During the manufacture of cheese or yogurt, addition of the starters and probiotic cultures usually result in a slower growth of the probiotic strains. This is possibly because the starter cultures produce substances that inhibit not only pathogens and spoilage microorganisms but also probiotics, and because of the rapid growth of starter cultures, the nutrients availability for probiotics decreases (Roy, 2005). Champagne et al. (2005) mentioned that very few strategies have been proposed to reduce the starters' negative effects on the probiotic cultures, and that the most common is reducing starter dose (entirely or partially). However, precautions must be taken when lowering the dose of the starter microorganisms, because probiotics can also show a negative effect on these cultures and this would slow their activity.

Environments with a rich concentration of oxygen due to transportation systems and stirring or whipping procedures are also commonly found in dairy processing, especially in ice creams and some types of yogurts and fermented milks. The exposure of cultures to dissolved oxygen causes the accumulation of toxic metabolites such as superoxide, hydroxyl radicals and hydrogen peroxide, which eventually lead to cell death of the probiotic microorganisms that partially or completely lack of an electrons transport system. Regarding this oxygen toxic effect on probiotics, there are variations between species. For example, *Bifidobacterium* spp., strictly anaerobic in nature, is generally considered more vulnerable than strains of *Lactobacillus acidophilus* (Talwaker and Kailasapathy, 2004).

Another important issue concerning the addition of probiotic strains into food is temperature. Heating temperatures below 45°C are usually compatible with the cultures, although this depends on the time and the specific strain. Processes that include heating steps above 45°C result in destruction of at least a portion of the probiotic population (Roy, 2005).

On the other hand, low temperatures are generally used to delay the chemical reactions and growth of microorganisms found in foods, therefore a lower temperature implies greater bacterial inhibition growth. A temperature low enough will inhibit the growth of all microorganisms including probiotics. Because of their nature, dairy products, fermented or not, require low storage temperature for preservation, and this fact determines the survival and development of probiotics in these products. It is believed that freezing also leads to a considerable reduction in the number of viable microorganisms in food, although this reduction would depend on the freezing rate and the specific strain tolerance to low temperature.

Corrales et al. (2007) evaluated the effect of the dynamic freezing operation on the viability of two different probiotic strains, *Lactobacillus acidophilus* and *Bifidobacterium lactis,* during ice cream production. It was found that the reduction rate of both strains during this operation was not significant ($P>0.05$), but throughout the whole process of elaboration of the ice cream (dynamic freezing and then hardening at -30°C) there was a significant reduction on both populations.

Other unit operations like pressing and draining could also affect the bacterial counts in the products. The effect of pressing and draining in a cheese probiotic cells is obviously a loss of these cells in the whey, so the final concentration in the pressed cheese is difficult to control (Heller, 1998). Segura (2005) evaluated the effect of the pressing operation in a Turrialba cheese (typical Costarican fresh cheese, >60% water, w/w) added whith *Bifidobacterium lactis*. Probiotic population was determined before and after the pressing operation, and significant differences were found ($P<0.05$). A loss of approximately two logarithms on probiotic population was reported after the pressing operation.

Despite the above results, it is believed that cheese could be a very good vehicle for delivering probiotic strains into the organism, since cheese has a stable structure and usually a high fat content (case of aged cheeses), factors that can help bacteria to survive during product storage and transit on the gastro-intestinal tract.

When comparing with yogurt, the problem for cheese (especially semi-hard and hard cheese) acting as carrier for probiotics results from the high fat and salt content and the relatively low recommended daily intake. Also the concentration of probiotics in cheese should be about four to five times higher than in yogurt. However, this does not apply to fresh cheese, which can easily be adjusted to low fat and salt contents, and for which recommended daily intake is rather high (Cruz et al., 2009a).

Figure 2 shows the growth of a strain of *L. paracasei* subsp. *paracasei* in a semi hard cheese during a ripening period of 45 days at 12°C and 85% RH (Boza et al., 2010). Probiotic population increased during the ripening period reaching interesting levels according with the high levels population goal.

Figure 3 shows the stationary behavior of the same bacteria viability in the ripened cheese kept under refrigeration for 49 days. It should be noted that strains of *Lactobacillus paracasei* have been isolated from naturally ripened cheeses and recognized as non starter lactic acid bacteria (Lynch et al., 1999), indicating that the matrix of the cheese is a good substrate for the growth of this bacterium.

The trend in cheeses, as in yogurt and fermented milks, is that probiotic bacteria populations remain stable or loose viability during ripening and storage (Kılıc et al., 2009; Ong et al., 2006; Songisepp et al., 2004; Vinderola et al., 2000; Yilmaztekin et al., 2004). There are also studies that have shown the growth of some probiotics in cheese during ripening periods or storage under refrigerated conditions (Boza et al., 2010; Buriti et al., 2005; Gardiner et al., 2002; Gardiner et al., 1998; Segura, 2005). However, growth and survival of probiotic microorganisms in ripened cheeses are believed to depend on many factors (like ripening temperature and the probiotic strain interactions with other microorganisms found in cheese) hence hard to generalize.

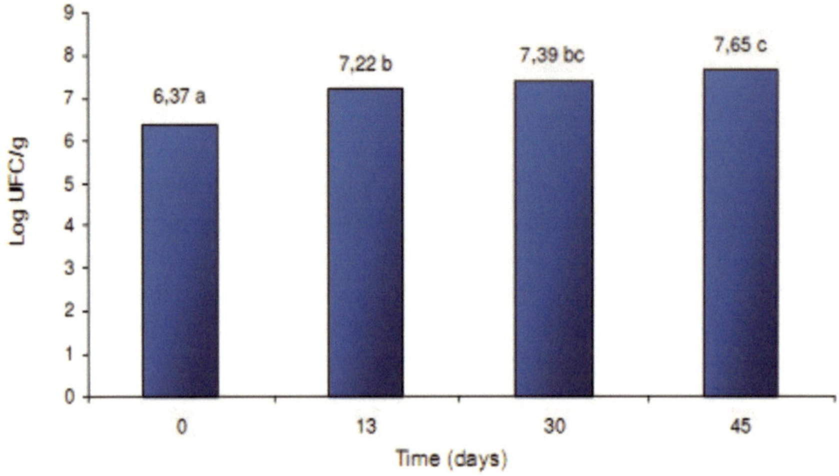

Figure 2. Logarithm of the number of colony forming units of *Lactobacillus paracasei* subsp. *paracasei* per gram of semi hard cheese for different time periods at 12°C and 85% RH. Different letters in the columns indicate significant differences (*P*<0,05).

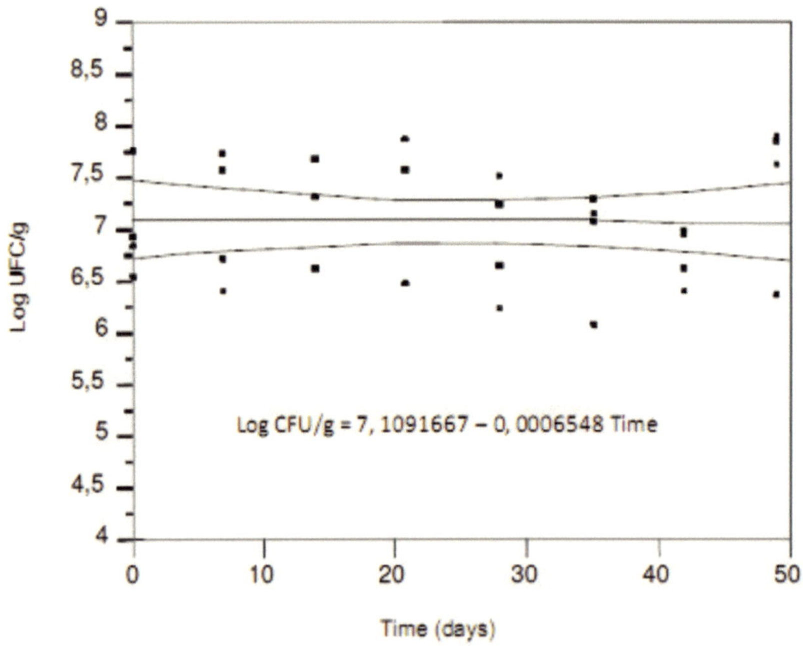

Figure 3. Logarithm of the number of colony forming units of *Lactobacillus paracasei* subsp. *paracasei* per gram of semi hard cheese vacuum packed and stored for 49 days at 5°C (Boza et al., 2010).

Indulgence products like ice-creams are potential probiotic vehicles as well, with the advantage of being appreciated by people belonging to all age groups and social levels (Cruz et al., 2009b). However, in these products, due to low storage temperatures and high concentration of dissolved oxygen, it is difficult for probiotic microorganisms to increase their number. The study conducted by Corrales et al. (2007) determined the behavior of two different probiotic strains, *L. acidophilus* and *B. lactis,* in ice cream throughout 85 days of storage at -30° C. Figure 4 (a and b) shows the behavior of probiotic strains.

The author found that freeze storage conditions affected significantly ($P<0.05$) the viability of the two microorganisms, and reported losses of 0.76 and 1.10 logarithmic units for *L. acidophilus* and *B. lactis* respectively. Functional shelf life (plate counts > 10^6 CFU/g) was found to be 90 days. An increase of 28% in variable costs was calculated for the product.

Salem et al. (2005) manufactured ice cream with different strains of *Lactobacilli* and *Bifidobacteria.* The probiotic ice cream was evaluated for cultures survival during 12 weeks of frozen storage at -26°C. Initial freezing of ice cream mix followed by hardening caused a reduction of less than one log cycle in viable counts of probiotics. The viable counts decreased during frozen storage by 2.23, 1.68, 1.54, 1.23 and 1.77 log for *Lactobacillus acidophilus, Bifidobacterium bifidum, Lactobacillus reuteri, Lactobacillus gasseri* and *Lactobacillus rhamnosus,* respectively. Although there was a decrease in the number of viable cells, the investigators considered the ice cream as a probiotic food during 12 weeks of storage, since the viable population remained above the recommended minimum limit of 1 x 10^6 CFU /g.

Feraz and colleagues (2012) investigated the survival of *L. acidophilus* in ice cream with different overrun levels during a 60 day storage period. All the ice creams presented a minimum count of 1 x 10^6 CFU/g at the end of 60 days of frozen storage.

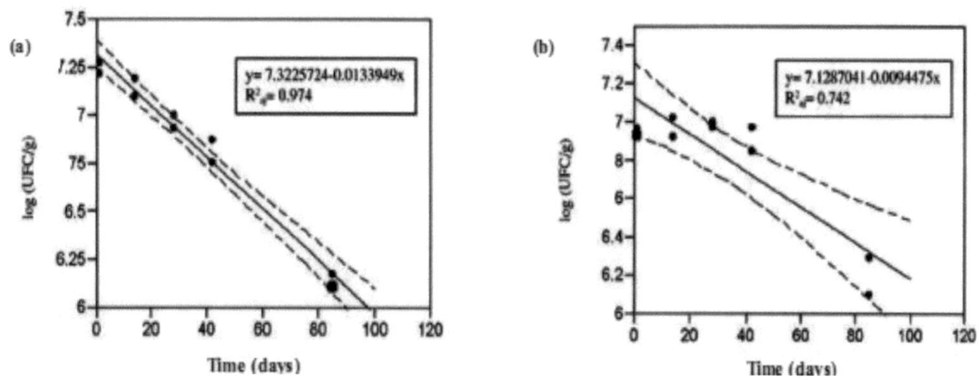

Figure 4. Behavior of *Lactobacillus acidophilus* (a) and *Bifidobacterium lactis* (b) during ice cream storage at -30°C (Corrales et al., 2007).

3. Techniques for the addition and protection of probiotics in dairy products

Controlled growth of probiotic bacteria in a dairy product during ripening or fermentation periods are desirable and interesting from a productive and economic point of view. This ideal situation may allow food producers to use a lower initial dose of inoculum, or may help to replace the microorganisms that could have been eliminated or destroyed during a specific step of the production process like thermal treatment, dynamic freezing or draining.

It has been already explained that probiotics generally do not grow well in milk, and in fact, as mentioned before, the populations of many probiotic bacteria are not even stable during storage of dairy products. However, it is possible to find variations among strains of the same species, and the current trend is the development of new dairy products by using new ingredients that favor the growth of these microorganisms, such as yeasts, tomato juice, rice and soy milk (Champagne et al., 2005; Liu and Tsao, 2009).

Champagne (2008) suggests some ways to address stability problems, and these include: strain selection, ingredients selection (flavours, enzymes, fruits or vegetables, prebiotics) and packaging. All these techniques can be used to innovate and develop new products. Other techniques may include the microencapsulation with lipid materials, alginate and prebiotics (Akhiar, 2010; Siuta-Cruce and Goulet, 2001), the addition of antioxidants such as ascorbate and L-Cysteine, and the elimination from the environment of strains producing hydrogen peroxide (Champagne et al. 2005).

It was mentioned (Cruz et al., 2009a) that one strategy for enhancing bacterial tolerance toward stresses such as temperature, pH or bile salts is prior exposure to sub-lethal levels of the given stress. Cruz et al. (2009a) proposed as alternative to avoid destruction by heat the addition of the probiotic after pasteurization, microencapsulation, pre-adaptation of cells to stress and changing technologies by a slight decrease in temperature.

In order to use probiotic bacteria with proven health benefits in the manufacture of dairy products, sometimes the process has to be modified and adapted for the strains, due to their high sensitivity. According to Cruz et al. (2009a) there are two options for the addition of probiotic bacteria during cheese processing which can directly affect the survival rate of these microorganisms: probiotic bacteria can be added before the fermentation (together with the starter culture), or after it.

Daigle et al. (1999) produced Cheddar cheese from microfiltered milk standardized with cream and fermented with *Bifidobacterium infantis*. In this case, bifidobacteria showed good survival (> 3×10^6 CFU/g) on cheese packaged under vacuum and kept at 4°C for 84 days. Cheddar cheese was also successfully produced with a spray dried adjunct of powder milk containing a strain of *Lactobacillus paracasei.* Data obtained demonstrated that probiotic spray-dried powder is a good option of probiotic addition to dairy products (Daigle et al, 1999).

Other research group (Songisepp et al., 2004) added *Lactobacillus fermentum* ME-3, which has been shown to possess antimicrobial and antioxidative properties, to a "Pikantne" cheese which is a semi-soft Estonian cheese with an open texture. They tested two different methods: adding the probiotic combination with the starter culture and adding the probiotic on the drained curd. The cheese produced using the first method showed better sensory characteristics and therefore was chosen to carry out stability tests of probiotic during ripening and storage. The results showed that the strain used was well suited to the process (levels of 5 x 10^7 CFU/g on ripened cheese) and maintained its probiotic effects.

Lactobacillus casei cells were immobilized on fruit pieces (apple and pear) and used them in the production of Feta cheese (Kourkoutas et al., 2005). Cheese was also produced with free cells of *L. casei*. At the end of the ripening period the authors concluded that the immobilized cells remained viable in the fruit, and in higher counts than in the cheese. Therefore, it is believed that these pieces of fruit were an effective support for the incorporation of probiotics in this type of product.

Ong and other researchers (2006) added combinations of *Lactobacillus acidophilus, L. casei* and *Bifidobacterium longum*; and *L. acidophilus, L. paracasei and B. Lactis* to Cheddar cheese. In this case cheese was produce following a standard procedure, in which milk, after being standardized was tempered to 31°C before inoculation with cheese starter culture and probiotic bacteria. All probiotic adjuncts survived manufacturing process and maintained their viability until the end of the ripening process.

Segura (2005) elaborated a probiotic fresh cheese (>60% water), adding *Bifidobacterium lactis* either to the milk before fermentation or to the curd (mixed with salt). It was found that a large number of bacteria were lost in subsequent operations such as pressing, but this phenomenon was lower when the probiotic culture was added to the curd (see Table 3).

Boza et al. (2010) modified the traditional process of semi hard cheese to avoid larger losses of probiotic in the whey. They added a strain of *Lactobacillus paracasei* mixed with salt after a preliminary pressing of the curd, wherein a major portion of whey was removed, obtaining a cheese with a viable probiotic cell number greater than 1 x 10^6 CFU/g.

Logarithm of the population of *B. lactis*			
Inoculation technique	Before pressing the curd	After pressing the curd	Variation in the logarithm of the probiotic population
Addition after pasteurization	8.51 ^{a1}	2.95 ^{b1}	5.56
Addition to the curd	9.81 ^{a2}	6.09 ^{b2}	3.72

$^{a, b}$... Different letters between columns indicate significant differences (*P*<0,05).
$^{1, 2}$... Different numbers between rows indicate significant differences (*P*<0,05).

Table 3. *Bifidobacterium lactis* population logarithmic variation before and after the pressing stage of a fresh cheese using two inoculation techniques.

Evaluation of the effect of inoculation time of the probiotics on viable counts of five bacteria in curds and whey during Cheddar cheese manufacture was performed (Fortin et al., 2011). These authors found that inoculation of probiotics in milk before renneting resulted in almost half the cell losses in whey compared with the addition just before the cheddarization step, and they also discovered that addition of probiotics in milk improved their subsequent stability by about 1 log over the 20 days storage period as compared with cells added at cheddarization. Specifically, significantly higher populations of *Bifidobacteria* in curds were detected when the probiotic culture was added to milk. They found that although the quantity of whey generated during cheddarization is much lower than that obtained after the first cutting, the population of probiotics in the whey was ten times higher than after the first cutting when probiotics were added to milk. The authors proposed that cells were not as well entrapped in the curd mass at cheddarization than at renneting.

Arguedas (2010) added *L. paracasei* subesp.*paracasei* in a Philadelphia type cheese (24% fat, w/w) and evaluated their survival behavior during 40 days at 5ºC. This author found that it was possible to reach a population around 7 x 10^6 CFU/g after 40 days of storage, and this cheese could be considered a functional product along the shelf life. Considering that during the Philadelphia type cheese production there is a pasteurization step followed by homogenization and fermentation, probiotic culture was added during the stirring step just before packaging. Figure 5 presents the modified production process. The author reported an increase of 11% on the final cost of the probiotic cream cheese when compared with the regular product.

When producing ice cream with probiotics, cultures may be added in two ways, considering that they are of the DVS (Direct Vat Set) type for direct addition to the product during its manufacture: either adding them directly to the pasteurized mix or using the milk as a substrate for fermentation, producing frozen yoghurt ice cream (Cruz et al., 2009b).

Corrales et al. (2007) developed a process of ice cream adding *Bifidobacterium lactis* and *Lactobacillus acidophilus*. Figure 6 presents the followed steps for the product preparation. The frozen bacteria was dispersed in 1 L of pasteurized milk (2% fat content), and then added the milk to the ice cream mix with constant stirring.

In a similar way, free and encapsulated cells of *L.casei* and *B.lactis* were added to ice cream to evaluate the effect of microencapsulation and resistant starch on the probiotic survival (Homayouni et al., 2008). In general, the results indicated that encapsulation can significantly increase the survival rate of probiotic bacteria on ice cream over an extended shelf-life.

Functional ice creams have been produced by mixing fortified milk fermented with probiotic strains with an ice cream mix, followed by freezing (Salem et al., 2005). Probiotic ice cream has been also produced by the addition of probiotic yogurt to the mix prior the dynamic freezing-step (Soukoulis et al., 2010).

More recently, the effect of different overrun levels on probiotics survival on ice cream has been studied by Ferraz et al. (2012), incorporating *Lactobacillus acidophilus* into a vanilla flavored product. *L. acidophilus* was added to the mix with constant stirring just before freezing. Ice creams were processed with overruns of 45%, 60%, and 90%. Although all presented a minimum count of 1 x 10^6 CFU/g at the end of 60 days of frozen storage, higher overrun levels negatively influenced cell viability, being reported a decrease of 2 log units for the 90% overrun treatment. The authors suggest that lower overrun levels should be adopted during the manufacture of ice cream with probiotics in order to maintain its functional status through the shelf life.

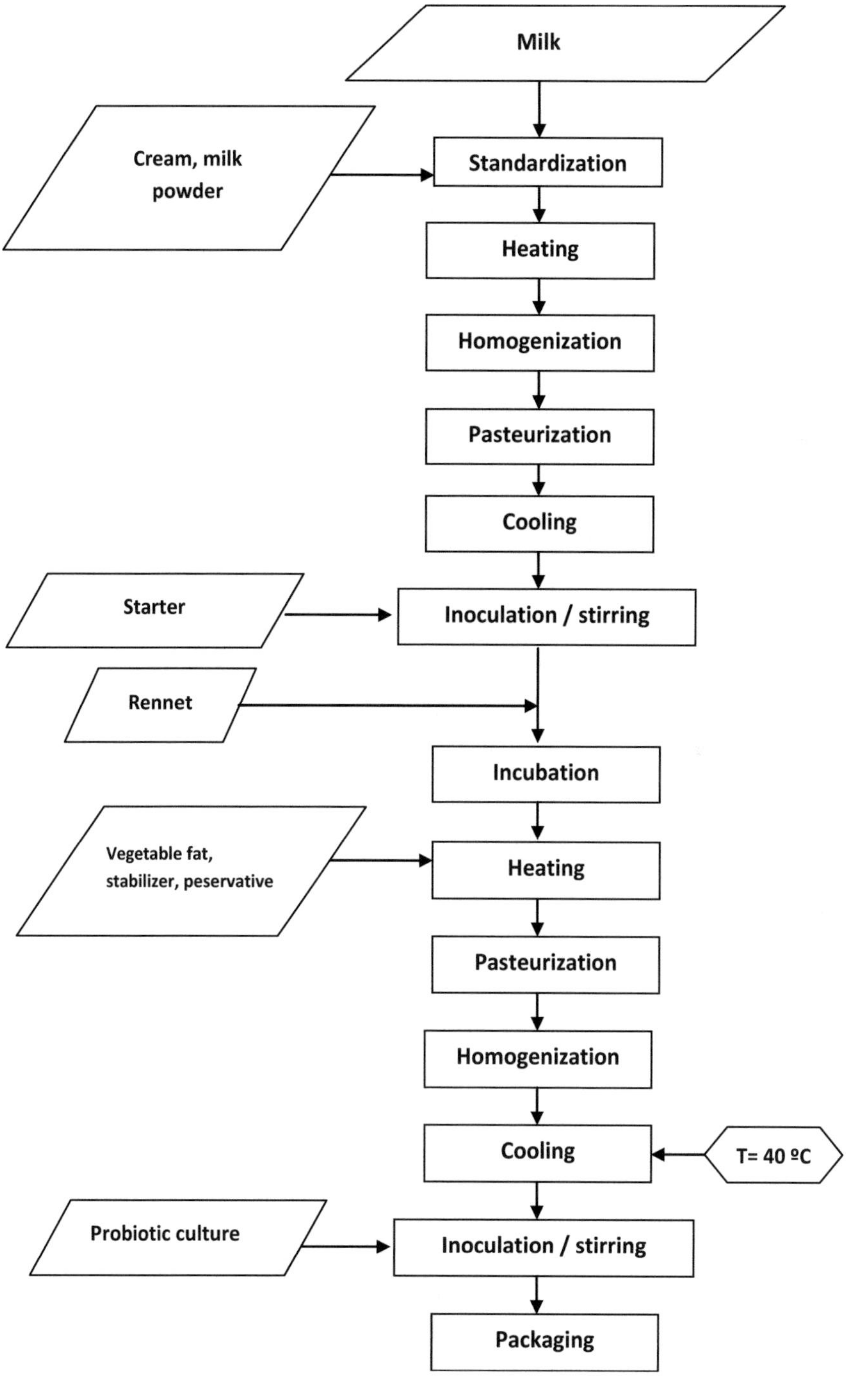

Figure 5. Production flow chart for Philadelphia type cheese with probiotics.

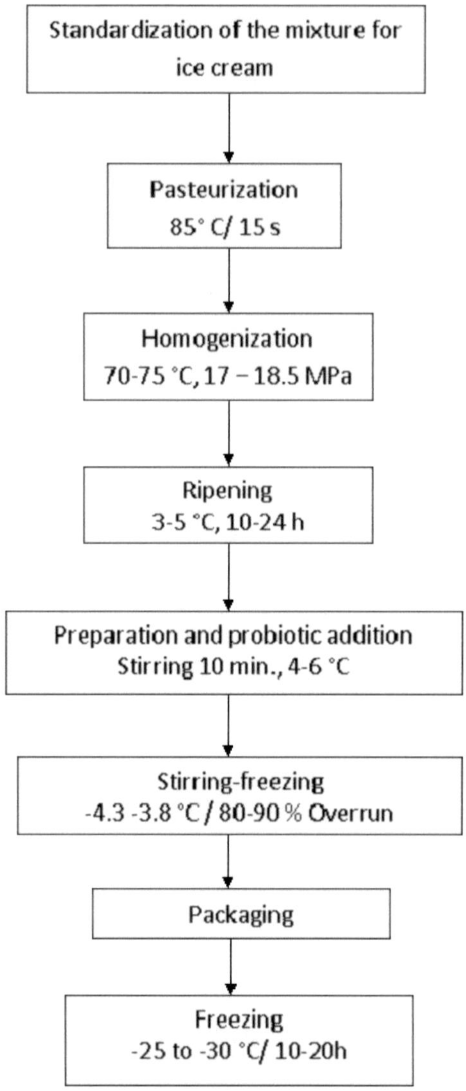

Figure 6. Production flow chart for ice cream with probiotics.

4. Quality modifications of products and sensory analysis

The products chosen for probiotic incorporation must be carefully studied, since the addition and/or multiplication of probiotic microorganisms could produce undesirable characteristics in the products (Dias and Mix, 2008; Komatsu et al., 2008). For many products the addition of probiotics may represent changes that significantly impact its physico-chemical properties, due to the metabolic activity of these living microorganisms and/or changes made on standard food processing procedures. Hence, careful selection of strains is necessary to minimize quality losses caused by alterations to flavor and texture of foods.

According to Champagne et al. (2005) many studies have shown that for some products the addition of probiotics do not lead to significant differences in the sensory properties, although changes in chemical composition and texture may occur these do not necessary have a relevant effect on flavor for some foods (depending on the extent of probiotic growth). This seems to be the case for fermented cheeses.

Natural cheeses are known for their complex microbial ecosystem which is in a constant state of flux as the cheese ages (Dias and mix, 2008). In general, a probiotic cheese should have the same acceptance as a conventional cheese: the incorporation of probiotic bacteria should not imply a loss of quality of the product. In this context, the level of proteolysis and lipolysis must be the same or even greater than cheese which does not have this functional status (Cruz et al., 2009a).

Buriti et al. (2005) evaluated the effect of *Lactobacillus acidophilus* on the instrumental texture profile and related properties of Minas fresh cheese (>65% water, w/w) during storage at 5°C up to 21 days. Parameters measured included hardness, elasticity, cohesiveness, chewiness and gumminess. Four cheese-making trials (T) were prepared, two supplemented with a mesophilic type O culture (T1, T2) and two with lactic acid (T3, T4). *L. acidophilus* was added in T2 and T3. Probiotic cheeses T3 were firmer by the end of storage, due to higher values of pH and hardness, and according to the authors also had better results in the sensory evaluation (preference-ranking test). Differences detected were attributed to the starter, rather than to *L. acidophilus*. In this study percentage of syneresis and the proteolytic index were also determined after the different storage times, finding no relevant differences.

For this same type of cheese, it was proved that the use of a probiotic culture (containing *L. acidophilus*, *B. animalis* and *S. thermophilus*) complementary to lactic acid, aiming to substitute tradicionally employed culture for Minas cheese production, is advantageous (Buriti et al., 2007). Cheeses with added probiotic culture showed to be less brittle and with more favorable sensory characteristics than those made with the traditional lactic acid culture. Researchers conducted an instrumental texture profile analysis of cheeses and a preference-ranking test.

In other study the influence of probiotic bacteria on proteolytic patterns and production of organic acid during ripening period of 6 months on Cheddar cheese at 4°C was evaluated (Ong et al., 2006). No significant differences (P>0.05) were observed in composition (fat, protein, moisture, salt content), but acetic acid concentration was higher in probiotic cheeses. The assessment of proteolysis during ripening showed no significant differences in the level of water-soluble nitrogen (primary proteolysis), but the concentration of free amino acids were significantly higher in probiotic cheeses (secondary proteolysis).

More recently, the survival and influence on sensory characteristics of probiotic strains of *Lactobacillus fermentum* and *Lactobacillus plantarum*, all derived from human faces, were investigated in Turkish Beyaz cheese production. Quantification of volatile aroma components by gas chromatography was performed as well as sensory evaluation. The results showed that tested probiotic culture mix was successfully used in cheese production without adversely affecting cheese quality during ripening. The chemical composition and sensory quality of probiotic cheeses were also comparable with traditional cheeses (Kılıc et al., 2009).

Arguedas (2010) analyzed the effect of adding *L. paracasei* subesp.*paracasei* in a Philadelphia type cheese (24% fat, w/w) on product texture during the shelf life. Table 4 shows the results obtained on hardness, cohesivity, adhesivity and gumminess (instrumental analysis) at day 2 and 44 for samples of regular and probiotic cheese at refrigerated storage (5ºC).

There was no significant difference (P> 0.05) in any parameter between regular and probiotic cream cheese although there was a variation as a function of time on hardness, cohesivity and gumminess for the samples analyzed. In general, these three parameters decreased along storage probably due to syneresis. Since there was no interaction between the time effect and the type of product effect, the decrease on these parameters is not related with the probiotic presence.

Treatment		Hardness (N)	Cohesivity	Adhesivity (erg)	Gumminess (N)
With probiotics	2 days	7,9970	0,3194	-141475,0	2,5964
	44 days	5,6058	0,2115	-120637,5	1,1735
Without probiotics	2 days	6,5627	0,2584	-139880,0	1,6967
	44 days	6,0673	0,2285	-115408,3	1,3882

Table 4. Philadelphia type cheese texture average values obtained during refrigerated storage at days 2 and 44 (Arguedas, 2010).

There was no significant difference (*P*> 0.05) in any parameter between regular and probiotic cream cheese although there was a variation as a function of time on hardness, cohesivity and gumminess for the samples analyzed. In general, these three parameters decreased along storage probably due to syneresis. Since there was no interaction between the time effect and the type of product effect, decreased on these parameters is not related with the probiotic presence.

Consumers rated taste liking degree for cheese during refrigerated storage (5ºC) at days 2, 16, 30 and 44. Figure 7 shows the average results for probiotic Philadelphia cheese type during this period of time. No significant differences (*P*>0.05) were found along shelf life considering taste liking degree for Philadelphia cheese type with *Lactobacillus paracasei* subsp. *paracase.* Average liking degree was 6.5.

Ice cream and ice milk appear to be good products for the delivery of probiotic bacteria. When the cream blend is prepared by adding a fermented milk, the resulting flavor of the product can be affected (Champagne et al., 2005; Cruz et al., 2009b). However, when small quantities of concentrated cultures are introduced, the sensory properties are not affected. Strain or species do seem to be important, since ice creams manufactured with *L. reuteri* cultures have shown to be "more sour" than those made from corresponding cultures of *L. acidophilus, L. rhamnosus,* or *B. bifidum* (Champagne et al., 2005). Also, products like non-fermented probiotic ice-cream will not normally present problems resulting from the microbial metabolism, since they are stored at very low temperatures, minimizing the probiotic microorganisms' biochemical reactions (Cruz et al., 2009b).

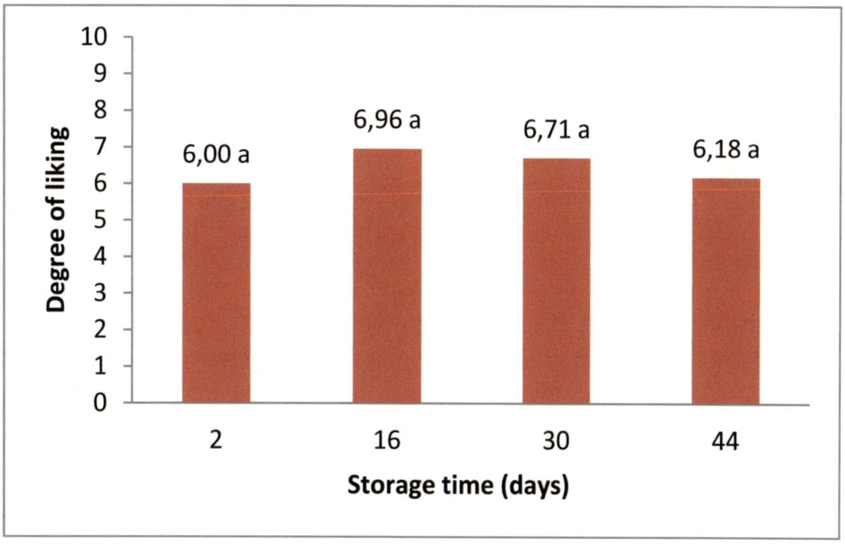

Figure 7. Consumers average taste liking degree of Philadelphia cheese type with *Lactobacillus paracasei* subsp. *paracasei* during storage (Arguedas, 2010). Different letters in the columns indicate significant differences (P<0,05).

Corrales et al. (2007) conducted a sensory evaluation of the ice cream flavor, using the duo-trio differentiation technique with 30 semi-trained panelists. It was found that 17 of the 30 semi-trained panelists were able to detect the sample that was equal to the pattern, indicating that no significant difference ($P > 0.05$) was found in the ice creams flavor with and without probiotics. This result supports the conclusion that the consumer did not detect changes in the flavor of ice cream, contributing to the product acceptance.

According to Soukoulis et al. (2010), probiotic ice cream is a functional frozen dairy dessert with particular sensory characteristics combining the flavor and taste of fermented milks with the texture of ice cream. In their study, the effects of compositional parameters (hydrocolloids type and amount, yogurt and milk fat content) on texture and flavor of a probiotic ice cream were evaluated. In such a product, the use of hydrocolloids like xanthan gum and low acidified formulations are recommended to improved creamy sensation, high textural quality and enhanced flavor. They found that based on hedonic and descriptive evaluation, consumers' acceptability of probiotic ice cream is mainly affected by ten sensory drivers including "sweet", "sour", "astringent", "vanilla flavor", "gummy", "coarse", "watery", "creamy", and "foamy".

The effect of several probiotic strains on the sensory acceptance of ice cream was evaluated by Salem et al. (2005). Probiotic ice cream was manufactured by mixing fortified milk fermented with probiotic strains with an ice cream mix. They found that all the ice cream samples received a high score in the sensory evaluation. Ice cream containing *Lactobacillus reuteri* was judged to be sourer and reached a higher score for "probiotic" flavor.

Two types of synbiotic ice cream containing 1% (w/w) resistant starch with free and encapsulated *Lactobacillus casei* and *Bifidobacterium lactis* were manufactured by Homayouni et al. (2008). The synbiotic ice cream samples were sensory assessed by 32 panelists. According to the authors, total evaluations in term of color, texture and taste of all samples were positive and did not have any marked off-flavor during the storage period. None of the ice creams were judged to be crumbly, weak, fluffy or sandy.

Finally, Ferraz et al. (2012) supplemented a vanilla ice cream with *Lactobacillus acidophilus* at different overrun levels (45%, 60%, and 90%). They did not report an influence for any overrun level (P>0.05) on acceptability regarding appearance, aroma, and taste of the ice creams.

Performing sensory evaluation is certainly an important step in probiotic dairy products development before the launch of the product into the market. As new products with probiotics may change some characteristics studying the behavior of trained panelists and consumers toward the developed product is a key factor and might represent a powerful tool to recover information that could support a product launch.

Another central issue in new probiotic products is to guarantee enough microorganism population in order to allow consumers to experience the beneficial effects described before. Probiotic quantification with an appropriate technique is a must in the product process development.

5. Probiotic quantification techniques

Proper selection of an analytical method for the probiotic microorganism's enumeration in food is critical since confirmation of whether the product has the minimum required amount of bacteria to provide the health benefits associated will depend on the result obtained.

The choice of culture medium and methodology for selective enumeration of commercial probiotic strains in combination with starters depends strongly on the product matrix, the target group and the taxonomic diversity of the bacterial background flora in the product (Van de Casteele et al., 2006). There is a wide variety of analysis methods that consider all these aspects and are extensively documented by various authors.

Several media have been suggested for the enumeration of probiotic bacteria alone or in combination in commercial cultures or products (Vinderola and Reinheimer, 2000). MRS agar is the media most commonly used and is normally supplemented with different sugars as maltose or glucose and with antibiotics solutions such as dicloxacillin, clindamycin, vancomycin, nalidixic acid, among many others. It is also common to add inhibitory agents as LiCl, NaCl, acids, bile salts and sorbitol. Supplements selection is made depending on the microorganism of interest and strains that wanted to be inhibited, for this purpose combination of both is very common. RCA agar with different antibiotics and salts is likewise used.

For *Bifidobacterium* sp. count, an incubation of plates under anaerobic conditions is required while Lactobacillus sp. strains can be recover both aerobically and anaerobically. Therefore one criterion for selecting the correct method is not only the strain of interest oxygen requirement but also accompanying flora characteristics. Similarly, temperature and incubation time varies between methods. Most of probiotic cultures are recovered at 37°C but increasing incubation temperature at 43°C is often use to inhibit mesophilic flora. Incubation times typically range from three to six days.

An important aspect to consider is that probiotic microorganisms viable cells amount should be kept at the minimum accepted level in order to be considered as a functional food during its entire shelf life. Therefore, in new product development probiotic bacteria count should be performed in fresh product and throughout shelf life. In many cases, shelf life of such products is determined as a function of time in which availability of minimum required concentration of probiotics can be guarantee.

In the scientific literature, populations of 10^6 - 10^7 CFU/g in the final product are established as therapeutic quantities of probiotic cultures in processed foods (Talwaker et al., 2004), reaching 10^8 - 10^9 CFU, provided by a daily consumption of 100 g or 100 ml of food, hence benefiting human health (Jayamanne and Adams, 2006). For example, in Brazil, the present legislation states that the minimum viable quantity of probiotic cultura should be between 10^8 and 10^9 CFU per daily portion of product and that the probiotic population should be stated on the product label (Brazilian Agency of Sanitary Surveillance, 2012).

6. Conclusion

The use of products like yogurt, fermented milks, different cheeses and ice cream as probiotic food carrier opened a valuable alternative for dairy industry. To meet consumers demand for probiotic foods in different countries, different types of products are needed. Research has demonstrated that is possible to incorporate successfully probiotics reaching the recommended amounts in order for consumers to experience the described health benefits. It is also possible to reach a reasonable shelf life according to the expected product characteristics.

From a technological point of view adding probiotics into dairy products could represent a difficult task depending on the type of product or microorganisms. Knowledge of all unit operations involved in processing and adaptations in traditional dairy process are helpful. Preliminary test to follow product and bacteria behavior provide useful information and sometimes it is necessary to change process parameters or inoculation step.

Proper techniques for population determination must be used to follow probiotic behavior during production and storage time and correctly predict shelf life. Performing physico-chemical analysis is decisive since characterization of product gives important information of probiotic effects and finally appropriate sensory techniques help to determine if attributes may have an influence on consumer acceptance. Since final product quality modifications could occur it is important to perform sensorial test with trained, semi-trained judges or directly with consumers at this stage. Results obtained in a product developing process are indeed specific for the product, microorganism or mixture of microorganisms and technology involved. It is not possible to generalize them to other products, strains or elaboration techniques.

Developing successful functional dairy food requires to be supported by scientific research. Product development in this field should consider knowing the consumer expectations, the technological process, the appropriate analyzing techniques and marketing. Nutrition advantages of dairy products need to be emphasized and information should be focused on consumers but also need to consider health care professionals.

Industry needs relevant regulation of physiological claims and health claims and nowadays some companies are performing clinical studies with particular strains to prove specific benefits but it is clear that production of functional dairy foods following the rules of medicine production is hardly of interest.

Considering the healthy population there may be potential to develop targeted products for different age groups. In the reduction of risk and treatments of various diseases, probiotics have resulting in promising benefits. However, it is important to understand the mechanisms behind the effects on our well-being. Information regarding the interaction between bacteria and dairy is focused on growth and survival of probiotics during production, storage and gastric transit therefore more research is needed to determine the effect of food substrate on metabolic activities of probiotics associated with their beneficial properties.

Author details

Esteban Boza-Méndez, Rebeca López-Calvo and Marianela Cortés-Muñoz*
National Research Center of Food Science and Technology (CITA),
University of Costa Rica, San José, Costa Rica

7. References

Akhiar, N.S. (2010). Enhancement of probiotics survival by microencapsulation with alginate and prebiotics. *Basic Biotechnology*, Vol.6, pp.13-18.

Arguedas, N. (2010). Determinación de la sobrevivencia del cultivo probiótico *Lactobacillus paracasei* subesp.*paracasei* (LC-01®) en un queso crema durante su almacenamiento y su influencia sobre la aceptación por consumidores, el pH, la textura y los costos variables. Thesis Lic. in Food Technology. Universidad de Costa Rica.

Aureli, P.; Capurso, L.; Castellazzi, A.M.; Clerici, M.; Giovannini, M.; Morelli, L.; Poli A.; Pregliasco, F.; Salvini, F.; Zuccotti, G.V. (2011). Probiotic and health: an evidence-based review. *Pharmacological Research*, Vol.63, No.5, pp. 366-376.

* Corresponding Author

Boza, E.; Morales, I.; Henderson, M. (2010). Development of mature cheese with the addition of the probiotic culture *Lactobacillus paracasei* subsp. *paracasei* Lc-01. *Revista Chilena de Nutrición*, Vol.37, No.2, pp.215-223.

Brazilian Agency of Sanitary Surveillance. (May 2012). Food with health claims, new foods/ingredients, bioactive compounds and probiotics. In: *Alimentos*, 05. 5. 2012, Available from http://www.anvisa.gov.br/alimentos/comissoes/tecno_lista_alega.htm

Buriti, F.; Da Rocha, J.; Saad, S. (2005). Incorporation of Lactobacillus acidophilus in Minas fresh cheese and its implications for textural and sensorial properties during storage. *International Dairy Journal*; Vol.15, pp.1279-1288.

Buriti, F.; Okazaki, T.; Alegro, J.; Saad, S. (2007). Effect of a probiotic mixed culture on texture profile and sensory performance of Minas fresh cheese in comparison with the traditional products. *Archivos Latinoamericanos de Nutrición*, Vol.57, No.2, pp. 179-185.

Champagne, C. (2008) Development of yoghurt and specialty milks containing probiotics. *Journal of Animal Science*, Vol.86, E-Suppl.2, pp.367-368.

Champagne, C.; Gardner, N.; Roy, D. (2005). Challenges in the addition of probiotic cultures to foods. *Critical Reviews in Food Science and Nutrition*, Vol.45, No.1, pp. 61-84.

Corrales, A.; Henderson, M.; Morales, I. (2007). Survival of probiotic microorganisms *Lactobacillus acidophilus* and *Bifidobacterium lactis* in whipped ice cream. Revista Chilena de Nutrición, Vol.34, No.2, pp.157-163.

Corriols, M. (2004). Determinación de la sobrevivencia del *Bifidobacterium lactis* en una natilla liviana durante el período de almacenamiento. Thesis Lic. in Food Technology. Universidad de Costa Rica.

Cruz, A.; Buriti, F.; Souza, C.; Faria, J.; Saad, S. (2009a). Probiotic cheese: health benefits, technological and stability aspects. *Trends in Food Science & Technology*, Vol.20, pp.344-354.

Cruz, A.; Antunes, A.; Sousa, A.; Faria, J.; Saad, S. (2009b). Ice-cream as a probiotic food carrier. *Food Research International*, Vol.42, pp.1233–1239.

Daigle, A.; Roy, D.; Belanger, D.; Vuillemard, J. (1999) Production of Probiotic Cheese (Cheddar Like Cheese) using enriched cream fermented by *Bifidobacterium infantis. Journal of Dairy Science*, Vol.82, pp.1081-1091.

Dias, B.; Mix, N. (2008) Probiotics in natural cheese. *Journal of Animal Science*, Vol.86, E-Suppl.2, pp.367-368.

Donkor, O.; Henriksson, A.; Vasiljevic, J.; Shah, N. (2006). Effect of acidification on the activity of probiotics in yoghurt during cold storage. *International Dairy Journal*, Vol.16, pp.1181-1189.

Ferraz, J.; Cruz, A.; Cadena, R.; Freitas, M.; Pinto, U.; Carvalho, C.; Faria, J.; Bolini, H. (2012). Sensory acceptance and survival of probiotic bacteria in ice cream produced with different overrun levels. *Journal of Food Science*, Vol. 71, No. 1, pp. 524-528.

Fortin, M.; Champagne, C.; St-Gelais, D.; Britten, M.; Fustier, P.; Lacroix, M. (2011). Effect of time of inoculation, starter addition, oxygen level and salting on the viability of probiotic cultures during Cheddar cheese production. *International Dairy Journal*, Vol.21, pp.75-82.

Gardiner, G.; Roos, R.P.; Collins, J.; Fitzgerald, G.; Stanton, C. (1998). Development of a probiotic Cheddar cheese containing human-derived *Lactobacillus paracasei* strains. *Applied and Environmental Microbiology*, Vol.64, No.6, pp.2192-2199.

Gardiner, G.; Bouchier, P.; O'Sullivan, E.; Kelly, J.; Collins, K.; Fitzgerald, G.; Ross, R.P.; Stanton, C. (2002). A spray-dried culture for probiotic Cheddar cheese manufacture. *International Dairy Journal*, Vol.12, pp.749-756.

Heller, J. (1998). Starterorganismen und Produktcharakteristika. *Deutsche Molkereizeitung*, Vol.12, pp.598-604.

Homayouni, A.; Azizi, A.; Ehsani, M.; Yarmand, M.; Razavi, S. (2008). Effect of microencapsulation and resistant starch on the probiotic survival and sensory properties of synbiotic ice cream. *Food Chemistry*, Vol.111, pp.50–55.

Jayamanne, V. S.; Adams, M. R. (2006). Determination of survival, identity, and stress resistance of probiotic bifidobacteria in bio-yoghurts. *Letters in Applied Microbiology*, Vol.42, No.3, pp. 189 -194.

Kılıc G.B, Kuleaşan H, Eralp I, Karahan A.G. (2009). Manufacture of Turkish Beyaz cheese added with probiotic strains. *LWT - Food Science and Technology*, Vol.42, pp.1003-1008.

Komatsu, T.; Buriti, F.; Saad, S. (2008). Inovação, persistência e criatividade superando barreiras no desenvolvimento de alimentos probióticos. *Brazilian Journal of Pharmaceutical Sciences*, Vol.44, No.3, pp. 329-347.

Kourkoutas, Y.; Xolias, V.; Kallis, M.; Bezirtzoglou, E.; Kanellaki, M. (2005) *Lactobacillus casei* cell immobilization on fruit pieces for probiotic additive, fermented milk and lactic acid production. *Process Biochemistry*, Vol.40, pp.411-416.

Liu, S-Q.; Tsao, M. (2009). Enhancement of survival of probiotic and non-probiotic lactic acid bacteria by yeasts in fermented milk under non-refrigerated conditions. *International Journal of Food Microbiology*, Vol.135, pp.34–38.

Lynch, C.; Muir, D.; Banks, J.; McSweeney, P.; Fox, P. (1999). Influence of adjunct cultures of *Lactobacillus paracasei* ssp. *paracasei* or *Lactobacillus plantarum* on Cheddar cheese ripening. *Journal of Dairy Science*, Vol.82, pp.1618-1628.

Ong, L.; Henriksson, A.; Shah, N. (2006). Development of probiotic Cheddar cheese containing *L. acidophilus, L. casei, L. paracasei* and *Bifidobacterium* sp. and the influence of these bacteria in proteolytic patterns and organic acid production. *International Dairy Journal*, Vol.16, pp.446-456.

Ranadheera R.D.C.S.; Baines, S.K.; Adams, M.C. (2010). Importance of food in probiotic efficacy. *Food Research International*, Vol.43, pp.1-7.

Rastall, R. A.; Fuller, R., Gaskins, H. R. (2000). Colonic functional foods. In: *Dairy Processing*, Gibson G.R., Williams C.M. (eds.), pp. 71-89, CRC Press, Boca Raton.

Ross, R.P.; Fitzgerald, G.F.; Collins, J.K.; Stanton, C. (2002). Cheese delivering biocultures-probiotic cheese. *Australian Journal of Dairy Technology*, Vol.57, No.2, pp.71-78.

Roy, D. (2005). Technological aspects related to the use of bifidobacteria in dairy products. *Lait*, Vol.85, pp.39-56.

Salem, M.; Fathi, F.; Awad, R. Production of probiotic ice cream. (2005). Polish Journal of Food and Nutrition Sciences, Vol.14, No.3, pp.267-271.

Saxelin, M.; Korpela, R.; Mäyrä-Mäkinen, A. Functional dairy products. In: *Dairy Processing. Improving Quality*, Smit, G. (ed.), pp.229-245, CRC Press, Boca Raton.

Saxelin, M. (2008). Probiotic Formulations and Applications, the Current Probiotics Market, and Changes in the Marketplace: A European Perspective. *Clinical Infectious Diseases*, Vol.46, pp.76-79.

Segura, N. (2005). Elaboración de queso fresco tipo Turrialba con probióticos. Thesis Lic. in Food Technology. Universidad de Costa Rica.

Shortt, C.; Shaw, D.; Mazza, G. (2003). Overview of opportunities for health-enhancing functional dairy products. In: *Handbook of Functional Dairy Products*, Shortt, C.; O'Brien, J. (eds), pp. 1-13, CR Press.

Siuta-Cruce, P.; Goulet, J. (2001). Improving probiotic survival rates. *Food Technology*, Vol.55, No.10, pp.36-42.

Songisepp, E.; Kullisaar, T.; Elias, P.; Hütt, P.; Zilmer, M.; Mikelsaar, M. (2004). A new probiotic cheese with antioxidative and antimicrobial activity. Journal of Dairy Science, Vol.87, pp.2017-2023.

Soukoulis, C.; Lyroni, E., Tzia, C. (2010). Sensory profiling and hedonic judgement of probiotic ice cream as a function of hydrocolloids, yogurt and milk fat content. *LWT-Food Science and Technology*, Vol.43, pp.1351-1358.

Stanton, C., Colette, D.; Coakley, M., Collins, K., Fitzgerald, G., Ross, P. (2003). Challenges facing development of probiotic-containing functional foods. In: *Handbook of Fermented Functional Foods*, Farnworth, E. (ed), pp.27-58, CR Press.

Talwaker, A.; Kailasapathy, K. (2004). A review of oxygen toxicity in probiotic yogurts: Influence on the survival of probiotic bacteria and protective techniques. *Comprehensive Reviews in Food Science and Food Safety*, Vol.3, pp.117-123.

Talwalkar, A.; Miller, C. W.; Kailasapathy, K.; Nguyen, M. H. (2004). Effect of packaging materials and dissolved oxygen on the survival of probiotic bacteria in yoghurt. *International Journal of Food Science and Technology*, Vol.39, No.6, pp.605 - 611.

Van de Casteele, S.; Vanheuverzwijn, T.; Ruyssen, T.; Van Assche, P.; Swings, J.; Huy, G. (2006). Evaluation of cultura media for selective enumeration of probiotic strains of lactobacilli and bifidobacteria in combination with yogurt or cheese starters. *International Dairy Journal*, Vol.16, pp. 1470-1476.

Vasiljevic, T., Shah N P. (2008). Probiotics –From Metchnikoff to bioactives. *International Dairy Journal*, Vol.18, pp.714-728.

Vinderola, C.; Prosello, W.; Ghiberto, D.; Reinheimer J. (2000). Viability of probiotic (Bifidobacterium, Lactobacillus acidophilus and Lactobacillus casei) and nonprobiotic microflora in Argentinian Fresco cheese. *Journal of Dairy Science*, Vol.83, pp.1905-1911.

Vinderola, C. G.; Reinheimer, J. A. (2000). Enumeration of Lactobacillus casei in the presence of *L. acidophilus*, bifidobacteria and lactic starter bacteria in fermented dairy products. *International Dairy Journal*, Vol.10, pp. 271-275.

Yilmaztekin, M.; Ozer, B.; Atasoy, F. (2004). Survival of *Lactobacillus acidophilus* LA-5 and *Bifidobacterium bifidum* BB-02 in white-brined cheese. *International Journal of Food Sciences and Nutrition*, Vol.55, No.1, pp.53-60.

Chapter 11

Milk and Dairy Products: Vectors to Create Probiotic Products

Gabriel-Danut Mocanu and Elisabeta Botez

Additional information is available at the end of the chapter

http://dx.doi.org/10.5772/50044

1. Introduction

The most important function of alimentation is represented by the assurance of human metabolic needs as well as wellbeing and satisfaction induced by sensorial characteristics of food. In the same time, by modulating some target functions of the body, the food components might have benefic psychological and physiological effects, beside the nutritional ones, already accepted.

In fact, food must contribute to health improving/protection and sustain systems of defence against different aggressions. We are situated at a new frontier of nutrition, in which the foods are evaluated by their biological potential and by their ability to reduce the risk of developing certain diseases. We can talk today about the fact that food for health represent an expanding field: *probiotic functional food*.

In essence, probiotic functional food are products that, by their biological active compounds and consumed in current diets, contribute to optimal human physical and psihycal health.

The appearance and development of functional probiotic food are the response of production field to the results of cellular and molecular biology field research, which demonstrates the implication of food components in proper functioning cellules and subcelular structures. The importance of these studies is essential in contemporaneous context in which the environment assaults by many ways the human body, fully stressing it's protection, adaption and equilibrium maintenance systems. By their specific action, the food components might contribute to the maintain the normal parameters of cellular edificium and of the human body equilibrium.

Nowadays we are assisting to an intensification of research in food – alimentation – health relationship field. The ideea that food might increase/defend health due to active biological components from it's composition conquers more and more acceptability in the scientific community and there are many publication in this field. Unlike the last years, the customers from many countries become more and more interested in health beneficial determined by alimentation, including probiotic functional food. In Romania, even before the adherence to UE, there were registered studies concerning manufacturing of probiotic functional foods, especially in dairy industry and explaining the induced benefits for health.

In this trend of food science are included some of the studies developed over the years by researchers from Galati Food Science and Engineering Faculty.

2. Probiotics: What are they?

2.1. Definitions

The name probiotic comes from the Greek *„pro bios"* which means *„for life"*. The history of probiotics began with the history of man; cheese and fermented milk were well known to the Greeks and Romans, who recommended their consumption, especially for children and convalescents. Probiotics are defined as the living microorganisms administered in a sufficient number to survive in the intestinal ecosystem. They must have a positive effect on the host [1].

The term „probiotic" was first used by [2] in 1965 to describe the „substances secreted by one microorganism that stimulate the growth of another". A powerful evolution of this definition was coined by [3] in 1974, who proposed that probiotics are „organisms and substances which contribute to intestinal microbial balance" [4]. In more modern definitions, the concept of an action on the gut microflora, and even that of live microorganisms disappeared [5] in 1998 defined probiotics as the „food which contains live bacteria beneficial to health", whereas [6] in 2001 defined them as „microbial cell preparations or components of microbial cells that have a beneficial effect on the health and well-being".

Some modern definitions include more precisely a preventive or therapeutic action of probiotics. [7] in 1997 for example, defined probiotics as „microorganisms which, when ingested, may have a positive effect in the prevention and treatment of a specific pathologic condition". Finally, since probiotics have been found to be effective in the treatment of some gastrointestinal diseases [6], they can be considered to be therapeutic agents. It is clear that a number of definitions of the term „probiotic" have been used over the years but the one derived by the Food and Agriculture Organization of the United Nations/World Health Organization [8] and endorsed by the International Scientific Association for Probiotics and Prebiotics [9] best exemplifies the breadth and scope of probiotics as they are known today:"live microorganisms which, when administered in adequate amounts, confer a health benefit on the host".

This definition retains historical elements of the use of living organisms for health purposes but does not restrict the application of the term only to oral probiotics with intestinal outcomes [10]. Despite these numerous theoretical definitions, however, the practical question arises whether a given microorganism can be considered to be a probiotic or not.

Some strict criteria have been proposed. [11] in 1992, for example, proposed the following parameters to select a probiotic: total safety for the host, resistance to gastric acidity and pancreatic secretions, adhesion to epithelial cells, antimicrobial activity, inhibition of adhesion of pathogenic bacteria, evaluation of resistance to antibiotics, tolerance to food additives and stability in the food matrix.

The probiotics in use today have not been selected on the basis of all these criteria, but the most commonly used probiotics are the strains of lactic acid bacteria such as *Lactobacillus, Bifidobacterium* and *Streptococcus* (*S. thermophilus*); the first two are known to resist gastric acid, bile salts and pancreatic enzymes, to adhere to colonic mucosa and readily colonize the intestinal tract [4, 12].

2.2. Properties of lactic acid bacteria

The lactic acid bacteria are generally defined as a cluster of lactic acid-producing, low %G+C, non-spore-forming, Gram-positive rods and cocci that share many biochemical, physiological, and genetic properties. They are distinguished from other Gram positive bacteria that also produce lactic acid (e.g., *Bacillus, Listeria,* and *Bifidobacterium*) by virtue of numerous phenotypic and genotypic differences. According to current taxonomy, the lactic acid bacteria group consists of twelve genera (table 1). All are in the phylum *Firmicutes,* Order, *Lactobacillales.* Based on 16S rRNA sequencing and other molecular techniques, the lactic acid bacteria can be grouped into a broad phylogenetic cluster, positioned not far from other low G +C Gram positive bacteria.

Five sub-clusters are evident from this tree, including: (1) a *Streptococcus-Lactococcus* branch (Family *Streptococcaceae*), (2) a *Lactobacillus* branch (Family *Lactobacillaceae*), (3) a separate *Lactobacillus-Pediococcus* branch (Family *Lactobacillaceae*); (4) an *Oenococcus-Leuconostoc-Weisella* branch (Family *Leuconostocaceae*), and (5) a *Carnobacterium-Aerococcus-Enterococcus-Tetragenococcus-Vagococcus* branch (Families *Carnobacteriaceae, Aerococcaceae,* and *Enterococcaceae*).

Seven of the twelve genera of lactic acid bacteria, *Lactobacillus, Lactococcus, Leuconostoc, Oenococcus, Pediococcus, Streptococcus,* and *Tetragenococcus,* are used directly in food fermentations. Although *Enterococcus* sp. Are often found in fermented foods (e.g., cheese, sausage, fermented vegetables), except for a few occasions, they are not added directly. In fact, their presence is often undesirable, in part, because they are sometimes used as indicators of fecal contamination and also because some strains may harbor mobile antibioticresistance genes.

Importantly, some strains of *Enterococcus* are capable of causing infections in humans. Likewise, *Carnobacterium* are also undesirable, mainly because they are considered as spoilage organisms in fermented meat products. Finally, species of *Aerococcus, Vagococcus,* and *Weisella* are not widely found in foods, and their overall significance in food is unclear.

2.3. Probiotics as functional foods

In the last decades consumer demands in the field of food production has changed considerably. Consumers more and more believe that foods contribute directly to their health [13, 14]. Today foods are not intended to only satisfy hunger and to provide necessary nutrients for humans but also to prevent nutrition-related diseases and improve physical and mental well-being of the consumers [15, 16].

Genus	Cell morphology	Fermentation route	Growth at		Growth in NaCl at		Growth at pH		Lactic acid isomer
			10ºC	45ºC	6.5%	18%	4.4	9.6	
Lactobacillus	rods	homo/hetero[4]	±[5]	±	±	-	±	-	D, L, DL[6]
Lactococcus	cocci	homo	+	-	-	-	±	-	L
Leuconostoc	cocci	hetero	+	-	±	-	±	-	D
Oenococcus	cocci	hetero	+	+	±	-	±	-	D
Pediococcus	cocci (tetrads)	homo	±	±	±	-	+	-	D, L, DL
Streptococcus	cocci	homo	-	+	-	-	-	-	L
Tetragenococcus	cocci (tetrads)	homo	+	-	+	+	-	+	L
Aerococcus	cocci (tetrads)	homo	+	-	+	-	-	+	L
Carnobacterium	rods	hetero	+	-	-	-	-	-	L
Enterococcus	cocci	homo	+	+	+	-	+	+	L
Vagococcus	cocci	homo	+	-	-	-	±	-	L
Weisella	coccoid	hetero	+	-	±	-	±	-	D, L, DL

[1]Adapted from [17]
[2]Adapted from [18]
[3]Refers to the general properties of the genus; some exceptions may exist
[4]Species of *Lactobacillus* may be homofermentative, heterofermentative, or both
[5]This phenotype is variable, depending on the species
[6]Some species produce D-, L-, or a mixture of D- and L-lactic acid.

Table 1. Genera of lactic acid bacteria and their properties [1, 2, 3]

In this regard, functional foods play an outstanding role. The increasing demand on such foods can be explained by the increasing cost of healthcare, the steady increase in life expectancy, and the desire of older people for improved quality of their later years [19, 15, 20].

The term "functional food" itself was first used in Japan, in the 1980s, for food products fortified with special constituents that possess advantageous physiological effects [21, 22]. Functional foods may improve the general conditions of the body (e.g. pre- and probiotics), decrease the risk of some diseases (e.g. cholesterol-lowering products), and could even be used for curing some illnesses.

The European Commission's Concerted Action on Functional Food Science in Europe (FuFoSE), coordinated by International Life Science Institute (ILSI) Europe defined functional food as follows: "a food product can only be considered functional if together with the basic nutritional impact it has beneficial effects on one or more functions of the human organism thus either improving the general and physical conditions or/and decreasing the risk of the evolution of diseases. The amount of intake and form of the functional food should be as it is normally expected for dietary purposes. Therefore, it could not be in the form of pill or capsule just as normal food form" [23].

European legislation however, does not consider functional foods as specific food categories, but rather a concept [22, 24]. Therefore, the rules to be applied are numerous and depend on the nature of the foodstuff. Functional foods have been developed in virtually all food categories. From a product point of view, the functional property can be included in numerous different ways as it can be seen in table 2.

Type of functional food	Definition	Example
Fortified product	A food fortified with additional nutrients	Fruit juices fortified with vitamin C
Enriched products	A food with added new nutrients or components not normally found in a particular food	Margarine with plant sterol ester, probiotics, prebiotics
Altered products	A food from which a deleterious component has been removed, reduced or replaced with another substance with beneficial effects	Fibers as fat releasers in meat or ice cream products
Enhanced commodities	A food in which one of the components has been naturally enhanced through special growing conditions, new feed composition, genetic manipulation, or otherwise	Eggs with increased omega-3 content achieved by altered chicken feed

Table 2. Prominent types of functional food [20, 25, 26]

It should be emphasized however, that this is just one of the possible classifications. According to alternative classification, some functional products are (1) "add good to your life", e.g. improve the regular stomach and colon functions (pre- and probiotics) or "improve children's life" by supporting their learning capability and behaviour. It is difficult, however to find good biomarkers for cognitive, behavioural and psychological, functions. Other group (2) of functional food is designed for reducing an existing health risk problem such as high cholesterol or high blood pressure. A third group (3) consists of those products, which "makes your life easier" (e.g. lactose-free, gluten-free products) [27].

These products have been mainly launched in the dairy-, confectionery-, soft-drinks-, bakery- and baby-food market [16, 20, 26].

3. Health benefits of probiotics

Since Metchnikoff's era, a number of health benefits have been contributed to products containing probiotic organisms. While some of these benefits have been well documented and established, others have shown a promising potential in animal models, with human studies required to substantiate these claims. More importantly, health benefits imparted by probiotic bacteria are very strain specific; therefore, there is no universal strain that would provide all proposed benefits, not even strains of the same species. Moreover, not all the strains of the same species are effective against defined health conditions. Some of these strain specific health effects are presented in figure 1.

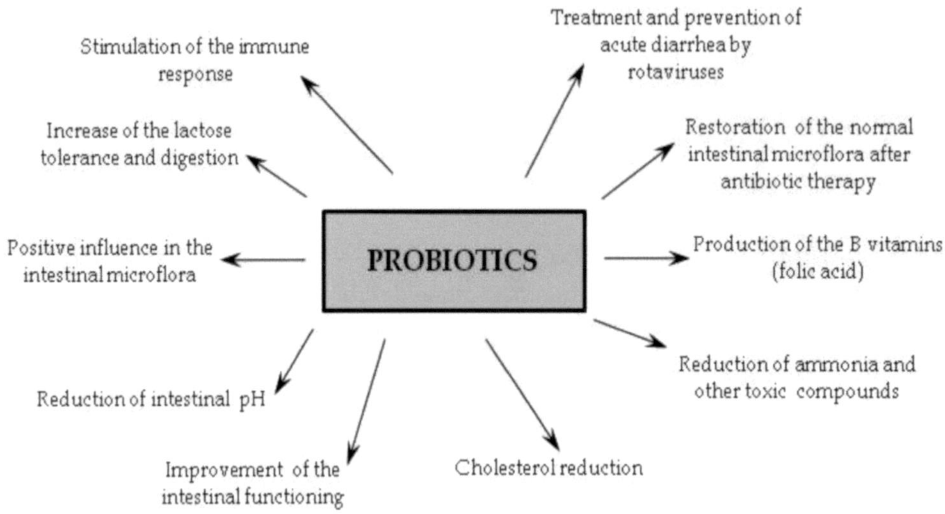

Figure 1. Probiotic beneficial effects on human health [28, 29]

4. Probiotic dairy products

Foods that affect specific functions or systems in the human body, providing health benefits beyond energy and nutrients—functional foods—have experienced rapid market growth in recent years. This growth is fueled by technological innovations, development of new products, and the increasing number of health-conscious consumers interested in products that improve life quality. Since the global market of functional foods is increasing annually, food product development is a key research priority and a challenge for both the industry and science sectors. Probiotics show considerable promise for the expansion of the dairy industry, especially in such specific sectors as yogurts, cheeses, beverages, ice creams, and other desserts. This book chapter presents an overview of functional foods and strategies for their development, with particular attention to probiotic dairy products.

4.1. Types of probiotic dairy product

The most common probiotic dairy products worldwide are various types of yogurt, other fermented dairy product, various lactic acid bacteria drinks and mixture of probiotic (fermented) milks and fruit juice. Probiotic cheese, both fresh and ripened, have also been launched recently. In table 3 are listed some dairy functional food products that have been developed recently in Faculty of Food Science and Engineering.

4.1.1. Fermented milks and beverages

Fermented milks and beverages make up an important contribution to the human diet in many countries because fermentation is an inexpensive technology, which preserves the food, improves its nutritional value and enhances its sensory properties.

Type of dairy functional food products	Description/Name
Fermented dairy products	Drink yogurts with La-5 and carrot juice/ BIOCOV
	Yogurt with La-5 and biomass of *Spirulina platensis*/YLaSP
	Yogurt with BB 12 and biomass of *Spirulina platensis*/YBbSP
	Yogurt with ABY 3/ABT 5 and medicinal plant extracts/ AFINOLACT
	Yogurt with ABY 3/ABT 5 and medicinal plant extracts/ CATINOLACT
	Yogurt with ABY 3/ABT 5 and medicinal plant extracts/ ROSALACT
Cheeses	Dessert based on fresh cheese and some fruit pulp
	Appetizer – type fresh cheese
	Probiotic Telemea cheese

Table 3. Some examples of dairy probiotic products developed

[30, 31] in 2011, proposed the realization of a probiotic dairy drink with added carrot juice. This probiotic product was obtained using goat milk (fat = 3.63%, proteins = 3.05%, lactose = 4.55%, dry matter = 12.05% and density = 1.030 g·mL^{-1}) which has been pasteurized at a temperature of 72ºC, for 20 minutes, a probiotic culture type Nutrish containing *Lactobacillus acidophilus* La-5 and carrot juice (dry matter = 9.35%, pH = 6.23, titratable acidity = 0.14 malic acid/100g, ash = 0.7%). After pasteurization, milk was quickly cooled to inoculation temperature at 37°C. The incubation of obtained fermented dairy drink was made at 37°C for 5 hours.

The addition of carrot juice (at a percentage of 10%) had a positive effect on physical – chemical and microbiological parameters of fermented dairy drink. Combining goat milk with carrot juice can get some food with potential therapeutic role.

As a result of the lactose fermentation, the titratable acidity increased fast during the incubation period. At the end of the storage period (after 5 days), the highest value of titratable acidity was 61 ºT. The pH of the obtained new product decreased during incubation period, and will stabilize during storage period, pH = 5.1 after 5 days of storage. The evolution of the number of microorganisms was analyzed for each sample during incubation and storage period. It was observed that the fermented dairy drink with added carrot juice product had been preserving its functional properties during storage (over 10^8 cfu·mL^{-1} probiotic bacteria).

The products were analyzed in terms of fluid flow thus establishing their rheological behavior. The literature shows that the rheological properties of fermented dairy products depend on the development of lactic bacteria as a consequence of metabolic changes leading physicochemical substrate in milk.

In figure 2 is presented the variation of shearing stress (τ, Pa) according to the shearing rate ($\dot{\gamma}$, s^{-1}). There was determined that samples have a rheological behavior similar with the one of the non-Newtonian fluids, time independent, therefore a pseudoplastic behavior. Specific for a fluid with this type of behavior is the flow resistance decrease as a result of the fluid shearing rate increase.

For all samples, it was noted that for low values of shear rate, tangential shear stress variation depending on shear rate was increasing (regression coefficient R^2 values varies from 0.962 and 0.995).

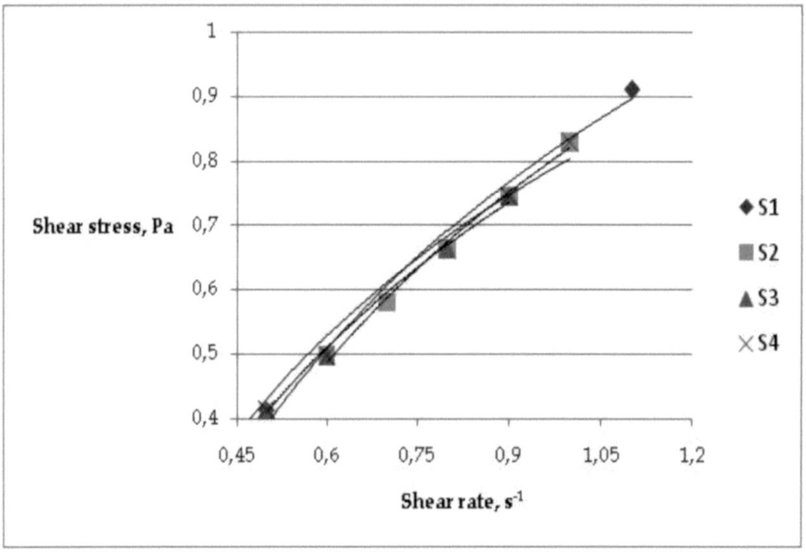

Figure 2. The shearing stress variation according to the shearing rate

To obtain yoghurt with *Spirulina platensis* biomass was used pasteurized cow milk (non fat dry matter = 9.08%, fat = 1.5%, proteins = 3.52%, lactose = 4.32%, mineral salts = 0.72%). Pasteurization of milk is achieved by maintaining standardized milk at 95 °C for 5 minutes. After pasteurization, milk was cooled to inoculation temperature at 42°C.

The inoculation of milk for obtaining these fermented dairy products is with a probiotic culture containing *Lactobacillus acidophilus* La-5 respectively *Bifidobacterium lactis* BB 12, at this time was added and biomass of *Spirulina platensis* (0.5 – 1% according to [32]).

After inoculation follows the distribution and packaging and incubation was made at 42°C for 6 hours in the thermostats set at the optimal temperature for the development of these bacteria. Meanwhile yoghurt gel gets a specific consistency. Cooling and storage of obtained yoghurts is performed at 6 °C for 15 days. In this storage period, coagulum is more compact, the flavor and taste become more pleasant. As a result of the lactose fermentation, the titratable acidity increased. This is slightly higher for the samples with La 5 from those with BB12.

All products with Spirulina platensis biomass have titratable acidity higher than control sample (1.1 times higher for samples with BB 12 and 1.2 times higher for samples with La 5).

The evolution of pH is correlated with lactose fermentation intensity and increased with titratable acidity, but in the same time it is influenced by the buffer substances that are found in *Spirulina platensis* biomass or formed during the manufacture of yoghurt. The pH of fermented dairy products fall between the values 4.11 and 4.53, values considered normal for such products.

The addition of *Spirulina platensis* biomass (figure 3) has positively influenced the number of viable probiotic microorganisms.

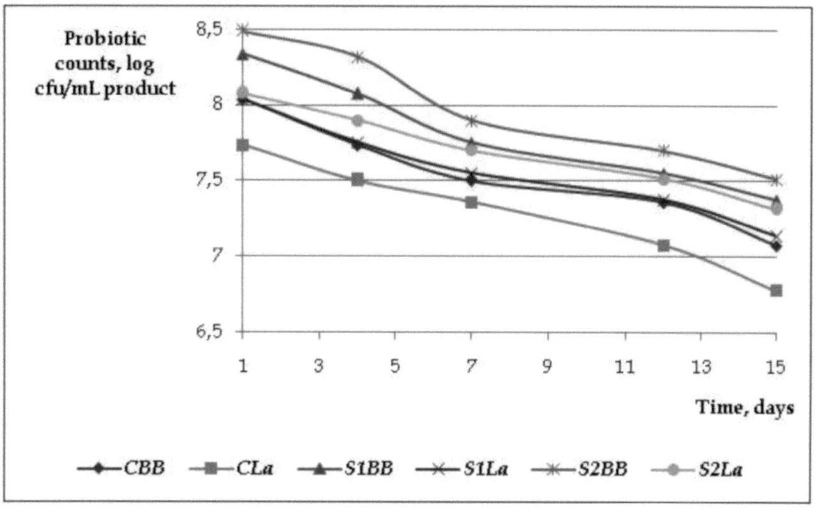

Figure 3. Viable counts variation during storage period

At the end of the storage period (after 15th days) the number of probiotic lactic bacteria for both, control samples and samples with *Spirulina platensis* biomass is still high, which shows that the product with *Spirulina platensis* biomass has been preserving its functional properties during storage period.

4.1.2. Cheeses

Perhaps no other fermented food starts with such a simple raw material and ends up with products having such an incredible diversity of color, flavor, texture, and appearance as does cheese. It is even more remarkable that milk, pale in color and bland in flavor, can be transformed into literally hundreds of different types of flavorful, colorful cheeses by manipulating just a few critical steps.

Just what happened to cause the milk to become transformed into a product with such a decidedly different appearance, texture, and flavor? To answer that question, it is first necessary to compare the composition of the starting material, milk, to that of the product, the finished cheese (figure 4).

In an attempt to diversify the range of probiotic dairy products, there has been made a series of research on the introduction of probiotic bacteria in cheese. According to [33], cheese is an

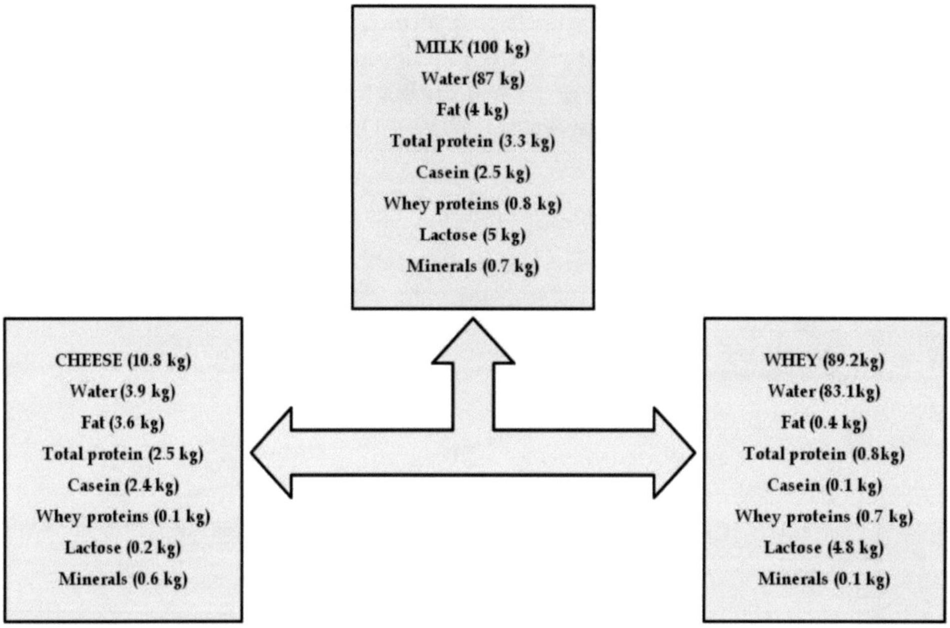

Figure 4. Partition of milk into cheese and whey (adapted from [18]).

interesting way of supplying probiotic bacteria due to the chemical composition of the raw milk that encourages their growth, metabolism and viability and also due to their relatively low acidity compared to other food products. The most of research has been focused on fresh cheese, but there are published some results on probiotic brined or ripened cheese, too.

Fresh cheese, mixt coagulated, is the most suitable cheese to carry probiotic bacteria, due to the high composition of nutrients, low acidity and low salt content. In 2009, [34] used probiotic fresh cheese and peach pulp in order to obtain a dessert, according to figure 5. Probiotic bacteria, *Lactobacillus acidophilus* La 5, was introduced in the fresh cheese as an agent of milk maturation, during coagulation stage. The product was rich in nutritive components (proteins: 10.9...11.3%; fat: 9.1...10.4% and minerals: 2...2.3%) and has a pseudoplastic rheological behaviour. This influenced the sensorial properties of the product, which achieved a creamy texture including in its structure the minced peach pulp and fat globules from the cream.

The research of the above mentioned authors continued, in the attempt to obtain a similar product using goat milk [35]. The amount of nutrients increased, comparing to the previous product (proteins: 12.4...12.5%; fat: 10.1...12.2% and minerals: 2.1...2.4%) but the rheological behaviour was not affected. Although there was expected a reserved attitude of the consumer because of the unpleasant flavour of goat milk, this was not observed.

In 2010 a new probiotic product based on fresh cheese was obtained, by mixing fresh cheese with caraway, cream and salt. The probiotic bacteria (*Bifidobacterium lactis* BB 12) were introduced in cheese at milk maturation stage. In figure 6 it can be observed that the caraway favourised the development of probiotic bacteria.

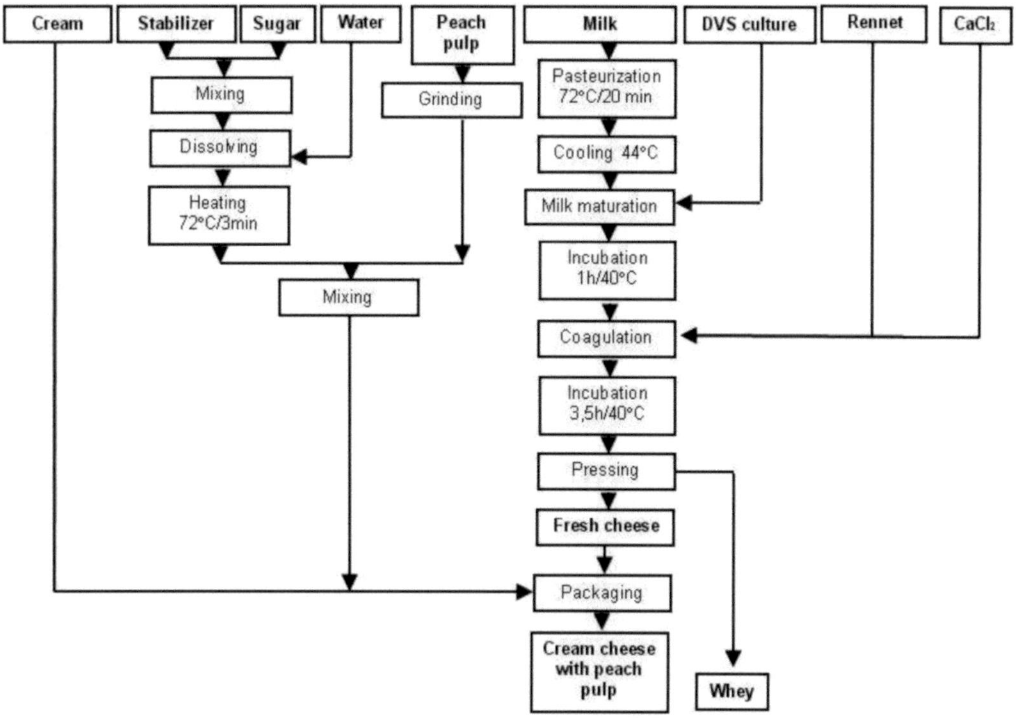

Figure 5. Technological flowchart for manufacturing the new product – Dessert based on fresh cheese and peach pulp

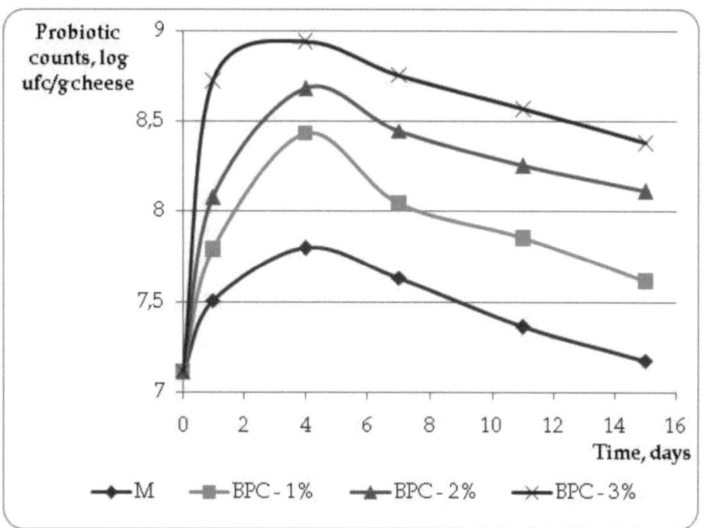

Figure 6. Evolution of bacteria during storage period

[36] and [37] studied the viability of probiotic bacteria *Bifidobacterium lactis, Lactobacillus acidophilus* and *Strepococcus thermophilus* in Telemea cheese during ripening and storage time. Telemea is a cheese variety originated in Romania, from where its manufacture spread

to other Balkan countries and Turkey [38]. The specific of this variety of cheese is ripening in brine. Evolution of probiotic bacteria during different stages of manufacturing process is presented in table 4. Conclusion of the study is that Telemea cheese can be considered a probiotic product, even if the high salt concentration disadvantages probiotic bacteria growth, as long as the number of viable cells remains above 10^7 cfu·g^{-1}.

Stage	*Bifidobacterium lactis*	*Lactobacillus acidophilus*	*Lactobacillus bulgaricus*	*Streptococcus thermophilus*
Inoculated milk	2.71	2.50	2.90	8
Milk, 10 minutes after renneting	2.68	3.21	4	12
Coagulum, before pressing	2.60	7.49	9.40	24.9
Pressed coagulum	2.40	7.29	9.50	24.3
Salted coagulum	2.20	7.20	9.60	22.1
Cheese ripened for 5 days	1.84	6.90	8.90	15
Cheese ripened for 10 days	1.75	6.36	8.40	14.2
Cheese ripened for 15 days	1.65	4.67	7.30	13.4
Cheese ripened for 20 days	1.57	4.21	6.70	12.6
Cheese ripened for 25 days	1.48	3.48	5.30	11.3
Cheese ripened for 30 days	1.41	2.90	4.20	9.1
Cheese ripened for 35 days	0.99	2.31	3.90	7.2
Cheese ripened for 40 days	0.60	2.21	3.50	6

Table 4. Evolution of probiotic bacteria during manufacturing of Telemea cheese (10^7 cfu·g^{-1})

There are registered many other studies about probiotic cheese and methods of manufacturing probiotic cheese. Most of them introduce probiotic bacteria in the milk maturation stage, but there are reports about introducing them after pressing [39] or immobilized on fruit pieces [40].

5. Improvement of benefical effect of probiotic dairy products through the use of bioactive compounds from plants

By sensorial analysis of several combinations milk-medicinal plants, as well as by physical and chemical analysis, there were selected the following medicinal plants: bilberry, sea-buckthorn, rosehip, liquorice, plants rich in active principles considered important for their pharmacological profile.

The research presented in this subchapter was realised on 14 variants of probiotic products (encoded according to table 5), manufactured from cow milk and medicinal plant extracts (bilberry, seabuckthorn, rosehip and liquorice), fermented by two types of probiotic cultures: ABY 3 (*Bifidobacterium lactis, Lactobacillus acidophilus, Lactobacillus delbrueckii* subsp. *bulgaricus* and *Streptococcus thermophilus*) and ABT 5 (*Bifidobacterium lactis, Lactobacillus acidophilus* and *Streptococcus thermophilus*).

Crt. No.	Product	Code	Culture	Description
1.	Control	M – 3	ABY 3	Milk + 0.02% DVS culture
2.	Afinolact	A – 3	ABY 3	Milk + 0.02% DVS culture + 6% bilberry extract
3.	Afinolact	LD+A – 3	ABY 3	Milk + 0.02% DVS culture + 6% bilberry extract + 6% liquorice extract
4.	Cătinolact	C – 3	ABY 3	Milk + 0.02% DVS culture + 6% seabuckthorn extract
5.	Cătinolact	LD+C – 3	ABY 3	Milk + 0.02% DVS culture + 6% seabuckthorn extract + 6% liquorice extract
6.	Rosalact	Mă– 3	ABY 3	Milk + 0.02% DVS culture + 6% rosehip extract
7.	Rosalact	LD+Mă – 3	ABY 3	Milk + 0.02% DVS culture + 6% rosehip extract + 6% liquorice extract
8.	Control	M – 5	ABT 5	Milk + 0.02% DVS culture
9.	Afinolact	A – 5	ABT 5	Milk + 0.02% DVS culture + 6% bilberry extract
10.	Afinolact	LD+A – 5	ABT 5	Milk + 0.02% DVS culture + 6% bilberry extract + 6% liquorice extract
11.	Cătinolact	C – 5	ABT 5	Milk + 0.02% DVS culture + 6% seabuckthorn extract
12.	Cătinolact	LD+C – 5	ABT 5	Milk + 0.02% DVS culture + 6% seabuckthorn extract + 6% liquorice extract
13.	Rosalact	Mă – 5	ABT 5	Milk + 0.02% DVS culture + 6% rosehip extract
14.	Rosalact	LD+Mă – 5	ABT 5	Milk + 0.02% DVS culture + 6% rosehip extract + 6% liquorice extract

Table 5. Experimental variants

The addition of aqueous medicinal plants extract positively influenced the number of viable probiotic microorganisms. At the end of storage period the number of probiotic lactic bacteria is high for both control samples and samples with medicinal plants, meaning that the products maintain their functional character [31, 41-46].

For the obtain products there was demonstrated the cytoprotective character, by studing the total antioxidant capacity, the total content of polyphenols, the superoxiddismutasic (SOD) activity, the minerals content, the ascorbic acid and anthocyaninis.

The results of the study reveal that probiotic dairy products with added medicinal plants contain a high level of total polyphenols and a high total antioxidant capacity. All these products are an excellent source of minerals with high biodisponibility in human diet. The addition of medicinal plants extract improved the SOD activity.

5.1. Identification of bioactive compounds from plants

Medicinal plants are extremely valuable biological currants. The valorification of this potential represents a never-ending source of raw materials for pharmaceutics and food industry. World Health Organisation has recently announced that 75-80% of world's population is treated with natural remedies.

The plants do not cure all the diseases but they might be extremely helpful in rational treating of some diseases and are not to dangerous. The plants have favourable effect to human body and unfavourable effect to some pathogen agents due to certain substances from their composition. In every plant species there must be known that substance or substances which assure them the therapeutic effect (the active principles).

In order to test the chemical composition of studied plants (bilberry, sea-buckthorn, rosehip and liquorice) there were determined by chemical analysis: ascorbic acid (for seabuckthorn and rosehip), glycyrrhizic acid (for liquorice) and anthocyaninis (for bilberry). The concentrations of the active principles in medicinal plants samples are reported in table 6.

Medicinal plants / Active principles	Bilberry (*Vaccinium myrtillus L.*)	Sea-buckthorn (*Hippophaë rhamnoides L.*)	Liquorice (*Glycyrrhiza glabra L.*)	Rosehip (*Rosa canina L.*)
Anthocyanins expressed as cyanidin-3-glucoside chloride	0.38 ± 0.06* (0.32÷0.47)	-	-	-
Ascorbic acid	-	0.8 ± 0.09* (0.66÷0.89)	-	1.24 ± 0.06* (1.18÷1.32)
Glycyrrhizic acid	-	-	5.03 ± 0.32* (4.6÷5.32)	-

The values were expressed in mean ± standard errors of regression and values in parenthesis indicate minimum and maximum level recorded.

Table 6. Active principles in medicinal plants

Regarding the active principles content there was demonstrated that all the analysed medicinal plants respect the values presented in European Pharmacopee, V edition: minimum 0.3% cyanidin–3–glucozide chloride in bilberry, the analysed probes having a maximum content of 0.47%. The ascorbic acid in seabuckthorn must be minimum in 0.5% and in rosehip minimum 1%. The analysed samples registered values of 0.66÷0.98% for seabuckthorn and 1,18% for rosehip. The glycyrrhizic acid, the main active principle in liquorice, must be minimum 4% (according to European Pharmacopee) and the determined values varied between 4.6 and 5.38% [31, 41-46].

5.2. The action of bioactive compounds from plants on probiotic bacteria metabolism

To have a probiotic effect, the strains of probiotic bacteria must be present in the product enough number. It is generally considered that the daily dose of probiotic strains must be between $1 \cdot 10^8$ and $1 \cdot 10^9$ cells. A portion of 100 g, the probiotic product should contain between 10^6 and 10^7 cfu·mL^{-1} product. The addition of aqueous medicinal plant extracts has positively influenced the number of viable probiotic microorganisms due to the presence of fermentable sugars and some growth factors (mineral salts, non-protein nitrogen). At the end of incubation period (after 5 hours), for the samples with medicinal plant extracts the lowest number of microorganisms has established for the samples: Mă–3 ($7.8 \cdot 10^8$ cfu·mL^{-1} probiotic bacteria) or A–5 and C–5 ($1.8 \cdot 10^9$ cfu·mL^{-1} probiotic bacteria) instead the higher number of probiotic bacteria was registered for LD+Mă–3 ($4.5 \cdot 10^9$ cfu·mL^{-1} probiotic bacteria) and LD+Mă–5 ($5.4 \cdot 10^9$ cfu·mL^{-1} probiotic bacteria).

After 8th days of storage period (table 7) the higher number of viable microorganisms was found in the sample with ABT 5 culture (LD+A–5: $2.6 \cdot 10^8$ cfu·mL^{-1} probiotic bacteria), and lowest number of probiotic bacteria cells was recorded for sample with ABY 3 culture (C–3: $0.9 \cdot 10^8$ cfu·mL^{-1} probiotic bacteria). The storage in refrigerated conditions causes a reduction of the number of probiotic bacteria up to 4.37 times in the products obtained with ABY 3 culture and 3.62 times for those with ABT 5.

Microbiological characteristics			
Sample code	cfu·mL^{-1} product	Sample code	cfu·mL^{-1} product
M – 3	$3 \cdot 10^7$	M – 5	$8 \cdot 10^7$
A – 3	$1 \cdot 10^8$	A – 5	$1.4 \cdot 10^8$
LD+A – 3	$1.8 \cdot 10^8$	LD+A – 5	$2.6 \cdot 10^8$
C – 3	$9 \cdot 10^7$	C – 5	$1.1 \cdot 10^8$
LD+C – 3	$1.9 \cdot 10^8$	LD+C – 5	$1.8 \cdot 10^8$
Mă – 3	$1.2 \cdot 10^8$	Mă – 5	$1.3 \cdot 10^8$
LD+Mă – 3	$2.2 \cdot 10^8$	LD+Mă – 5	$2.1 \cdot 10^8$

Table 7. Microbiological characteristics of fermented dairy products wit ABY 3 or ABT 5 after 8th days of storage

At the end of storage period, the number of probiotic lactic acid bacteria for both control samples and for samples with medicinal plant extracts is still high ($1{\cdot}10^7{\div}1{\cdot}10^8$ $cfu{\cdot}mL^{-1}$ probiotic bacteria), which shows that the products has been preserving its functional properties during storage period. Both cultures can be used in the production of probiotic products [31, 41-46].

Besides the cytoprotective effect conferred by the presence of probiotic bacteria, research has shown that products with added medicinal plants have a increased cytoprotective nature and because the content of biologically active compounds. Experimental results showed that the probiotic fermented dairy product with added medicinal plant extracts have a high content of total polyphenols with beneficial effects on human health, which help to prevent various diseases, such as cardiovascular disease, diabetes [47, 48] and consequently a higher total antioxidant capacity.

The higher amount of total polyphenols (table 8) was determined for samples: LD+Mă–3 (280.78 $\mu g{\cdot}mL^{-1}$) or LD+Mă–5 (285.56 $\mu g{\cdot}mL^{-1}$).

Total polyphenols expressed as catechin, $\mu g{\cdot}mL^{-1}$			
Sample code	ABY 3	Sample code	ABT 5
M – 3	62.086	M – 5	82.086
A – 3	99.91	A – 5	106
LD+A – 3	152.95	LD+A – 5	158.6
C – 3	72.086	C – 5	87.73
LD+C – 3	135.56	LD+C – 5	147.3
Mă – 3	262.95	Mă – 5	239.47
LD+Mă – 3	280.78	LD+Mă – 5	285.56

Table 8. The total polyphenols content for samples with ABY 3 or ABT 5 culture

Compared with control samples (not containing medicinal plant extracts) total antioxidant capacity (Table 9) increased by 3.25-9.94 times in products made with ABY 3 culture and 2.1-8.3 times the ABT 5 products.

TEAC, $mM{\cdot}L^{-1}$			
Sample code	ABY 3	Sample code	ABT 5
M – 3	0.16	M – 5	0.2
A – 3	0.57	A – 5	0.43
LD+A – 3	0.70	LD+A – 5	0.73
C – 3	0.52	C – 5	0.48
LD+C – 3	0.81	LD+C – 5	0.62
Mă – 3	1.19	Mă – 5	1.27
LD+Mă – 3	1.59	LD+Mă – 5	1.66

Table 9. Total antioxidant capacity for sample with ABY 3 or ABT 5 culture

For products manufactured with ABY 3 culture and those with ABT 5 was observed that the total antioxidant capacity and total polyphenols content is higher for mixtures with liquorice extract, from the rest of the samples tested, except for samples Mă–3 şi Mă–5.

In addition to being an excellent source of protein, probiotic dairy products based on milk and medicinal plant extracts are a good source of minerals, calcium, potassium, phosphorus, magnesium, zinc. The minerals in these products are from raw milk, and medicinal plant extracts. Because the extracts of bilberry, sea-buckthorn, rosehip and liquorice have different mineral content, the products made have a different content in some microelements.

Distribution of broad in probiotic dairy products based on milk and medicinal plant extracts depends on the content of plant extracts and reactions/associations that occur during the technological process. The results of measurements are presented in Tables 10 and 11.

Crt. No.	Sample name	Microelements content, mg/100g product									
		Ca	Mg	Na	K	Mn	Fe	Zn	Cu	Pb	Cd
1.	M – 3	130	6	36	130	-	0.1	0.4	ND	ND	ND
2.	A – 3	135	7	39	115	0.1	0.15	0.45	ND	ND	ND
3.	C – 3	135	7.5	37.5	125	0.1	0.2	0.5	ND	ND	ND
4.	Mă – 3	137.5	9	41	135	0.1	0.27	0.5	ND	ND	ND
5.	LD+A – 3	140	7.8	47.5	135	0.1	0.21	0.5	ND	ND	ND
6.	LD+C – 3	141	7.8	40	140	0.1	0.25	0.5	ND	ND	ND
7.	LD+Mă - 3	139.5	9.3	46	150	0.1	0.3	0.5	ND	ND	ND

Table 10. Mineral concentration of fermented dairy products with ABY 3 culture

Crt. No.	Sample name	Microelements content, mg/100g product									
		Ca	Mg	Na	K	Mn	Fe	Zn	Cu	Pb	Cd
1.	M – 5	130	6	36	130	-	0.1	0.4	ND	ND	ND
2.	A – 5	135	7	34.5	110	0.1	0.14	0.45	ND	ND	ND
3.	C – 5	140	7.8	32.5	115	0.1	0.21	0.5	ND	ND	ND
4.	Mă – 5	140	8	35	120	0.1	0.25	0.5	ND	ND	ND
5.	LD+A – 5	150	7.5	39.5	140	0.1	0.17	0.5	ND	ND	ND
6.	LD+C – 5	145	8	38	135	0.1	0.23	0.5	ND	ND	ND
7.	LD+Mă–5	147	9	41	140	0.1	0.3	0.5	ND	ND	ND

Table 11. Mineral concentration of fermented dairy products with ABY 5 culture

Minerals in fermented dairy products based on milk and medicinal plant extract fulfill in human body the following functions:

- Are composed of hard tissue: Ca and Mg contribute in a major portion at the formation of the skeleton and teeth. Ca is also one of the most sensitive elements that regulate cellular functions. Is the regulator of enzymes involved in carbohydrate, lipid and protein metabolism, is also involved in important physiological processes such as muscle contraction, blood coagulation, apoptosis and necrosis;

- Are components of soft tissue: Fe and K in the form of organic compounds contribute to muscles, organs and blood. Fe are component of hemoglobin involved in oxygen transport, the of myoglobin, the body's oxygen tank. Fe is considered a major potential prooxidant metals from the human body;
- Are regulators of biological functions: as solubilized salts in body fluids contribute to sensitivity of nervous stimulus, maintain muscle elasticity, adjustment of pH digestive fluids and other secretions, maintaining of osmotic pressure.

Fermented dairy product based on milk and medicinal plant extracts had a higher superoxiddismutase activity. The relationship between the iron and SOD activity is presented in figures 7 and 8.

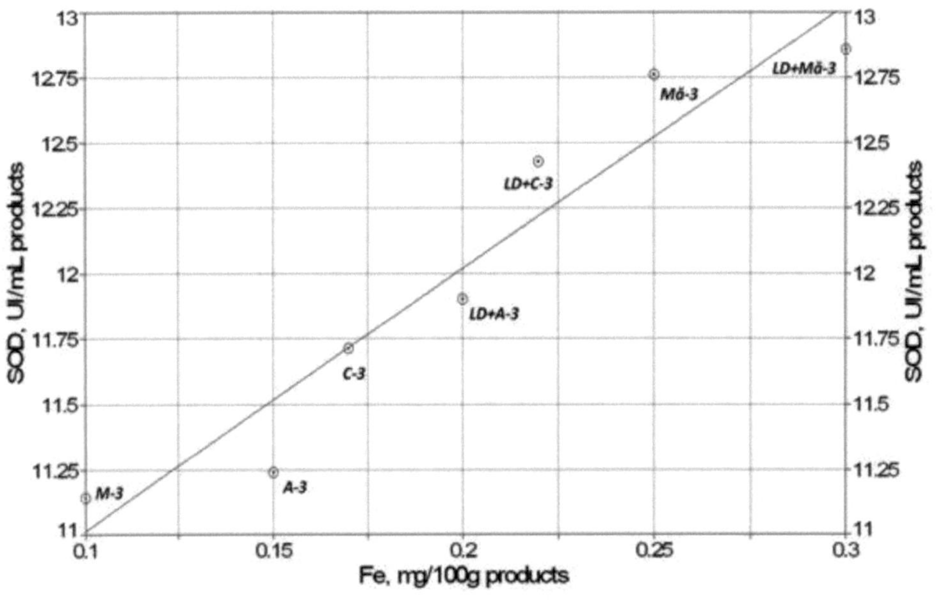

Figure 7. The relationship between SOD activity and iron content of products obtained with ABY 3 culture

For all samples of fermented dairy products with medicinal plant extracts is an increase in SOD activity compared with the control sample. Measured activity is total SOD-like activity (which contributes enzyme as such and superoxiddismutase-like activity of polyphenols and iron or zinc). SOD activity ranged from 11.142 to 12.857 IU·mL^{-1} product; it was maximum for the sample LD+Mă–3. Samples obtained with ABY 3 culture had a higher SOD activity than samples with ABT 5.

5.3. Probiotic dairy products with added plant extracts

To obtain the probiotic dairy products with medicinal plant extracts was used standardized cow milk to 1.5% fat. The technological process for production of fermented dairy products

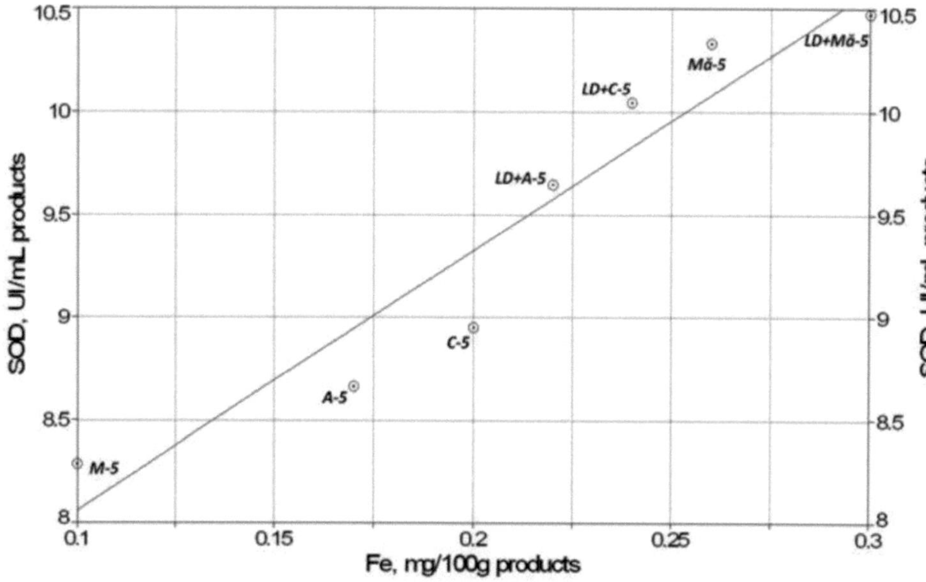

Figure 8. The relationship between SOD activity and iron content of products obtained with ABY 5 culture

Sample / Characteristics	Bilberries extract	Sea-buckthorn extract	Rosehip extract	Liquorice extract
Dry matter, g/100g	4.84	4.84	5.29	4
Ash insoluble in hydrochloric acid, g/100g	0.74	0.83	0.67	0.81
Total carbohydrate, g/100g	4.69	0.2	7.19	7.29
Total proteins, g/100g	0.21	0.41	0.62	2.15
Calcium, mg/100g product	13.2	4.4	34.5	30
Magnesium, mg/100g product	15.4	4.4	13.8	60
Sodium, mg/100g product	6.6	4.4	8.05	9
Potassium, mg/100g product	132	110	300	230
Manganese, mg/100g product	2.2	2.2	1.38	0.4
Iron, mg/100g product	1.1	2.2	0.46	0.6
Zinc, mg/100g product	0.66	0.44	0.69	0.6
Copper, mg/100g product	0.22	0.22	0.23	0.6
Lead, mg/100g product	ND	ND	ND	ND
Cadmium, mg/100g product	ND	ND	ND	ND
Caffeic Acid, g/100g	1.47	1.69	1.63	0.54
Cyanidin-3-glucoside chloride, g/100g	0.55	-	-	-
Ascorbic acid, g/100g	-	0.26	0.18	-
Glycyrrhizic acid, g/100g	-	-	-	1.96

Table 12. Characteristics of concentrated medicinal plant extracts

with medicinal plant extracts is presented in figure 8. The pasteurization of milk is achieved by maintaining standardized milk at 95 °C for 5 minutes. After pasteurization, milk is cooled to a temperature of 42 °C. Milk inoculation for these probiotic dairy products was made with two Probio-Tec probiotic cultures type: ABY 3 respectively ABT 5, at this time were added and aqueous extracts of medicinal plants (bilberries, sea-buckthorn, rosehip and liquorice) that have a number of characteristics presented in table 12.

After inoculation follows the distribution and packaging and incubation was made at 42°C for 6 hours in the thermostats set at the optimal temperature for the development of these bacteria. Meanwhile yoghurt gel gets a specific consistency. Cooling and storage of obtained yoghurts is performed at 6 °C for 8 days. In this storage period, coagulum is more compact, the flavor and taste become more pleasant.

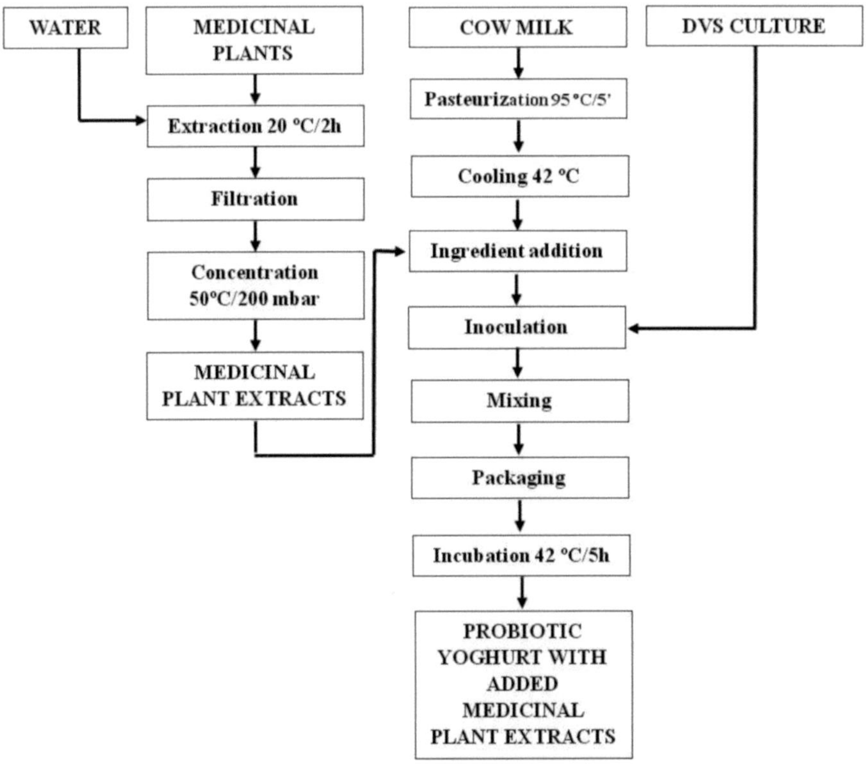

Figure 9. Technological flowchart for manufacturing the new product – Probiotic yoghurt with added medicinal plant extracts

The characteristics of fermented dairy products studied, in terms of chemical properties are:

- Total dry matter have values between 12.05% and 12.5% (lowest for products that contain liquorice extract), exceeding the minimum specified in Romanian standard for fermented dairy products (12%);
- The fat content of the samples vary between 0.6% and 1.3% lowest in products with liquorice compared with other, because smaller proportions of milk of these products;

- Lactic fermentation is faster for the samples with added plant extracts because of monosaccharides content (glucose, fructose, arabinose, xylose) and oligosaccharides (sucrose, raffinose, maltose, xiloglucan) from medicinal plants, which are fermented faster than lactose;
- Titratable acidity at the end of incubation period is between 67ºT and 78ºT, with higher values for products liquorice extracts. After 8th days of storage period, the titratable acidity is between 84ºT and 97ºT;
- The pH of products after incubation period varies between 5.035 and 5.287. After 8th days of storage it reaches values of 4.225-4.553, lowest value was obtained for the products with ABT 5 culture;
- ABT 5 probiotic culture which contains *Bifidobacterium lactis, Lactobacillus acidophilus* and *Strepococcus thermophilus* is more active than ABY 3 consists of *Bifidobacterium lactis, Lactobacillus acidophilus, Lactobacillus delbrueckii* subsp. *bulgaricus* and *Strepococcus thermophilus*;
- The established technological flowchart leads to obtaining some appropriate products in terms of physical-chemical characterization.

6. Conclusions

The researchers team of the Faculty of Food Science and Engineering, with many researchers in the scientific world, were concerned to investigate the possibility of obtaining probiotic products based on milk. Use milk as a vehicle for creating probiotic product was a constant concern of the staff of the Faculty of Food Science and Engineering in recent years. Probiotic character and functional role of probiotic products was obtained by adding fruit and vegetable juices, medicinal plant extracts, *Spirulina platensis* biomass, etc. We plan to continue research in this direction by investigating other products that may stimulate growth of probiotic bacteria.

Author details

Gabriel-Danut Mocanu and Elisabeta Botez
Department of Food Science, Food Engineering and Applied Biotechnology, Faculty of Food Science and Engineering, „Dunarea de Jos" University of Galati, Romania

Acknowledgement

Authors are grateful to the S.C. Hofigal Export – Import S.A. Bucharest for the material support of this work (medicinal plants, biomass of *Spirulina platensis*).

7. References

[1] Gismondo M.R. Drago L. Lombardi A. Review of probiotics available to modify gastrointestinal flora. International Journal of Antimicrobial Agents 1999; 12, 287–292.

[2] Lilly D.M. Stillwell R.H. Probiotics: Growth-promoting factors produced by microorganisms. Science 1965; 147, 747–748.

[3] Parker R.B. Probiotics, the other half of the antibiotic story. Animal Nutrition & Health 1974; 29, 4–8.

[4] Fioramonti J. Theodorou V. Bueno L. Probiotics: What are they? What are their effects on gut physiology?. Best Practice & Research Clinical Gastroenterology 2003; 17, 711–724.

[5] Salminen S. von Wright A. Morelli L. Marteau P. Brassart D. de Vos W.M. Fondén R. Saxelin M. Collins K. Mogensen G. Birkeland S.E. Mattila-Sandholm T. Demonstration of safety of probiotics – A review. International Journal of Food Microbiology 1998; 44, 93–106.

[6] Marteau P.R. de Vrese M. Cellier C.J. Schrezenmeir J. Protection from gastrointestinal diseases with the use of probiotics. The American Journal of Clinical Nutrition (Suppl.) 2001; 73, 430–436.

[7] Charteris W.P. Kelly P.M. Morelli L. Collins J.K. Selective detection, enumeration and identification of potentially probiotic *Lactobacillus* and *Bifidobacterium* species in mixed bacterial populations. International Journal of Food Microbiology 1997; 35,1–27.

[8] Food and Agriculture Organization/World Health Organization (FAO/WHO), (2001). Health and nutritional properties of probiotics in food including powder milk with live lactic acid bacteria, Report of a Joint FAO/WHO Expert Consultation on Evaluation of Health and Nutritional Properties of Probiotics in Food including Powder Milk with Live Lactic Acid Bacteria, pp. 1-28, ISBN: 92-5-105513-0 Córdoba, Argentina, October 1-4, 2001

[9] Reid G. Sanders M.E. Gaskins H.R. Gibson G.R. Mercenier A. Rastall R. New scientific paradigms for probiotics and prebiotics. Journal of Clinical Gastroenterology 2003; 37, 105–118

[10] Reid G. Safe and efficacious probiotics: What are they?. Trends in Microbiology 2006; 14, 348–352.

[11] Havenaar R. Ten Brink B. Huis in't Veld J.H.J. Selection of Strains for Probiotic Use. In: R. Fuller (ed.) Probiotics: The Scientific Basis. Chapman & Hall; 1992. p151-170.

[12] Soccol C.R. de Souza Vandenberghe L.P. Spier M.R. Pedroni Medeiros A.B. Yamaguishi C.T. De Dea Lindner J. Pandey A. Soccol V.T. The Potential of Probiotics. Food Technology and Biotechnology 2010;48(4) 413–434.

[13] Young Y. Functional foods and the European consumer. In: J. Buttriss & M. Saltmarsh (ed.) Functional foods. II. Claims and evidence. The Royal Society of Chemistry; 2000.

[14] Mollet B. Rowland I. Functional foods: At the frontier between food and pharma. Current Opinion in Biotechnology 2002;13, 483–485.

[15] Roberfroid M.B. An European consensus of scientific concepts of functional foods. The American Journal of Clinical Nutrition 2000b;16, 689–691.

[16] Menrad K. Market and marketing of functional food in Europe. Journal of Food Engineering 2003;56, 181–188.

[17] Axelsson L. Lactic acid bacteria: classification and physiology. In: Salminen, S., A. von Write, and A. Ouwehand, M. Dekker (ed.) Lactic Acid Bacteria Microbiological and Functional Aspects, Third Edition, Inc. New York; 2004. p1—66

[18] Hutkins R.W. Microorganisms and Metabolism. In: Microbiology and technology of fermented foods. Blackwell Publishing Professional; 2006. P.15-66.
[19] Roberfroid M.B. Concepts and strategy of functional food science: The European perspective. The American Journal of Clinical Nutrition 2000a;71, S1660–S1664.
[20] Kotilainen L. Rajalahti R. Ragasa C. Pehu E. Health enhancing foods: Opportunities for strengthening the sector in developing countries. Agriculture and Rural Development Discussion 2006;Paper 30.
[21] Kwak N.S. Jukes D.J. Functional foods. Part 1. The development of a regulatory concept. Food Control 2001a;12, 99–107.
[22] Stanton C. Ross R.P. Fitzgerald G.F. Van Sinderen D. Fermented functional foods based on probiotics and their biogenic metabolites. Current Opinion in Biotechnology 2005;16, 198–203.
[23] Diplock A.T. Aggett P.J. Ashwell M. Bornet F. Fern E.B. Roberfroid M.B. Scientific concepts of functional foods in Europe: Concensus document. British Journal of Nutrition 1999;81 (suppl. 1), S1–S27.
[24] Coppens P. Fernandes Da Silva M. Pettman S. European regulations on nutraceuticals, dietary supplements and functional foods: A framework based on safety. Toxicology 2006;221, 59–74.
[25] Spence J.T. Challenges related to the composition of functional foods. Journal of Food Composition and Analysis 2006;19, S4–S6.
[26] Siró I. Kápolna E. Kápolna B. Lugasi A. Functional food. Product development, marketing and consumer acceptance—A review. Appetite 2008;51, 456–467.
[27] Mäkinen-Aakula M. Trends in functional foods dairy market. In Proceedings of the third functional food net meeting 2006.
[28] Gibson G.R. Roberfroid M.B. Dietary modulation of the human colonic microbiota: Introducing the concept of prebiotics. Journal of Nutrition 1995;125, 1401–1412.
[29] Prado F.C. Parada J.L. Pandey A. Soccol C.R. Trends in non-dairy probiotic beverages – review. Food Research International 2008;41, 111-123.
[30] Mocanu G.D. Botez E. Andronoiu D.G. Gîtin L. Researches concerning the production of a fermented dairy drink with added carrot juice. Journal of Environmental Protection and Ecology 2011;12, 718-727.
[31] Mocanu G.D. Botez E. Nistor O.V. Andronoiu D.G. Characterization of probiotic yoghurt obtained with medicinal plant extracts and modelling of bacteria cell growth during its production. Journal of Agroalimentary Processes and Technologies 2011;17, 65-71.
[32] Guldas M. Irkin R. Influence of Spirulina platensis powder on the microflora of yoghurt and acidophilus milk. Mljekarstvo 2010;60, 237-243.
[33] Gomes M. Malcata F.X. Development of Probiotic Cheese Manufactured from Goat Milk: Response Surface Analysis via Technological Manipulation. Journal of Dairy Science, 1998;81, 1492–1507.
[34] Andronoiu D.G. Botez E. Nistor O.V. Mocanu G.D. Technological variants for obtaining of a dessert based on fresh probiotic chees. Journal of Agroalimentary Processes and Technologies 2009;XV, 102 – 106.

[35] Andronoiu D.G. Gîtin L. Botez E. Mocanu G.D. Researches Concerning the Production and Characterisation of a Dessert Based on Fresh Cheese and Peach Pulp. Journal of Environmental Protection and Ecology 2011;12, 502-508.

[36] Uliescu M. Rotaru G. Mocanu D. Stanciu V. Study of the probiotic telemea cheese maturation. The Annals of the University „Dunarea de Jos" Galaţi, Fascicle VI-Food Technology 2007;XXX, 92–99.

[37] Rotaru G. Mocanu D. Uliescu M. Andronoiu D. Research studies on cheese brine ripening. Innovative Romanian Food Biotechnology 2008;2, 30-39.

[38] Abd El-Salam M.H. Alichanidis E. Cheese Varieties Ripened in Brine. In: Fox, P.F.; P.L.H. McSweeney, T. M. Cogan, and T. P. Guinee (ed.) Cheese: chemistry, physics and microbiology, Third edition, volume 2, Elsevier Academic Press, 2004.

[39] Songisepp E. Kullisaar T. Hutt P. Elias P. Brilene T. Zilmer M. Mikelsaar M. A New Probiotic Cheese with Antioxidative and Antimicrobial Activity. Journal of Dairy Science 2004;87, 2017–2023.

[40] Kourkoutas Y. Bosnea L. Taboukos S. Baras C. Lambrou D. Kanellaki M. Probiotic Cheese Production Using Lactobacillus casei Cells Immobilized on Fruit Pieces. Journal of Dairy Science 2006;89, 1439–1451/

[41] Mocanu D.G. Rotaru G. Vasile A. Botez E. Andronoiu D.G. Nistor O. Vlăsceanu G. Dune A. Researches on maintaining functionality in the storage of probiotic dairy product – Afinolact. Journal of Agroalimentary Processes and Technologies 2009;15, 229 – 233.

[42] Mocanu D.G. Rotaru G. Vasile A. Botez E. Andronoiu D.G. Nistor O. Vlăsceanu G. Dune A. Studies on the production of probiotic dairy products based on milk and medicinal plant extracts. Journal of Agroalimentary Processes and Technologies 2009;15, 234 – 238.

[43] Mocanu G.D. Rotaru G. Botez E. Vasile A. Andronoiu D. Nistor O. Gîtin L. Vlăsceanu G. Dune A. Research concerning the production of a probiotic dairy product with added medicinal plant extracts. The Annals of the University „Dunărea de Jos" of Galaţi. Fascicle VI – Food Technology 2009;XXXII, 37 – 44.

[44] Mocanu G.D. Rotaru G. Botez E. Vasile A. Andronoiu D. Nistor O. Gîtin L. Vlăsceanu G. Dune A. Sensorial characteristics and rheological properties of probiotic product Cătinolact. The Annals of the University „Dunărea de Jos" of Galaţi. Fascicle VI – Food Technology 2009;XXXII, 64 – 69.

[45] Mocanu G.D. Rotaru G. Botez E. Gîtin L. Andronoiu D. Nistor O. Vlăsceanu G. Dune A. Sensory evaluation and rheological behavior of probiotic dairy products with Rosa canina L. and Glycyrriza glabra L. extracts. Innovative Romanian Food Biotechnology 2009;4, 32 – 39.

[46] Mocanu D. Rotaru G. Botez E. Andronoiu D. Nistor O. Probiotic yogurt with medicinal plants extract: Physical–chemical, microbiological and rheological characteristics. Journal of Agroalimentary Processes and Technologies 2010;16, 469-476.

[47] Manach C. Mazur A. Scalbert A. Polyphenols and prevention of cardiovascular diseases. Current Opinion in Lipidology. Nutrition and metabolism 2005;16, 1–7.

[48] Hărmănescu M. Moisuc A. Radu F. Drăgan S. Gergen I. Total polyphenols content determination in complex matrix of medical plants from Romania by NIR spectroscopy. Bulletin UASVM, Agriculture 2008;65, 123–128.

Chapter 12

Probiotic Confectionery Products – Preparation and Properties

Dorota Żyżelewicz, Ilona Motyl, Ewa Nebesny, Grażyna Budryn, Wiesława Krysiak, Justyna Rosicka-Kaczmarek and Zdzisława Libudzisz

Additional information is available at the end of the chapter

http://dx.doi.org/10.5772/50047

1. Introduction

Proper orientation of the gastrointestinal tract biocenosis and consumption of probiotic products is becoming more and more important in the industrialized world as problems such as civilization diseases and population aging are spreading.

The word "probiotic" is derived from the Greek "pro bios" and means "for life". As defined by FAO /WHO, probiotics are specific strains of microorganisms, which when served to human in proper amount, have a beneficial effect on our body (improve health or reduce risk of getting sick) [1, 2]. Probiotic bacteria most commonly belong to *Lactobacillus* and *Bifidobacterium* species.

However, not all bacteria have equally strong effect on human health improvement. Activity of probiotic bacteria is a specific feature of the strain.

The effect of improving human health depends not only on strain (its probiotic activity) but also on media (a matrix on which bacteria are carried). The media should provide probiotic bacteria with a high viability and activity during transit through intestinal tract and at their final destination.

Probiotic bacteria support both, specific and nonspecific human and animal defense mechanisms.

Probiotics improve digestion of lactose in subjects suffering from disorders in its absorption and relieve symptoms of the gastrointestinal tract disorders. Additionally they may contribute to lowering of cholesterol as well as reduce adherence, and thereby prevent translocation of pathogenic microorganisms into the intestinal lumen. There are many different evidence that prove ability of probiotic bacteria to prevent or slow down the processes leading to colorectal cancer. Lactic acid bacteria are also able to use (or bond) carcinogenic compounds derived from diet or produced by pathogenic bacteria in the intestines, such as nitrosamines, azo dyes, mycotoxins or amino acids pyrolisates. However, the strongest clinical evidence demonstrating the beneficial effect of probiotics on human health is immunity increase (immunomodulation) [4-9].

Probiotics may be consumed in the form of pharmaceutical preparations, food supplements or food additives.

LAB probiotic bacteria may play a role of a supplement in: vegetable, fruit and fruit and vegetable juices, breakfast cereals, different kinds of chips, mousses and creams, ice creams and fruit jellies. They may also serve as supplement when properly selected probiotic strain is added to fermented meats, vegetable silages and not soured dairy products, cottage and ripened cheeses as well as many other products. Probiotic bacteria are also used as an additive in nutrition products for children. Most commonly, however, they are used in process of manufacturing fermented dairy products such as yogurts or probiotic kefirs.

Fixation of lactic acid bacteria with the use of innovative processes, thanks to the elimination of characteristic sour taste allows to extend its possible application to a whole new group of products. LAB viability in this type of products is often caused by low water content and water activity, as well as leaving LAB in the state of anabiosis without performing fermentation. This criteria is met by a certain number of semi-finished and final products in confectionery industry. Under polish research projects no. 3 P06T 054 24 and no. R12 018 01 attempts were made to include LAB into the composition of such products as: chocolate and chocolate products, raisins coated with chocolate (dragees), confectionery cores from fatty masses, biscuits coated with chocolate couverture, interleaved wafers and bread spreads. In these products the number of bacteria as CFU $\cdot$ g^{-1} and LAB survival rate during several months of storage (depending on the type of testing material) was determined. This chapter describes the technology of manufacturing products such as interleaved wafers and chocolate covered raisins, biscuits and cores from peanut fatty masses, supplemented with lyophilized live bacterial cultures of lactic acid bacteria from *Lactobacillus* group [10-16]. The lyophilized preparation of LAB contained 3 strains:

- *Lactobacillus casei* strain no ŁOCK 0900 B/00019,
- *Lactobacillus casei* strain no ŁOCK 0908 B/00020,
- *Lactobacillus paracasei* strain no ŁOCK 0919 B/00021.

All these strains were derived from the Collection of Pure Industrial Microbial Cultures at the Lodz University of Technology ŁOCK 105. These strains were deposited in the Institute of Immunology and Experimental Therapy of the Polish Academy of Sciences in Wroclaw.

The strains were selected on the basis of results of *in vitro* studies. They were resistant to the acidity of gastric juice, resistant to the bile, adhered to epithelial cells and displayed an antimicrobial activity. The studies were carried out according to FAO/WHO recommendations [1, 2]. On the basis of the sequence of the gene encoding 16S rRNA, the examined bacterial strains were classified as *Lactobacillus casei* and *Lactobacillus paracasei* (97 ÷ 99% similarity). Both these species rank among the typical microflora of human intestines and can be safely used for production of fermented milk products and preparations of probiotics. The examined strains were tolerant to pH 3.5. Almost all cells survived 3 h incubation at pH 3.5 and at neutral pH (6.5) while 80 ÷ 100% cells survived at pH 2.5 (it depended on a strain) while in the presence of 4% bile salts only 60% cells survived. All the examined LAB strains exerted an inhibitory effect on pathogenic bacteria, both gram-negative and gram-positive. The *in vivo* studies employing 2-month old, immunocompetent mice Balb/c revealed no translocation of these bacteria to the blood and other internal organs. Minor amounts of these bacteria in mesenteric lymph nodes could be an evidence of activation of immune system. The safety of application of these strains was also proved through *in vivo* studies employing children suffering from the atopic skin inflammation [17].

2. Methods

Obtained probiotic confectionery products, namely: interleaved wafers, raisins coated in chocolate, as well as confectionery cores such as biscuits and peanut fatty masses were analyzed with the use of following methods:

- Casson viscosity and Casson yield value of couverture according to Casson method, with the use of digital rheoviscosimeter HADV – III+ from Brookfield Engineering Laboratories Inc. (USA), with co-axially arranged rotor SC4–27 (11.75 mm diameter), stator and an attachment for small volume samples (cylinder with diameter – 25.13 mm) [15, 18-20],
- percentage of couverture content established by a difference in weight of confectionery cores (biscuits, cores from peanut fatty masses, wafers, raisins) with coating and before coating,
- dry mass content by drying a sample with sand at a temperature 102 – 105°C,
- water activity with the use of a measuring instrument HYGROPALM AW 1 from Rotronic (Switzerland) with a digital probe AW-DIO at a temperature T=23 ± 1°C,
- total acidity by potentiometric titration to a pH value of 8.2 with the use of pH-meter from SCHOTT CG 843 with combined electrode – BlueLine 11 from SCHOTT GERÄTE GmbH (Germany),
- texture analysis at a temperature of 20°C, with the use of digital texture analyzer TA.XT Plus from Stable Micro Systems (UK) with driver and software, probes used: A/CKB – chocolate coated raisins, HDP/90 (heavy duty platform) – biscuits and peanut fatty masses coated with couverture, HDP/SR – wafer filling (consistency masses – spreadability), HDP/VB – wafer cores (hardness – crunchiness),
- changes in fat by DSC method with the use of DSC 111 apparatus from Setaram (France), according to the following procedure:
    - cooling a sample (with an initial room temperature) to a temperature of 10°C with cooling speed of 1°C·min^{-1} to obtain a complete crystallization of fat,
    - leveling initial conditions by keeping a sample at a temperature of 10°C for 2 min,

- heating a sample to a temperature of 55°C with a heating speed of 3°C·min^{-1}, during which melting of fat in a sample occurred. Changes occurring during this stage were presented as melting curves. Maximum of peak created on developed curves describes as the melting point (Tm), meanwhile from a peak area melting enthalpy (ΔH) was calculated. Heating temperatures of samples were chosen from a range of melting temperatures of fat present in a product [21],

- organoleptic analysis covered evaluation of color, exterior surface, interior of product, consistency, as well as taste and smell with a hedonic 5-point scale [22],
- viability of lactic acid bacteria during storage of products at temperatures of 4, 18 and 30°C. The amount of *Lactobacillus* bacteria was determined by Koch's plate-cultivating method with the use of MRS growth medium. Products were suspended in a solution of physiological saline and peptone. With this manner first dilution was obtained. Prepared with this method samples were incubated in a water bath at a temperature of 37°C for 30 min, afterward samples were homogenized for 1 min. Next step included preparing serial decimal dilutions from which an inoculation of 1 ml of samples onto a Petri dish was performed (each dilution in triplicate). Plates were incubated for 48 h at a temperature of 37°C in a CO_2 WT3 Binder incubator (anaerobic conditions with an addition of 5% volume of CO_2).

Bacteria viability in studied confectionery products was calculated according to following formula:

$$\text{Viability}\left[\%\right] = \frac{N}{N_O} \times 100\%$$

N – log CFU · g^{-1} after a certain period of storage
N_0 – log CFU · g^{-1} directly after product preparation,

- statistical analysis, including a calculation of arithmetical average and standard deviation, was performed with a Microsoft Excel software. Results were obtained from at least three replicates.

3. Products coated with chocolate couverture supplemented with live cultures of lactic acid bacteria

3.1. Chocolate couverture as a media for lactic acid bacteria

Chocolate couvertures contain usually 30-40% of fat. Primary components of chocolate couverture are: cocoa fat, sugar, powder milk (in milk and white couvertures), cocoa liquor and lecithin. It has a fluid consistency during tempering and a solid form in a final product. Couverture can include bigger, possible to sense particles of additives, such as fragmented nuts, which can be found in couverture in a final product, although they were put on a product during processing before or after coating with couverture. Shelf life of couverture is usually 3 to 12 months, depending on its type, but ultimately shelf life of a couverture coated product depends on the kind of used filling. The content of chocolate couverture in products consist at least 15% of products mass. The process of obtaining chocolate couverture, as an exterior layer on products, include: conching of couverture components, tempering of couverture, coating with a tempered couverture.

Conching takes place at a temperature of at least 40°C (in most of the times 60-70°C) and lasts for up to 48 h. In these conditions it is impossible to maintain high LAB viability, when they are introduced to a chocolate mass in form of a preparation. Tempering of milk chocolate couverture is performed at a temperature of 28°C, and dark chocolate couverture at 30°C [23]. Thus, temperatures used during tempering allow the possibility to introduce to the product probiotic additive in form of fixated LAB preparation. Additionally, low water activity of couverture – on a level of 0.3 – 0.5, allows quite high viability of LAB in products [15, 16]. In this studies LAB preparation fixated by freeze-drying on a powder milk as a carrier media. Obtained this way cultures of lactic acid bacteria, which in lyophilized preparation as well as in a final products, namely chocolate couvertures and chocolate products, were in a state of anabiosis, ready to return to normal life functions when found in proper environment, such as human digestive system.

The aim of this part of the study regarding supplementing of chocolate couverture, used for coating confectionery products, with a lactic acid bacteria preparation was to establish the possibility of obtaining such confectionery products with functional properties in the whole time of shelf life. Furthermore, to establish a minimal level of supplementation to maintain functional properties, for products with significant differences in used confectionery core. Finally, to study the most important properties of used couverture itself, as well as the whole coated with couverture product, which could lower the quality of final product, despite it maintaining full functional properties throughout the whole shelf life.

3.2. Obtaining chocolate couverture supplemented with live cultures of lactic acid bacteria used for coating of various confectionery cores

Obtaining chocolate couvertures enriched with live cultures of lactic acid bacteria was performed by adding lyophilized LAB to a industrially obtained chocolate couverture. Dark chocolate couverture produced by Union Chocolate Sp. z o. o. (Żychlin, Poland) and a preparation of live cultures of lactic acid bacteria (LAB) with a concentration of live bacterial cells from *Lactobacillus* species on a level of 9×10^{10} CFU · g^{-1} from Institute of Fermentation Technology and Microbiology, Lodz University of Technology (Poland) were used. Couverture to LAB preparation ratio was 96:4 (w/w).

In couverture supplemented with LAB, and in a control couverture, rheological properties were established (Table 1), which are extremely important from a technological standpoint, because an eventual increase in a couverture viscosity caused by LAB addition could significantly hinder latter stage of coating [15, 18-20].

Rheological properties analysis have shown an increase of viscosity caused by addition of LAB by about 5% (Table 1). From technological point of view this change is not big enough to cause any repercussions in a form of incomplete product coating. In this regard LAB addition didn't cause any significant changes in a couverture. Thus, couverture without any further modifications (e.g. content of fat or emulsifier) can be used to selected confectionery products.

Type of couverture	Casson viscosity (Pa · s)	Casson yield value (Pa)
Dark	**1.33** ± 0.02	**8.86** ± 0.28
Dark +LAB	**1.40** ± 0.01	**8.47** ± 0.09

Table 1. Casson viscosity and yield value of couvertures used for confectionery cores coating.

3.3. Obtaining confectionery products coated with couverture supplemented with live cultures of lactic acid bacteria

For couverture coating, as cores, industrially produced biscuits and cores from peanut fatty masses, obtained in a laboratory, were used. Both these products significantly differed, both, in chemical composition, as well as an area to volume ratio (thus the development of couverture surface). Both factors could significantly influence the viability of bacteria present in LAB preparations during products manufacture and storage.

Obtaining confectionery cores used for couverture coating

As cores for couverture coating (with various thickness of its layer) Petit Beurre biscuits were used (Z.P.C. Piast Sp. z o. o., Głogówek, Poland), they contain of: wheat flour, sugar, eggs, confectionery fat and a raising agent in the mass ratio of 100:30:20:10:1. The second type of confectionery cores were candies from peanut fatty mass obtained in a laboratory. Raw materials used for obtaining this product originated from: sugar from Promyk Cukrohurt Sp. z o. o. (Siedlce, Poland) – 17 g · 100 g^{-1}, confectionery fat Efekt 40 MT "middle-tans" from Z.P.T. Kruszwica S.A. (Kruszwica, Poland) – 27 g · 100 g^{-1}, powdered skim milk from S.M. Spomlek (Radzyń Podlaski, Poland) – 17 g · 100 g^{-1}, peanut mash from Plus (Łódź, Poland) – 20 g · 100 g^{-1}, wafer production discards from Dybalski-Cukiernie (Łódź, Poland) – 19 g · 100 g^{-1}. Due to nutritional policy fat used in a recipe had a decreased amount of trans fatty acids [24]. Fat completely devoid of trans fatty acids didn't maintain proper rheological properties in the whole time of storage.

Confectionery fat was grind to a paste in a mixer with single work-load of 3 kg with a hook stirrer. Friable components i.e. powdered sugar, peanut mash, powdered milk and ground wafer discards were all mixed with each other in amounts featured in a recipe. To a ground to a paste fat, prepared mixture of components was gradually added. The pulp was mixed to obtain homogeneous consistency. Prepared pulp was carried to a rectangular mold. The surface of the pulp was leveled. Molds with pulp were cooled to a temperature 8-10°C, and then cut to single pieces with a size of 25×20×45 mm.

Obtaining confectionery cores for coating and the process of coating with a chocolate couverture

Peanut fatty mass cut to a shape of candies was lead to obtain a temperature of 15-18°C (to obtain a solid consistency). Biscuits were coated without cooling them beforehand. Prepared cores were placed on a grid of coating machine and coated with a previously tempered couverture. Planned percentage of couverture layer on cores, i.e. 30%, 35% and 40% for biscuits and 16%, 25% and 30% for peanut fatty mass, was obtained by regulating the speed of movement of coating machines grid (Promet, Łódź, Polska) on which a layer of couverture poured on cores was blown away to a proper thickness by a stream of air. Chocolate couverture was heated to a temperature of 45-50°C. After the bulk liquidated, it was tempered to a temperature of 28-30°C, and next it was slowly heated to 31-32°C and finally then measured amount of LAB was added. The amounts of couverture on cores ware picked experimentally, to obtain a proper level of CFU of LAB per 1 gram of a whole coated product during storage time, with a possibly thinnest layer of couvurture. For couverture coated biscuits the amount of lyophilized LAB amounted a least 0.5% in relation to a weight of a product. This amount corresponded to 10^7 CFU of LAB per 1 gram of fresh product. To couverture used for peanut fatty mass coating the amount of lyophilized LAB was increased to 0.55% per mass of product to provide probiotic properties during the whole storage time. It corresponded to 10^8 CFU of LAB per 1 gram of fresh product.

Storage of coated biscuits and candy from peanut fatty mass

Finished products were left at a temperature of 6-8°C to cool down and solidify. Next, products were wrapped in aluminum foil and stored at 4, 18 and 30°C for a period of time predicted as a suitable shelf life for given product, that is for 4 months in case of biscuits, and for 3 months for candy from peanut fatty masses.

3.4. Results

Water activity in couverture coated cores from biscuits and peanut fatty mass

Changes in water activity were presented only for fresh products and after the full period of storage, because of very small variation of this parameter (Table 2 and 3).

Storage temp.	Content of couverture on biscuits (%)					
	30	35	40	30	35	40
	Biscuits coated with couverture **supplemented** with LAB			Biscuits coated with couverture non-supplemented with LAB		
	Water activity					
Fresh product	**0.229** ± 0.010	**0.336** ± 0.026	**0.275** ± 0.007	**0.233** ± 0.012	**0.338** ± 0.060	**0.235** ± 0.070
4°C	**0.282** ± 0.007	**0.266** ± 0.002	**0.307** ± 0.004	**0.314** ± 0.005	**0.292** ± 0.005	**0.307** ± 0.012
18°C	**0.314** ± 0.012	**0.319** ± 0.005	**0.303** ± 0.004	**0.302** ± 0.002	**0.316** ± 0.004	**0.300** ± 0.018
30°C	**0.303** ± 0.008	**0.308** ± 0.001	**0.300** ± 0.006	**0.294** ± 0.006	**0.308** ± 0.002	**0.312** ± 0.012

Table 2. Water activity in biscuits coated with various amount of couverture supplemented and non-supplemented with LAB in fresh product and after storage for 4 months at temperatures 4, 18 or 30°C.

Storage temp.	Content of couverture on candy (%)					
	16	25	30	16	25	30
	Candy coated with couverture **supplemented** with LAB			Candy coated with couverture non-supplemented with LAB		
	Water activity					
Fresh product	**0.221** ± 0.008	**0.365** ± 0.026	**0.210** ± 0.090	**0.221** ± 0.003	**0.323** ± 0.005	**0.281** ± 0.002
4°C	**0.346** ± 0.002	**0.328** ± 0.002	**0.339** ± 0.013	**0.329** ± 0001	**0.322** ± 0.002	**0.315** ± 0.006
18°C	**0.342** ± 0.001	**0.340** ± 0.004	**0.309** ± 0.005	**0.315** ± 0.004	**0.342** ± 0.003	**0.340** ± 0.003
30°C	**0.306** ± 0.004	**0.309** ± 0.003	**0.304** ± 0.001	**0.287** ± 0.002	**0.329** ± 0.001	**0.338** ± 0.002

Table 3. Water activity in peanut fatty mass candy coated with various amount of couverture supplemented and non-supplemented with LAB in fresh product and after storage for 3 months at temperatures 4, 18 or 30°C.

An increase in water activity was observed, with an exception of 35% of couverture on biscuits and 25% of couverture on candy (both couvertures supplemented with LAB). It can be explained by re-crystallization of saccharose during storage, which is linked to releasing of water and increasing its activity, although in both candies and biscuits, water activity of products coated with couverture of middle thickness was relatively high [25]. Water activity in coated biscuits and candy in the whole time of storage was in a range of 0.210 – 0.340 and 0.229 – 0.338, respectively. The level of water activity allowed LAB to stay in a state of anabiosis, which provided stability and high viability of probiotic microorganisms [26].

Total acidity of couverture coated cores from biscuits and peanut fatty mass

Total acidity of couverture coated cores from biscuits and peanut fatty mass is presented in Table 4 and 5.

Storage temp.	Content of couverture on biscuits (%)					
	30	35	40	30	35	40
	Biscuits coated with couverture **supplemented** with LAB			Biscuits coated with couverture non-supplemented with LAB		
	Total acidity (ml 1 M NaOH · 100 g^{-1})					
Fresh product	**2.67** ± 0.04	**2.74** ± 0.07	**2.82** ± 0.04	**2.64** ± 0.09	**2.70** ± 0.02	**3.06** ± 0.08
4°C	**3.02** ± 0.07	**3.14** ± 0.12	**3.18** ± 0.11	**3.00** ± 0.08	**3.10** ± 0.09	**3.12** ± 0.20
18°C	**3.12** ± 0.07	**3.20** ± 0.04	**3.22** ± 0.09	**3.08** ± 0.02	**3.26** ± 0.14	**3.28** ± 0.12
30°C	**3.19** ± 0.14	**3.39** ± 0.06	**3.43** ± 0.06	**3.17** ± 0.09	**3.32** ± 0.08	**3.38** ± 0.11

Table 4. Total acidity of biscuits coated with various amount of couverture supplemented and non-supplemented with LAB in fresh product and after storage for 4 months at temperatures 4, 18 or 30°C.

Storage temperature	Content of couverture on candy (%)					
	16	25	30	16	25	30
	Candy coated with couverture **supplemented** with LAB			Candy coated with couverture non-supplemented with LAB		
	Total acidity (ml 1 M NaOH · 100 g^{-1})					
Fresh product	**2.20** ± 0.07	**2.32** ± 0.09	**2.38** ± 0.04	**2.16** ± 0.02	**2.30** ± 0.11	**2.38** ± 0.08
4°C	**2.34** ± 0.03	**2.35** ± 0.02	**2.39** ± 0.07	**2.29** ± 0.12	**2.42** ± 0.03	**2.52** ± 0.08
18°C	**2.46** ± 0.09	**2.48** ± 0.05	**2.54** ± 0.07	**2.52** ± 0.04	**2.56** ± 0.06	**2.58** ± 0.03
30°C	**2.50** ± 0.03	**2.56** ± 0.07	**2.64** ± 0.06	**2.43** ± 0.12	**2.50** ± 0.12	**2.64** ± 0.03

Table 5. Total acidity of peanut fatty mass candy coated with various amount of couverture supplemented and non-supplemented with LAB in fresh product and after storage for 3 months at temperatures 4, 18 or 30°C.

Both, coated biscuits and candy from peanut fatty mass directly after preparation showed an increase in total acidity along an increase in a amount of couverture on products. It indicates that a presence of couverture caused an increase in an amount of components with acidic properties. Couverture contains cocoa liquor, which is rich in volatile and non-volatile organic acids, thus acidity of couverture alone can amount to 8 ml 1 M NaOH·100 g^{-1}. Meanwhile coated biscuits and candy had total acidity in range of 2.7 – 3.4 ml 1 M NaOH · 100 g^{-1} and 2.2 – 2.6 ml 1 M NaOH · 100 g^{-1}, respectively. Higher values of total acidity in coated biscuits result from bigger amounts of couverture, comparing to candy. No noticeable influence of LAB addition on total acidity of products was observed. 3 months of storage caused total acidity to increase, more the higher temperature of storage was used. Furthermore, bigger increase of this parameter was observed in biscuits, which could be caused by two factors. Firstly, by longer storage time, which was dictated by normative requirements, and secondly by bigger area of surface of biscuits in relation to their weight. Because of that they had greater contact with external agents causing degradation changes, such as releasing of free fatty acids. In studied storage period LAB addition didn't cause any changes in total acidity, both in biscuits and candy. It can be considered to be a marker of keeping of probiotic microorganisms in a state of anabiosis, because their activity would cause a lactic acid production and it would influence the acidity of product, and consequently lead to its deterioration.

Hardness of couverture coated cores from biscuits and peanut fatty mass

In fresh biscuits an increase of hardness caused by the content of couverture was observed (Table 6). Biscuit itself was fresh, tender and rather brittle, but properly tempered couverture, with properly crystallized fat in its V polymorphic form, formed a hard surface, which decided about product hardness. Higher hardness values of biscuits coated with couverture supplemented with LAB, testify that it had good textural properties, which means that the addition of LAB didn't hinder cocoa fat crystallization in couverture. During biscuits storage decreasing of hardness was noticed. However no definite correlation between hardness changes and LAB supplementation, storage temperature of couverture content was observed. It could be probably caused by the fact that changes in hardness are quite complex and a few factors affect it, including softening of cocoa fat in a couverture at temperatures above 15°C (especially at 30°C), an increase of water content in a couverture resulting from water diffusion from product, drying of biscuit core, re-crystallization of saccharose and retrogradation of starch in biscuits. More precise image of hardness changes in coated biscuits can be observed in a chart showing a cutting force (Figure 1).

Storage temp.	**Content of couverture on biscuits (%)**					
	30	35	40	30	35	40
	Biscuits coated with couverture **supplemented** with LAB			Biscuits coated with couverture non-supplemented with LAB		
	Hardness (kg)					
Fresh product	**5.30** ± 0.09	**7.18** ± 0.02	**8.41** ± 0.08	**5.37** ± 0.04	**6.68** ± 0.07	**6.96** ± 0.04
4°C	**5.66** ± 0.08	**6.00** ± 0.09	**6.57** ± 0.20	**5.85** ± 0.07	**5.66** ± 0.12	**5.66** ± 0.11
18°C	**4.57** ± 0.02	**4.45** ± 0.14	**4.25** ± 0.12	**4.59** ± 0.07	**6.01** ± 0.04	**6.55** ± 0.09
30°C	**5.21** ± 0.09	**4.51** ± 0.08	**6.98** ± 0.11	**4.35** ± 0.14	**4.65** ± 0.06	**4.99** ± 0.06

Table 6. Hardness of biscuits coated with various amount of couverture supplemented and non-supplemented with LAB in fresh product and after storage for 4 months at temperatures 4, 18 or 30°C.

During cutting of fresh biscuit the biggest action was observed after around ¾ of a second, that corresponds to a depth of about 1.5 mm. Thickness of couverture layer measured for this amount of couverture in a product amounted to 1 mm from each side. The biscuit was brittle, instantly breaking under the pressure of a cutting probe, and the couverture was less hard than the biscuit, and was gently cut by the blade. Cutting profile of a biscuit stored for 4 months at a temperature of 4°C was quite similar to the fresh biscuit, only with lower hardness peak, which was caused by a smoother cut caused by leveling of moisture in a whole product and by declining parts of tensions created during baking. Biscuits stored for 4 months at a temperature of 30°C showed a highest hardness after 1.5 s of the test, thus in deeper parts of the product. However, earlier in a cutting profile a local maximum with a lower values of hardness can be observed. This indicates that biscuit core dried to some degree, it crumbled not in a whole cut but in several layers. Overall hardness value was lower, however it was probably caused by a lower hardness of couverture. It can be observed that the beginning of diagram progresses with a slope at a lower angle comparing to fresh product, and the one stored at refrigeration conditions. After 4 months of storage at a temperature of 18°C similar tendency can be noticed, meaning a couverture is softer than on a fresh biscuit, a biscuit is dried and it crumbles unevenly. Above considerations lead to a conclusion that high biscuit hardness stored at a relatively high temperature is caused by its drying. From all samples stored for 4 month of biscuits coated with couverture supplemented with LAB in an amount of 40% showed statistically significantly higher hardness (about 30%) comparing to analogous samples in non-supplemented couverture, stored at temperatures of 18 and 30°C. Hardness of other samples coated in supplemented couverture, comparing to non-supplemented ones didn't differ by more than 20%.

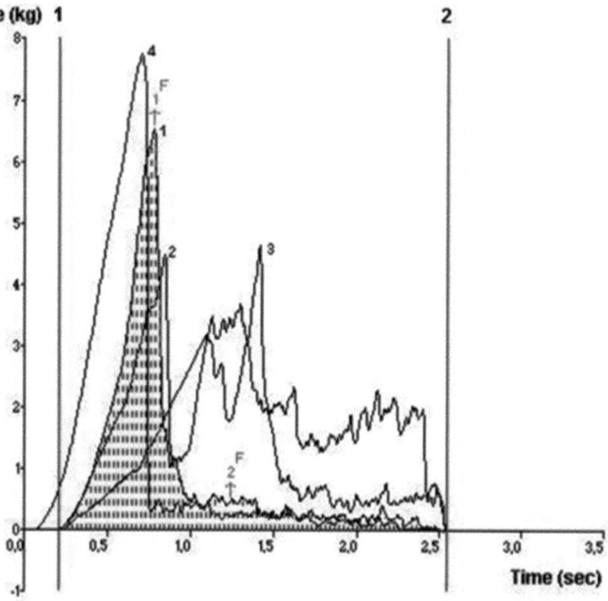

Figure 1. Exemplary profile of texture of fresh and stored for 4 months biscuits coated with couverture (35%) supplemented with LAB, maximal used force is the hardness of product; 1 – fresh product, 2 - stored at 4°C, 3 – stored at 18°C, 4 – stored at 30°C.

Fresh peanut fatty mass candy showed statistically similar hardness regardless of couverture content on cores or supplementation with LAB (Table 7). Noticeable decrease in hardness of candy stored in a period of 3 months at temperatures of 18 and 30°C was observed. Especially at the highest temperature, which resulted from plasticizing both, of cocoa butter in couverture and confectionery fat in candy core. Statistically higher hardness after storage was observed in candy coated with supplemented couverture. It can indicate that LAB preparation gives couverture additional rigidity, as well as makes couverture less susceptible to melting.

Storage temp.	**Content of couverture on biscuits (%)**					
	16	25	30	16	25	30
	Candy coated with couverture **supplemented** with LAB			Candy coated with couverture non-supplemented with LAB		
	Hardness (kg)					
Fresh product	**0.66** ± 0.02	**0.62** ± 0.11	**0.63** ± 0.08	**0.66** ± 0.07	**0.66** ± 0.09	**0.66** ± 0.04
4°C	**0.53** ± 0.12	**0.65** ± 0.03	**0.69** ± 0.08	**0.52** ± 0.03	**0.56** ± 0.02	**0.55** ± 0.07
18°C	**0.35** ± 0.04	**0.35** ± 0.06	**0.36** ± 0.03	**0.28** ± 0.09	**0.28** ± 0.05	**0.35** ± 0.07
30°C	**0.12** ± 0.12	**0.17** ± 0.12	**0.14** ± 0.03	**0.13** ± 0.03	**0.14** ± 0.07	**2.64** ± 0.03

Table 7. Hardness of peanut fatty mass candy coated with various amount of couverture supplemented and non-supplemented with LAB in fresh product and after storage for 3 months at temperatures 4, 18 or 30°C.

Cutting profile of coated candy shows that in fresh product the biggest hardness was present in a layer of couverture at a depth of almost 1 mm (Figure 2). Storage at a temperature of 4°C caused a lowering of hardness of couverture and an increase in deeper layers – at half-height, where its partial fracture took place [25]. In products stored at 18°C a lowering of hardness of both, couverture and core was observed. They showed local maximum of hardness on a similar level. Storage at a temperature of 30°C caused a significant softening of both, couverture and core. Cutting curve did not show any local hardness maximum. The blade evenly and gently delved into candy.

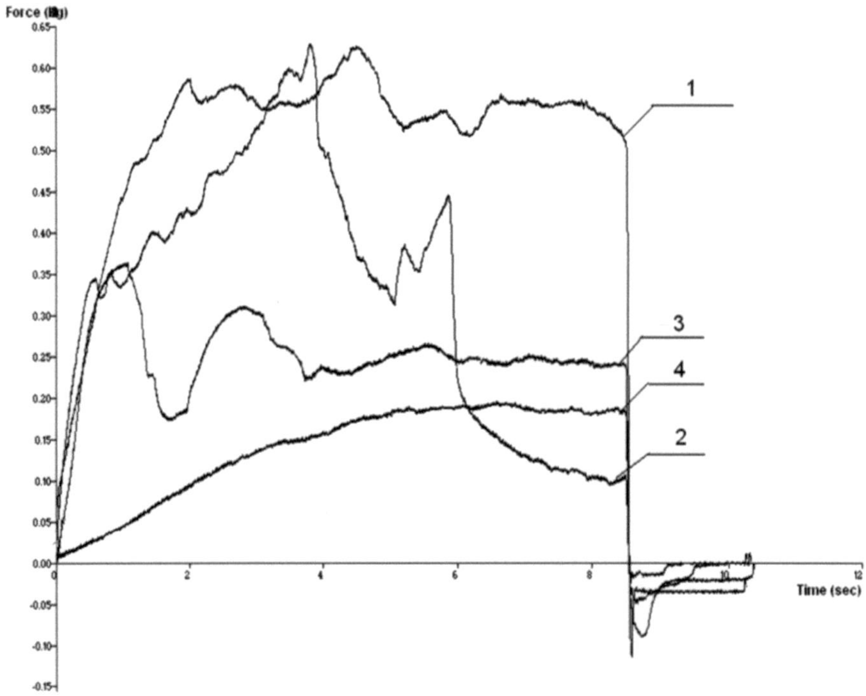

Figure 2. Exemplary profile of texture of fresh and stored for 3 months peanut fatty mass candy coated with couverture (25%) supplemented with LAB, maximal used force is the hardness of product; 1 – fresh product, 2 - stored at 4°C, 3 – stored at 18°C, 4 – stored at 30°C.

Thermal profile of fat from chocolate couverture from biscuits and peanut fatty mass candy

Melting enthalpy of cocoa butter from couverture, which coated biscuits increased with an increase of couverture content in a product (Table 8 and 9).

Furthermore, an increase of melting temperature with an increase of couverture thickness was observed, regardless if it was supplemented of non-supplemented with LAB. Supplemented couverture showed bigger values of melting enthalpy comparing to non-supplemented couverture. During 4 months of storage of coated biscuits melting enthalpy of couverture decreased both, in supplemented and non-supplemented product. This decrease was bigger when storage temperature increased, furthermore, bigger decrease was observed in couverture supplemented with LAB. In supplemented biscuits melting temperature of cocoa butter remained at the same level during whole storage. In non-supplemented couverture a decrease of melting temperature during storage was noticed. Summarizing these changes it can be observed, that in both, supplemented and non-supplemented couverture fat remained in its stable V polymorphic form only when biscuits were stored at refrigeration temperature. At other temperatures it was partly in amorphous form with a lower melting temperature. Supplementing of couverture with LAB influenced positively maintaining of fats crystalline form. Exemplary thermogram can be seen in Figure 3.

Storage temp.	Content of couverture on biscuits (%)					
	30		35		40	
	Enthalpy ΔH (J · g-1)	Temperature Tm (°C)	Enthalpy ΔH (J · g-1)	Temperature Tm (°C)	Enthalpy ΔH (J · g-1)	Temperature Tm (°C)
Fresh product	**36.05** ± 0.18	**30.68** ± 0.09	**37.71** ± 0.28	**32.02** ± 0.12	**42.16** ± 0.37	**33.08** ± 1.06
4°C	**32.83** ± 0.78	**32.35** ± 0.55	**34.03** ± 0.46	**32.89** ± 0.25	**37.39** ± 0.71	**34.04** ± 0.69
18°C	**24.23** ± 0.65	**32.02** ± 0.37	**24.92** ± 0.41	**32.43** ± 0.55	**25.63** ± 0.18	**33.82** ± 0.91
18°C	**22.21** ± 0.72	**31.78** ± 0.75	**23.36** ± 0.61	**31.58** ± 0.38	**23.64** ± 0.58	**32.50** ± 0.29

Table 8. Enthalpy and maximal melting temperature of cocoa fat in dark couverture supplemented with LAB used for coating of biscuits during 4 months of storage at temperatures of 4, 18 and 30°C.

Storage temp.	Content of couverture on biscuits (%)					
	30		35		40	
	Enthalpy ΔH (J ·g-1)	Temperature Tm (°C)	Enthalpy ΔH (J ·g-1)	Temperature Tm (°C)	Enthalpy ΔH (J ·g-1)	Temperature T=(°C)
Fresh product	**35.07** ± 0.34	**32.92** ± 0.27	**36.55** ± 0.15	**33.35** ± 0.24	**37.29** ± 0.07	**33.35** ± 0.40
4°C	**31.09** ± 0.61	**31.34** ± 0.37	**31.66** ± 0.62	**32.50** ± 0.78	**31.71** ± 0.54	**33.76** ± 0.95
18°C	**28.22** ± 1.12	**26.53** ± 0.87	**28.26** ± 0.67	**33.27** ± 0.58	**29.30** ± 0.82	**33.85** ± 1.05
18°C	**23.77** ± 0.76	**25.28** ± 0.60	**26.64** ± 1.10	**27.99** ± 0.85	**28.35** ± 0.63	**31.99** ± 0.61

Table 9. Enthalpy and maximal melting temperature of cocoa butter in dark couverture non-supplemented with LAB used for coating of biscuits during 4 months of storage at temperatures of 4, 18 and 30°C.

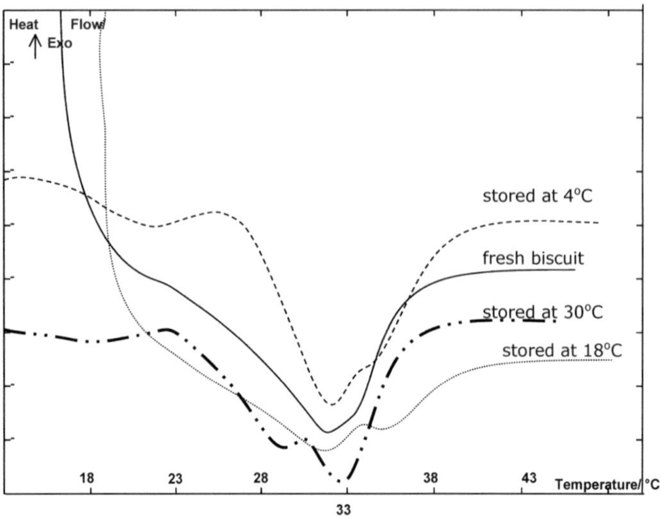

Figure 3. Exemplary DSC thermogram of cocoa butter in the couverture (in the amount of 35%) supplemented with LAB used for biscuit coating in fresh biscuit and biscuits stored for 4 months at temperatures of 4, 18 and 30°C.

In case of peanut fatty mass candy similar tendencies were observed.

Organoleptic analysis of couverture coated cores from biscuits and peanut fatty mass

The best rating of 5.00 obtained biscuits containing the least, namely 30% of couverture (Table 10). With increasing amounts of couverture on cores products obtained lower ratings. Similarly, the best rating 4.80 obtained cores from peanut fatty mass with the lowest amount of couverture, namely 16% (Table 11). Storage of products caused a decrease in organoleptic evaluation, especially those stored at 30°C, which were practically disqualified because of to soft consistency of couverture, and in case of candy also to soft consistency of cores. Organoleptic evaluation of products in supplemented and non-supplemented couverture was statistically on the same level.

Storage temperature	**Content of couverture on biscuits (%)**					
	30	35	40	30	35	40
	Biscuits coated with couverture **supplemented** with LAB			Biscuits coated with couverture non-supplemented with LAB		
Organoleptic rating (points 1- 5)						
Fresh product	**5.00** ± 0.11	**4.85** ± 0.16	**4.75** ± 0.14	**5.00** ± 0.18	**4.85** ± 0.17	**4.75** ± 0.30
4°C	**4.90** ± 0.27	**4.75** ± 0.10	**4.75** ± 0.11	**4.90** ± 0.14	**4.85** ± 0.09	**4.75** ± 0.24
18°C	**4.90** ± 0.24	**4.75** ± 0.31	**4.75** ± 0.24	**4.85** ± 0.11	**4.90** ± 0.14	**4.90** ± 0.31
30°C	4.00 ± 0.17	**4.00** ± 0.17	**3.50** ± 0.23	**4.00** ± 0.30	**3.90**± 0.19	**3.60** ± 0.08

Table 10. Organoleptic analysis of biscuits coated with various amount of couverture supplemented and non-supplemented with LAB in fresh product and after storage for 4 months at temperatures 4, 18 or 30°C.

Storage temperature	Content of couverture on candy (%)					
	16	25	30	16	25	30
	Candy coated with couverture **supplemented** with LAB			Candy coated with couverture non-supplemented with LAB		
	Organoleptic rating (points 1- 5)					
Fresh product	**4.80** ± 0.12	**4.80** ± 0.17	**4.95** ± 0.09	**4.70** ± 0.19	**4.75** ± 0.22	**4.75** ± 0.11
4°C	**4.75** ± 0.12	**4.75** ± 0.08	**4.70** ± 0.28	**4.75** ± 0.30	4.60 ± 0.17	**4.75** ± 0.24
18°C	**4.55** ± 0.13	**4.75** ± 0.14	**4.75** ± 0.27	**4.80** ± 0.19	**4.70** ± 0.18	**4.75** ± 0.14
30°C	**3.90** ± 0.19	**3.70** ± 0.17	**3.60** ± 0.22	**3.60** ± 0.18	**3.50** ± 0.18	**3.50** ± 0.17

Table 11. Organoleptic analysis of peanut fatty mass candy coated with various amount of couverture supplemented and non-supplemented with LAB in fresh product and after storage for 3 months at temperatures 4, 18 or 30°C.

Viability of Lactobacillus bacteria in couverture coated biscuits

Viability of bacteria from *Lactobacillus* species was established in biscuits coated with various amounts of couverture – 30%, 35% and 40%. Biscuits were stored at temperatures of 4, 18 and 30°C for a period of 3 months. The content of *Lactobacillus* bacteria in all products directly after their production amounted from 6.80×10^{7} CFU · g^{-1} (30% of couverture) to 1.74×10^{8} CFU · g^{-1} (35% of couverture). The amount of probiotic bacteria in biscuits stored for a period of 3 moths varied, depending on a storage temperature. After 4 months of storage at a temperature of 4°C of couverture coated biscuits, amount of probiotic bacteria in all studied products was 10^{7} CFU · g^{-1}. Viability of *Lactobacillus* bacteria in coated biscuits after 4 month storage period at 4°C was at a level of 92.6% (30% of couverture) to 96.9% (35% of couverture) (Table 12). Storage at a temperature of 18°C caused a decrease of the amount of live probiotic bacteria in biscuits coated with couverture in amounts of 30% and 35% by two orders of magnitude, comparing to initial amounts. Only in biscuits coated with 40% of couverture bacteria amount maintained on the same level, and after 4 months amounted 2.3×10^{7} CFU · g^{-1}. Viability of probiotic bacteria in a product stored at a temperature of 18°C after 4 months was lower than when stored at 4°C, and ranged from 75.7% (35% of couverture) to 92.6% (40% of couverture). The use of temperature of 30°C during storage caused a significant decrease in an amount of bacteria in a product, comparing to initial level of bacteria as well as to products stored at other temperatures. The content of probiotic bacteria, after 4 months of storage, lowered by 3 - 4 orders of magnitude – from10^{8} CFU · g^{-1} to 10^{3}-10^{4} CFU · g^{-1}. The highest viability of bacteria showed biscuits coated with 30% of couverture (64.9%), and the lowest biscuits coated with 40% of couverture (40.2%). On the basis of performed analyses it can be noticed that probiotic bacteria *L. casei* and *L. paracasei* show the best viability, in couverture coated biscuits stored for during 4 months, when kept at a temperature of 4°C. Confectionery products stored at this temperature also don't change their consistency and organoleptic properties. In case of products stored at temperatures of 18 and 30°C, obtained low amounts of live bacterial cells from *Lactobacillus* species, is not high enough to establish a product to be functional, with an exception of biscuits coated with couverture in an amount of 40%, stored at 18°C.

Couverture content	Storage temperature		
	4°C	18°C	30°C
	Viability of bacteria (%)		
30%	94.4 ± 3.7	78.5 ± 3.4	64.9 ± 4.1
35%	96.9 ± 4.1	75.7 ± 4.1	40.2 ± 4.0
40%	92.6 ± 2.8	92.6 ± 6.2	62.0 ± 3.9

Table 12. Viability of *Lactobacillus* bacteria in biscuits coated with various amounts of couverture after 4 months of storage.

Viability of Lactobacillus bacteria in candy from peanut fatty mass

Directly after product manufacture the amount of live cells of *Lactobacillus* bacteria in couverture amounted 1.6×10^8 CFU · g^{-1} and 1.4×10^8 CFU · g^{-1}, respectively. After 3 month storage period at a temperature of 4°C a slight decrease in an amount of live cells was observed, on average by 2.5%. Lactic bacilli in a couverture, coating candy from peanut fatty mass, in an amount of 16% and 30% maintained the highest viability, after 3 months of storage, at refrigeration temperature (4°C) and was 95.2% and 96.4%, respectively. At a temperature of 18°C after 3 month storage period amount of bacteria decreased by two orders of magnitude (from 10^8 CFU · g^{-1} to 10^6 CFU · g^{-1}), whereas storing at 30°C caused a decrease of three orders of magnitude – from 10^8 CFU · g^{-1} to 10^5 CFU · g^{-1} (Table 13). Increasing the amount of couverture of products slightly improved viability of bacteria, however these changes are not statistically significant. From performed experiments it can be concluded, that probiotic bacteria maintain the highest viability, after 3 month storage period, both at 4 and 18°C. However, the best temperature for storage of candy from peanut fatty mass coated with couverture with an addition of probiotic bacteria, was at the refrigeration temperature (4°C). At these conditions, after 3 months of storage, bacteria viability was the highest and amounted from 95.2% to 96.4%. High viability of bacteria, above 76%, was achieved during storage of candy at a temperature of 18°C. On the other hand, the lowest viability, from 68.1% to 67.8% was observed in products stored at 30°C.

Couverture content in candy from peanut fatty mass	Storage temperature		
	4°C	18°C	30°C
	Viability of bacteria (%)		
16%	95.2 ± 3.3	82.1 ± 4.3	68.1 ± 3.0
30%	96.4 ± 2.4	83.6 ± 3.3	67.8 ± 4.3

Table 13. Viability of *Lactobacillus* bacteria in candy from peanut fatty mass after 3 months of storage.

In Table 14 the amounts of live bacterial cells, after storage for 3 months at different temperatures are presented. Results are calculated per final product, namely per a single candy from peanut fatty mass coated with couverture with a weight of 15 g. Storing this product at temperatures of 4 and 18°C, provides a high level of live *Lactobacillus* bacterial cells, above 10^7 CFU · 15 g^{-1}. Consumed with a confectionery product amount of lactic bacilli is high enough, to provide a beneficial effect of health and well-being of a consumer. During final product storage at a temperature of 30°C level of live bacterial cells (10^6 CFU · 15 g^{-1}) is not high enough, for a product to become functional.

Couverture content	**Storage temperature**		
	4°C	18°C	30°C
	The amount of live bacterial cells in a single piece of candy (CFU · 15 g^{-1})		
16%	9.6×10^8	8.0×10^7	5.8×10^6
30%	1.1×10^9	9.8×10^7	5.1×10^6

Table 14. The amount of live bacterial cells of *Lactobacillus* species in candy from peanut fatty mass, with a weight of 15g, after 3 months of storage.

Full summary of results of analysis regarding all stages of storage can be found in a report from research project supported by Polish Ministry of Science and High Education within development project [11].

Possibility of application of live bacterial cultures of lactic acid preparation for supplementation of chocolate couverture used for confectionery cores coating

Biscuits and cores from fatty masses coated with chocolate couverture supplemented with cultures of lactic acid bacteria, with various percentage content on cores were characterized by correct physicochemical and organoleptic properties for this kind of products. Couverture supplementation with LAB didn't cause any deterioration of physicochemical and organoleptic properties of coated candy and biscuits. For both products, temperatures of 4 and 18°C were proper to achieve high viability of LAB and to classify them as functional food during the whole storage time.

4. Wafers supplemented with lactic acid bacteria

4.1. Wafers

Wafer cream, as an environment for LAB, is a confectionery semi-product with a moisture content below 3%, obtained by aerating of fillings such as: praline, sugar-fat, received from oil seeds, and others (e.g. sugar-protein). Consistency of cream is a sticky and smooth paste, it gives wafer products their characteristic taste. Main components of creams are fat and powdered sugar. The amount of fat in cream depends on relative costs of sugar and fat and on the nutritional purpose of a product. Most of the times 30% of fat are used, but this amount can vary between 23 and 45% of fat in a cream. A certain content of sugar is not exceeded, because it weakens cohesion of a cream. Cream consistency depends on the amount of fat used in a recipe. Other components of creams are: powdered milk, organic acids, flavoring and coloring agents. The addition of wafer production discards gives creams brown color and lowers the concentration of water in the mass. As a partial saccharose substitute glucose can also be used. As a thickening agent and a stabilizer of consistency starch can be added, however it can hinder mass aeration. Water present in a mass causes an increase of viscosity, which can be lowered by an addition of lecithin in an amount of 0.2% per products mass. When producing a cream, to the sticky fat all friable components predicted with a recipe are added, which lowers the temperature of fat. Later, while mixing and aeration the temperature of the mass increases again. After finished mixing process, cream with a definite temperature, density and consistency is received. Density of cream varies between 0.75 and 1.15 g · ml^{-1}. Latter squeezing of cream, under increased pressure, onto a product provides further aeration. In case of wafer creams, it is necessary for them to have high nutritional value, proper taste, flavor and color, smooth and spongy consistency, low water content, or to have bounded structure so they won't soften the wafer. Creams should provide good adhesion to wafers, plasticity and an ability to harden after cooling down of final products. It is important for a cream to be stable at room temperature and to have certain melting characteristics, namely to be solid at a temperature of 20°C, and to melt quickly in the mouth [27-30].

To a cream, used for interleaving wafers, lactic acid bacteria were introduced, making it a product with probiotic properties [31]. Certain physicochemical, organoleptic and textural properties of this cream, compared to non-supplemented cream are described. Research-development works in this subject area were conducted under project no. R12 018 01 [11].

4.2. Obtaining interleaved wafers supplemented with lactic acid bacteria

Production of wafer product includes following steps: preparation and measurement of raw materials, mixing and graining of components, grinding, pouring semi-fluid mass onto individual wafers, sticking of wafer with one another, cutting wafer to required size, optional coating with tempered couverture and finally storage. Main, and the longest process from all mentioned above is the process of mass grinding. It is performed until solid phase particles do not exceed 30 μm. In case of probiotic creams, lyophilized lactic acid bacteria preparation is added to the mass (with a temperature of 40°C) in the final stage of its grinding. Interleaved wafers were received by sticking together three individual wafers (Wafer factory "MIRAN WAFEL" Sp. z .o. o., Poland) with a filling consisting 70% of core mass. Final products comprised of wafer cores coated and non-coated with a dark couverture (Union Chocolate Sp. z o. o., Żychlin, Poland). In coated wafers couverture was added in an amount of 30% per final product mass.

Wafer fillings differed by the type of used fat and its concentration. For human health it is preferable that fats used in confectionery industry have as few trans-configured fatty acids as possible. To realize the idea of nutritional policy for producing wafer fillings transless fats: Akomic 2000 (AarhusKarlshamn, Sweeden) and Akotres S30 (AarhusKarlshamn, Sweeden) and medium-trans fat Efekt 40 (Z.T. Kruszwica S.A., Kruszwica, Polska) were used, in amounts of 34.44, 37.44 and 40.44%, respectively. In supplemented fillings amount of added powdered milk was lowered proportionally to the amount of added lactic acid bacteria lyophilisate, with a concentration of live bacterial cells of *Lactobacillus* species on a level of 9×10^{10} CFU · g^{-1}. Initially 3.5% of lyophilisate was used, however bacteria content was so high, that for economic reasons, this amount was lowered to 0.5% per products weight. This amount, with ease provided a probiotic character of wafer products during whole storage time. Other materials used in wafer filling production are: powdered sugar (Promyk Cukrohurt Sp. z o.o., Siedlce, Poland), wafer discards (Dybalski-Cukiernie, Łódź, Poland), powdered skim milk (S.M. Spomlek, Radzyń Podlaski, Poland), lecithin (Cargill S.A., Bielany Wrocławskie, Poland), ethyl vanillin (Plus, Łódź, Poland). In Table 15 whole recipe for obtaining probiotic wafer fillings in presented. Finished wafer cores were stored at temperatures of 4, 18 and 30°C for a period of 3 months, during which the changes in physicochemical properties and LAB viability were established.

Raw material	Concentration of raw material (%)		
Fat	40.44	37.44	34.44
Sugar	25.71	28.71	31.71
Powdered milk	27.60	27.60	27.60
Production wafer discards	6.18	6.18	6.18
Lecithin	0.05	0.05	0.05
Ethyl vanillin	0.02	0.02	0.02

Table 15. Recipe for obtaining probiotic creams, used as a filling for interleaving wafers, with the use of Efekt 40, Akomic 2000 and Akotres S30 fats.

Considering required physicochemical and organoleptic properties of wafer products for it to be a probiotic product, i.e. proper texture (mainly crunchiness of final product and spreadability of filling), right amount of LAB and unchanged sensory properties, comparing to product without LAB, products were analyzed to establish following parameters: water activity, spreadability of cream, hardness (crunchiness) of product and finally organoleptic evaluation.

4.3. Physicochemical analysis of interleaved wafers supplemented with lactic acid bacteria

Considering the great amount of obtained results of physicochemical analyses of wafers only selected were chosen and presented in a following chapter, namely only those for products stored at 18°C. This temperature was chosen because of the fact that confectionery products are stored at it most of the times on a store shelf. Full results are presented in a report from a research-development project no. R12 018 01 [11].

Water activity

The content of easily accessible water in food as well as the amount and type of solute influences microorganism development in a product. Most of microorganisms prefer a_w from 0.9 to almost 1. Xerophilic mold on a solid surface are able to develop still when a_w values 0.65, and osmophilic yeast are able to expand with a_w of 0.61. Most microorganisms can endure conditions, when water activity of environment is lower than needed for their development. This is the case with lactic bacteria, which during the state of anabiosis, while in lyophilized form were added to wafer fillings. In Figures 4 and 5 variability of water activity of wafer products stored at 18°C for 3 months, which fillings were supplemented with lactic acid bacteria.

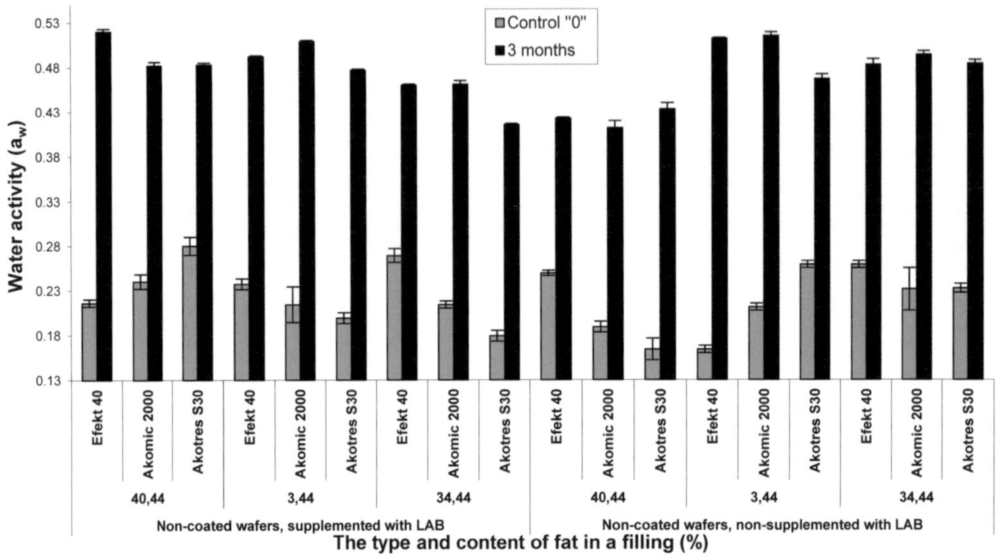

Figure 4. Water activity in **non-coated wafers interleaved with cream supplemented and non-supplemented with LAB**, stored at a temperature of **18°C** for a period of **3 months,** depending on the type and the amount of used fat.

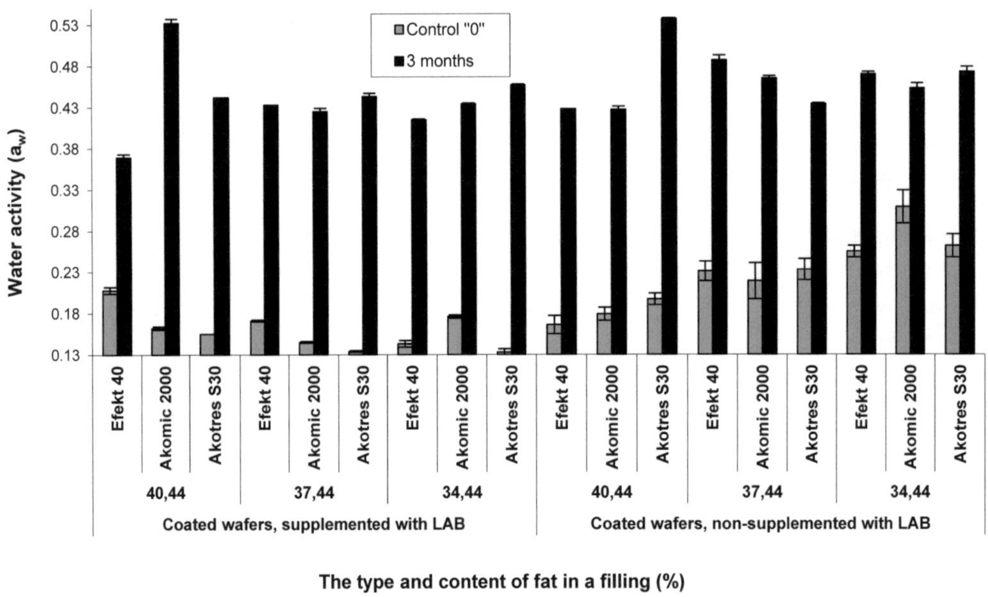

Figure 5. Water activity in **coated wafers interleaved with cream supplemented and non-supplemented with LAB**, stored at a temperature of **18°C** for a period of **3 months**, depending on the type and the amount of used fat.

Initial samples of studied wafers had low water activity, and it ranged between 0.133 and 0.280. The lowest a_w value had wafers coated with couverture supplemented with LAB. During storage wafers showed an overall tendency to absorb moisture from the environment, especially when stored at 18°C. Final values of a_w of products stored in these conditions were in a range of 0.405 – 0.520. It is a level at which LAB are still in a state of anabiosis. Those conditions guaranteed high viability of probiotic bacteria. Judging by obtained results of water activity in wafers: coated and non-coated, supplemented and non-supplemented with LAB, it can be concluded that 3 month storage period at a temperature of 18°C won't have any negative influence on microbiological stability of studied product. Protective role of couverture on wafer core clearly showed in couverture coated wafers, both supplemented and non-supplemented with LAB, when stored at 18°C.

No noticeable correlation between a_w and the type and amount of used fat was observed.

Consistency (spreadability) of confectionery fillings derived from fatty masses

For wafers to be properly, evenly glued together the filling used for their interleaving must have proper consistency (spreadability). This parameter was expressed as a force necessary to immerse in a mass (spreadability) and emerge (adhesiveness) from a mass a conical probe, moving with a constant speed. In Tables 16 and 17 spreadability of masses used as wafer fillings, depending on the type and amount of used fat, is presented. Obtained results for spreadability of fatty masses used as fillings for interleaving wafers show a distinct dependency from the type and amount of used fat.

The biggest hardness, which is equal to the worst spreadablity showed masses received with 34.44% of fat (in this case force values often were non-determinable, above 40.000 g). With an increasing content of fat in a filling its spreadability was improving (force values necessary for immersing the probe averaged between 5.000 and 37.500 g). On the other hand, considering the type of fat used in masses, and the spreadability of those fillings, it can be observed that the least hard, which is equal to the best spreading fillings were obtained with the use of Akotres S30 fat, regardless of the amount of fat. Even in a concentration of 34.44% masses were quite spreadable (force value ranged between 5.000 and 28.000 g), while for fats Efekt 40 and Akomic 2000 in this amount spreadability could not be determined.

When observing obtained results of consistency measurement of filling used for interleaving wafer, no significant influence of LAB supplementation was noticed. It can be concluded that supplementation with LAB of masses used for interleaving wafers won't hinder the ability to properly bind them together. Obtained results of hardness analysis of wafer cores, showed a lack of clear dependency of this parameter from the type and amount of fat used for obtaining creams, as well as from its supplementation with LAB.

Comparing force values from a "biting test" of control samples, it can be noticed that the least amount of force necessary to break, showed wafers interleaved with masses obtained with Akotres S30 fat.

In case of non-coated wafers it can also be noticed, that harder values were obtained by products received with Akomic 2000, comparing to wafers with Efekt 40 fat in its material composition. Similar dependency from the type of used fat was observed for consistency (spreadability) studies of fillings (Table 16).

Type of mass	Type and amount of fat (%)								
	Efekt 40			Akomic 2000			Akotres S30		
	40.44	37.44	34.44	40.44	37.44	34.44	40.44	37.44	34.44
1. Supplemented masses; non-coated	**7.064** ± 0.162	**19.456** ± 1.032	*	**35.116** ± 0.777	**23.491** ± 0.330	*	**11.892** ± 1.245	**17.505** ± 0.411	**25.110** ± 0.078
2. Non-supplemented masses; non-coated	**37.434** ± 0.177	**37.505** ± 0.093	*	**30.811** ± 0.202	**36.438** ± 0.459	**37.531** ± 0.074	**16.149** ± 0.351	**21.036** ± 0.073	**18.904** ± 0.181
3. Supplemented masses; coated	**9.123** ± 0.329	**14.957** ± 1.374	**37.549** ± 0.050	**32.507** ± 0.549	**37.559** ± 0,230	**37.559** ±0.557	**4.399** ± 0.623	**6.656** ± 0.473	**5.822** ± 0.814
4. Non-supplemented masses; non-coated	**34.594** ± 0.529	**22.505** ± 0.220	**37.559** ± 0.042	**26.870** ± 0.395	**34.489** ± 0.304	**37.561** ± 0,050	**13.171** ± 0.048	**4.065** ± 0.375	**28.120** ± 0.294

Table 16. Consistency (spreadability) of cream used for interleaving wafers, expressed as force (g) necessary to immerse a conical probe in a mass, depending on material composition.
* Masses, in which consistency could not be determined, the value of applied force above 38 000 g

Type of mass	Type and amount of fat (%)								
	Efekt 40			Akomic 2000			Akotres S30		
	40.44	37.44	34.44	40.44	37.44	34.44	40.44	37.44	34.44
1. Supplemented masses; non-coated	**-4.952** ± 0.156	**-6.206** ± 0.351	*	**-7.471** ± 0.252	**-8.776** ± 0.080	*	**-6.074** ± 0.682	**-5.674** ± 0.174	**-5.769** ± 0.029
2. Non-supplemented masses; non-coated	**-4.847** ± 0.444	**-3.938** ± 0.109	*	**-7.266** ± 0.548	**-6.451** ± 0.505	**-6.821** ± 0.626	**-6.487** ± 0.201	**-5.761** ± 0.095	**-6.489** ± 0.197
3. Supplemented masses; coated	**-6.205** ± 0.242	**-6.274** ± 0.519	**-0.002** ± 0.650	**-7.529** ± 0.267	**-0.003** ± 0.001	**-3.858** ± 0.342	**-4.849** ± 0.282	**-4.489** ± 0.030	**-4.276** ± 0.012
4. Non-supplemented masses; non-coated	**-5.347** ± 0.036	**-6.447** ± 0.257	**-3.619** ± 0.022	**-8.737** ± 0.172	**-6.850** ± 0.631	**-0.002** ± 0.000	**-7.882** ± 0.151	**-4.195** ± 0.424	**-5.075** ± 0.162

Table 17. Consistency (adhesiveness) of cream used for interleaving wafers, expressed as force (g) necessary to emerge a conical probe from a mass, depending on material composition.
* Masses, in which consistency could not be determined

An increase of hardness of wafer products during storage, might be caused by an increase of individual wafers hardness caused by moisture absorption as well as changes occurring in consistency of filling resulting from shifting proportion between the content of solid to liquid phase. Polymorphic form, in which initially fat components crystallized (change of melting temperature – DSC measuring) could change, and an altering of crystalline network structure of masses used for interleaving wafers could occur. As a result of new crystals emergence from already present crystal germ, the filling could harden, or as a result of crystal aggregation – soften (an increase of a decrease of solid phase surface). It seems probable, that those changes in quite significant degree, could be the reason of hardness changes in final products.

Hardness (crunchiness) of wafer cores, interleaved with LAB supplemented cream

In case of wafer type products, one the most important organoleptic property, which consumer pays close attention to when choosing his favorite product, is wafer hardness (crunchiness). This parameter in established in a "biting test". It is expressed as a force value necessary to fully cut the wafer core. In Figures 6 and 7 hardness values of wafers: supplemented and non-supplemented with LAB, coated and non-coated with couverture, after 3 months of storage at a temperature of 18°C are presented.

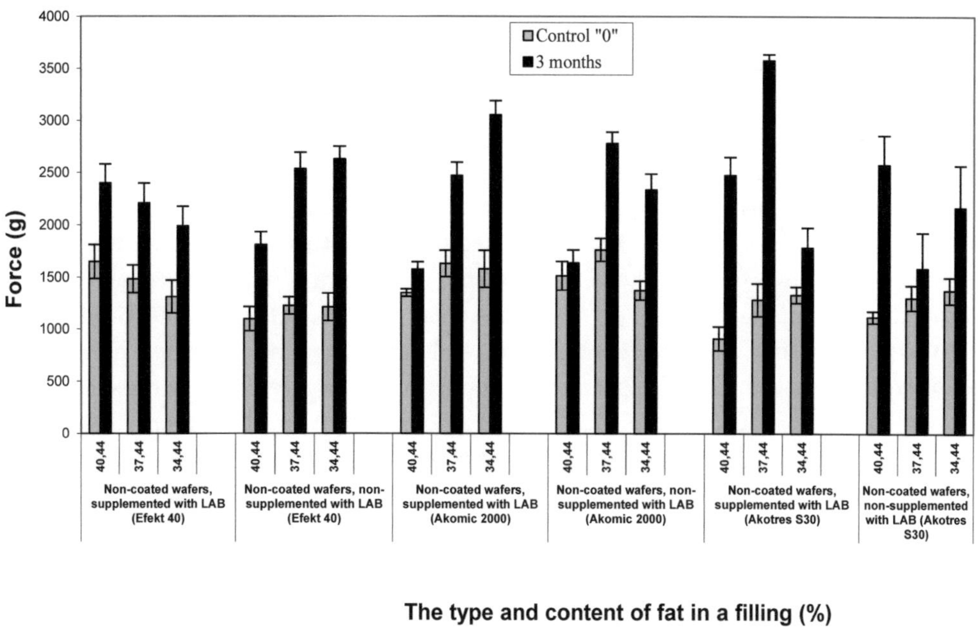

Figure 6. Comparison of hardness values of **non-coated wafers interleaved with cream supplemented and non-supplemented with LAB**, stored for **3 months** at a temperature of **18°C**, depending on the type and amount of used fat.

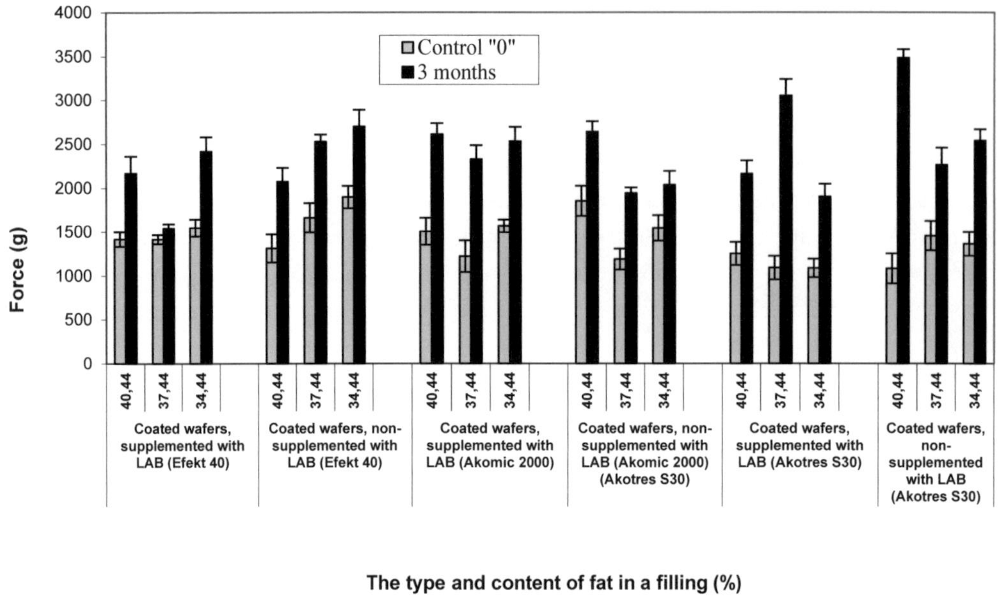

Figure 7. Comparison of hardness values of **coated wafers interleaved with cream supplemented and non-supplemented with LAB**, stored for **3 months** at a temperature of **18°C**, depending on the type and amount of used fat.

Polymorphic changes of fats

Thermal analysis of fatty mass fillings used for interleaving wafers indicated differences in polymorphism of fats used for obtaining products, mainly depending on temperature and storage period. No significant influence of LAB supplementation on the amounts, or temperatures of disintegration of polymorphic forms of used fats was observed. On average, 3 polymorphic forms of used fats occured, regardless of their type. In filling received with Akotres S30 fat (transless) 3 peaks on an endothermic curve were observed, there were temperatures of 12.50, 24.8 and 34.6°C, for Akomic 2000 fat (transless) those temperatures were: 13.98, 26.7 and 33.34°C. The highest melting temperatures of polymorphic forms were obtained for fillings with Efekt 40 fat (trans-containing), namely 16.78, 27.95 and 34.9°C. With an increasing storage temperature and storage time changes occurring in polymorphism of fats used for producing fillings were observed. An increase of melting temperatures was observed, also a new polymorphic form of fat in products stored at 30°C was noticed. The biggest changes occurred in products made with medium-trans Efekt 40 fat, and the least significant ones in products with transless Akotres S30 fat.

The lower melting temperatures of polymorphic forms of fats are, the more it is possible for it to contain a significant amount of unsaturated fatty acids. Whereas, the higher melting temperature of a polymorphic form of fat, the more saturated fatty acids can be found in its composition. Taking this criteria into account, the most beneficial it is to use Akotres S30 fat for obtaining fatty mass fillings. Endotherms obtained for melting of this fat in control samples didn't indicate any significant changes in shape of values of melting temperatures and enthalpies, compared to samples obtained from products stored for 6 and 12 weeks. It indicates that studied wafers are suitable for consumption during a whole 3 month period of storage. This tendency was noticed for wafers both, supplemented and non-supplemented with LAB. DSC analysis of fillings used for interleaving wafers also revealed that additional supplementation with LAB didn't influence significantly physicochemical properties of final product.

Organoleptic evaluation

The best flavor and appearance properties had product both, supplemented and non-supplemented directly after production. During storage in all products crunchiness parameter decreased, also in case of coated wafers, an appearance of couverture was changing (grey coating on a surface of couverture appeared). In Tables 18 and 19 organoleptic rating of coated wafers supplemented and non-supplemented with LAB is presented. Rating of non-coated wafer cores and a wide description of all types of products can be read in a report from a project [11].

Storage time / **Fat concentration**	Efekt 40			Akomic 2000			Akotres S30		
	40.44%	37.44%	34.44%	40.44%	37.44%	34.44%	40.44%	37.44%	34.44%
Coated wafers supplemented with LAB lyophilisate at concentation of 0.5%									
Control „0"	**3.9**±0.3	**3.9**±0.1	**3.8**±0.1	**3.8**±0.2	**3.9**±0.1	**3.9**±0.3	**3.9**±0.1	**3.9**±0.1	**3.8**±0.1
Stored at 4°C									
6 weeks	**3.8**±0.2	**3.8**±0.1	**3.8**±0.1	**3.8**±0.1	**3.9**±0.2	**3.9**±0.2	**3.8**±0.1	**3.8**±0.1	**3.8**±0.1
12 weeks	**3.7**±0.2	**3.7**±0.1	**3.7**±0.2	**3.7**±0.2	**3.7**±0.2	**3.7**±0.2	**3.7**±0.2	**3.7**±0.2	**3.7**±0.3
Stored at 18°C									
6 weeks	**3.5**±0.1	**3.4**±0.2	**3.4**±0.1	**3.4**±0.1	3.5±0.3	**3.5**±0.1	**3.3**±0.2	**3.5**±0.1	**3.5**±0.1
12 weeks	**3.0**±0.2	**2.7**±0.3	**2.7**±0.2	**2.7**±0.2	**3.0**±0.3	**3.0**±0.2	**3.7**±0.1	**3.0**±0.2	**3.0**±0.1
Stored at 30°C									
6 weeks	**3.0**±0.0	**3.0**±0.1	**2.7**±0.3	**3.0**±0.1	**3.0**±0.3	**2.7**±0.1	**3.0**±0.2	**3.0**±0.2	**2.7**±0.2
12 weeks	**2.7**±0.1	**2.5**±0.2	**2.6**±0.1	**2.5**±0.2	**2.4**±0.3	**2.7**±0.1	**2.6**±0.1	**2.6**±0.2	**2.7**±0.1

Table 18. Organoleptic rating of **coated wafers supplemented with LAB**, differing by a material composition, depending on the storage time and temperature.

In case of non-coated wafers, regardless of storage period and temperature, a few parameters remained at the same level, namely: wafer color, filling color and filling consistency at room temperature. Tastiness of products didn't change, but overall taste impressions were worse than in control samples, resulting from changes which occurred in products. Wafers stored at a temperature of 4°C lost their crispiness. Products stored at 18°C dried, or lost their crispiness, and in some cases became harder than control samples. Wafers stored at 30°C showed good crunchiness and crispiness, but at the same time were very fragile. Coated wafers, even then freshly made were not evenly coated with couverture, "overcoatings" were observed, and because of that organoleptic rating of those products suffered, receiving grades below 4 (desirable quality). Coated wafers stored at 18 and 30°C received in a rating values of "tolerable" or below. Factor that disqualified wafers stored at a temperature of 30°C were changes of color and consistency of couverture and taste of a whole product.

Storage time / Fat concentration	Efekt 40			Akomic 2000			Akotres S30		
	40.44%	37.44%	34.44%	40.44%	37.44%	34.44%	40.44%	37.44%	34.44%
Model wafers – coated, non-supplemented with LAB									
Control „0"	**3.9**±0.2	**3.9**±0.1	**3.8**±0.2	**3.9**±0.2	**3.9**±0.1	**3.8**±0.3	**3.8**±0.2	**3.9**±0.1	**3.9**±0.1
Stored at 4°C									
6 weeks	**3.9**±0.1	**3.9**±0.1	**3.8**±0.1	**3.8**±0.2	**3.8**±0.1	**3.8**±0.2	**3.8**±0.2	**3.9**±0.2	**3.8**±0.1
12 weeks	**3.7**±0.1	**3.9**±0.1	**3.7**±0.2	**3.7**±0.1	**3.7**±0.1	**3.7**±0.3	**3.7**±0.1	**3.9**±0.1	**3.7**±0.2
Stored at 18°C									
6 weeks	**3.7**±0.1	**3.5**±0.3	**3.5**±0.1	**3.4**±0.1	**3.7**±0.2	**3.5**±0.1	**3.5**±0.2	**3.7**±0.1	**3.4**±0.3
12 weeks	**3.0**±0.1	**3.0**±0.2	**3.0**±0.1	**2.7**±0.2	**2.7**±0.1	**3.0**±0.1	**3.0**±0.2	**3.0**±0.1	**3.0**±0.1
Stored at 30°C									
6 weeks	**3.0**±0.2	**3.0**±0.1	**3.0**±0.1	**3.0**±0.2	**2.7**±0.2	**3.0**±0.1	**3.0**±0.1	**3.0**±0.3	**3.0**±0.2
12 weeks	**2.5**±0.2	**2.7**±0.2	**2.5**±0.3	**2.7**±0.1	**2.4**±0.1	**2.6**±0.2	**2.7**±0.1	**2.5**±0.1	**2.7**±0.1

Table 19. Organoleptic rating of **coated wafers non-supplemented with LAB**, differing by a material composition, depending on the storage time and temperature.

No noticeable influence of the type and amount of fat used in fillings on changes occurring during storage of products was observed. In no way, in products supplemented with LAB, the presence of lactic acid bacteria was noticed during organoleptic evaluation.

Viability of Lactobacillus bacteria in wafers

Similar to previously described products, wafers were also stored at temperatures of 4, 18 and 30°C. 27 types of wafers with various composition were examined, including 18 wafers non-coated with couverture (W1 - W18) and 9 types of wafers with a couverture coating (W19 - W27).

Lactic acid bacteria in a form of lyophilisate were introduced into the filling of wafers. Initially 3.5% of lyophilisate was used (wafers W1 - W5), however bacteria level proved to be so high, that for economic reasons the amount of added lyophilisate was reduced to a range of 0.71% (W10 - W18) to 0.5% (wafers W6 - W9 and W19 - W27).

Initial level of LAB in non-coated wafers W1 – W5 ranged between 4.0×10^{8} CFU · g^{-1} and 7.0×10^{8} CFU · g^{-1}, in wafers W6 - W18 it ranged from 2.8×10^{7} CFU · g^{-1} to 1.7×10^{8} CFU · g^{-1}. Storage of non-coated wafers at a refrigeration temperature (4°C) allowed to maintain a high viability of probiotic bacteria from *Lactobacillus* species. In all types of wafers, regardless of its composition, after three months of storage viability was quite high and ranged from 92.2% (W6) to 98.3% , which is equal to an amount of live cells from 3.8×10^{7} CFU · g^{-1} to

8.4×10^7 CFU · g^{-1}. Very high viability of bacteria was maintained in wafers non-coated with chocolate, with 40.44% and 37.44% of Akomic 2000 fat in its composition (W13 – 98.1%, W14 – 98.0%), and in non-coated wafers with 34.44% of Akotres S30 fat (W18 – 99.4%) (Table 20).

During storage at a temperature of 18°C lactic acid bacteria viability was at a similar level in all studied types of wafers. The biggest amount of live cells of probiotic bacilli was observed in non-coated wafers, to which 3.5% of lyophilisate was added (W1 – W5), and in non-coated wafers with Akotres S30 fat in a concentrations of 34.44% and 37.44% (W8 – W9), ranging from 80.3% (W8) to 87.9% (W9) (Table 20). In other wafers viability of probiotic bacteria kept on a level of 72-7 – 77.8% (Table 23). Low level of live cells of *Lactobacillus* bacteria was observed after storage of wafers at a temperature of 30°C. Beside wafers W1 – W5, in which viability of bacteria amounted from 70.4% (W1) to 72.2% (W5), in other non-coated wafers viability of lactic bacilli ranged from 62.5% (W18) to 69.4% (W9) (Table 20). The rest of the products, i.e. W19 – W27 contained of wafers with chocolate coating, in which initial levels of probiotic bacteria ranged from 2.5×10^7 CFU · g^{-1} (W25) to 8.1×10^7 CFU · g^{-1} (W23). The addition of couverture was performed to hinder oxygen access to the filling, containing live cells of probiotic bacteria, and therefore to improve the viability of bacteria in the product. The biggest viability, from 95.69% (W23) to 98.64% (W21), was achieved by storing couverture coated wafers at a temperature of 4°C. In couverture coated wafers containing Efekt 40 fat, at a concentration of 37.44% and 34.44% (W20, W21) and in coated wafers containing Acomic 2000 fat at a concentration of 40.44% (W22), bacteria viability after 3 months of storage was the biggest and amounted 98.2%, 98.6% and 98.0%, respectively (Table 20). Also on a high level was viability of live cells of probiotic bacteria in wafers stored at 18°C. At this temperature, after 3 months of storage viability of bacteria ranged between 72.2% (W23) and 83.6% (W19). A temperature of 30°C proved to be the least desirable for couverture coated wafers storage. After 3 months of storage at this temperature, the amount of live cells of *Lactobacillus* bacteria ranged from 7.1×10^4 CFU · g^{-1} to 1.6×10^5 CFU · g^{-1}, which equaled to a viability range from 61.6% (W26) to 67.2% (W20) (Table 20).

Examined wafers differed not only in the amounts of added lyophilisate, but also in the content of fat and sugar. Fat can be a substance protecting cells, and sugar participates in lactic fermentation. However, no influence of those constituents on viability of *Lactobacillus* bacteria in wafers was observed. Similarly, coating wafers with couverture did not influence bacteria viability in examined wafers in a significant manner.

Obtained results allowed to conclude, that preferable temperatures for storage of wafers, both coated and non-coated with courertuve, which provides high probiotic bacteria viability are two temperatures, i.e. refrigeration temperature (4°C) as well as a temperature suggested by normative legislations for storage of this type of products, namely 18°C.

The amount of live cells of *Lactobacillus* bacteria consumed with one piece of wafer, stored for 3 months at a temperature of 4°C reached 10^9 CFU per wafer (Table 21). It was the highest in wafers non-coated with chocolate, in which lyophilisate content was 3.5% (W1 – W5) and amounted from 5.3×10^9 CFU · 28 g^{-1} to 9.8×10^9 CFU · 28 g^{-1} (Table 21). High level of live cells of probiotic bacteria, i.e. from 10^7 CFU to 10^8 CFU per individual product, is also maintained when products are stored for 3 months at a temperature of 18°C (Table 21). In case of wafers stored at a temperature of 30°C, the amount of live and active cells consumed by a potential buyer, would be lower than 10^7 CFU per wafer, for most of examined products. With an exception of wafers W1 – W5, in which this level was from 4.3×10^7 CFU · 28 g^{-1} to 5.8×10^7 CFU · 28 g^{-1}, but this requires an addition of 3.5% of lyophilisate to the product.

Couverture	Type of fat	Fat content (%)	Sugar content (%)	Symbol of product	Lyophilisate content in final product (%)	Storage temperature		
						4°C	18°C	30°C
						Viability of bacteria (%)*		
Non-coated	Efekt 40	40.44	25.71	W1	3.50	97.2	81.6	70.4
	Efekt 40	37.44	28.71	W2	3.50	96.9	80.7	70.7
	Efekt 40	34.44	31.71	W3	3.50	96.6	80.5	72.1
	Akomic 2000	40.44	25.71	W4	3.50	94.1	84.5	70.8
	Akomic 2000	37.44	28.71	W5	3.50	96.1	85.7	72.2
	Akomic 2000	34.44	31.71	W6	0.50	92.2	77.5	63.6
	Akotres S30	40.44	25.71	W7	0.50	93.,5	76.2	63.9
	Akotres S30	37.44	28.71	W8	0.50	95.5	80.3	65.9
	Akotres S30	34.44	31.71	W9	0.50	97.2	87.9	67.4
	Efekt 40	40.44	25.71	W10	0.71	95.7	75.7	64.7
	Efekt 40	37.44	28.71	W11	0.71	94.5	77.4	65.6
	Efekt 40	34.44	31.71	W12	0.71	97.7	75.0	65.1
	Akomic 2000	40.44	25.71	W13	0.71	98.1	77.8	65.1
	Akomic 2000	37.44	28.71	W14	0.71	98.0	76.5	64.6
	Akomic 2000	34.44	31.71	W15	0.71	96.6	74.2	64.1
	Akotres S30	40.44	25.71	W16	0.71	97.3	75.0	64.4
	Akotres S30	37.44	28.71	W17	0.71	98.0	72.7	63.0
	Akotres S30	34.44	31.71	W18	0.71	98.4	76.7	62.5
Coated	Efekt 40	40.44	25.71	W19	0.50	97.6	82.6	65.4
	Efekt 40	37.44	28.71	W20	0.50	98.2	81.4	67.2
	Efekt 40	34.44	31.71	W21	0.50	98.6	77.6	66.9
	Akomic 2000	40.44	25.71	W22	0.50	98.0	72.6	62.7
	Akomic 2000	37.44	28.71	W23	0.50	95.7	72.2	63.0
	Akomic 2000	34.44	31.71	W24	0.50	97.2	73.4	63.0
	Akotres S30	40.44	25.71	W25	0.50	96.3	74.1	64.0
	Akotres S30	37.44	28.71	W26	0.50	96.1	74.7	61.6
	Akotres S30	34.44	31.71	W27	0.50	98.0	75.2	63.2

Table 20. Viability of *Lactobacillus* bacteria in wafers after 3 months of storage.
*Weight of an average wafer without couverture coating is 28 g, and with couverture coating - 39 g.

Couverture	Type of fat	Fat content (%)	Sugar content (%)	Symbol of product	Lyophilisate content in final product (%)	Storage temperature		
						4°C	18°C	30°C
						The amount of live bacterial cells		
Non-coated	Efekt 40	40.44	25.71	W1	3.50	9.8×10^9	4.2×10^8	4.3×10^7
	Efekt 40	37.44	28.71	W2	3.50	8.6×10^9	3.3×10^8	4.4×10^7
	Efekt 40	34.44	31.71	W3	3.50	8.3×10^9	3.2×10^8	5.8×10^7
	Akomic 2000	40.44	25.71	W4	3.50	5.9×10^9	8.4×10^8	5.2×10^7
	Akomic 2000	37.44	28.71	W5	3.50	5.3×10^9	6.7×10^8	4.6×10^7
	Akomic 2000	34.44	31.71	W6	0.50	1.1×10^9	6.6×10^7	4.8×10^6
	Akotres S30	40.44	25.71	W7	0.50	1.2×10^9	4.6×10^7	4.5×10^6
	Akotres S30	37.44	28.71	W8	0.50	1.1×10^9	6.6×10^7	4.8×10^6
	Akotres S30	34.44	31.71	W9	0.50	1.4×10^9	5.2×10^7	5.4×10^6
	Efekt 40	40.44	25.71	W10	0.71	1.4×10^9	3.5×10^7	4.6×10^6
	Efekt 40	37.44	28.71	W11	0.71	1.7×10^9	3.5×10^7	4.1×10^6
	Efekt 40	34.44	31.71	W12	0.71	1.7×10^9	2.6×10^7	4.2×10^6
	Akomic 2000	40.44	25.71	W13	0.71	1.9×10^9	4.5×10^7	4.4×10^6
	Akomic 2000	37.44	28.71	W14	0.71	1.9×10^9	3.6×10^7	4.0×10^6
	Akomic 2000	34.44	31.71	W15	0.71	1.6×10^9	2.5×10^7	4.1×10^6
	Akotres S30	40.44	25.71	W16	0.71	1.3×10^9	2.2×10^7	3.1×10^6
	Akotres S30	37.44	28.71	W17	0.71	1.8×10^9	1.8×10^7	3.0×10^6
	Akotres S30	34.44	31.71	W18	0.71	1.8×10^9	3.4×10^7	2.6×10^6
Coated	Efekt 40	40.44	25.71	W19	0.50	1.9×10^9	1.0×10^8	4.8×10^6
	Efekt 40	37.44	28.71	W20	0.50	1.9×10^9	8.0×10^7	6.3×10^6
	Efekt 40	34.44	31.71	W21	0.50	2.0×10^9	4.1×10^7	6.0×10^6
	Akomic 2000	40.44	25.71	W22	0.50	1.5×10^9	1.7×10^7	2.8×10^6
	Akomic 2000	37.44	28.71	W23	0.50	1.4×10^9	2.0×10^7	3.8×10^6

	Akomic 2000	34.44	31.71	W24	0.50	1.5×10^9	2.1×10^7	3.2×10^6
	Akotres S30	40.44	25.71	W25	0.50	1.3×10^9	1.9×10^7	3.2×10^6
	Akotres S30	37.44	28.71	W26	0.50	1.4×10^9	3.0×10^7	2.8×10^6
	Akotres S30	34.44	31.71	W27	0.50	1.5×10^9	2.6×10^7	3.1×10^6

Table 21. The amount of live bacterial cells of Lactobacillus species in non-coated wafer, with a weight of 28 g, and in couverture coated wafer, with a weight of 39 g, after 3 months of storage.

Applications

Received probiotic product in a form of wafers interleaved with a mass supplemented with LAB had similar organoleptic properties, i.e. color, structure, exterior appearance, consistency, balanced taste and smell to wafers produced with a mass without lactic bacteria lyophilisate. Additional presence of bacterial preparation didn't influence in any significant manner water activity in products. Supplemented wafer products maintained proper conditions, which provided lactic acid bacteria with an environment, and allowed it to stay on a level, so that it can be considered to be a product with functional properties. In case creams used for interleaving wafers, besides proper organoleptic rating, they have to have certain textural properties, such as: adhesiveness, hardness and spreadability. In this regard supplemented masses were very similar to non-supplemented ones. According to above observations it can be concluded that it is safe to use masses supplemented with lactic acid bacteria for interleaving wafers, increasing this way health benefits of final wafer products.

5. Raisins coated with chocolate supplemented with live cultures of lactic acid bacteria

To receive raisins coated with chocolate sultana raisins from Iran and chocolate couvertures (dark, milk and white) from Union Chocolate (Żychlin, Poland) supplemented with lyophilized live cultures of lactic acid bacteria from *Lactobacillus* species on a level of 9×10^{10} CFU · g^{-1} were used. For polishing raisins in chocolate polishing agent was used, prepared according to a recipe: distilled water (56.4% w/w), citric acid (0.35% w/w), glucose-fructose syrup (5.275% w/w), saccharose (16.082% w/w), acacia gum (47.47% w/w), edible oil (0.3% w/w) and soy lecithin (0.1% w/w) [32].

In a table below results of analysis of Casson viscosity and yield value of couverture are presented.

Using chocolate couverture supplemented with LAB in an amount of 0.5% based on the weight of the product, at a level of about 50% in relation to raisin core, caused an increase in couverture viscosity. The increase was the highest when white couverture was LAB supplemented. Addition of lactic acid bacteria preparation also influenced the yield value of couvertures. A significant drop of yield value was observed in white couverture supplemented with LAB. Smaller changes were noticed in dark couverture. Whereas in milk couverture yield value was practically the same.

Type of couverture	η_{CA} (Pa·s)	τ_{CA} (Pa)
Dark	**1.333** ± 0.023	**8.86** ± 0.28
Dark + LAB	**1.398** ± 0.011	**8.47** ± 0.09
White	**1.913** ± 0.008	**4.31** ± 0.19
White + LAB	**2.616** ± 0.032	**2.45** ± 0.02
Milk	**2.189** ± 0.009	**1.02** ± 0.02
Milk + LAB	**2.586** ± 0.047	**1.08** ± 0.01

Table 22. Casson viscosity (η_{CA}) and yield value (τ_{CA}) of couvertures.

5.1. Obtaining chocolate coated raisins

To receive chocolate coated raisins following procedure was used. Raisins were washed, dried at a temperature of 30°C, sorted according to size and placed in a coating drum heated previously to 25°C. Temperature was kept constant during the whole process of coating. Onto rotating in a drum raisins subsequent portions of tempered couverture with a temperature of 33°C were poured, total in an amount of 50% in relation to the weight of the product. In raisins in chocolate supplemented with lactic acid bacteria, before coating, to a couverture LAB lyophilisate in an amount of 0.8% based on the weight of the product, was added and stirred for 5 min to provide a full distribution. First portion of couverture was laid on raisins without the use of cool air stream. Latter layers of couverture were placed on raisins with cool air blowing on it while coating. The time of coating of one layer of couverture was 30 s. Total time of coating amounted to 65-150 min, depending on a temperature and air humidity. After coating raisins in chocolate were placed on sieves (in a single layer) and left for 24 hours to obtain a full solidification and consolidation of chocolate couverture structure. After 24 hours, ready product was polished in a spinning coating machine by gradually pouring portions of polishing agent onto it. After each polishing layer product was left in a spinning coating machine for 2 min with cool air blower turned off. After this period cool air was turned on again. Next layer of polishing agent was used when polished product was dry. After finished polishing process, dry chocolate coated raisins were placed on sieves for at least 2 hour period. Afterward, product was packed into plastic bags and kept for storage at temperatures of 4, 18 and 30°C for a period of 3 months. Analyses were performed at monthly intervals [11, 32].

In obtained raisins coated with chocolate supplemented with LAB and with normal (control) chocolate couverture content was established to verify the degree of stratification (which should be about 50%). Obtained results are presented in Table 23.

From obtained results of average percentage content of couverture in received raisins coated with chocolate it can be noticed, that this parameter was at a level of about 50% (w/w), as planned. Raisins coated with supplemented white chocolate were coated in the smallest degree. It was probably caused by bigger losses in a coating machine.

Average content of couverture (%) in chocolate coated raisins					
Dark + LAB	Dark	White + LAB	White	Milk + LAB	Milk
50.47 ± 2.21	**49.68** ± 0.75	**46.36** ± 2.80	**49.99** ± 1.59	**50.55** ± 0.04	**50.64** ± 1.32

Table 23. Average percentage content of couverture in raisins with diffetent types of couverture.

5.2. Physicochemical analysis of chocolate coated raisins

Dry mass content in chocolate coated raisins

Results of dry mass analysis in chocolate coated raisins are placed in Table 24.

Storage time	Dry mass content (%) in chocolate coated raisins:					
	Dark + LAB	Milk + LAB	White + LAB	Dark	Milk	White
Control „0"	**92.85** ± 0.04	**92.72** ± 0.07	**92.45** ± 0.03	**92.15** ± 0.02	**91.80** ± 0.04	**92.76** ± 0.03
Stored at 4°C						
1 month	**91.56** ± 0.03	**91.80** ± 0.09	**91.43** ± 0.07	**92.23** ± 0.04	**91.85** ± 0.07	**91.76** ± 0.07
2 months	**92.38** ± 0.02	**91.24** ± 0.06	**92.33** ± 0.03	**92.69** ± 0.03	**92.33** ± 0.09	**92.03** ± 0.04
3 months	**92.36** ± 0.06	**92.84** ± 0.07	**92.93** ± 0.04	**92.69** ± 0.07	**93.20** ± 0.10	**92.23** ± 0.06
Stored at 18°C						
1 month	**91.70** ± 0.02	**91.90** ± 0.03	**90.64** ± 0.04	**93.04** ± 0.04	**91.56** ± 0.04	**92.19** ± 0.02
2 months	**92.70** ± 0.06	**92.53** ± 0.09	**92.56** ± 0.03	**92.52** ± 0.03	**91.95** ± 0.07	**92.60** ± 0.07
3 months	**93.12** ± 0.05	**92.46** ± 0.02	**92.97** ± 0.02	**92.91** ±0.06	**93.10** ± 0.03	**92.47** ± 0.05
Stored at 30°C						
1 month	**92.23** ± 0.02	**93.00** ± 0.07	**92.28** ± 0.02	**92.93** ± 0.09	**92.44**± 0.06	**92.08** ± 0.03
2 months	**93.89** ± 0.04	**93.66** ± 0.04	**94.47** ± 0.04	**92.99** ± 0.04	**93.66** ± 0.02	**93.76** ± 0.05
3 months	**94.41** ± 0.03	**94.25** ± 0.02	**94.73** ± 0.03	**94.47** ± 0.07	**94.95** ± 0.05	**94.18** ± 0.04

Table 24. Dry mass content (%) in chocolate coated raisins, received with the use of different types of couverture **supplemented**, and as a comparison **non-supplemented** with LAB, stored at temperatures of 4, 18 and 30°C for 3 months.

Directly after obtaining the biggest dry mass content was noticed in raisins coated with dark chocolate supplemented with LAB.

Supplementation of dark and milk couvertures caused an increase of dry mass content in final products, comparing to analogous products with non-supplemented couvertures. In case of raisins coated with white couverture supplemented with LAB, dry mass content was slightly smaller than in non-supplemented one, namely by about 0.3% percentage point. During storage dry mass content in chocolate coated raisins changed. Usually after slight decrease after the first month, dry mass content increased during latter storage month. Higher temperature used during storage of supplemented raisins in chocolate caused an increase of dry mass content, regardless of the type of couverture used for coating (dark, white, milk).

Water activity in chocolate coated raisins

Results of water activity (a_w) in chocolate coated raisins are presented in Table 25.

Storage time		Water activity in chocolate coated raisins:					
		Dark + LAB	**Milk + LAB**	**White + LAB**	**Dark**	**Milk**	**White**
Control „0"	whole	**0.427 ±** 0.001	**0.408 ±** 0.001	**0.414 ±** 0.001	**0.414 ±** 0.002	**0.420 ±** 0.011	**0.386 ±** 0.003
	crushed	**0.472 ±** 0.007	**0.474 ±** 0.003	**0.507 ±** 0.004	**0.486 ±** 0.003	**0.510 ±** 0.020	**0.389 ±** 0.007
Stored at 4°C							
1 month	whole	**0.491 ±** 0.002	**0.457 ±** 0.002	**0.480 ±** 0.001	**0.476 ±** 0.003	**0.429 ±** 0.001	**0.390 ±** 0.001
	crushed	**0.535 ±** 0.009	**0.496 ±** 0.004	**0.490 ±** 0.007	**0.512 ±** 0.012	**0.512 ±** 0.006	**0.550 ±** 0.004
2 months	whole	**0.540 ±** 0.001	**0.433 ±** 0.001	**0.499 ±** 0.001	**0.523 ±** 0.001	**0.533 ±** 0.004	**0.490 ±** 0.003
	crushed	**0.527 ±** 0.004	**0.524 ±** 0.009	**0.522 ±** 0.004	**0.533 ±** 0.007	**0.563 ±** 0.008	**0.537 ±** 0.013
3 months	whole	**0.545 ±** 0.001	**0.541 ±** 0.001	**0.514 ±** 0.004	**0.550 ±** 0.001	**0.490 ±** 0.005	**0.510 ±** 0.001
	crushed	**0.521 ±** 0.003	**0.565 ±** 0.004	**0.547 ±** 0.009	**0.545 ±** 0.009	**0.524 ±** 0.011	**0.527 ±** 0.019
Stored at 18°C							
1 month	whole	**0.494 ±** 0.001	**0.460 ±** 0.001	**0.483 ±** 0.002	**0.416 ±** 0.001	**0.426 ±** 0.004	**0.358 ±** 0.003
	crushed	**0.546 ±** 0.001	**0.488 ±** 0.003	**0.493 ±** 0.007	**0.504 ±** 0.014	**0.513 ±** 0.009	**0.532 ±** 0.004
2 months	whole	**0.514 ±** 0.002	**0.416 ±** 0.001	**0.497 ±** 0.001	**0.505 ±** 0.001	**0.532 ±** 0.001	**0.498 ±** 0.001
	crushed	**0.520 ±** 0.003	**0.511 ±** 0.002	**0.522 ±** 0.011	**0.521 ±** 0.006	**0.541 ±** 0.017	**0.529 ±** 0.007
3 months	whole	**0.535 ±** 0.001	**0.555 ±** 0.002	**0.551 ±** 0.001	**0.566 ±** 0.001	**0.539 ±** 0.001	**0.555 ±** 0.004
	crushed	**0.516 ±** 0.007	**0.548 ±** 0.009	**0.552 ±** 0.007	**0.541 ±** 0.008	**0.528 ±** 0.011	**0.521 ±** 0.003
Stored at 30°C							
1 month	whole	**0.397 ±**	**0.396 ±**	**0.400 ±**	**0.415 ±** 0.06	**0.358 ±**	**0.345 ±**

		0.007	0.009	0.001		0.001	0.004
	crushed	**0.499** ± 0.011	**0.463** ± 0.003	**0.477** ± 0.006	**0.495** ± 0.014	**0.472** ± 0.003	**0.513** ± 0.002
2 months	whole	**0.406** ± 0.003	**0.295** ± 0.004	**0.404** ± 0.004	**0.476** ± 0.050	**0.481** ± 0.005	**0.493** ± 0.001
	crushed	**0.495** ± 0.007	**0.451** ± 0.004	**0.461** ± 0.002	**0.483** ± 0.020	**0.490** ± 0.005	**0.474** ± 0.011
3 months	whole	**0.433** ± 0.002	**0.478** ± 0.002	**0.469** ± 0.002	**0.472** ± 0.070	**0.457** ± 0.002	**0.462** ± 0.001
	crushed	**0.456** ± 0.004	**0.487** ± 0.007	**0.488** ± 0.005	**0.480** ± 0.009	**0.469** ± 0.007	**0.469** ± 0.009

Table 25. Water activity in chocolate coated raisins, received with the use of different types of couverture **supplemented**, and as a comparison **non-supplemented** with LAB, stored at temperatures of 4, 18 and 30°C for 3 months.

Raisins coated with dark and milk couverture without LAB addition showed similar water activity values (for whole chocolate coated raisins). Slightly lower value of a_w had raisins coated with white couverture. Supplementation of couvertures with lactic acid bacteria only very slightly increased the values of a_w in final products, obtained with the use of dark and milk couvertures. More noticeable increase of a_w – from 0.389 to 0.414 was observed for raisins coated with white couverture.

During storage of raisins coated with all types of couverture supplemented with LAB at refrigeration and room temperatures water activity increased (whole raisins in chocolate). Only at higher storage temperature of 30°C water activity was decreasing for 2 months of storage to finally increase during third month. Similar changes of a_w during storage were observed for raisins coated with non-supplemented couverture. A difference was noticed for a_w changes of chocolate coated raisins stored at 30°C, in which during first month of storage a_w decreased, and during following months of storage rose to values higher than in initial samples (directly after production).

Water activity in crushed products was generally higher comparing to the values of this parameter analyzed in a whole product. Comparing water activity values in whole and crushed raisins coated with chocolate, obtained with the use of supplemented with LAB and non-supplemented couvertures – dark, milk and white, directly after production and during 3 months of storage at temperatures of 4, 18 and 30°C it can be concluded that they kept under the value of 0.6. Due to that fact, it is probable that during the whole time of storage no bacterial activity in both, supplemented and non-supplemented with LAB, will be maintained.

Total acidity in chocolate coated raisins

Total acidity changes of chocolate coated raisins during 3 months of storage in various temperatures is presented in Table 26.

Storage time	Total acidity (ml 1 M NaOH · 100 g^{-1}) in chocolate coated raisins:					
	Dark + LAB	Milk + LAB	White + LAB	Dark	Milk	White
Control „0"	**17.0** ± 0.1	**15.0** ± 0.2	**14.2** ± 0.2	**17.7** ± 0.8	**14.9** ± 0.5	**13.6** ± 0.1
Stored at 4°C						
1 month	**15.6** ± 0.4	**14.8** ± 0.2	**14.2** ± 0.1	**16.2** ± 0.4	**15.7** ± 0.1	**14.0** ± 0.2
2 months	**16.1** ± 0.2	**13.9** ± 0.2	**14.1** ± 0.1	**15.5** ± 0.2	**15.1** ± 0.2	**14.0** ± 0.1
3 months	**15.3** ± 0.1	**15.1** ± 0.2	**14.2** ± 0.2	**14.9** ± 0.1	**16.8** ± 0.2	**13.5** ± 0.2
Stored at 18°C						
1 month	**16.6** ± 0.1	**14.8** ± 0.3	**13.2** ± 0.3	**16.7** ± 0.1	**14.7** ± 0.1	**13.8** ± 0.3
2 months	**15.7** ± 0.3	**13.6** ± 0.3	**14.1** ± 0.2	**14.8** ± 0.2	**14.5** ± 0.1	**13.4** ± 0.1
3 months	**16.0** ± 0.1	**13.1** ± 0.3	**13.1** ± 0.1	**15.5** ± 0.2	**15.2** ± 0.1	**14.4** ± 0.2
Stored at 30°C						
1 month	**15.2** ± 0.2	**14.6** ± 0.1	**14.6** ± 0.2	**15.8** ± 0.2	**15.1** ± 0.1	**13.8** ± 0.3
2 months	**14.3** ± 0.2	**14.1** ± 0.2	**15.1** ± 0.2	**14.0** ± 0.2	**15.7** ± 0.1	**13.8** ± 0.2
3 months	**14.8** ± 0.2	**12.7** ± 0.1	**13.9** ± 0.1	**15.0** ± 0.1	**15.9** ± 0.1	**13.6** ± 0.1

Table 26. Total acidity in chocolate coated raisins, received with the use of different types of couverture **supplemented**, and as a comparison **non-supplemented** with LAB, stored at temperatures of 4, 18 and 30°C for 3 months.

Directly after obtaining the highest total acidity, amounting 17.7 ml 1 M NaOH · 100 g^{-1}, had raisins coated with dark couverture non-supplemented with LAB, and the lowest, amounting 13.6 ml 1 M NaOH · 100 g^{-1}, was observed in raisins coated with white couverture.

Supplementation of dark, milk and white couvertures with bacteria from *Lactobacillus* species didn't influence significantly the total acidity of products after production. The biggest changes of this parameter after supplementation of couverture with LAB, namely by 0.7 ml 1 M NaOH · 100 g^{-1}, were noticed in raisins coated with dark couverture.

In raisins coated with white and milk couvertures total acidity after supplementation with LAB increased by 0.1 and 0.6 ml 1 M NaOH · 100 g^{-1}, respectively.

During storage of chocolate coated raisins only slight decrease of total acidity was observed. With an exception in product containing milk couverture non-supplemented with LAB, in which total acidity increased by 1 ml 1 M NaOH · 100 g^{-1} after 3 months of storage at a temperature of 30°C. The magnitude of total acidity changes in chocolate coated raisins depended on storage temperature.

In raisins coated with dark, white and milk couverture supplemented with LAB total acidity decrease during storage. The biggest decrease of this parameter was noticed in chocolate coated raisins stored at 30°C and in products in dark, milk and white couverture supplemented with LAB was 2.2, 2.3 and 0.3 ml 1 M NaOH · 100 g^{-1}, respectively.

Analysis of fat quality in a coating of chocolate coated raisins by DSC method

In obtained chocolate coated raisins, analyses of changes occurring in fat from a couverture (which is a coating of the product), by differential scanning calorimetry method were performed. These changes are presented in tables 30 and 31. Exemplary thermograms of fats from dark couverture supplemented and non-supplemented with LAB, which are a coating of chocolate coated raisins, stored during 3 months period at a temperature of 18°C are presented in Figures 8 and 9, respectively.

Storage time	**Enthalpy and melting temperature of fat from coatings of chocolate coated raisins:**					
	Dark + LAB		**Milk + LAB**		**White + LAB**	
	ΔH (J/g)	Tm (°C)	ΔH (J/g)	Tm (°C)	ΔH (J/g)	Tm (°C)
Control „0"	23.014	34.30	22.906	34.30	24.020	34.20
			Stored at 4°C			
1 month	35.480	34.78	16.530	34.51	22.357	34.47
2 months	32.550	35.00	20.794	32.03	27.496 4.487 + 23.009	Tm_1=30.58 Tm_2=34.54
3 months	25.261	34.51	25.218	34.03	33.653	34.15
			Stored at 18°C			
1 month	35.742	34.54	24.230	33.94	17.124	34.73
2 months	34.312	34.77	22.906	33.46	25.591	34.00
3 months	29.820 16.564 + 13.255	Tm_1=32.47 Tm_2=34.10	34.744	32.48	33.698	34.74
			Stored at 30°C			
1 month	34.000	36.45	16.353	34.96	29.053	35.08
2 months	32.475 2.440 + 30.034	Tm_1=29.57 Tm_2=35.82	18.786 7.561+11.229	Tm_1=31.48 Tm_2=35.12	23.137	35.60
3 months	36.578 22.555+14.022	Tm_1=33.01 Tm_2= 4.57	24.102 5.988+18.113	Tm_1=29.67 Tm_2=34.42	35.622 8.977+26.645	Tm_1=31.11 Tm_2=35.34

Table 27. Enthalpy (ΔH) and melting temperature (Tm) of fat from coatings of chocolate coated raisins obtained with the use of different types of couvertures **supplemented** with LAB, stored at temperatures of 4, 18 and 30°C during 3 months.

In samples of chocolate coated raisins directly after production the value of melting enthalpy of fat extracted from product coating, in all types of couvertures supplemented with lactic acid bacteria was lower than in analogous products coated with non-supplemented couverture. This phenomena can be explained by the fact, that LAB preparation influenced fat crystallization, namely in supplemented couverture more liquid phase of fat was present than in analogous products without LAB.

During storage of chocolate coated raisins melting enthalpy value of fat from coatings of dark and white chocolate coated raisins, increased regardless of storage conditions. It can be caused by crystallization of previously present sources of crystallization or by increasing the area of already existing fat crystals.

Storage time	Enthalpy and melting temperature of fat from coatings of chocolate coated raisins:					
	Dark		Milk		White	
	ΔH (J/g)	Tm (°C)	ΔH (J/g)	Tm (°C)	ΔH (J/g)	Tm (°C)
„0" control	30.157	35.51	27.559	34.34	34.93	34.25
Stored at 4°C						
1 month	31.568	35.90	25.400	33.94	22.026	34.61
2 months	33.064 0.911+32.153	T_{m1}= 28.64 T_{m2}= 35.56	33.148	34.57	31.872	35.03
3 months	37.718	34.25	23.874 3.744+20.129	T_{m1}=30.28 T_{m2}=33.69	37.027	34.93
Stored at 18°C						
1 month	36.765	34.54	26.622	33.57	32.684	34.86
2 months	31.847	36.20	28.173 1.580+26.59	T_{m1}=26.84 T_{m2}=34.17	33.455	34.61
3 months	38.068 25.119+12.917	T_{m1}= 32.32 T_{m2}= 34.54	34.592	32.48	34.568 13.273+21.294	T_{m1}= 31.64 T_{m2}= 34.75
Stored at 30°C						
1 month	36.373	35.35	24.059 5.225+18.834	T_{m1}=31.29 T_{m2}=34.86	21.614	35.31
2 months	29.962 0.741+29.22	T_{m1}=29.14 T_{m2}=35.94	18.764 5.244+13.519	T_{m1}=32.39 T_{m2}=35.06	31.425	36.07
3 months	33.30 4.558+28.743	Tm=31.04 Tm=35.95	24.259 6.087+18.171	T_{m1}=29.69 T_{m2}=34.43	26.953 3.650+23.302	T_{m1}= 31.65 T_{m2}= 35.71

Table 28. Enthalpy (ΔH) and melting temperature (Tm) of fat from coatings of chocolate coated raisins obtained with the use of different types of couvertures **non-supplemented** with LAB, stored at temperatures of 4, 18 and 30°C during 3 months.

Temperature and storage time of chocolate coated raisins had an influence on changes of polymorphic forms of fat from couverture coating products. In coatings without LAB addition, bigger tendency to two polymorphic forms creation was observed, mainly in raisins stored at temperatures of 18 and 30°C. The range of maximal melting temperatures of first polymorphic form of fat from couverture from chocolate coated raisins non-supplemented with LAB was from 26.80 to 32.30°C. Whereas for second polymorphic form it ranged from 33.69 to 35.95°C. Range of melting temperatures of fat from couverture from chocolate coated raisins supplemented with LAB, for first polymorphic form was from 29.57 to 33.01°C, and for second form, from 34.10 to 38.82°C. Range of melting temperatures of first and second polymorphic forms of fat was similar in raisins coated with both, supplemented and non-supplemented couvertures. At a temperature of 30°C in dark and milk couvertures both, supplemented and non-supplemented, appearance of second polymorphic form of fat was observed after 2 months of storage. In white couverture appearance of second polymorphic form was noticed after 3 month storage of product. In can concluded that lack of cocoa liquor and bigger content of milk in a white couverture delayed polymorphic changes of fats in this couverture.

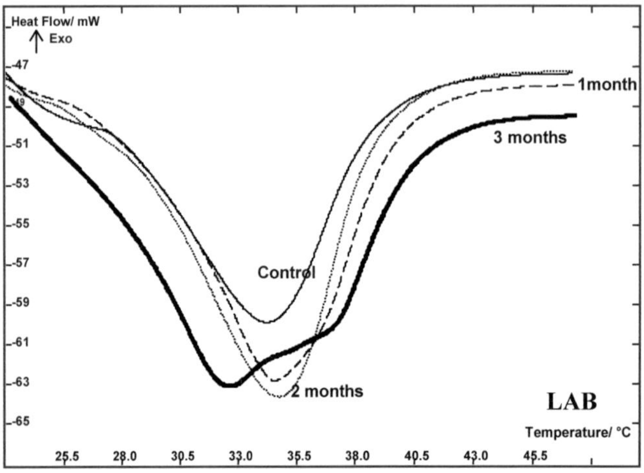

Figure 8. Thermogram of fat from coatings of chocolate coated raisins obtained from dark couverture **supplemented** with LAB stored at a temperature of 18°C during 3 months.

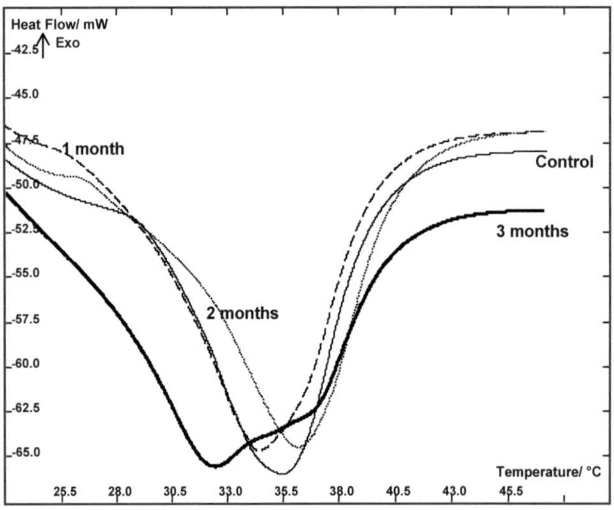

Figure 9. Thermogram of fat from coatings of chocolate coated raisins obtained from dark couverture **non-supplemented** with LAB stored at a temperature of 18°C during 3 months.

Texture of chocolate coated raisins

In Table 29 results of cutting test of chocolate coated raisins are presented.

Dark and milk chocolate coated raisins supplemented with lactic acid bacteria became harder. On the other hand raisins in white chocolate after supplementation with LAB softened.

During storage, from all supplemented products, raisins coated with white couverture became the hardest. During storage of all studied chocolate coated raisins (with and without LAB addition) at temperatures of 4 and 18°C hardness gradually decreased, which can be associated with water diffusion. During storage of supplemented and non-supplemented chocolate coated raisins at temperature of 30°C hardness initially rose (drying of surface), next it decreased (water diffusion from raisin to coating and from environment into product), and finally to increase after third month. Additionally hardness of supplemented products was higher than hardness analyzed directly after production. With an exception of raisins coated with white chocolate, in which hardness was significantly lower than in fresh products.

Storage time	Force (kg) required to cut chocolate coated raisins:					
	Dark + LAB	Milk + LAB	White + LAB	Dark	Milk	White
Control „0"	3.002	2.949	3.781	2.777	2.850	3.861
Stored at 4°C						
1 month	3.005	3.125	3.531	2.461	2.596	2.344
2 months	2.821	2.731	2.424	2.247	2.492	2.067
3 months	2.078	2.068	2.329	2.227	2.271	1.838
Stored at 18°C						
1 month	2.712	3.007	3.319	2.023	3.110	1.898
2 months	2.512	2.597	2.785	1.772	2.316	1.723
3 months	2.125	2.136	2.015	2.180	2.527	2.037
Stored at 30°C						
1 month	3.010	3.154	3.337	2.832	3.480	2.156
2 months	2.786	2.727	2.751	2.314	3.321	2.139
3 months	3.068	3.404	3.608	2.649	2.641	2.413

Table 29. Force required to cut chocolate coated raisins received with the use of different types of couverture **supplemented**, and as a comparison **non-supplemented** with LAB, stored at temperatures of 4, 18 and 30°C for 3 months.

Organoleptic evaluation of chocolate coated raisins

In received raisins coated with couverture supplemented with LAB organoleptic analysis was performed, according to a 5-point scale, and it was compared to products obtained with non-supplemented courevture (Table 30).

The highest note in a 5-point scale received raisins coated with white chocolate and then with milk couverture. Lowest ratings (below 4 points) received raisins coated with dark chocolate both, fresh and during the whole storage period, regardless of LAB supplementation. To high grade of raisins coated with white couverture was caused by their delicate, gentle taste, and soft, elastic consistency. Addition of lactic acid bacteria to couvertures coating raisins didn't influence significantly sensory properties of products, which is favorable from the point of view of a consumer, who highly appreciates sensory quality of chocolate. However, although in first month of storage of chocolate coated raisins stored at all temperatures organoleptic rating didn't change, in latter months this parameter degraded, especially when stored at 30°C. It was caused by the changes occurring in products during storage. Most noticeably these changes were observed in chocolate coated raisins stored at 30°C. They included changes of taste, caused by modifications of fat in a coating, an increase of dry mass content in cores (raisins), increase of hardness of raisins coated with dark and milk chocolates, connected to an increase of dry mass content in chocolate couvertures and surfaces of products, which became less shiny with time.

Organoleptic ratings of raisins coated with dark, milk and white chocolates stored at temperatures of 4 and 18°C were practically identical in a first month of storage (differences of 0.0 – 0.1 points) comparing to fresh product, they were slightly different after second month (by 0.0 – 0.2 points) and third month (0.0 – 0.1 points) of storage. In case of 3 month storage period of non-supplemented products, differences in organoleptic evaluation between fresh product and product stored for a 3 month period were more noticeable and reached 0.7 points. Considering similar organoleptic evaluation of raisins coated with chocolate stored at 4 and 18°C it can be concluded, that examined raisins coated with chocolate don't have to be kept at refrigeration conditions and can be stored at a store shelf as well, where they can easily be found by a consumer next to analogous traditional products.

Storage time	Grades (points) of chocolate coated raisins:					
	Dark + LAB	Milk + LAB	White + LAB	Dark	Milk	White
Control „0"	4.0 ±0.1	4.2 ± 0.1	4.7 ± 0.2	3.9 ± 0.0	4.0 ± 0.1	4.6 ± 0.1
Stored at4°C						
1 month	3.7 ± 0.1	4.1 ± 0.1	4.6 ± 0.3	3.9 ± 0.2	3.9 ± 0.2	4.5 ± 0.3
2 months	3.6 ± 0.2	4.1 ± 0.1	4.6 ± 0.1	3.8 ± 0.2	3.9 ± 0.1	4.6 ± 0.1
3 months	3.1 ± 0.0	3.9 ± 0.1	4.5 ± 0.2	3.6 ± 0.1	3.8 ± 0.2	4.5 ± 0.2
Stored at18°C						
1 month	3.8 ± 0.2	4.0 ± 0.2	4.6 ± 0.2	3.8 ± 0.2	3.9 ± 0.1	4.5 ± 0.3
2 months	3.7 ± 0.2	4.0 ± 0.1	4.5 ± 0.2	3.7 ± 0.2	3.8 ± 0.2	4.3 ± 0.2
3 months	3.0 ± 0.1	3.7 ± 0.2	4.4 ± 0.3	3.4 ± 0.1	3.1 ± 0.3	4.0 ± 0.2
Stored at30°C						
1 month	3.9 ± 0.2	4.1 ± 0.2	4.6 ± 0.2	3.9 ± 0.1	3.8 ± 0.2	4.5 ± 0.1
2 months	3.7 ± 0.2	3.9 ± 0.3	4.4 ± 0.2	3.4 ± 0.2	3.7 ± 0.1	4.2 ± 0.2
3 months	2.8 ± 0.3	3.6 ± 0.2	3.5 ± 0.2	2.9 ± 0.2	3.0 ± 0.1	3.4 ± 0.2

Table 30. Organoleptic evaluation of chocolate coated raisins received with the use of different types of couverture **supplemented**, and as a comparison **non-supplemented** with LAB, stored at temperatures of 4, 18 and 30°C for 3 months.

Viability of LAB in a product

The biggest viability of *Lactobacillus* bacteria in products after 3 months of storage was observed when stored at a refrigeration temperature (4°C). The highest viability was observed in raisins coated with dark (88.9%) and white (88.0%) chocolate, slightly lower was noticed in raisins coated with milk chocolate (86.5%) (Table 31). In products stored at a temperature of 18°C amount of live bacterial cells was lower by two orders of magnitude, amounting 10^5 CFU · g^{-1}. When products were stored at a stress temperature (30°C) a severe decrease in an amount of live cells was observed, even just after one month of storage. The biggest drop in LAB viability was observed in raisins coated with white chocolate, to 58.6%, next in raisins coated with milk chocolate, to 59.7%, and finally in raisins coated with dark chocolate viability lowered to 61.3%.

Sample	**Storage temperature**		
	4°C	18°C	30°C
	Viability of bacteria (%)		
Raisins coated with milk chocolate	**88.0** ± 2.3	**73.1** ± 3.0	**59.7** ± 3.9
Raisins coated with dark chocolate	**88.9** ± 3.0	**72.7** ± 3.8	**61.3** ± 4.0
Raisins coated with white chocolate	**86.5** ± 2.7	**66.3** ± 2.0	**58.6** ± 4.0

Table 31. Viability of *Lactobacillus* bacteria in chocolate coated raisins after 3 months of storage.

Sample	**Storage temperature**		
	4°C	18°C	30°C
	The amount of live bacterial cells in the product (CFU · 80 g^{-1})		
Raisins coated with milk chocolate	3.4×10^8	2.5×10^7	2.5×10^6
Raisins coated with dark chocolate	3.8×10^8	2.3×10^7	3.2×10^6
Raisins coated with white chocolate	2.7×10^8	8.0×10^6	2.1×10^6

Table 32. The amount of live bacterial cells of *Lactobacillus* species in chocolate coated raisins, with a weight of 80g, after 3 months of storage.

The worst *Lactobacillus* bacteria viability, at all storage temperatures, was observed in raisins coated with white couverture. Storage of raisins coated with dark, white and milk couvertures supplemented with *Lactobacillus* bacteria, at a temperature of 4°C provides a maintenance of probiotic properties of these products. Temperature of 18°C, in case of raisins coated with dark and milk couvertures, also prevents them from loosing probiotic properties during storage. Storage of those products in this temperature allows to maintain high lactic bacteria viability during the whole storage period, namely 3 months.

Consuming a package of chocolate coated raisins, with a weight of 80 g, stored at 4°C provides a consumer with 10^8 CFU of probiotic bacteria. The same package of raisins coated with dark and milk chocolate stored at a temperature of 18°C contains 2.3×10^7 CFU · 80 g^{-1} and 2.5×10^7 CFU · 80 g^{-1}, respectively, while raisins coated with white chocolate an amount of 8.02×10^6 CFU · 80 g^{-1} of final product (Table 32). After storage of chocolate coated raisins at 30°C consumed amount of lactic acid bacteria would amount to a level of 10^6 CFU · 80 g^{-1} of final product, and would be below recommended level (10^7 CFU · g^{-1}) for functional food.

6. Summary

Proposed technology enables to introduce to dark, white and milk couvertures, live cultures of lactic acid bacteria (as a lyophilisate) and to use them for obtaining raisins coated with chocolate, characterized by soft consistency.

Results of research and development project indicated what follows:

- Addition of live cultures of lactic acid bacteria to dark and milk couvertures caused a slight increase of dry mass if products, and in raisins coated with white couverture a slight lowering of this parameter.
- LAB supplementation of couvertures used for coating of raisins, only mildly increased the value of water activity in products coated with dark and milk couverture. Bigger increase of a_w was noticed for raisins coated with white couverture.
- Supplementation with lactic acid bacteria of dark, white and milk couvertures didn't influence acidity of fresh products, and during storage this parameter decreased only slightly.
- Temperature and storage time of chocolate coated raisins influenced changes of polymorphic forms of fat from couvertures used for coating of raisins. Fat from supplemented and non-supplemented with LAB couvertures was characterized by similar ranges of melting temperatures of both, first and second polymorphic form.
- Raisins coated with dark and milk couvertures after supplementation with LAB had higher hardness values than raisins coated with analogous non-supplemented couvertures. While raisins coated with white couverture after LAB supplementation

became softer. During storage at temperatures of 4 and 18°C raisins coated with dark, milk and white couvertures supplemented with LAB gradual decrease in hardness was observed. After 2 months of storage products were softer than when they were fresh.

- LAB supplementation of couvertures used for coating raisins practically didn't affect organoleptic properties of received products. The best rating in a 5-point scale received raisins coated with white couverture, next notes belonged to products in milk and dark couverture.
- Raisins coated in chocolate supplemented with LAB can be stored and exhibited in a store at room temperature. Number of live bacterial cells in products during whole storage period remained at a functional level.

7. Conclusion

Viability of lactic acid bacteria in some confectionery products appears to be unexpectedly high. It is caused by low moisture content in products mentioned in this chapter, as well as reqiured water activity (below 0.6), high concentration of carbohydrates, mainly saccharose and limited access of oxygen. However, viability depends mainly on recipe of product (mainly the type of fat), technological processes used for obtaining products and the time of these processes, and finally storage conditions.

Many solutions for application of lactic acid bacteria to confectionery, pastry and other kinds of products, cited in this chapter, is the subject of patent protection.

Application of bacteria in a form of preserved preparation, in which live cells are put in a state of anabiosis, allows to maintain high viability of LAB in confectionery products during storage. LAB addition to confectionery products - a type of food often consumed by kids and youth, allows to enrich the diet of this group of consumers with probiotic products with taste similar to traditional products, which are also ready for distribution and sale analogous to products without addition of bacterial preparations, not requiring refrigeration temperatures and hence being always "within reach".

Author details

Dorota Żyżelewicz, Ilona Motyl, Ewa Nebesny, Grażyna Budryn, Wiesława Krysiak, Justyna Rosicka-Kaczmarek and Zdzisława Libudzisz
University of Technology, Faculty of Biotechnology and Food Sciences, Lodz, Poland

Acknowledgement

Authors wish to thank Polish Ministry of Science and High Education for financial support of research and development project No. R12 018 01 about: "Semi-products and Products Suplemented with Viable Lactic Acid Bacteria" in which presented studies were performed.

We also would like to acknowledge help received from companies: ZPT "Kruszwica" S.A. (Kruszwica, Poland) and AARHUSKARLSHAMN SWEDEN AB (Karlshamn, Sweden) for transferring a portion of fats used for obtaining nutty fatty mass and wafer creams.

8. References

[1] FAO/WHO Report. Health and nutritional properties of probiotics in food including milk with live lactic acid bacteria. Report of a Joint FAO/WHO Expert Consultation. 2001. *www.who.int/foodsafety/.../fs.../probiotics.pdf (accessed 11 June 2012).*

[2] FAO/WHO Report. Guidelines for the Evaluation of Probiotics in Food. Report a Joint FAO/WHO Working Group. 2002. ftp://ftp.fao.org/es/esn/food/wgreport2.pdf *(accessed 11 June 2012).*

[3] Burns AJ, Rowland IR. Anti–carcinogenicity of probiotics and prebiotics. Current Issues of Intestinal Microbiology 2000;1 13-24.

[4] Rafter J. Probiotics and colon cancer. Best Practice and Research Clinical Gastroenterology 2003;17 849-859.

[5] Doron S, Gorbach SL. Probiotics: Their role in the treatment and prevention of disease. Expert Review of Anti-infective Therapy 2006;4: 261-275.

[6] Boutron-Ruault MC. Probiotics and colorectal cancer. Clinical Nutrition and Metabolism 2007;21: 85-88.

[7] Nowak A, Libudzisz Z. Ability of intestinal lactic acid bacteria to bind or/and metabolize phenol and p-cresol. Annals of Microbiology 2007;57(3) 329-335.

[8] Nowak A, Arabski M, Libudzisz Z. Ability of intestinal lactic acid bacteria to bind or/and metabolise indole. Food Technology and Biotechnology 2008;46(3) 299-304.

[9] Dicks LMT, Botes M. Probiotic lactic acid bacteria in the gastro-intestinal tract: Health Benefits, Safety and Mode of Action. Beneficial microbes 2010;1 11-29.

[10] Nebesny E, Żyżelewicz D, Krysiak W, Libudzisz Z, Motyl I. Physico-chemical, Microbiological and Organoleptic Properties of Sugar Free Chocolates Enriched with Viable Lactic Acid Bacteria. Grant's report No 3 P06T 054 24; 2005 (in polish).

[11] Nebesny E, Żyżelewicz D, Budryn G, Krysiak W, Libudzisz Z, Motyl I. Confectionery Semi-products and Products Suplemented with Viable Lactic Acid Bacteria. Grant's report No R12 018 01; 2008 (in polish).

[12] Krysiak W, Nebesny E, Żyżelewicz D, Budryn G, Motyl I, Libudzisz Z. Method of obtaining of pralines with sugar-fat filling of improved healthy properties. Polish patent application: P-384866; 2008 (in polish).

[13] Żyżelewicz D, Nebesny E, Krysiak W, Budryn G, Motyl I, Libudzisz Z. Spreadable product for bread of probiotic properties and preparation thereof. Polish patent application: P-388158; 2009 (in polish).

[14] Nebesny E, Żyżelewicz D, Motyl I, Libudzisz Z. Chocolate and preparation thereof. Polish Patent: P- 366273; 2012 (Patent Application, 2004) (in polish) – in press.

[15] Żyżelewicz D, Nebesny E, Motyl I, Libudzisz Z. Effect of milk chocolate supplementation with lyophilized *Lactobacillus* cells on its attributes. Czech Journal of Food Sciences 2010;28(5) 392-406.

[16] Nebesny E, Żyżelewicz D, Motyl I, Libudzisz Z. Dark chocolates supplemented with *Lactobacillus* strains. European Food Research and Technology 2007;225(5) 33-42.

[17] Cukrowska B, Ceregra A, Rosiak I, Libudzisz Z, Motyl I, Klewicka E. Effect of Oral Dosage of Probiotic Bacteria *Lactobacillus* on Composition of Gut Ecosystem, Immune System and Course of the Food Allergy in Infants. Grant's rapport No KBN PB 777/P05/2004/26; 2007 (in polish).

[18] Sokmen A, Gunes G. Influence of some bulk sweeteners on rheological properties of chocolate. LWT 2006;39 1053-1058.

[19] Chevalley J. An adaptation of the Casson equation for the rheology of chocolate. Journal of Texture Studies 1991; 22 219-229.

[20] Steiner EH. A new rheological relationship to express the flow properties of melted chocolates. International Chocolate Review 1958;13 290.

[21] Foubert I, Vanrolleghem PA, Dewettinck K. A differential scanning calorimetry method to determine the isothermal crystallization kinetics of cocoa butter. Thermochimica Acta 2003;400 131-142.

[22] Polish Standard. Confectionery products. Organoleptic evaluation. PN-A-88032 1998; (in polish).

[23] Loisel C, Kelle G, Lecq G, Launay B, Ollivon M. Tempering of chocolate in a scraped surface heat exchanger. Journal of Food Science 1997;62 773-780.

[24] Maat J, Rossi D, Babuchowski A, Beekmans F, Castenmiller J, Fenwick R, Haber J, Hogg T, Israelachwili D, Kettlitz B, Kohnke J, Lienemann K, Majou D, Petersen B, Schiefer G, Tomás-Barberán F. European Technology Platform on Food for Life. The Vision for 2020 and Beyond. Available; 2007 http://etp.ciaa.be/documents/SRA_2007_2010.pdf *(accessed 11 June 2012).*

[25] Budryn G, Nebesny E, Żyżelewicz D, Krysiak W, Motyl I, Libudzisz Z. Confectionery product of sugar-fat cores. Polish patent application: P-384154; 2007 (in polish).

[26] Saarela M, Mogensen G, Fondén R, Mättö J, Mattila-Sandholm T. Probiotic bacteria: Safety, functional and technological properties. Journal of Biotechnology 2000;84 197-215.

[27] Besselich N. Die Keks-, Biskuit- und Waffelfabrikation. Verlag der Konditor-Zeitung, Trewir 1950;177-184, 188-190.

[28] Manley D. Technology of Biscuits, Crackers and Cookies. Cambridge: Woodhead Publishing Limited, Boca Raton, CRC Press; 2000 pp. 430-436.

[29] Warsza H. Wafer Products Manufacturing. Warsaw: Science-Technical Publications; 1970 pp. 5-7, 71-87, 93, 97, 101, 121-122 (in polish).

[30] Wyczański S. Confectionery Technology. Part 3. Warsaw: Ligot and Food Publications; 1965 pp. 177-179 (in polish).

[31] Żyżelewicz D, Nebesny E, Motyl I, Libudzisz Z, Budryn G, Krysiak W, Rosicka-Kaczmarek J. Confectionery product with functional properties. Wafers coated and non-coated with chocolate couverture, interleaved with filling supplemented with lactic acid bacteria (LAB). Patent application: P-393270; 2010 (in polish).
[32] Nebesny E, Żyżelewicz D, Krysiak W, Budryn G, Motyl I, Libudzisz Z. Raisins in chocolate coating. Polish patent application: P-384153; 2007 (in polish).

Section 2

Probiotics in Health

Chapter 13

Probiotics in Pediatrics – Properties, Mechanisms of Action, and Indications

Antigoni Mavroudi

Additional information is available at the end of the chapter

http://dx.doi.org/10.5772/50043

1. Introduction

Probiotics have been the topic of many studies over the past 20 years. Metchnikoff and Tissier (Metchnikoff 1907, Tissier, 1906) were the first to make scientific suggestions concerning the probiotic use of bacteria. They suggested that these bacteria could be administered to patients with diarrhea to help restore a healthy gut flora. Fuller (1989) in order to point out the microbial nature of probiotics redefined the word as "A live microbial feed supplement which beneficially affects the host animal by improving its intestinal balance. " The most recent but probably not the last definition is "live microorganisms, which when consumed in adequate amounts, confer a health effect on the host"(Guarner and Schaafsma,1998). In the last 20 years however, research in the probiotic area has progressed considerably and significant advances have been made in selection and characterization of specific probiotic cultures. Most of the studies aim to investigate the physiological and functional properties of various probiotic strains, the mechanisms of action and the indications for human use and health benefits.

Probiotic bacteria are a subset of specific organisms, which, when ingested, transiently occupy the gastrointestinal tract and lead to documented health benefits. Lactic-acid-producing bacteria (LAB), particularly members of the genus Lactobacilli, Bifidobacteria, non pathogenic gram positive bacteria and non bacterial microorganisms (for example certain yeasts, such as Saccharomyces boulardii) have been used as probiotic agents. [1] The use of specific probiotic bacteria has been shown to enhance host defense mechanisms. [2] Prebiotics are non-digestable food ingredients that beneficially affect the host by stimulating the growth and/or activity of a limited number of bacterial species in the colon. Compounds most commonly studied for their prebiotic nature are non-digestable carbohydrates. In particular, oligosaccharides are considered the main units among prebiotics, which include fructooligosaccharides (FOS), inulin, lactulose and galactooligosaccharides (GOS). Synbiotics are a combination of probiotics and prebiotics and it is the synergy between these two substances that becomes known as synbiotics.

Several clinical benefits have been reported as a result of interaction between host and becteria ,such as for treatment and prevention of viral diarrhea [3] and reducing the risk of necrotizing enterocolitis (NEC), mitigating antibiotic associated diarrhea ,and modulating host immune response (such as in allergic disease).

2. Properties

Intestinal microflora is composed of both well-established resident microbes and those ingested orally which transiently occupy the gastrointestinal (GI) tract. Probiotics are generally defined as non pathogenic organisms in food supply (ingested microbes) that are capable of conferring a health benefit to the host by modifying gut microbial ecology.

Probiotics are live microorganisms which when ingested in adequate amounts confer a health effect on the host by enhancing host defense mechanisms. Several clinical benefits have been reported with various specific microbes in pediatric populations. It is increasingly clear that these benefits to the host are mostly mediated by the profound effect that intestinal microflora (microbiota) have on gut barrier function and host immune response. The most frequently used probiotic agents are the lactic acid producing bacteria, such as Lactobacilli and Bifidobacteria, non pathogenic strains of Gram positive bacteria, such as Streptococcus, E. Coli and non bacterial microorganisms, such as Saccharomyces Bulardii

There are several generally accepted characteristics that define probiotic bacteria. [4-6]

- They are microbial organisms
- They remain viable and stable after culture manipulation, and storage before consumption
- They survive gastric, biliary, and pancreatic digestion.
- They are able to induce a host response once they enter the intestinal microbial ecosystem (by adhering to gut epithelium or other mechanisms) and they yield a functional and clinical benefit to the host when consumed.
- It has been suggested that probiotic bacteria should be of "human origin" and that they should "colonize" the intestine. [5,6]

Probiotics can be found in certain foods, such as yogurts, fruit juices and soy beverages. They are also found in supplements that come in liquid, capsule and powdered forms. It is believed that a concentration of 10 live microorganisms per gram or ml of product is required in order to exert a health benefit on the host.

Probiotics have a wide range of beneficial effects and numerous indications of use in pediatric populations, such as:

- Acute diarrhea
- Antibiotic-Associated Diarrhea
- Allergy prevention
- Necrotizing enterocolitis

3. Mechanisms of action

The intestine of the newborn is essentially sterile. During the birthing process and during the first days of life, the gut is inoculated with bacteria. In the first two days of life, an infant's intestinal tract is rapidly colonized with bacteria consisting mainly of Enterobacteria. In most breastfed infants, the Bifidobacteria counts increase rapidly to constitute 80-90% of the total flora. Formula-fed infants, on the other hand, tent to have a flora that is more complex, consisting mostly of coliforms and Bacteroides with significantly lower prevalence of Bifidobacteria. [7] Although the composition of the microflora varies among individuals, the composition within each individual remains stable over prolonged periods. [8] A normal microbial flora is necessary for the development of gut associated lymphoid tissue (GALT). The gut luminal microbes are responsible for mucosal immune system development in healthy infants. Signaling through specific receptors, particularly toll-like receptors, intestinal bacteria affect epithelium cell function, which determines T-cell differentiation and antibody responses to T-cell-dependent antigens, regulating immune gut response for IgA responses to luminal antigens. [9] Resident bacteria, particularly Lactobacilli and Bifidobacteria, can exert antimicrobial activities influencing both local and systemic immunity. [10]

Intestinal bacteria have a major effect on enhancing secretory immune function. Among the more consistently found effects of specific Bifidobacteria and Lactobacilli in pediatric populations is the effect on humoral immunity, particularly on secretory IgA(s IgA) and other immunoglobulins. An increase in IgA-, IgM-,and IgG-secreting cells in circulation ,as well as fecal IgA concentrations ,has been reported. During the neonatal period, s IgA in the stool of formula-fed infants is essentially undetectable. [11, 12] Bifidobacteria and Lactobacillus given orally have been shown to influence s IgA in a number of animal trials [13] Infant studies that investigated the effects of specific Lactobacilli and Bifidobacteria supplementation on stimulating the mucosal immune response have reported similar positive results. Breast milk contains significant levels of sIgA that are transferred to the infant. Bifidobacteria, which predominate in breast-fed infants, have shown to stimulate the synthesis and secretion of IgA. Recent reports indicate similar IgA increases in premature infants receiving B lactis. [14] sIgA, the most important and predominant immunoglobulin in mucosal surfaces, provides protection against antigens, potential pathogens, toxins, and virulence factors. [15]

The resident Bifidobacteria and Lactobacilli in the gut can offer resistance to colonization by other potentially pathogenic microbes, thereby functioning as part of the gut defense barrier. They have also been associated with the secretion of substrates that have antimicrobial properties [16] and the secretion of mucins via activation of MUC2 and MUC3 genes, part of the intestinal barrier that can inhibit adherence of pathogenic bacteria. [17]

An increasing number of clinical trials have documented effects of ingestion of specific probiotic bacteria on gut barrier function and immunity. For example in both animal and human models, ingestion of L casei, L bulgaricus, and L acidophilus has been shown to activate production of macrophages and enhance phagocytosis. [8] Serum sCD14, a marker of immunologic maturation in the neonate, was significantly greater than placebo in infants provided probiotics. Additionally, decreased gut permeability with Lactobacilli [18] , and recently in premature infants receiving Bifidobacteria [19] , is another mechanism by which probiotics may function.

Some probiotic bacteria have been shown to exert beneficial effects on pro- and anti-inflammatory cytokine secretion [8]. Decreases in fecal 1 antitrypsin, urinary protein eosinophil X, tumor necrosis factor (TNF)-α [20,21] have been reported as a result of down-regulation of the inflammatory immune response by probiotic agents.

It is being recognized that host-microbe interactions have an effect on atopic disease. Alterations in the balance of intestinal microflora, particularly in immune and inflammatory-related diseases coupled with significant reduction in the oral ingestion and exposure to a microbe that has led to postulation of the "hygiene hypothesis". This theory suggests that a lower exposure in early childhood to bacterial and other antigens in industrialized societies has led to inadequate development and maturation of immune responses and appears responsible for the increased prevalence of asthma and allergies due to inadequate defensive and immune-modulating gut immune diseases. [22, 23, 24] Infants are born with a predominance of Th2 (T helper 2) lymphocyte activity ,which predisposes them to an exaggerated response to allergens ,with increased IgE production. Exposure to intestinal bacteria ,on the other hand ,stimulates Th1 (T helper 1) activity (which primarily reacts defensively to bacterial stimuli as part of the protective immune response). As a consequence ,intestinal microbes (resident and ingested)can redirect immune balance from a Th2-predominant response to a balanced Th1/Th2 response ,decreasing the changes for a potential exaggerated allergic response. Finally, TReg (regulatory) cells release cytokines such as transforming growth factor β(TGF-β) ,which can inhibit Th1 or Th2 overexpression and also play a role in adequate balancing the host response to bacterial food antigens ,and their activity seems to be increased by luminal microbes [25,26,27,28] Some Bifidobacteria and Lactobacilli given orally may enhance the production of a balanced T-helper-cell response [29,30] and stimulate production of interleukin (IL)-10, and TGF-β [21,31,32] both of which have a role in the development of immunologic tolerance to antigens and can decrease allergic type immune responses.

Bifidobacteria supplementation in premature infants has been shown to positively modify the microflora of the intestines. [33] Beneficial increases in stool, short-chain fatty acids, reductions in stool pH, and decreases in fecal ammonia and indoles [34, 35] and concentrations of Bacteroides and E. Coli have been documented [36, 37] with Bifidobacteria supplementation. Specific probiotic bacteria positively affect the ratio of favorable to unfavorable in the gut luminal environment. Lactobacilli and Bifidobacteria when ingested they are not part of the resident microflora of the host, and their counts typically decrease or disappear once ingestion stops. Specific Lactobacilli and Bifidobacteria, when ingested, can modify the composition of intestinal microbial ecology. They are not typically pathogenic and seem beneficial in fostering host immune development and response. These ingested organisms have the potential of further supporting gut barrier function and appropriate host immune system development and immune response.

In summary effects have been documented supported by a large body of evidence from in vitro and animal studies. These include effects on innate (nonspecific immune defense) and adaptive immunity (responses that require exposure to pathogens or antigens that the immune system recognizes and "remembers"). Adequate adaptive responses are important for host defense, as well as to develop immune tolerance, which decreases chances for abnormal immune hyperreactivity or inflammation. The following effects on innate and adaptive immunity have been reported:

Effects on innate immunity

- Compete with and inhibit growth of potential pathogens
- Promote mucin production
- Decrease gut permeability
- Enhance natural killer cell activity, macrophage stimulation, and phagocytosis

Effects on adaptive immunity

- Increase total and specific s IgA in serum and intestinal lumen
- Increase IgA-, IgG-, and IgM- secreting cells
- Modulate inflammatory gut immune responses [5]

4. Indications

Clinical benefits with specific probiotic bacteria by enhancing defense mechanisms, as well as by modulating host immune response include prevention and treatment of acute infectious diarrhea and antibiotic-associated diarrhea, modulating allergic immune response, prevention of NEC and treating constipation.

4.1. Acute infectious diarrhea

The clinical outcome with the use of probiotic bacteria in order to treat or prevent acute diarrheal diseases has been supported by a large and growing body of evidence. The larger number of trials documents therapeutic use of probiotics as supplements early in the course of the disease. The rationale of using probiotics to treat and prevent diarrheal diseases is based on the assumption that they modify the composition of colonic microflora and act against enteric pathogens. The majority of studies have included various species of Lactobacilli, and by far the most used has been L rhamnosus (GG). This specific strain has shown efficacy when given as a supplement early in the course of rotaviral diarrhea. The most consistent effect reported is a reduction in duration of illness (0, 5 to 1, 5 days). While for the individual infant the effect may be modest, the effect on the population may be significant. [38]

A reduction in incidence of acute diarrheal disease has been reported by another body of literature. Several studies have documented a reduction in incidence or severity of acute diarrhea with Bifidobacteria mainly B. lactis [39, 40] and with Lactobacilli, mainly L rhamnosus (GG) [41, 42] though protection is not always significant. [43] Both L rhamnosus

(GG) and L reuteri (during treatment) [44] and B lactis (used prophylactically) [45] have documented reduced rotaviral shedding. Thirty-four randomized clinical trials reviewed by a meta-analysis evaluated the efficacy of probiotics in the prevention of acute diarrhea. Probiotics significantly reduced the risk of diarrhea developing in infants and children by 57%. The protective effect did not significantly vary among the probiotic strains used, including B lactis, L rhamnosus GG, L acidophilus, S bouladrii, and other agents used alone or in combination with 2 or more strains. [46] Decreased hospitalization [47] and reduced duration of hospitalization were also confirmed. All studies suggested that the effect occurs on both the manifestations of the disease and on the course of the infection. There has been no study so far documenting an increase in diarrheal disease with probiotic use. These findings suggest that specific probiotics may be used in a long-term and prophylactic manner, particularly in infancy.

Several mechanisms have been proposed in order to explain the efficacy of probiotics in preventing or treating acute diarrhea. It has been shown that probiotics stimulate or modify nonspecific and specific immune responses to pathogens. Probiotics have been shown to enhance mucosal immune defenses and protect structural and functional damage promoted by enteric pathogens in the brush border of enterocytes, probably by interfering with the cross-talk between the pathogen and host cells. [48] It has been shown that L rhamnosus (GG) and Lactobacillus plantarum 299v inhibit, in a dose-dependent manner, the binding of E.coli to intestinal-derived epithelial cells grown in tissue culture by stimulation of synthesis and secretion of mucins. [49] Certain probiotics increase the number of circulating lymphocytes [50] and lymphocyte proliferation [51,] stimulate phagocytosis, increase specific antibody responses to rotavirus vaccine strain [52] , and increase cytokine secretion, including interferon-γ. [51] L rhamnosus GG and Lactobacillus acidophilus have been shown to produce antimicrobial substances against some gram-positive and gram-negative pathogens. [53, 54] Other mechanisms proposed by which probiotics might exert their activity against pathogens are competition for nutrients required for growth of pathogens [55,56] ,competitive inhibition of adhesion of pathogens [57-60] ,and modification of toxins and toxin receptors. [61,62]

4.2. Antibiotic-associated diarrhea

Antibiotic-associated diarrhea (AAD) is defined as an acute inflammation of the intestinal mucosa caused by the administration of a broad spectrum of antibiotics. The single bacterial agent most commonly associated with AAD is Clostiridium difficile. However, when the normal fecal gram-negative organisms are absent, overgrowth by staphylococci, yeasts and fungi has been implicated. [63] In fact, most episodes of AAD in childhood are not due to C. difficile. [64] The rationale for the use of probiotics in AAD is based on the assumption that the key factor in the pathogenesis of AAD is a disturbance in normal intestinal flora.

Several probiotic bacteria have proved to be beneficial in reducing the risk of antibiotic-associated diarrhea in infants and children. [65-67] Six randomized controlled trials that collectively assessed 766 children for the efficacy of probiotics in the prevention of AAD indicated that concomitant treatment with probiotics, compared with placebo reduced the risk of diarrhea from 28, 5% to 11, 9%. [67] Beneficial effects were strongest for B lactis and S thermophilus given in infant formula and L rhamnosus (GG) as a supplement.

In conclusion, Randomized Controlled Trials (RCTs) in children have provided so far evidence of a moderate beneficial effect of L rhamnosus (GG), B. lactis, S. thermophilus and S. boulardii in preventing AAD. No data on efficacy of other probiotic strains are available in children. Based on the previously reported evidence probiotics have been shown capable of providing reasonable protection against the development of AAD. Their use is probably warranted whenever the physician feels that preventing this usually self-limited complication is important.

4.3. Nosocomial diarrhea

Nosocomial diarrhea refers to any diarrhea contracted in a health care institution and is more commonly caused by enteric pathogens especially rotavirus. [68] The reported incidence ranges from 4, 5 to 22, 6 episodes per 100 admissions. It may prolong hospital stays and increase medical costs. Although hand washing is the essential infection control measure, other cost-effective approaches to prevent nosocomial diarrhea are being evaluated.

Two RCTs evaluated the use of L rhamnosus G [69, 70] on nosocomial diarrhea prevention. The first study showed that L rhamnosus G administered orally twice daily significantly reduced the risk of diarrhea compared with placebo (6, 7% vs 33, 3%; p=0,002) [69]. The second RCT evaluating L rhamnosus G in the prevention of diarrhea involved 220 children. L rhamnosus (GG) was administered orally once daily and did not prevent nosocomial rotavirus infections compared with placebo (25, 4% vs 30, 2%; p=0, 4). However, the rate of symptomatic rotavirus enteritis was lower in children receiving L rhamnosus (GG) compared with placebo (13% vs 21%; p=0, 13). [70]

The available data do not provide strong evidence for the routine use of L rhamnosus (GG) to prevent nosocomial rotavirus diarrhea in infants and toddlers.

Two other RCTs evaluated the efficacy of B. bifidum and S. thermophilus in preventing nosocomial diarrhea. The first study showed that the administration of standard infant formula supplemented with B. bifidum and S. thermophilus reduced the prevalence of nosocomial diarrhea compared with placebo. The risk of rotavirus gastroenteritis was significantly lower in those receiving probiotic-supplemented formula [71]. The second RCT showed that infants living in residential care settings, although they differ from hospital settings are also at increased risk for diarrheal illnesses, and the mode of acquiring diarrhea is similar. The infants received milk formula supplemented with viable B. lactis strain Bb12. It was shown that the previous intervention did not reduce the prevalence of diarrhea compared to placebo. [72]

In conclusion there is conflicting evidence on the efficacy of L rhamnosus (GG) provided by 2 RCTs in preventing nosocomial diarrhea. One small RCT suggests a benefit of B. bifidum and S. thermophilus in sick infants admitted to the hospital, but no such benefit has been identified in healthy children in residential care settings. The already mentioned studies suggest that there is currently not enough evidence to recommend the routine use of probiotics to prevent nosocomial diarrhea. There is a need for large and well-designed RCTs.

4.4. Allergy

The rationale for using probiotics in prevention and treatment of allergic disorders is based on the concept that appropriate microbial stimuli are required for normal early immunological development. Microbial-gut interactions can improve the integrity of the gut barrier by decreasing intestinal permeability, reducing both adherence of potential antigens and their systemic effect, and by modulating GALT immune response toward antigen tolerance. The intestinal microflora interacts with the mucosal immune system. It has been found that different strains of commercial bacteria vary in the cytokine response they generate. The Th1/Th2 imbalance is crucial to the clinical expression of allergy. Probiotic bacteria can produce significant antiallergenic effects by intricate interactions inducing Th1 cytokines, such as interferon-γ [73] , T-regulatory cytokines, such as IL-10 and TGF-β [74] , and mucosal immunoglobulin A production [75].

Three species of Lactobacillus were shown to modulate the phenotype and functions of human myeloid dendritic cells (DCs). Lactobacillus-exposed myeloid DCs up-regulated HLA-DR, CD83, CD40, CD80, and CD86, and secreted high levels of IL-12 and IL-18, but not IL-10. [76]

Infants with atopic dermatitis who received hydrolyzed whey formula supplemented with L rhamnosus (GG) showed greater clinical improvement than those who received the hydrolyzed formula alone. They also excreted less TNF-α and α-1-antitrypsin in their stool suggesting that the probiotics decreased gut inflammation. [77] Atopic infants treated with extensively hydrolyzed whey-based formula with L rhamnosus (GG) or B lactis showed greater improvement in severity of skin manifestations than with hydrolysate formula alone. The probiotic-supplemented group also demonstrated a reduction in serum soluble CD4 (a marker of T-cell activation) and an increase in serum TGF-β1 involved in suppressing the inflammatory response via IgE production and oral tolerance induction. [21] These studies suggest that regular probiotic supplementation may stabilize intestinal barrier function and play a role in modulating allergic responses leading to a decreased severity of atopic symptoms, particularly atopic dermatitis associated with cow's milk protein [21,29,78].

A marked anti-inflammatory effect was produced by bifidobacteria with an IL-10 induction by dendritic cells and consequent inhibition of Th1 activation with decreased interferon-γ production [79]. In atopic infants supplemented with B lactis a decrease of Bacteroides and E coli in the stool was shown. Most interestingly, serum IgE correlated with E coli counts, and in highly sensitized infants correlated with Bacteroides counts. Thus, certain probiotics seem to influence the gut's allergen-stimulated inflammatory response and provide a barrier effect against antigens that might otherwise ultimately lead to systemic allergic symptoms such as eczema. [37]

Proliferation and growth of beneficial bacteria in the digestive system is being promoted with the use of prebiotics. Prebiotics are generally considered to be safe and they are naturally present in several kinds of food. A food ingredient must fulfill the following criteria to be considered a prebiotic: it should be hydrolyzed or absorbed in the upper part of the gastrointestinal tract, it has to be a selective substrate for beneficial bacteria in the colon for example bifidobacteria, and it must be able to alter the intestinal microflora towards a healthier composition [80].

In regards to the immunomodulatory effect of prebiotics, the proposed mechanisms of action are the following: They are thought to stimulate the activity of lactic acid bacteria, such as lactobacilli and bifidobacteria, which have immunomodulatory qualities. A second mechanism of action is that fermentation of prebiotics by lactic acid bacteria enhances Short Chain Fatty Acids (SCFA) that they act as energy substrate for colonocytes. It has been shown that SCFA stimulate Interferon-γ and IL-10 production. [81-84]

The immunomodulatory effect of prebiotics on the prevention of atopic dermatitis has been evaluated by several studies. A study by Moro et al showed that a mixture of prebiotic oligosaccharides reduces the incidence of atopic dermatitis during the first six months of age [85]. A study by van der Aa et al determined the effect of Bifidobacterium breve M-16V combined with a prebiotic oligosaccharide mixture (synbiotic) on atopic markers. The synbiotic mixture had no detectable effect on plasma levels of the analysed atopic disease markers in vivo [86]. Another study by de Kivit S, et al investigated the effect of prebiotic galacto- and fructo-oligosaccharides (scGOS/lcFOS) in combination with Bifidobacterium breve M-16V (GF/Bb) on atopy. The study showed that dietary supplementation with GF/Bb enhances serum galectin-9 levels, which associates with the prevention of the allergic symptoms. [87]

In conclusion, although theoretically pro-, pre and synbiotics are promising candidates to prevent or treat AD, results of the clinical trials performed so far are not conclusive. Prevention trials show promising but heterogenic results. Therefore at this moment there is not enough evidence to support the use of pro-, pre-, or synbiotics for prevention of AD in clinical practice. Results of treatment trials are not very convincing, however pro- or synbiotics could possibly play a role in the treatment of IgE-associated AD, which should be elucidated in future prospective trials.

4.5. Necrotizing enterocolitis

Microflora establishment and composition in premature infants is a major determinant in the pathophysiology of NEC. The premature infant is exposed to a variety of factors that negatively affect their possibilities of attaining an appropriate colonization. These factors include increasing exposure to potential delayed colonization, colonization with "neonatal intensive care unit environmental microbes", use of antibiotics, lack of exposure to maternal flora and breast milk.

Mechanisms by which probiotics could prevent NEC include increase in favorable type microflora with reduced colonization by pathogens, increased intestinal barrier to translocation of bacteria into the bloodstream, modification of the host response to microbial products by sensitization and immunization, and enhanced tolerance and advancement of enteral nutrition [88-91.]

Several RCTs have assessed the efficacy of probiotics in preventing NEC. In a preospective, double-blind study premature infants (n=585) were randomized to receive standard milk formula supplemented with L rhamnosus G, or placebo. The group supplemented with L rhamnosus GG was found to have lower incidence of urinary tract infections and lower, but not statistically significant, incidence of NEC [92]. Two other trials have shown various degrees of reduction in relative risk of NEC with probiotics. The first study compared the incidence of NEC and the mortality of very-low-birth-weight (VLBW) infants fed breast milk with or without added probiotics. Infants (n=187) were randomized to receive breast milk or breast milk with L. acidophilus and B. infantis. In the intervention group the incidence of NEC was significantly decreased compared with the incidence in infants given breast milk alone [93]. The second study compared neonates receiving B infantis, S thermophilus, and B bifidus with neonates receiving no probiotic supplement. The incidence of NEC was 4% in 72 supplemented infants versus 16, 4% in 73 controls. The severity of NEC was less severe in the probiotic group. Three of 15 infants with NEC died, all in the control group [94].

A meta-analysis of RCTs evaluated if probiotic supplementation in preterm (<34 weeks gestation) VLBW(< 1500 gr) neonates could prevent NEC. The risk for NEC and death was significantly lower in the intervention group, while the risk for sepsis was not significantly different between the intervention group and the placebo. No significant adverse effects were reported [95].

In conclusion, specific clinical benefits are increasingly demonstrated for specific bacteria, which determine their probiotic capability. The protective and immune modulatory mechanisms that explain these effects are increasingly being documented.

5. Safety concerns of probiotics use

Newborn infants can develop infection from many species of resident microflora. The mechanisms for these infections and route of contamination are unclear. Many strains of Lactobacilli and Bifidobacteria are generally recognized as safe for use in the food supply. Documented correlations between systemic infections and probiotic consumption are few, and they have all occurred in patients with underlying medical conditions. Sporadic lactobacillemia from environmental, dietary, or fecal lactobacilli has been very rarely reported. Case reports of L rhamnosus (GG) infections possibly associated with probiotic consumption, in immunocompromised patients have been even less common [96, 97].

As opposed to the rarely reported episodes of lactobacillemia (some associated to ingested Lactobacilli), bifidobacteremia has not been sporadically reported, whether associated with consumption of commercial products containing Bifidobacteria or not. Bifidobacteria have also been consumed in infant formulas for more than 15 years worldwide and have not been associated with any pathologic or adverse event. Studies so far have documented safety and adequate growth with B. lactis in infants from birth [39] and in vulnerable populations, including preterm infants, [33, 19] malnourished infants, [98] and infants born to mothers with HIV disease [99]

From the safety point of view, according to current available information, Bifidobacteria, particularly B lactis, has a uniquely strong safety profile, making it a good probiotic candidate for newborns and young infants. Lactobacilli, particularly L rhamnosus (GG), also seems generally safe and be appropriate for older infants and children. Until adequate data are available for each specific probiotic bacterium, use of probiotics in general cannot be recommended in immunocompromised populations. However, as safety is better documented for specific bacteria, we may be able to use them in certain populations that may benefit the most from probiotic use.

Author details

Antigoni Mavroudi
Aristotle University of Thessaloniki, Greece

6. References

[1] Gaurner F, Schaafsma GJ. Probiotics. Int J Food Microbiol 1998; 39:237-238.

[2] Maldonado J, Caňabade F,Sempere L,et al. Human milk probiotic Lactobacillus fermentum CECT 5716 reduces the incidence of gastrointestinal and upper respiratory tract infections in infants. J Pediatr Gastroenterol Nutr 2012; 54:56-62.

[3] Corrêa N, Penna F,Lima F,et al. Treatment of acute diarrhea with Saccharomyces boulardii in infants. J Pediatr Gastroenterol Nutr 2011; 53: 497-501.

[4] Food and Agriculture Organization, World Health Organization. The Food and Agriculture Organization of the United Nations and the World Health Organization Joint FAO/WHO expert consultation on evaluation of health and nutritional properties of probiotics in food including powder milk with live lactic acid bacteria. FAO/WHO Report No. 10-1-2001.

[5] Saavedra JM. Clinical applications of probiotic agents. Am J Clin Nutr. 2001; 73: 1147S-1151S.

[6] Isolauri E. Probiotics in human disease. Am J Clin Nutr. 2001; 73:1142S-1146S.

[7] Harmsen HJ,Wildeboer-Veloo AC,Raangs GC et al. Analysis of intestinal flora development in breast-fed infants and formula-fed infants by using molecular identification and detection methods. J Pediatr Gastroenterol Nutr. 2000; 30:61-67.

[8] Isolauri E,Sutas Y,Kankaanpaa P,Arvilommi H,Salminen S. Probiotics :effects on immunity. Am J Clin Nutr. 2001 ;73(2 Suppl): 444S-450S.

[9] Saavedra JM. Use of Probiotics in Pediatrics : Rationale ,Mechanisms of Action ,and Practical Aspects Nutr Clin Pract 2007 22: 351.

[10] Servin AL. Antagonistic activities of Lactobacilli and Bifidobacteria against microbial pathogens. FEMS Microbiol Rev. 2004; 28:405-440.

[11] Bakker-Zierikzee AM, Tol EA ,Kroes H ,Alles MS ,Kok FJ ,Bindels JG. Faecal s IgA secretion in infants fed on pre- or probiotic infant formula. Pediatr Allergy Immunol. 2006; 17:134-140.

[12] Kohler H, Donarski S ,Stocks B ,Parret A ,Edwards C ,Schroten H. Antibacterial characteristics in the feces of breast fed and formula-fed infants during the first year of life. J Pediatr Gastroenterol Nutr 2002 ; 34 :188-193.

[13] Roller M ,Rechkemmer G ,Watzl B. Prebiotic inulin enriched with oligofructose in combination with the probiotics Lactobacilus rhamnosus and Bifidobacterium lactis modulates intestinal immune functions in rats. J Nutr. 2004 ;134 :153-156.

[14] Mohan R , Koebnick C, Radke M, Schildt J , Possner M,Blaut M. Microbial colonization of the gastrointestinal tract of preterm infants :diversity and new ways for prevention of infactions (abstract). European Academy of Pediatrics ,Barcelona ,Spain ,October 2006. Abstract No. PG3-07

[15] Forchielli ML, Walker WA. The role of gut-associated lymphoid tissues and mucosal defense. Br J Nutr. 2005; 93 (suppl 1): S41-S48.

[16] Silva M, Jacobus NV, Deneke C, Gorbach SL. Antimicrobial substance from a human Lactobacillus strain. Antimicrob Agents Chemother. 1987;31:1231-1233.

[17] Mack DR, Michail S, Wei S, McDougall L, Hollingsworth MA. Probiotics inhibit enteropathogenic E. coli adherence in vitro by inducing intestinal mucin gene expression. Am J Physiol. 1999; 276:G941-G950.

[18] Gupta P,Andrew H,Kirschner BS,Guandalini S. Is Lactobacilli helpful in children with Crohn's disease? Results of a preliminary open-label study. J Pediatr Gastroenterol Nutr 2000; 31:453-457.

[19] Stratiki Z,Kostalos C,Sevastiadou S,et al. The effect of a bifidobacter supplemented bovine milk on intestinal permeability of preterm infants. Early hum Dev. Availible at: http://dx. doi. org/10. 1016/j. earlhumdev. 2006. 12. 002. Accessed February 25, 2007.

[20] Majamaa H, Isolauri E. Probiotics: a novel approach in the management of food allergy. J Allergy Clin Immunol. 1997; 99:179-185.

[21] Isolauri E ,Arvola T ,Sutas Y, Moilanen E ,Salminen S. Probiotics in the management of atopic eczema. Clin Exp Allergy 2000; 30:1604-1610.

[22] Bufford JD, Gern JE. The hygiene hypothesis revisited. Immunol Allergy Clin North Am. 2005; 25:247-262.

[23] Schaub B,Lauener R,von Mutius E. The many faces of the hygiene hypothesis. J Allergy Clin Immunol. 2006; 117:S969-S977.

[24] Weng M, Walker WA. Bacterial colonization, probiotics and clinical disease. J Pediatr. 2006; 149:S107-S114.

[25] Bjorksten B. Allergy prevention: interventions during pregnancy and early infancy. Clin Rev Allergy Immunol. 2004; 26:129-138.

[26] Becker AB, Chan –Yeung M. Primary prevention of asthma. Curr Opin Pulm Med. 2002; 8:16-24.

[27] Rook GA, Brunet LR. Microbes, immunoregulation, and the gut. Gut. 2005; 54:317-320.

[28] Rautava S,Ruuskanen O,Ouwehand A,Salminen S,Isolauri E. The hygiene hypothesis of atopic disease :an extended version. J Pediatr Gastroenterol Nutr. 2004; 38:378-388.

[29] Aldinucci C, Bellussi L, Monciatti G, et al. Effects of dietary yoghurt on immunological and clinical parameters of rhinopathic patients. Eur J Clin Nutr. 2002; 56: 1155-1161.

[30] Arunachalam K, Gill HS, Chandra RK. Enhancement of natural immune function by dietary consumption of Bifidobacterium lactis (HN019). Eur J Clin Nutr. 2000; 54:263-267.

[31] Kalliomaki M ,Salminen S, Poussa T, Arvilommi H, Isolauri E. Probiotics and prevention of atopic disease :4-year follow-up of a randomized placebo-controlled trial. Lancet. 2003;361:1869-1871.
[32] Pessi T, Sutas Y, Hurme M, Isolauri E. Interleukin-10 generation in atopic children following oral Lactobacillus rhamnosus GG. Clin Exp Allergy. 2000;30:1804-1808.
[33] Mohan R ,Koebnick C ,Schildt J, et al. Effects of Bifidobacterium lactis Bb12 supplementation on intestinal microbiota of preterm infants :a double-blind placebo-controlled ,randomized study. J Clin Microbiol. 2006;44:4025-4031.
[34] Langhendries JP, Detry J,Van HJ, et at. Effect of a fermented infant formula containing viable Bifidobacteria on the fecal flora composition and pH of healthy full-term infants. J Pediatr Gastroenterol Nutr. 1995; 21:177-181.
[35] Bakker-Zierikzee AM, Alles MS, Knol J, Kok FJ, Tolboom JJ,Bindels JG. Effects of infant formula containing a mixture of galacto- and fructo-oligosaccharides or viable Bifidobacterium animalis on the intestinal microflora during the first 4 months of life Br J Nutr. 2005; 94:783-790.
[36] Fukushima Y,Li S-T, Hara H,Terada A,Mitsuoka T. Effect of follow-up formula containing Bifidobacteria (NAN BF) on fecal flora and fecal metabolites in healthy children. Biosci Microflora. 1997;16:65-72.
[37] Kirjavainen PV, Arvola T, Salminen SJ, Isolauri E. Aberrant composition of gut microbiota of allergic infants: a target of bifidobacterial therapy at weaning? Gut. 2002; 51:51-55.
[38] Szajewska H, Setty M, Mrukowicz J, Guandalini S. Probiotics in gastrointestinal diseases in children : hard and not-so-hard evidence of efficacy. J Pediatr Gastroenterol Nutr. 2006; 42:454-457.
[39] Weizman Z, Asli G, Alsheikh A. Effect of a probiotic infant formula on infections in child care centers: comparison of two probiotic agents. Pediatrics. 2005; 115: 5-9.
[40] Chouraqui JP, Van Ergoo LD, Fichot MC. Acidified milk formula supplemented with Bifidobacterium lactis: impact on infant diarrhea in residential care settings. J Pediatr Gastroenterol Nutr. 2004; 38:288-292.
[41] Swajewska H, Mrukowicz JZ. Probiotics in the treatment and prevention of acute infectious diarrhea in infants and children: a systematic review of published randomized, double-blind, placebo controlled trials. J Pediatr Gastroenterol Nutr. 2001; 33(suppl 2):S17-S25.
[42] Oberhelman RA, Gilman RH, Sheen P, et al. A placebo-controlled trial of Lactobacillus GG to prevent diarrhea in undernourished Peruvian children. 1999; 134:15-20.
[43] Mastretta E, Longo P, Laccisaglia A, et al. Effect of Lactobacillus GG and breast-feeding in the prevention of rotavirus nosocomial infection. J Pediatr Gastroenterol Nutr 2002; 35: 527-531.
[44] Rosenfeldt V, Michaelson KF, Jakobsen M, et al. Effect of probiotic Lactobacillus strains in young children hospitalized with acute diarrhea. Pediatr Infect Dis J. 2002; 21:411-416.
[45] Saavedra JM, Bauman NA, Oung I, Perman JA, Yolken RH. Feeding of Bifidobacterium bifidum and Streptococcus thermophilus to infants in hospital for prevention of diarrhoea and shedding of rotavirus. Lancet 1994; 344: 1046-1049.
[46] Sazawal S, Hiremath G, Dhingra U, Malik P, Deb S, Black RE. Efficacy of probiotics in prevention of acute diarrhoea: a meta-analysis of masked, randomized, placebo-controlled trials. Lancet Infect Dis. 2006; 6: 374-382.

[47] Guandalini S, Pensabene L, Zikri MA, et al. Lactobacillus GG administered in oral rehydration solution to children with acute diarrhea: a multicenter European trial. J Pediatr Gastroenterol Nutr. 2000; 30: 54-60.
[48] Lievin-Le Moal V, Amsellem R, Servin AL, et al. Lactobacillus acidophilus (strain LB) from the resident adult human gastrointestinal microflora exerts activity against brush border damage promoted by a diarrhealgenic Escherichia coli in human enterocyte –like cells. Gut 2002; 50: 803-11.
[49] Mack DR, Michail S, Wei S, et al. Probiotics inhibit enteropathogenic E. coli adherence in vitro by inducing intestinal mucin gene expression. Am J Physiol 1999; 276: G941-G50.
[50] De Simone C, Ciardi A, Grassi A, et al. Effect of bifidobacterium bifidum and Lactobacillus acidophilus on gut mucosa and peripheral blood B lymphocytes. Immunopharmacol Immunotoxicol 1992; 14: 331-340.
[51] Aattour N, Bouras M, Tome D, et al. Oral ingestion of lactic-acid bacteria by rats increases lymphocyte proliferation and interferon-gamma production. Br J Nutr 2002; 87: 376-373.
[52] Kaila M, Isolauri E, Soppi E, et al. Enhancement of the circulating antibody secreting cell response in human diarrhea by a human Lactobacillus strain. Pediatr Res 1992; 32:141-144.
[53] Silva M, Jacobus NV, Deneke C, et al. Antimicrobial substance from a human Lactobacillus strain. Antimicrob Agents Chemother 1987; 31: 1231-1233.
[54] Cocinnier MH, Lievin V, Bernet-Camard MF, et al. Antibacterial effect of the adhering human Lactobacillus acidophilus strain LB. Antimicrob Agents Chemother 1997; 41: 1046-1052.
[55] Wilson KH, Perini I. Role of competition for nutrients in suppression of Clostiridium difficile by the colonic microflora. Infect Immun 1988; 56: 2610-2614.
[56] Walker WA. Role of nutrients and bacterial colonization in the development of intestinal host defense. J Pediatr Gastroenterol Nutr 2000; 30 (suppl): S2-S7.
[57] Bernet MF, Brassat D, Nesser JR, et al. Lactobacillus acidophilus LA1 binds to human intestinal cell lines and inhibits cell attachment and cell invasion by enterovirulent bacteria. Gut 1994; 35: 483-439.
[58] Davidson JN, Hirsch DC. Bacterial competition as a mean of preventing diarrhea in pigs. Infect Immun 1976; 13: 1773-1774.
[59] Rigothier MC, Maccanio J, Gayral P. Inhibitory activity of Saccharomyces yeasts of adhesion of Entamoeba histolytica trophozoites to human erythrocytes in vitro. Parasitol Res 1994; 80: 10-15.
[60] Michail S, Abernathy F. Lactobacillus plantarum reduces the in vitro secretory response of intestinal epithelial cells to enteropathogenic Escherichia coli infection. J Pediatr Gastroenterol Nutr 2002; 35: 350-355.
[61] Pothoulakis C, Kelly CP, Joshi MA, et al. Saccharomyces boulardii inhibits Clostiridium difficile toxin A binding and enterotoxicity in rat ileum. Gastroenterology 1994; 104: 1108-1115.
[62] Czerucka D, Roux I, Rampal P. Saccharomyces boulardii inhibits secretagogue-mediated adenosine 3, 5-cyclic monophosphate induction in intestinal cells. Gastroenterology 1994; 106: 65-72.
[63] Hogenauuer C, Hammer HF, Krejs GJ, et al. Mechanisms and management of antibiotic-associated diarrhea. Clin Infect Dis 1998; 27: 702-710.

[64] McFarland LV, Brandmarker SA, Guandalini S. Pediatric Clostiridium difficile: a phantom menace or clinical reality? J Pediatr Gastroenterol Nutr 2000; 31: 220-231.
[65] Correa NB, Peret Filho LA, Penna FJ, Nicoli JR. A randomized formula controlled trial of Bifidobacterium lactis and Streptococcus thermophilus for prevention of antibiotic-associated diarrhea in infants. J Clin Gastroenterol. 2005; 39: 385-389.
[66] Kotowska M, Albrecht P, Szajewsks H. Saccharomyces boulardii in the prevention of antibiotic-associated diarrhea in children: a randomized double-blind placebo-controlled trial. Aliment Pharmacol Ther. 2005; 21: 583-590.
[67] Szajewska H, Ruszczynski M, Radzikowski A. Probiotics in the prevention of antibiotic-associated diarrhea in children a meta-analysis of randomized controlled trials. J Pediatr. 2006; 367-373.
[68] Matson DO, Estes MK. Impact of rotavirus infection at a large pediatric hospital. J Infect Dis 1990; 162:598-604.
[69] Szajewska H, Kotowska M, Mrukowicz J, et al. Lactobacillus GG in prevention of diarrhea in hospitalized children. J Pediatr 2001; 138:361-365.
[70] Mastretta E, Longo P, Laccisaglia A, et al. Lactobacillus GG and breast feeding in the prevention of rotavirus nosocomial infection. J Pediatr Gastroenterol Nutr 2002; 35:527-531.
[71] Saavedra JM, Bauman NA, Oung I, et al. Feeding of Bididobacterium bifidum and Streptococcus thermophilus to infants in hospital for prevention of diarrhea and shedding of rotavirus. Lancet 1994; 344: 1046-1049.
[72] Chouraqui JP, Van Ergoo LD, Fichot MC. Acidified milk formula supplemented with Bifidobacterium lactis: impact on infant diarrhea in residential care settings. J Pediatr Gastroenterol Nutr 2004; 38:288-292.
[73] He F, Morita H, Hashimoto H, et al. Intestinal Bifidobacterium species induce varying cytokine production. J Allergy Clin Immunol 2002; 109:1035-1036.
[74] Kalliomaki M, Ouwehand A, Arvilommi H, et al. Transforming growth factor-beta in human breast milk: a potential regulator of atopic disease at early age. J Allergy Clin Immunol 1999; 104: 1251-1257.
[75] Park JH, Um JI, Lee BJ, et al. Encapsulated Bifidobacterium bifidum potentiates intestinal IgA production. Cell Immunol 2002; 219: 22-27.
[76] Mohamadzadeh M, Olson S, Kalina WV, et al. Lactobacilli activate human dendritic cells that skew T cells toward T helper 1 polarization. Proc Natl Acad Sci USA 2005; 102: 2880-2885.
[77] Isolauri E. Studies on Lactobacillus GG in food hypersensitivity disorders. Nutr Today Suppl. 1996; 31: 285-315.
[78] Pohjavuori E, Viljanen M, Korpela R, et al. Lactobacillus GG effect in increasing IFN-γ production in infants with cow's milk allergy. J Allergy Clin Immunol. 2004; 114: 131-136.
[79] Hart AL, Lammers K, Brigidi P, et al. Modulation of human dendritic cell phenotype and function by probiotic bacteria. Gut 2004; 53: 1602-1609.
[80] Collins MD, Gibson GR. Probiotics, prebiotics and synbiotics: approaches for modulating the microbial ecology of the gut. Am J Clin Nutr 1999: 69: 1052S-7S.
[81] Gibson GR, Roberfroid MB. Dietary modulation of the human colonic microbiota: introducing the concept of prebiotics. J Nutr. 1995; 125: 1401-1402.
[82] Wong JM, De SR, Kedall CW, Emam A, Jenkins DJ. Colonic health: fermentation and short chain fatty acids. J Clin Gastroenterol. 2006; 40: 235-243.

[83] Cavaglieri CR, Nishiyama A, Fernandes LC, Curi R, Miles EA, Calder PC. Differential effects of short-chain fatty acids on proliferation and production of pro- and anti-inflammatory cytokines by cultured lymphocytes. Life Sci. 2003; 73: 1683-1690.

[84] Mavroudi A, Xinias I. Dietary interventions for primary allergy prevention in infants. Hippokratia 2011; 15: 216-222.

[85] Moro G, Aslanoglu S, Stahl B, Jelinek J, Wahn U, Boehm G. A mixture of prebiotic oligosaccharides reduces the incidence of atopic dermatitis during the first six months of age. Arch Dis Child. 2006; 91: 814-819.

[86] van der Aa L. B, Lutter R, Heymans H. S. A, Smids B. S, Dekker T, van Aalderen W. M. C, Sillevis Smitt J. H, Knippels L. M. J, Garssen J, Nauta A. J, Sprikkelman A. B and the Synbad Study Group. No detectable beneficial systemic immunomodulatory effects of a specific synbiotic mixture in infants with atopic dermatitis. Clinical & Experimental Allergy 2012; 42: 531-539.

[87] de Kivit S, Saeland E, Kraneveld A. D, van de Kant H. J. G, Schouten B, van Esch B. C. A. M, Knol J, Sprikkelman A. B, van der Aa L. B, Knippels L. M. J, Garssen J, van Kooyk Y, Willemsen L. E. M. Galectin-9 induced by dietary synbiotics is involved in suppression of allergic symptoms in mice and humans. Allergy 2012; 67: 343-352.

[88] Magne F, Suan A, Pochart P, Desjeux JF. Fecal microbial community in preterm infants. J Pediatr Gastroenterol Nutr. 2005; 41: 386-392.

[89] Millar M, Wilks M, Costeloe K. Probiotics for preterm infants? Arch Dis Child Fetal Neonatal Ed. 2003; 88: F354-F358.

[90] Agostini C, Axelsson I, Braegger C, et al. Probiotic bacteria in dietetic products for infants: a commentary by the ESPGHAN Committee on Nutrition. J Pediatr Gastroenterol Nutr. 2004: 38: 365-374.

[91] Vanderhoof JA, Young RJ. Pediatric applications of probiotics. Gastroenterol Clin North Am. 2005; 34: 451-454, ix.

[92] Dani C, Biadaioli R, Bertini G, Martelli E, Rubaltelli FF. Probiotics feeding in prevention of urinary tract infection, bacterial sepsis and necrotizing enterocolitis in preterm infants: a prospective double-blind study. Biol Neocate. 2002; 82: 103-108.

[93] Lin HC, Su BH, Chen AC, et al. Oral probiotics reduce the incidence and severity of necrotizing enterocolitis in very low birth weight infants. Pediatrics. 2005; 115: 1-4.

[94] Bin-Nun A, Bromiker R, Wilschanski M, et al. Oral probiotics prevent necrotizing enterocolitis in very low birth weight neonates. Pediatrics. 2005; 192-196.

[95] Deshpande G, Rao S, Patole S and Bulsara M. Updated Meta-analysis of Probiotics for Preventing Necrotizing Enterocolitis in Preterm Neonates. Pediatrics. 2010; 125: 921-930.

[96] Land MH, Rouster-Stevens K, Woods CR, Cannon ML, Cnota J, Shetty AK. Lactobacillus sepsis associated with probiotic therapy. Pediatrics. 2005; 115: 178-181.

[97] Kunz AN, Noel JM, Fairchok MP. Two cases of Lactobacillus bacteremia during probiotic treatment of short gut syndrome. J Pediatr Gastroenterol Nutr. 2004; 38: 457-458.

[98] Nopchinda S, Varavithya W, Phuapradit P, et al. Effect of Bifidobacterium Bb12 with or without Streptococcus thermophilus supplemented formula on nutritional status. J Med Assoc Thai. 2002; 85(suppl 4): S1225-S1231.

[99] Cooper PA, Mokhachane M, Bolton KD, Steenhout P, Hager C. Growth of infants born from HIV positive mothers fed with acidified starter formula containing Bifidobacterium lactis (abstract). Eur J Peds. 2006; 165(Suppl 13): 114.

Chapter 14

Probiotics Applications in Autoimmune Diseases

Hani Al-Salami, Rima Caccetta,
Svetlana Golocorbin-Kon and Momir Mikov

Additional information is available at the end of the chapter

http://dx.doi.org/10.5772/50463

1. Introduction

An autoimmune disorder (AD) is a condition in which the immune system mistakenly attacks its own body cells through the production of antibodies that target certain tissues. Such attack triggers further inflammation that result in more attacks and a significant inflammatory response leading to tissue destruction and cessation of functionality [1]. ADs include diabetes, rheumatoid arthritis, Graves' disease, systemic lupus and inflammatory bowel disease (IBD) [2]. ADs are on the rise worldwide and have major health implications from the diseases themselves as well as complications. Even though the causes of AD have been postulated to be genetic and environmental, the actual triggers remain poorly defined [3]. Genetic predisposition contribute to about 30% of AD while 70% to environmental factors such as infections (e.g., virus, bacteria) and lifestyle-associated factors such as food.

Recent data show that AD has prevalence of 6-8% and are currently affecting 400 million people worldwide, with the majority of all those affected being women. Previous figures underestimated the scope of the problem, while even the most pessimistic predictions fell short of the current figure. It is predicted that the total number of people living with AD will increase drastically within the coming thirty years if no new and substantially more effective drugs are produced [4]. On 2009, estimated health costs of autoimmune disorders have exceeded 100 billion dollars only in the US. This adds to the cost generated from higher rate of hospitalization, higher mortality rate, and impaired performance of workers with the disease [5]. AD is a condition that incorporates various metabolic disturbances and inflammatory physiological and biochemical reactions including blood dyscrasias and endocronological and pathophysiological imbalances. Of recently, gastrointestinal abnormalities have been directly linked to the initiation and progression of autoimmune diseases especially slower gut movement (gastroparesis) and microfloral overgrowth (especially of fermentation bacteria and yeasts due to the slightly more acidic gut contents). Improving AD complications, reducing prevalence and restoring normal physiological patterns should significantly optimise treatment outcomes and the quality of life for patients.

In healthy individuals, the immune system prevents self-attack by two main routes. Firstly, by neutralizing dysfunctional lymphocytes in the thymus before they start attacking own body cells. This results in preventing the initiation of inflammation and progression of the autoimmune symptoms. Secondly, when dysfunctional lymphocytes are released into the mainstream, the immune system minimizes their ability to interact with triggers (antigens) through direct and indirect effects [6-8]. This results in a significant reduction in the severity of potential inflammatory response. Accordingly, treating AD can be achieved by either replacing the function of the damaged tissues (e.g. through injecting insulin when treating Type 1 diabetes, T1D) or suppressing the dysfunctional immune cells (e.g. through steroid therapy) [9-11].

Generally, clinical and laboratory research has suggested that certain immune cells called B-cells may have a stronger influence on the development and progression of various autoimmune diseases than previously thought [12]. Inflammatory cells attack different organs in different autoimmune disorders. In T1D, the autoimmune system attacks the β-cells of the pancreas triggering an inflammatory reaction, which results in the destruction of these cells and the cessation of insulin production [13]. In rheumatoid arthritis, rheumatoid factor antibodies are produced by the immune system and are interact with γ globulin (blood proteins) forming a complex that triggers inflammation that targets muscles and bones [14]. In Graves's diseases, an autoimmune disease of the thyroid gland, antibodies are produced against the thyroid protein thyroglobulin. These antibodies are called Thyroid Stimulating Hormones Receptors (TSHR) antibodies results in the increase in thyroid synthesis and section and thyroid growth as well as all accompanying symptoms [15-17]. In some autoimmune blood disorders, antibodies are produced against the body red and white blood cells, while in other autoimmune disorders, antibodies attack a wide range of tissues and organs resulting in more debilitating symptoms [18]. In systemic lupus, antibodies target antigens that are present in nucleic acids and cell organelles such as ribosomes and mitochondria. Lupus can cause dysfunction of many organs, including the heart, kidneys, and joints [19]. IBDs include two main conditions, ulcerative colitis and Crohn's disease. The inflammation in both conditions can affect the small and large intestine and sometimes other parts of the digestive system. Generally, ulcerative colitis is limited to the colon, primarily affecting the mucosa and the lining of the colon. Extensive inflammation gives rise to small ulcerations and microscopic abscesses that produce bleeding which exacerbate further the inflammatory response and worsen symptoms. Crohn's disease affects the small and large intestine, and rarely the stomach or oesophagus.

Many ADs have been characterized by a compromised gut movement which has been linked to the disturbed immune system and can result in substantial gut bacterial and yeast overgrowth [20-24]. Such an overgrowth is postulated to disturb body physiological and biochemical reactions and exacerbate the autoimmune-associated inflammation. This has also been linked to long term complications and weaker prognosis resulting in poor drug response and worsening quality of life [25, 26].

Diagnosing autoimmune diseases can be particularly difficult, because these disorders can affect any organ or tissue in the body and produce a wide variety of signs and symptoms. Many early symptoms of these disorders — such as fatigue, joint and muscle pain, fever or weight change — are nonspecific. Symptoms are often not apparent until the disease has reached a relatively advanced stage. Accordingly, prevention in most susceptible individuals and early diagnosis are two most important approaches, when researching the future therapy for autoimmune diseases.

ADs include wide range of inflammatory disease models that are characterized by the presence of a colossal inflammatory response. The trigger of the inflammation is versatile and complex with many hypotheses ranging from ingested toxins to idiopathic causes [9, 18, 27]. However, genetic influence remains a strong cause and is considered a contributing factor for the development and progression of these diseases. AD-associated inflammation can cause chemical unbalance that has been linked to poor tissue sensitivity to drug stimulation, rise in the levels of reactive radicals in the blood, poor enterohepatic recirculation and negatively affecting liver detoxification and performance. The level and extent of tissue damage depend on the severity of the inflammatory response and varies in different disease models. Accordingly, future therapy should focus not only on symptomatic relief, but also on rectifying the disturbances in body physiology and associated short and long term complications. These disturbances may affect the whole body and have been strongly linked to inflammatory lymph nodes in the gut walls. Thus, future therapy should also focus on normalizing gut disturbed immune response, which can be achieved through normalizing the composition of bile acids and microflora, gut immune-response and microflora-epithelial interactions towards maintaining normal biochemical reactions and healthy body physiology.

Of recently, the applications of probiotics in autoimmune diseases have gained great interest due to the feasibility of their administration and also their safety. A good example is hypoglycemic effect of probiotics in a rat model of Type 1 diabetes [28]. Possible mechanisms of actions include their anti-inflammatory effect resulting in a significant reduction in diabetes progression and complications [24]. This can be brought about through the normalization of gut disturbed-microflora by the administered probiotic-bacteria. Interesting, probiotic co-administration with a sulphonylureas antidiabetic drug has shown to reduce inflammation and ameliorate diabetes complications suggesting a significant role and great potential of probiotic applications as anti-inflammatory adjunct therapy.

Probiotics are dietary supplements containing bacteria which, when administered in adequate amounts, confer a health benefit on the host. Combinations of different bacterial strains can be used but a mixture of Lactobacilli and Bifidobacteria is a common choice. Probiotics have been shown to be beneficial in a wide range of conditions including infections, allergies, metabolic disorders such as diabetes mellitus, ulcerative colitis and Crohn's disease.

This chapter aims to explore the changes in gut microflora, physiology and metabolic pathways which are associated with the autoimmune diseases. A great focus will be on the potential application of probiotics on rectifying the disturbed gut composition associated with these diseases and whether such intervention can prevent or even treat these diseases.

2. Autoimmune-associated disturbances in gut microflora

The initial set of gut microfloral composition in human starts during birth. The physical structure of the gut is altered by the presence of microorganisms during growth. Once matured, the integrity of the epithelial barrier is maintained by the presence of these same microbes. Accordingly, the mother's microflora is considered a source of the infant own initial gut bacterial colonization, which is then influenced by the mother's milk, tissues' growth, the maturation of the immune system, as well as other factors. Gut motility and contents have been emerging as an important area of research when investigating the origin and potential therapeutics of autoimmune disease. Many patients with autoimmune disease have shown strong evidence of disturbances in the composition of gut microflora and the subsequent toxin buildup and other associated physiological and biochemical abnormalities [29]. A good example is Type 1 diabetic patients. Although the pathogenesis of T1D remains unclear, there is strong evidence supporting the hypothesis that the trigger leading to T1D, starts in the gut of genetically susceptible individuals [30, 31]. This inflammation causes major disturbances in both, the gut microfloral composition and bile acids ratios. This results in ongoing inflammatory response that brings about the destruction of pancreatic tissues and subsequent cessation of insulin production leading to clinical signs and symptoms of Type 1 diabetes. Another good example showing disturbed microfloral composition is IBD. Patients with IBD have shown clear shift of the gut microfloral composition towards less lactic acid-producing bacteria. In addition, the relative load of some species of colon-associated bacteria such as Bifidobacteria shows little presence in the gut of IBD patients indicating less bacterial-synchronization and disturbed quorum sensing processes in such patients. Interestingly, antibiotics are used in IBD to treat infective complications and to improve symptoms through altering the gut microfloral composition [32].

Maintenance of the physical integrity of the gut is essential to limit penetration of harmful bacteria. Dorsal to the epithelial layer in the gastrointestinal tract is a protective mucous gel layer which is altered by the existing microbial colonies. The neutral pH of the epithelium is preserved by the mucin, which creates a gradient to the acidic contents of the gut. It acts as a physical barrier to block microorganisms from adhering to the underlying epithelium and prevents sheer stress on the gut. The spread of harmful xenobiotics through the gut is limited by the mucin, which is normally a thick and viscous layer. In a germ-free environment the mucous layer is thinner and has a different mucin content and composition. Recent literature has shown that in ulcerative colitis and, to a lesser extent, Crohn's disease are associated with a significant reduction of the protective gut-mucus layer, however, the role of this alteration in the pathogenesis of both diseases remain unclear [33].

Localized inflammatory responses are modulated by the gut microfloral bacteria that seek to establish an ideal environment for their growth. The gut microfloral bacteria also alter inflammatory mediators which utilize the lymphatic system for transport, altering sites of inflammation outside the gut.

Intercellular interactions can also change gut permeability and are influenced by gut microflora. Zonula occludens are proteins that provide a structural framework to cells and seal the space between them, preventing the movement of ions across the barrier. A number of pathogenic bacteria and parasites target these epithelial cell membranes to increase the gut vulnerability to penetration. Comparatively, the presence of some beneficial bacteria can increase the expression of zonula occludens at tight junctions, improving epithelial integrity and cell-cell adhesiveness.

It is important to stress the fact that both, the complexity and versatility of gut microflora, remain major challenges to precisely be able to measure the changes in bacterial composition in diseases patients and compare that to healthy ones. In addition, the effect of food, drug consumption, gender and age may also influence gut microfloral composition adding complexity when comparing healthy versus disease states. To complicate this further, the interaction between bile acids and gut microflora has a significant effect on the density, composition, type, colonization and quorum sensing processes of various strains of gut bacteria, thus, making bile acids (BA) a major component of the bacterial-ecosystem that exists in the gut. This necessitates including bile acids, with when investigating autoimmune-associated disturbances in gut microbiota.

BAs are naturally produced in human. They are known to provide human with health benefits through their endocronological, microfloral, metabolic and other known and unknown effects. Disturbances in bile acids composition and functionality may cause tissue damage and eventual necrosis due to higher than normal concentrations of potent bile acids such as lithocholic acid compared with less potent bile acids such as chenodeoxycholic acid [34]. The nature of the interaction between gut microflora and bile acids is based on the fact that secondary bile acids are solely produced by the action of gut microflora. Gut microflora activates primary bile acids to secondary bile acids. This interaction between bile acid composition and the composition of gut microflora represents the base of the hypothesized linking between bile acid, gut microflora and energy balance. However, even though the compositions of bile acids and gut microflora are reported to be different in diabetic patients [35], it is still not clear how these changes directly affect the development and progression of diabetes or its complications. These complications include cardiovascular, tissue necrosis and ulcerations, and metabolic disturbances.

T1D is a good example of a common autoimmune disease which is on the rise worldwide. Even though the composition of gut microflora has been reported to be different in T1D patients, it may be difficult to quantify or qualify such a difference. Gut microflora interacts closely with the body immune system and has shown to control the immune response to various inflammatory stimuli. The mechanism of action of probiotics could be one or more of the following. Firstly, by competitive exclusion, where gut microfloral bacteria resist colonization of other 'foreign' bacteria. Secondly, by barrier formation where the microflora forms a physical barrier reducing bacterial translocation by forming a wall surrounding the outside part of the gut enterocytes. Thirdly, gut bacteria can produce bacteriocins and change the pH to create a harsher environment for other invading bacteria to settle in the gut. Fourthly, gut microflora can influence the immune system through its effect on gut enterocytes (quorum sensing) and the innate and adaptive immune system [36, 37]. To understand better the autoimmune-associated disturbances in the gut microflora, there is a definite need to understand the mechanism by which gut microflora interacts with the epithelial mucosa lining up the intestinal tract. Over the last decade, there have been growing interests in studying the mechanism by which enterocytes interact with gut microflora.

The epithelial mucosa is inhabited by significant number of various immune cells that work as a link between the gut epithelia and lumen-contents [38]. One of these immune cells is lymphocytes such as T helper cells. These cells play an important role in the adaptive immune response. Thus, T helper cells have a more administrative role where it comes to neutralizing infected cells. Accordingly, they do not have direct cytotoxic or phagocytic effect. This role covers activating and directing other immune cells to destroy xenobiotics. They are essential in B cell antibody class switching, in the activation and growth of cytotoxic T cells, and in maximizing the antibacterial activity of phagocytes such as macrophages [39-41]. After a period of time, T helper cells start expressing CD4 which is a specialized surface protein. So when a body-cell is infected with an antigen, and this cell expresses this antigen on MHC class 2, a CD4 cell will promote the cell interactions and elimination. The lamina propria is a layer of connective tissue that lies adjacent to the epithelium of a mucous membrane. The intestinal epithelial mucosa consists of the lamina propria and the mucus. Many T helper cells, macrophages and IgA-producing plasma cells are present in the lamina propria [4].

Specialized microfold (M) cells of the lymph tissues can be found in the epithelial mucosa in the gut. M cells play a crucial role in the genesis of systemic immune response by delivering antigenic substrate to the underlying lymphoid tissue where immune responses start. Although it has been shown that dendritic cells also have the ability to sample antigens directly from the gut lumen, M cells certainly remain the most important antigen-sampling cell and are affected in the autoimmune diseases. M cells transport bacteria and antigen to the lymphatic tissue. Dendritic cells are bone marrow-derived antigen-presenting cells that essentially influence all aspects of innate and acquired immunity (Figure 2). These cells sense the microbes in their milieu through TLRs, and by signalling via different TLRs, generate biological reactions which produce variable responses from excitatory to suppressive. Dendritic cells are heterogeneous inhabitants of the intestine found scattered in all lymphoid compartments and can enter between epithelial cells to taster lumenal bacteria which they can then present to immune cells in the mucosa.

In healthy individuals, cytokines and mature T cells suppress 'exaggerated' T cell response that may result in unwanted cell damage, apoptosis and death. Thus, gut microflora in each individual, works as a finger print and exerts a significant control over the immune response to various 'antigenic' stimuli. In addition to the gut microfloral control on the intestinal immunoregulatory system and the mucosal barrier, it is also involved in the pathogenesis of symptoms related to metabolic interactions of the microflora with intestinal contents or intestinal functions such as peristaltic movement [25, 26, 42-44]. As a result, many gastrointestinal disorders can be benefited from probiotic treatments. This includes travel diarrhoea, bloating and irritable bowel disease. Changes in the permeation of the intestine have been strongly associated with various autoimmune diseases such as T1D and IBD. However, the efficacy of probiotic treatment in autoimmune diseases is still under scrutiny and despite excellent progress in studying changes in gut microfloral composition associated with many autoimmune diseases, probiotic therapy has still not shown clear clinical efficacy in treating such conditions. The reported changes of intestinal permeation seem to indicate weakness of enterocytic tight junctions as well as the integrity of the epithelial mucosa as a whole. During the autoimmune process, inflammation becomes sound resulting in increased mucosal permeability (**Figure 1**). This may result in antigens reaching the lamina propria (from the lumen) triggering an autoimmune response. This starts through activation of the T cells and proinflammatory cytokines release. This results in further increase to the mucosal permeability and exacerbates the immune response [45-48].

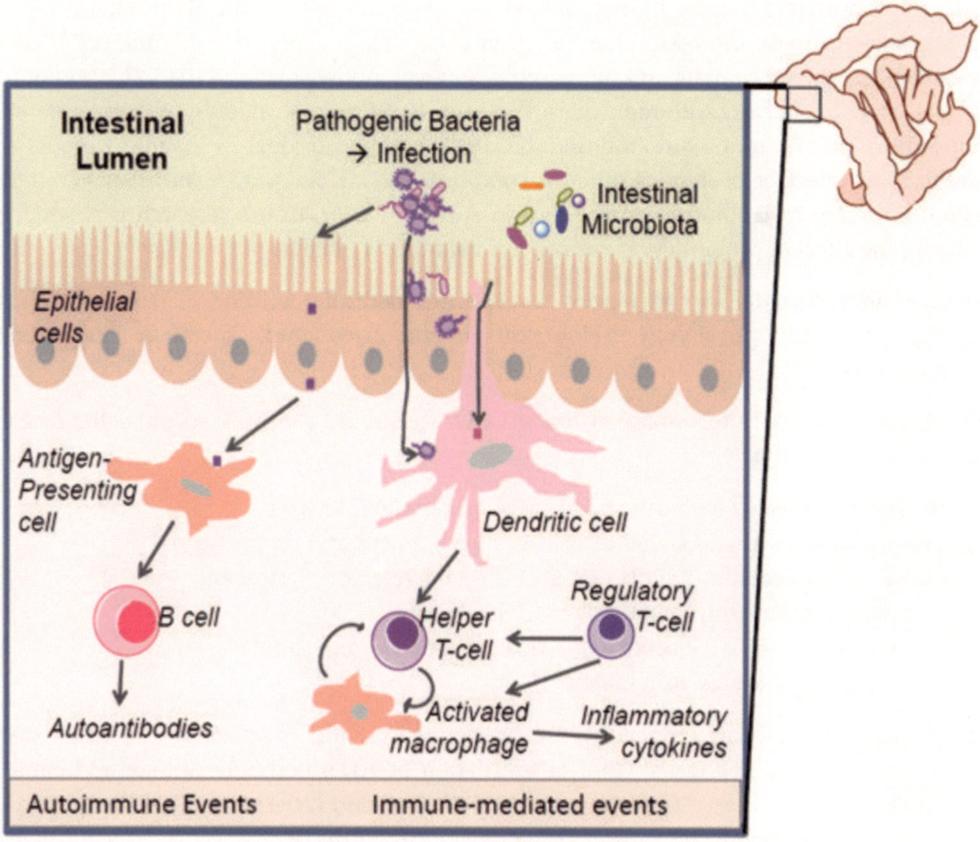

Figure 1. Intestinal permeability during an autoimmune response

3. Animal models suitable for investigating probiotic applications in autoimmune diseases

During the process of drug development, various *in vivo, ex vivo, in situ* and *in silico* methods can be used. Each method has advantages and disadvantages, and so using more than one method can provide better confirmation of findings. *In silico* methods can provide an initial insight into a potential drug candidate with predicted high pharmacological activity and good stability, while *ex vivo* methods can provide more information about a drug's interaction with living tissue, and are more cost-effective compared with *in vivo* animal models [49]. *In situ* methods can better predict drug absorption compared with *ex vivo* models but *in vivo* models can provide more comprehensive pharmacokinetic profiles and give a better understanding of drug-tissue interactions [50]. I*n vivo* studies are usually carried out where drug therapeutic formulations are administered to animals in order to investigate short and long term safety, to explore various clinical effects and to study different physicochemical parameters before confirming suitability of the formulation to a disease condition(s). Various animal models are used to represent various diseases [51].

In vivo studies on specialized animal models have allowed a great progress in tailoring research questions towards individualized gene contributions and their effect on the pathogenesis of these diseases. This has been done using standard inflammatory disease models in transgenic animals and by identifying novel models through the induction of the disease using chemicals. Although there is a surplus of animal models (spontaneous and induced) to study various autoimmune diseases, there is no ideal or standard model for studying the effect of probiotics on each condition [52-55]. Rats, mice and hamsters have been used to study probiotics applications in Ads. However, future research is needed, to compare the effect of probiotics on various animal models of ADs.

An ideal animal model should represent a specific medical condition in terms of disease development, pathophysiology, biological disturbances and short & long term complications [56-58].

If we are to create a better model of human AD, we should carefully consider the disease effect on the following:

- Relevant end points including primary, secondary and tertiary.
- The relevant speed and stages of disease development and progression.
- Disease complications, their progression and the relevant clinical end point(s).
- Symptomatic/nonsymptomatic signs of the disease.
- Feasibility of sample collections in terms of tissue site and sample volume.
- The incidence in males vs. females.

The current therapeutics for ADs are inadequate, which necessitates further drug development and *in vivo* trials. Clinical translation of AD's pathophysiology and clinical manifestations, from animal to human, has been limited and rather difficult. This is because very little is known about the pathophysiology and prognosis of such conditions; the extent of heterogeneity, polymorphism, genetic distance, the exact site of initial immune response (gut, lymph nodes, blood, brain or?), and 'potential' triggering antigens. To complicate this further, different Ads have different signs and symptoms and thus, one animal model is unlikely to be always suitable for all conditions. Creating a suitable animal model for ADs requires the ability to accurately translate the findings to human. These findings include therapeutic efficacy (prevention/treatment), safety and PK/PD profiles. With regards to different ADs, various animal models have been proposed. In fact, many ADs have more than one animal model representing the disease. For example, T1D has many animal models. The nonobese diabetic (NOD) mouse is considered the 'standard' animal model of the disease. Other models are induction models of rats, mice and hamsters using alloxan or streptozotocin to destroy pancreatic beta cells and induce T1D. The NOD mouse represents the best spontaneous model for a human autoimmune disease, in particular, T1D. NOD mouse model allows the investigation of various immunointerventions that can be used in human T1D. Similar to T1D in human, NOD mice have higher levels of macrophages, dendritic cells, CD4+ and B cells. The induction of T1D in NOD mouse can be achieved through environmental conditions, mimicking the development of T1D in human. However, the development of T1D in NOD mouse takes place quickly and can produce a significant inflammatory condition that may over-respond to immunomanipulation and exaggerate the effect of a treatment. Also, the incidence of T1D is different between males and females in this model while the incidence is the same in males and females in human. This can further limit the applications and the findings of this animal model [59]. Many therapeutics that showed good efficacy in this model failed to achieve similar results in T1D human subjects [60]. Having said that and regardless of how different this model is, from the 'true' human TID, NOD mouse remains the most representative of human T1D. Interestingly, in a recently published study, the incidence of T1D was much higher, when the mice were maintained in a germ-free environment suggesting direct connection between gut microflora and the development of T1D [61, 62].

Overall, a suitable animal model for human AD should ideally be easy to breed and handle, and can accommodate various medical conditions that may come about or be associated with the condition it is representing. Thus, extrapolation of its findings to human should be easily done, and with great accuracy and precision.

4. The influence of gut microflora on the development of autoimmune diseases

In many autoimmune diseases, the gut microfloral composition is different than that of healthy individuals. However, the cause of this change of composition and whether this change is a contributing factor to the development of the disease remain unclear. Probiotic treatment has demonstrated potential benefits in many Ads, assumingly, through normalizing such changes in the gut microfloral composition. Interestingly, the literature suggests that the effect of probiotic treatment on ADs' development and progression may be brought about through the effect on the expression and functionality of certain protein transporters. Recent publications suggest that many transporters have their expression and functionality altered in the autoimmune disease; T1D [23, 27, 72]. The exact mechanism associating the change in transporters and diabetes' development is still unknown but there are few assumptions to explain such an interaction. The first assumption is that some ADs, start with a direct insult in the gut, initiating a disturbance in the gut microflora and a consequent disturbed bile flow. This results in an altered bile feedback mechanisms and a change in the expression of protein transporters responsible for bile enterohepatic recirculation. The second assumption is that disturbance in protein transporters expression and functionality, caused by a genetic mutation, produces a disturbance in enterocytic-microfloral interactions triggering an inflammatory response. This response is further exacerbated by the resulted increase in gut permeability and ileal lymph/tissue necrosis. The third assumption is that the functionality of the immune system is altered (due to either an insult in the gut or genetic mutation). This alters the composition of gut microflora resulting in initiating of inflammation reaching various body tissues causing systemic inflammatory response triggering an autoimmune disorder and eventuating in autoimmune systematic response. In all these assumptions, genetic susceptibility is expected, and contributes further to the disease development and progression. The above assumptions were based on the work of the authors as well as careful evaluation of the literature.

In recent publications, alterations in the functionality of some transporters have been linked directly to the development of some autoimmune diseases such as diabetes. In addition, the enterohepatic recirculation of bile acids has also been related, by association, since secondary bile acids are solely produced by the action of gut microflora [13]. Bile salts' output in diabetic animals was high compared with healthy, and the expression of Mdr2 was also high after STZ treatment [63]. In another study, a mutation in Zinc transporter 8 (ZT8) located in beta cells, is implicated in the dysregulation of insulin transport and release, and an exacerbation of the inflammatory response leading to T1D. In this study, ZT8 was considered as an autoantigen resulting in the stimulation and production of beta cells autoantibodies and T1D development [64]. Moreover, streptozotocin (STZ) had different but significant effect on the expression of Na/Cl/glucose cotransporters, and the administration of insulin reduced such an effect [65]. Hyperglyemia itself directly reduced the activity of Mdr1 suggesting a clear association between pre-T1D hyperglycemia and disturbances in protein transporters [66]. In another recent study, the effect of STZ on cation protein transporters was reported, interestingly, at different levels of protein synthesis; transcriptional and posttranscriptional depending on the type of the transporters affected [67]. However, some studies suggest a diabetic influence is stronger on enzymatic activities than on protein transporters with the enzymatic influence being the cause of exacerbation of inflammation and development of the disease [68]. The impairment of protein transporters functionality, reported in the diabetic animals can take place either by reduced protein expression or reduced action. When glucose protein transporters in the blood brain barrier were studied under chronic hyperglycemia, their concentrations remain constant but functionality and glucose intake were impaired [69]. However, under acute hyperglycemia induced by STZ, their concentration decreased suggesting different response at different stages of the disease [70-72]. Accordingly, protein transporters have shown strong association with diabetes development and progression as well as diabetic complications.

Although there is some evidence suggesting that unrelated infections can result in the induction of organ specific autoimmunity [73], there is abundant epidemiological, clinical, and experimental evidence linking similar and closely related infectious agents with autoimmune diseases. Accordingly, the most acceptable hypothesis explaining how infectious agents cause autoimmunity is "molecular mimicry". Molecular mimicry directly invokes the specificity of the immune response to the resultant breakdown of tolerance. It proposes that microbial peptides have structural similarities to self-peptides and are therefore involved in the activation of autoreactive immune cells [74, 75]. Peptides, primarily, heat shock proteins (HSPs), have been implicated in autoimmunity [76, 77].

HSPs are a highly conserved family of proteins with significant structural homology between humans and bacteria. HSPs are located on almost all subcellular and cellular membranes and their numbers are induced in response to high temperatures and stress. HSPs function as molecular chaperons which are instrumental for signalling and protein trafficking. HSPs induced synthesis is implicated in autoimmunity. HSPs are believed to act through the activation of Toll-like receptors (TLRs) which trigger the expression of several genes that are involved in immune responses.

TLRs are only present in vertebrates and at least 11 TLRs are currently known. Distinct TLRs are differentially distributed within cells: TLR1, TLR2, TLR4, TLR5, TLR6, TLR10 and TLR11 are transmembrane proteins expressed on cell surfaces that contain extracellular domains rich in leucine that interact with pathogenic peptides, whereas TLR3, TLR7, TLR8 and TLR9 are primarily distributed on the membranes of intracellular compartments such as endosomes [78, 79]. Accordingly, TLRs are another potential target to bacterial manipulation. They are proteins on intestinal membranes that bind to pathogen-associated molecular patterns (PAMPs). After binding they release nuclear factor-kappa B (NF-kB) which moves into the cell nucleus and stimulates the release of pro-inflammatory mediators to target pathogens [80, 81]. Gut microfloral bacteria can directly trigger TLRs through adhering to the epithelial mucosa. As the human gut contains such large volumes of beneficial bacteria, they constantly trigger the TLRs. This leads to an eventual attenuation in the TLR response [82-84], (see **Figure 2**).

Although both pathogenic and probiotic bacteria regulate immunity via activation of TLRs, they do not usually trigger the same pathogenic inflammatory responses. Different probiotic bacteria stimulate distinct TLRs on host cells. Therefore, it is of biological and clinical importance to understand how very similar molecular proteins (HSPs) released by both commensal and pathogenic bacteria can trigger different responses by stimulating the same cellular receptors. One of the reasons for this may be that although the proteins are very similar they are not identical and thus they may stimulate the receptors in different ways to either produce a pro-inflammatory or an anti-inflammatory response. Another possibility is that the slight differences in the peptides allow them to bind to different TLRs leading to dissimilar responses. A third reason might be that more than one TLR is involved and that the effects seen are a synergistic effect depending on which TLRs are involved. TLR2 recognizes a variety of microbial components which include lipopeptides and peptidoglycan as well as lipopolysaccharides (LPS) from non-enterobacteria. TLR4 is an essential receptor for (LPS) recognition [85-87] and it has been shown to be involved in the recognition of endogenous heat shock proteins, eg HSP60 and HSP70. Microbial recognition by TLRs facilitates dimerization of these receptors. TLR2 appears to form a heterophilic dimer with TLR1 or TLR6 but other TLRs are believed to form homodimers. TLR1 and TLR6 that are functionally associated with TLR2 allow for the discrimination between diacyl and triacyl lipopeptides. Dimerisation of TLRs triggers activation of signalling pathways through the cell and into the nucleus. However, different gene expression profiles are triggered depending on which TLRs and TLR combinations are activated.

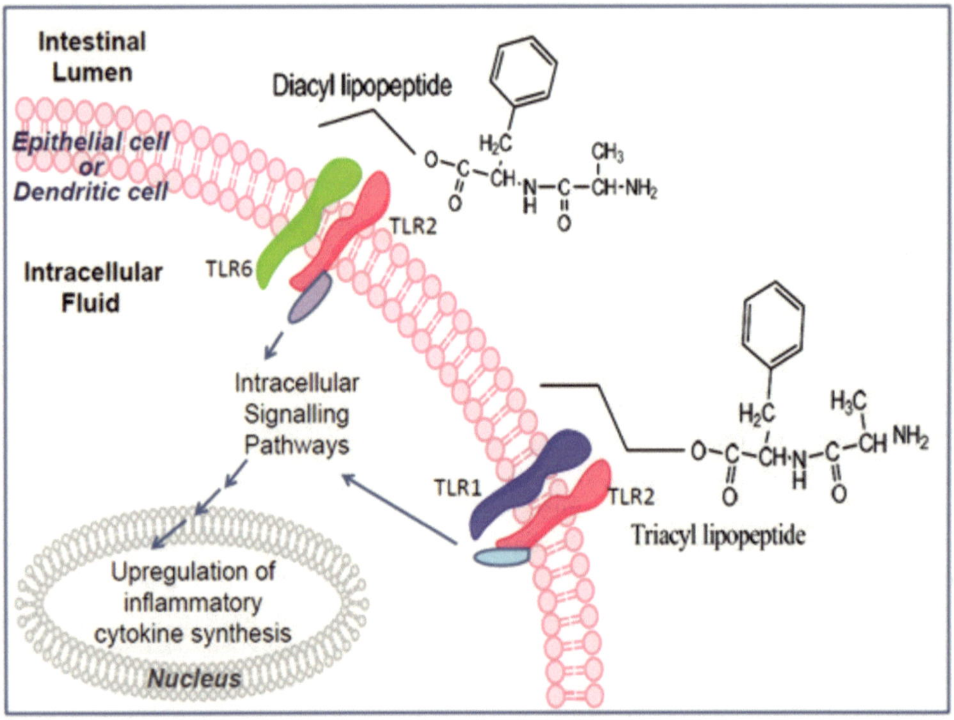

Figure 2. Molecular mimicry as a proposed cause of autoimmune diseases through the induction of 'mistaken-identity' immune response.

Loss of tolerance of the immune system to the body's own tissues can be caused by a number of factors including infection, excessive dendritic cell stimulation by intestinal microbiota, inadequate regulatory T-cell function or genetic factors. Dendritic cells are believed to be critical to the balance between tolerance and active immunity. Intestinal Dendritic cells are excessively activated in IBD as well as other autoimmune diseases which indirectly links the gut microfloral disturbances with the initiation or the progression of the disease (see Figure 2). Thus, the influence of disturbances in normal gut microflora may be indirectly linked to the initiation, development, progression and prognosis of many of the autoimmune disease. Such disturbances have been linked to changes in the expression and functionality of protein transporters in and outside the gastrointestinal tract. These disturbances have also been linked to changes in the composition and functionality of bile acids and many physiological and biochemical feedback mechanisms that showed clear impact on the stability, performance and efficiency of the immune system and its associated lymph tissues. However, many studies may show a significant impact or the lack of it, when trying to rectify these disturbances through the treatment with probiotics, making the influence of gut microflora on the development and progress of autoimmune disease difficult to clearly explain. Consequently, a direct influence of normal microfloral composition on the body's inflammatory response has been demonstrated in the literature. This directs further research towards investigating how the gut microflora can potentially control the immune system to the extent where its manipulation may delay or even prevent the initiation of the inflammatory response leading to the clinical signs and symptoms of the immune disease.

5. The effect of probiotics on autoimmune-associated inflammation

Bacterial gut-microflora live in an ecosystem, where each bacterial colony is part of a bacterial strain that colonizes the gut, and interacts with each other, as well as, with other gut-bacterial strains. The nature of this interaction is being currently studied at many scientific labs worldwide, and evidence of cross-talking continues to emerge. Bacterial cross-talking process involves polypeptide-based signals being secreted by various bacteria that influence the protein expression and functionality in other bacteria [25, 88]. This means that bacteria can influence the expressions and functionality of various proteins and membrane-transporters of other bacteria, via changing the gut concentrations of certain polypeptides. This can be brought about through the induction or suppression of membrane-transporters or through the process of direct-signalling [38]. In matter of fact, sequencing of human faecal samples has identified over 5000 different active gut-bacteria, with known metabolic activities [24]. This exceeds the average number of mammalian cells present in the body! Infants in the womb are mainly germ-free with the exception of some microbes that may be acquired through the swallowing of the amniotic fluid. The type and variance of these microbes and the role each gut-bacterial strain plays in initial gut-ecosystem development is still not completely understood. The next exposure to microflora takes place during birth when infants inherit a bacterial profile from their mother that shapes the composition of the matured gut. This profile of bacteria differs with type of delivery (vaginal or caesarean), time taken for the membrane of the amniotic sac to rupture, gestational age and use of antibiotics during labour. The human gut undergoes continuous maturation over many years, and has a shifting microbe population that varies between individuals and their exposure to family members, especially siblings, the sanitation of living conditions, and food and drink. The balance of different bacteria stabilises as people age but is still affected by factors including diet, location, antibiotic use and radiation exposure in adults. Gut composition seems to become more unstable again as people age, as the faecal microbial profiles of those 65 years and older show considerably more variability between individuals [89].

Compromised gut movement associated with autoimmune disease can result in substantial bacterial and yeast overgrowth which is postulated to disturb bile acids composition and exacerbate the disease-associated inflammation [105-107]. Autoimmune disease such as diabetes, show substantial inflammatory response, and bile acids disturbances can cause chemical unbalance that has been linked to poor tissue sensitivity to insulin [108], rise in the levels of reactive radicals in the blood [109], poor enterohepatic recirculation and dysfunctional protein-transporters in the gut that is negatively affecting liver detoxification and performance [110]. Accordingly, future AD-therapy should not only focus on rectifying physiological imbalance but also in targeting the disturbances in bile acids composition, protein transporters and overall the inflammation cascade initiated in the gut. This can be achieved through normalizing the composition of gut microflora and bile acids, gut immune-response and microflora-epithelial interactions towards maintaining normal biochemical reactions and healthy body physiology. Physiological features of human development including the innate and adaptive immunity, immune tolerance, bioavailability of nutrients, and intestinal barrier functions, are directly related to the composition and functionality of the human microflora. This includes the percentages of what is currently known as good and bad gut microflora. Good microflora includes two main species, Lactobacillus and Bifidobacteria. Microflora modifications may take place due to antibiotics consumption, prebiotic and probiotics administration and the use of drugs which affect gastric motility resulting in changes in gastric pH and gut-emptying rate. These modifications have been shown to be significantly profound in diabetic subjects resulting in the reduction of the percentage of good bacteria, the increase of the percentage of bad bacteria and yeasts and the consequent increase in the percentage of toxic bile salts such as lithocholic acid. This can also contribute to the higher incidence of gall stones and liver necrosis reported in diabetic patients. Accordingly, probiotics can introduce missing microbial components with known beneficial functions for the human host, while prebiotics can enhance the proliferation of beneficial microbes or probiotics, resulting in sustainable changes in the human microflora. Symbiotic relationship between probiotics and prebiotic administration is expected to exert a synergistic effect and in the right dose, may normalize and even reverse dysbiosis-associated complications.

Continuous exposure to bacteria can induce mucin secretion and change the structure of the mucous layer which can play a role in maintaining mucus thickness and its protective effects. In a recent *in vivo* study, Wistar rats were administered a probiotic formulation (VSL#3) daily for seven days. After probiotic treatment, basal luminal mucin content increased by 60% which has been linked to better protective effect and substantial stimulation of mucin secretion at the level of DNA-gene expression [90-93].

The significance and magnitude of the effect of host genetics on gut microfloral composition and functionality is difficult to accurately determine [94, 95]. It is generally agreed on that initial colonisation has the greatest effect on the lifelong bacterial types and functionality. Accordingly, it is expected that family members with shared genetic factors are likely to share the same initial colonisation similarities between their bacterial types. However, when the similarity of bacterial populations was compared between identical twins, non-identical twins and siblings, it was found that identical twins had significantly closer microflora compositions while others did not [96]. Other studies have observed bacteria modification after changes in host allele types, which also indicates some genetic effects but evidence remains controversial. Thus, it is clear that genetics do influence bacterial types in the gut, as does diet, environment and a multitude of other factors. Accurate definition to the contribution of each factor to the types and functionality of gut microflora remains to be studied. Microfloral bacteria in the gut play a number of beneficial roles [97]. They ferment and break down otherwise indigestible food components, thus, making additional nutrients available to the human host. The presence of gut bacteria is protective against pathogens; the multitude of bacteria reduce the amount of available nutrients for invading pathogens, adhesion of pathogens to epithelial walls is restricted and commensal bacteria may produce bacteriocins that have an inhibitory effect of pathogenic bacterial growth.

Gut microflora is reported to influence the formation of cells essential to the immune system. Gut-associated lymphoid tissues are collections of immune cells in lymphoid tissue in the gastrointestinal tract [98]. They play an essential role in the localised immune defence of the gut. While small accumulations of lymphoid tissue occur throughout the gastrointestinal tract, the majority is found in Peyer's patches, mesenteric lymph nodes and dendritic cells [99] (see Figure 3).

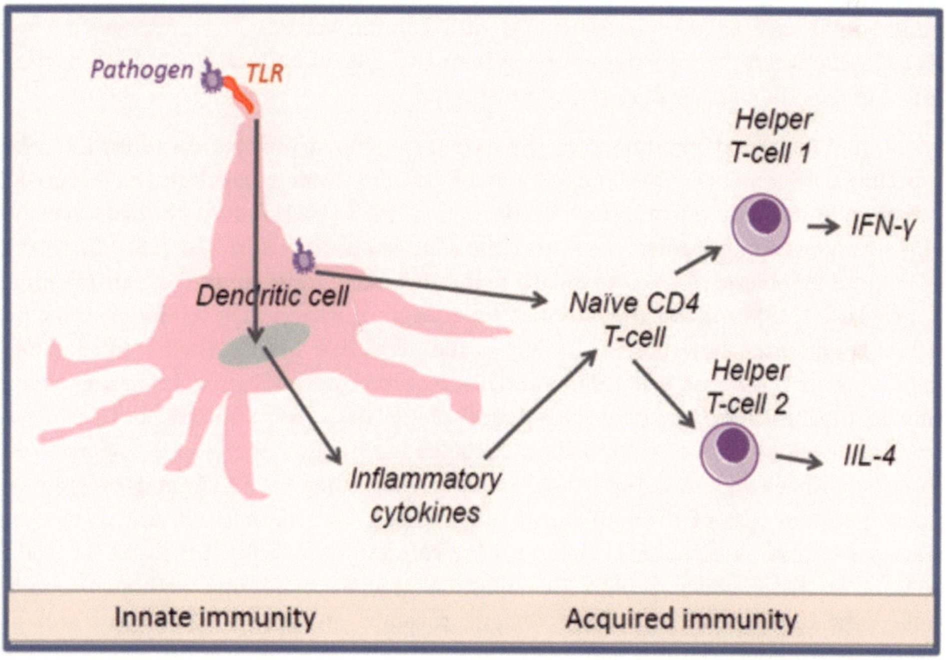

Figure 3. The influence of gut microflora on the activation of intestinal epithelial immune cells.

Peyer's patches store the inflammatory mediators, of a localised immune response including naive T-cells. Dendritic cells function as messengers which present endocytosed antigens to the Peyer's patches or mesenteric lymph nodes to prime T-cells into effector cells [100]. If the antigens are presented to the mesenteric lymph nodes, the effector cells are released into systemic circulation via the efferent lymphatic system, leading to an inflammatory response from central lymph nodes. Through effects on the dendritic cell intermediary, bacteria can modulate T-cell regulators which can lead to alter systemic inflammation via lymphatic systems. Gut growth in animal studies where mice are raised in a microbe free environment shows a different intestinal structure compared to normal gut growth and the amount of gut-associated lymphoid tissue is reduced [101, 102]. This results in reduced gut microfloral differentiation between beneficial and pathogenic bacteria, bringing about a significant reduction in the area of the gut which can launch an innate immune response and decreases the communication of antigen information to central lymph nodes. This makes the entire body more vulnerable to harmful bacteria passing through the gut epithelium unnoticed [103-105].

In mice, a disturbed TLR-pathway results in compromised TLR signalling which results in any intestinal injury being met with an exaggerated response [81, 106-108]. A down-regulated TLR pathway caused by dysbiosis could cause a similar inflammatory process, making commensal bacteria potentially protective against IBD [109, 110]. This indicates the necessity of the TLR conditioning to develop an immune tolerance to bacterial threats in the gut. Bacteria in the gut can also bind to PAMPs to deliberately initiate an inflammatory response to signal the presence of invading pathogens.

Overall, these changes to inflammatory signalling and response based on interactions with gut microfloral bacteria are numerous and varied in mechanism. This indicates a complex relationship between the innate immune system and gut microflora where both parties are adaptive to the other, rather than static in response.

Many autoimmune and inflammatory diseases have shown positive response to probiotic and prebiotic treatments. The composition of the intestinal microflora may even affect mammalian physiology outside the gastrointestinal tract [111]. Recent studies have shown significant changes in gut microfloral and bile acid compositions in T1D [28, 43]. Thus, it is clear that our symbiotic microflora award many metabolic capabilities that our mammalian genomes lack [112], and so therapeutics that target microfloral modulation may prove rewarding. When the new born baby leaves the germ free uterus, she/he enters a highly contaminated extra-uterus environment. This requires the activation of her/his immune system to prevent infection. Over the period of the first year, the new born's intestinal microflora develops and its composition becomes her/his gut microfloral fingerprint! Gut microflora has been shown to play a major rule in controlling the inflammatory response of the host immune system through direct and indirect bacteria-bacteria and bacteria-host interactions. These interactions include physical and metabolic functions of the gut microfloral bacteria, which protect the intestinal tract from foreign pathogenic bacteria, eliminate the presence of unwanted bacteria through producing bacteriocins and other chemicals, and inform the gut epithelium and the host immune system about whether a local inflammatory response is needed [37, 113]. Gut microflora can control the host immune system through four main actions. The induction of IgA secretion to protect against infection, triggers localized inflammatory responses, neutralizing T-helper (Th) cell response and also contributing to the induction or inhibition of generalized mucosal immune responses. Recent studies have shown that in autoimmune diseases and gut inflammation disorders, there is a significant disturbances in the ratios of Th cells such as the increase in the Th-2/Th-1 ratio associated with inflammatory bowel diseases, which has been linked to exacerbation of the gut inflammation and the development of the disease. In recent studies, gut-associated dendritic cells in the lamina propria can extend their appendices reaching the gut mucosa and using their Toll-like receptors (TLR) 2 and 4, to sample bacterial metabolites [114, 115]. This may result in dendritic cells releasing certain cytokines that stimulate the activation of naive Th-0 into active Th- cells such as 1, 2 and 3/1 [115]. Interestingly, some microfloral bacteria can actually cross enterocytic microfolds and interact with antigen presenting immune cells in mesenteric lymph nodes to activate naive plasma cells into IgA-producing B cells [116]. IgA coats the intestinal mucosa and control further bacterial penetration thus protecting the host from potential pathogenic bacteria. Even more interestingly, gut microflora bacteria have shown ability to not only initiate an inflammatory response but also to control and inhibit such a response. Some microfloral bacteria or their metabolites can interact with the intracellular receptor TLR-9, to which the bacteria activates T cells through the production of potent anti-inflammatory cytokines such as IL-10 [117, 118]. Microfloral bacteria can also produce small molecules that can enter intestinal epithelial cells to inhibit activation of nuclear factor kappa-light-chain-enhancer of activated B cells (NFkB) [119]. Moreover, prolonged exposure to bacterial endotoxins, in particular, LPS (which interacts with TLR 2 and 4) can activate intracellular anti-inflammatory associated proteins that result in an overall anti-inflammatory effect [120]. Such gut bacterial-host interactions are critical in maintaining a balanced and effective immune response to various infections while maintaining control over prolonged or chronic inflammation and reducing the overstimulation of the host immune system.

Recent evidence suggests that a particular gut microfloral community may favour occurrence of the metabolic diseases. It is well know that the composition of gut microflora changes with diet and also as we age [121, 122]. In one study, a high fat diet was associated with higher endotoxaemia and a lowering of bifidobacterium species in mice cecum [123-125]. In a follow up study, the administration of prebiotics, in particular, oligofructose, to mice given high fat diet, restored the reduced quantity of bifidobacterium. This also resulted in reducing metabolic endotoxaemia, the inflammatory tone and slowing the development of diabetes. In this study and compared with control mice on chow diet, high fat diet significantly reduced intestinal Gram negative and Gram positive gut bacteria, increased endotoxaemia and diabetes-associated inflammation. However, when diabetic mice on high fat diet were given oligofructose, metabolic normalization took place including the quantity of gut bifidobacteria. In these mice, multiple correlation analyses showed that endotoxaemia negatively correlated with bifidobacteria quantity [126, 127]. By the same token, bifidobacterium quantity significantly and positively correlated with improved glucose tolerance, glucose-induced insulin secretion and normalised inflammatory tone (decreased endotoxaemia and plasma and adipose tissue proinflammatory cytokines) [123-125]. In general, the level of microfloral diversity and gut bifidobacteria in human, relate to health status and both decrease with age [128, 129].

6. The potential applications of probiotics in autoimmune diseases

Probiotics have been shown to be beneficial in wide range of conditions including infections, allergies, and metabolic disorders such as diabetes mellitus, ulcerative colitis and Crohn's disease [130-132].

When discussing therapeutic applications in AD, the use of probiotics is an area of growing interest, not just as an adjunct therapy but also as a mainstream treatment aiming at normalizing the disturbed gut-microfloral composition, as well as, directly relieving signs and symptoms of the disease. In order to design a probiotic formulation that targets disease-associated disturbances in gut microflora, a better and more detailed understanding of these disturbances is necessary. Better understanding of microfloral composition in the gut can be achieved through cell-culturing and protein-based assays that analyse the nature, type and quantity of various bacteria that exist in the gut.

However, beneficial effects of probiotics in ADs are modest, bacterial-strain and disease-state specific and limited to certain manifestations of disease and duration of use of the probiotic.

6.1. Type 1 diabetes and probiotics

Probiotic administration in animal models of Type 1 diabetes has shown great potentials. Combinations of different bacterial strains can be used [133] but a mixture of *Lactobacilli* and *Bifidobacteria* is a common choice [20-23, 26, 42, 92, 134-136]

There are reports in the literature that probiotic treatment can be useful in diabetes [28] but there is little explanation of the mechanisms involved. The initial site of diabetogenic cells has been hypothesized to be in the gut whereas pancreatic lymph nodes serve as the site of amplification of the autoimmune response [137]. This autoimmune response may disturb the composition of the normal gut flora. Treatment with *Bifidobacteria* and *Lactobacilli* has been shown to normalize the composition of the gut flora in children with T1D [131, 138]. In addition, the administration of *Lactobacilli* to alloxan-induced diabetic mice prolonged their survival [139, 140] and administration to non-obese diabetic (NOD, a rodent model of T1D) mice inhibited diabetes development possibly by the regulation of the host immune response and reduction of nitric oxide production [140]. Furthermore, the administration of a mixture of *Bifidobacteria, Lactobacilli* and *Streptococci* to NOD mice was protective against T1D development postulated to be through induction of interleukins IL4 and IL10 [141].

Slowing of peristalsis (gastroparesis) has been reported in T1D patients. This can result in a bigger population of bacteria in the gut and a subsequent rise in the concentration of secondary bile acids [142, 143] such as lithocholic acid [144, 145]. In addition, the disturbed bile acid composition in T1D (8) is strongly linked with autoimmune and liver diseases. The administration of *Lactobacilli* and *Bifidobacteria* may restore the bile acid composition [146, 147]. It is important to select the right probiotic species based on efficacy, stability in the gut (bile and pH tolerability) and long term safety. For example, some probiotic-bacterial cells have been examined for stability as well as efficacy in various autoimmune diseases. *Lactobacillus rhamnosus, Lactobacillus acidophilus* and *Bifidobacterium lactis* show good bile and pH tolerability under normal conditions of pH (1.5-8) and bile acid concentration (0.8 – 3 %), in addition to long term safety [148-150].

6.2. Inflammatory bowel diseases and probiotics

In IBD such as UC colitis, there is a substantial inflammatory component with atypical type 2 T-helper cell (Th2) activation. Th2 are activated by the presence of antigens and then direct other immune cells in the body. In UC they can become overly sensitised and secrete interleukin-13, an inflammatory mediator [151]. This drives T-cells not normally present in the colon to migrate there and makes the colon mucosa more sensitive to commensal bacteria which drives further inflammatory responses [152].

Naïve CD4 T cells differentiate into Th1 or Th2 effector T cells on activation by antigen-presenting cells (see Figure 4). Th1 and Th2 cells carry out distinct antigen specific adaptive immune functions; Th1 cells mediate cellular immunity against intracellular pathogens, whereas Th2 cells enable humoral immunity and immunity against extracellular pathogens. The effector functions of Th1 cells are exerted in part by production of interferon (IFN)-γ and those of Th2 cells by interleukins including IL4. Inappropriate regulation of Th1 and Th2 cell functions can cause autoimmune diseases.

In IBD, UC in particular, as with other inflammatory conditions, the production of immunoglobulins is elevated. Immunoglobulins, or antigens, bind to antibodies to encourage an immune response to the antigen while limiting the harm the antigen can do. UC displays an increased production of IgA, IgM, IgF but also has a disproportionately high level of IgG1. IgG1 binds to a colonic epithelial antigen in an autoimmune response. That antigen is also present in the eyes, skin and joints and inflammatory responses there can cause the extraintestinal symptoms associated with UC, including peripheral arthritis, erythema nodosum, iritis, uveitis and thromboembolism [153].

The identification of a causative UC pathogen would greatly simplify diagnosis and new treatment identification. Three broad studies used sequenced bacteria from the human gut to try and identify a healthy gut microbial profile. When the bacteria strains were divided by phylogenetic type it was found that 98% of bacteria were part of four phyla [154-156]. Another study compared this control data to samples from patients with Crohn's disease and UC. Two-thirds and three-quarters of the diseased samples, respectively, had the same bacterial balance as healthy controls. In the other IBD samples there was no consistency in the atypical bacterial groups, indicating that although dysbiosis is present there are no single causative bacteria [154]. Unfortunately, it is still unknown whether the dysbiosis precipitates gut inflammation or if another cause initiates the disease and dysbiosis occurs due to the inflammatory changes [157]

It has been shown that patients with UC display an increased microflora density [151] meaning the total population of bacteria in the colon is increased. In one study the number of bacteria in colon biopsies taken during endoscopy from newly diagnosed and untreated UC patients was double that of healthy controls [158]. The samples from UC patients also showed a thinner and less sulphated mucosal layer of the gut epithelium [159] which could support the increased bacterial levels through a lessened mucus flow to dislodge bacteria or an improved nutritional role from less sulphate.

VSL#3 is a high dose probiotic mixture that shows how information from multiple trials and *in vitro* studies can be brought together. Considering how new data fits into the probiotic profile established from previous investigations can help highlight any challenges to existing assumptions. Alternatively, when study results are replicated by different research centres the significance of the findings is increased. This reflective process should develop an understanding of the probiotic that is based on clinical evidence. VSL#3 contains a combination of three strains of bifidobacterium, four strains of lactobacilli and one strain of streptococcus salivarius. A trial in 1999, shortly after the probiotic was developed, tested faecal samples of 20 UC patients to determine changes in bacterial concentrations when VSL#3 was administered with no other treatment. An increase in the bacterial numbers of strains found in the probiotic was observed in all patients from the 20th day of treatment and remained stable. This established that the probiotic could colonise the gut and encouraged further clinical trials [160]. VSL#3 was then trialled repeatedly in small studies which had similar conclusions regarding safety and efficacy. The studies showed a low number of reported side effects which were consistently mild, so safety in the trialled patient types was assumed. The outcomes from the trials were encouraging as the probiotic treated groups usually showed an improvement in disease state [92, 161-166]. This identified VSL#3 as a feasible new UC treatment but a large, randomised, placebo controlled study was needed to verify results [167]. Two studies have provided the additional clinical evidence needed to substantiate the conclusions from earlier trials. The first was conducted on patients in India in 2009 over a 12 week treatment regime. The second trial, in 2010, had a shorter treatment time of 8 weeks and was carried out in Italy. Both trials were multicentre, randomised and placebo controlled and were conducted on 144 patients. Information on the safety of VSL#3 was definitely supported by both trials. The only side effects reported by the probiotic treatment group were mild, primarily abdominal bloating and discomfort. Additionally, there were no patient withdrawals from the VSL#3 group due to worsening of symptoms [167-169]. As both trials were on patients with mild to moderate UC as determined by the Ulcerative Colitis Disease Activity Index (UCDAI) score, safety in this demographic can be seen to have been established. The safety of VSL#3 in more severe disease stages were not assessed by these trials and remains unknown. The primary outcome from both trials was a 50% reduction in the patient UCDAI score. When the results of the group receiving probiotics were compared to the group not receiving probiotics it was shown that a significantly greater the percentage of VSL#3 treated patients achieved the outcome compared to the placebo. This was consistent between the two trials. One of the secondary outcomes was the achievement of disease remission, which was the reduction in UCDAI to 2 or less. It is interesting that this was only a secondary outcome as remission is often considered the main goal of treatment of UC by patients. Both trials achieved remission in approximately 50% of patients on VSL#3. This was statistically significant in the 2009 Indian trial as the placebo remission rate was only 15% [168]. The second trial, based in Italy, had an unusually high placebo remission rate of 40% which meant that 50% remission in the VSL#3 was not significant [169]. This placebo rate weakens the evidence for VSL#3 inducing disease remission when adjunctive treatments are unchanged. However, these results do support the role of VSL#3 as an effective UC treatment to reduce symptom severity.

Despite promising treatment outcomes with VSL#3, exact mechanisms of action and the extent and significance of synergism remain to be clearly identified. The mechanism of action has been investigated a number of times and these studies suggest alteration of intestinal integrity is likely to be central to VSL#3 activity. Intestinal epithelial cells incubated in media with VSL#3 show increased transepithelial resistance. This may be mediated by specific elements of the Mitogen-activated protein kinase (MAPK) pathway, which was activated by VSL#3. Pathogen-induced reduction in transepithelial resistance was diminished by VSL#3, probably due to the prevention of cell structure dysfunction at tight junctions [170]. VSL#3 may also alter mucin secretion, which makes up the mucous layer in the gastrointestinal tract. Of the nine identified genes, MUC2 is the predominant gel-forming mucin. MUC2 was induced in a concentration dependant manner by the exposure of the probiotic mixture to cells in media. It was postulated that this would correlate with an increase in mucin secretion. Rats fed with VSL#3 for seven days had an increase in MUC2 gene expression leading to an increase in the total mucin pool [159] When rat colonic loops were exposed to live VSL#3 an increase in mucin secretion was observed immediately without the need for a change in the mucin pool. Separate colonisation of the bacterial strains in VSL#3 identified that Lactobacilli is most likely to be responsible for mucin changes. Mucin secretion is known to effect bacterial adhesion and colonisation, so lactobacilli may upregulate MUC2 to improve colonisation. This implies that the benefits to intestinal structure are coincidental. One murine model of colitis, dextran-sodium sulphate-induced colitis, showed no mucin response to VSL#3 treatment. Mucous barrier thickness and expression of mucin genes were unchanged and inflammation did not decrease. The inactivity of VSL#3 may be a result of the colitis model used, which may have altered probiotic mediated effects as VSL#3 did adhere and change the microflora population. Trials on intestinal biopsies with ulcerative colitis could aid in supporting or invalidating the effect of VSL#3 on mucin.

Inflammatory mediators also play an important role in the reduced inflammation reported after treatment with VSL#3. The expression of TLR2 by dendritic cells is down regulated, which lessens the potential for TLR signalling for pro-inflammatory processes. An increase in production of IL-10, an anti-inflammatory cytokine, was also observed. This may be as a result of the changes to TLR2 or the overall reduction in inflammation. VSL#3 exerts multiple direct and indirect effects on gut inflammation which have not been fully elucidated, but can be observed in patient trials. While some studies suggest limitations to VSL#3 usefulness in UC treatment, further research is needed before they can be confirmed. Current information suggests that VSL#3 holds great promise as a low risk adjunctive treatment for mild to moderate UC to reduce symptom severity.

Strains that are identified for use as probiotics should not be pathogenic or carry antibiotic resistance as their use would be potentially harmful. There may be other consequences from treatment that can lead to physiological harm. As probiotic treatments often utilise bacterial strains found in the healthy human gut there is an assumption that probiotic treatment is without risks. Low withdrawal rates due to side effects from clinical trials support this notion, even in critically ill patients [171]. However, probiotic sepsis, a potentially deadly complication, has occasionally been reported [172]. Sepsis may be more likely in individuals with severe illness as they may be immunologically compromised.

HLA-DR is a MHC class 2 surface receptor responsible for identifying and binding to an antigen before presenting to the immune system to educate T and B-cells. There are more than a dozen major subtypes of HLA-DR, some of which have been associated with specific diseases. The prevalence of serotypes DR2, DR9, and DRB1*0103 is significantly higher in people with active UC when compared to a healthy population. This could be a genetic factor that indicates a susceptibility to UC [173].Alternatively, the more common strains may be created by the body in response to the mucosal damage in the colon as a reparative effort [174]. As the prevalence of HLA-DR subtypes differs between populations the implications of these results are complex to apply. For example, the DR2 subtype showed a definite increased occurrence in UC patients from Japanese, Finn and Siscilian populations. In other culturally heterogenous populations the association is less strong or even absent, even though the association with DR2 is still significant when considered over all populations. DR9 is also more prevalent in Japanese populations, so it may be more important when assessing factors of disease susceptibility then in other ethnic groups. DRB1*0103 may be applied more specifically as it may be an indicator for how extensive UC could be. DR4, though, seems to be protective against UC, as the frequency that is occurs at is much lower in people with UC [173].

Another potential genetic factor in the development of UC is the expression of transcription factor XPB1 which regulates secretory and other stress-responsive cells in the endoplasmic reticulum stress response. In mice where the factor is absent, intestinal epithelial cells are more susceptible to potential colitis inducers and displayed spontaneous enteritis [175]. In humans, a variance in XPB1 has been associated with both Crohn's disease and UC. The activity of peroxisome proliferator-activated receptor-gamma (ppar-gamma) is an inflammatory system change that is unique to ulcerative colitis. In healthy individuals ppar-gamma modulates inflammation by attenuating nuclear factor-kappa B (NF-kB), a protein present in almost all cells that responds to harmful cell stimuli. Ppar-gamma activity in colonic epithelial cells of UC patients is reduced, but gene expression of ppar-gamma is normal. This indicates that bacteria present in the gut affect the activity of ppar-gamma in UC [176].

Bacterial imbalance may indicate more aggressive disease progression. The intestinal samples for the study were taken during surgery required to treat IBD or other conditions (primarily colonic cancer), not especially for the study. The age of the patients with atypical bacterial balances was on average 8 years younger than that of the control group. The need for surgery at a younger age could demonstrate a more aggressive disease. Alternatively, the changes in bacteria may be secondary to (not causative of) severe disease. The samples with Crohn's disease in the atypical group were also more likely to have abscesses [154]. Whether an imbalanced gut microflora was a contributing factor to the development of the abscess, or if the development of the abscess encouraged the growth of bacteria normally atypical to the human gut is difficult to discern.

When the microbial composition in the rectum was compared between patients with UC and normal patients, it was found that levels of Bifidobacterium were reduced in the samples with the inflammatory disease [177]. This is in keeping with a theory that post-operative pouchitis after surgical resection of the colon to manage UC is linked to a reduction in levels of Lactobacillus lactis and Bifidobacterium *[178]* Pouchitis occurs when the illeoanal pouchy becomes inflamed and passes diarrhoea, sometimes bloody, and causes fever. After up to 10% of surgeries pouchitis becomes recurrent although the cause is unknown [179].

Even with these changes in microbial balance it has been found that use of antibiotics has no effect on the development or progression of UC. This is a marked point of difference compared to Crohn's disease where certain antibiotic therapies have been known to complete remission [180]. This may be associated with the absence of serum bacterial antibodies in patients with UC. While Crohn's disease has numerous elevated bacterial antibodies, indicating that particular bacteria may play a specific role in the disease, there is only one that has been identified in UC; perinuclear antineutrophil antibody. This antibody identifies bacterial antigens that have cross-reacted with nuclear antigens and it responds in tests to enteric bacterial antigens [181]. This shows a generalized overactive immune response targeting much of the gut bacteria resulting in wide spread exacerbation of the immune system and damaging further the intestinal tissues including the gut-associated lymphoid system. Thus, probiotic treatment poses great potential in treating IBD and further research is needed to investigate whether normalizing the gut microfloral composition will result in preventing the disease or ameliorating its severity and long term complications.

7. Lupus and probiotics

Systemic Lupus (SL) is an autoimmune disease which shares a significant inflammatory response and overactive and hypersensitive Th2 cells. A study of the autoimmune response in SL has found that one type of T cells is commonly found among SL patients. Cytotoxic CD8+ T-cell is found to be initially activated at the early stages of the disease and results in wide spread generalized activation of a long inflammatory cascade that brings about a full SL symptoms.

Similar to that of T1D, there are clear disturbances in gut microflora in SL, and, similar to other autoimmune diseases, a direct link between such changes and the initiation of the disease remains unclear. The literature suggests that gut microflora participates in the progression and complications of SL. This is brought about through an initial antigenic trigger that results in immune system 'confusion' which brings about an inflammatory response that attacks and destroys body's own tissues. The role of gut microflora in the initiation and development of SL is complex. This starts with a trigger that initiates a shift in gut microfloral composition which results in a formation of specific DNA-targeting antibodies directed towards specific pathogenic bacterial cells e.g. burkholderia bacteria [182]. This antibodies production is exacerbated through wider inflammatory response which brings about symptomatic SL and further complications of the disease. In theory and similar to the potential beneficial effect of probiotic administration on other autoimmune diseases, probiotic treatment, in particular, long term, is anticipated to neutralize gut-microfloral disturbances that brings about a stabilization of antibody production and eventual cessation of the inflammatory response which results in less severity and reduced signs and symptoms of the disease. In one study, authors measured the resistance of normal gut microflora to the colonization of pathogenic bacteria. This was done by a comprehensive biotyping technique in healthy individuals and patients with inactive and active SL. Colonization resistance was found to be lower in active SL patients than in healthy individuals ($P = 0.09$, Wilcoxon one sided, with correction for ties) suggesting that in patients with SL, various types and more bacteria are translocating across the gut wall than in healthy individuals, due to lower colonization resistances in these patients. Some of these may serve as polyclonal B cell activators or as antigens cross-reacting with DNA [183]. Thus, administering probiotic bacteria such as bifidobacteria which may restore normal gut-microflora and reduce the inflammatory response and production of such antibodies should be beneficial. However, the use of probiotics in the prevention or treatment of SL remains doubtable due to many challenges including dose and frequency required to exert a clinical beneficial effect, targeted delivery to live bacteria to the large intestine, bacterial loading and bacterial interaction with other drugs.

Overall, the therapeutic applications of probiotics in autoimmune diseases can be summarized in three main mechanisms covering preventative measures as well reliving the signs and symptoms of the diseases. This focuses on the role of probiotic 'long-term' treatment of the gut aiming at manipulating and neutralizing the gut-microfloral bacteria to restore healthy body physiology and biochemical reactions, as well as minimizing symptoms through ameliorating the inflammatory response. In addition, probiotics have been shown to increase non-specific host resistance to pathogenic bacteria. Probiotics are believed to deliver their effects via three main mechanisms: (1) competitive exclusion, (2) production of anti-bacterial substances and (3) regulation of immune responses.

7.1. Competitive exclusion

Probiotics compete with pathogens and toxins for adherence to the intestinal epithelium. This concept describes the manner by which probiotic bacteria populate, overtake the pathogenic bacteria and go on to completely colonize and 'crowd' the gut.

7.2. Production of anti-bacterial substances

Probiotics exert anti-bacterial effects on pathogenic bacteria by producing bactericidal substances including bacteriocins and acid which work synergistically or alone to inhibit pathogenic bacterial growth. Bacteriocins are antimicrobial peptides which are produced by some gram positive bacteria while acetic, lactic and propionic acid are produced by a wide range of probiotic bacteria leading to a decrease in pH and inhibition of growth of many pathogenic gram negative bacteria.

7.3. Regulation of immune responses

Infections can disrupt T-cell tolerance [Rocken et al, 1992] due to the enormous bacterial load of the intestinal lumen. It appears that sustained exposure to bacterial antigens can result in impaired T-cell function [Bronstein-Sitton et al, 2003]. An inadequate function of immunoregulatory cells can lead to loss of tolerance.

Probiotics regulate immune responses by modulating pathogen induced inflammation caused by TLR-mediated signalling pathways. Probiotic bacteria have been shown to skew the Th1/Th2 balance toward Th1, which helps down-regulate overactive Th2-mediated allergic responses. Effects on the Th1/Th2 balance have been observed in some animal models of allergy [184]; however not all strains stimulated Th1 immunity [185, 186]. Nonetheless, stimulation of Th1 immunity has been reported in clinical trials [187-191] and clinical efficacy has been demonstrated in adults, children and infants for diseases including IBS and IBD [192, 193], see Figure 4.

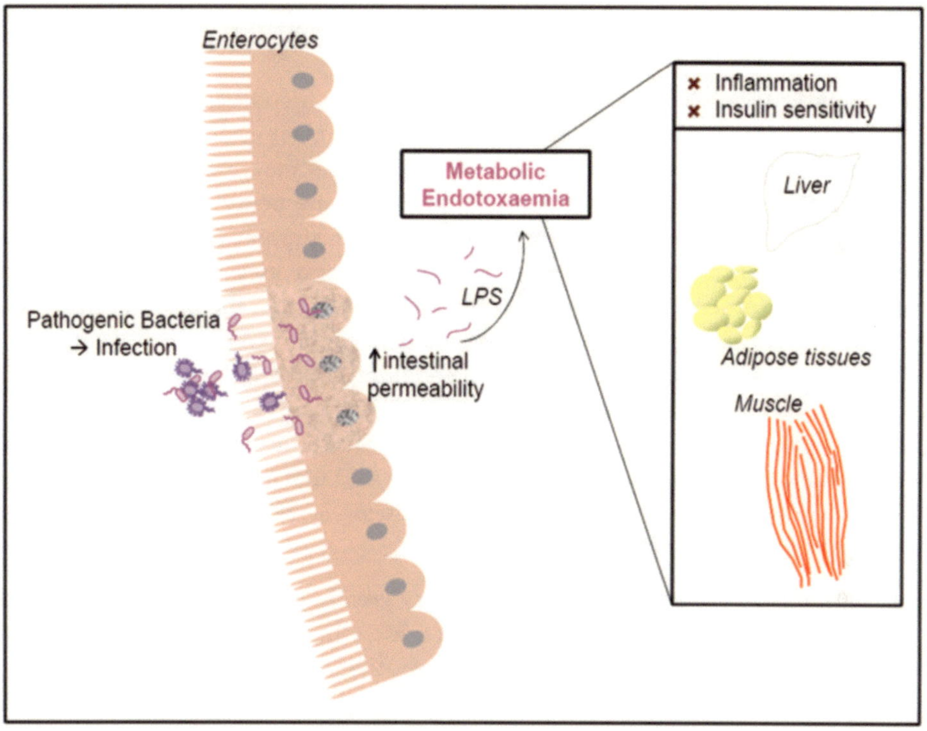

Figure 4. The relationship between LPS endotoxins and inflammation pathology in some autoimmune diseases. This figure adapted with modification from Cani P & Delzenne NM [105].

8. Safety and toxicology of probiotics

The World Health Organisation has guidelines for the evaluation of probiotic health claims. The guidelines begin by emphasising the importance of identifying the genus and species of the probiotic bacteria, as effects are strain specific. The WHO report also outlines assessment of probiotic storage, safety and evidence used to substantiate health claims [194].

Strains that are identified for use as probiotics should not be pathogenic or carry antibiotic resistance as their use would be potentially harmful. There may be other consequences from treatment that can lead to physiological harm. As probiotic treatments often utilise bacterial strains found in the healthy human gut there is an assumption that probiotic treatment is without risks. Low withdrawal rates due to side effects from clinical trials support this notion, even in critically ill patients [171]. However, probiotic sepsis, a potentially deadly complication, has occasionally been reported [172]. Sepsis may be more likely in individuals with severe illness as they may be immunologically compromised.

The mechanism of immune system modulation through gut microflora may change during certain disease states. A large trial on patients with acute pancreatitis found that 16% patients in the probiotic group died compared with 6% of the placebo group, indicating an increase in mortality with prophylactic probiotic treatment in such immunocompromised patients [195]. This highlights the need for caution when treating a disease state or severity that safety has not been established with.

A range of probiotics have been used to treat mild to moderate UC without severe side effects. However, probiotic safety in severe UC has not been established. While patients with symptoms that are unresponsive to current therapies may benefit greatly from new treatments, until the mechanisms of action of probiotics are better understood the risk to patients is also unknown. Accordingly, probiotic administration has shown good safety profile in individuals with overall good health status, and may be suffering from mild infections or GI disorders [196, 197]. Probiotic safety stems from the fact that many strains are of human origin and present in large numbers in human GIT [131]. Accordingly, the reported incidences of probiotics inducing bacterial infection and bacteremia are very low (18). The only major concern with probiotic administration is the potential of bacterial translocation resulting in the induction of antibiotic-resistance strains that may lead to pathogenesis and haemodyscrasia [198, 199]. Having said that and as previously explained, the risks of infections caused by probiotic treatment is expected to be significant in immunocompromised patients [200-204].

Clinical trials of new treatments for many Ads vary greatly in trial length, inclusion criteria and *in vivo* models used. The diversity of these trials makes meaningful comparison of probiotic treatments difficult. For example there is no standard index for UC, with variety of different symptom based evaluations, composite scores and patient evaluated scoring systems used in clinical trials [205]. Patient inclusion in the trial, response to a treatment, and whether remission is induced, is usually determined by a disease activity index score of a pre-specified value being met. Comparison of different definitions of success is complex, as a patient could be considered in remission by one trial but in a state of active disease by another. In addition, clinical trials of treatments of UC are known to have a diverse and unpredictable placebo response rate [206]. A 2007 meta-analysis of 40 clinical trials found that placebo induced remission rates ranged from 0-40% while placebo response was as high as 67% [207]. An unpredictable placebo response can interfere with the perceived usefulness of new treatments making findings hard to interpret. On the other hand, clinical trials that evaluated outcomes based on subjective scores (physician impression of disease severity, patient reported quality of life, etc.) were associated with higher placebo rates of response and remission. Use of objective assessments, e.g. the presence of inflammatory markers or sigmoidoscopy score, can reduce placebo values and make comparison of clinical trials simpler. The patient acceptability and cost of invasive tests like colonoscopies and blood sampling limit their use. Objective scores also do not quantify changes in time off work and symptoms like urgency and tenesmus, which are reported to be most important to patients.

The length of the clinical trial can change both rates of success and placebo responses. Shorter trials with fewer study visits lessen the cost of the study and reduce placebo values [206]. Long term trials may document a decrease in clinical effectiveness as relapses occur, the treatment ceases working and symptoms return. This may be due to the nature of disease rather than the treatment, as e.g. 67% of UC patients experience a relapse within the first ten years [208].

Risk of relapse makes withdrawal of existing therapy prior to commencing clinical trials undesirable. As a result, most probiotic treatments are initiated as adjunctive therapy to a stable oral dose of 5-aminosalicylic acid or an immunosuppressant. The period of time the dosage of other medications must have been stable for prior to the trial varies. The effect of these existing medications on the mechanism and efficacy of probiotics is unknown.

The adoption of a standardised disease activity index and trial endpoints would allow for comparison and combination of data from multiple trials. Until then, the value of an individual probiotic trial should be assessed with an understanding of how the trial characteristics may have influenced the reported results.

Commercially available probiotics often contain more than one bacterial type. The careful selection and administration of multiple strains of bacteria in combination has the potential to be more effective than any strain on its own. This concept is supported by a small review of 16 studies which found the multiple strain products was more effective than the composite single strains 75% of the time. Additionally, a study that did ex vivo screening of probiotic strains for beneficial changes in the regulation of T-cells and pro-inflammatory cytokines identified that multistrain combinations were more potent, adding to the theory that the use of multiple bacterial strains allows for better therapeutic effects.(37)

Doses may play a role in the comparative effectiveness of a probiotic mixture. The number of bacteria in a dose can be as high as the combined quantity from a therapeutically effective dose of each composite strain assuming no synergism. The higher combined dose may have a greater effect, making the multistrain probiotic therapy more likely to be effective especially if synergistic interaction exists between used bacterial strains [209]. Countering this as the sole mechanism influencing efficacy are studies where animals were administered single strain and multiple strain probiotics to protect against pathogens. Although the total dose of each probiotic was the same, the mixtures still had a greater protective effect or survival rate, indicating the presence of bacterial synergism [210-212].

A number of potential mechanisms for additive and synergistic interactions between probiotic strains exist. Some are probably the result of fortunate coincidence, while others are likely to be due to bacterial adaptation. The mechanism for the synergy may be simple, e.g. a byproduct of one bacteria increasing another strains' rate of growth. Other mechanisms may be more complex, involving more than two strains or using intermediaries to alter signalling pathways. The potential intricacy of these bacterial interactions prevents any single strain from a multi strain probiotic being identified as the sole cause of a therapeutic effect without detailed additional research. Using more strains of bacteria in a probiotic preparation does not guarantee a better therapeutic response. Multiple strains of bacteria can have an antagonistic effect on each other through the production of agents that inhibit growth or competition for resources and adhesion sites. Other bacterial interactions could mask the influence of the antagonism on patient response, to the point where it may not be identified at all. This means bacteria with no clinical benefit could be included in probiotics unnecessarily.

Given that the effects of probiotics are strain specific, it is not possible to determine whether multiple strain probiotics are 'better' than single strain probiotics or vice versa. It does seem that some bacterial strains do have an increased clinical efficacy in one preparation over the other. Additional strain specific research could develop a reference to aid in determining if a probiotic bacterial strain is likely to benefit more from the reduced competition when administered alone or the potential synergism when multiple strains interact.

The mechanism of immune modulation through gut microfloral bacteria change during certain disease states. A large trial on patients with acute pancreatitis found that 16% patients in the probiotic group died compared with 6% of the placebo group, indicating an increase in mortality with prophylactic probiotic treatment [195]. This highlights the need for caution when treating a disease state or severity that safety has not been established with.

If the use of probiotics is to become part of autoimmune disease therapy, their safety concerns may be overcome by thoroughly studying appropriate dosing and frequency, their short and long term effect on mucosal membranes and the variation of their effect in different populations.

9. Conclusion

It is becoming more evident that the initiation, modulation and exacerbation of the inflammatory response resulting in ADs, is associated with disturbances of the gut microflora, as well as other biophysiological and biochemical processes inside and outside the gastrointestinal tract. *In vitro* studies have elucidated some of the complex proposed mechanisms associating gut microfloral disturbances with the development and progress of many ADs. Clinical trials have also provided evidence implicating probiotic intake to some health benefits noticed in ADs such as UC and T1D. However, significant clinical applications of probiotics as first line treatment for ADs have not been demonstrated or clearly proven, despite limited success in alleviating signs and symptoms of the diseases. As they are safe, probiotics are easily available to patients interested in trialling their effects. Many probiotics can be taken only once or twice a day which makes dosing convenient. Human trials have, so far, had a low incidence and severity of side effects. However, until trials are done using a broader range of disease severities with multiple bacterial strains, probiotic use may be limited to mild to moderate disease state and efficacy remains limited and at times controversial.

Main limitations to probiotic efficacies include formulation challenges, survival rate, cell-forming-bacterial-units required to exert a clinical effect and the versatility of gut microflora in different individuals and different stages of the disease. This makes selection of the bacterial strains, dosing volume and frequency and safety of AD patients, challenging. In addition, direct comparison of multiple clinical trials is complicated by the variability in study endpoints, disease severity assessment and other medication usage.

Ultimately, the primary treating physician, alongside the patient and the health care team, needs to assess whether a patient may benefit from probiotic treatment. If probiotics are to be used, trials on populations with a similar disease state to the patient can provide some guidance in strain selection. Clinical evidence should be used to determine if probiotic treatment is to be adjunctive or not, whether remission or symptom improvement is possible and to manage expectations. Disease state activity index scoring can monitor patient improvement or deterioration. For the patient, though, it is likely that the only monitoring that is meaningful is whether probiotic treatment has improved their perceived quality of life, thus, patient perception should always be taken into account when probiotic intake is considered.

Author details

Hani Al-Salami and Rima Caccetta
School of Pharmacy, Curtin Health Innovation Research Institute, Curtin University of Technology, Perth WA, Australia

Svetlana Golocorbin-Kon and Momir Mikov
Pharmacy Faculty, University of Montenegro, Podgorica, Montenegro

Acknowledgement

This work has been supported by the School of Pharmacy, Curtin University, Perth WA, Australia

10. References

[1] Tlaskalova-Hogenova, H., et al., *The role of gut microbiota (commensal bacteria) and the mucosal barrier in the pathogenesis of inflammatory and autoimmune diseases and cancer: contribution of germ-free and gnotobiotic animal models of human diseases.* Cell Mol.Immunol., 2011. 8(2): p. 110-120.

[2] Bach, J.F., *The effect of infections on susceptibility to autoimmune and allergic diseases.* N Engl J Med, 2002. 347(12): p. 911-920.

[3] Ebringer, A., et al., *Bovine spongiform encephalopathy: is it an autoimmune disease due to bacteria showing molecular mimicry with brain antigens?* Environmental health perspectives, 1997. 105(11): p. 1172-4.

[4] Paccagnini, D., et al., *Linking chronic infection and autoimmune diseases: Mycobacterium avium subspecies paratuberculosis, SLC11A1 polymorphisms and type-1 diabetes mellitus.* Plos One, 2009. 4(9): p. e7109.

[5] Waterhouse, J.C., T.H. Perez, and P.J. Albert, *Reversing bacteria-induced vitamin D receptor dysfunction is key to autoimmune disease.* Annals of the New York Academy of Sciences, 2009. 1173: p. 757-65.

[6] Cheng, G., et al., *Pharmacologic activation of the innate immune system to prevent respiratory viral infections.* American journal of respiratory cell and molecular biology, 2011. 45(3): p. 480-8.

[7] Hausmann, J., et al., *CD8 T cells require gamma interferon to clear borna disease virus from the brain and prevent immune system-mediated neuronal damage.* Journal of virology, 2005. 79(21): p. 13509-18.

[8] Singh, B., *Stimulation of the developing immune system can prevent autoimmunity.* Journal of autoimmunity, 2000. 14(1): p. 15-22.

[9] Alzabin, S. and P.J. Venables, *Etiology of autoimmune disease: past, present and future.* Expert review of clinical immunology, 2012. 8(2): p. 111-3.

[10] Brix, T.H. and L. Hegedus, *The complexity of the etiology of autoimmune thyroid disease is gravely underestimated.* Thyroid : official journal of the American Thyroid Association, 2011. 21(12): p. 1289-92.

[11] Cooper, G.S., *Unraveling the etiology of systemic autoimmune diseases: peering into the preclinical phase of disease.* The Journal of rheumatology, 2009. 36(9): p. 1853-5.

[12] Iwatani, Y., et al., *Intrathyroidal lymphocyte subsets, including unusual CD4+ CD8+ cells and CD3loTCR alpha beta lo/-CD4-CD8- cells, in autoimmune thyroid disease.* Clinical and experimental immunology, 1993. 93(3): p. 430-6.

[13] Al-Salami, H., et al., *Bile acids: a bitter sweet remedy for diabetes.* The New Zealand Pharmacy Journal, 2007. 27(10): p. 17-20.

[14] Goebel, K.M., A. Krause, and F. Neurath, *Acquired transient autoimmune reactions in Lyme arthritis: correlation between rheumatoid factor and disease activity.* Scandinavian journal of rheumatology. Supplement, 1988. 75: p. 314-7.

[15] Fukushima, H., et al., *Diagnosis and discrimination of autoimmune Graves' disease and Hashimoto's disease using thyroid-stimulating hormone receptor-containing recombinant proteoliposomes.* Journal of bioscience and bioengineering, 2009. 108(6): p. 551-6.

[16] Miyamoto, S., et al., *Assessment of thyroid growth stimulating activity of immunoglobulins from patients with autoimmune thyroid disease by cytokinesis arrest assay.* European journal of endocrinology / European Federation of Endocrine Societies, 1997. 136(5): p. 499-507.

[17] Fujihira, T., *[Significance of circulating anti-thyroid stimulating hormone (TSH) receptor antibodies in patients with autoimmune thyroid diseases--thyroid stimulation blocking antibody in patients with Graves' disease].* Journal of UOEH, 1989. 11(4): p. 393-401.

[18] Peakman, M. and D. Vergani, *Autoimmune disease: etiology, therapy and regeneration.* Immunology today, 1994. 15(8): p. 345-7.

[19] Lipsky, P.E., *Systemic lupus erythematosus: an autoimmune disease of B cell hyperactivity.* Nature immunology, 2001. 2(9): p. 764-6.

[20] Al-Salami, H., et al., *Probiotic treatment reduces blood glucose levels and increases systemic absorption of gliclazide in diabetic rats.* European Journal of Drug Metabolism and Pharmacokinetics, 2008. 33(2): p. 101-6.

[21] Al-Salami, H., et al., *Gliclazide reduces MKC intestinal transport in healthy but not diabetic rats.* European Journal of Drug Metabolism and Pharmacokinetics, 2009. 34(1): p. 43-50.

[22] Al-Salami, H., et al., *Influence of the semisynthetic bile acid MKC on the ileal permeation of gliclazide ex vivo in healthy and diabetic rats treated with probiotics.* Methods and Findings in Experimental and Clinical Pharmacology, 2008. 30(2): p. 107-113.

[23] Al-Salami, H., et al., *Probiotic treatment proceeded by a single dose of bile acid and gliclazide exert the most hypoglycemic effect in Type 1 diabetic rats.* Medical Hypothesis Research, 2008. 4(2): p. 93-101.

[24] Rodes, L., et al., *Transit time affects the community stability of lactobacillus and bifidobacterium species in an in vitro model of human colonic microbiotia.* Artificial cells, blood substitutes, and immobilization biotechnology, 2011. 39(6): p. 351-6.

[25] Al-Salami, H., et al., *Probiotic Pre-treatment Reduces Gliclazide Permeation (ex vivo) in Healthy Rats but Increases It in Diabetic Rats to the Level Seen in Untreated Healthy Rats.* Arch.Drug Inf., 2008. 1(1): p. 35-41.

[26] Al-Salami, H., et al., *Probiotic treatment decreases the oral absorption of the semisynthetic bile acid, MKC, in healthy and diabetic rats.* The European journal of drug metabolism and pharmacokinetics, 2009.

[27] Bernard, C.C. and N. Kerlero de Rosbo, *Multiple sclerosis: an autoimmune disease of multifactorial etiology.* Current opinion in immunology, 1992. 4(6): p. 760-5.

[28] Al-Salami, H., et al., *Probiotic treatment reduces blood glucose levels and increases systemic absorption of gliclazide in diabetic rats.* Eur.J.Drug Metab Pharmacokinet., 2008. 33(2): p. 101-106.

[29] Tlaskalova-Hogenova, H., et al., *Commensal bacteria (normal microflora), mucosal immunity and chronic inflammatory and autoimmune diseases.* Immunology letters, 2004. 93(2-3): p. 97-108.

[30] Ghosh, S., D. van Heel, and R.J. Playford, *Probiotics in inflammatory bowel disease: is it all gut flora modulation?* Gut, 2004. 53(5): p. 620-622.

[31] Bourlioux, P., et al., *The intestine and its microflora are partners for the protection of the host: report on the Danone Symposium "The Intelligent Intestine," held in Paris, June 14, 2002.* Am J Clin Nutr, 2003. 78(4): p. 675-683.

[32] Hammer, H.F., *Gut Microbiota and Inflammatory Bowel Disease.* Digestive Diseases, 2011. 29(6): p. 550-553.

[33] Hanski, C., et al., *Defective post-transcriptional processing of MUC2 mucin in ulcerative colitis and in Crohn's disease increases detectability of the MUC2 protein core.* The Journal of pathology, 1999. 188(3): p. 304-11.

[34] Mikov, M., et al., *Pharmacology of bile acids and their derivatives: absorption promoters and therapeutic agents.* Eur J Drug Metab Pharmacokinet, 2006. 31(3): p. 237-251.

[35] Duan, F., K.L. Curtis, and J.C. March, *Secretion of insulinotropic proteins by commensal bacteria: rewiring the gut to treat diabetes.* Appl.Environ.Microbiol., 2008. 74(23): p. 7437-7438.

[36] Gareau, M.G., P.M. Sherman, and W.A. Walker, *Probiotics and the gut microbiota in intestinal health and disease.* Nature Reviews Gastroenterology & Hepatology, 2010. 7(9): p. 503-514.

[37] Walker, W.A., *Mechanisms of action of probiotics.* Clinical Infectious Diseases, 2008. 46: p. S87-S91.

[38] Falk, P.G., et al., *Creating and maintaining the gastrointestinal ecosystem: what we know and need to know from gnotobiology.* Microbiol Mol Biol Rev, 1998. 62(4): p. 1157-1170.

[39] Isolauri, E., *Probiotics in human disease.* The American journal of clinical nutrition, 2001. 73(6): p. 1142S-1146S.

[40] Macfarlane, G.T., et al., *The Gut Microbiota in Inflammatory Bowel Disease.* Current Pharmaceutical Design, 2009. 15(13): p. 1528-1536.

[41] Ouwehand, A., E. Isolauri, and S. Salminen, *The role of the intestinal microflora for the development of the immune system in early childhood.* Eur J Nutr, 2002. 41 Suppl 1: p. I32-I37.

[42] Al-Salami, H., et al., *Influence of the semisynthetic bile acid MKC on the ileal permeation of gliclazide in vitro in healthy and diabetic rats treated with probiotics.* Methods and Findings in Experimental and Clinical Pharmacology, 2008. 30(2): p. 107-13.

[43] Al-Salami, H., et al., *Probiotic Pre-treatment Reduces Gliclazide Permeation (ex vivo) in Healthy Rats but Increases It in Diabetic Rats to the Level Seen in Untreated Healthy Rats.* Archives of drug information, 2008. 1(1): p. 35-41.

[44] Al-Salami, H., et al., *The influence of probiotics pre-treatment, on the ileal permeation of gliclazide, in healthy and diabetic rats.* The archives of drug information, 2008. 1(1): p. 35-41.

[45] Collett, A., et al., *Modulation of the permeability of H2 receptor antagonists cimetidine and ranitidine by P-glycoprotein in rat intestine and the human colonic cell line Caco-2.* J Pharmacol Exp Ther, 1999. 288(1): p. 171-178.

[46] Legen, I. and A. Kristl, *D-glucose triggers multidrug resistance-associated protein (MRP)-mediated secretion of fluorescein across rat jejunum in vitro.* Pharm Res, 2004. 21(4): p. 635-640.

[47] Linskens, R.K., et al., *The bacterial flora in inflammatory bowel disease: current insights in pathogenesis and the influence of antibiotics and probiotics.* Scandinavian journal of gastroenterology. Supplement, 2001(234): p. 29-40.

[48] Sun, Y.Q., et al., *Long-standing gastric mucosal barrier dysfunction in Helicobacter pylori-induced gastritis in mongolian gerbils.* Helicobacter, 2004. 9(3): p. 217-227.

[49] Qin, X., et al., *Oral characteristics of bergenin and the effect of absorption enhancers in situ, in vitro and in vivo.* Arzneimittelforschung., 2010. 60(4): p. 198-204.

[50] Zanchi, A., et al., *Differences in the mechanical properties of the rat carotid artery in vivo, in situ, and in vitro.* Hypertension, 1998. 32(1): p. 180-185.

[51] Adorini, L., S. Gregori, and L.C. Harrison, *Understanding autoimmune diabetes: insights from mouse models.* Trends Mol Med, 2002. 8(1): p. 31-38.

[52] Leung, P.S., et al., *Development and validation of gene therapies in autoimmune diseases: Epidemiology to animal models.* Autoimmunity reviews, 2010. 9(5): p. A400-5.

[53] Lam-Tse, W.K., A. Lernmark, and H.A. Drexhage, *Animal models of endocrine/organ-specific autoimmune diseases: do they really help us to understand human autoimmunity?* Springer seminars in immunopathology, 2002. 24(3): p. 297-321.

[54] Ludewig, B., R.M. Zinkernagel, and H. Hengartner, *Transgenic animal models for virus-induced autoimmune diseases.* Experimental physiology, 2000. 85(6): p. 653-9.

[55] Korganow, A.S., J.C. Weber, and T. Martin, *[Animal models and autoimmune diseases].* La Revue de medecine interne / fondee ... par la Societe nationale francaise de medecine interne, 1999. 20(3): p. 283-6.

[56] Burkhardt, H. and J.R. Kalden, *Xenobiotic immunosuppressive agents: therapeutic effects in animal models of autoimmune diseases.* Rheumatology international, 1997. 17(3): p. 85-90.

[57] Kano, K., *[Study of autoimmune diseases using animal disease models. 2) Confronting the clinical situation].* Nihon rinsho. Japanese journal of clinical medicine, 1988. 46(4): p. 850-9.

[58] Nagasawa, R. and N. Maruyama, *[Study of autoimmune diseases using animal disease models. 1) Molecular genetic analysis of autoimmune disease models].* Nihon rinsho. Japanese journal of clinical medicine, 1988. 46(4): p. 845-9.

[59] Dieleman, L.A., et al., *Role of animal models for the pathogenesis and treatment of inflammatory bowel disease.* Scand J Gastroenterol Suppl, 1997. 223: p. 99-104.

[60] Srinivasan, K. and P. Ramarao, *Animal models in type 2 diabetes research: an overview.* Indian J Med Res, 2007. 125(3): p. 451-472.

[61] Alam, C., et al., *Effects of a germ-free environment on gut immune regulation and diabetes progression in non-obese diabetic (NOD) mice.* Diabetologia, 2011. 54(6): p. 1398-406.

[62] King, C. and N. Sarvetnick, *The incidence of type-1 diabetes in NOD mice is modulated by restricted flora not germ-free conditions.* Plos One, 2011. 6(2): p. e17049.

[63] van Waarde, W.M., et al., *Differential effects of streptozotocin-induced diabetes on expression of hepatic ABC-transporters in rats.* Gastroenterology, 2002. 122(7): p. 1842-1852.

[64] Rungby, J., *Zinc, zinc transporters and diabetes.* Diabetologia, 2010. 53(8): p. 1549-1551.

[65] Vidotti, D.B., et al., *Effect of long-term type 1 diabetes on renal sodium and water transporters in rats.* Am J Nephrol, 2008. 28(1): p. 107-114.

[66] Tramonti, G., et al., *Expression and functional characteristics of tubular transporters: P-glycoprotein, PEPT1, and PEPT2 in renal mass reduction and diabetes.* Am.J.Physiol Renal Physiol, 2006. 291(5): p. F972-F980.

[67] Grover, B., et al., *Reduced expression of organic cation transporters rOCT1 and rOCT2 in experimental diabetes.* J.Pharmacol.Exp Ther., 2004. 308(3): p. 949-956.

[68] Py, G., et al., *Effects of streptozotocin-induced diabetes on markers of skeletal muscle metabolism and monocarboxylate transporter 1 to monocarboxylate transporter 4 transporters.* Metabolism, 2002. 51(7): p. 807-813.

[69] Mooradian, A.D. and A.M. Morin, *Brain uptake of glucose in diabetes mellitus: the role of glucose transporters.* Am.J.Med.Sci., 1991. 301(3): p. 173-177.

[70] Zilberstein, D., et al., *Identification and biochemical characterization of the plasma membrane glucose transporter of Leishmania donovani.* The Journal of Biological Chemistry, 1986. 261(32): p. 15053-7.

[71] Matthaei, S., D.L. Baly, and R. Horuk, *Rapid and effective transfer of integral membrane proteins from isoelectric focusing gels to nitrocellulose membranes.* Analytical biochemistry, 1986. 157(1): p. 123-8.

[72] Matthaei, D., et al., *Multipurpose NMR imaging using stimulated echoes.* Magnetic resonance in medicine : official journal of the Society of Magnetic Resonance in Medicine / Society of Magnetic Resonance in Medicine, 1986. 3(4): p. 554-61.

[73] Vezys, V. and L. Lefrancois, *Cutting edge: inflammatory signals drive organ-specific autoimmunity to normally cross-tolerizing endogenous antigen.* Journal of immunology, 2002. 169(12): p. 6677-80.

[74] Wucherpfennig, K.W., *Mechanisms for the induction of autoimmunity by infectious agents.* The Journal of clinical investigation, 2001. 108(8): p. 1097-104.

[75] Wucherpfennig, K.W., *Structural basis of molecular mimicry.* Journal of autoimmunity, 2001. 16(3): p. 293-302.

[76] Srivastava, P., *Roles of heat-shock proteins in innate and adaptive immunity.* Nature reviews. Immunology, 2002. 2(3): p. 185-94.

[77] Li, Z., A. Menoret, and P. Srivastava, *Roles of heat-shock proteins in antigen presentation and cross-presentation.* Current opinion in immunology, 2002. 14(1): p. 45-51.

[78] Heil, F., et al., *The Toll-like receptor 7 (TLR7)-specific stimulus loxoribine uncovers a strong relationship within the TLR7, 8 and 9 subfamily.* European journal of immunology, 2003. 33(11): p. 2987-97.

[79] Latz, E., et al., *Mechanisms of TLR9 activation.* Journal of endotoxin research, 2004. 10(6): p. 406-12.

[80] Rifkin, I.R., et al., *Toll-like receptors, endogenous ligands, and systemic autoimmune disease.* Immunological reviews, 2005. 204: p. 27-42.

[81] Toubi, E. and Y. Shoenfeld, *Toll-like receptors and their role in the development of autoimmune diseases.* Autoimmunity, 2004. 37(3): p. 183-8.

[82] Shimizu, S., et al., *Involvement of toll-like receptors in autoimmune sialoadenitis of the non-obese diabetic mouse.* Journal of oral pathology & medicine : official publication of the International Association of Oral Pathologists and the American Academy of Oral Pathology, 2012.

[83] Watanabe, T., et al., *Involvement of activation of toll-like receptors and nucleotide-binding oligomerization domain-like receptors in enhanced IgG4 responses in autoimmune pancreatitis.* Arthritis and rheumatism, 2012. 64(3): p. 914-24.

[84] Lampropoulou, V., et al., *Suppressive functions of activated B cells in autoimmune diseases reveal the dual roles of Toll-like receptors in immunity.* Immunological reviews, 2010. 233(1): p. 146-61.

[85] Takeuchi, O., et al., *Differential roles of TLR2 and TLR4 in recognition of gram-negative and gram-positive bacterial cell wall components.* Immunity, 1999. 11(4): p. 443-51.

[86] Shimizu, N., et al., *Changes in and discrepancies between cell tropisms and coreceptor uses of human immunodeficiency virus type 1 induced by single point mutations at the V3 tip of the env protein.* Virology, 1999. 259(2): p. 324-33.

[87] Hoshino, K., et al., *Cutting edge: Toll-like receptor 4 (TLR4)-deficient mice are hyporesponsive to lipopolysaccharide: evidence for TLR4 as the Lps gene product.* Journal of immunology, 1999. 162(7): p. 3749-52.

[88] Al-Salami, H., et al., *Probiotics decreased the bioavailability of the bile acid analog, monoketocholic acid, when coadministered with gliclazide, in healthy but not diabetic rats.* European Journal of Drug Metabolism and Pharmacokinetics, 2011.

[89] Collado, M.C., M. Hernandez, and Y. Sanz, *Production of bacteriocin-like inhibitory compounds by human fecal Bifidobacterium strains.* J Food Prot, 2005. 68(5): p. 1034-1040.

[90] Caballero-Franco, C., et al., *The VSL#3 probiotic formula induces mucin gene expression and secretion in colonic epithelial cells.* American Journal of Physiology-Gastrointestinal and Liver Physiology, 2007. 292(1): p. G315-G322.

[91] Chapman, T.M., G.L. Plosker, and D.P. Figgitt, *VSL#3 probiotic mixture: a review of its use in chronic inflammatory bowel diseases.* Drugs, 2006. 66(10): p. 1371-87.

[92] Karimi, O., A.S. Pena, and A.A. van Bodegraven, *Probiotics (VSL#3) in arthralgia in patients with ulcerative colitis and Crohn's disease: a pilot study.* Drugs of today, 2005. 41(7): p. 453-9.

[93] Rachmilewitz, D., et al., *Toll-like receptor 9 signaling mediates the anti-inflammatory effects of probiotics in murine experimental colitis.* Gastroenterology, 2004. 126(2): p. 520-528.

[94] Savard, J. and J.A. Sawatzky, *The use of a nursing model to understand diarrhea and the role of probiotics in patients with inflammatory bowel disease.* Gastroenterology nursing : the official journal of the Society of Gastroenterology Nurses and Associates, 2007. 30(6): p. 418-23; quiz 424-5.

[95] Tamboli, C.P., et al., *Probiotics in inflammatory bowel disease: a critical review.* Best practice & research. Clinical gastroenterology, 2003. 17(5): p. 805-20.

[96] Michalowicz, B.S., et al., *Periodontal bacteria in adult twins.* Journal of periodontology, 1999. 70(3): p. 263-73.

[97] Buhnik-Rosenblau, K., Y. Danin-Poleg, and Y. Kashi, *Predominant effect of host genetics on levels of Lactobacillus johnsonii bacteria in the mouse gut.* Applied and environmental microbiology, 2011. 77(18): p. 6531-8.

[98] Brisbin, J.T., J. Gong, and S. Sharif, *Interactions between commensal bacteria and the gut-associated immune system of the chicken.* Animal health research reviews / Conference of Research Workers in Animal Diseases, 2008. 9(1): p. 101-10.

[99] Talham, G.L., et al., *Segmented filamentous bacteria are potent stimuli of a physiologically normal state of the murine gut mucosal immune system.* Infection and immunity, 1999. 67(4): p. 1992-2000.

[100] Yoshioka, H., et al., *Immunohistochemical examination of Peyer's patches in autoimmune mice.* Histochemistry, 1988. 90(2): p. 145-50.

[101] Buchman, A.L. and S.M. Rao, *A patient with a polyglandular autoimmune syndrome involving the salivary glands, thyroid, intestine, and pancreas.* Digestive Diseases and Sciences, 2004. 49(4): p. 590-3.

[102] Atserova, I.S., S.E. Makievskaia, and A.A. Misautova, *[Microflora of the intestine and autoimmune reactions in chronic colitis].* Terapevticheskii arkhiv, 1970. 42(5): p. 36-9.

[103] Wu, H.J., et al., *Gut-residing segmented filamentous bacteria drive autoimmune arthritis via T helper 17 cells.* Immunity, 2010. 32(6): p. 815-27.

[104] Sblattero, D., et al., *The gut as site of production of autoimmune antibodies.* Journal of Pediatric Gastroenterology and Nutrition, 2004. 39 Suppl 3: p. S730-1.

[105] Hill, S.M., et al., *Autoimmune enteropathy and colitis: is there a generalised autoimmune gut disorder?* Gut, 1991. 32(1): p. 36-42.

[106] Zhu, W., et al., *Overexpressing autoimmune regulator regulates the expression of toll-like receptors by interacting with their promoters in RAW264.7 cells.* Cellular immunology, 2011. 270(2): p. 156-63.

[107] Marshak-Rothstein, A., *Toll-like receptors in systemic autoimmune disease.* Nature reviews. Immunology, 2006. 6(11): p. 823-35.

[108] Wen, L., et al., *The effect of innate immunity on autoimmune diabetes and the expression of Toll-like receptors on pancreatic islets.* Journal of immunology, 2004. 172(5): p. 3173-80.

[109] Tanaka, K., *Expression of Toll-like receptors in the intestinal mucosa of patients with inflammatory bowel disease.* Expert review of gastroenterology & hepatology, 2008. 2(2): p. 193-6.

[110] Himmel, M.E., et al., *The role of T-regulatory cells and Toll-like receptors in the pathogenesis of human inflammatory bowel disease.* Immunology, 2008. 125(2): p. 145-53.

[111] Macfarlane, G.T. and S. Macfarlane, *Human colonic microbiota: ecology, physiology and metabolic potential of intestinal bacteria.* Scandinavian journal of gastroenterology. Supplement, 1997. 222: p. 3-9.

[112] Zaneveld, J., et al., *Host-bacterial coevolution and the search for new drug targets.* Curr Opin Chem Biol, 2008. 12(1): p. 109-114.

[113] Shi, H.N. and A. Walker, *Bacterial colonization and the development of intestinal defences.* Can.J.Gastroenterol., 2004. 18(8): p. 493-500.

[114] Rescigno, M., et al., *Dendritic cells express tight junction proteins and penetrate gut epithelial monolayers to sample bacteria.* Nat.Immunol., 2001. 2(4): p. 361-367.

[115] von, H.M. and G.T. Nepom, *Animal models of human type 1 diabetes.* Nat.Immunol., 2009. 10(2): p. 129-132.

[116] Macpherson, A.J. and T. Uhr, *Induction of protective IgA by intestinal dendritic cells carrying commensal bacteria.* Science, 2004. 303(5664): p. 1662-1665.

[117] Mishan-Eisenberg, G., et al., *Differential regulation of Th1/Th2 cytokine responses by placental protein 14.* Journal of immunology, 2004. 173(9): p. 5524-30.

[118] Tukel, T., et al., *Crohn disease: frequency and nature of CARD15 mutations in Ashkenazi and Sephardi/Oriental Jewish families.* American journal of human genetics, 2004. 74(4): p. 623-36.

[119] Neish, A.S., et al., *Prokaryotic regulation of epithelial responses by inhibition of IkappaB-alpha ubiquitination.* Science, 2000. 289(5484): p. 1560-1563.

[120] Otte, J.M. and D.K. Podolsky, *Functional modulation of enterocytes by gram-positive and gram-negative microorganisms.* Am.J.Physiol Gastrointest.Liver Physiol, 2004. 286(4): p. G613-G626.

[121] Rebole, A., et al., *Effects of inulin and enzyme complex, individually or in combination, on growth performance, intestinal microflora, cecal fermentation characteristics, and jejunal*

histomorphology in broiler chickens fed a wheat- and barley-based diet. Poult.Sci., 2010. 89(2): p. 276-286.

[122] Yen, C.H., et al., *Long-term supplementation of isomalto-oligosaccharides improved colonic microflora profile, bowel function, and blood cholesterol levels in constipated elderly people--a placebo-controlled, diet-controlled trial.* Nutrition, 2011. 27(4): p. 445-450.

[123] Cani, P.D., et al., *Changes in gut microbiota control metabolic endotoxemia-induced inflammation in high-fat diet-induced obesity and diabetes in mice.* Diabetes, 2008. 57(6): p. 1470-1481.

[124] Cani, P.D., et al., *Selective increases of bifidobacteria in gut microflora improve high-fat-diet-induced diabetes in mice through a mechanism associated with endotoxaemia.* Diabetologia, 2007. 50(11): p. 2374-2383.

[125] Cani, P.D., et al., *Changes in gut microbiota control inflammation in obese mice through a mechanism involving GLP-2-driven improvement of gut permeability.* Gut, 2009. 58(8): p. 1091-1103.

[126] Goris, H., F. de Boer, and D. van der Waaij, *Kinetics of endotoxin release by gram-negative bacteria in the intestinal tract of mice during oral administration of bacitracin and during in vitro growth.* Scand J Infect Dis, 1988. 20(2): p. 213-219.

[127] Li, L.J., et al., *Changes of gut flora and endotoxin in rats with D-galactosamine-induced acute liver failure.* World J Gastroenterol, 2004. 10(14): p. 2087-2090.

[128] Sazawal, S., et al., *Effects of Bifidobacterium lactis HN019 and prebiotic oligosaccharide added to milk on iron status, anemia, and growth among children 1 to 4 years old.* Journal of pediatric gastroenterology and nutrition, 2010. 51(3): p. 341-6.

[129] Sazawal, S., et al., *Efficacy of probiotics in prevention of acute diarrhoea: a meta-analysis of masked, randomised, placebo-controlled trials.* The Lancet infectious diseases, 2006. 6(6): p. 374-82.

[130] Nagy, G., et al., *Both alpha-haemolysin determinants contribute to full virulence of uropathogenic Escherichia coli strain 536.* Microbes and infection / Institut Pasteur, 2006. 8(8): p. 2006-12.

[131] Rozanova, G.N. and D.A. Voevodin, *[A case of an effective application of probiotics in the complex therapy of severe type 1 diabetes mellitus and intestinal disbacteriosis].* Klin.Med (Mosk), 2008. 86(1): p. 67-68.

[132] Ziegler, A.G., et al., *Early infant feeding and risk of developing type 1 diabetes-associated autoantibodies.* JAMA, 2003. 290(13): p. 1721-1728.

[133] Bezkorovainy, A., *Probiotics: determinants of survival and growth in the gut.* Am J Clin Nutr, 2001. 73((2 Suppl)): p. 399S-405S.

[134] Karimi, O. and A.S. Pena, *Probiotics: Isolated bacteria strain or mixtures of different strains? Two different approaches in the use of probiotics as therapeutics.* Drugs Today (Barc), 2003. 39(8): p. 565-597.

[135] Karimi, O. and A.S. Pena, *Indications and challenges of probiotics, prebiotics, and synbiotics in the management of arthralgias and spondyloarthropathies in inflammatory bowel disease.* Journal of clinical gastroenterology, 2008. 42 Suppl 3 Pt 1: p. S136-41.

[136] Altenhoefer, A., et al., *The probiotic Escherichia coli strain Nissle 1917 interferes with invasion of human intestinal epithelial cells by different enteroinvasive bacterial pathogens.* FEMS Immunol Med Microbiol, 2004. 40(3): p. 223-229.

[137] Jacobs, D.B., G.R. Hayes, and D.H. Lockwood, *In vitro effects of sulfonylurea on glucose transport and translocation of glucose transporters in adipocytes from streptozocin-induced diabetic rats.* Diabetes, 1989. 38(2): p. 205-211.

[138] Rozanova, G.N., et al., *Pathogenetic role of dysbacteriosis in the development of complications of type 1 diabetes mellitus in children.* Bull Exp Biol Med, 2002. 133(2): p. 164-166.

[139] Matsuzaki, T., et al., *Effect of oral administration of Lactobacillus casei on alloxan-induced diabetes in mice.* APMIS, 1997. 105(8): p. 637-642.

[140] Matsuzaki, T., et al., *Prevention of onset in an insulin-dependent diabetes mellitus model, NOD mice, by oral feeding of Lactobacillus casei.* APMIS, 1997. 105(8): p. 643-649.

[141] Calcinaro, F., et al., *Oral probiotic administration induces interleukin-10 production and prevents spontaneous autoimmune diabetes in the non-obese diabetic mouse.* Diabetologia, 2005. 48(8): p. 1565-1575.

[142] Meinders, A.E., et al., *Biliary lipid and bile acid composition in insulin-dependent diabetes mellitus.* Dig Dis Sci, 1981. 26(3): p. 402-408.

[143] Meinders, A.E., et al., *Biliary Lipid and Bile-Acid Composition in Insulin-Dependent Diabetes-Mellitus - Arguments for Increased Intestinal Bacterial Bile-Acid Degradation.* Digestive Diseases and Sciences, 1981. 26(5): p. 402-408.

[144] Malavolti, M., et al., *Interaction of Potentially Toxic Bile-Acids with Human-Plasma Proteins - Binding of Lithocholic (3-Alpha-Hydroxy-5-Beta-Cholan-24-Oic) Acid to Lipoproteins and Albumin.* Lipids, 1989. 24(7): p. 673-676.

[145] Miyai, K., et al., *Hepatotoxicity of Bile-Acids in Rabbits - Ursodeoxycholic Acid Is Less Toxic Than Chenodeoxycholic Acid.* Laboratory Investigation, 1982. 46(4): p. 428-437.

[146] Kurdi, P., et al., *Cholic acid is accumulated spontaneously, driven by membrane deltapH, in many lactobacilli.* J Bacteriol, 2000. 182(22): p. 6525-6528.

[147] Kurdi, P., et al., *Mechanism of growth inhibition by free bile acids in lactobacilli and bifidobacteria.* J Bacteriol, 2006. 188(5): p. 1979-1986.

[148] Franz, B. and J.C. Bode, *[Plasma bile acid concentration (PGK): fasting values, daily fluctuations and effect of intraduodenal bile acid administration in healthy subjects and patients with chronic liver diseases].* Z Gastroenterol, 1973. 11(2): p. 131-134.

[149] Hedenborg, G. and A. Norman, *The nature of urinary bile acid conjugates in patients with extrahepatic cholestasis.* Scand J Clin Lab Invest, 1984. 44(8): p. 725-733.

[150] Hedenborg, G. and A. Norman, *Fasting and postprandial serum bile acid concentration with special reference to variations in the conjugate profile.* Scand J Clin Lab Invest, 1985. 45(2): p. 151-156.

[151] Danese, S. and C. Fiocchi, *Ulcerative Colitis.* New England Journal of Medicine, 2011. 365(18): p. 1713-1725.

[152] Ho, G.-T., C. Lees, and J. Satsangi, *Ulcerative colitis.* Medicine, 2007. 35(5): p. 277-282.

[153] Ghosh, S., A. Shand, and A. Ferguson, *Ulcerative colitis.* BMJ, 2000. 320(7242): p. 1119-1123.

[154] Frank, D., *Molecular-Phylogenetic Characterization of Microbial Community Imbalances in Human Inflammatory Bowel Diseases.* Proceedings of the National Academy of Sciences of the United States of America, 2007. 104(34): p. 13780-13785.

[155] Eckburg, P.B., et al., *Diversity of the Human Intestinal Microbial Flora.* Science, 2005. 308(5728): p. 1635-1638.

[156] Ley, R.E., et al., *Microbial ecology: human gut microbes associated with obesity.* Nature, 2006. 444(7122): p. 1022-1023.

[157] CAMPIERI, M. and P. GIONCHETTI, *Bacteria as the cause of ulcerative colitis.* Gut, 2001. 48(1): p. 132-135.

[158] Bibiloni, R., *The bacteriology of biopsies differs between newly diagnosed, untreated, Crohn's disease and ulcerative colitis patients.* Journal of Medical Microbiology, 2006. 55(8): p. 1141-9.

[159] Atuma, C., et al., *The adherent gastrointestinal mucus gel layer: thickness and physical state in vivo.* American Journal of Physiology - Gastrointestinal and Liver Physiology, 2001. 280(5): p. G922-G929.

[160] Venturi, A., et al., *Impact on the composition of the faecal flora by a new probiotic preparation: preliminary data on maintenance treatment of patients with ulcerative colitis.* Alimentary Pharmacology & Therapeutics, 1999. 13(8): p. 1103-8.

[161] Tursi, A., et al., *Low-dose balsalazide plus a high-potency probiotic preparation is more effective than balsalazide alone or mesalazine in the treatment of acute mild-to-moderate ulcerative colitis.* Medical Science Monitor, 2004. 10(11): p. PI126-31.

[162] Bibiloni, R., et al., *VSL#3 probiotic-mixture induces remission in patients with active ulcerative colitis.* Am J Gastroenterol, 2005. 100(7): p. 1539-46.

[163] Soo, I., et al., *VSL#3 probiotic upregulates intestinal mucosal alkaline sphingomyelinase and reduces inflammation.* Canadian Journal of Gastroenterology, 2008. 22(3): p. 237-42.

[164] Tursi, A., *Balsalazide plus high-potency probiotic preparation (VSL[sharp]3) in the treatment of acute mild-to-moderate ulcerative colitis and uncomplicated diverticulitis of the colon.* J Clin Gastroenterol, 2008. 42 Suppl 3 Pt 1: p. S119-22.

[165] Huynh, H.Q., et al., *Probiotic preparation VSL#3 induces remission in children with mild to moderate acute ulcerative colitis: a pilot study.* Inflammatory Bowel Diseases, 2009. 15(5): p. 760-8.

[166] Miele, E., et al., *Effect of a probiotic preparation (VSL#3) on induction and maintenance of remission in children with ulcerative colitis.* American Journal of Gastroenterology, 2009. 104(2): p. 437-43.

[167] Turcotte, J.F. and H.Q. Huynh, *Treatment with the probiotic VSL#3 as an adjunctive therapy in relapsing mild-to-moderate ulcerative colitis significantly reduces ulcerative colitis disease activity.* Evid Based Med, 2011. 16(4): p. 108-9.

[168] Sood, A., et al., *The probiotic preparation, VSL#3 induces remission in patients with mild-to-moderately active ulcerative colitis.* Clin Gastroenterol Hepatol, 2009. 7(11): p. 1202-9, 1209 e1.

[169] Tursi, A., *Treatment of Relapsing Mild-to-Moderate Ulcerative Colitis With the Probiotic VSL&3 as Adjunctive to a Standard Pharmaceutical Treatment: A Double-Blind, Randomized, Placebo-Controlled Study.* The American journal of gastroenterology, 2010. 105(10): p. 2218-2227.

[170] Otte J, P.D.K., *Functional modulation of enterocytes by gram-positive and gram-negative microorganisms.* Am J Physiol Gastrointest Liver Physiol, 2003. 286(4): p. G613-G626.

[171] Jacobi, C.A., C. Schulz, and P. Malfertheiner, *Treating critically ill patients with probiotics: Beneficial or dangerous?* Gut Pathog, 2011. 3(1): p. 2.

[172] Verna, E.C. and S. Lucak, *Use of probiotics in gastrointestinal disorders: what to recommend?* Therap Adv Gastroenterol, 2010. 3(5): p. 307-19.

[173] P C F Stokkers, P.H.R., G N J Tytgat, S J H van Deventer, *HLA-DR and -DQ phenotypes in inflammatory bowel disease: A meta-analysis.* Gut, London, 1999. 45(3): p. 395-402.

[174] S C Ng, S.P., H O Al-Hassi, N English, N Gellatly, M A Kamm, S C Knight, A J Stagg, *A novel population of human CD56+ human leucocyte antigen D-related (HLA-DR+) colonic lamina propria cells is associated with inflammation in ulcerative colitis.* Clinical & Experimental Immunology, 2009. 158(2): p. 205–218.

[175] Kaser, A., et al., *XBP1 links ER stress to intestinal inflammation and confers genetic risk for human inflammatory bowel disease.* Cell, 2008. 134(5): p. 743-56.

[176] Dubuquoy, L., et al., *Impaired expression of peroxisome proliferator-activated receptor gamma in ulcerative colitis.* Gastroenterology, 2003. 124(5): p. 1265-76.

[177] Macfarlane, S.F., E ; Kennedy, A ; Cummings, JH ; Macfarlane, GT, *Mucosal bacteria in ulcerative colitis.* British journal of nutrition, 2005. 93: p. S67-S72.

[178] Li-Xuan Sang, B.C., Wen-Liang Zhang, Xiao-Mei Wu, Xiao-Hang Li, and Min Jiang, *Remission induction and maintenance effect of probiotics on ulcerative colitis: A meta-analysis.* World journal of Gastroenterology, December 14, 2009. 16(15): p. 1908–1915.

[179] Sanjay Chaudhri, P.R., *Surgical management of inflammatory bowel disease.* Surgery (Oxford), 2005. 23(10): p. 373-376.

[180] Chamberlin W., B.T., Campbell J.. *Primary treatment of Crohn's disease: combined antibiotics taking center stage.* Expert Review of Clinical Immunology, 2011. 7(6): p. 751-760.

[181] Seibold, F., et al., *pANCA represents a cross-reactivity to enteric bacterial antigens.* J Clin Immunol, 1998. 18(2): p. 153-60.

[182] Zhang, W. and M. Reichlin, *A possible link between infection with burkholderia bacteria and systemic lupus erythematosus based on epitope mimicry.* Clinical & Developmental Immunology, 2008.

[183] Apperloo-Renkema, H.Z., et al., *Host-microflora interaction in systemic lupus erythematosus (SLE): colonization resistance of the indigenous bacteria of the intestinal tract.* Epidemiology and infection, 1994. 112(2): p. 367-73.

[184] Hirose, Y., et al., *Safety studies of LP20 powder produced from heat-killed Lactobacillus plantarum L-137.* Regulatory toxicology and pharmacology : RTP, 2009. 54(3): p. 214-20.

[185] Cross, M.L., *Microbes versus microbes: immune signals generated by probiotic lactobacilli and their role in protection against microbial pathogens.* FEMS immunology and medical microbiology, 2002. 34(4): p. 245-53.

[186] Joshi, V.D., et al., *IL-18 levels and the outcome of innate immune response to lipopolysaccharide: importance of a positive feedback loop with caspase-1 in IL-18 expression.* Journal of immunology, 2002. 169(5): p. 2536-44.

[187] Wheeler, J.G., et al., *Immune and clinical impact of Lactobacillus acidophilus on asthma.* Annals of allergy, asthma & immunology : official publication of the American College of Allergy, Asthma, & Immunology, 1997. 79(3): p. 229-33.

[188] Wheeler, J.G., et al., *Impact of dietary yogurt on immune function.* The American journal of the medical sciences, 1997. 313(2): p. 120-3.

[189] Aldinucci, C., et al., *Effects of dietary yoghurt on immunological and clinical parameters of rhinopathic patients.* European journal of clinical nutrition, 2002. 56(12): p. 1155-61.

[190] Aldinucci, D., et al., *CD40L induces proliferation, self-renewal, rescue from apoptosis, and production of cytokines by CD40-expressing AML blasts.* Experimental hematology, 2002. 30(11): p. 1283-92.

[191] Aldinucci, D., et al., *Expression of functional interleukin-3 receptors on Hodgkin and Reed-Sternberg cells.* The American journal of pathology, 2002. 160(2): p. 585-96.

[192] Schultz, M., et al., *Immunomodulatory consequences of oral administration of Lactobacillus rhamnosus strain GG in healthy volunteers.* J Dairy Res, 2003. 70(2): p. 165-173.

[193] Schultz, M., et al., *Preventive effects of Escherichia coli strain Nissle 1917 on acute and chronic intestinal inflammation in two different murine models of colitis.* Clin Diagn Lab Immunol, 2004. 11(2): p. 372-378.

[194] Araya M, M.L., Reid G, Sanders M E, Stanton C, *Guidelines for the Evaluation of Probiotics in Food*, in *Joint FAO/WHO Working Group meeting*2002: London Ontario, Canada.

[195] Besselink, M.G., et al., *Probiotic prophylaxis in predicted severe acute pancreatitis: a randomised, double-blind, placebo-controlled trial.* Lancet, 2008. 371(9613): p. 651-9.

[196] Luoto, R., E. Isolauri, and L. Lehtonen, *Safety of Lactobacillus GG probiotic in infants with very low birth weight: twelve years of experience.* Clinical infectious diseases : an official publication of the Infectious Diseases Society of America, 2010. 50(9): p. 1327-8.

[197] Luoto, R., et al., *Impact of maternal probiotic-supplemented dietary counselling on pregnancy outcome and prenatal and postnatal growth: a double-blind, placebo-controlled study.* Br J Nutr, 2010. 103(12): p. 1792-1799.

[198] Snydman, D.R., *The safety of probiotics.* Clin.Infect.Dis., 2008. 46 Suppl 2: p. S104-S111.

[199] Liong, M.T., *Safety of probiotics: translocation and infection.* Nutr.Rev., 2008. 66(4): p. 192-202.

[200] Rayes, N., et al., *Supply of pre- and probiotics reduces bacterial infection rates after liver transplantation--a randomized, double-blind trial.* Am.J.Transplant., 2005. 5(1): p. 125-130.

[201] Marteau, P. and P. Seksik, *[Role of prebiotics and probiotics in therapeutic management of cryptogenetic inflammatory bowel disease].* Gastroenterologie clinique et biologique, 2001. 25(2 Pt 2): p. C94-7.

[202] Marteau, P. and F. Shanahan, *Basic aspects and pharmacology of probiotics: an overview of pharmacokinetics, mechanisms of action and side-effects.* Best Pract Res Clin Gastroenterol, 2003. 17(5): p. 725-740.

[203] Marteau, P., et al., *Review article: gut flora and inflammatory bowel disease.* Aliment Pharmacol Ther, 2004. 20 Suppl 4: p. 18-23.

[204] Marteau, P., *Probiotics, prebiotics, synbiotics: ecological treatment for inflammatory bowel disease?* Gut, 2006. 55(12): p. 1692-3.

[205] Cooney, R., et al., *Outcome measurement in clinical trials for ulcerative colitis: towards standardisation.* Trials, 2007. 8(1): p. 17.

[206] Sands, B.E., *The Placebo Response Rate in Irritable Bowel Syndrome and Inflammatory Bowel Disease.* Digestive Diseases, 2009. 27(Suppl. 1): p. 68-75.

[207] Su, C., et al., *A Meta-Analysis of the Placebo Rates of Remission and Response in Clinical Trials of Active Ulcerative Colitis.* Gastroenterology, 2007. 132(2): p. 516-526.

[208] Hoie, O., et al., *Ulcerative colitis: patient characteristics may predict 10-yr disease recurrence in a European-wide population-based cohort.* Am J Gastroenterol, 2007. 102(8): p. 1692-701.

[209] Dunne, C., et al., *In vitro selection criteria for probiotic bacteria of human origin: correlation with in vivo findings.* Am J Clin Nutr, 2001. 73(2 Suppl): p. 386S-392S.

[210] PaubertBraquet, M., *Enhancement of host resistance against Salmonella typhimurium in mice fed a diet supplemented with yogurt or milks fermented with various Lactobacillus casei strains.* International journal of immunotherapy, 1995. 11(4): p. 9-9.

[211] Lema, M., L. Williams, and D.R. Rao, *Reduction of fecal shedding of enterohemorrhagic Escherichia coli O157:H7 in lambs by feeding microbial feed supplement.* Small Ruminant Research, 2001. 39(1): p. 31-39.

[212] Perdigon, G. and Perdigon, *Prevention of gastrointestinal infection using immunobiological methods with milk fermented with and.* The Journal of Dairy Research, 1990. 57(02): p. 255-264.

Chapter 15

Probiotics: The Effects on Human Health and Current Prospects

Giselle Nobre Costa and Lucia Helena S. Miglioranza

Additional information is available at the end of the chapter

http://dx.doi.org/10.5772/50048

1. Introduction

Studies employing genes sequence for genotyping analysis of microorganisms, are allowing the knowledge expansion about the microbiota of the human gastrointestinal tract (GIT). Only in the last decade, the number of species detected molecularly has exceeded on a large scale the number of species accessible by cultivation-dependent methods.

The molecular techniques ranging from the identification of intestinal microbiota, particularly probiotic microorganism in different environments, detection of pathogenicity genes in foods, identification and quantification using real-time polimerase chain reaction (PCR), till studies with proteomics approach, which evaluate the expression of genes of interest or the changes in the host due to the microorganisms impact, have providing new perspectives in the investigation of diversity, abundance and dynamics of the intestinal ecosystem.

Research on probiotics microorganisms has focused on methods of evaluating the GIT microbiota survival and function, cross-talk between the intestinal microbiota and the host and the probiotic interactions with the immune system. Actually, the data generated by clinical studies reinforces the effect of this microbiota on the human health.

A substantial number of clinical studies have supported the idea that health can be affected by the daily consumption of probiotics. The exploitation of these data allows understanding the mechanisms by which probiotic microorganisms survive the passage through the GI tract to interact with the resident microbiota, and affect physiological functions in the host. Thus the probiotics have been extensively studied and commercially explored in many different products in the world.

2. The gastrointestinal microbiota

The human gastrointestinal tract (GIT) is composed of several connected organs that are involved in nutrient conversion and providing energy sources from the food absorbed. This complex system has a well-known anatomical architecture that is approximately 7 m long, comprising a 300 m^2 surface area in adults. From the mouth to the colon, there exists a complex microbiota consisting of facultative and strict anaerobes, including streptococci, bacteroides, lactobacilli and yeasts. The microbial community, inhabitants of these organs, is collectively called the gut microbiota and is composed of a myriad of microbial cells that outnumber the cells number of our body by a factor of at least 10. In addition, there is a great diversity of species, some of which have not yet been identified or cultured, and understanding the dynamics of this population is a challenge to the TGI ecologist (Zoetendal, et al., 2008).

However, the development of molecular biology since the discovery of polymerase chain reaction (PCR) by Mullins and Fallona (1996) up to the current approaches "omics", have focused on molecular characterization of specific environments such as GIT, as well as their interactions with probiotic bacteria. The knowledge of this microbiota that is underway has increased our understanding of the beneficial effects of probiotics on the human and animal health.

Prior to birth, humans develop in a sterile environment, the womb. However, the rupture of the membranes at delivery exposes the neonate to a wide variety of microorganisms, especially those that colonize the GIT, forming its microbiota. Over the course of human development, this microbiota undergoes variations according to the stages of life and related to the habits and habitats to which the individual is exposed (Isolauri et al., 2004, Tiihonen et al., 2010).

The most dramatic changes in the composition of the intestinal microbiota occur during childhood. During the first days of life, the microorganism population is unstable and tends to stabilize with breastfeeding or the intake of breast milk substitutes. The greatest change in this composition, however, occurs through weaning and the introduction of solid foods (Favier, et al., 2002). Throughout adulthood, the intestinal microorganisms are relatively stable; however, this stability is reduced in the elderly (Tiihonen et al., 2010). These changes can be attributed to dietary restrictions, changes in eating habits and the increased incidence of diseases and concomitant medication use, all of which are found with increasing age (Gill, et al., 2001, Tiihonen et al., 2010).

Early studies focused on the changes in the human intestinal microbiota, reporting the reduction of anaerobes and bifidobacteria and an increase of enterobacteria in the elderly (Mitsuoka, 1990). However, recent studies suggest a lower stability and increased diversity of the intestinal microbiota with advancing age (Hopkins and Macfarlane, 2002; Maukonen, et al., 2008, Tiihonen et al., 2010).

The human GIT has a very complex microbial ecosystem that is based on competition and symbiosis (Mackie et al., 1999) and consists of at least 400 to 500 different bacterial species, approximately 10^{14} cells (Ott et al., 2004; Zoetendal, et al., 2004; Zoetendal, et al., 2008). This population, have the composition which differs both along the gastrointestinal tract as along the lumen to the mucosa (Tiihonen et al., 2010), is affected by several factors; some are determined by the interactions between genetic, environmental or disease factors to which the individual is exposed, the diet, the secretion of mucus, digestive enzymes and intestinal peristalsis. As a result, each individual has a unique characteristic microbiota (Isolauri, et al., 2004; Ley, et al., 2006).

The lack of bacteria in the upper GI tract (esophagus, stomach and duodenum) is related to the composition of the luminal medium (acid, bile and pancreatic secretions). In addition, the propulsive motor activity at the end of the ileum eliminates most of ingested microorganisms, preventing the stability of bacterial colonization in the lumen (Guarner and Malangelada, 2003). However, the lower portion of the GI tract, comprising the lower duodenum and small and large intestines, contains a complex and dynamic microbial ecosystem, with a high density of live bacteria reaching concentrations 10^{11}-10^{12} cells / g of luminal contents, which corresponds to 1.5 kg of microorganisms (Moore and Holdeman, 1974; Whitman et al., 1998; del Piano, 2006).

In this environment, the permanent organisms that colonize and grow in the place where they are found are considered to be autochthonous microbiota, whereas the non-native or transients are those that are vehicled by food, water and environmental components passing through the region (Ley, et al., 2006)

The TGI naturally has the function of protecting the body against pathogens and / or toxic metabolites. This protection is ensured by a number of factors, including saliva, gastric acids, peristalsis, mucus, intestinal proteolysis, intestinal microbiota balance and the epithelial membranes with intercellular junctional complexes (Ouwehand et al., 2002).

The intestinal mucosa forms an interface between the body and luminal environment, with the function of allowing the passage of nutrients and simultaneously acting as a barrier against microorganisms, toxins and other undesirable substances. The mucus produced by the goblet cells exerts this protective function; therefore, the barrier effect is guaranteed by the physical, chemical and functional epithelium integrity (Cenčič and Langerholc, 2010).

The balance of the microbiota has been gaining special attention from the scientific community for years, and many studies indicate and confirm a close relationship between intestinal disbioses and microbial imbalance in addition to intestinal homeostasis and the maintenance of the equilibrium of the intestinal microbiota. Some microorganisms, particularly the probiotics, have great importance in maintaining this balance.

Although feces are the most available sample to investigate the intestinal microbiota, it is questionable how well the fecal microorganisms represent the intestinal microbiota, as they originate from the lumen and the distal colon. Indeed, the composition of intestinal microbiota is different in the lumen and the distal colon and throughout the TGI and mucosa. Moreover, the TGI has large species diversity and consists of known species and those that have not yet been cultured.

Thus, for more precise information on the gut microbial population, appropriate samples should be collected during endoscopies or surgical procedures; however, such invasive procedures are rather unsuitable and rarely used in research. Moreover, the scarcity of information on the effects of anesthetics and disinfectants used in these procedures suggests the possibility that they may compromise the investigation (Isolauri et al., 2004, Ley, et al., 2006). Therefore, the approaches of studies on human intestinal microbiota are usually based on *in vitro* or animal models and in the evaluation of the fecal microbiota.

3. Probiotics and human health

Evidence derived from clinical and mechanistic studies indicate that the health benefits promoted by healthy lifestyle habits and the consumption of a balanced diet rich in bioactive ingredients are approaches that are increasingly attractive to the pharmaceuticals and food industries in addition to the general population.

Functional foods are defined as any substance or constituent of a food that, in addition to providing basic nutrition, promotes metabolic and / or physiological health benefits (Walker, et al., 2006). These foods are broadly grouped into conventional foods, bioactive substances and synthesized foods. In general, the term refers to a food that has been modified to become functional or that naturally contains bioactive compounds. Functional foods are also known as designer foods, medicinal foods, nutraceuticals, therapeutic foods, superfoods, foodiceuticals, and medifoods (Shah, 2007).

Thus, the probiotic microorganisms capable of promoting beneficial effects in a host for the production of bioactive compounds or the equilibrium of the intestinal tract are often associated with functional foods.

There is a long history of health claims concerning the beneficial effects of probiotic microorganisms in food, particularly lactic acid bacteria and bifidobacteria. Additionally, studies involving probiotic microorganisms have distinguished these microbes into different categories according to their mode of action, the aims of the administration of the probiotics and their mode of administration in addition to claims regarding legal regulations.

4. Probiotics: History and concepts

There is a long history of the beneficial effects that some microbes have on human health, with the effects of lactic acid bacteria, in particular, being the earliest record. In a Persian version of the Old Testament (Genesis 18:8), there is a statement that "Abraham owed his longevity to the consumption of sour milk." In 76 BC, the Roman historian Plinius recommended the administration of fermented dairy products for the treatment of gastroenteritis (Bottazzi, 1983; Schrezenmeir and de Vrese, 2001). However, studies involving these organisms and their clinical effects in animals and humans are contemporary and are based on the production of beneficial substances and / or the promotion of a balance that favors the microbial host.

The concept of beneficial microorganisms has been attributed to *Lactobacillus bulgaricus* when, more than a century ago, Elie Metchnikoff (1905) emphasized the importance of lactobacilli in the intestinal microbiota, providing the properties of health maintenance and longevity to the host. However, the term "probiotics" was proposed decades later by Lilly and Stillwell (1965) in reference to a substance secreted by protozoa in symbiosis. Parker (1974) first used the concept of combining the use of organisms or substances, as opposed to antibiotics, to contribute to the balance of intestinal microbiota. The term was later popularized by Fuller (1989) and defined as a probiotic food supplement based on live microorganisms with beneficial effects to the host in balancing the intestinal microbiota.

The term "probiotic" has been widely used, and according to research data, the general concept has experienced subtle changes. Schrezenmeyer and Vrese (2001) defined the term as a microorganism preparation or product containing viable microorganisms in sufficient numbers to change, through colonization, the host microbiota, thus promoting health benefits. Salminen and colleagues (1999) defined probiotics as microbial cell preparations (or components thereof), viable or inactive, with favorable effects on the health and welfare of the host. Clearly, the benefits must be evaluated in terms of the mechanisms and properly established and documented selection criteria.

Some authors also extend the action of probiotics to inactive cells and argue that both living and dead cells in probiotic products can produce beneficial biological responses (Havenaar et al., 1992; Adams, 2010). This approach will open new perspectives for research, for example, about the amount of cells needed and the proportion viable / non-viable cells required to obtain the desired effect. Furthermore, the use of inactivated probiotics has attractive advantages, such as consumption safety and the possibility of products with long shelf lives (Adams, 2010).

The WHO and FAO (World Health Organization and Food and Agriculture Organization of the United Nations) maintain the general concept that defines probiotics as live microorganisms that, when consumed in adequate amounts, confer benefits to the host (FAO / WHO, 2001). In Brazil, according to the currently enforced food legislation, the National Sanitary Surveillance Agency (ANVISA) has set forth that, to produce the claimed benefits of a probiotic food, the product should contain a minimum number of viable probiotic cells between 10^8 and 10^9 Colony-former unit (CFU) per day (BRAZIL, 2008).

However, the scientific community agrees that the effects of probiotic microorganisms can vary depending on the species, the quantity ingested and the physiologic characteristics of the host. Furthermore, the current evidence suggests that the probiotic effects are species and even strain specific (FAO/WHO 2002, Isolauri et al., 2004, Tiihonem et al., 2010).

Although the *Lactobacillus* and *Bifidobacterium* have been predominantly used as commercial probiotic; the market is not exclusive to these genera. In fact, is growing the number of probiotic foods available to the consumer. Based in scientific studies, the regulatory agencies worldwide have characterized a broader number of microorganisms as probiotics. Because the technologic and functional characteristics, these strains have been used in food and pharmaceutical industry (Table 1).

Species	Strains
Bacillus lactis	DR10™
Bifidobacterium adolescentis	ATCC 15703, 94-BIM
B. animalis and subspecies *lactis*	BB-12™
B. breve	Yakult™, BB-03
B. bifidus	BB-11™
B. essencis	Danone™
B. infantis	Shirota™, Immunitas™, 744,
B. lactis	Bb-02, Lafti™, DSM-B94, DR10™
B. laterosporus	CRL431
B. longum	BB536, SBT2928, UCC 35624
Lactobacillus acidophilus	LA-1™, La-5™, NCFM, DDS-1, SBT-2062, La-14™
L. casei	Shirota™, LC™, DN1114001™, Immunitas™
L. casei shirota	Yakult™
L. casei ssp. *defensis*	Danone™
L. lactis	L1A,
L. fermentum	RC-14
L. helveticus	B02
L. johnsonii	La1™
L. paracasei	CRL 431™
L. plantarum	299 Probi™, LP115™, Lp01
L. rhamnosus	GG, GR-1, LB21, 271Probi™
L. reuteri	SD2112
L. salivarius	Ls-33
Sacharomyces cereviseae	NCYC Sc 47
S. boulardii	17™

Table 1. Some microorganisms used as probiotic cultures in commercial products.

The characterization of the probiotic species or strain is supported by the screening of resistance to the adverse conditions in the TGI. To survive passage through the TGI, microbes must exhibit a resistance to a low pH, bile and pancreatic enzymes. Moreover, it is desirable that these bacteria display adhesion to the intestinal mucosa and pathogen exclusion abilities and have positive effects on the immune system of the host; evidently, these bacteria should be non-pathogenic and have a GRAS (Generally Recognized as Safe) status. These effects are evaluated by intensive *in vitro* and *in vivo* approaches. The intestinal homeostasis relies upon the equilibrium between substance absorption, secretion and the barrier capacity of the digestive epithelium, and probiotic microorganisms are highly related to homeostasis.

The scientific literature reports sufficient data to demonstrate that the benefits attributed to probiotics are inherent to their population increase in a given environment, concomitant with a decrease in potentially pathogenic bacteria (Jankovic et al., 2010). In addition, it had been demonstrated for more than 20 years that the intestinal microbiota of healthy individuals is altered with the ingestion of probiotics in favor of lactobacilli and bifidobacteria species. Although such alterations and the beneficial effects in healthy populations remains a complex issue (Saxelin, et al., 1993; de Vrese, et al., 2006), there is a consensus on the association of disbioses with chronic inflammatory diseases (Manichanh, et al., 2006), obesity (Ley et al, 2006) and allergies (Penders et al., 2006).

There has been a substantial increase in the number of articles published in scientific journals and the lay press, focusing on the popularity of probiotic foods and their effects. Thus, the FAO and WHO (2001) established scientific committees, whose discussions have produced a document with guidelines designed to regulate the characterization of potentially probiotic microorganisms, ensure the security of the host, assess at the technological and commercial aspects of probiotics in food and evaluate the clinical proof of the expected effects on individuals (FAO / WHO, 2002).

Understanding the complex microbial system of the TGI will help to characterize the intestinal microbial community and recognize the mechanisms by which probiotics exert their effect on the health of humans and animals. Although the traditional culture-based and phenotypic techniques used to study this complex ecosystem are unfeasible, the current molecular approaches have increased our knowledge of the structure, diversity, interactions and mechanisms that influence the dynamics of the TGI microbial community.

5. Molecular approaches in the study of probiotic microorganisms

Studies of the gut microbiota that use traditional techniques for microbial cultivation are supported by phenotypic analysis based on morphological and biochemical characterization. These techniques are laborious, time consuming, subject to misinterpretation and identify only approximately 40% of the microbiota (Carey et al., 2007). The reasons for the deficiencies in microorganism cultivation by traditional methods include ignorance of the nutritional profile of the microorganism, culture medium selectivity, the stress imposed by cultivation procedures, the need to restrict the environmental conditions and difficulties in simulating the host interactions with microorganisms (Zoetendal, et al., 2004).

Research involving nucleic acid analysis indicated that the majority of the bacteria in a variety of ecosystems are different from those related on the cultivation methods. This idea led to the development and application of methods that are independent of the culture medium to study complex microbial ecosystems (Zoetendal, et al., 2004; Zoetendal, et al., 2008).

The polymerase chain reaction (PCR), developed by Kary Mullis in the 1980's, enabled the *in vitro* production of multiple copies of specific DNA sequences, without cloning (Alberts, et al. 1994). Variations of this technique have targeted the needs and advancement of biotechnology.

In addition, LAB and bifidobacteria have received much attention, especially since the creation of the consortium for sequencing the genome of these microorganisms (Lactic Acid Bacteria Genome Consortium - LABGC) in the U.S., which culminated in the genomic sequencing of industrial strains and many other relevant sequences that are ongoing. Currently, fourteen strains of *Lactobacillus* and ten strains of *Bifidobacterium* have been sequenced by the consortium (http://www.jgi.doe.gov/genome-projects/) or by private initiatives, such as *B. longum* NCC2705 in 2002, the first bifidobacteria to have its genome sequenced, and *L. plantarum* WCSF1 in 2003, the first *Lactobacillus* sequenced (O'Flaherty et al., 2009).

Molecular approaches to evaluate phylogeny and genetic and chemotaxonomic identification of the related species have been used successfully in the recent decades in studies. Additionally, the use of bioinformatics tools, along with access to available databases in the GenBank / NCBI (National Center for Biotechnology Information) has boosted research, aiming at the development of strategies for identifying target species (Costa, et al., 2011).

The significant increase in the knowledge of the structure, diversity and factors that influence the GIT microbial community dynamics and the mechanisms by which probiotics may influence intestinal homeostasis are due to ready access to their genomic data. Furthermore, the variety of *in vivo* immunoassays aimed at elucidating the physiological effects of probiotic therapies and the molecular approaches based on PCR, ribotyping and hybridization with probes have also contributed to the body of knowledge (Vaugh, et al., 2005; Walker, et al., 2006; Carey, et al., 2007).

Molecular markers are successfully employed in this environment favorable to the identification of probiotic microorganisms, and various molecular techniques have become powerful tools. Indeed, there are a large number of techniques that are useful for the identification of *Lactobacillus* in different environments (Moreira et al., 2005, , Costa, et al, 2011), the detection of pathogenicity genes in foods (Bottero, et al., 2004), the identification and quantification of bifidobacteria via real-time PCR (Masco, et al, 2007). In addition, proteomic approaches evaluates the expression of genes of interest or changes in the host related to the effects of the microorganisms (Yuan, et al. 2008; O'Flaherty, et al., 2010).

The use these of technologies associated with suitable choice of the molecular marker is very important to differentiate closely species. The *rec*A gene has provided a high discriminatory ability for the differentiation of the LAB species (Figure 1).

Furthermore, studies employing the sequence analysis of genes for microorganism genotyping, such as ribosomal small subunit rRNA (SSU rRNA), allow the expansion of the knowledge about the diversity of the gut microbiota. Only a decade after the introduction of genotyping, the number of species molecularly detected in the TGI has greatly exceeded the number of species accessible using cultivation-dependent methods (Zoetendal, et al., 2008).

One of the most increasingly used techniques is real-time PCR or quantitative PCR (qPCR), which identifies and quantifies organisms of interest. This technique, coupled with the use of specific primers, has proven to be an accurate method that is suitable for the identification and quantification of microorganisms (Matsuki, et al., 2004). Moreover, this tool provides new perspectives in the studies of the diversity, abundance and dynamics of the intestinal ecosystem (Walker, et al, 2006; Masco, et al., 2007, Zoetendal, et al., 2008). Thus, the qPCR has attracted attention for being a reliable method that is highly sensitive for the detection and quantification of many organisms in different environments.

The technique is based on the traditional technology of PCR in combination with compounds that fluoresce at certain wavelengths, making it possible to monitor the amount of PCR products generated in each reaction cycle (Wittwer et al., 1997; Vitali, et al., 2003).

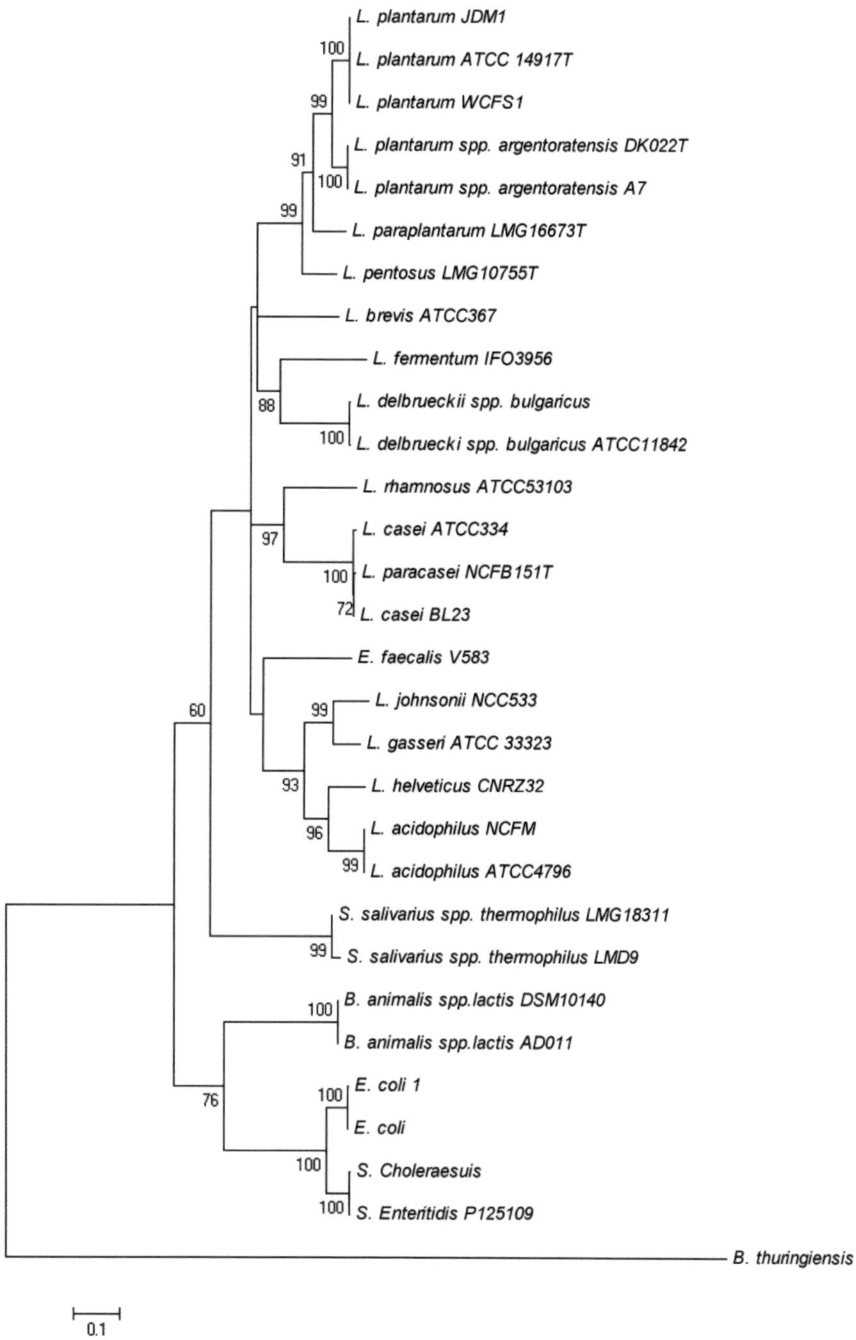

Figure 1. The phylogenetic tree *consensus* from the *rec*A gene sequence comparisons, demonstrating the relationship of closely related species of the BAL, *Bifidobacterium* and enteric bacteria. The tree was constructed with the Neighbor-Joining method and the Clustal W algorithm. Genetic distances were computed by using Nei's coefficient. Bootstrap values based on 1000 replicates are provided at branch nodes. The *B. thuringiensis* sequence was included as an out-group sequence.

The methods used for qPCR are based on the measurement of the fluorescence emitted as a function of the value of the cycle threshold (CT) or Crossing Point (CP), which is posteriorly related to mathematical expressions for absolute or relative quantification (Livak and Schmittgen, 2001; Pfaffl, 2001). The CT method is directly related to the quantity of the amplification product in the PCR reaction.

The normalization of the target gene using an endogenous standard is recommended (Pfaffl, 2001). The addition of a gene normalizer to the reaction is highly recommended and is intended to correct any concentration differences or defects in DNA extraction.

Normalization ensures that fluctuations in the signal strength due to impurities or amounts of target DNA below the detection limit are taken into account during the analysis. However, the uniformity of the normalizer gene during the entire process or the stability of the expression during the experimental treatment must be confirmed (Kubista, et al., 2006; Marcelino, 2009; Hofstätter, et al., 2010; Dang and Sun 2011).

In the development of these methodologies, some alternatives have emerged to further refine the technique. Thus, the application of qPCR to quantify only viable cells (vqPCR) has eliminated one of the common criticisms in the quantification of probiotic microorganisms because qPCR does not distinguish between viable and non-viable cells.

The approach of vqPCR is based on the differentiation between viable cells and non-viable cells based on the membrane integrity. Theoretically, the selective dye used can only penetrate the permeable membranes of dead cells and intercalate extracellular DNA. The dye makes the DNA unavailable for amplification due to the presence of an azide group, present in such substances as ethidium monoazide (EMA) or propidium monoazide (PMA), which allows cross-links between the dye and DNA after the exposure to high-intensity visible light. The photolysis of these substances (EMA and PMA) converts the azide group into a highly reactive nitrene radicals, which can react with any organic molecule in its vicinity, including DNA, which then cannot be amplified by PCR (Varma, et al., 2007; Fitipaldi, et al., 2010).

Unquestionably, the use of genetic tools has accelerated the knowledge and understanding of the complexities found in the intestinal microbiota and their interactions. It is now possible to gain a better comprehension of the role of these organisms, including the accurate analysis of the functionality of probiotics and to obtain strains lacking one or more proteins (O'Flaherty and Klaenhammer, 2010). Furthermore, it is obvious that an understanding of the interactions through the cross-talk between the intestinal microbiota and its host would expand the knowledge of the relationship between microbiota and their effects on health.

There is an increasing tendency of probiotic studies to focus on metagenomics (Ventura, et al, 2009), which is, which is defined as the study of the collection of genomes of an ecosystem and can be used to study the phylogenetic, physical and functional properties of microbial communities. From the point of view of functional genomics, the application of these technologies provides a wealth of information and fosters research aiming at a better understanding of probiotic microorganisms and their effects.

6. Market prospects

The interest in functional foods is directly related to the growing appreciation of the quality of life and disease prevention because these foods affect specific functions or systems in the human body and are intended to complement basic nutrition (Shah, 2007). The food industry has developed a variety of new products containing active ingredients that promote consumer health.

The global market for functional foods generated US$ 32.07 billion in 2000 and US$ 68.39 billion in 2005; in 2010, the total surpassed US$ 150 billion and continues to expand (Granato et al., 2010). Latin America is considered an emerging market, and despite the general lack of nutritional knowledge by the population, Brazil and Mexico are potential trade markets for probiotics (Granato et al., 2010). The probiotic market in Latin America grew 32% per year between 2005 and 2007 (Crowley, 2008), and the annual sales growth rate of probiotic drinks and yogurts was 5% between 2006 and 2011 (Özer and Kirmaci, 2010).

Among the functional foods, dairy products with functional claims accounted for almost 43% of the world market between 2005 and 2010 (Özer and Kirmaci, 2010). In this scenario, the use of probiotic microorganisms in foods and pharmaceuticals had such an increase in the world market, that the sales reached $ 15 billion in 2007, amounted to $21.6 billion in 2010 with the prospect of more than $ 31.1 billion by 2015 (Agheyisi, 2011).

Following the same trend, the sales of foods with functional claims reached $ 500,000 in 2007, representing 1% of the total spending on food in Brazil (Cruz, et al, 2007; Granato, et al., 2010). According to Euromonitor International Consulting data released in 2010, the market for products for intestinal microbiota balance had a 60% growth in Brazil in five years, from R $ 57 million in 2004 to $ 92 million in 2009 (Revista Fator, 2011).

Over the last two decades, a substantial number of research studies have supported the idea that health can be affected by the daily consumption of probiotic foods (Heyman and Menárd, 2002), with clinical evidence demonstrating the actual effect of these organisms to the host. These data provide an understanding of the mechanisms by which probiotic microorganisms survive the passage through the GI tract to interact with the resident microbiota and affect physiological functions in the host. In addition, there is much investigation into both the classification of probiotic strains and the production technologies and regulation of the products.

To assess the impact of scientific research in the dissemination and consolidation of the benefits of probiotics in the diet, a search was conducted using three major scientific databases (Isi Web of Knowledge, Pub Med and Scopus). The search was restricted to two periods, and the key word "probiotic" in the title of the publication was used as a selection parameter. On average, there were 410 publications from 1991 to 2001, whereas 2406 records were found in the 2002 to 2011 period. According to the database Isi Web of Knowledge, in a period of ten years (2001 to 2011), 2686 publications were available in the database, documenting 791 patents, and 100 records are related to reviews; all of the other publications are related to primary literature.

Clearly, in a market in which product development should meet the needs of the consumer, it is important that scientific research does not neglect the technology and logistical aspects or the regulations of each country. The market will continue to grow as consumers maintain an interest in the products offered; however, the credibility of the product is based on its effects, which are often supported by scientific studies and the "know-how" from the manufacturer.

The majority of probiotic products on the market includes *Lactobacillus* and/or *Bifidobacterium* species but also yeasts; *Bacillus* and *Enterococcus* are common in these products. (Shah, 2007; Gaggìa, et al.; 2010). Some probiotics marketed in food and pharmaceutical industries worldwide, the microorganisms involved, category of product, manufacturer and country from origin are listed in Table 2.

Country	**Category**	**Commercial brand**	**Manufacturer**	**Probiotic**
Australia	Ingredient	Probiomics	Bioxyne	*L. fermentum* VRI003 (PCC)
Brazil	Capsules	Floratil	Merck	*S. boulardii*
	Sachet	Fiber Mais Flora	Nestlé	*Lactobacillus reuteri*
		Activia	Danone	*B. animalis* DN173010
		Actimel	Danone	*L. casei defensis*
		Batavito	Batavo	*L. casei*
		Chamyto	Nestlé	*L. jonhsonii/ L. helveticus*
		Danito	Danone	*L. casei*
		Leite fermentado	Paulista	*L. casei*
		Leite fermentado	Parmalat	*L. acidophilus/L. casei/ B. animalis* subsp. *lactis*
		Sofyl	Yakult	*L casei shirota*
		Vigor club	Vigor	*L. acidophillus/L. casei*
		Yakult	Yakult	*L. casei shirota*
	Traditional yogurt or Drinking yogurt	Activia	Danone	*B. animalis* DN173010
		Biofibras	Batavo	*B. animalis/ L. acidophilus*
		Lective	Vigor	*B. animalis* subsp. *lactis*
		Nesvita	Nestlé	*B. animalis* subsp. *lactis*
	Cheese	Equilibra	Danubio	*B. animalis*
		SanBIOS	Coop. Santa Clara	*B. lactis*
Canadian/ USA	Capsules, Fermented Milk, Fermented soy and Fermented rice	Bio-K+ CL1285	Bio-K+ International	*L. acidophilus CL1285* & *L. casei LBC80R*
France	Fermented milk	DanActive	Danon	*L. casei* DN-114 001 ("L. casei Immunitas")
	Ingredient	Lacteol	Laboratory Houdan	*L. acidophilus* LB

Country	Category	Commercial brand	Manufacturer	Probiotic
Japan	Fermented milk	Yakult	Yakult	*L. casei* Shirota *B. breve* strain Yakult
	Ingredient		Morinaga Milk Industry Co Ltd	*B. longum* BB536
Sweden	Juice	GoodBelly ProbiMage	Probi	*L. plantarum 299v*
	Juice, Cultured Milk	ProViva	Probi	*L. plantarum 299v*
	Juice	Bravo Friscus/ ProbiFrisk	Probi	*L. plantarum HEAL 9* & *L. paracasei 8700*
	Probiotic chewable tablets or drops	Protectis	Biogaia	*L. reuteri* ATCC 55730
Switzerland	Fermented milk	LC1	Nestlé	*L. johnsonii* Lj-1 same as NCC533
UK		Floralfit (Blend strains)		*L. acidophilus* La14, *L. casei* Lc11, *L. salivarius* Ls-33, *L. plantarum* Lp115, *L. rhamnosus*, Lr-32, *B. lactis* Bl-04 & *B. longum* Bl05.
USA	Capsules	Align	Procter & Gamble	*Bifidobacterium infantis 35624*
	Supplement	GanedenBC	Ganeden Biotech	*Bacillus coagulans GBI-30, 6086*
	Capsules	Florajen products (blend or only strain)	American Lifeline, Inc	*L. acidophilus, Bifidobacterium lactis & B. longum*
	Yogurt	Activia	Dannon	*B. animalis* DN173 010
	Dairy products	Frozen Kefir, Milk cultured kefir, Traditional kefir	Lifeway Foods Inc.	*L. acidophilus, L. casei, L. lactis, L. rhamnosus, L. reuteri, B. breve, B. lactis, L. plantarum, B. longum, Leuconostoc cremoris, Sacharomyces florentinus* & *Streptococcus diacetylactis.*
USA/ Finland	Supplements or Chewable for kids	Culturelle, Dannon Danimals	Valio and Dannon	*L. rhamnosus* GG ("LGG")

Table 2. Foods and pharmaceuticals probiotics products, marketed worldwide, manufacturer and microorganism in use.

The probiotic market is constantly changing. Within this context, many innovations that direct studies and the functional microorganism market are being applied, and there are prospects of many other approaches because this branch of science is challenging.

What factors are predominant in the probiotics development? From the standpoint of marketing, the factors are a fully expanding open field, and the numbers reflect this scenario. From a scientific standpoint, many studies are aimed at the selection of strains with desirable and efficient characteristics, invoking the research of new effects and the elucidation of the mechanisms of action. The application of techniques for the functional genomics of probiotic bacteria certainly will accelerate the development of such products (de Vos, et al., 2004).

Furthermore, advances in the "genomic era" will increasingly be used to answer questions related to interactions between organisms. Molecular biology and its tools, the access to molecular databases, and the speed with which information is disclosed are essential for accurate identification of the benefits attributed to probiotics.

Most of the probiotic bacteria currently marketed were selected on basis on their technological properties, but not for their ability to confer health benefits. However, is evident that the use and development of novel technologies aiming products that meet the nutritional and physiological requirements desired by the target population is a priority among research and Industries. Additionally, the "feedback" among science, industry and the market is extremely important, and is desired that there is dynamism between these sectors.

Author details

Lucia Helena S. Miglioranza*
Department of Food Science and Technology – UEL, Brazil

Giselle Nobre Costa **
Graduate Program in Food Science
Master's Degree in Dairy Technology – UNOPAR, Brazil

7. References

Adams, C.A. (2010). The probiotic paradox: live and dead cells are biological response modifiers. N*utrition Research Reviews*, 23 (1): 37-46.

Agheyisi, R. (2011). The probiotics market: Ingredients, supplements, foods - FOD035B. Food & Beverage, June 2008. Available from http://www.bccresearch.com/report/probiotics-market-ingredients-foods-fod035c.html. Acessed 05 May 2012.

Alberts, B.; Lewis, J.; Ralf, M.; Roberts, K.; Watson, J.D. (1994). *Molecular Biology of the Cell*. 3.ed. New York : Garland Publishing; 1294p.

* Corresponding Author

** *Financial Support from CNPq Research Agency*

Bottazzi, V. (1983). Food and feed production with microorganisms. *Biotechnology*,5: 315-63.

Bottero, M.T.; Dalmasso, A.; Soglia, D.; Rosati, S.; Decastelli, L.; Civera, T. (2004). Development of a multiplex PCR assay for the identification of pathogenic genes of *Escherichia coli* in milk and milk products. *Molecular and Cellular Probes*, 18: 283-288.

BRAZIL, 2008. Agência Nacional de Vigilância Sanitária/ANVISA. Resolução RDC nº 278, de 22 de setembro de 2005, atualizada em Julho de 2008. Legislação para alimentos com alegações de propriedades funcionais e ou de saúde, novos alimentos/ingredientes, substâncias bioativas e probióticos. Available in: http://www.anvisa.gov.br/alimentos/comi ssoes/tecno_lista_alega.htm. Acessed 20 December 2011.

Carey, C.M.; Kirk, J.L.; Ojha, S. and Kostrzynska, M. (2007). Current and future uses of real time polymerase chain reaction and microarrays in the study of intestinal microbiota, and probiotic use and effectiveness. *Canadian Journal of Microbiology*, 53: 537-550.

Cencič, A. and Langerholc, T. (2010). Functional cell models of the gut and their applications in food microbiology - A review. *International Journal of Food Microbiology* 141: S4-S14.

Costa, G. N; Vilas-Bôas, G. T.; Vilas-Boas, L. A; Miglioranza, L. H. S. (2011). *In silico* phylogenetic analysis of lactic acid bacteria and new primer set for identification of *Lactobacillus plantarum* in food samples. *European Food Research Technology*, 233 (2): 233-241.

Crowley, L. (2008). Danisco meets growing South American demand for probiotics.

Cruz, A.G.; Faria, A.F.J.; Van Dender, A.G.F. (2007). Packaging system and probiotic dairy foods. *Food Research International*, 40:951-956.

Dang, W. and Sun, L. (2011). Determination of internal controls for quantitative real time RT-PCR analysis of the effect of *Edwardsiella tarda* infection on gene expression in turbot (Scophthalmus maximus). *Fish & Shellfish Immunology*, 30: 720 -728.

de Vos, W.M.; Bron, P.A. and Kleerebezem, M. (2004).Post-genomics of lactic acid bacteria and other bacteria to discover gut functionality. *Current Opinion and Biotechnology*

de Vrese, M.; Winkler, P.; Rautenberg, P.; Harder, T.; Noah, C.; Laue, C.; Ott, S.; Hampe, J.; Schreiber, S.; Heller, K.; Schrezenmeir, J. (2006). Probiotic bacteria reduced duration and severity but not the incidence of common cold episodes in a double blind, randomized, controlled trial. *Vaccine*, 24:6670-6674.

Del Piano, M.; Morelli, L.; Strozzi, G.P.; Allesina, Barba, S.; Deidda, F.; Lorenzini, P.; Ballar′e, M.; Montino, F.; Orsello, M.; Sartori, M.; Garello, E.; Carmagnola, S.; Pagliarulo, M.; Capurso, L. (2006). Probiotics: from research to consumer. *Digestive and Liver Disease*, 38 Suppl. (2): 248-255.

FAO - Food and Nutrition paper 85. Probiotics in food - Health and nutritional properties and guidelines for evaluation. (2006). Report of a Joint FAO-WHO expert consultation on evaluation of health and nutritional properties of probiotics in food including powder milk with live lactic acid bacteria. Cordoba, Argentina, 2001.

FAO - Food and Nutrition paper 85. Probiotics in food - Health and nutritional properties and guidelines for evaluation. (2006). Report of a Joint FAO-WHO working group on drafting guidelines for the evaluation of probiotics in food. Ontario, Canadá, 2002.

Favier, C.F.; Vaughan, E.E.; De Vos, W.M.; Akkermans, A.D.L. (2002). Molecular monitoring of succession of bacterial communities in human neonates. *Applied and Environonmental Microbiology*, 68 (1): 219 -226.

Fittipaldi, M.; Codony, F.; Adrados, B.; Camper, A.K. and Morató, J. (2010). Viable Real-Time PCR in environmental samples: Can all data be interpreted directly? *Microbiology and Ecology*, 61:7-12.

Fuller, R. (1989). Probiotics in man and animals. *Journal Applied Bacteriology*, 66: 365-378.

Gaggìa, F.; Mattarelli, P. and Biavati. B. (2010). Probiotics and prebiotics in animal feeding for safe food production. *International Journal of Food Microbiology*, v. 141, p. S15-S28.

Gill, H.S.; Darragh, A.J.; Cross, M.L. (2001). Optimizing immunity and gut function in the elderly. *The Journal of Nutrition Health & Aging*, 5 (2): 80-91.

Granato, D.; Branco, G.F.; Cruz, A.G.; Faria, J.A.F. and Shah, N.P. (2010). Probiotic dairy products as functional foods. Comprehensive Reviews in Food Science and Food Safety, v. 9, p. 455-470.

Guarner, F. and Malangelada, J.R. (2003). Gut flora in health and disease. *Lancet*, 361: 512-519.

Havenaar, R.; Ten Brink, B. and Huis In't Veld, J.H.J. (1992). Selection of strains for probiotic use. In: Fuller, R (Ed.), Probiotics: the scientific basis. (1st Edn.), Chapman and Hall, London, p: 209-224.

Heyman, M. and MÉNARD, S. (2002). Probiotic microorganisms: How they affect intestinal athophysiology. *Cellular and Molecular Life Sciences*, 59: 1151 - 1165

Hofstätter, H.B.; Tschernutter, M. and Kunert, R. (2010). Comparison of hybridization methods and real-time PCR: their value in animal cell line characterization. *Applied and Environmental Microbiology*, 87:419-425.

Hopkins, M.J.; Macfarlane, G.T. (2002). Changes in predominant bacterial populations in human faeces with age and with *Clostridium difficile* infection. *Journal of Medical Microbiology*, 51: 448-454.

Isolauri, E.; Salminen, S.; Ouwehand, A.C. (2004). Probiotics. *Best Practice & Research Clinical Gastroenterology*, 18 (2): .299-313.

Jankovic, I.; Sybesma, W.; Phothirath, P.; Ananta, E.; Mercenier, A. (2010). Application of probiotics in food productos – Challenges and new approaches. *Current Opinion in Biotechnology*, 21: 175-181.

Kubista, M.; Andrade, J.M.; Bengtsson, M.; Forootan, A.; Jonák, J.; Lind, K.; Sindelka, R.; Sjöback, R.; Sjögreen, B.; Strömbom, L.; Ståhlberg, A.; Zoric, N. (2006). The real-time polymerase chain reaction. *Molecular Aspects of Medicine*, 27: 95-125.

Ley, R.E.; Peterson, D.A. and Gordon, J.I. (2006). Ecological and evolutionary forces shaping microbial diversity in the human intestine. *Cell*, 124: 837-848.

Lilly, D.M.E.; Stillwell, R.H. (1965). Probiotics: Growth promoting factors produced by microorganisms. *Science*, 147: 747-748.

Livak, K.J. and Schmittgen, T.D. (2001). Analysis of relative gene expression data using real-time quantitative PCR and the $2^{-\Delta\Delta CT}$ method. *Methods* 25, 402- 408.

Mackie, R.I.; Sghir, A.; GASKINS, H.R. (1999). Developmental microbial ecology of the neonatal gastrointestinal tract. *American Journal of Clinical Nutrition*, 69: 1035S-1045S.

Manichanh, C.; Rigottier-Gois, L.; Bonnaud, E.; Gloux, K.; Pelletier, E.; Frangeul, L.; Nalin, R.; Jarrin, C.; Chardon, P.; Marteau, P.;Roca, J. and Dore, J. (2006). Reduced diversity of faecal microbiota in Crohn's disease revealed by a metagenomic approach. *Gut*, 55: 205-211.

Marcelino, F.C. (2009). PCR quantitative em tempo real: Metodologias e aplicações. *Curso*: Embrapa Soja, Londrina. pg 1-29.

Masco, L.; Vanhoutte, L.; Temmerman, R.; Swings, J.; Huys, G. (2007). Evaluation of real-time PCR targeting the 16S rRNA and *rec*A genes for the enumeration of bifidobacteria in probiotic products. *International Journal of Food Microbiology*, 113 : 351-357.

Matsuki, T.; Watanabe, K.; Fujimoto, J.; Kado, Y.; Takada, T.; Matsumoto, K.; and Tanaka, R.. (2004). Quantitative PCR with 16S rRNA-Gene-Targeted Species-Specific Primers for Analysis of Human Intestinal Bifidobacteria. *Applied and Environmental Microbiology*, 39: 167-173.

Maukonen, J.; Mätö, J.; Kajander, K.; Mattila-Sandholm, T.; Saarela, M. (2008). Diversity and temporal stability of fecal bacterial populations in elderly subjects consuming galacto-oligosaccharide containing probiotic yogurt. *International Dairy Journal*, 18: 386-395.

Metchnikoff, E. (1905). Immunity in infective diseases. Cambridge University Press, New York.

Mitsuoka, T. (1990). Bifidobacteria and their role in human health. *Journal of Industrial Microbiology*, 6: 263 -267.

Moore, W. E. C.; Holdeman, L.V. (1974). Human fecal flora: The normal flora of 20 Japanese–Hawaiians. Applied Microbiology. 27 (5): 961-979.

Moreira, J.L.S.; Mota, R.M.; Horta, M.F.; Teixeira, S.M.R.; Neumann, E.; Nicoli, J.R. and Nunes, A.C. (2005). Identification to the species level of *Lactobacillus* isolated in probiotic prospecting studies of human, animal or food origin by 16S-23S rRNA restriction profiling. *BMC Microbiology*, 5(15): 1-9.

Mullins, K. B. and Faloona, F. A. (1987). Specific synthesis of DNA in vitro via a polymerase-catalyzed chain reaction. Methods in Enzymology, 155: 335-350.

O'Flaherty, S. and Klaenhammer, T.R. (2010). The role and potential of probiotic bacteria in the gut, and the communication between gut microflora and gut/host. *International Dairy Journal*, 20: 262-268.

O'Flaherty, S.; Goh, Y.J. and Klaenhammer, T.R. (2009). Genomics of probiotic bacteria, In: Prebiotics and Probiotics Science and Technology. Charalampopoulos, D.; Rastall, R.A (Eds). 1st Ed. Springer-Verlag New York INC, New York, 1265p.

Ott, SJ.; Musfeldta, M.; Timmisb, KN.; Hampea,J.; Wenderothb, DF.; Schreiber, S. (2004). *In vitro* alterations of intestinal bacterial microbiota in fecal samples during storage. *Diagnostic Microbiology and Infectious Disease*, 50: 237-245.

Ouwehand, A.C.; Salminen, S. and Isolauri, E. (2002). Probiotics: an overview of beneficial effects. *Antonie van Leeuwenhoek*, 82: 279-289.

Özer, B.H. and Kirmaci, H.A.(2010). Functional milks and dairy beverages. *International Journal of Dairy Technology*, 63 (1): 1-15.

Parker, R.B. (1974). Probiotics, the other half of the antibiotic story. *Animal Nutrition Health*, 29: 4-8.

Penders, J.; Stobberingh, E.E.; Van Den Brandt, P.A.; Thijs, C. (2007). The role of the intestinal microbiota in the development of atopic disorders. *Allergy*, 62:1223-1236.

Pfaffl, M.W. (2001). A new mathematical model for relative quantification in real-time RT-PCR. *Nucleic Acids Research*, 29 (9): 2002-2007.

REVISTA FATOR. Conteúdo online, http://www.revistafator.com.br/ver_noticia.php?not=142778. Acessed 05 April 2012.

Salminen, S.; Ouwehand, A.C.; Benno, Y. and Lee, Y.K. (1999). Probiotics: how should they be defined? *Trends in Food Science and Technology*, 10: 107-110.

Saxelin M.; Ahokas, M.; Salminen, S. (1993). Dose-response on the fecal colonization of *Lactobacillus* strainGG administered in 2 different formulations. *Microbial Ecology in Health and Disease*, 6(3): 119-122.

Schrezenmeir, J.; de Vrese, M. (2001). Probiotics, prebiotics, and synbiotics – Approaching a definition. *American Journal Clinical Nutrition*, 73: 361-364.

Shah, N.P. (2007). Functional cultures and health benefits. *International of Dairy Journal*, 17: 1262-1277.

Tiihonen, K.; Ouwehand, A.C.; and Rautonen, N. (2010). Human intestinal microbiota and healthy ageing. *Ageing Research Reviews*, 9: 107-116.

Varma, M.; Field, R.; Stinson, M.; Rukovets, B.; Wymer, L. and Haugland, R. (2009). Quantitative real-time PCR analysis of total and propidium monoazide resistant fecal indicator bacteria in wastewater. *Water Research*, 43:4790-4801.

Vaughan, E. E.; Heilig, H.G.H.J. ; Ben-Amor, K. and de Vos, W.M. (2005). Diversity, vitality and activities of intestinal lactic acid bacteria and bifidobacteria assessed by molecular approaches. *FEMS Microbiology Reviews*, 29: 477-490.

Ventura, M.; O'Flaherty, S.; Claesson, M.J.; Turroni, F. Klaenhammer, T.R.; Van Sinderen, D.; O'Toole, P.W. (2009). Genome-scale analyses of health promoting bacteria: probiogenomics. *Nature Reviews*, 7: 61-71.

Vitali, B.; Candela, M.; Matteuzzi, D.; Brigidi P. (2003). Quantitative Detection of Probiotic *Bifidobacterium* Strains in Bacterial Mixtures by Using Real-time PCR. *Systematic Applied Microbiology* 26: 269-276.

Walker, W.A.; Goulet, O.; Morelli, L.O.; Antoine, J.M. (2006). Progress in the science of probiotics. *European Journal of Nutrition*, 45, (Suppl 1):1-18.

Whitman, W.B.; Coleman, D.C. and WIEBE, W.J. (1998). Prokaryotes: the unseen majority. *Proceedings of the National Academy of Sciences of the United States of America*, 95: 6578-6583.

Wittwer, C. T., Herrmann, M.G., Moss, A.A., Rasmussen, R. P. (1997). Continuous fluorescence monitoring of rapid cycle DNA amplification. *BioTechniques* 22, 130 -138.

Yuan, J.; Wang, B.; Sun, Z.; Bo, X.; Yuan, X.; He, X.; He, X.; Zhao, H.; Du, X.; Wang, F.; Jiang, Z.; Zhang, L.; Jia, L.; Wang, Y.; Wei, K.; Wang, J.; Zhang, X.; Sun, Y.; Huang, L. And Zeng, M. (2008). Analysis of hostinducing proteome changes in *Bifidobacterium longum* NCC2705 grown *in vivo*. *Journal of Proteome Research*, 7: 375-385.

Zoetendal, E.G.; Rajilić-Stojanović, M. and de Vos, W.M. (2008). High-throughput diversity and functionality analysis of the gastrointestinal tract microbiota. *Gut*, 57(11): 1605-1615.

Zoetendal, Eg.; Collier, Ct.; Koike, S.; Gaskins, Hr.; Mackie, Ri.; Gaskins, HR. (2004). Molecular ecological analysis of the gastrointestinal microbiota: A review. *The Journal of Nutrition*, 134: 465-472.

Chapter 16

Saccharomyces cerevisiae var. *boulardii* – Probiotic Yeast

Marcin Łukaszewicz

Additional information is available at the end of the chapter

http://dx.doi.org/10.5772/50105

1. Introduction

The discovery and study of the budding yeast *Saccharomyces cerevisiae var. boulardii* (*Sb*) is strictly related to the concept of health promoting microorganisms from food. The first most well-known and popularized throughout Europe assumption of health promoting food containing living microorganisms was yogurt. Appointed in 1888 by Louis Pasteur, Ilya Ilyich Metchnikov working in Paris developed a theory that aging is caused by toxic bacteria in the gut and that lactic acid could prolong life which resulted in popularization of yogurt consumption. Metchnikov received with Paul Ehrlich the Nobel Prize in Medicine in 1908 for his previous work on phagocytosis, which probably promoted his idea of today's so called functional food further and triggered subsequent research on this subject. Scientists started to look for traditional, regional food products considered good for health. One of them was French scientist Henri Boulard who was in IndoChina in 1920 during cholera outbreak. He observed that some people chewing the skin of lychee and mangosteen or preparing special tea did not develop the symptoms of cholera. This observation lead Henri Boulard to the isolation of a tropical strain of yeast named *Saccharomyces boulardii* (*Sb*) from lychee and mangosteen fruit, which is nowadays the only commercialized probiotic yeast.

At the beginning Metchnikov's theory that lactic acid bacteria (LAB) can prolong life was disputable and some researchers doubted it. For example, Cheplin and Rettger (1920)[1] demonstrated that Metchnikov's strain, today called *Lactobacillus delbrueckii subsp. bulgaricus,* could not live in the human intestine. A scientific discussion to be constructive needs to forge and define new argued ideas. Such a new term was probiotic (*pro* Lat. "for" and *biotic* Greek adjective from *bios* "life") used by Werner Kollath [2] in 1953 to denote, in contrast to harmful antibiotics, all good organic and inorganic complexes. It is attributed to Lilly and Stillwel [3] who in 1965 defined the probiotic as "a substance produced by one microorganism stimulating the growth of another microorganism". The significance of probiotics evolved over time. In 1974 Parker [4] defined it as "organisms and substances which contribute to intestinal microbial balance", in 1989 Fuller [5] defined it as "a live microbial feed supplement, which beneficially affects the host animal by improving its intestinal microbial balance", in 1996 Sanders [6] wrote "Probiotics, simply defined, are microbes consumed for health effect. The term probiotic is used in food applications. The term biotherapeutic is used in clinical applications". To distinguish between the beneficial effect of living microorganism from organic compounds the term prebiotic was introduced for the latter. However, living microorganism during their growth always affect the chemical composition of the environment, thus it is very difficult to differentiate the influence of microorganisms alone from the impact of organic compounds resulting from microorganisms metabolism. Unfortunately, there is still no general agreement to clear-cut definition of the probiotic.

Irrespectively of the assumed probiotic definition, during over half of the last century the conducted research showed that *Sb* may be beneficial for human health [7]. As mentioned before, the history of probiotic strain started in 1920. Henri Boulard after his return to France patented isolated strain and in 1947 sold it to Biocodex company created for its production. *Sb* was registered as a drug for the first time in 1953 and so far it is the only registered eukaryotic probiotic microorganism.

While commercial application of *Sb* in diarrhea treatment has been steadily growing since 1953, the scientific interest measured in number of publications was in a "lag phase" during next 30-40 years. While searching year by year Scholar Google for "boulardii" it has been found out that there were no articles after 1953, with the first appearing in 1977. From 1977 to 1986 only 17 publications were found.

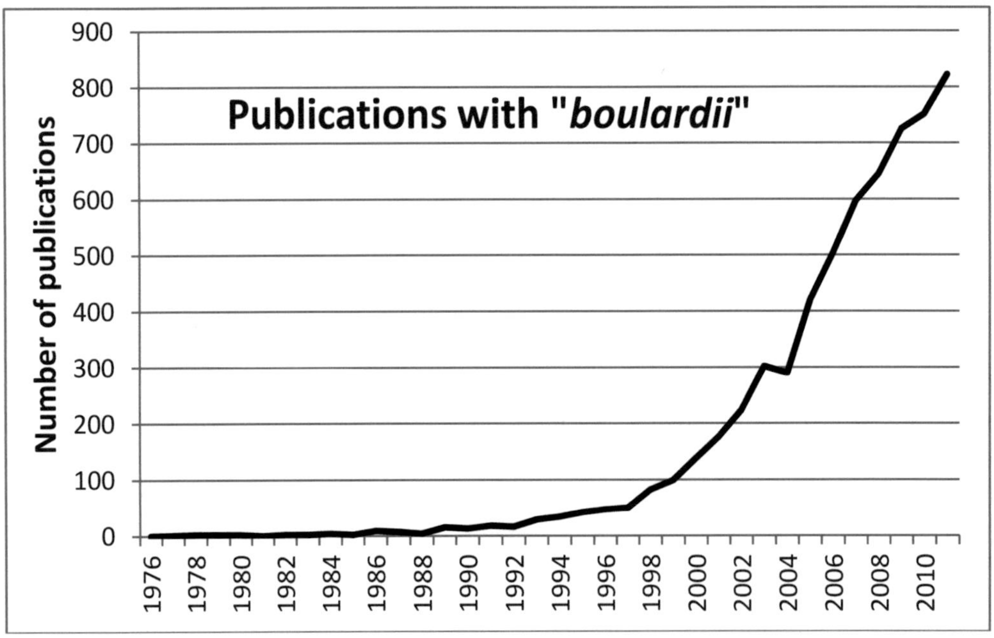

Figure 1. Number of peer-reviewed publications mentioning *Sb* from 1976 to 2010.

The publication of two successive patents in 1986 "Method for preventing or treating pseudo-membranous colitis" [8] and in 1987 "Method for the treatment of amoebiasis" [9], was probably the turning point. Thus, while in 1987 there were only 7 publication in 2011 there were already 822.

Why has *S. boulrdii* been so extensively studied in recent years? Diarrheal diseases are of various origin and continue to represent a major threat to global health. In developing countries, mortality due to acute diarrhea, especially in children, is alarmingly high. In contrast, in developed countries, mortality caused by diarrheal diseases may be considered marginal, yet these disorders are burdensome and widespread, having important economic impact on the society. While the majority of physicians regard probiotics as a very effective therapy they still criticize the lack of useful clinical guidelines [10]. Indeed, beside various origins of diarrheal diseases there are various mechanisms of action of *Sb* and the fields of its potential application are growing.

2. Systematic classification

Sb is a close relative to baker's yeast *Saccharomyces cerevisiae,* the most wildly used organism in industrial microbiology for various foodstuff products. The most obvious difference between them is unusually high optimal growth of *Sb* in the temperature (37 °C) which fits very well with the temperature of human body. Another important feature is better survival at acid pH. Yeast classification was traditionally based on their physiological and biochemical profiles. However, it fails to distinguish between several yeast species or cultivars and it resulted in a discussion whether *Sb* should remain as species or subspecies of *S. cerevisiae*. Thus, molecular methods have been developed and applied to yeast strain typing and identification.

Table 1. summarizes some results of the investigation on differences and similarities between *Sb* and *S. cerevisiae*. Although *S. boulardii* strains differ significantly from laboratory strains of *S. cerevisiae* [11], finally according to current nomenclature like International Code of Botanical Nomenclature (ICBN) *Sb* yeasts should be referred to as *S. cerevisiae var. boulardii* [16]. It should, however, be pointed out, that strongly reduces ability to mate with other strains puts *Sb* on the evolutionary way of becoming a separate species.

Taxonomy attempts to achieve two aims: first the classification that reflects the evolution and phylogenetic relationships and second the development of procedures enabling identification of individual species. Thus, independently of discussion on the systematic classification, very important issue concerns identification of species which affect human health. *S. cerevisiae* appears to be an emerging pathogen [17-19]. Thus, recent research concentrates on unravelling features determining the pathogenicity. It has been shown that yeast pathogenicity correlates with survival in oxidative stress [20] which could be triggered by transcription factor Rds2 [21] or activation of MAP kinases and variability in the polyglutamine tract of Slt2 [22]. Probiotic properties are also strain specific, which is the case for *Sb* used as probiotic [11, 12]. Thus there is a need for a valuable molecular markers able to distinguish among strains and establish appropriate methods for the identification of probiotic strains of the *Sb*. *Such a method could be, for example,* microsatellite length polymorphism, having a discriminatory power of 99% [15, 23], restriction fragment length polymorphism [24], full genome hybridization [14], randomly amplified polymorphic DNA [25], GeneChip hybridization [11], artificial neural network–assisted Fourier-transform infrared spectroscopy [26] or multilocus enzyme electrophoresis [27]. These identification methods enable the discrimination between various strains but are not necessarily related to mechanisms of probiotic activity. Metabolic footprinting using mass spectrometry may be useful in this regard. Using gas chromatography–time of flight–mass spectrometry there was good correlation with genetic method of strains classification. Probiotic strains of *Sb* showed tight clustering both genetically and metabolically. The major discriminatory metabolites were: trehalose, myo-inositol, lactic acid, fumaric acid and glycerol 3-phosphate [28]. Next very important step is very to find out a functional relationship between specific DNA and probiotic action.

Sb	***S. cerevisiae***
Higher optimal growth temperature (~37 °C)	Lower optimal growth temperature (~30 °C)
Higher resistance to low pH [11]	Lower resistance to low pH [11]
The karyotypes of *Sb* are very similar to those of *S. cerevisiae*	Typing RFLPs or PCR- (ex 5.8S rDNA) failed to distinguish *Sb* from *S. cerevisiae* [12]
Do not use galactose [13]	Use galactose
Asporogenous in contrast to S. cerevisiae but may produce fertile hybrids with of *S. cerevisiae* strains [11]	Sporogenous
Lost all intact Ty1/2 elements [14].	Contains several Ty1/2 elements
Microsatellite typing shows genotypic differences [15]	
Trisomic for chromosome IX	There are stable strains with various ploidy

Table 1. Summary of some differences and similarities between *Sb* and *S. cerevisiae.*

3. Medical applications of *Sb*

Several published medical studies have shown the efficacy and safety of *Sb* for various disease indications both in adults and children. Regarding the medical use, different indications of *Sb* could be listed: prevention of antibiotic-associated diarrhea, recurrent *Clostridium difficile*-associated diarrhea and colitis, Travellers' diarrhea, acute bacterial and viral diarrhea, diarrhea in patients with total enteral feeling, anti-inflammatory bowel diseases, supplement to hydration in adults and children, against diarrhea associated with the use of antibiotics. [29-32]. There is an increasing number of publications showing the results of double blind clinical trials, clinical guidelines including new applications of the usage of *Sb* and new potential fields. While the number of different possible application of *Sb* in prevention and treatment of health disorders is growing, it is crucial to determine mechanisms of its action. This is an extremely difficult task due to a high number of factors involved in the observed health benefits.

Use for disease	Dose (mg/d)	Duration	Adjunct to
Prevention of antibiotic associated diarrhea	500-1000	During antibiotics with additional 3 days to 2 weeks after	Nothing
Prevention of Traveller's diarrhea	250-1000	Duration of trip (3 weeks)	Nothing
Enteral nutrition-related diarrhea	2000	8-28 days	Nothing
H. pylori symptoms	1000	2 weeks	Standard triple therapy
Treatment of *Clostridium difficile* infections	1000	4 weeks	Vancomycin or metronidazole
Acute adult diarrhea	500 - 750	8-10 days	Nothing
Inflammatory bowel disease	750-1000	7 weeks to 6 months	Mesalamine
Irritable bowel syndrome	500	4 weeks	Nothing
Giardiasis	500	4 weeks	Metronidazole
HIV-related diarrhea	3000	7 days	Nothing

Table 2. Summary of recommendations for clinical use of *Sb* in adults [7]

Mechanisms of action of *Sb*

While *Sb* has been proven effective in several double-blind studies and yeast preparation is sold in several countries as both a preventive and therapeutic agent, not all mechanisms of its action have been studied [7, 33] and the new ones are still being discovered. Figure 2 summarizes most of the postulated mechanism of *Sb* activity which are :

a. antimicrobial effect,
b. nutritional effect,
c. inactivation of bacterial toxins,
d. quorum sensing,
e. trophic effects,
f. immuno-modulatory effects
g. anti-inflammatory effects,
h. cell restitution and maintenance of epithelial barrier integrity.

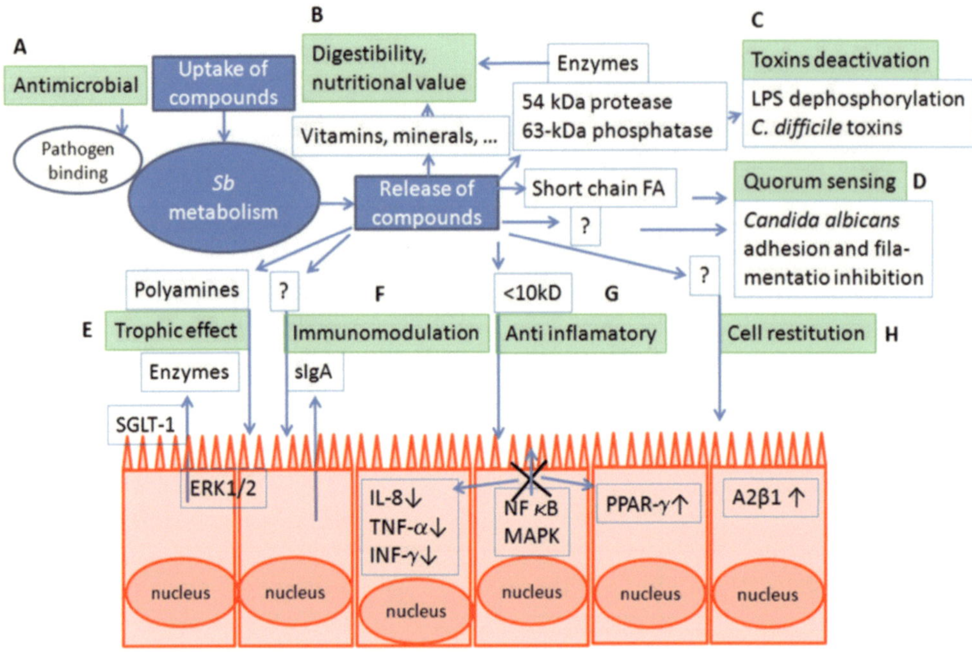

Figure 2. *Sb* possible probiotic mechanisms of action.

This enumeration is somehow artificial because one factor may play multiple roles and various processes may act synergistically.

Antimicrobial effect may be exerted through several mechanisms. One of them is irreversible binding of bacteria to the yeast surface, preventing their adhesion to the mucous membranes and subsequent elimination by the flow Fig. 2A. It has been shown that *Sb* has the ability to bind enteric pathogens to mannose as a receptor [34]. That yeast viability was not necessary for the adhesion phenomenon. Furthermore it has been shown that in the binding beside process beside mannose-containing glycoprotein other proteins are involved [35]. On the other hand, Tasteyre et al. [36] showed that the yeast could inhibit adherence of *C. difficile* to cells, thanks to its proteolytic activity and steric hindrance. This is exerted trough the modification the eukaryotic cell surface receptors involved in adhesion of *C. difficile.* Other mechanisms exerting antimicrobial effect are utilization of substrates, modification of the environment and release of various compounds.

Some of the released compounds are **quorum sensing** molecules Fig. 2D. They influence metabolism and properties of microorganisms, for example, reducing the ability to adhesion or filamentation, which are both important factors of strains pathogenicity [37, 38].

Sb may inhibit pathogens through action on microbial virulence factors. Invasive properties of *Salmonella enterica* serovar Typhimurium is closely related to the flagellum-associated motility. Study performed on human colonic cells infected by the *S. enterica* showed that in in presence of *Sb* the pathogen motility was reduced [39]. *Sb* also acts by **inactivation of**

bacterial toxins (Fig. 2C). For example, it has been shown that the 63-kDa protein phosphatase from *Sb* is able to dephosphorylate and partially inactivate the endotoxin (LPS) of *Escherichia coli*. Furthermore, *Sb* releases *in vivo* a 54-kDa serine protease that digests toxins A and B of *Clostridium difficile* and the BBM receptor of toxin A [40].

Sb also influences the growth of gut microflora and the host by its metabolism (uptake of substrates and release of products or multitude of cell components by dying cells). Yeast from *Saccharomyces* genus has been used in human and animal **nutrition** (Fig 2 B) for many centuries and new applications in agro-industries are being developed [41]. They are of high nutritional value and are used as food additive or to obtain some products such as white or "living" beer. Yeast cells are also a well-known source of proteins, B-complex vitamins, nucleic acids, vitamins and minerals, including a biologically active form of chromium known as glucose tolerance factor [42]. In some countries a mixture of a small amount of baker yeast with water and sugar was prepared as a drink for children as supplementation with B-complex vitamins. *Sb* releases during its passage through gastrointestinal track at least 1500 various compounds [43]. While vitamins are necessary exogenous organic compound which must be ingested, enzymes may help to transform bigger to smaller compounds which may be absorbed by brush border. The brush border is the structure formed by microvilli increasing the cellular surface area responsible for secretion, absorption, adhesion and transduction of signals. Within the gastrointestinal tract brush border is crucial for digestion and nutrient absorption. It has been shown that oral administration of probiotic strain of *Sb* enhanced the activities of the brush border ectomembrane enzymes (ex. sucrase, maltase, trehalase, lactase, aminopeptidase, alkaline phosphatase), carriers (sodium glucose cotransporter-1) receptors of immunoglobulins (the secretory component) or secretory immunoglobulin A [44-48]. *Sb* cells contain substantial amounts of polyamines (spermidine and spermine) which are known to affect cell maturation, enzyme expression and membrane transport, thus polyamines were suggested as mediators in the intestinal trophic response [45]. **Trophic effect** Fig 2E has been recently reviewed by Buts [33, 43]. It was postulated that *Sb* upgraded intestinal function by at least three mechanisms:

- The endoluminal secretion of various compounds by yeast
- The secretion of polyamines triggering transduction trophic signals and resulting in enhanced synthesis of brush border membrane proteins (enzymes and carriers).

Clinical studies have shown that oral administration of *Sb* is effective in treatment of inflammatory bowel diseases and control of irritable bowel syndrome. There are several possible mechanisms of **anti-inflammatory effect** (Fig 2G) recently reviewed by Pothoulakis [49], Vandenplas [50] or Vohra [51]. The activity may be exerted through released compounds which modifies epithelial cell and mucosal immune system signaling pathways. One mechanism of anti-inflammatory effect could be exerted by producing by *Sb* a heat stable low molecular weight (<1 kDa) soluble factor [52]. The mechanism is based on blocking activation of nuclear factor-kappa B (NF-◎B) and mitogen activated protein kinase (MAPK). As a result, pro-inflammatory compounds such as interleukin 8 (IL-8), tumor necrosis factor alpha (TNF-◎) and interferon gamma (IFN- ◎) are down regulated. *Sb* and *Sb* secreted-protein(s) inhibit production of pro-inflammatory cytokines by interfering with the global mediator of inflammation nuclear factor ◎B, and modulating the activity of the mitogen-activated protein kinases ERK1/2 [53] and p38 [54]. *Sb* activates expression of peroxisome proliferator-activated receptor-gamma (PPAR-γ) that protects the digestive track from inflammation. *Sb* also suppresses 'bacteria overgrowth' and host cell adherence as described before.

Another mechanism mutually related to inflammation and synergistically acting with antimicrobial and anti-inflammatory effect [55] is **immunomodulation** Fig 2F. Sb has been shown to increase secretion of immunoglobulin A [48]. Immunomodulation could be exerted by *Sb* interactions with mucosal dendritic cells. Dendritic cells discriminate commensal microorganisms from potential pathogens and take part in maintaining the balance between tolerance and active immunity. They respond to intestinal inflammation and thus are potential target in inflammatory bowel disease [56]. Dendritic cells produce regulatory cytokines and induce T cells. *Sb* inhibits dendritic cell-induced activation of naïve T cells [57] and may interfere with IBD pathogenesis by trapping T cells in mesenteric lymph nodes [58].

Bacterial infections leading to inflammatory bowel diseases results in intestinal epithelial cell damage. Thus, remission of these diseases requires both the cessation of inflammation and the **cell restitution** Fig. 2H within the damaged epithelium, which is effected by enterocyte migration. It has been recently shown that *Sb* accelerate enterocyte migration by secretion of motogenic factors that enhance cell restitution through the dynamic regulation of $\alpha2\beta1$ integrin activity [59].

4. Effect of *Sb* on the virulence factors of *Candida albicans*

While there is quickly increasing information on the influence of *Sb* on the bacterial origin diseases the interaction between *Sb* and *Candida albicans* is much less studied filed. *C. albicans* is a dimorphic fungus growing commensally in the gastrointestinal tract of healthy humans. Switching between morphotypes is a striking feature enabling the growth as budding yeast or as filamentous forms. It also enables in formation of complicate biofilm structures [60]. The transition between morphotypes contributes to the overall virulence and constitutes potential target for development of antifungal drugs.

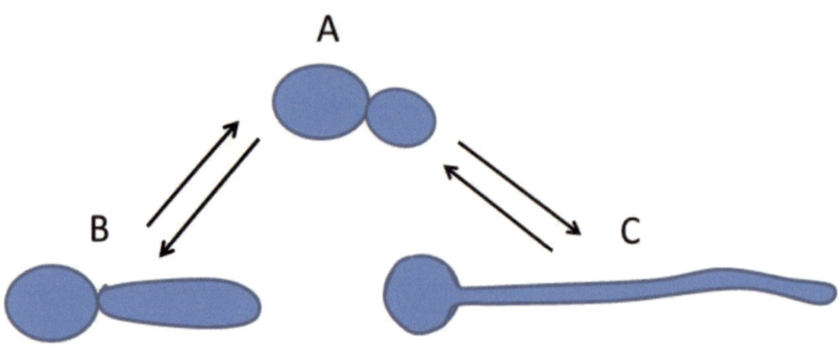

Figure 3. Phenotypic switching of *C. albicans*. (A) budding yeast, (B) pseudohyphal growth, (C) hyphal growth.

Pathogenicity of *C. albicans*, like all pathogens, is conditioned by their virulence. All the features that improve microbial colonization of host cells, multiplication and spread within organism or toxins production, which in turn leads to the development of the disease are called virulence factors. The virulence of *C. albicans* include: the ability to adhesion, biofilm formation and production of coatings, as well as morphological transformation, phenotypic switching and secretion of proteases, phospholipases and endotoxin [61]. Morphogenesis in *C. albicans* can be impaired by various small molecules such as farnesol, fatty acids, sugars, rapamycin, geldanamycin, histone deacetylase inhibitors, and cell cycle inhibitors recently reviewed by Shareck [62]. Affecting metabolism of the C. albicans may also have indirect effect as for example synergism with the antifungal drugs. Indeed metabolic state of the cell greatly affects activity of the PDR pump activity [63].

It has been shown that both live *Sb* cells and the extract from *Sb* culture filtrate diminish *C. albicans* adhesion to and subsequent biofilm formation [38]. Thus, independently of the trophic relationships, for example, elimination from the medium of carbon source (sugars) or polyunsaturated fatty acids [64], *Sb* releases to the medium active compounds. These compounds in dose dependent manners are able to inhibit switching from budding yeast to hyphae growth. The extract prepared from *Sb* culture filtrate was showed to contain 2-phenylethanol, caproic, caprylic and capric acid. The highest activity reducing candidal virulence factors was capric acid (C10:0), which is responsible for inhibition of hyphae formation. It also reduced candidal adhesion and biofilm formation, though three times less than the extract. Thus *Sb* release to the medium other factors, not yet identified, suppressing *C. albicans* adherence [37]. Capric acid acts through the activation of cAMP pathways and Hog1 kinase cascade, reducing the expression of genes of *C. albicans* virulence. Capric acid reduces *CSH1*, *INO1*, *HWP1* transcripts. *CSH1* encodes a protein related to the hydrophobicity surface of the fungal cell wall and is involved in adhesion. *INO1* encodes an enzyme involved in the biosynthesis of inositol, which is a precursor components on the surface of the cell wall of *C. albicans* involved in the virulence. *HWP1* (Hyphal Wall Protein) encodes protein present in hyphae and pseudogyphae and involved in adhesion and biofilm formation. Besides inhibition of *C. albicans* adhesion to epithelial cell lines, *Sb* living cells and compounds released to the medium, reduced cytokine-mediated inflammatory host response. In fact the IL-8 gene expression was suppressed in *C. albicans*-infected epithelial cells by the compounds released to the medium by *Sb* [65].

It is clear that *Sb* secretes many compounds and some of them may act as quorum sensing and modulate growth of other microorganisms including other eukaryotes such as *C. albicans*. Besides identified compounds and their activity it is clear that there are still other biologically active compounds produced by *Sb* which remain to be discovered [65].

5. Conclusions and future perspectives

A century after publication of the Metchnikov's theory there is no more doubt concerning potential positive influence of selected strains of living microorganisms in the ingested food on human health. Nevertheless, the discussion has been even more turbulent and the topic is "hot", as seen from increasing number of scientific publications. In contrast to most of the registered drugs which are single, pure compounds, *Sb* has been shown to be beneficial through various mechanism. Thus, due to very complex and various interactions it is exiting research area with a lot of things to discover, but it is also extremely laborious, costly and time consuming. There is a number of organisms in traditional fermented food that has been shown to be potentially beneficial for human health. However, probiotic properties are strain specific and very often not well characterized. Properties of strains from the same species may be very different, thus for human health benefits potential probiotic strain should be very well characterized. It is clear that microflora of the human body is very complex and it is important to maintain appropriate homeostasis, which may be unbalanced by use of antibiotics. This can be prevented or regained by use of appropriate probiotics. However, due to the complexity of the possible interactions and various mechanisms of actions it is very difficult to register and commercialize a new probiotic. It is a great challenge to resolve this bottleneck in the future.

Author details

Marcin Łukaszewicz
Faculty of Biotechnology, University of Wrocław, Wrocław, Poland

6. References

[1] Cheplin HA, Rettger LF. Studies on the Transformation of the Intestinal Flora, with Special Reference to the Implantation of Bacillus Acidophilus: II. Feeding Experiments on Man. Proceedings of the National Academy of Sciences of the United States of America. 1920;6(12):704-5. Epub 1920/12/01.

[2] Kollath W. [Nutrition and the tooth system; general review with special reference to vitamins]. Deutsche zahnarztliche Zeitschrift. 1953;8(11):Suppl 7-16. Epub 1953/06/01. Ernahrung und Zahnsystem; Ubersichtsreferat mit besonderer Berucksichtigung der Vitamine.

[3] Lilly DM, Stillwell RH. Probiotics: Growth-Promoting Factors Produced by Microorganisms. Science. 1965;147(3659):747-8. Epub 1965/02/12.

[4] Parker RB. Probiotics, the other half of the antibiotic story. Animal Nutr Health. 1974;29:4-8.

[5] Fuller R. Probiotics in man and animals. The Journal of applied bacteriology. 1989;66(5):365-78. Epub 1989/05/01.

[6] Sanders ME. Probiotic cultures and human health. In: Germfree life and its ramifications Proceedings of the XIIth International Symposium on Gnotobiology Honolulu USA, June 24-28, 1996 (Eds: Hashimoto, K, Sakakibara, B, Tazume, S, and Shi-mizu, K) XIIth ISG Publishing Committee, Shiozawa. 1996:91-5.

[7] McFarland LV. Systematic review and meta-analysis of *Saccharomyces boulardii* in adult patients. World journal of gastroenterology : WJG. 2010;16(18):2202-22. Epub 2010/05/12.

[8] Hublot B, Levy RH, inventors; Method for preventing or treating pseudo-membranous colitis patent 4595590. 1986.

[9] Gayral PG, Hublot BM, inventors; Method for the treatment of amoebiasis patent 4643897. 1987.

[10] Guandalini S. Commentary on 'Probiotics for treating acute infectious diarrhoea'. Evidence-Based Child Health: A Cochrane Review Journal. 2011;6(6):2024-5.

[11] Edwards-Ingram L, Gitsham P, Burton N, Warhurst G, Clarke I, Hoyle D, et al. Genotypic and physiological characterization of *Saccharomyces boulardii*, the probiotic strain of *Saccharomyces cerevisiae*. Applied and Environmental Microbiology. 2007;73(8):2458-67.

[12] McCullough MJ, Clemons KV, McCusker JH, Stevens DA. Species identification and virulence attributes of *Saccharomyces boulardii* (nom. inval.). Journal of Clinical Microbiology. 1998;36(9):2613-7.

[13] McFarland LV. *Saccharomyces boulardii* is not *Saccharomyces cerevisiae*. Clinical Infectious Diseases. 1996;22(1):200-1.

[14] Edwards-Ingram LC, Gent ME, Hoyle DC, Hayes A, Stateva LI, Oliver SG. Comparative genomic hybridization provides new insights into the molecular taxonomy of the *Saccharomyces* sensu stricto complex. Genome Research. 2004;14(6):1043-51.

[15] Malgoire JY, Bertout S, Renaud F, Bastide JM, Mallié M. Typing of *Saccharomyces cerevisiae* clinical strains by using microsatellite sequence polymorphism. Journal of Clinical Microbiology. 2005;43(3):1133-7.

[16] Rajkowska K, Kunicka-Styczyńska A. Phenotypic and genotypic characterization of probiotic yeasts. Biotechnology & Biotechnological Equipment. 2009;23(2):662-5.

[17] Heitman J. *Saccharomyces cerevisiae*: an emerging and model pathogenic fungus. : ASM Press; 2006.

[18] McCusker JH, Clemons KV, Stevens DA, Davis RW. Genetic characterization of pathogenic *Saccharomyces cerevisiae* isolates. Genetics. 1994;136(4):1261-9.

[19] Skovgaard N. New trends in emerging pathogens. International Journal of Food Microbiology. 2007;120(3):217-24.

[20] Diezmann S, Dietrich FS. *Saccharomyces cerevisiae*: Population divergence and resistance to oxidative stress in clinical, domesticated and wild isolates. PloS one. 2009;4(4):e5317.

[21] Diezmann S, Dietrich FS. Oxidative stress survival in a clinical *Saccharomyces cerevisiae* isolate is Influenced by a major quantitative trait nucleotide. Genetics. 2011;188(3):709-22.

[22] de Llanos R, Hernández-Haro C, Barrio E, Querol A, Fernández-Espinar MT, Molina M. Differences in activation of MAP kinases and variability in the polyglutamine tract of Slt2 in clinical and non-clinical isolates of *Saccharomyces cerevisiae*. Yeast. 2010;27(8):549-61.

[23] Hennequin C, Thierry A, Richard GF, Lecointre G, Nguyen HV, Gaillardin C, et al. Microsatellite typing as a new tool for identification of *Saccharomyces cerevisiae* strains. Journal of Clinical Microbiology. 2001;39(2):551-9.

[24] Zerva L, Hollis RJ, Pfaller MA. *In vitro* susceptibility testing and DNA typing of *Saccharomyces cerevisiae* clinical isolates. Journal of Clinical Microbiology. 1996;34(12):3031-4.

[25] Mitterdorfer G, Mayer HK, Kneifel W, Viernstein H. Clustering of *Saccharomyces boulardii* strains within the species *S. cerevisiae* using molecular typing techniques. Journal of Applied Microbiology. 2002;93(4):521-30.

[26] Büchl NR, Hutzler M, Mietke-Hofmann H, Wenning M, Scherer S. Differentiation of probiotic and environmental *Saccharomyces cerevisiae* strains in animal feed. Journal of Applied Microbiology. 2010;109(3):783-91.

[27] Duarte FL, Pais C, Spencer-Martins I, Leäo C. Distinctive electrophoretic isoenzyme profiles in *Saccharomyces* sensu stricto. International Journal of Systematic and Evolutionary Microbiology. 1999;49(4):1907-13.

[28] MacKenzie DA, Defernez M, Dunn WB, Brown M, Fuller LJ, de Herrera SRMS, et al. Relatedness of medically important strains of *Saccharomyces cerevisiae* as revealed by phylogenetics and metabolomics. Yeast. 2008;25(7):501-12.

[29] Surawicz CM. [The microbiota and infectious diarrhea]. Gastroenterologie clinique et biologique. 2010;34 Suppl 1:S29-36. Epub 2010/10/05. Le microbiote dans les diarrhees infectieuses.

[30] Czerucka D, Piche T, Rampal P. Review article: yeast as probiotics -- *Saccharomyces boulardii*. Alimentary pharmacology & therapeutics. 2007;26(6):767-78. Epub 2007/09/05.

[31] Im E, Pothoulakis C. [Recent advances in *Saccharomyces boulardii* research]. Gastroenterologie clinique et biologique. 2010;34 Suppl 1:S62-70. Epub 2010/10/05. Progres recents dans la recherche sur *Saccharomyces boulardii*.

[32] Szajewska H, Horvath A, Piwowarczyk A. Meta-analysis: the effects of *Saccharomyces boulardii* supplementation on *Helicobacter pylori* eradication rates and side effects during treatment. Alimentary pharmacology & therapeutics. 2010;32(9):1069-79.

[33] Buts JP, De Keyser N. Interaction of *Saccharomyces boulardii* with intestinal brush border membranes: key to probiotic effects? Journal of pediatric gastroenterology and nutrition. 2010;51(4):532-3. Epub 2010/08/14.

[34] Gedek. Adherence of *Escherichia coli* serogroup 0 157 and the *Salmonella* Typhimurium mutant DT 104 to the surface of *Saccharomyces boulardii*. Mycoses. 1999;42(4):261-4.

[35] Tiago FC, Martins FS, Souza EL, Pimenta PF, Araujo HR, Castro IM, et al. Adhesion on yeast cell surface as a trapping mechanism of pathogenic bacteria by *Saccharomyces* probiotics. Journal of Medical Microbiology. 2012.

[36] Tasteyre A, Barc M-C, Karjalainen T, Bourlioux P, Collignon A. Inhibition of *in vitro* cell adherence of *Clostridium difficile* by *Saccharomyces boulardii*. Microbial Pathogenesis. 2002;32(5):219-25.

[37] Murzyn A, Krasowska A, Stefanowicz P, Dziadkowiec D, Lukaszewicz M. Capric acid secreted by *S. boulardii* inhibits *C. albicans* filamentous growth, adhesion and biofilm formation. PloS one. 2010;5(8):e12050. Epub 2010/08/14.

[38] Krasowska A, Murzyn A, Dyjankiewicz A, Lukaszewicz M, Dziadkowiec D. The antagonistic effect of Saccharomyces boulardii on Candida albicans filamentation,

adhesion and biofilm formation. FEMS yeast research. 2009;9(8):1312-21. Epub 2009/09/08.

[39] Pontier-Bres R, Prodon F, Munro P, Rampal P, Lemichez E, Peyron JF, et al. Modification of *Salmonella* Typhimurium Motility by the Probiotic Yeast Strain *Saccharomyces boulardii*. PloS one. 2012;7(3):e33796.

[40] Castagliuolo I, LaMont JT, Nikulasson ST, Pothoulakis C. *Saccharomyces boulardii* protease inhibits *Clostridium difficile* toxin A effects in the rat ileum. Infection and Immunity. 1996;64(12):5225-32.

[41] Ferreira IMPLVO, Pinho O, Vieira E, Tavarela JG. Brewer's *Saccharomyces* yeast biomass: characteristics and potential applications. Trends in Food Science & Technology. 2010;21(2):77-84.

[42] Schwarz K, Mertz W. Chromium(III) and the glucose tolerance factor. Archives of Biochemistry and Biophysics. 1959;85(1):292-5.

[43] Buts J-P. Twenty-five years of research on *Saccharomyces boulardii* trophic effects: updates and perspectives. Digestive Diseases and Sciences. 2009;54(1):15-8.

[44] Buts J-P, De Keyser N. Effects of *Saccharomyces boulardii* on Intestinal Mucosa. Digestive Diseases and Sciences. 2006;51(8):1485-92.

[45] Buts JP, De Keyser N, De Raedemaeker L. *Saccharomyces boulardii* enhances rat intestinal enzyme expression by endoluminal release of polyamines. Pediatric Research. 1994;36(4):522-7.

[46] Jahn H-U, Ullrich R, Schneider T, Liehr R-M, Schieferdecker HL, Holst H, et al. Immunological and trophical effects of *Saccharomyces boulardii* on the small Intestine in healthy human volunteers. Digestion. 1996;57(2):95-104.

[47] Buts J-P, Bernasconi P, Van Craynest M-P, Maldague P, De Meyer R. Response of human and rat small intestinal mucosa to oral administration of *Saccharomyces boulardii*. Pediatr Res. 1986;20(2):192-6.

[48] Buts J-P, Bernasconi P, Vaerman J-P, Dive C. Stimulation of secretory IgA and secretory component of immunoglobulins in small intestine of rats treated with *Saccharomyces boulardii*. Digestive Diseases and Sciences. 1990;35(2):251-6.

[49] Pothoulakis C. Review article: anti-inflammatory mechanisms of action of *Saccharomyces boulardii*. Alimentary pharmacology & therapeutics. 2009;30(8):826-33.

[50] Vandenplas Y, Brunser O, Szajewska H. *Saccharomyces boulardii* in childhood. European Journal of Pediatrics. 2009;168(3):253-65.

[51] Vohra A, Satyanarayana T. Probiotic yeasts microorganisms in sustainable agriculture and biotechnology. In: Satyanarayana T, Johri BN, Prakash A, editors.: Springer Netherlands; 2012. p. 411-33.

[52] Sougioultzis S, Simeonidis S, Bhaskar KR, Chen X, Anton PM, Keates S, et al. *Saccharomyces boulardii* produces a soluble anti-inflammatory factor that inhibits NF-κB-mediated IL-8 gene expression. Biochemical and Biophysical Research Communications. 2006;343(1):69-76.

[53] Chen X, Kokkotou EG, Mustafa N, Bhaskar KR, Sougioultzis S, O'Brien M, et al. *Saccharomyces boulardii* inhibits ERK1/2 mitogen-activated protein kinase activation both

in vitro and in vivo and protects against *Clostridium difficile* toxin A-induced enteritis. Journal of Biological Chemistry. 2006;281(34):24449-54.

[54] Zanello G, Berri M, Dupont J, Sizaret P-Y, D'Inca R, Salmon H, et al. *Saccharomyces cerevisiae* modulates immune gene expressions and inhibits ETEC-mediated ERK1/2 and p38 signaling pathways in intestinal epithelial cells. PloS one. 2011;6(4):e18573.

[55] Thomas S, Metzke D, Schmitz J, Dörffel Y, Baumgart DC. Anti-inflammatory effects of *Saccharomyces boulardii* mediated by myeloid dendritic cells from patients with Crohn's disease and ulcerative colitis. American Journal of Physiology - Gastrointestinal and Liver Physiology. 2011;301(6):G1083-G92.

[56] Ng SC, Kamm MA, Stagg AJ, Knight SC. Intestinal dendritic cells: Their role in bacterial recognition, lymphocyte homing, and intestinal inflammation. Inflammatory Bowel Diseases. 2010;16(10):1787-807.

[57] Baumgart D. The probiotic yeast *Saccharomyces boulardii* inhibits DC-induced activation of naïve T-cells. Gastroenterology. 2007;135(4):A-559 (sup 1).

[58] Dalmasso G, Cottrez F, Imbert V, Lagadec P, Peyron J-F, Rampal P, et al. *Saccharomyces boulardii* inhibits inflammatory bowel disease by trapping T cells in mesenteric lymph nodes. Gastroenterology. 2006;131(6):1812-25.

[59] Canonici A, Siret C, Pellegrino E, Pontier-Bres R, Pouyet L, Montero MP, et al. *Saccharomyces boulardii* Improves Intestinal Cell Restitution through Activation of the $\alpha2\beta1$ Integrin Collagen Receptor. PloS one. 2011;6(3):e18427.

[60] Whiteway M, Oberholzer U. Candida morphogenesis and host–pathogen interactions. Current Opinion in Microbiology. 2004;7(4):350-7.

[61] Calderone RA, Fonzi WA. Virulence factors of Candida albicans. Trends in Microbiology. 2001;9(7):327-35.

[62] Shareck J, Belhumeur P. Modulation of Morphogenesis in Candida albicans by Various Small Molecules. Eukaryotic Cell. 2011;10(8):1004-12.

[63] Krasowska A, Łukaszewicz M, Bartosiewicz D, Sigler K. Cell ATP level of *Saccharomyces cerevisiae* sensitively responds to culture growth and drug-inflicted variations in membrane integrity and PDR pump activity. Biochemical and Biophysical Research Communications. 2010;395(1):51-5.

[64] Krasowska A, Kubik A, Prescha A, Lukaszewicz M. Assimilation of omega 3 and omega 6 fatty acids and removing of cholesterol from environment by *Saccharomyces cerevisiae* and *Saccharomyces boulardii* strains. Journal of Biotechnology. 2007;131(2):S63-S4.

[65] Murzyn A, Krasowska A, Augustyniak D, Majkowska-Skrobek G, Lukaszewicz M, Dziadkowiec D. The effect of *Saccharomyces boulardii* on *Candida albicans*-infected human intestinal cell lines Caco-2 and Intestin 407. FEMS microbiology letters. 2010;310(1):17-23. Epub 2010/07/16.

Chapter 17

Microbial Interactions in the Gut: The Role of Bioactive Components in Milk and Honey

Rosa Helena Luchese

Additional information is available at the end of the chapter

http://dx.doi.org/10.5772/50122

1. Introduction

The fact that living organisms play a key role on health, was put on a scientific basis at the beginning of the last century by Elie Metchinikoff, when working at the Pasteur Institute in Paris. The findings that Bulgarian peasants, who ingested large amounts of soured milks, also lived to a ripe old age led him to conclude about the beneficial effects of fermented milks.

One of the most convincing demonstrations of the role of the gut microbiota in resistance to disease was provided by Collins and Carter [1]. These authors proved that germ-free guinea–pig was killed by 10 cells of *Salmonella Enteritidis,* but it required 10^9 cells to kill a conventional animal with a complete gut microbiota.

Probiotic was initially defined by Parker [2] as "Organisms and substances which contributes to intestinal microbial balance". Fuller [3] redefined probiotics as "A live microbial feed supplement which beneficially affects the host animal by improving its intestinal microbial balance". This definition clarifies the need for a probiotic to be viable.

The term prebiotic was subsequently adopted to define "non-digestible food ingredients that beneficially affect the host by selectively stimulating the growth and/or activity of one or a limited number of bacteria in the colon that improve host health"[4] Modification by prebiotics of the composition of the colonic microbiota leads to the predominance of a few of the potentially health-promoting bacteria, especially, but not exclusively, lactobacilli and bifidobacteria. Much of the work on prebiotics deals with the use of oligosaccharides, although the first demonstration of this type of effect was observed with a disaccharide, lactulose. Gibson and Roberfroid [4] also launched the concept of symbiotic by combining the rationale of pro- and prebiotics, is proposed to characterize some colonic foods with interesting nutritional properties that make these compounds candidates for classification as health-enhancing functional food ingredients.

The bacterial genera most often used as probiotics are lactobacilli and bifidobacteria. At present, probiotics are almost exclusively consumed as fermented dairy products such as yogurt or freeze-dried cultures, but in the future they may also be found in fermented vegetables and meats [5].

The microbial community inhabiting the gastrointestinal tract is characterized by its high population density, wide diversity, and complexity of interactions. Bacteria are predominant but a variety of protozoans, yeasts and bacteriophages are also found. Bacteria are not distributed randomly throughout the gastrointestinal tract but instead are found at population levels and species distributions that are characteristic of specific regions of the tract. The stomach and proximal small intestine contain relatively low numbers of microorganisms. Acid- tolerant lactobacilli and streptocococci predominate in the upper smal intestine. The distal small intestine (ileum) maintains a more diverse microbiota and higher bacterial numbers. The large intestine (colon) is characterized by large numbers of bacteria, low redox potential, and relatively high short-chain fatty acid concentrations. The prominent role played by anaerobic bacteria in this dynamic ecosystem is evident from the finding that more than 99% of the bacteria isolated from human fecal specimens are anaerobic or aerotolerant [6].

The intestinal tract is a dynamic ecosystem that is influenced by host, intrinsic, and environmental factors. Thus, our undestanding of gut microbial interactions and how the gastrointestinal activity is modulated, might help on establishing screening criteria to identify potentially probiotic bacteria suitable for human or animal use.

2. Microbial interactions in the gut

The nature of the microbial interaction can be predominantly by competition or mutualism [7]. In the gut they can affect either the population level of a given strain or the metabolic activity of that strain. In addition, genetic transfers can occur between strains within the gut. The host and the diet cam modulate the expression of the microbial interactions. These interactions involve multiple mechanisms that are poorly understood. Such mechanisms are involved either in the size of subdominant microbial populations or in the metabolic activities of predominant populations. Diet and perhaps other environmental factors, such as stress, can modify their expression.

The gastrointestinal tract of neonates becomes colonized immediately after birth with environmental microorganisms, mainly from the mother by several processes including sucking, kissing, and caressing. The proximity of the birth canal and the anus, as well as parental expression of neonatal care, are effective methods of ensuring transmission of microbes from one generation to the next [6].The pattern and level of exposure during the neonatal period is likely to influence the microbial succession and colonization in the gastrointestinal tract. Infants from developing countries have an early colonization with enterobacteria whereas those born in countries with good obstetric and hygienic procedures, may result in a delayed development pattern or even the absence of certain groups of intestinal bacteria during succession [8].

After the birth process, neonates are continuously exposed to new microbes that enter the gastrointestinal tract with food. This begins with breast milk, which contains up to 10^6 microbes/mL in healthy mothers. The most frequently encountered bacterial groups include staphylococci, streptococci, corynebacteria, lactobacilli, micrococci, propionibacteria and bifidobacteria originated from the nipple and surrounding skin as well as the milk ducts in the breast [6, 9, 10].

A pronounced dominance of bifidobacteria was observed over the entire breast-feeding period, with a corresponding reduction in facultative bacteria [11, 12]. There is a strong evidence suggests that the early composition of the microbiota of neonates plays an important role for the postnatal development of the immune system [13, 14].

Both adults and neonates are regularly exposed to microorganisms via the diet, but are affected differently. The microorganisms entering newborns via milk are more likely to colonize than are those entering healthy adults [6, 15].

Bacterial species or strains that will be established in the infant bowel might be capable to utilize the substrates provided by the diet and the particular human host. *Bifidobacteria, E. coli* and enterococci can utilize a wide range of monosaccharides and oligosaccharides which would be provided by the diet. Once established the range of fermentable substrates available to the bacteria changes from mono and oligosaccharides to complex plant polymers (dietary fibre) that pass undigested through to the small bowel. The other major complex carbohydrates is provided by the mucins that are continuously secreted into the bowel by the goblet cells present in the mucosal lining. Strict regulations of catabolic pathways must be an extremely important attribute in a habitat where the nutritional profile will vary from day to day according to the omnivorous and varied dietary preferences of the human host and help [16]

Protection against colonization of the intestinal tract by potentially pathogenic microorganisms, due to the gut microbiota, was called competitive exclusion [17], whose pioneering evidence had been obtained by Nurmi and Rantala [18], with birds. When these, soon after birth, were inoculated with cecal material of an adult bird, the frequency of Salmonella infections was significantly reduced.

Undoubtedly the main benefit attributed to probiotics is the competitive exclusion of pathogens that occurs by different mechanisms including: a) competition for receptors in the intestinal epithelium as occurs with lactobacilli that directly inhibits the binding of *Salmonella, E. coli* and other foodborne pathogens b) secretion of factors that inhibit internalization and adhesion of pathogens, as well as increased secretion of mucin as with lactobacilli which stimulate the secretion of MUC2 and MUC3 2 which inhibits the adherence of enteropathogenic *E. coli* c) stimulating the mucosal barrier effect, such as the lactobacilli and bifidobacteria which helps to prevent pathogens from inducing an increase in intestinal permeability; d) production of volatile fatty acids and / or other antibacterial substances, by the anaerobic microbiota besides nutrient competition [19, 20].

Constituents of the normal microbiota and some pathogenic bacteria have the ability to colonize the mucosal surfaces [21] Some microorganisms seem to be able to securely attach to the intestinal epithelium [22], and is thought to be this an important prerequisite for probiotics in a long-term survival during competition against other microorganisms for specific niches and subsequent multiplication. However, no consensus among researchers exists about the fact that a probiotic should or should not adhere to mucosal surfaces, colonize and then exert a probiotic effect, being an alternative its regular consumption to maintain the levels needed to promote the effect, forming a transient microbiota [23].

Another desired effect of a probiotic includes altered metabolism of the intestinal microbiota as the reduction in the synthesis of toxins or carcinogenic substances or an increased production of short-chain fatty acids or other substances that improve the condition of the mucosa. Prebiotics may also be given to augment immune reaction, preferably those that have a protective effect without causing overt inflammation . The ability of lactic bacteria to inactivate mutagenic compounds, such as dyes and N-nitrosamines, has been attributed to cell wall components, such as peptidoglycan and polysaccharides [24].. The lactic acid bacteria also may mediate anticarcinogenic activities by reducing the activity of fecal bacterial enzymes such as nitroredutases, azoredutases and β glucuronidase (EC 3.2.1.31) that convert procarcinogenic to carcinogenic compounds in the colon [14]

The ability to sense other bacteria may have important consequences for competitive and nutritional strategies controlling for example, entry into stationary phase, dispersal and the production of antimicrobial compounds. The ability to interfere with the signalling of bacteria will determine the fitness of the given organism to survive in the gut and may also have therapeutic potential. The study of cell-to-cell communication in gastrointestinal(GI) tract bacteria is not as advanced as it is for bacteria from other ecosystems. In Gram-negative bacteria the best-characterized systems involve *N*-acylhomoserine lactone (acyl-HSL) signals, LuxI family signal synthases and LuxR family response regulators. It appears that Gram-positive bacteria prefer peptide signals, also termed peptide pheromones [25].

Probiotics may play an active role inflammatory bowel diseases by enhancing the intestinal barrier at the mucosal surface. Caballero-Franco et al. [26] investigated whether the clinically tested VSL#3 probiotic formula and/or its secreted components could augment the protective mucus layer in vivo and in vitro. For in vivo studies, Wistar rats were orally administered the probiotic mixture VSL#3 on a daily basis for seven days. After treatment, basal luminal mucin content increased by 60%. In contrast to the animal studies, cultured cells incubated with VSL#3 bacteria did not exhibit increased mucin secretion. However, the bacterial secreted products contained in the conditioned media stimulated a remarkable mucin secretion effect. Among the three bacterial groups (*Lactobacilli, Bifidobacteria,* and *Streptococci*) contained in VSL#3, the *Lactobacillus* species were the strongest potentiator of mucin secretion in vitro.

The competitive exclusion of pathogens mediated by lactobacilli is usually performed by two mechanisms: (i) production of antimicrobial substances such as lactic acid and bacteriocins, and (ii) adhesion to the mucosa and coaggregation which can form a barrier which prevents colonization by pathogenic microorganisms [27].

Three mechanisms of aggregation have been reported so far. The first is related to the interaction between the components of the cell surface, as in the oral cavity with *Streptococcus sanguis* and *Prevotella locscheii* in which adhesins are protein-type lectins. Adlerberth et al. [28] observed that the adhesion of *Lactobacillus plantarum* to human colonic cells HT-29 was due to mannose-sensitive attaching mecanism. As the cell walls of the yeast *Saccharomyces cerevisiae* consists polysaccharide containing mannose (mannans), *Escherichia coli* and other enterobacteria containing mannose-specific adhesin receptors agglutinate yeast cells. The ability of binding yeast cells may therefore be an indication of mannose specific activity [29].

Autoaggregation has been correlated with adhesion, which is known to be a prerequisite for colonization and infection of the gastrointestinal tract by many pathogens. Adherence to the epithelium is therefore a prerequisite for enterotoxigenic *Escherichia coli* both to colonize the small intestine and to cause diarrhea, since adherence targets toxins directly onto the epithelial cell [30].

Coaggregation is a process by which genetically distinct bacteria become attached to one another via specific molecules. Cumulative evidence suggests that such adhesion influences the development of complex multi-species biofilms. The coaggregation properties of probiotic strains with pathogens as well as their ability to displace pathogens are of importance for therapeutic manipulation of the aberrant intestinal microbiota. Aggregation abilities of a probiotic with the pathogen strains were strain-specific and dependent on time and incubation conditions [31]

Recently, the complement protein mannose-binding lectin (MBL) has been shown to play a role in the first line of defense against *Candida albicans*. MBL binds to a wide variety of microorganisms through a carbohydrate recognition domain, exhibiting strong binding to *Candida* and other yeast species. The complement system is activated via this lectin pathway, causing opsonization and direct lysis of microorganisms[32]. A number of probiotic bacteria contact recognition proteins, including lectins, enzymes and other factors involved in carbohydrate metablolism , are involved in microbe-microbe host interactions [33].

In other cases, the adhesins are not lectins, such as in the case of *Streptococcus sanguis* and *Streptococccus gordonii* [34].

The second mechanism, described in lactobacilli, is dependent upon secretion of a protein of 32 kDa that promotes aggregation and a high frequency of conjugation [35] According to Collado, Meriluoto and Salminen [31] the ability to autoaggregate, together with cell-surface hydrophobicity and coaggregation abilities with pathogen strains can be used for preliminary screening in order to identify potentially probiotic bacteria suitable for human or animal use.

Finally, in *Enterococcus faecalis*, the ability to promote aggregation is due to secretion of small hydrophobic peptides called sex pheromone with consequent increase of the frequency combination [36, 37]. Pheromones appear to induce the synthesis surface proteins encoded by the plasmid, which mediate cell-cell contact.The sex pheromone system of *Enterococcus faecalis* is responsible for the clumping response of a plasmid carrying donor strain with a corresponding plasmid free recipient strain due to the production of sex pheromones by the recipient strain. The clumping response is mediated by a surface material (called aggregation substance) which is synthesized upon addition of sex pheromones to the cultures. After induction a dense layer of "hairlike" structures is formed on the cell wall of the bacteria that are responsible for the cell-cell contact which leads to the aggregation of cells [38]

Boris et al. [39] have characterized a peptide produced by *Lactobacillus gasseri* (previously classified as plantarum), which promotes the aggregation of cells of *L. plantarum* and *Enterococcus* spp. The authors hypothesize that these aggregates could mediate protection of the mucosa by the formation of a bacterial film that prevents access of undesirable microorganisms in the vaginal mucosa.

3. Bioactive prebiotic components in milk

Many components of human milk are multifunctional, providing antimicrobial, antiinflammatory, antioxidant effect besides being growth factors [40].

Breast milk not only provides a range of substrates for bacterial growth, but it also appears to be a reservoir for some of the bacteria we inherit, including Lactobacillus sp. and *Bifidobacteria* [41] Breast milk contains viable lactobacilli and bifidobacteria that might contribute to the initial establishment of the microbiota in the new born [10]. Although this needs to be verified and an explanation given with mechanism uncovered as to how lactobacilli reach the mammary gland and if other bacteria do likewise, the end result is that infants are colonized predominantly by lactic acid bacteria [20].

Although it is likely that antimicrobial components in human milk inhibit the growth of pathogenic bacteria, it is also likely that some substances stimulate the growth of beneficial bacteria, *ie*, they have prebiotic activity. This factor, originally called the bifidus factor, may promote the growth of *Lactobacilli* and *Bifidobacteria*, which can limit the growth of several pathogens by decreasing intestinal pH. One possible substance identified was *N*-acetyl-glucosamine [42]. Subsequently, several oligosaccharides have been shown to have this activity, but it is also possible that milk proteins also have such prebiotic activity . Increasing the lactobacilli and bifidobacteria levels is a target for infant formulas and the most common approach to this end has been to include prebiotic compounds [10].

The gut microbiota of breastfed infants is different from that of formula-fed infants. According to Penders [43], exclusively formula-fed infants were more often colonized with *E coli, C difficile, Bacteroides,* and lactobacilli, compared with breastfed infants. Although Penders et al. [44] showed that formula-fed infants have similar counts of bifidobacteria compared with breast-fed infants, most reports found that breast-fed infants have higher number of bifidobacteria, whereas formula-fed infants develop a mixed flora with a lower level of bifidobacteria [45].

Oliveira [12] studied the influence of diet and type of delivery in 68 neonates aged between seven and 21 days on both composition and evolution of the gut *Bifidobacterium spp., Lactobacillus spp.* microbiota. Gut colonization by bifidobacteria was not influenced by the type of delivery but the counts of lactobacilli were higher in those born vaginally as shown in table 1. Lactobacilli numbers in infants fed formula and human milk and born vaginally were significantly higher ($p<0.05$) than those born by caesarean, suggesting a possible microbiota transference from mother to the child. Similar results were reported by Biasucci [46] that demonstrated significant retarded colonization by lactobacilli at 10 days of age in babies delivered by cesarean section. Differently, Martin et al. [47] found that lactic acid bacteria colonization was not significantly related to the delivery method.

Oliveira [12] also found that bifidobacteria numbers in infants born vaginally and fed with breast milk (BM) were higher than the others, while those who received pasteurized human milk from milk banks (HMB) showed a significant lower number of *Bifidobacterium* as compared to other types of feeding (Table 1). No significant differences were observed on infants born by cesarean. These *in vivo* results corroborate with previously, *in vitro* observed data, by Borba and Ferreira [48], who evaluated the effect of human milk pasteurization on growth of different species of *Bifidobacterium*. It was demonstrated that pasteurization of human milk affected the growth of bifidobacteria, indicating that, somehow, the pasteurization process (65°C/30minutos) inhibits bifidogenic factors, or results in the production of inhibitory compounds to this microbial group

The same negative pasteurization effect was observed by Oliveira [12] on the growth of lactobacilli (Table 1). Although breast-milk contains viable lactobacilli and bifidobacteria that might contribute to the initial establishment of the microbiota in the newborn, the negative effect of human milk pasteurization on the lactobacilli and bifidobacteria gut population, cannot be explained solely on the destruction of those bacteria by the pasteurization process. Milk formulas do not contain these bacteria, but favored the development of bifidobacteria and lactobacilli in the intestine reaching a number significantly higher, as compared to the gut microbiota of pasteurized human milk fed infants.

Indeed, the health-promoting effects of breast-milk have been linked partly to the presence of lactobacilli and bifidobacteria in breast-milk [10, 47], but clearly also to different milk bifidogenic components.

Both lactotobacilli and bifidobacteria benefit in environments with low redox potential and the presence of antioxidant compounds present in human milk. Anti-oxidants such as lactoferrin, α-tocopherol, β carotene, cysteine, ascorbic acid, uric acid, catalase and glutathione peroxidase are present in human milk [40]. Most of these compounds are thermo-labile and might have been destroyed during milk pasteurization process. Whey protein is rich in *cysteine*, the thermo-labile amino acid which represents an effective *cysteine* delivery system for the cellular synthesis of glutathione. In addition, the ability of cysteine and cysteine to lower redox potential stimulates de growth of anaerobic or anaero-tolerant bacteria. The repeated processes that donor human milk is submitted before delivery to newborn infants cause a reduction in the fat and protein concentration. The magnitude of this decrease is higher on the fat concentration and it needs to be considered when this processed milk is used to feed preterm infants [49].

	Cesarean	Vaginally
	Lactobacillus	
HMB	2,4 a A	3,3 b A
FM	2,8 a B	5,7 a A
BM	3,8 a B	5,6 a A
	Bifidobacterium	
HMB	5,6 a A	3,7 b A
FM	5,7 a A	6,5 ab A
BM	6,2 a A	7,4 a A

Treatments with the same small letters in columns and capital letters in rows do not differ significantly by Tukey test (P> 0.05)

Table 1. Averages of the Lactobacilli and Bifidobacteria log numbers, in babies born by cesarean section and vaginally delivery, fed with pasteurized milk from human milk banks (HMB), formula (FM) and breast milk (BM).

3.1. Milk oligosaccharides

For many years, the oligosaccharides were considered for his role in the modulation of intestinal microbiota of infants. Currently, there is strong evidence that free oligosaccharides as well as glycoproteins are potent inhibitors of bacterial adhesion on the surface of the epithelium in the early stages of the infectious process. Therefore, the milk oligosaccharides have two important functions. The first as a source prebiotic stimulating the growth of probiotic bacteria and a second, operating in a non-specific defense mechanism inhibiting pathogens from adhering to the gastrointestinal mucosa. Although the exact pathophysiological mechanism of diarrhea is not yet fully elucidated, it seems that the ability of microorganisms to adhere to the mucosal surface is essential for spreading diarrheagenic bacteria in the duodenum [50].

Concentrations of total oligosaccharides in human milk (HMO) is 5,0-8,0 g per liter whereas just traces are found in cow's milk. In cow's milk, only small amounts of oligosaccharides are detectable, with sialyllactose being the major component [51].

Differences in the qualitative or quantitative aspects of term and preterm milk have not been observed, but compositional changes of oligosaccharides in term milk occurs during lactation with the largest amounts being found at early stages. The highest concentrations of HMOs can be found in colostrum (20 g/L), but even mature milk contains oligosaccharides in concentrations up to 13 g/L [52]. Coppa [11] reported that lactose concentration (±SD) in human milk increased from 56 ± 6.06 g/L on day 4 to 68.9 ± 8.16 g/L on day 120. Oligosaccharide level decreased from 20.9 ± 4.81 g/L to 12.9 ± 3.30 gIL, respectively. Monosaccharides represented only 1.2% of total carbohydrates.

Although intact HMOs may be absorbed, ENGFER et al. [52] postulate that a majority of HOs reach the large intestine, where they serve as substrates for bacterial metabolism. Therefore, HMOs might be considered the soluble fiber fraction of human milk

Human milk compared with other milk species, is considered unique in terms of its complex oligosaccharides content. With few exceptions, HMOs have a core structure consisting of a lactose unit at the reducing end linked to *N*-acetyllactosamine units (type 1 and 2), with branching occurring frequently Residues of L-fucose, sialic acid [N-acetylneuraminic acid (NeuAc), or both can be found linked to the core without further elongation. An elongation is achieved by an enzymatic attachment of GlcNAc residues linked in ß1-3 or in ß1-6 linkage to a Gal residue followed by further addition of Gal in a ß-1-3 or ß-1-4 bond. Thus, a large number of core structures can be formed. Further variations occur due to the attachment of lactosamine, Fuc, and/or NeuAc residues at different positions of the core region and of the core elongation chain (10, 50). The addition of Fuc is dependent on the actions of at least three different fucosyltransferases in a genetically determined process.[51, 52]..

Within human milk oligosaccharides at least 10 containing GlcNAc are known as growth factors for a so-called bifidus biota in breastfed infants. Dietary modulation of the intestinal microflora is today one of the main topics of interest in the nutritional sciences. Fructo-oligosaccharides (FOS) and galacto-oligosaccharides (GOS) are prebiotics whose bifidogenic activity has been proven in adults. Moro and Arslanoglu [19] demonstrated that supplementation of infant formulas with a mixture of GOS and FOS modified the fecal flora of term and preterm infants, stimulating the growth of Bifidobacteria. In the trial with term infants, the bifidogenic effect of the prebiotic mixture was dose dependent and there was also a significant increase in the number of Lactobacilli in the supplemented group.

The similarities between epithelial cell surface carbohydrates and oligosaccharides in human milk strengthen the idea that specific interactions of those oligosaccharides with pathogenic microorganisms do occur preventing the attachment of microbes to epithelial cells. HMOs may act as soluble receptors for different pathogens, thus increasing the resistance of breast-fed infants. Some of the best-characterized adhesins of bacteria are those of *E. coli*, which possesses type 1 fimbriae (mannose sensitive), S fimbriae (sensitive to sialylated galactosides), or colonization factors [a heterogeneous group with various receptor specificities. The various ligand specificities of *E. coli* strains could explain the differences in intestinal colonization of breastfed versus formula-fed newborns: The free oligosaccharides and glycoproteins of human milk, which are present in large amounts and great variety, might prevent intestinal attachment of microorganisms by acting as receptor analogs competing with epithelial ligands for bacterial binding [51]

Rockova et al. [53] reported that two strains of *B. animalis* were unable to grow on a medium containing human oligosaccharides as the sole carbon source in contrast of bifidobacteria from human origin. On the other hand human oligosaccharides seem to be more specific for human origin bifidobacteria compared with fructooligosaccharides. Hence, new prebiotics with similar bifidogenic properties like human oligosaccharides should be developed.

3.2. Milk proteins

Whey proteins constitute about 60-80% of the total protein content of human milk, but only 18% of bovine milk. Furthermore, the composition of whey proteins is different for each of the milks: beta-lactoglobulin, that is not found in human milk, predominates in bovine milk, while alfalactalbumin and lactoferrin predominate in human milk. The alfalactalbumin is necessary for the synthesis of lactose in the mammary gland, through the action of the lactose synthetase enzyme, their concentration in human milk ranges from 0.22 to 0.46 g/dl. The betalactoglobulin has been blamed for allergies to bovine milk [54].

Undenatured whey protein is rich in *cysteine*, the thermo-labile amino acid which represents an effective *cysteine* delivery system for the cellular synthesis of glutathione. Both cysteine and glutamine, along with glycine, are necessary the synthesis of the tri-peptide *glutathione* (GSH), one of the major detoxifiers (Phase II sulfonation) and antioxidants of the body. Enhancing glutathione levels also helps reduce the risk of infections by improving white blood cell functions. However, the unique disulfide cystine bonds of whey are heat sensitive (thermo-labile) so only carefully processed, undenatured whey proteins deliver bioavailable cystine di-peptides for intracellular conversion to cysteine, thus maximizing glutathione levels with its important immune, antioxidant, and detoxification benefits. [55].

3.2.1. Lactoferrin

Whey proteins present in human milk, such as secretory IgA, lactoferrin and lysozyme are very stable in acid medium, and reasonably resistant the action of proteolytic enzymes, it is believed, therefore, that over three quarters of these proteins appear intact in the feces of infants. Approximately 6-10% of lactoferrin is not digested by the intestinal tract, assuming that it can reach the colon and play prebiotic activities [56]

Lactoferrin, a glyco-protein, is a major protein in human milk (1.3-2.8 g/L) while it is present only in traces in cow´s milk. Lactoferrin inhibits the growth of bacteria and fungi due to its ability to bind iron, a function known as *ferro-privation*. Iron is a nutrient usually required for bacterial growth. In this way the effect of lactoferrin can be ascribed to an inhibitory effect against a pathogens rather than a direct stimulus to the development of Bifidobacteria [11].

In addition, lactoferrin also promotes the growth of beneficial bacteria such as *L. bifidus*, helping infants establish good microbial conditions in their intestines, described as "*eubiosis*". It is also an antioxidant that naturally occurs in many body secretions such as tears, blood, breast milk, saliva and mucus. Lactoferrin has anti-viral, anti-tumor activity, anti-infl ammatory / anti-oxidant activity, and immuno-modulating activity [57] Lactoferrin is also a cystine rich sub fraction.

3.2.2. Lisozime

Lysozyme is an antimicrobial enzyme (EC 3.2.1.17) found in tears, saliva, human milk whey, mucus, neutrophil granules and egg- white. It hydrolyses b (1,4) linkage between N acetylglucosamine and N-acetylmuramic acid in bacterial cell wall. Gram positive bacteria are more susceptible to lysozyme than Gram negative. The enzyme synergistically interacts with other immunoprotective components like IgA, C3 complement components and lactoferin. Human milk contains up to 400 mg/mL of lysozyme, which is a concentration approx. 3000 times higher than in bovine milk.[58]

Resistance to lysozyme and the ability to utilize human milk oligosaccharides (HMOs) were identified as the most important factors affecting the growth of bifidobacteria in human milk. Four out of 5 strains of human origin were resistant to lysozyme and utilized HMOs. In contrast, *B. animalis* was susceptible to lysozyme and did not utilize HMOs [53]

According to Rockova et al. [58] the lysozyme-resistant *Bifidobacterium bifidum* and *Bifidobacterium longum* strains exhibited excellent growth in human milk. In contrast, most of non-indigenous species, such as *C. butyricum*, did not grow in human milk oligosaccharides together with lysozyme may act as prebiotic-bifidogenic compounds inhibiting intestinal clostridia.

3.2.3. Lactoperoxidase

Lactoperoxidase makes up approximately 0.5% of the whey protein. In the presence of hydrogen peroxide (formed in small quantities by cells), catalyzes the oxidation of thiocyanate (part of saliva), forming hypothiocyanate, which can kill both gram-positive and gram-negative bacteria. Thus, lactoperoxidase in human milk may contribute to the defense against infection already in the mouth and upper gastrointestinal tract. Human milk contains active lactoperoxidase, but its physiologic significance is not yet known.[42]

3.2.4. κ-Casein and glycomacropeptide

κ-Casein, a minor casein subunit in human milk, is a glycoprotein with charged sialic acid residues. The heavily glycosylated k-casein molecule has been shown to inhibit the adhesion of *Helicobacter pylori* to human gastric mucosa. K-Casein has been shown to prevent the attachment of bacteria to the mucosal lining by acting as a receptor analogue [42].

Glycomacropeptide is resultant from the tryptic hydrolysis of human k-casein, containing sugars glucosamine and galactosamine. The molecular weight of intact human *k*-casein was estimated to be approximately 33,000. The human *k*-casein contained about 40% carbohydrate (15% galactose, 3% fucose, 15% hexosamines, and 5% sialic acid) and 0.10% (1 mol/mol) phosphorus. Its amino acid composition was similar to that of bovine *k*-casein except for serine, glutamic acid, and lysine contents [59]

Glycomacropeptide helps control appetite and inhibit the formation of dental plaque and dental cavities. It is a growth factor for bifidobacteria (bifidogenic factor 1) Levels of glycomacropeptide may range from 1% to 18% [40]

3.3. Milk fat

The main fatty acids present in human milk are restricted to those with 12-18 carbon atoms chains,namely lauric, myristic, palmitic, palmitoleic, stearic, oleic, linoleic and linolenic. Some of the long chain polyunsaturated acids such as arachidonic and others are derived from essential fatty acids linoleic and linolenic acids, totaling together with their precursors, about 15% of fat of human milk. This percentage is much higher than that found in bovine milk. Palmitic, oleic and linoleic add up together about 70% of total fatty acids of colostrum and 74% of that of mature milk [54]

Corcoran et al. [60] studied the effect of inclusion of various C18 fatty acids with 0–2 double bonds in either *cis* or *trans* configuration on *Lactobacillus rhamnosus* GG survival in simulated gastric juice at pH 2.5. Overall, the data suggest that probiotic lactobacilli can use an exogenous oleic acid source to increase their acid survival and the underlying mechanism most likely involves the ability of increased membrane oleic acid to be reduced by H^+ to stearic acid.

Rosberg-Cody et al. [61] isolate different strains of the genus *Bifidobacterium* from the fecal material of neonates and assessed their ability to produce the cis-9, trans-11 conjugated linoleic acid (CLA) isomer from free linoleic acid. The most efficient producers belonged to the species *Bifidobacterium breve*, of which two different strains converted 29 and 27% of the free linoleic acid to the cis-9, trans-11 isomer per microgram of dry cells, respectively. In addition, a strain of *Bifidobacterium bifidum* showed a conversion rate of 18%/μg dry cells. The ability of some *Bifidobacterium* strains to produce CLA could be another human health-promoting property linked to members of the genus, given that this metabolite has demonstrated anticarcinogenic activity in vitro and in vivo.

4. Bioactive prebiotic components in honey

Most of the honey in the world is produced by bees from the nectar. Nectar is a sugar solution and water, may contain pure sucrose, a mixture of sucrose, glucose and fructose, or glucose and fructose only. The nectar is transported to the combs of the hive, where they will undergo physical and chemical changes responsible for their maturation (Crane, 1983). The chemical composition of honey, as well as aroma, color and medicinal properties, are directly related to the nectar source that originated with the bee species that produced it, with their geographic and climatic conditions. All these factors contribute to the wide variation found in honey [62].

Shin and Ustunol [63] defines honey as natural syrup containing mainly fructose (38.5%) and glucose (31.3%). Other sugars in honey include maltose (7.2%), sucrose (1,5%) and a variety of oligosaccharides (4.2%). In addition to the complex mixture of carbohydrates, are enzymes, minerals, pigments, waxes and pollen. More than one hundred eighty substances have been found in different honey types.

Honey is a complex product of easy digestion and assimilation, constituting a source of energy that contributes to the balance of biological processes in that it contains suitable proportions, enzymes, vitamins, fatty acids, amino acids, phenolic and aromatic substances [64]. In addition contains oligosaccharides which stimulates the growth of probiotic bacteria in the gut [65, 66].

Leite et al. [65], found in various di-and trisaccharides in Brazilian honeys. Maltose showed up in higher levels in honeys surveyed followed by other five disaccharides, turanose, nigerose, melibiose, sucrose, isomaltose and four trisaccharides, maltotriose, panose, melezitose and raffinose..

Cellobiose, gentiobiose, isomaltose, kojibiose, laminaribiose, maltose, maltulose, melibiose, nigerose, palatinose, trehalose, trehalulose, turanose, and sucrose are the main disaccharides found in honey [66, 67]. However, it would be rather difficult to identify the predominant disaccharide or certain combinations in the previously studied honey types. For example, maltulose and turanose were found in many honey samples, however their concentrations varied to a wide extent. Thus, Sanz and others [66] found the highest amounts of maltulose and turanose (0.66 to 3.52 and 0.72 to 2.87 g/100 g of honey, respectively) in 10 samples of honey from different regions of Spain and commercially available nectar and honeydew honeys.

Carbohydrate degradation has been extensively studied in a variety of different *Bifidobacterium* species. Various α- and β-galactosidases, α- and β-glucosidase and β-fructofuranosidases during growth on fructooligosaccharides activities have been characterized in *Bifidobacterium species.* Additionally, starch-, amylopectin-, and pullulan-degrading activities in bifidobacteria have been investigated [68]

Pokusaeva et al. [68] describe the identification of two genes, *agl1* and *agl2*, present in the genome of *B. breve* UCC2003 and responsible for the hydrolysis of α-glycosidic linkages, such as those present in palatinose. The preferred substrates for both enzymes were panose, isomaltose, and trehalulose. The two purified α-1,6-glucosidases were also shown to have transglycosylation activity, synthesizing oligosaccharides from palatinose, trehalulose, trehalose, panose, and isomaltotriose.

Proline is the main amino acid present in honey; it is added by the bee and its amount varies depending on the floral source.[67].

Macedo et al. [69] studied the effect of the *Apis mellifera* honey on growth and viability of commercial strains of lactobacilli and bifidobacteria in fermented milk. Milk was inoculated with 2% of each probiotic separately and added with 3% of honey. After fermentation, were stored at 7 º C for up to 46 days and were evaluated periodically. The honey did not affect the growth or activity of lactobacilli, but exerted significant positive effect ($p<0.05$) on *Bifidobacterium* cultures assisting in maintaining the viability and stimulating metabolic activity of these bacteria, with increased pH reduction.

5. Conclusion

It is well stablished the role of several oligosaccharides as prebiotic substances. The prebiotic effect of human milk, however, is not related to a single growth-promoting substance, but rather to a complex of interacting factors. In particular the prebiotic effect has been ascribed to several oligosaccharides, that is clearly proved. The role and the way milk fat and proteins such as lactoferrin, lysozyme stimulate the growth of probiotic bacteria is not yet clearly defined.

Author details

Rosa Helena Luchese
Food Microbiology Laboratory, Department of Food Technology,
UFRRJ-Federal Rural University of Rio de Janeiro, Rio de Janeiro, Brazil

6. References

[1] Collins, F.M.; Carter, P.B. Growth of Salmonellae in orally infected germfree mice. Infect. Immun.1978; 21: 41-47.
[2] Parker, R.B. Probiotics, the other half of the antibiotic story. Anim. Nutr. Health 1974; 29: 4-8.
[3] Fuller, R. Probiotics in man and animals. 1989; 66:365-78.
[4] Gibson GR, Roberfroid MB. Dietary modulation of the human colonic microflora: introducing the concept of prebiotics. J Nutr 1995;125:1401–12.
[5] Marcel B Roberfroid MB. Prebiotics and probiotics: are they functional foods? American Journal of Clinical Nutrition 2000; 71(6): 1682S-1687s.
[6] Mackie RI, Sghir A, Gaskins HR. Developmental Microbial Ecology of the Neonatal Gastrointestinal Tract. Am. J. Clin. Nutr. 1999; 69: 1035S-45S.
[7] Boddy L, Wimpenny JWT. Ecological Concepts in Food Microbiology 1992; 73:23S-38S.
[8] Allerberth I, Carlsson B, Man P. Intestinal Colonization with Enterobacteriaceae in Pakistan and Swedish Hospital Delivered Infants. Acta Pediatr. Scand. 1991; 80: 602-10.
[9] Almeida, J. A. G.; Guimarães, V.; Novak, F. R. Normas técnicas para bancos de leite humano. Fiocruz/IFF-BLH. Rio de Janeiro, 2005.
[10] Solís G,.de los Reyes-Gavilan CG, Fernández N, Margolles A, Gueimonde M. Establishment and development of lactic acid bactéria and bifidobacteria microbiota in breast-milk and the infant gut. Anaerobe 2010; 16: 307-10.
[11] Coppa G V, Zampini L., Galeazzi T, Gabrielli, O. Prebiotics in human milk: a review. Digestive and Liver Disease. 2006; 38(Suppl. 2): 291-94.
[12] Oliveira GS. Modulação da Microbiota Colônica e Sanidade de Lactentes: Fatores Prébióticos de Leite e de Virulência de microrganismos. 2011, 122p. Tese (Doutorado) – Programa de Ciência e Tecnologia de Alimentos, Universidade Federal Rural do Rio de Janeiro, Seropédica, 2011.

[13] Fooks LJ, Fuller R, Gibson GR. Probiotics, prebiotics and human gut microbiology. Int Dairy J 1999; 9: 53–61.
[14] Dunne C. Adaptation of bacteria to the intestinal niche: Probiotics and gut disorder. Inflammatory Bowel Diseases. 2001; 7(2): 136-45.
[15] Marini A, Negretti F, Boehm G, Destri ML, Clerici-Bagozzi D, Mosca F, Agosti M. Pro- and Pre-biotics administration in preterm infants: colonization and influence on faecal flora. 2003; Acta Paediatr. Suppl. 441:80-81.
[16] Tannock GW. What pediatricians need to know about the analysis of the gut microbiota. In Michail S, Sherman PM. (ed.) Probiotics in Pediatric Medicine. Humana Press; 2009. p17-28.
[17] Sanders M.E. Probiotics: considerations for human health. Nutrition Reviews. 2003; 61(3): 91-99.
[18] Nurmi IE, Rantala M. New aspects of Salmonella infection in broiler production. Nature.19731978; 21: 41-47.
[19] Moro G.E, Arslanoglu S. Reproducing the bifidogenic effect of human milk in formula-fed infants: Why and how? Acta Pediatrica. 2005; 94 (449): 14-17.
[20] Reid G. Probiotics and prebiotics – Progress and challenges. International Dairy Journal, 2008; 18(10-11): 969-75.
[21] Goldin B. R., Gorbach SL. 1992. Probiotics for humans. In: Probiotics: the scientific basis. (R.Fuller, ed.) pp. 355-376. Chapman and Hall, London, UK.
[22] Kleeman, E. G., and T. R. Klaenhammer. Adherence of *Lactobacillus* species to human fetal intestinal cells. J. Dairy Science. 1982; 65:2063-2069.
[23] Saloff-Coste, C. J. De. La microflora gastrointestinal y lãs leches fermentadas. Danone World Newsletter, n. 14, 22p. maio, 1997. Disponível em: <http://www.danonevitapole.com/ nutri_views/newsletter/esp/news_14/ref.htm.> (accessed 13 march. 2002)
[24] Zhang XB, Ohta Y. Binding of mutagens by fractions of the cell wall skeleton of lactic acid bacteria on mutagens. Journal of Dairy Science. 1991;74:1477–1481.
[25] Simon Swift, Elaine E. Vaughan, Willem M. de Vos. Quorum Sensing within the Gut Ecosystem. Microbial Ecology in Health and Disease 2000; 12 (2): 81-92.
[26] Caballero-Franco C, Keller K, De Simone, Chadee C . The VSL#3 probiotic formula induces mucin gene expression and secretion in colonic epithelial cells. Mucosal Biology 2007; 292(1) G315-G322.
[27] Redondo-Lopez V, Cook RL, Sobel JD. Emerging role of lactobacilli in the control and maintenance of the vaginal bacterial microflora. Reviews of Infectious Diseases. 1990; 12: 856-72.
[28] Adlerbert HI, Ahrné S, Johansson ML, Molin G. A mannose-specific adherence mechanism in *Lactobacillus plantarum* coferring binding to the human colonic cell line HT-29. Applied and Environmental Microbiology. 1996; 62(7): 2244-51.
[29] Mirelman D, Altmann G, Eshdat Y Screening of bacterial isolates for mannose-specific lectin activity by agglutination of yeast. Journal of Clinical Microbiology. 1980; 11: 328-31.

[30] Zafiri D, Oron Y, Eisenstein BI, Ofek I. Growth advantage and enhanced toxicity of *Escherichia coli* adherent to tissue culture cells due to restricted diffusion of products secreted by the cells. J. Clin. Invest. 1987; 79: 1210-16.

[31] Collado C , Meriluoto J, Salminen S. Measurement of aggregation properties between probiotics and pathogens: *In vitro* evaluation of different methods. Journal of Microbiological Methods. 2007, 71: 71–4.

[32] Olivier van Till1 JW, Modderman PW, de Boer M, Hart MHL, Beld MGHM, Boermeester MA. Mannose-Binding Lectin Deficiency Facilitates Abdominal *Candida* Infections in Patients with Secondary Peritonitis. Clin Vaccine Immunol 2008;15 (1): 65-70.

[33] Lakhtin M, Alyoshkin V, Lakhtin V, Afanasyev S, Pozhalostina L. Probiotic Lactobacillus and Bifidobacterial Lectins Against *Candida albicans* and *Staphylococcus aureus* Clinical Strains: New Class of the Pathogen Biofilm Destructors Probiotics and Antimicrobial Proteins. 2010; 2(3): 186-96.

[34] Kolenbrander PE, London J. Adhere today, here tomorrow: oral bacterial adherence. Journal of Bacteriology. 1993; 175: 3247-52.

[35] Reniero R., Cocconcelli P, Botazzi V, Morelli L. High frequency of conjugation in *Lactobacillus* mediated by an aggregation-promoting factor. Journal of General Microbiology. 1991; 138: 763-68.

[36] Ehrenfeld EE, Kessler RE, Clewell DB. Identification of pherormone-induced surface proteins in *Streptococcus faecalis* and evidence of a role for lipoteichoic acid in formation of mating aggregates. Journal of Bacteriology.1986; 168: 6-12.

[37] Mori M, Tanaka H, Sakagami Y. Isolation and structure of the *Streptococcus faecalis* sex pherormone, CAM 373. FEBS Letters. 1986; 206: 69-72.

[38] Chandler JR, Dunny GM. Characterization of the Sequence Specificity Determinants Required for Processing and Control of Sex Pheromone by the Intramembrane Protease Eep and the Plasmid-Encoded Protein PrgY▿ .J. Bacteriol. 2008; 190(4): 1172-83.

[39] Boris S, Suarez JE, Barbés C. Characterization of the aggregation promoting factor from *Lactobacillus gasseri,* a vaginal isolate. Journal of Applied Microbiology. 1997; 83: 413-20.

[40] Neto, M.T. Aleitamento materno e infecção ou da importância do mesmo na sua prevenção. Act Pediatr Port. 2006; 37(1): 23-26.

[41] Martín R, Olivares M, Marín ML, Fernández L. Probiotic Potential of 3 Lactobacilli Strains Isolated From Breast Milk. J Hum Lact 2005; 21(1): 8-17.

[42] Lönnerdal Bo. Nutritional and physiologic significance of human milk proteins. Am J Clin Nutr 2003; 77(suppl):1537S–43S.

[43] Penders J, Thijs C, Vink C, Stelma FF, Snijders B; Kummeling L, Van den Brant PA, Stobbering EE. Factors influencing the composition of the intestinal microbiota in early infancy. Pediatrics, 2006; 118: 511-20.

[44] Penders J, Vink C, Driessen C, London N, Thijs, C, Stobberingh EE. Quantification of *Bifidobateriem ssp, Escherichia coli, and Clostridium dficile* in faecal samples of breast-fed and formula-fed enfants by real-time PCR. Femms Microbilology Letters. 2005; 243:141-47.

[45] Coppa GV, Zampini L, Galeazzi T, Gabrielli O Prebiotics in human milk: a review. Digestive and liver disease official journal of the Italian Society of Gastroenterology and the Italian Association for the Study of the Liver. 2006; 38(2): S291-S294.

[46] Biasucci G, Rubini M, Riboni S, Morelli L, Bessi E, Retetangos C.. Mode of delivery affects the bacterial community in the newborn gut. Early Human Development. 2010; 86(1): 20113–15

[47] Martın R, Hans GHJ, Heilig EG, Zoetendal E, Jiménez L-F, Smidt H, Rodríguez JM. Cultivation-independent assessment of the bacterial diversity of breast milk among healthy women. Research in Microbiology 2007; 158: 31-37.

[48] Borba L, Ferreira CLLF. Probióticos e Prebióticos em Bancos de Leite Humano. In: Ferreira,C.L.L.F. Ed. Prebióticos e Probióticos: .Suprema Gráfica e Editora, Rio Branco, MG p.103-121, 2003

[49] Vieira AA, Soares FVM, Pimenta HP, Abranches AD, Moreira MEL. Analysis of the influence of pasteurization, freezing/thawing, and offet processes on human milk's macronutrient concentrations. Early Human Development. 2011; 87:577-580.

[50] Mirelman D, Altmann G, Eshdat Y. Screening of bacterial isolates for mannose-specific lectin activity by agglutination of yeast. Journal of Clinical Microbiology.1986; 11: 328-331

[51] Kunz C, Rudloff S, Baier W, Klein N, Strobel S. Oligosaccharides in human milk: Structural, Functional, and Metabolic Aspects. Annu. Rev. Nutr. 2000; 20: 699–722.

[52] Engfer M B, Stahl B, Finke B. Human milk oligossacharides are resistant to enzymatic hydrolysis in the upper gastrointestinal tract. Am J Clim, 2000.

[53] Rockova S, Nevoral J, Rada V, Marsik P, Sklenar J, Hinkova A, Vlkova E, Marounek M. Factors affecting the growth of bifidobacteria in human milk. International Dairy Journal 2011; 21: 504-508.

[54] Laurindo VM, Calil T, Leone CN, Ramos JL. Composição nutricional do colostro de mães de recém nascidos de termo adequados e pequenos para a idade gestacional. II – Composição nutricional do leite humano nos diversos estágios da lactação. Vantagens em relação ao leite de vaca. Revisões e Ensaios; 1991; 14 – 23.

[55] Douglas Jr. FW, Greenberg R, Farrell Jr. HM, Edmondson LF. Effects of Ultra-High-Temperature Pasteurization on Milk Proteins; J. Agric. Food Chem. 1981, 29, 11-15

[56] Davidson LA, Lönnerdal B. Persistence of human milk proteins in the breast fed infant. Acta Paediatr Scand 1987;76:733–40.

[57] Arnold D, Di Biase AM, Marchetti M, Pietrantoni A, Valenti P, Seganti L, Superti F. Anti-adenovirus activity of milk proteins: lactoferrin prevents viral infection. Antiviral Res, 2002, 53, 153-8.

[58] Rockova S, Rada V, Marsik P, Vlkova E, Bunesova V, Sklenar J, Splichal I. Growth of bifidobacteria and clostridia on human and cow milk saccharides. Anaerobe 2011; 17: 223-225.

[59] Yamauchi K, Azuma N, KOBAYASHI Y , KAMINOGAWA S Isolation and Properties of Human *k*-Casein J Biochem 1981; 90 (4): 1005-1012.

[60] Corcoran B. M, Stanton C., Fitzgerald GF, Ross R P. Growth of probiotic lactobacilli in the presence of oleic acid enhances subsequent survival in gastric juice. Microbiology 2007; 153 (1) 291-299.

[61] Rosberg-Cody E, Ross RP, Hussey S, Ryan CA, Murphy BP, Fitzgerald GF, Devery R, Stanton C. Mining the microbiota of the neonatal gastrointestinal tract for conjugated linoleic acid-producing bifidobacteria. Appl Environ Microbiol. 2004; 70(8):4635-41.

[62] Silva, C.L.; Queiroz, A.J.M.; Figueirêdo, R.M.F. Caracterização físico-química de méis produzidos no estado do Piauí para as diferentes floradas. Revista Brasileira de Engenharia Agrícola e Ambiental 2004; 8 (2/3):260-65.

[63] Shin, H.S. Ustunol, Z. Carbohydrate composition of honey from different floral sources and their influence on growth of selected intestinal bacteria: An in vitro comparison. Food Research International. 2005; 38:721-728.

[64] Komatsu, S.S.; Marchini, L.C.; Moreti, A.C.C.C. Análises físico-químicas de amostras de méis de flores silvestres, de eucalipto e de laranjeira, produzidos por *Apis mellifera* L., 1758 (Hymenoptera, apidae) no estado de São Paulo. Conteúdo de açúcares e de proteína. Ciência e Tecnologia de Alimentos 2002; 22(2): 143-46.

[65] Leite, J.M, C. Trugo, L.C.; Costa, L.S.M.; Quinteiro, L.M.C.; Barth, O.M.; Dutra, V.M.L.; Maria, C.A.B. Determination of oligosaccharides in Brazilian honeys of different botanical origin. Food Chemistry 2000; 70: 93-98.

[66] Sanz, M.L.; Sanz, J.; Martinez, C.I. Gás chromatographic-mass spectrometric method for the qualitative and quantitative determination of disaccharides and trisaccharides in honey. Journal of Chromatography A. 2004; 1059(1-2): 143-148.

[67] Kaškonienė V, Venskutonis PR. Floral Markers in Honey of Various Botanical and Geographic Origins: A Review. Comprehensive Reviews in Food Science and Food Safety 2010; 9 (6): 620–34.

[68] Pokusaeva K, O'Connell-Motherway M, Zomer A, Fitzgerald GF. Douwe van Sinderen Characterization of Two Novel α-Glucosidases from *Bifidobacterium breve* UCC2003 Appl Environ Microbiol. 2009 75(4): 1135–1143.

[69] Macedo, L.N.; Luchese, R.H.; Guerra, A.F.; Barbosa, C.G. Efeito prebiótico do mel sobre o crescimento e viabilidade de *Bifidobacterium* spp. e *Lactobacillus* spp. em leite. Ciência e Tecnologia de Alimentos, 2008; 28(4): 935-942.

Chapter 18

Lectin Systems Imitating Probiotics: Potential and Prospects for Biotechnology and Medical Microbiology

Mikhail Lakhtin, Vladimir Lakhtin, Alexandra Bajrakova, Andrey Aleshkin, Stanislav Afanasiev and Vladimir Aleshkin

Additional information is available at the end of the chapter

http://dx.doi.org/10.5772/47828

1. Introduction

On the one hand, probiotics as microbial cellular preparations of usefulness for human include a lot of examples of successful applications supporting healthy status of organism. Majority of probiotics are represented by lactobacilli, bifidobacteria, and their mixtures [1]. Among them Acilact (consortium *Lactobacillus acidophilus* NK1 + 100_{ash} + K_3III_{24}), Lactobacterin (*L. plantarum* 8RA-3), Bifidin (*Bifidobacterium adolescentis* MC-42), Bifidumbacterin (*B. bifidum* N1), Biovestin (*B. adolescentis* + *B. bifidum*) and others are well-known probiotics produced and used in Russia (**Table 1**). These probiotics are based on probiotic strains from healthy adults gut (Collection of microorganism at G.N. Gabrichevsky Research Institute for Epidemiology & Microbiology [2]). However being of live cell origin, survival and metabolism of probiotics could not be reliably controlled, and theoretically in some cases originally probiotic bacteria have some risk to be changed towards decreasing useful activities and revealing negative features similarly to some relative pathogens. So search of non-cellular types of natural agents imitating probiotics is really important.

On the other hand, lectins as carbohydrate-binding/recognizing/sensitive proteins of non-immunoglobulin nature are multifunctional and multidomain (at least one type domain is CRD: carbohydrate binding at the level of aminoacid sequence), widely occur in nature [3 - 9], and can be specifically assembled to different soluble or not glycans, polysaccharides or glycoconjugates (GC) [glycoproteins, glycolipids, other glycol-non-proteins, any targets with exposed GC] in selected directions especially on solid or cell surfaces [10 - 15]. During assembling, lectin complexes: a) increase their multivalent and multifunctional recognition (more CBS: carbohydrate binding sites [CRD or epitopes in space], appearance of new types of CBS and new targets are reached), b) form a dynamic partially reversible net system of lectin associates revealing carbohydrate recognition (the relatively changeable vector of resulting recognition by such a system can be evaluated by ordering a panel of carbohydrate targets according to their affinity to lectins). As a result, any lectin molecule in biological surroundings can be theoretically represented as: a) a lectin system (LS) of complexes and ensembles, b) a cascade of the directed assembling reactions, and c) a cascade system [16]. For example, complexes or oligomers of lectins or lectin-GC may be able to reveal new or modified carbohydrate/GC specificity, for example, in locations between subunits [14]. So lectin type cascades involving changeable originally the same molecules of lectins are possible.

Lectins are represented by more than 20 families and large groups involving in regulation of metabolism and widely used in biotechnology [5, 7, 8, 13, 15, 17]. Symbiotic microbial lectins are important regulators of relationships between microbes and eukaryotic macroorganisms [16]. However, among symbiotic lectins, PBL are the least studied recognition factors [8, 16, 18].

In 2004 probiotic bacterial lectins (PBL) including lactobacillar and bifidobacterial lectins (LL and BL) of human origin were firstly identified and preliminarily characterized by us [19]. The present study extend our knowledge concerning PBL as new class of natural symbiotic compounds. Such lectins may play important role in human superorganism in the regulation of inter- and intrapopulation relationships between bacteria and between bacteria and the host [20]. The data concerning lectins allow evaluation of important potential of PBL as cofunctioning factors produced by probiotics. The aim was to review our current study of PBL in aspects of their prospects for biotechnology and medicine.

2. Isolation and characterization of PBL

Criteria of choice of bacterial sources of PBL were:: a) probiotic lactobacilli and bifidobacteria, b) industrial strains, and c) consortium variants of increased antagonistic activities against reference microbial diagnosticums. Acilact corresponded to all these criteria. So LL isolated were represented as a combination of lectins of all ingredient strains of Acilact. Analogously, BL isolated included combination of lectins of strains MC-42 and N1. We studied lectins from probiotical lactobacilli and bifidobacteria, originally isolated from the healthy adults gut (**Table 1**).

Identification of PBL [20] was performed using a panel of biotinylated artificial polymeric linear water-soluble GC (www.lectinity.com). Advantages of such GC were homogenecity, multiple carbohydrate residues in side exposed positions (on polyacrylamide chain) similar to mucin glycan clusters or to simple carbohydrate antigen organization, and increased affinity of interaction due to polyvalent carbohydrate targets. The combined scheme of identification and isolation of PBL is presented in **Fig. 1**. The critical step of identification is isoelectric focusing of protein fractions in the slab of polyacrylamide gel followed by gel electric blotting to membrane. Immobilized lectins treated with biotinyl-GC were visualized by streptavidin-peroxidase conjugate in the presence of chemiluminescent substrate of peroxidase using Dark Room of the BioChemi System (UVP, Calif.). Chemiluminescence kinetics was registered to optimise regime of PBL registration. The main positions of PBL were established (see **Table 2**). Lectins revealed in acidic region (within pI 4-4.5) of pH gradient were combined as acidic PBL (preparations aLL and aBL), and lectins revealed in basic region (within pI 7.6-8) were combined as basic PBL (preparations bLL or bBL). Additional PBL were identified as slightly acidic (within pI 5.1-6) [21] or approximately neutral. Artificial Mannan [GC as polyMan]- or (Mucin-like[GC as polyGalNAc])-binding PBL were rerpresented by LL (preferentially Mucin-like binding) and BL (preferentially Mannan-like binding). Combined preparations of LL (aLL or bLL) of Acilact were represented by contributions of the corresponding aLL or bLL of Acilact ingredient strains. Similarly, combined LB of strains MC-42 and N1 completed each other.

PBL were localized on the surface of bacteria (lactobacilli) within complexes which can be simply desorbed in the presence of LiCl (not NaCl). System of cell surface LL (as more protected) was represented by more extended panel of forms compared to secreted LL (as more dissociated and available to hydrolases of surrounding) into cultural fluid. Maximal forms of LL were obtained when boiled in the presence of sodium dodecyl sulfate (SDS) and 2-mercaptoethanol (ME) (**Table 3**).

2.1. Isolation of PBL

Scheme of isolation of PBL is presented in **Fig. 1**. Being on the bacterial cell surface in complexes, PBL can be esially desorpted in vitro or in cultural fluids in the presence of chaotropic agents in combinations with surfactants (endogenic or exogenic) and chelate compounds. The way of isolation of active PBL is protected by the patent (in process). Procedure of PBL isolation needed approximately 3 days. As a result, PBL preparations were characterized as uncolor, transparent fluids, without smell, resistant to freezing.

1. Growth of bacteria in fluid medium.
2. Microfiltration and sterilization in *Steriflip* (Millipore).
3. Concentration and concentrate washing in *Centricon Plus-20* (Millipore).
4. Precipitation of concentrate with ice acetone.
5. Solubilization of precipitate in small volume.
6. Isoelectric focusing in slab of polyacrylamide gel in the presence of urea and saccharose.
7. Cutting out of lectins from the gel regions where acidic or basic PBL were identified*.
8. Extraction of PBL from gel.
9. Concentration and concentrate washing in phosphate buffer saline pH 7 (PBS).
10. Freezing and storing aliquots of PBL.

*Simultaneous identification of GC-binding PBL by blotting of a part of gel plate to membrane followed by membrane treating with GC-biotin and Streptavidin-Peroxidase.

Figure 1. Scheme of identification and isolation of PBL [5, 20].

No	Species*, strains	Previous names	Probiotics in Russia including s train as ingredient
1	*L. helveticus* NK1	*L. acidophilus* NK1	Acilact, Normospectrum, Polybacterin
2	*L. casei/paracasei* K_3III_{24}	*L. acidophilus* K_3III_{24}	Acilact, Normospectrum
3	*L. helveticus* 100_{ash}	*L. acidophilus* 100_{ash}	Acilact
4	*L. plantarum 8RA-3*	*L. plantarum 8RA-3*	Lactobacterin
5	*B. longum* MC-42	*B. adolescentis* MC-42	Bifidin
6	*B. bifidum* N1	*B. bifidum* N1	Bifidok, Bifidumbacterin

*[46, 47].

Table 1. Probiotic lactobacillar and bifidobacterial strains (ingredients of probiotics) used in our work

<table>
<tr><td colspan="2">General properties:
• Original localization in ordered complexes within cell surface layers; facilitated desorption into surroundings
• Molecular masses within 52-80 kD
• System forms: acidic [a] (within pI 3.7-4.5), slowly acidic (within pI 5.1-6), neutral (within pI 6.5-7.5) and basic [b] (pI 7.6-8)
• Contain exposed aromatic aminoacids: Tyr (partially masked in different erxtent in aLL and aBL), Trp (preferentially in BL, Phe (some differences between aLL and aBL)
• Aggregation state (preferentially for aL)
• Sensitivity to detergents (preferentially for bL)
• Capability to adhesion on hydrophobic surfaces like polysterene and immobillon P (aL > bL)
• Contain ions Ca, Mg</td></tr>
<tr><td>Acidic LL:
major 58-59 kD
minors 60-62 and 53-55 kD
pI 3.8-4 (2 bands)
$(D_{350} - D_{400})/D_{240} = 46.3$</td><td>Acidic BL:
Majors and minors 56-57, 53-54, 60-64 kD

pI 3.7-4.2 (1 band + 2 dublet bands)
$(D_{350} - D_{400})/D_{240} = 66.7$</td></tr>
<tr><td>Basic LL:
62-80 kD , pI 7.6-8
$(D_{350} - D_{400})/D_{240} = 33.8$</td><td>Basic BL:
58-62; 52-54 kD; pI 7.6-8
$(D_{350} - D_{400})/D_{240} = 33.2$</td></tr>
</table>

D= optical density.

Table 2. Physicochemical and biochemical properties of PBL [5, 20, 22]

Sources of PBL	Specificity to polymeric GC	Positions of PBL bands, pI*	Intensity of PBL**
Protein concentrates			
B. adolescentis MC-42	GalNAc- Man(6-P)- Gal(3-Sulfate)-	7.5-8 6; 8 4-5; 7	4+/3+ 3+/+ 4+/+
B. gallinarum GB***	GalNAc- Man(6-P)- Gal(3-Sulfate)-	8 6.5; 7.5-8 7	2+ 2+/+/3+ 2+
B. bifidum N1	GalNAc- Man(6-P)- Gal(3-Sulfate)-	7-7.5 8 4-5	2+ 2+ 4+
LiCl-cell surface extracted protein concentrates			
L. acidophilus 100_{ash}	GalNAc- Man(6-P)- Gal(3-Sulfate)-	5.1-7-7.5;8 6.5 5; 7	+/4+/3+ 2+ +/3+
L. acidophilus NK1	GalNAc- Man(6-P)-	8 6;6.5;7	2+ +/2+/+
(SDS+ME)-treated proteins of concentrate fractions			
Acilact	GalNAc- Man(6-P)-	4.5-5.5;5.8;6.3 5.5-6.5	3+/3+2+ 2+/3+/2+
L. acidophilus 100_{ash}	GalNAc- Man(6-P)- Gal(3-Sulfate)-	5.8;6.2 6-7 5; 7	+/+ 2+/+ +/3+
L. acidophilus NK1	GalNAc- Man(6-P)-	4.5-5;5.7;6.2 5;6-6.8;7-8	3+/2+/+ 2+/2+/4+

*Isoelectric points (pI) according to isoelectric focusing in PAA gel in gradient of pH 4 - 8; ** in scale "+" - "4+" (relative chemiluminescence of complex PBL-b—Streptavidin-Peroxidase in the presence of chemiluminescent substrate of peroxidase). *** strain from chicken gut. SDS: sodium dodecyl sulfate, ME: 2-mercaptoethanol.

Table 3. Identified PBL of different types [20].

The main physicochemical and biochemical properties of PBL are presented in **Table 2**. As it can be seen from the **Table 2**, PBL are relatively hydrophobic proteins and can be presented in aggregated forms with partially exposed aromatic aminoacid residues (especially controlled for Tyr and Trp). Protein stability of PBL needed the presence of cocktail of protease inhibitors (*"Complete"*, R & D). Increased disappearance of bBL upon storing in glass tubes (compared to polypropylene tubes) for a long time was observed (increased sorption on glass walls is possible). PBL contained cations of metals. For example, major forms of PBL of *L. helveticus* NK1 (strain as dominated contributors of LL into Acilact) contain approximately 2 Ca^{2+} in molecule. Fluorescent properties of PBL (especially in case of BL) are increased in PBL complexes including endogenic exopolymers.

Aforementioned data allow preliminary classifications of PBL [5, 7, 16]. Currently, PBL can be considered as: originally surface proteins of recognition, Ca^{2+} (and other metal cation)-containing and binding proteins, relatively (random structure)-organized (decreasing of randomly ordered structure in complexes as refolding recognition process), preferentially originally mono- or bivalent (one CBS in polypeptide) low sensitive haemagglutinins (similar to pan-agglutinins), with capability to create complexes, oligomeres and aggregated particles, members of functional superfamilies.

2.2. Biological properties of PBL

PBL imitate the following general main activities of probiotics: antimicrobial, immunocorrecting, ssupporting consortium, stabilizing healthy status in communicative directions "Microbes - Microbes" and "Microbes - Host". In addition, PBL reveal unique properties which complete probiotics to synbiotics and extend spectrum of useful activities in combinations "Probiotics + PBL" (see below).

PBL are represented by four LS (**Table 1).** Among them LL and LB (acidic and basic) were isolated and studied by us in detail. In addition, in case of slowly acidic LL it was suggested their potential cofunctioning to oxidase-reductase system within potential lactobacillus consortium of Acilact strains and *L. plantarum* 8RA-3 [21]. The role of such LL may be in regulation of protection of probiotic consortium in biotopes against peroxide stress. Examples of regulation of oxidoreductases with lectins are well documented [15]. Mean time, the role of neutral LL is still unclear.

2.2.1. Interactions between PBL and GC [14, 17, 19, 20, 22-24]

Major forms of soluble PBL are represented mainly as molecules and their complexes with one CBS. Such PBL forms needed hydrolase treated red cells for visualization of haemagglutination reaction. In haemagglutination reaction (*Clostridium perfringens* sialidase-treated human AII-blood group erythrocytes) interaction between PBL and GC was as approximately equimolar (1 : 1, M/M).

We identified different lectins secreted by lactobacilli and bifidobacteria using a panel of GC and mainly three methods including: a) dot-blotted supernatant concentrates on Immobillon-P membrane (Millipore), b) proteins blotted after isoelectric focusing supernatant concentrated protein fractions in polyacrylamide plate, c) proteins sorpted on sialidase (or trypsin)-treated human AII-red cells [5, 22 - 25].

For identification of lectins among extended panel of lactobacilli and bifidobacteria strains we used GC (0.5-5 mkg/ml, PBS) containing multiply exposed side carbohydrate residues on biotynylated (b) or not polyacrylamide (PAA) chain (www.lectinity.com):

- Fucα1- [α-L-Fucan-like],
- Galβ1- [β-D-Galactan-like],
- Gal(3-Sulfate)β1- [3-HSO_3Galβ1- ;β-D-Galactan-3-Sulfate polymer],
- GaNAcα1- [Tn-like antigen containing polymer],
- GalNAcα1,3Galβ1- [A_{di} as (AII-blood group substance)-like containing polymer],
- GalNAcα1,3GalNAcβ1- [Fs as (Forssman antigen)-like containing polymer],
- GalNAcα1,3GalNAcα1- ,
- Galα1,3GalNAcα1- [$T_{\alpha\alpha}$-like antigen containing polymer],
- GalNAcβ1- [desialylated Mucin-like],
- Galβ1,4GlcNAcβ1- [poly(LacNAc)-containing mucin-like],
- GlcNAcβ1- [soluble linear Chitin-like],
- Manα1- [α-D-Mannan-like],
- Man(6-phosphate)α1- [6-H_2PO_3Manα1-polymer; α-D-PhosphoMannan],
- (MurNAc-L-Ala-D-isoGln)β1- [MDP-; Muramyldipeptide containing polymer; bacterial Peptidoglycan-like],
- Rhaα1- [α-L-Rhamnan-like].

The whole resulted chemiluminescent pictures of LL and BL separated by isoelectric focusing followed by blotting were distinct and needed individual optimized regimes of registration. It is seen from the **Table 3** that: a) the pictures of PBL are unigue and depended on strain origin, b) dominated PBL types are revealed as mucin- and/or Mannan-binding; b) PBL of probiotic consortium include PBL of ingredient strains. Mannan-binding lectins of *L. plantarum* 8RA-3 possessed increased intensities of chemiluminescence [19]. These data were supported by study of PBL specificity to GC in haemagglutination reaction [5, 23]. Dissociation of PBL-(hydrolase-treated human AII-red cells) agglutinates was observed in the presence of 0.5-1 mkg/ml of GC. Effectivenes of GC was decreased in the order: poly(GalNAc) or Mannan > Galactan >> Chitin-like polymer (no influence).

In other seria of experiments we extended panel of probiotic bacteria and extended panel of GC to identify new PBL types using dot-blotting technique [24, 25]. It was shown that PBL of lactobacilli and bifidobacteria are capable to discriminate GalNAc-containing GC (GalNAc residues as exposed, internal/masked, or dublicated) glycoantigens A_{di}-, Fs-, or T_n-depending on strain origin. No binding of PBL to $T_{\alpha\alpha}$ was observed. PBL also discriminated artificial peptidoglycan, mannans and mucins. Due to PBL revealing as LS [16] when two or more PBL forms (major and minor ones) vary on specificity, similarity (identical part of mosaic of the same specificity) and differences (the whole mosaic as unique, ranging intensity of components with the same specificity, some components which simultaneously recognize two types of target GC) between recognizing potential of species and genus of lactobacilli and bifidobacteria can be established.

Using dot-blotting technique, at least 7 types of LS were identified for extended panel of lactobacilli and bifidobacteria which occur in human gut. Among these, LS were represented by lectins which especially significantly recognized α–D-Mannan (phosphorylated or not; yeast-like), α-L-Fucan (algal-like), peptidoglycan (bacterial-like), mucins (mammalian gut-like); antigens T_n and Forssman, blood group AII substance. Such lectins were identified as mosaic within bacterial mainly acidic protein massive.

Aforementioned data on interaction between PBL and GC indicate that PBL may serve as additional important functional characteristic. The latter can serve the basis to study biotope metabolic relationships involving probiotic bacteria as antagonistic to opportunistic microorganisms in keeping healthy biotope status; and to construct cofunctioning systems of PBL together with yeast and higher plant ingredients.

Antimicrobial activities of PBL against clinical microbial strains [21, 26-32] included:

- Growth inhibition;
- Involving biodegradation (proteolysis) (LL > BL);
- Synergistic action (LL + BL: against staphylococci [effectiveness: LL > BL]; BL + LL: against microfungi [effectiveness: BL > LL]; BL + antibiotics: against *Candida* species [possibility to decrease effective work doses of antibiotics]);
- Action as cascades (action of aPBL followed by action of bPBL);
- Concurent use of resources of pathogens during their different live cycle steps (wrong essembling of biofilms of pathogens, choice and switching of metabolome nets, increased degradation of pathogen constructions including their lysis).

The following general comments on antimicrobial action of PBL should be noted. The action of PBL is directed against colorectal and urogenital clinical strains from human biotopes. PBL act as the members of new class of biofilm destructors [27]. Anti-*Staphylococcus* and anti-*Candida* action reveal multistep synergism in space (different regions of action of aLL and bLL, aBL and bBL, aLL and aBL, antibiotic-like and lytic actions) and in time (earlier action as antibiotic-like, later lytic action of aPBL followed by lysis by bPBL). It takes place multisynergism of anti-*Candida* action between PBL and antibiotics (azoles, amphotericin B, nystatin). Taken together, PBL imitate anti-*Staphylococcus* and anti-*Candida* activities of probiotic lactobacilli and bifidobacteria [33, 34] and can be potentially used for treatment of candidoses and staphylococcoses.

It should be also noted that PBL possess advantages compared to other antimicrobials: prolonged action; cascade synergistic action, low subcytoagglutinating doses; non-dependence on antibiotic types (upon therapy) [probiotics delivered can be inactivated by some antibiotics]. In addition, ketokonazol and some other antibiotics are poorly soluble in PBS that decreases their effectiveness and control.

2.2.2. Activities of PBL in respect of potential probiotic compartment of biotope [35, 37, 38]

PBL reveal a spectrum of activities in respect of populations of lactobacilli isolated from the same biotope. Results indicate that LL support healthy status of normoflora in biotope due to realization of supervisor signal functions of PBL. It is expected that when delivered, PBL increase synbiotic compartment of biotope against potential pathogenic compartment (in addition to other positive events in direction "Microbiocenoses - Host").

2.2.3. *Other biological activities of PBL*

Activities of PBL in respect of cells of mammalian protection systems [22, 27]:

- Inducing production of TNF-α by human periphery blood lymphocytes;
- Modulation of peritoneal macrophage migration in manner which is differed from action of GC;

Predicted PBL activities based on similarities to symbiotic bacterial lectins [16], other lectins and probiotics:

- As possible ingredients of drug forms in cases of colorectal alterations (potential effectiveness: BL > LL), or urogenital alterations (potential effectiveness: LL > BL).
- Direct antitumor action: against changed human cell systems similarly to PBL action against eukaryotic pathogens as xenoagents in organism (potential effectiveness: BL > LL); througph increased affinity of PBL to Poly(LacNac) as potential tumor antigens [14] (potential effectiveness: LL > BL);
- Against protozoan pathogens (like action against another type of eukaryotic pathogens - microfungi);
- Against viruses (like Acilact action against rotavirus infection of children; similar to Mannan-binding phytolectins possessing activity against HIV-1);
- Intracellular sorting into organells and vesicules [due to capability of PBL to recognize poly(Man-6-P) within targets, similar to animal Man-6-P-binding lectins)];
- Biocompatibility and synergism of LL and/or BL together with other probiotic microbial lectins as antimicrobials;
- Biocompatibility and synergism of PBL to other type antimicrobials possessing distinct mechanisms of action;
- The possible forming additional antimicrobial pool as PBL fragments in the presence of host and microbial hydrolases of surroundings in biotopes.

3. Conclusion

All aforementioned data support wide potential of PBL for industrial and medical biotechnology.

The following main prospects of applications of PBL can be underlined:

For cell cultures [autostimulators, supporting probiotic bacterial cultures: mixed or not, in the presence of pathogen, etc.],

In constructing bioadditives, anti-infectives and drug forms [system drugs of synergistic and selective action as antipathogenic agents, and as factors supporting probiotic compartment in biotopes];

In diagnostics [microassays; for typing clinical pathogen strains; for detecting altered anormal surface and metabolome net of pathogenic significance] [28, 29, 32],

In constructing of cascade biosensors based on LS-organization of PBL [for monitoring biotope healthy balance, for screening strains and their mixtures especially on solid surfaces like sensibilized membranes, polysterol or polypropelene;

In constructing predictable lactobacilli- and bifidobacteria-based consortia as potentially probiotics-like, constructing synbiotic consortia [37, 38]:

- keeping metabolome status;
- switching (on/off) microbial nets and cascades;
- controlling microbiocenosis functioning; providing cell teaching;
- factors in constructing of beneficial microbiocenoses;
- directed antipathogen action by changing ontogenesis of pathogens) [27];
- helpers in building cell and cytokine-like gradients [23];
- synergistic and synbiotic factors in mixed cultures of microorganisms or in host biotopes [32];
- stabilizers of poorly growth probiotic microorganisms [35, 36];
- co-functioning with other PBL-like and non-PBL antimicrobials produced by probiotic compartment of biotope (bacteriocins, antimicrobial peptides and biosurfactants) [22, 39]
- co-functioning with human cytokines (PBL as cytokine inducers), defensins, antibiotics (synergism), antibodies [9, 39, 40];
- synergism of PBL signaling and signal proteinases/oligopeptidases and (oligo)peptides of surrounding;
- screening, selection and typing of strains;
- ingredients of both free cell drug ointments and cosmetic creams (improving formulas) [39, 41];
- in recombinant lectin technologies [17];
- carriers for drug delivery, carriers of low molecular weight highly hydrophobic heterocyclic effectors (some antibiotics, chemotherapeutic antitumor agents, apoptose inducers, etc.) [3, 17];
- ingredients of functional bioadditives [41];

Upon chemotherapy and radiotherapy of tumors to support healthy status of organism [36, 42]

In system drug therapy when added PBL (LL and/or BL) will modulate whole spectrum of system drug activities;

In landscape microecology and architectuire of microbiocenoses (PBL as the direct participants and organizers of landscapes) [32].

It is clear that solid or cell surfaces are of preferential importance for any directed assembling initiated by PBL (increased accumulated interphased concentrations of reactants, initiating or triggering assembling on immobilized first components of cascades, achievement of maximally long and asymmetric products). That is why PBL within pore PAA hydrophilic gels or membranes (Durapore membranes as [multi]layer microaccumulators), immobilized PBL on PVDF membranes (Immobillon P) or polysterene microplates and latex particles are of especial perspectiveness.

Author details

Mikhail Lakhtin, Vladimir Lakhtin*, Alexandra Bajrakova, Andrey Aleshkin,
Stanislav Afanasiev and Vladimir Aleshkin
*Department of Medical Biotechnology,
G.N. Gabrichevsky Research Institute for Epidemiology & Microbiology, Moscow, Russia*

Appendix

List of abbreviations

a – acidic

b – basic

CBD – carbohydrate binding domain

CBS – carbohydrate binding site

D – optical density

BL bifidobacterial lectins

LL lactobacillar lectins

LS lectin system(s)

PBL – probiotic bacterial lectins

GC – glycoconjugate(s)

PBS – phosphate buffer saline pH 7

4. References

[1] Shenderov BA (2008) [Functional Foods and their role in prophylaxis of metabolic syndrome (in Russian)]. Moscow: DeLi Print, 319 pp. ISBN 978-5-94343-166-1.

[2] Aleshkin VA, Amerhanova AM, Pospelova VV, Afanasyev SS, Shenderov BA (2008) History, Present Situation and Prospects of Probiotic Research Conducted in the G.N. Gabrichevsky Institute for Epidemiology and Microbiology. Microb Ecol Health & Dis 20: 113-115.

[3] Lakhtin VM (1987) [Lectins for Investigation of Proteins and Carbohydrates (in Russian)]. In: Vsesoyuzniy Institut Nautchnoy I Tehnitcheskoy Informatsii, Moskva, Itogi Nauki I Tehniki, Seriya Biotehnologiya, Vol 2 (Klyosov AA, ed) [VINITI, Moscow, Reviews of Science and Technique, Series Biotechnology, Vol 2 (Klyosov AA, ed) (in Russian)]: 290 p. ISSN 0208-2330.

* Corresponding Author

[4] Lakhtin VM (1989) [Lectins and Aspects of Their Study (in Russian)]. Mikrobiologitcheskiy Zhurnal (Kiev) 51; No 3: 69 – 74. ISSN 0201-8462.

[5] Lakhtin VM, Lakhtin MV, Pospelova VV, Shenderov BA (2007) Lectins of Lactobacilli and Bifidobacteria. II. Probiotic Lectins of Lactobacilli and Bifidobacteria as Possible Signal Molecules Regulating Inter- and Intrapopulation Relationships Between Bacteria and Between Bacteria and the Host. Microb Ecol Health & Dis 19: 153 - 157.

[6] Korsun VF, Lakhtin VM, Korsun EV, Mitskonas A (2007) [Phytolectins (in Russian)]. Moscow: Practical Medicine, 288 p. ISBN 5-98811-077-0.

[7] Lakhtin VM, Afanasiev SS, Aleshkin VA, Nesvizskyi UV, Lakhtin MV, Shubin VV, Cherepanova YV, Pospelova VV (2009) [Classification of lectins as universal regulator molecules of biological systems (in Russian)]. Vestnik Rossiiskoy Academii Meditsinskih Nauk No 3: 36-43. ISSN 0869-6047.

[8] Lakhtin V, Lakhtin M, Aleshkin V (2011) Lectins of Living Organisms. The Overview. Anaerobe 17: 452 - 455. DOI: 10.1016/j.anaerobe.2011.06.004.

[9] Lakhtin MV, Karaulov AV, Lakhtin VM, Alyoshkin VA, Afanasyev SS, Nesvizskyi UV, Afanasyev MS, Vorapaeva EA, Alyoshkin AV (2012) [Lectin – glycoconjugate systems in human organism (in Russian)]. Immunopathology, Allergology, Infectology No 1: 27-36. ISSN 0236-297X.

[10] Lakhtin VM (1991) [Lectins in the Investigation of Receptors (in Russian)]. Uspekhi Khimii (Moskva) 60: 1777 - 1816. ISSN 0044-460X.

[11] Lakhtin VM (1992) [Specificity of Microbial Lectins (in Russian)]. Prikladnaya Biokhimiya i Mikrobiologiya (Moskva) 28: 483 - 501. ISSN 0555-1099.

[12] Lakhtin VM (1994) Lectin Sorbents in Microbiology. In: "Lectin - Microorganism Interactions". Doyle RJ, Slifkin M, eds. New York: Marcel Dekker. 1994. pp. 249 - 298. ISBN 0-8247-9113-4.

[13] Lakhtin VM (1994) [Molecular organization of lectins (in Russian)]. Molekulyarnaya Biologiya (Moskva) 28: 245-273. ISSN 0026-8984.

[14] Lakhtin VM (1995) [Use of Lectins in the Analysis of Carbohydrate Moieties of Glycoproteins and Other Natural Glycoconjugates (in Russian)]. Biokhimiya (Moskva) 60: 187-217. ISSN 0320-5725.

[15] Lakhtin MV, Lakhtin VM, Aleshkin VA, Afanasiev SS, Aleshkin AV (2010) [Lectins and Enzymes in Biology and Medicine (in Russian)]. Moscow: "Dynasty" Publishing House, 496 pp. ISBN 978-5-98125-076-7.

[16] Lakhtin M, Lakhtin V, Aleshkin V, Afanasiev S (2011) Lectins of Beneficial Microbes: System Organization, Functioning and Functional Superfamily. Beneficial Microbes 2: 155 – 165. DOI: 10.3920/BM2010.0014.

[17] Lakhtin VM (1989) [Biotechnology of Lectins (in Russian)]. Biotechnologiya (Moskva) No 6: 676-691.

[18] Lakhtin VM, Aleshkin VA, Lakhtin MV, Afanasiev SS, Pospelova VV, Shenderov BA (2006) [Lectins, Adhesins and Lectin-like Substances of Lactobacilli and Bifidobacteria

(in Russian)]. Vestnik Rossiiskoy Academii Meditsinskih Nauk 1: 28-34. ISSN 0869-6047.

[19] Lakhtin VM, Pospelova VV, Lakhtin MV, Medvedkova NA, Bovin NV (2004) [Interaction of components of cultural fluids of lactobacilli and bifidobacteria to synthetic analogs of polysaccharides (in Russian)]. Proceedings of the International Conference "Probiotics, Prebiotics, Synbiotics and Functional Foods. Modern State and Prospects" (June 2-4, 2004, Moscow) Moscow, 2004: 28.

[20] Lakhtin VM, Lakhtin MV, Pospelova VV, Shenderov BA (2006) Lactobacillus and Bifidobacterial Lectins as Possible Signal Molecules Regulating Intra- and Interpopulation Bacteria - Bacteria and Host - Bacteria Relationships. Part I. Methods of Bacterial Lectin Isolation, Physicochemical Characterization and Some Biological Activity Investigation. Microb Ecol Health & Dis 18: 55 - 60.

[21] Lakhtin M, Aleshkin V, Afanasiev S, Pospelova V (2011) Three Protective Lectin Systems In Probiotic Lactobacillus Consortium (poster 360). Proceedings of the 4th Congress of European Microbiologists (26-30 June, 2011, Geneva, Switzerland). Program Book: P. 94. CD-ROM Abstracts, "L" Names.

[22] Lakhtin MV, Alyoshkin VA, Lakhtin VM, Nesvizhsky YV, Afanasyev SS & Pospelova VV (2010) [The Role of Lectins from Probiotic Microorganisms in Sustaining the Macroorganism (in Russian)]. Vestnik Rossiiskoy Academii Meditsinskih Nauk 2: 3-8. ISSN 0869-6047.

[23] Lakhtin VM, Lakhtin MV, Korsun VF & Shenderov BA (2008) [Mutual Potential of Lectins of Probiotic Microorganisms and Fungi in Conditions of Organization and Functioning of the Model Eukaryotic Cell Biofilms - for Further Use in Clinical Practice within Phytocompositions (in Russian)]. Praktitcheskaya Fitoterapiya 2: 11-17. ISBN5-88010-096-0.

[24] Lakhtin VM, Dykhal YI, Lakhtin MV, Cherepanova YV, Pospelova VV, Shenderov BA (2009) New Lectin Systems in Cultural Fluids of Probiotic Strains of Lactobacilli and Bifidobacteria Capable to Discriminate Glycoantigens Containing GalNAc. Gastroenterologia Sankt-Peterburga [Gastroenterology of Sankt-Petersburg] 4: A14-A15. ISSN 1727-7906.

[25] Lakhtin V, Alyoshkin V, Lakhtin M, Afanasyev S (2009) Glycoconjugates in Discrimination of Glycoconjugate Recognition Systems of Probiotic Microorganisms. New Potential Keys for Strains and Glycometabolome Typing. Glycoconjugate Journal. 26: 876.

[26] Lakhtin MV, Lakhtin VM, Aleshkin VA, Afanasiev SS, Korsun VF (2010) [Phyto- and Probiotic Lectins – Synergistic Antipathogens (in Russian)]. Praktitcheskaya Fitoterapiya 1: 5-11. ISBN5-88010-096-0.

[27] Lakhtin M, Aleshkin V, Lakhtin V, Afanasiev S, Pozhalostina L, Pospelova V (2010) Probiotic Lactobacillus and Bifidobacterial Lectins Against *Candida albicans* and *Staphylococcus aureus* Clinical Strains: New Class of Pathogen Biofilm Destructors. Probiotics & Antimicrobial Proteins 2: 186-196. DOI: 10.1007/s12602-010-9046-3.

[28] Lakhtin MV, Bajrakova AL, Lakhtin VM, Alyoshkin AV, Afanasiev SS, Aleshkin VA (2011) [Microbiocenosis Model "Pathogen – Probiotic lectins" for Monitoring Primate Biotope Dysbioses (in Russian)]. Proceedings of the 2nd International Scientific Conference "Fundamental and Applied Aspects of Medical Primatology" (August 8-10, 2011, Sochi, Russia). Vol 1. Sochi: 88-92.

[29] Lakhtin MV, Bajrakova AL, Lakhtin VM, Afanasiev SS, Aleshkin VA, Korsun VF (2011) [Probiotic Lectins of Human in Protection against Dysbioses in Different Human Biotopes (in Russian)]. Praktitcheskaya Fitoterapiya (Moskva) 1: 4-13. ISBN5-88010-096-0.

[30] Lakhtin MV, Lakhtin VM, Aleshkin VA, Bajrakova AL, Afanasiev SS, Korsun VF (2011) [Phyto- and Probiotic-Analog Therapy of Microfungal Infections: Theory and Practice Potential in the Development (in Russian)]. ARS MEDICA 15: 183-187. ISSN 2220-5497.

[31] Lakhtin M, Lakhtin V, Bajrakova A, Aleshkin A, Afanasiev S, Aleshkin V (2012) Interaction of Probiotic Bacterial Lectins to *Candida* Species. Proceedings of the VIII International Conference "Sciences and Technology: Step in Future (Feb. 27 – March 5, Prague, Chechia)". – Vol. 29: 34 - 41. Materiály VIII mezinárodní vědecko - praktická conference «Věda a technologie: krok do budoucnosti - 2012». - Díl 29: 34 – 41. Biologické vědy: Praha. Publishing House «Education and Science». ISBN 978-966-8736-05-6.

[32] Lakhtin VM, Lakhtin MV, Afanasiev SS (2012) [The fight for the space and resources between probiotic and relatively pathogenic compartments in potential biotope: lectin type imitators of probiotics against eukaryotic pathogens – importance for biotechnology (in Russian)]. Proceedings of the VIII International Conference "Key questions in Modern Science" (April 17 – 25, Sofia, Bulgaria). Sofia, Vol. 28: 28 – 33. Материали за 8-а международна научна практична конференция, «Ключови въпроси в съвременната наука», - 2012. Том 28: 28 – 33. Биологии. Селско стопанство. София. «Бял ГРАД-БГ» ООД. ISBN 978-966-8736-05-6.

[33] Krisenko OV, Sclyar TV, Vinnikov AI, Kiryukhantseva IM (2012) [Features of antagonistic activity of lactic acid bacteria in respect of microfungi (in Ukrainian)]. Proceedings of the VIII International Conference "Key questions in Modern Science" (April 17 – 25, Sofia, Bulgaria). Sofia, Vol. 28: 18-23. [Материали за 8-а международна научна практична конференция, «Ключови въпроси в съвременната наука», - 2012. Том 28. Биологии. Селско стопанство. София. «Бял ГРАД-БГ» ООД]. ISBN 978-966-8736-05-6.

[34] Lazarenko L, Babenko L, Shynkarenko-Sichel L, Pidgorskyi V, Mokrozub V, Voronkova O, Spivak M (2012) Antagonistic Action of Lactobacilli and Bifidobacteria in Relation to *Staphylococcus aureus* and Their Influence on the Immune Response in Cases of Intravaginal Staphylococcosis in Mice. Probiotics & Antimicro Prot 4: 78 – 89. DOI 10.1007/s12602-012-9093-z.

[35] Lakhtin VM, Bajrakova AL, Lakhtin MV, Belikova YV, Afanasyev SS, Aleshkin VA (2012) [Regulating properties of the human probiotic bacterial consortium lectins: Screening, modulation and selection of normoflora of *Lactobacillus* populations from the same biotope (in Russian)]. Proceedings of the VIII International Conference "Days of Science". Prague, Vol. 74: 38 – 44. Materiály VIII mezinárodní vědecko - praktická conference «Dny vědy - 2012». - Díl 74: 38 – 44. Biologické vědy: Praha. Publishing House «Education and Science» s.r.o. ISBN 978-966-8736-05-6.

[36] Lakhtin M, Lakhtin V, Bajrakova A, Afanasyev S, Alyoshkin V (2011) Human Biotope Probiotic Bacterial Lectins as Signal System Supporting Biotope Healthy Balance. Immunology 135, Suppl. 1:111.

[37] Lakhtin MV, Lakhtin VM, Cherepanova YV, Pospelova VV, Afanasyev SS, Alyoshkin VA (2011) Ranging qualities of industrial ingredient probiotic strains of human bifidobacteria and lactobacilli to predict new probiotic formulas. Proceedings of International Conference "Modern Achievements of Biotechnology" (June 21-23, 2011, Stavropol, Russia). Part 2. Division "Bioadditives". Moscow: NOU "Education Scientidic Technical Center of Milk Industry", 49 – 51.

[38] Lakhtin VM, Lakhtin MV, Agapova YV, Belikova YV, Kulakova YV, Afanasyev SS, Alyoshkin VA (2012) Advantadges of the probiotic "Acilact" compared to ingredient strains using algorithmic ranges of qualities. Proceedings of the VIII International Conference "Scientific Space of Europe - 2012" (April 7-15, Przemyśl. Poland). Przemyśl, Nauka i studia. Vol. 32: 50 – 57. Materiały VIII Międzynarodowej naukowi-praktycznej konferencji «Naukowa przestrzeń Europy - 2012» Vol. 32. – P. 50-57. Nauk biologicznych.: Przemyśl. Nauka i studia. ISBN 978-966-8736-05-6.

[39] Lakhtin VM, Afanasiev SS, Aleshkin VA, Nesvizhsky YV, Pospelova VV, Lakhtin MV, Cherepanova YV, Agapova YV (2008) [Strategical Aspects of Constructing Probiotics of the Future (in Russian)]. Vestnik Rossiiskoy Academii Meditsinskih Nauk 2: 33-44. ISSN 0869-6047.

[40] Lakhtin VM, Afanasyev SS, Aleshkin VA, Nesvizhsky YV, Pospelova VA (2008) [Nanotechnologies and perspectives of their application in medicine and biotechnology (in Russian)]. Vestnik Rossiiskoy Academii Meditsinskih Nauk No 4, 50-55. ISSN 0869-6047.

[41] Lakhtin MV, Lakhtin VM, Aleshkin AV, Bajrakova AL, Afanasiev SS, Aleshkin VA, Korsun VF (2012) [Probiotic Lectins – Ingredients of Biopreparations, Bioadditives and Drug Forms (in Russian)]. Praktitcheskaya Fitoterapiya (Moskva) No 1: ISBN5-88010-096-0.

[42] Lakhtin MV, Aleshkin VA, Lakhtin VM, Nesvizhsky YV, Afanasiev SS, Pospelova VV (2012) [Prospects of Lectins of Probiotics in Chemotherapy (in Russian)]. Proceedings of the VIII International Conference "Days of Science" (March 27 – April 5, Prague, Chechia). Prague, Publishing House «Education and Science» s.r.o. Vol. 74: 6 – 10. Materiály VIII mezinárodní vědecko - praktická conference «Dny vědy - 2012». - Díl 74:

6 – 10. Biologické vědy: Praha. Publishing House «Education and Science» s.r.o. ISBN 978-966-8736-05-6.

[43] Botina SG (2011) [Molecular biological approaches in selection of bacterial cultures upon creation of inoculi for biotechnology (in Russian)]. Thesis, DSc (Biotechnology). Moscow: 280 pp.

[44] Subbotina ME (2009) [Development of method of gene typing of bifidobacteria using bi-locus sequenation to identify species of strains (in Russian)]. Thesis, PhD (Genetics). Moscow: 122 pp.

Chapter 19

Probiotic Use for the Prevention of Necrotizing Enterocolitis in Preterm Infants

Fatma Nur Sari and Ugur Dilmen

Additional information is available at the end of the chapter

http://dx.doi.org/10.5772/50049

1. Introduction

Necrotizing enterocolitis (NEC) is among the most common and devastating diseases that primarily afflicts preterm infants in neonatal intensive care units (NICU) (1). Despite recent advances in neonatal care, the incidence of necrotizing enterocolitis and the associated morbidity and mortality have remained unchanged because of the improved survival for smaller, more premature infants (2). Both medical and surgical management play critical role in the treatment of NEC once it occurs, but prevention is likely to have the most dramatic impact on overall morbidity and mortality.

2. Epidemiology

The incidence of NEC varies among NICUs worldwide, but ranges 3% and 28% with an average of 7% in infants born weighing less than 1500 g (3). NEC occurs more commonly in the smallest and most immature infants, with the incidence increasing inversely to gestational age and birth weight among appropriately grown preterm infants. Although NEC is almost exclusively a disease of prematurity, 5 % to 10 % of cases occur in infants born greater than or equal to 37 weeks gestation (4). Most of the infants in whom NEC develops are previously fed and the disease usually occurs in the second week of life after the initiation of enteral feeding (5).

The estimated rate of death related with NEC ranges between 20 and 30 %, with the highest rate among infants requiring surgery (6). Beyond the mortality and gastrointestinal morbidities, NEC is also the harbinger of neurologic deficits and developmental delay (7).

3. Pathophysiology

NEC is a disease with a multifactorial etiology leading to the one common final pathway of necrosis and inflammation of the neonatal intestine (8). Although the pathophysiology of NEC is incompletely understood, epidemiologic studies have identified multiple factors that increase an infant's risk for the development of NEC, although prematurity, enteral feeding, intestinal ischemia/asphyxia and bacterial colonization are thought to play central roles in disease pathogenesis (9).

Prematurity is the most consistent and important risk factor for NEC. Anand et al. (10) proposed major altered components of the intestinal barrier of preterm neonates such as disruption of the integrity of epithelial tight junctions, impaired peristalsis and deficiencies in components of the mucous coat that may contribute to the onset of NEC (11-13).

Enteral feeding is a significant risk factor for disease in preterm infants, because most cases of NEC occur after feedings have been introduced. Although the precise relationship between enteral feeding and NEC remain poorly understood, studies have identified the importance of breast milk as opposed to formula, osmolality, volume and rate of feeding as important factors (14, 15). Breast milk appears to reduce the incidence of NEC in human studies and controlled animal models (16, 17).

Intestinal ischemia is another risk factor in the development of NEC. There is a delicate balance between vasodilatation and vasoconstriction in neonatal circulation, mediated formerly by nitric oxide and the latter by endothelin-1. The basal intestinal vascular resistance is decreased by the predominance of nitric oxide. Pathologic states cause endothelial dysfunction which leads to endothelin-1 activation and resultant vasoconstriction, intestinal ischemia, and cellular injury (18).

GI tract of the preterm infants are susceptible to abnormal bacterial colonization because of the immature immunologic defenses (9). The intestinal flora that is normally populated plays an important role in maintaining the intestinal barrier, and also has the ability to dampen the inflammatory response. Colonization of the intestine with pathogenic microorganisms, depending on the exposition to a variety of nosocomial bacteria in the NICUs and immature immune systems, may serve as predisposing factors in development of NEC in preterm infants (19).

4. Diagnosis and management

The clinical syndrome associated with NEC is nonspecific. Infants with NEC may exhibit several gastrointestinal signs including abdominal distention, increased gastric residuals, occult or gross blood in the stool, and abdominal wall erythema or ecchymosis. In addition to GI-specific signs, NEC infants may exhibit systemic signs such as lethargy, apnea, bradycardia and temperature instability (19).

The diagnosis of disease continues to be made with the use of pathognomonic radiographic findings. The most specific signs, which still are the only "signs" that allow the diagnosis to be confirmed prior to surgical inspection of the intestine, are pneumatosis intestinalis in most cases, hepatic portal venous gas or pneumoperitoneum in a minority of cases. A diagnosis requires one of the specific radiographic findings or direct inspection of the intestine in the clinical context (5, 20). As soon as the diagnosis of NEC is suspected, initial management should include bowel rest, decompression, cultures of blood, urine and sputum, administration of broad-spectrum antibiotics, appropriate fluid resuscitation, serial abdominal examinations and radiographs. Surgical intervention for NEC is required in 30% to 50% of cases reported; therefore close observation with serial examinations and radiographs is essential. Surgical intervention involving primer peritoneal drainage or laparatomy with the resection of affected bowel are generally required in infants with intestinal perforation or deteriorating clinical condition (20).

5. Preventative strategies for NEC

Based on the epidemiologic studies and understanding of the pathophysiology there have been several approaches attempted to prevent NEC in animal and human studies. Reduction of NEC has been shown with breast milk feeding, antibiotic prophylaxis, steroids, IgA supplementation, probiotics, epidermal growth factor, polyunsaturated fatty acids, platelet activating factor (PAF) antagonists, PAF-acetylhydrolase, trefoil factor, leukocyte depletion, and oxygen radical scavengers in animal models. In human studies, there remains no standard effective alternative for NEC prevention, although breast milk feeding is the best option that neonatologists have to offer. Besides breast milk feeding, strategies with the most evidence supporting their effectiveness are careful feeding advancement and prophylactic probiotics supplementation in at-risk neonates (4, 5).

6. Probiotic prophylaxis in NEC

The intestine of the newborn is devoid of bacterial flora at birth but is rapidly colonized thereafter (9). Although the maternal flora constitutes the main source of intestinal colonization, gestational age, the mode of delivery, the neonatal diet and genetic factors also influence the colonization (21).

Colonization by commensal bacteria is required for the normal development and maturation of the newborn intestine. Lactobacilli and Bifidobacteria that are the principal kinds of probiotics bacteria predominate in the normal gut flora of healthy, breastfed, term neonates (22). In contrast, the intestine of the preterm infant tends to be colonized by different microorganisms, predominantly coliforms, enterococci and bacteroides species (23). Even among VLBW infants receiving breast milk, Sakata et al. (24) found that the Bifidobacteria were undetectable in the intestinal flora during the first 1 to 2 weeks after birth and did not predominate until after the third week of life. Hoy et al. (25) and Millar et al. (26) observed a decline in the variety of species and shift to a predominance of Enterobacteriaceae before the onset of NEC.

Intestinal microbiological flora is an important factor in the host-defense mechanism against bacterial infections. The combination of an increase in potentially pathogenic microorganisms together with a decrease "in normal flora" found in preterm infants is one of the factors that render these infants at increased risk of developing NEC (23, 27). It has been suggested that the growth of pathogens might be prevented by inducing the colonization of the intestine non-pathogenic bacteria (probiotics) of species normally resident in the gut of preterm and term infants (28).

The identification of probiotics bacterial species involved in gut homeostasis and potential therapeutic benefits of probiotics have led to interest in their use in the prevention of NEC (29, 30). Probiotics compete with other microbes for binding sites and substrates in the bowel, enhance the IgA mucosal response, improve the mucosal barrier, reduce mucosal permeability, stimulate intestinal mucosal lactase activity, increase anti-inflammatory cytokines, and produce a wide range of antimicrobial substances such as bacteriocins, microcins, reuterin, hydrogen peroxide and hydrogen ions (20, 28).

Gastrointestinal mucosa is the primary interface between the external environment and the immune system. Whenever intestinal microflora reduces, antigen transport is increased indicating that the normal gut microflora maintains gut defenses (31). The non-pathogenic probiotic bacteria interact with the gut epithelial cells and the immune cells to start the immune signals. These bacteria must interact with M cells in the Peyer's patches, with gut epithelial cells, and with associated immune cells. Probiotic bacteria have been shown to modulate immunoglobulin production. Secretory IgA plays an important role in mucosal immunity, contributing to the barrier against pathogenic bacteria and viruses. The increase in the number of IgA producing cells was the most remarkable property induced by probiotic organisms (32, 33).

Probiotic supplementation has resulted in a reduction in the incidence of NEC-like intestinal lesions in several animal models. Caplan et al. (34) demonstrated that Bifidobacteria supplementation resulted in intestinal colonization and subsequent reduction in NEC-like lesions in a neonatal rat model of intestinal ischemia/reperfusion. Butel et al. (35) showed in a NEC model in quail, that supplementation with Bifidobacteria prevented the development of cecal lesions reminiscent of NEC.

Several studies have specially assessed the colonization pattern and the incidence of NEC in preterm infants supplemented with various probiotics (19, 36-38). (Table 1)

A randomized controlled trial found that infants whose feed was supplemented with Bifidobacterium breve had higher rates of fecal bifidobacterial colonization at 2 weeks of age (73 vs. 12 %), improved weight gain and had feeding tolerance. However, the incidence and severity of NEC were not reported in this study (39).

In a multicenter double-blind study, preterm infants with a gestational age of <33 weeks or birth weight of <1500 g, who survived 42 weeks, were randomized to receive either placebo or L. rhamnosus GG (LB-GG) once a day, starting with the first fed until discharged. The incidence of urinary tract infection, bacterial sepsis and NEC were examined as outcome measures. There were no significant differences between the probiotics and placebo groups with regard to any of the outcome variables (28).

Another study performed by Bin-Nun et al. (40) was conducted to test the hypothesis that normalizing the intestinal flora by administration of prophylactic probiotics would provide a natural defense, thereby reducing both the incidence and severity of NEC in preterm infants. Preterm infants ≤1500g birth weight were randomized to either receive a daily feeding supplementation with a probiotic mixture (Bifidobacteria infantis, Bifidobacteria bifidus, and Streptococcus thermophilus) of 10^9 cfu/day or to not receive feed supplements. In this study, probiotic supplementation had resulted in a reduction in the incidence and the severity of NEC in very low birth weight (VLBW) infants.

In addition, Lin et al. (19) reported a decrease in NEC, NEC plus mortality and severity of NEC, following probiotics L. acidophilus and B. infantis (Infloran), prophylaxis in a prospective, randomized blinded study. They also recently reported a multicenter-blinded trial regarding who were randomized to receive Bifidobacterium bifidum and L. acidophilus for 6 weeks. The results showed a significant reduction in the incidence of death or NEC and no adverse effect, such as sepsis, flatulence or diarrhea (37).

Similarly, Hoyos (36) reported a significant reduction in the incidence of NEC and NEC-associated death in infants in the NICU after the prophylactic administration of probiotics in the form of Infloran-supplemented enteral feeding. However, infants were more mature and generally had higher birth weights; it is not a blinded trial and comparison was made with historical controls.

The results of the study performed by Sari et al (38) suggested a trend toward lower incidence of NEC and, death or NEC, although the difference was not statistically significant. None of the L. sporogenes-supplemented fed infants died from NEC; they could not find significant difference in severity of NEC or in mortality rate attributable to NEC between the probiotics and control groups. The use of a single probiotics agent rather than two agents and utility of a relatively low dose of L. sporogenes may explain, at least in part, the smaller treatment effect in their study. Longer duration of umbilical venous catheterization in probiotics group also may be another cause in the lesser effect of L. sporogenes on NEC prevention.

In a recent report by Manzoni et al. (41) routinely supplementation of probiotic LB-GG in a large, 6-year VLBW infants cohort was proved microbiologically safe and clinically well tolerated.

Although some of the studies (19) predicated that probiotics may reduce the incidence of sepsis; literature did not confirm this association (42, 43). Sari et al. (38) also did not show that L. sporogenes reduced the incidence of sepsis in VLBW infants. Sepsis has a complex pathogenesis that is favored by many factors (that is, immune deficiencies of preterm infants, type and frequency of invasive procedures and so on) that cannot be influenced by probiotic administration. The main effect of orally administered probiotics is in the gastrointestinal tract, and so probiotics alone cannot overcome the invasive procedures including infection.

Lactobacilli and Bifidobacteria are generally regarded as non-pathogenic, except a few reported cases of Lactobacillus bacteremia that seemed to occur in immunocompromised or extremely sick infants receiving high doses of Lactobacillus (44). Kunz et al. (45) described L. bacteremia in two preterm infants who received LB- GG, and both of those infants had short-gut syndrome. The other authors did not observe sepsis attributable to probiotics in the studies (28, 44, 45). Sari et al (38) observed no cases of sepsis or other adverse effects, such as diarrhea, flatulence attributable to probiotic supplementation.

In 2011, Alfaleh et al. (29) performed a meta-analysis of randomized controlled trials, including some of those discussed here, to evaluate the efficacy of probiotics in the prevention of severe NEC and/or sepsis in preterm infants. Sixteen eligible trials randomizing 2842 infants were included. Included trials were highly variable with regard to enrollment criteria (i.e. birth weight and gestational age), baseline risk of NEC in the control groups, timing, dose, formulation of the probiotics, and feeding regimens. In a meta-analysis of trial data, enteral probiotics supplementation significantly reduced the incidence of severe NEC (stage ≥2) (typical RR 0.35, 95% CI 0.24 to 0.52) and mortality (typical RR 0.40, 95% CI 0.27 to 0.60). There was no evidence of significant reduction of nosocomial sepsis (typical RR 0.90, 95% CI 0.76 to 1.07). The included trials reported no systematic infection with probiotic supplemental organism. Author concluded that enteral supplementation of probiotics prevents severe NEC and all cause mortality in preterm infants.

A recent systematic review performed by Mihatsch et al. (46) reported that there is insufficient evidence to recommend routine probiotics. However there is encouraging data which justifies the further investigation regarding the efficacy and safety of specific probiotics in circumstances of high local incidence of severe NEC.

There is limited information about the long-term effects of probiotics supplementation in neonates. Chou et al. (47) reported the long-term neurodevelopmental outcomes of preterm infants in their trial of oral probiotics for NEC. A total of 83.1% of infants from their trial were assessed by Bayley infant developmental assessment tool (BSID-II) at 3 years' corrected age; 1 of 153 and 4 of 148 had died after discharge. There were no significant differences in growth, neurodevelopmental and sensory outcomes at 3 years' corrected age. Recently, a prospective follow-up study was conducted to evaluate growth and neurodevelopmental outcomes in a cohort of infants enrolled in a randomized controlled trial of oral probiotics for the prevention of NEC in VLBW infants. The authors concluded that administration of oral probiotics to VLBW infants in the early neonatal period had no adverse effects on growth, neuromotor, neurosensory, and cognitive outcomes at 18-22 months' corrected age (48). Given the importance of this issue, it is critical that authors of all trials in this field report long-term neurodevelopmental outcomes of the enrolled infants.

Source	GA/BW	Probiotic Agent(s)	Dosage and Duration	Type of milk	Results
Hoyos et al, (36) 1999	<37 wk	LB-A,BI	LB-A $0.25x10^9$ CFU, BI $0.25x10^9$ CFU, once daily from first feed until discharge	MM, DM, or FM	**Significant decrease in NEC and NEC associated mortality**
Dani et al, (28) 2002	<33 wk or <1500 g	LB-GG	$6x10^9$ CFU once daily from first feed until discharge	MM, DM, or FM	Non-significant decrease in NEC, UTI and sepsis
Bin-Nun et al, (40) 2005	<1500 g	BI, ST, BBB	BI $0.35x10^9$ CFU, ST $0.35x10^9$ CFU, BBB $0.35x10^9$ CFU once daily from first feed to 36 wk corrected age	MM or FM	**Significant decrease in NEC**
Lin et al, (19) 2005	<1500 g	LB-A, BI	LB-A 1004356 and BI 1015697 organisms twice daily from day 7 until discharge	MM or DM	**Significant decrease in NEC or death**
Lee et al, (49) 2007	<37 wk	LB-A	10^8 CFU from first feed for 14 d	MM or FM	**Non-significant decrease in NEC, improved feeding tolerance**
Lin et al, (37) 2008	<34 wk and <1500 g	LB-A, BBB	2 x 10^9 CFU/d for 6 wk	MM or FM	**Significant decrease in NEC or death**
Samanta et al (50), 2009	<34 wk and <1500 g	BBB, BB-L, BI, LB-A	$2.5x10^9$ CFU/d until discharge	MM or FM	**Significant decrease in NEC, death or sepsis**
Sari et al (38), 2011	<33 wk or <1500 g	LB-S	$0.35x10^9$ CFU/d from first feed until discharge	MM or FM	**Non-significant decrease in NEC or death, improved feeding tolerance**

Table 1. Studies examining effect of probiotic supplementation on incidence of NEC

GA indicates gestational age; BW, birth weight; LB-A, Lactobacillus acidophilus; BI,Bifidobacteria infantis; LB-GG, Lactobacillus GG; ST, Streptococcus thermophilus; BBB, Bifidobacterium bifidus; BB-L, Bifidobacteria longum; LB-S, Lactobacillus sporogenes; CFU, colony forming unit; MM, mother's milk; DM, donor milk; FM, formula milk; UTI, urinary tract infection

7. Summary

NEC is one of the commonest causes of acute morbidity and mortality in preterm infants as well as a cause of long term disability for older children. The pathogenesis is multifactorial but probably requires the classic triad of injury to the intestinal mucosa, presence of enteral food substrate and the presence of bacteria and bacterial products. Recent advances in neonatology have led to improved survival for younger and smaller infants, and a resultant increase in the disease burden of NEC. The morbidity and mortality rates for NEC have still remained constant, by contrast with the improvement in outcomes for many prematurity-related diseases. There are several prosperous researches that could ultimately result in novel preventative or therapeutic options but there is currently no effective preventive strategy, and treatment options are limited.

Although probiotics may be a promising approach for prevention and decreased severity of NEC, issues exist regarding the standardization of an appropriate probiotic supplement for neonates. Most studies have utilized various combinations of probiotic bacteria and amounts of culture-forming units for different lengths of time. These differences in methodology have created difficulties in elucidating the most beneficial probiotic supplement for the preterm population. Questions remain concerning the strains or combinations of strains that offer the best benefit. Potential exists for a significant difference in the magnitude of the benefit when administered to formula versus breast-fed neonates. There are also uncertainties over the optimal time to start probiotics in order to confer maximal benefit, possible adverse effects including probiotic-associated sepsis and tolerance of milk feeding and the long-term consequences of probiotic supplementation. So, before routine probiotic prophylaxis could be recommended to neonatologists, it would be important to have evidence in support of such use from large, prospective, single-protocol, randomized, double-blind trials.

Author details

Fatma Nur Sari* and Ugur Dilmen
Neonatal Intensive Care Unit in Zekai Tahir Burak Maternity and Teaching Hospital, Ankara, Turkey

* Corresponding Author

8. References

[1] Neu J, Walker WA. Necrotizing enterocolitis. N Engl J Med 2011;364:255-64.

[2] Henry MC, Moss RL. Neonatal necrotizing enterocolitis. Semin Pediatr Surg 2008;17:98-109.

[3] Uauy RD, Fanaroff AA, Korones SB, et al. Necrotizing enterocolitis in very low birth weight infants: biodemographic and clinical correlates. National Institute of Child Health and Human Development Neonatal Research Network. J Pediatr 1991;119:630-8.

[4] Caplan MS. Neonatal Necrotizing Enterocolitis: Clinical Observations, Pathophysiology, and Prevention. In: Martin RJ, Fanaroff AA, Walsh MC, editors. Fanaroff and Martin's Neonatal-Perinatal Medicine: Diseases of the Fetus and Infant. 9th Ed. ed. St. Louis, Missouri: Mosby; 2011. p. 1431-1442.

[5] Berman L, Moss RL. Necrotizing enterocolitis: an update. Semin Fetal Neonatal Med 2011;16:145-50.

[6] Fitzgibbons SC, Ching Y, Yu D, et al. Mortality of necrotizing enterocolitis expressed by birth weight categories. J Pediatr Surg 2009;44:1072-5; discussion 1075-6.

[7] Bedrick AD. Necrotizing enterocolitis: neurodevelopmental "risky business". J Perinatol 2004;24:531-3.

[8] Neu J. Necrotizing enterocolitis: the search for a unifying pathogenic theory leading to prevention. Pediatr Clin North Am 1996;43:409-32.

[9] Hunter CJ, Upperman JS, Ford HR, et al. Understanding the susceptibility of the preterm infant to necrotizing enterocolitis (NEC). Pediatr Res 2008;63:117-23.

[10] Anand RJ, Leaphart CL, Mollen KP, et al. The role of the intestinal barrier in the pathogenesis of necrotizing enterocolitis. Shock 2007;27:124-33.

[11] Allen A, Bell A, Mantle M, et al. The structure and physiology of gastrointestinal mucus. Adv Exp Med Biol 1982;144:115-33.

[12] Muresan Z, Paul DL, Goodenough DA. Occludin 1B, a variant of the tight junction protein occludin. Mol Biol Cell 2000;11:627-34.

[13] Cetin S, Ford HR, Sysko LR, et al. Endotoxin inhibits intestinal epithelial restitution through activation of Rho-GTPase and increased focal adhesions. J Biol Chem 2004;279:24592-600.

[14] Di Lorenzo M, Bass J, Krantis A. An intraluminal model of necrotizing enterocolitis in the developing neonatal piglet. J Pediatr Surg 1995;30:1138-42.

[15] Kamitsuka MD, Horton MK, Williams MA. The incidence of necrotizing enterocolitis after introducing standardized feeding schedules for infants between 1250 and 2500 grams and less than 35 weeks of gestation. Pediatrics 2000;105:379-84.

[16] Caplan MS, Hedlund E, Adler L, et al. Role of asphyxia and feeding in a neonatal rat model of necrotizing enterocolitis. Pediatr Pathol 1994;14:1017-28.

[17] Lucas A, Cole TJ. Breast milk and neonatal necrotising enterocolitis. Lancet 1990;336:1519-23.
[18] Nowicki PT. Ischemia and necrotizing enterocolitis: where, when, and how. Semin Pediatr Surg 2005;14:152-8.
[19] Lin HC, Su BH, Chen AC, et al. Oral probiotics reduce the incidence and severity of necrotizing enterocolitis in very low birth weight infants. Pediatrics 2005; 115:1-4.
[20] Thompson AM, Bizzarro MJ. Necrotizing enterocolitis in newborns: pathogenesis, prevention and management. Drugs 2008;68:1227-38.
[21] Ruemmele FM, Bier D, Marteau P, et al. Clinical evidence for immunomodulatory effects of probiotic bacteria. J Pediatr Gastroenterol Nutr 2009;48:126-41.
[22] Orrhage K, Nord CE. Factors controlling the bacterial colonization of the intestine in breastfed infants. Acta Paediatr Suppl 1999;88:47-57.
[23] Claud EC, Walker WA. Hypothesis: inappropriate colonization of the preterm intestine can cause neonatal necrotizing enterocolitis. FASEB J 2001;15:1398-403.
[24] Sakata H, Yoshioka H, Fujita K. Development of the intestinal flora in very low birth weight infants compared to normal full-term newborns. Eur J Pediatr 1985;144:186-90.
[25] Hoy C, Millar MR, MacKay P, et al. Quantitative changes in faecal microflora preceding necrotising enterocolitis in preterm neonates. Arch Dis Child 1990;65:1057-9.
[26] Millar MR, MacKay P, Levene M, et al. Enterobacteriaceae and neonatal necrotising enterocolitis. Arch Dis Child 1992;67:53-6.
[27] Hall MA, Cole CB, Smith SL, et al. Factors influencing the presence of faecal lactobacilli in early infancy. Arch Dis Child 1990;65:185-8.
[28] Dani C, Biadaioli R, Bertini G, et al. Probiotics feeding in prevention of urinary tract infection, bacterial sepsis and necrotizing enterocolitis in preterm infants. A prospective double-blind study. Biol Neonate 2002;82:103-8.
[29] Alfaleh K, Anabrees J, Bassler D, et al. Probiotics for prevention of necrotizing enterocolitis in preterm infants. Cochrane Database Syst Rev 2011:CD005496.
[30] Embleton ND, Yates R. Probiotics and other preventative strategies for necrotising enterocolitis. Semin Fetal Neonatal Med 2008;13:35-43.
[31] Madsen K, Cornish A, Soper P, et al. Probiotic bacteria enhance murine and human intestinal epithelial barrier function. Gastroenterology 2001;121:580-91.
[32] Isolauri E, Kaila M, Mykkanen H, et al. Oral bacteriotherapy for viral gastroenteritis. Dig Dis Sci 1994;39:2595-600.
[33] Szajewska H, Kotowska M, Mrukowicz JZ, et al. Efficacy of Lactobacillus GG in prevention of nosocomial diarrhea in infants. J Pediatr 2001;138:361-5.
[34] Caplan MS, Miller-Catchpole R, Kaup S, et al. Bifidobacterial supplementation reduces the incidence of necrotizing enterocolitis in a neonatal rat model. Gastroenterology 1999;117:577-83.

[35] Butel MJ, Waligora-Dupriet AJ, Szylit O. Oligofructose and experimental model of neonatal necrotising enterocolitis. Br J Nutr 2002;87 Suppl 2:S213-9.

[36] Hoyos AB. Reduced incidence of necrotizing enterocolitis associated with enteral administration of Lactobacillus acidophilus and Bifidobacterium infantis to neonates in an intensive care unit. Int J Infect Dis 1999;3:197-202.

[37] Lin HC, Hsu CH, Chen HL, et al. Oral probiotics prevent necrotizing enterocolitis in very low birth weight preterm infants: a multicenter, randomized, controlled trial. Pediatrics 2008;122:693-700.

[38] Sari FN, Dizdar EA, Oguz S, et al. Oral probiotics: Lactobacillus sporogenes for prevention of necrotizing enterocolitis in very low-birth weight infants: a randomized, controlled trial. Eur J Clin Nutr 2011;65:434-9.

[39] Kitajima H, Sumida Y, Tanaka R, et al. Early administration of Bifidobacterium breve to preterm infants: randomised controlled trial. Arch Dis Child Fetal Neonatal Ed 1997;76:F101-7.

[40] Bin-Nun A, Bromiker R, Wilschanski M, et al. Oral probiotics prevent necrotizing enterocolitis in very low birth weight neonates. J Pediatr 2005;147:192-6.

[41] Manzoni P, Lista G, Gallo E, et al. Routine Lactobacillus rhamnosus GG administration in VLBW infants: a retrospective, 6-year cohort study. Early Hum Dev 2011;87 Suppl 1:S35-8.

[42] Schanler RJ. Probiotics and necrotising enterocolitis in preterm infants. Arch Dis Child Fetal Neonatal Ed 2006;91:F395-7.

[43] Deshpande G, Rao S, Patole S, et al. Updated meta-analysis of probiotics for preventing necrotizing enterocolitis in preterm neonates. Pediatrics 2010;125:921-30.

[44] Land MH, Rouster-Stevens K, Woods CR, et al. Lactobacillus sepsis associated with probiotic therapy. Pediatrics 2005;115:178-81.

[45] Kunz AN, Noel JM, Fairchok MP. Two cases of Lactobacillus bacteremia during probiotic treatment of short gut syndrome. J Pediatr Gastroenterol Nutr 2004;38:457-8.

[46] Mihatsch WA, Braegger CP, Decsi T, et al. Critical systematic review of the level of evidence for routine use of probiotics for reduction of mortality and prevention of necrotizing enterocolitis and sepsis in preterm infants. Clin Nutr 2012;31:6-15.

[47] Chou IC, Kuo HT, Chang JS, et al. Lack of effects of oral probiotics on growth and neurodevelopmental outcomes in preterm very low birth weight infants. J Pediatr 2010;156:393-6.

[48] Sari FN, Eras Z, Dizdar EA, et al. Do Oral Probiotics Affect Growth and Neurodevelopmental Outcomes in Very Low-Birth-Weight Preterm Infants? Am J Perinatol 2012.

[49] Lee SJ, Cho SJ, Park EA. Effects of probiotics on enteric flora and feeding tolerance in preterm infants. Neonatology 2007;91:174-9.

[50] Samanta M, Sarkar M, Ghosh P, et al. Prophylactic probiotics for prevention of necrotizing enterocolitis in very low birth weight newborns. J Trop Pediatr 2009;55:128-31.

Chapter 20

Dairy Probiotic Foods and Bacterial Vaginosis: A Review on Mechanism of Action

Parvin Bastani, Aziz Homayouni,
Violet Gasemnezhad Tabrizian and Somayeh Ziyadi

Additional information is available at the end of the chapter

http://dx.doi.org/10.5772/50083

1. Introduction

Bacterial vaginosis (BV) is the most common urogenital disease in women, affecting about 19-24% of them in reproductive ages. 10-26% of pregnant women in the United States have been reported to suffer from BV. The prevalence of BV varies in different parts of the world and is higher in developing countries. The disease has been found in 12 to 25 percent of women in routine clinic populations, and accounts for 32 to 64 percent of women in clinics for sexually transmitted diseases; however, there is still some controversy about whether or not BV is a sexually transmitted disease (STD) in the "traditional" sense. Current data indicate that the overall prevalence of BV is much higher among STD clinic attendees and commercial sex workers [1]. BV is believed to occur as a result of an imbalance in the normal vaginal microbiota [2] when the normal Lactobacillus bacteria in the vagina are disrupted and subsequently replaced by predominantly anaerobic bacteria including Gardnerella vaginalis, Mycoplasma hominis, Prevotella, and Peptostreptococcus [3]. Other bacteria such as Escherichia coli from the rectum have also been shown to cause the disease. Lactobacilli bacteria, by producing a natural antibacterial, hydrogen peroxide, keep the healthy normal balance of vaginal microorganisms. Factors that upset this balance in the vagina are not well-understood. However, the activities or behaviours that have been related with BV incidence include having a new sex partner or multiple sex partners and douching [4, 5]. BV is mainly followed by irritating symptoms mainly foul, fish-like or musty odor which is sometimes stronger after a woman has sex, watery or foamy, white (milky) or gray vaginal secretions, itching on the outside of the vagina and Burning or discomfort during urination [6]. It is also known that BV is associated with potentially severe gynaecological and obstetric complications. Current data suggest a causal association between BV, pelvic inflammatory disease and tubal factor infertility [7]. Pregnant women with BV have a higher risk of adverse outcomes such as late miscarriage, chorioamnionitis, premature rupture of membranes, preterm birth and postpartum endometritis; they are more susceptible to having babies of low birth weight as well [8, 9]. BV has been identified as a risk factor for herpes virus type 2 infections and increased viral shedding in infected women [10, 11]. It has also been suggested that the presence of BV increases the risk for human immunodeficiency virus infection [12]. It is noteworthy that many women with BV do not show any symptoms [13], pelvic inflammatory disease [14], infections following gynecological surgery [15] and pre-term birth. BV is not transmitted through toilet seats, bedding, swimming pools, or touching of objects. Women, who have not had sexual intercourse, hardly develop BV [16].

Typically, a cure for BV refers to resolution of symptoms and maybe a repeat BV-negative screen. We know from clinical studies that BV has both an unprompted resolution and repetition [13]. As many as 30 percent of women relapse within 1 month of treatment, with unprompted relapse occurring more commonly among women treated with topical compared with systemic antibiotics [17]. The most common oral treatment for BV in both pregnant and non-pregnant women is metronidazole and clindamycin [18]. The individual cure rate given a 7-day, twice-daily course of 500 mg of metronidazole ranges from 84 percent to 96 percent, and the cure rate given a 2 g single dose of metronidazole is 54-62 percent [19]. The second systemic treatment for BV is oral clindamycin. The one known clinical trial conducted describing the efficacy of oral clindamycin reported that a 300-mg, twice-daily course of clindamycin for 7 days resulted in a 94 percent cure rate [15]. The two topical treatments for BV include metronidazole 0.75 percent vaginal gel and clindamycin 2 percent vaginal cream [5].

Probiotics have been documented to be beneficial in curing BV as well as reducing its recurrence and have been administered both orally and vaginally [20]. Oral administration introduces the beneficial bacteria directly into the vagina; probiotics consumed orally are believed to ascend to the vaginal tract after they are excreted from the rectum. Mechanism through which probiotics play a role in BV treatment include: [1] occupation of specific adhesion sites at the epithelial surface of the urinary tract; [2] maintenance of a low pH and production of antimicrobial substances like acids, hydrogen peroxide and bacteriocins; [3] degradation of polyamines; and [4] the production of surfactants with antiadhesive properties [21]. Probiotics have been shown to exert the beneficial effects both in foods such as yoghurt [22], ice cream [23, 24], and supplements [25]. However, foods may be preferred by patients since BV is not considered a disease by public and the affected women may not want to be prescribed supplements.

The purpose of the present chapter is to review recent research into aspects influencing the impact probiotics have on bacterial vaginosis. All papers published between 1990 and 2011 were searched in Pubmed and Science Direct, using probiotic, bacterial vaginosis and urinary tract infections (UTI) as key words; only clinical trials were included.

2. Probiotics

2.1. History

The expression "probiotic" was probably first defined by Kollath in 1953, when he suggested the term to denote all organic and inorganic food complexes as "probiotics" in contrast to harmful antibiotics, for the purpose of upgrading such food complexes as supplements [26]. In 1998, probiotics were described as "live microorganisms which, when ingested in adequate amounts, confer a health benefit". The term "probiotic" is an etymological hybrid derived from Greek and Latin meaning "for life" [27]. The original observation of the positive role played by some selected bacteria is attributed to Eli Metchnikoff, who extolled the virtues of consuming fermented dairy products and postulated his "Longevity without aging" theory, in which he claimed that lactic bacteria by replacing the harmful bacteria indigenous to the intestines, prolong life. The Russian born Nobel Prize recipient, working at the Pasteur Institute at the beginning of the last century suggested that the dependence of the intestinal microbes on the food makes it possible to adopt measures to modify the flora in our bodies and to replace the harmful microbes by useful microbes [28].

2.2. Definition

Presently, there is general agreement that a "probiotic" refers to viable microorganisms that promote or support a beneficial balance of the autochthonous microbial population of the gastrointestinal tract [29, 30]. Probiotics are defined as live microorganisms which, when consumed in appropriate amounts, confer a health benefit on the host, by FAO/WHO [31]. When ingested, some of these probiotic microorganisms are able to resist the physicochemical conditions prevailing in the digestive tract [32]. The strains most frequently used as probiotics belong to the genera bifidobacterium and Lactobacillus [33]. Some of the species used in probiotic products are: 1) Lactic acid producing bacteria (LAB): Lactobacillus, bifidobacterium, streptococcus; 2) Non-lactic acid producing bacterial species: Bacillus, propionibacterium; 3) Nonpathogenic yeasts: Saccharomyces; 4) Non-spore forming and non-flagellated rod or coccobacilli [31].

2.3. Health benefits

Some mostly documented health effects of probiotics are: relieving diarrhea, improving lactose intolerance, relief of respiratory and urinary tract infections and its immunomodulatory, anticarcinogenic, antidiabetic, hypocholesterolemic and hypotensive properties [25, 34, 35]. LAB also have some other advantageous effects such as vitamin synthesis, improvement of mineral and nutrient absorption, deprivation of antinutritional factors, and/or modulation of GI physiology and reduction of pain perception. Special probiotic strains may induce the expression of receptors on epithelial cells that locally control the transmission of nociceptive information to the GI nervous system [36]. By reducing inflammatory responses, probiotics have been shown to correct insulin sensitivity and reduce development of diabetes mellitus [34]. A beneficial effect of "lactic acid producing" microorganisms on vaginal microflora has also been suggested more than 100 years ago [37]. There are differing degrees of evidence supporting the verification of such effects, and the consultation recognizes that there are reports showing no clinical effects of certain probiotic strains in specific situations [38].

3. Probiotic and bacterial vaginosis

Since antimicrobial treatment of urogenital infections is not constantly effectual and problems remain due to bacterial and yeast resistance, recurrent infections as well as side effects, it is no wonder why alternative remedies are sought for, by patients and their caregivers [39, 40]. The basis for use of probiotics in BV treatment emerged in 1973, when healthy women with no history of UTI were reported to have lactobacilli in their vagina [39]. Lactobacillus organisms that predominate in the vagina of healthy women spread from their rectum and perineum and form a barrier to the entry of uropathogens from vagina into the bladder [41].

Probiotics are believed to protect the host against infections by means of several mechanisms including: [1] occupation of specific adhesion sites at the epithelial surface of the urinary tract; [2] maintenance of a low pH and production of antimicrobial substances like acids, hydrogen peroxide, and bacteriocins; [3] degradation of polyamines; and [4] the production of surfactants with antiadhesive properties [21, 42].

There are important issues to which a great attention must be paid regarding the effects of probiotics on BV treatment and prevention. Probiotics have been administered both orally and vaginally; however it is still not clear as to which route is more efficient. Foods and supplements have been used as carriers when oral administration was aimed; no studies have compared the efficacy of these two vehicles. Not all strains have exerted the desired effects in the patients; poor colonization of some strains in the vagina could be a reason [39, 40, 43]. The most profitable dose and treatment duration must be taken into consideration as well.

3.1. Route of administration

Probiotics must colonize the vagina to confer the benefits claimed for them; therefore they have to reach the organ intact. Vaginal probiotic capsules have widely been used, by the means of which, the probiotic bacteria are directly introduced into the vagina; however, in an attempt to come up with a more practical route which could also prevent BV in healthy women as well as presenting the consumer with other health benefits of these beneficial microorganisms, probiotics were administered orally [41, 43]. Researchers assumed that, similar to pathogenic bacteria with colonic origin which cause urogenital disorders, probiotic bacteria must be capable of ascending to the vaginal tract after being excreted from the rectum (Figure 1). This application is also justified by observations that the normal vaginal microflora colonizes from an intestinal origin which means that microbial ascension is a natural process actually contributing to a the development of a healthy vaginal microflora in the host [39]; this has been shown by a number of clinical trials as well [41, 44]. Thus far, no clinical trials have compared the efficacy of probiotics when administered vaginally versus orally. In tables 1-2, clinical trials performed in this regard have been summarized. It appears that vaginal administration has no predominance to oral consumption of probiotics, when it comes to treating BV.

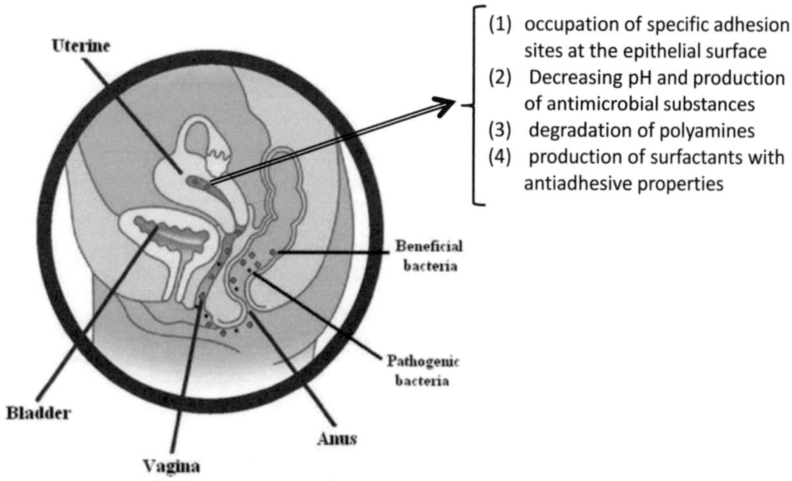

Figure 1. Capability of pathogenic and probiotic bacteria to ascend the vagina after being excreted from rectum

Type	Strain	Dose	Period	Heath condition	Effect	Ref.
Yoghurt	Lactobacillus acidophilus	1.0×10^8 CFU	Once daily for 2 month	Bacterial vaginosis, candidiasis	Reduction in BV episodes at 1 mo was 60% for probiotic yoghurt vs 25% for pasteurized	[45]
Capsules	Lactobacillus rhamnosus GR-1 plus Lactobacillus fermentum RC-14 or L. rhamnosus	10^8 CFU	Each day for 28 days	Bacterial vaginosis	Normal vaginal flora was restored using specific probiotic strains administered orally	[41]
Skim milk	Lactobacillus rhamnosus GR-1 and Lactobacillus fermentum RC-14	10^9 CFU	Given twice daily for 14 days	Bacterial vaginosis	Treatment correlated with a healthy vaginal flora in up to 90% of patients	[46]
Capsule	1- L. rhamnosus GR-1/ L. fermentum RC-14 2-L. rhamnosus GR-1/L. fermentum RC-14 3-L. rhamnosus GR-1/L. fermentum RC-14	8×10^8 CFU 1.6×10^9 CFU 6×10^9 CFU	Day orally for 28 days	History of BV	Through 6 weeks after treatment with probiotics, Nugent score decreased, indicative of BV resolution	[43]
Capsule	L. rhamnosus GR-1 + L. fermentum RC-14	$>10^9$ CFU	Once-daily for 60 days	Bacterial vaginosis	Probiotics colonized the vagina properly and the Nugent score normalized after the treatment	[47]
Capsules	Lactobacillus rhamnosus GR-1 and Lactobacillus fermentum RC-14	10^9 CFU	60 days	Urogenital infections	Lactobacilli counts increased while yeast and coliforms decreased significantly after supplementation	[48]
Capsule	Lactobacillus reuteri RC-14 Lactobacillus rhamnosus GR-1	10^9 CFU	Twice daily from days 1 to 30	Bacterial vaginosis	88% were cured in the antibiotic/probiotic group compared to 40% in the antibiotic/placebo group [$p < 0.001$]. High counts of Lactobacillus sp. Colonized the vagina properly	[49]
Capsules	Lactobacillus rhamnosus GR-1 and Lactobacillus reuteri RC-14	1.0×10^9 CFU	BID for 30, after 500 mg metronidazole BID PO for 7 d	Bacterial vaginosis	BV cure rate was 88% in probiotic group vs. 40% in placebo group	[42]
Capsules	Lactobacillus rhamnosus GR-1 and Lactobacillus reuteri RC-14	2.5×10^9 CFU	14 days	Bacterial vaginosis	The median difference in Nugent scores between baseline and the end of the study was 3 in the intervention group and 0 in the control group	[50]

Table 1. The effects of oral administration of probiotics on BV, performed between 1990 and 2011

Type	Strain	Dose	Period	Heath condition	Effect	Ref.
10–15 mL yoghurt, vaginal douche	L acidophilus	1.0×10^{8} CFU	BID for 7 d	First trimester of pregnancy with BV diagnosis	BV cure rate was 88% probiotic group at 4 and 8 w and 38% in control group	[51]
Vaginal tablets	L acidophilus and oestriol 0.03 mg	10^{6} CFU	Once daily or twice daily for 6 days	Bacterial vaginosis	Microbiological cure [Nugent criteria] and clinical cure were observed on days 15 and 28 post intervention	[1]
Tampons	L.gasseri, L casei var rhamnosus & L fermentum	10^{8} CFU	5 tampons during menstruation	Bacterial vaginosis	Microbiological cure was observed based on Nugent score and Amsel criteria	[2]
Vaginal tablet	Lactobacillus acidophilus, 0.03 mg oestriol and 600 mg lactose.	> 10^{7} CFU	Daily for 6 days	Vaginal infections	Vaginal flora was enhanced significantly by the probiotic administration in combination with low dose oestriol	[52]
Capsules	L rhamnosus GR1, L reuteri RC14	1× 10^{9} CFU	Bedtime for 5 consecutive days	Bacterial vaginosis	Microbiological cure at days 6, 15 and 30 and clinical cure at days 6, 15, and 30 were reported	[42]
Vaginal tablet	Lactobacillus rhamnosus	> 4×10^{4} CFU	Once a week at bedtime for two months	Bacterial vaginosis	Significant difference between the two treatment groups were seen at day 90	[53]
Vaginal tablets	L. brevis L. salivarius subsp salicinius , and L.plantarum	10^{9} CFU	7 days	Bacterial vaginosis	All of the patients in the probiotic group were free of BV, showing a normal or intermediate vaginal flora	[4]
Vaginal application	40 mg of Lactobacillus rhamnosus	> 4×10^{4} CFU	for 6 months	Prevent the recurrence of bacterial vaginosis	The vaginal administration of the probiotic allows stabilization of the vaginal flora and reduces BV recurrence	[54]
Vaginal Capsules	L gasseri LN40, Lactobacillus fermentum LN99, L. casei subsp. rhamnosus LN113 and P. acidilactici LN23	Between 10^{8} and 10^{10} CFU	Five days, after conventional treatment of bacterial vaginosis	Bacterial vaginosis, vulvovaginal candidiasis	LN had a good colonization rate in the vagina BV patients and women receiving LN were cured 2-3 days after Administration	[55]
Vaginal capsule	Lactobacillus rhamnosus, L acidophilus, and Streptococcus thermophiles	10^{8} CFU	21 days, for 7 days on 7 days off, and 7 days on.	Prophylaxis bacterial vaginosis	Probiotic prophylaxis resulted in lower recurrence rates for BV women	[3]

Table 2. The effects of vaginal administration of probiotics on BV, performed between 1990 and 2011

3.2. Administration vehicles

As for administration route, no studies by now have investigated the efficacy of foods versus supplements in exerting the benefits expected from the probiotics. Supplements have been used in a greater number of studies in BV patients and the number of studies in which foods were opted as probiotic vehicles are limited. Consumption of fermented milk containing lactobacilli has been found to reduce BV episodes [45]. Supplements have been used in a variety of forms including oral capsules, vaginal tablets and vaginal capsules. Clinical trials in which patients were administered oral capsules, reported a positive effect of the treatment on BV [39, 41-43, 49, 56]. Vaginal probiotic tablets were reported to be effective in alleviating BV symptoms and decreasing its recurrence [1, 4, 52, 53]. Vaginal capsules have also been reported to efficiently ease BV symptoms in some studies [3, 42, 50, 55].

3.3. Appropriate strains for treatment of bacterial vaginosis

Various in-vitro studies have shown that specific strains of lactobacilli inhibit the growth of bacteria causing BV by producing H2O2, lactic acid, and/or bacteriocins and/or inhibit the adherence of G. vaginalis to the vaginal epithelium [52]. According to a general theory a probiotic must have two criteria to be selected as an efficient strain in the treatment of urogenital infections: 1] It must be able to colonize the host without any adverse side effects and 2] It must be capable of inhibiting urogenital pathogens [57]. According to Reid and colleagues [43] different probiotic bacteria have varying capabilities to colonize the vagina of different patients; this indicates the importance of using a combination of strains in probiotic products. Oral administration of L. acidophilus, or intra-vaginal administration of L. acidophilus or L. rhamnosus GR-1 and L. fermentum RC-14, have been documented to most efficiently increase the numbers of vaginal lactobacilli, restore the vaginal microbiota to normal, and cure women of BV [52].

3.4. Appropriate dose for treatment of bacterial vaginosis

Researchers have tried different dosages in their attempts to treat BV with probiotics, many of which have resulted in positive outcomes. There is strong evidence that BV is most appropriately treated when over 10^8 viable organisms per day is used [41]. However, the minimum dose which can generally confer the favored benefits in women must to be determined.

3.5. Effect of treatment duration

What BV patients and their caregivers are mostly looking for, is a treatment protocol to get them rid of the recurrence of the infection. Probiotics are a good option to fulfill this goal, provided that they are properly colonized in the vagina. Parent and colleagues [1] found that cure was more common, and the number of vaginal lactobacilli was significantly higher, in women with BV at both 2 and 4 weeks after the start of a 6-day treatment with L. acidophilus and oestriol, when compared to women with BV who received a placebo. However, most clinical trials have reported that 2 months of oral administration of L. acidophilus, Lactobacillus rhamnosus GR-1 and L. fermentum RC-14 can be more effective in preventing recurrences of BV and/or increasing vaginal colonization with lactobacilli, thus restoring the normal vaginal microbiota [58].

4. Conclusion

This study confirms the potential efficacy of lactobacilli as a non-chemotherapeutic means to restore and maintain a normal urogenital flora, and shows that probiotic bacteria especially L. acidophilus, L. rhamnosus GR-1 and L. fermentum RC-14 when administered over 10^8 CFU for 2 months can most appropriately normalize vaginal flora, help cure the existing infection and prevent recurrence of BV. Longer periods of probiotic administration may be useful for long term control of BV relapses after conventional therapy with metronidazole.

Probiotics have been reported useful when used either vaginally or orally; foods and supplements have both been shown to be efficient vehicles as well; however, since BV is a common disorder for the prevention of which, the vaginal flora needs to be normal and devoid of pathogens by the help of beneficial bacteria, suggesting women to consume probiotic foods will not only protect them against BV, but will also reward them with other health benefits of probiotics.

Author details

Parvin Bastani

Women's Reproductive Health Research Center, Tabriz University of Medical Sciences, I.R. Iran

Aziz Homayouni and Violet Gasemnezhad Tabrizian

Department of Food Science and Technology, Faculty of Health and Nutrition, Tabriz University of Medical Sciences, I.R. Iran

Somayeh Ziyadi*

Department of Midwifery, Faculty of Nursing and Midwifery, Tabriz University of Medical Sciences, I.R. Iran

Acknowledgment

The authors would like to express their thanks to Dr. Vahid Zijah, Head of Research and Science Department of Behboud Hospital for financial support of this study.

5. References

[1] Parent D, Bossens M, Bayot D, Kirkpatrick C, Graf F, Wilkinson FE, et al. Therapy of bacterial vaginosis using exogenously-applied Lactobacilli acidophili and a low dose of estriol: a placebo-controlled multicentric clinical trial. Arzneimittelforschung. 1996;46[1]:68-73.

[2] Eriksson K, Carlsson B, Forsum U, Larsson PG. A double-blind treatment study of bacterial vaginosis with normal vaginal lactobacilli after an open treatment with vaginal clindamycin ovules. Acta Dermato-Venereologica. 2005;85[1]:42-6.

[3] Ya W, Reifer C, Miller LE. Efficacy of vaginal probiotic capsules for recurrent bacterial vaginosis: a double-blind, randomized, placebocontrolled study. American Journal of Obstetrics and Gynecology. 2010;203[2]:120-5.

[4] Mastromarino P, Macchia S, Meggiorini L, Trinchieri V, Mosca L, Perluigi M, et al. Effectiveness of Lactobacillus-containing vaginal tablets in the treatment of symptomatic bacterial vaginosis. Clinical Microbiology and Infection 2009;15[1]:67-74.

* Corresponding Author

[5] Fethers K, Fairley CK, Hocking J, Gurrin LC, Bradshow CS. Sexual risk factors and bacterial vaginosis: a systematic review and meta-analysis. Clinical Infectious Diseases 2008;47[11]:1426-35.

[6] Razzak MSA, Al-Charrakh AH, Al-Greitty BH. Relationship between lactobacilli and opportunistic bacterial pathogens associated with vaginitis. North American Journal of Medical Sciences. 2011;3[4]:185-92.

[7] Wilson JD, Ralph SG, Rutherford AJ. Rates of bacterial vaginosis in women undergoing in vitro fertilisation for different types of infertility. British Journal of Obstetrics and Gynaecology. 2002;109[6]:714-7.

[8] Sweet RL. Role of bacterial vaginosis in pelvic inflammatory disease. Clinical Infectious Diseases. 1995;20:271-5.

[9] Hauth JC, Goldenberg RL, Andrews WW, DuBard MB, Copper RL. Reduced incidence of preterm delivery with metronidazole and erythromycin in women with bacterial vaginosis. The New England Journal of Medicine. 1995;333[26]:1732-6.

[10] Cherpes TL, Melan MA, Kant JA, Cosentino LA, Meyn LA, Hillier SL. Genital Tract Shedding of Herpes Simplex Virus Type 2 in Women: Effects of Hormonal Contraception, Bacterial Vaginosis, and Vaginal Group B Streptococcus Colonization. Clinical Infectious Diseases. 2005;40:1422–8.

[11] Marrazzo JM. Evolving issues in understanding and treating bacterial vaginosis. Expert Review of Antiinfective Therapy 2004;2[6]:913-22.

[12] Jamieson DJ, Duerr A, Klein RS, Paramsothy P, Brown W, Cu-Uvin S, et al. Longitudinal analysis of bacterial vaginosis: findings from the HIV epidemiology research study. Obstetrics & Gynecology. 2001;98[4]:656-63.

[13] Verstraelen H, Senok AC. Vaginal Lactobacilli, probiotics and IVF. Reproductive BioMedicine Online. 2005;11[6]:674-5.

[14] Ness RB, Kip KE, Hillier SL, Soper DE, Stamm CA, Sweet RL, et al. A cluster analysis of bacterial vaginosis associated microflora and pelvic inflammatory disease. 162. 2005;6[585-590].

[15] Larsson PG, Bergström M, Forsum U, Jacobsson B, Strand A, Wölner-Hanssen P. Bacterial vaginosis. Transmission, role in genital tract infection and pregnancy outcome: an enigma. Acta Pathologica, Microbiologica et Immunologica. 2005;113[4]:233-45.

[16] Reece EA, Barbieri R. Obstetric and Gynocology. The essential of clinical care. Ney York: Thieme; 2010.

[17] Joesoef MR, Schmid GP, Hillier SL. review of treatment options and potential clinical indications for therapy. Clin Infect Dis 1999;28:57-65.

[18] Sobel J, Peipert JF, McGregor JA, Livengood C, Martin M, Robbins J, et al. Efficacy of clindamycin vaginal ovule [3-day treatment] vs. clindamycin vaginal cream [7-day treatment] in bacterial vaginosis. Infectious Diseases in Obstetrics and Gynecology. 2001;9[1]:9-15.

[19] Rauh VA, Culhane JF, Hogan VK. Bacterial vaginosis: a public health problem for women. Journal of the American Medical Women's Association 2000;55[4]:220-4.

[20] Senok A, Verstraelen H, Temmerman M, Botta GA. Probiotics for the treatment of bacterial vaginosis. The Cochrane Library. [Review]. 2009[4].

[21] Goldin BR, Gorbach SL. Clinical indications for probiotics: An overview. Clinical Infectious Diseases. 2008;46:96-100.

[22] Ejtahed, H. S., Mohtadi-Nia, J., Homayouni-Rad, A., Niafar, M., Asghari-Jafarabadi, M., Mofid, V. and Akbarian-Moghari, A. (2011). Effect of probiotic yogurt containing Lactobacillus acidophilus and Bifidobacterium lactis on lipid profile in individuals with type 2 diabetes mellitus. Journal of Dairy Science, 94: 3288-3294.

[23] Homayouni A, Azizi A, Ehsani MR, Yarmand MS, Razavi SH. Effect of microencapsulation and resistant starch on the probiotic survival and sensory properties of synbiotic ice cream Food Chemistry. 2008;111:50-5.

[24] Homayouni A, Azizi A, Javadi M, Mahdipour S, Ejtahed H. Factors influencing probiotic survival in ice cream: A Review. International Journal of Dairy Science. 2012.

[25] Homayouni Rad A, Vaghef Mehrabany E, Alipoor B, Vaghef Mehrabany L, Javadi M. Do probiotics act more efficiently in foods than in supplements? Nutrition. 2012;28:733-6.

[26] Mitsuoka T. Intestinal flora and human health. Asia Pacific Journal of Clinical Nutrition. 1996;5:2-9.

[27] Screzenmeir J, Vrese M. Probiotics, prebiotics, and synbiotics-approaching a definition. American Journal of Clinical Nutrition. 2001;73:361-4.

[28] Lourens-Hattingh A, Viljoen BC. Yogurt as probiotic carrier food. International Dairy Journal 2001;11[1-2]:1-17.

[29] Holzapfel WH, Habere P, Geisen R, Björkroth J, Schillinger U. Taxonomy and important features of probiotic microorganisms in food and nutrition. American Journal Clinical Nutrition. 2001;73[2]:365-73.

[30] Homayouni, A., Akbarzadeh, F. and Vaghef Mehrabany E. Which are more important: Prebiotics or probiotics? Nutrition (2012), doi:10.1016/j.nut.2012.03.017.

[31] Champagne CP, Ross RP, Saarela M, Hansen KF, Charalampopoulos D. Recommendations for the viability assessment of probiotics as concentrated cultures and in food matrices. International journal of food microbiology. 2011;149:185-93.

[32] Salminen S, Bouley C, Boutron-Ruault MC, Cummings JH, Franck A, Gibson GR, et al. Functional food science and gastrointestinal physiology and function. British Journal of Nutrition. 1998;80:147-71.

[33] Heyman M, Menard S. Probiotic microorganisms: how they affect intestinal pathophysiology. Cellular and Molecular Life Sciences. 2002;59[7]:1151-65.

[34] Lye HS, Kuan CY, Ewe JA, Fung WY, Liong MT. The Improvement of Hypertension by Probiotics: Effects on Cholesterol, Diabetes, Renin, and Phytoestrogens. International Journal of Molecular Sciences. 2009;10:3755-75.

[35] Wolvers D, Antoine JM, Myllyluoma E, Schrezenmeir J, Szajewska H, Rijkers GT. Guidance for substantiating the evidence for beneficial effects of probiotics: prevention and management of infections by probiotics. Journal of Nutrition. 2010;140[3]:698-712.

[36] Rousseaux C, Thuru X, Gelot A, Barnich N, Neut C, Dubuquoy L, et al. Lactobacillus acidophilus modulates intestinal pain and induces opioid and cannabinoid receptors. Nature Medicine. 2007;13[1]:35-7.

[37] Döderlein A. Das Scheidensekret und seine Bedeutung für das Puerperalfieber. Carolina: Nabu Press; 1892.

[38] Andersson H, Asp NG, Bruce A, Roos S, Wadstrom T, Wold AE. Health effects of probiotics and prebiotics: a literature review on human studies. Food and Nutrition Research. 2001;45:58-75.

[39] Reid G, Bruce AW. Urogenital infections in women: can probiotics help? Postgraduate Medical Journal. 2003;79[934]:428-32.

[40] Cribby S, Taylor M, Reid G. VaginalMicrobiota and the Use of Probiotics. Interdisciplinary Perspectives on Infectious Diseases. 2008.

[41] Reid G, Beuerman D, Heinemann C, Bruce AW. Probiotic Lactobacillus dose required to restore and maintain a normal vaginal flora. FEMS Immunology and Medical Microbiology. 2001;32[1]:37-41.

[42] Anukam K, Osazuwa E, Ahonkhai I, Ngwu M, Osemene G, Bruce AW, et al. Augmentation of antimicrobial metronidazole therapy of bacterial vaginosis with oral probiotic Lactobacillus rhamnosus GR-1 and Lactobacillus reuteri RC-14: randomized, double-blind, placebo controlled trial. Microbes and Infection. 2006;8[6]:1450-4.

[43] Reid G, Bruce AW. Selection of lactobacillus strains for urogenital probiotic applications. Journal of Infectious Diseases. 2001;183:77-80.

[44] Antonio MA, Rabe LK, Hillier SL. Colonization of the rectum by Lactobacillus species and decreased risk of bacterial vaginosis. Journal of Infectious Diseases. 2005;192[3]:394-8.

[45] Shalev E, Battino S, Weiner E, Colodner R, Keness Y. Ingestion of yogurt containing Lactobacillus acidophilus compared with pasteurized yogurt as prophylaxis for recurrent candidal vaginitis and bacterial vaginosis. Archives of Family Medicine. 1996;5[10]:593-6.

[46] Gardiner GE, Heinemann C, Baroja ML, Bruce AW, Beuerman D, Madrenas J, et al. Oral administration of the probiotic combination Lactobacillus rhamnosus GR-1 and L. fermentum RC-14 for human intestinal applications. International Dairy Journal. 2002; 12:191–6.

[47] Reid G, Burton J, Hammond JA, Bruce AW. Nucleic acid based diagnosis of bacterial vaginosis and improved management using probiotic lactobacilli. Journal of Medicinal Food. 2004;7:223-8.

[48] Reid G, Charbonneau D, Erb J, kochanowski B, Beuerman D, Poehner R, et al. Oral use of lactobacillus rhamnosus GR-1 and L.fermentum RC-14 significantly alters vaginal flora: randomized placebo-controlled trial in 64 healthy women. FEMS Immunology & Medical Microbiology. 2003;35[2]:131-4.

[49] Anukam KC, Osazuwa E, Osemene GI, Ehigiagbe F, Bruce AW, Reid G. Clinical study comparing probiotic Lactobacillus GR-1 and RC-14 with metronidazole vaginal gel to treat symptomatic bacterial vaginosis. Microbes Infect. 2006;8[12-13]:2772-6.

[50] Petricevic L, Unger FM, Viernstein H, Kiss H. Randomized, double-blind, placebo-controlled study of oral lactobacilli to improve the vaginal flora of postmenopausal women. European Journal of Obstetrics & Gynecology and Reproductive Biology. 2008;141[1]:54-7.

[51] Neri A, Sabah G, Samra Z. Bacterial vaginosis in pregnancy treated with yoghurt. Acta Obstetricia et Gynecologica Scandinavica. 1993;72:17-9.

[52] Ozkinay E, Terek MC, Yayci M, Kaiser R, Grob P, Tuncay G. The effectiveness of live lactobacilli in combination with low dose oestriol [Gynoflor] to restore the vaginal flora after treatment of vaginal infections. British Journal of Obstetrics and Gynaecology. 2005;112[2]:234-40.

[53] Marcone V, Calzolari E, Bertini M. Effectiveness of vaginal administration of Lactobacillus rhamnosus following conventional metronidazole therapy: how to lower the rate of bacterial vaginosis recurrences. New Microbiologica. 2008;31[3]:429-33.

[54] Marcone V, Rocca G, Lichtner M, Calzolari E. Long-term vaginal administration of Lactobacillus rhamnosus as a complementary approach to management of bacterial vaginosis. International Journal of Gynecology and obstetrics. 2010;110[3]:223-6.

[55] Ehrström S, Daroczy K, Rylander E, Samuelsson C, Johannesson U, Anzén B, et al. Lactic acid bacteria colonization and clinical outcome after probiotic supplementation in conventionally treated bacterial vaginosis and vulvovaginal candidiasis. Microbes and Infection. 2010;12[10]:691-9.

[56] Reid G, Bruce AW, Taylor M. influence of 3-day antimicrobial therapy and Lactobacillus vaginal suppositories on recurrence of urinary tract infections. Clin Ther. 1992;14:11-6.

[57] Reid G, Bruce AW, Taylor M. Instillation of Lactobacillus and stimulation of indigenous organisms to prevent recurrence of urinary tract infections. Microecology and therapy. 1995;23:32-45.

[58] Alvarez-Olmos MI, Barousse MM, Rajan L, Van Der Pol BJ, Fortenberry D, Orr D, et al. Vaginal lactobacilli in adolescents: presence and relationship to local and systemic immunity, and to bacterial vaginosis. Sexually Transmited Diseases. 2004;31[7]:393-400.

Chapter 21

Usefulness of Probiotics for Neonates?

Marie-José Butel, Anne-Judith Waligora-Dupriet and Julio Aires

Additional information is available at the end of the chapter

http://dx.doi.org/10.5772/51265

1. Introduction

1.1. Gut microbiota, health and diseases

In humans there are a multitude of site-specific communities of bacteria localized on the skin, mucosal surfaces, and in the intestinal tract [1,2]. The total number of prokaryotic cells is estimated to be around 10^{14}, ten times more than the number of eukaryotic cells. These microbial communities interact extensively with the host, a process which is crucial for host development and homeostasis. Most of the microbiota is located in the gastrointestinal (GI) tract, and progressively increase in number from the jejunum to the colon. In the colon, the levels of bacteria are as high as 10^{11} microorganisms per gram of luminal content with a very wide diversity. The composition of gut microbial communities was originally known through culture-based studies, which estimated that 400 to 500 different species are present in the adult human intestinal tract [3]. Through the most recent culture-independent analyses, gut microbiota is thought to comprise up to 1000 bacterial species per individual and over 5000 species in total [4]. The gut microbiota is dominated by only four phyla, i.e. Firmicutes, Bacteroidetes, Actinobacteria, and Proteobacteria, although there are more than 50 bacterial phyla on Earth [1].

Although the gut microbiota community was mostly studied in terms of pathogenic relationships for several decades, it is now recognized that most microorganism-host interactions in the gut are, in fact, commensal or even mutualistic [1,2]. This complex ecosystem has many functions which contribute to major roles for the host, including metabolic functions, barrier effects, and maturation of the immune system [5,6]. Indeed, bacterial colonic fermentation of non-digestible dietary residues and endogenous mucus is an important metabolic process in humans. The metabolites produced by this bacterial fermentation are mostly short-chain fatty acids (SCFAs) which supply energy and nutritive products to the bacteria, and trophic functions on the intestinal epithelium [7]. However, bacterial fermentation of proteins and peptides can also generate potentially pathogenic metabolites, such as phenol, amines, indols, and thiols [8]. The barrier effect refers to a resistance to colonization by exogenous or opportunistic bacteria that are at a low level in the gut [9]. Many mechanisms are thought to be responsible for this effect, including secretion of antimicrobial molecules, competition for nutrients, and attachment to ecological niches. These mechanisms also contribute to maintaining equilibrium in the microbial population of the gut. Finally, the gut microbial community has a major immune function [10].The intestinal immune system is separated from the gut microbiota by a single epithelial layer, which allows cross-talk between bacteria and the host. The commensal gut microbiota therefore profoundly influences the development of the intestinal adaptative immune system, being crucial for the development of gastrointestinal lymphoid tissue (GALT), homeostasis between T-helper 1 (Th1) and T-helper 2 (Th2) cell activity , as well as the acquisition of oral tolerance [10].

As the gut microbiota is greatly involved in the intestinal homeostasis, any dysbiosis could lead to dysfunctions. Hence, several diseases have been associated with alterations in the composition of the gut microbiota such as inflammatory bowel diseases (IBD) [11,12], irritable bowel syndrome (IBS) [13], and allergic diseases [14].

As IBD is concerned, although a direct pathogenic role for a specific agent has not been shown, there is evidence that autochthonous intestinal microbiota is involved (for review, see [15]). Several studies through culture-dependent and –independent analyses have reported differences in microbiota in patients suffering from IBD compared to healthy ones with less diversity in fecal microbiota [11] and higher numbers of mucosa-associated bacteria [16] in IBD patients. Indeed, IBD patients have fewer bacteria with anti-inflammatory properties and/or more bacteria with proinflammatory properties [15]. Likewise, some clinical studies reported differences in the composition of bacterial communities compared to period without allergic symptoms [17,18].

Irritable bowel syndrome (IBS) is defined by functional recurrent abdominal pain associated with abdominal distension and changes in bowel habits (constipation, diarrhea, or both). The etiology remains elusive; however, there is growing evidence of the role of gut microbiota in IBS [19].

Some recent studies have also suggested that obese individuals have a higher abundance of *Firmicutes* at the expense of Bacteroidetes in their gut microbiota compared with lean people [20,21]. This increase was reversed by surgically-induced or diet-induced weight loss [20,22]. Type 2 diabetes seems also to be associated with changes in gut microbial composition, regardless of body weight [23,24]. However, such associations have not been found by all authors [25]. Differences in the composition of gut microbiota have also been linked with type 1 diabetes [26].

Lastly, antibiotic courses have been shown to impact the microbiota with long term alterations [27,28]. Few studies investigated the health consequences of such alterations, but for *Clostridium difficile* colonization, responsible for antibiotic-associated diarrhea or pseudomembranous colitis [29].

These associations need to be confirmed in large studies. Moreover, it is still unclear whether the altered microbiota composition is a consequence rather than a cause of these disorders. Moreover, microbiota could promote disease in genetically susceptible hosts. Nevertheless, studies conducted to identify relationships between gut microbiota and diseases are a prerequisite to new approaches of therapeutics.

2. Probiotics, prebiotics, tools for modulating the gut microbiota

The associations of gut microbiota and diseases have given rise to the interest in manipulating gut microbiota as a new means of prevention or therapy. Indeed, some bacteria, mainly bifidobacteria and lactobacilli, have for a long time been thought to have beneficial health effects. They were firstly described by a few visionary scientists like Metchnikoff, Nissle, and Shirota about a century ago. This concept of "useful microbes" as written by Metchnikoff in his publication "On the prolongation of life" in 1907 [30] has led many years later to the use of "probiotic" strains to deliberately manipulate the microbiota. This concept has been forgotten during the expansion of the era of antibiotics and vaccines. However, research on the roles of the commensal microbiota gave a renewed interest for these beneficial microorganisms. Currently, probiotics are defined as "live microorganisms which when administered in adequate amounts confer a health benefit on the host" [31,32]. The most widely used probiotics include lactic acid bacteria, specifically *Lactobacillus* and *Bifidobacterium* species [33]. Although the efficacy of probiotics is sometimes debatable, they offer great potential benefits to health and are safe for human use, and their areas of interest are wide [34]. Effectiveness has been reported in the treatment and/or prevention of various gastrointestinal diseases, such as acute viral gastroenteritidis, antibiotic-associated diarrhea, pouchitis, and irritable bowel syndrome [33,35,36]. Some beneficial effects have also been reported in ulcerative colitis, ventilator-associated pneumonia, functional constipation, and reduction of cholesterol (see [34] for review).

Their beneficial effects could be through the production of metabolites, such as short chain fatty acids or other small molecules, or the bacterial components, such as DNA or peptidoglycan. However, these effects are strain-specific and further work is still required to confirm their benefits to health.

Modulation of the gut microbiota can be also achieved by the use of prebiotics. Prebiotics are defined as non-digestible dietary components that beneficially affects the host by selectively stimulating the growth and/or the activity of one or a limited number of bacteria in the colon, and thus improves host health [37]. They are mainly oligosaccharides, and bacteria mainly enhanced are bifidobacteria. Their potential interest lies in the fact that their effect is linked to a modification of the equilibrium of the autochthonous gut microbiota and not to a single or a limited number of exogenous strain(s) as for probiotics. Moreover, in terms of safety, they have not the side effect of probiotic supplementation, for which systemic translocation of the ingested live bacteria has been reported in some cases during probiotic uses [38]. Prebiotic supplementation has been less studied than probiotic supplementation. Although prebiotic supplementation leads constantly to an increase in gut bifidobacteria levels, their effects in terms of health benefits of an early use of infant formula enriched with prebiotics appear with limited or unclear clinical significances [39]. Thus, the Committee on Nutrition of the European Society for Paediatric Gastroenterology, Hepatology, and Nutrition (ESPGHAN) did not recommend the routine use of prebiotic-supplemented formula [39]. However, no adverse effects have been observed.

The increase use of association of probiotics and prebiotics, named "synbiotic" is appealing. However, a very limited number of such supplementation has been studied in infants. An alternative option is the use formulas fermented with lactic acid-producing bacteria during the production process that are subsequently inactivated by heat or other means at the end of the process [40]. This leads to a probiotic/prebiotic activity likely related to both production of active bacterial metabolites such as transoligosaccharides and presence of bacterial components such as cell membrane and DNA [41,42]. The limited number of studies on this kind of formula does not allow general conclusions to be drawn on the use and effects of fermented formulae [40]. It is recommended that the observed effects should be assessed in further randomized controlled trials.

Both uses of prebiotics and synbiotics in neonates are not included in the present review.

3. Gut bacterial establishment

The formation of the intestinal ecosystem starts rapidly during the neonatal stage of life (see [43,44] for review). Colonizing bacteria originate mainly from the mother; the gut microbiota is a major source. Other sources include the microbiota of the vagina, perineum, skin, and even breast milk [45,46]. The first colonizing bacteria are facultative anaerobes due to the abundance of oxygen in the gut. This decreases the redox potential in the gut lumen, creating a reduced environment that favors the establishment of obligate anaerobes [43]. However, little is known about the factors that lead to the establishment of specific bacterial strains. Then, during the infant stage of life, numerous bacteria are encountered in the environment including the skin microbiota of parents, siblings, nurses, and foods. Hence, over time, successively larger numbers of bacteria are established in the infant gut, and these are mainly comprised of obligate anaerobes. This leads to a high interindividual variability in the composition and patterns of bacterial colonization during the first weeks of life. By the end of the first year of life, the gut bacterial composition converges toward an adult-like microbiota profile [47].

Various external factors can affect the pattern of bacterial colonization, i.e. mode of delivery, mode of infant feeding, and environment [43,44]. Infants born by cesarean section are deprived of contact with their mother's gut and vaginal microbiota, which decreases bacterial diversity and colonization by obligate anaerobes such as bifidobacteria and *Bacteroides* [48,49]. The mode of infant feeding also strongly influences bacterial establishment, the hallmark being a dominant colonization by bifidobacteria in breastfed infants compared with formula-fed ones. However, improvements in infant formulas have led to only minor differences in colonization following each feeding method [43,44].

Moreover, changes in the establishment of gut microbiota have been observed in modern Western infants, most likely due to improved hygiene and general cleanliness in Western countries, resulting in reduced bacterial exposure [43,44]. Finally, gestational age can also affect bacterial colonization. Preterm birth leads to a delayed and abnormal pattern of microbial colonization in the gut [50-53]. In particular, colonization by beneficial bacteria such as bifidobacteria, which are normally dominant in fullterm babies, is delayed especially in very and extremely preterm neonates [54].

4. Gut microbiota and pediatric diseases: a rational for probiotic use in neonates

The early bacterial pattern in the first weeks of life appears to be a crucial step in the establishment of the various functions of the gut microbiota. In fact, recognition of self– and non–self–antigens begins early in life, perhaps even *in utero* [55]. Maturation of the intestinal immune system is thought to be significantly affected by the sequential bacterial establishment [10,56]. Indeed, at birth, the lymphoid system is not yet mature even though it is developed and the fetus is in a Th2 immunological context, and Th1 responses are repressed in order to avoid its rejection [57]. Therefore, after birth, the newborn must quickly restore the Th1/Th2 balance. The existence of a rich microbial environment is thought to be important in this process, the first bacteria to colonize the infant's gut being the first stimuli for post-natal maturation of the T-helper balance. The immature Th2-dominant neonatal response undergoes environment-driven maturation via microbial contact during the early postnatal period resulting in a gradual inhibition of the Th2 response and an increase of the Th1 response and prevention of allergic diseases which are Th2 linked, a basis of the so-called "hygiene hypothesis" [56].

Late-onset diseases could be therefore associated with an impairment of this step, all the more as early impairment in bacterial establishment can have long term effects in terms of bacterial pattern [58] as well as in terms of immune maturation [49,59]. Indeed, a large number of studies have shown that an imbalance of the numbers of Th1 and Th2 cells may be at the origin of a great variety of disease processes.

The first disease associated to this imbalance is allergy. Thus, the initial composition of the infant gut microbiota may be a key determinant in the development of atopic disease [60]. This hypothesis is consistent with the delayed colonization of the digestive tract associated with changes in lifestyle over the last 15 years in Western countries [43,44], where incidence of allergic diseases had sharply increased since a decade. Moreover, factors known to modify establishment of the gut microbiota, e.g. birth through caesarian section [61,62], prematurity [63], and exposure to antibiotics during pregnancy [64] have been associated with a higher risk of atopic disease. This hygiene hypothesis implicating a relationship between allergic diseases and gut microbiota is supported by several clinical studies which reported differences in the composition of the fecal microbiota between infants who live in countries with high or low prevalence of allergy, as well between infants with or without allergic diseases. In fact, several reports have associated allergic diseases with abnormal bacterial pattern. Low diversity [65] and low levels of bifidobacteria have been associated with allergy development [66,67], as well as high levels of clostridia [14,66]. A recent study revealed differences in the abundance of *Bifidobacterium* and enterobacteria among 7 cesarean-delivered infants with and without eczema over a 2 year-follow-up and preceding the apparition of the symptoms [68].

Likewise, early alterations in the gut microbiota have been linked with the risk of later overweight or obesity associated with lower levels of bifidobacteria and higher levels of *Staphylocccus aureus* during the first year of life [69].

For many years, a number of studies have documented differences between patients suffering from inflammatory bowel diseases and healthy persons, even if there is still debate about whether changes precede or follow the development of IBD [70]. For instance, a decreased prevalence of dominant members of the human commensal microbiota, i.e. *Clostridium* IXa and IV groups, *Bacteroides*, bifidobacteria and a concomitant increase in detrimental bacteria, i.e. sulphate-reducing bacteria and *Escherichia coli* has been reported [71]. A pilot study found differences in mucosa-associated bacteria in duodenal mucosa with higher number of aerobic and facultative-anaerobic bacteria and a decrease in *Bacteroides*, a strictly anaerobic genus in pediatric IBD patients compared to control patients [72]. This peculiar microbial profile, with higher diversity in duodenal mucosa from children suffering from celiac disease and the specific harmful role of *Escherichia coli* supported the idea of a disease associated with the gut microbiota environment [73,74]. Other studies reported decrease in fecal and duodenal bifidobacteria populations in celiac patients [75].

Lastly, associations between intestinal microbiota and autism have been reported such as the overgrowth of neurotoxin-producing clostridia [76]. Several reports indicate that certain clusters of clostridia are present in higher levels in fecal microbiota from autistic infants [77,78]. Overgrowth of *Desulfovibrio* sp may also lead to direct damage through interaction between the host and lipopolysaccharide and sulfate reduction [79].

Hence, although a causal relationship has not been categorically established, there is emerging evidence that the initial gut bacterial colonization during the first weeks of life is of great importance for infant health. Perinatal determinants altering the colonization pattern could therefore lead to a higher risk of later diseases. For instance, as already mentioned, infants born through cesarean section and therefore colonized by an altered bacterial pattern as compared with vaginally delivered ones have been reported to be at higher risk of either allergic diseases [80-82], or celiac disease [83], or obesity [84-86], or type 1 diabetes [87]. A prolonged breast-feeding over one year has been linked to a lower risk of overweight or obesity [88]. Likewise, changes in the establishment of gut microbiota observed in modern Western infants result in reduced bacterial exposure [43,44]. Thus, these infants lack of adequate bacterial stimuli, leading to a deviated maturation of their immune system likely responsible for a higher risk of allergic disease development or inflammatory bowel diseases [56].

5. Probiotics in fullterm neonates

The potential benefits of the use of probiotics in pediatrics have been recently reviewed [89,90]. It mainly includes treatment acute viral gastroenteritis [91], prevention of antibiotic-associated diarrhea [92,93], reduction of the inflammatory response in IBD patients [11]. Limited effects have been observed in colicky infants [94]. However, a recent study reported a clear improvement of the symptoms of colic within one week of *Lactobacillus reuteri* administration as compared with simethicone treated infants [95] linked to an antimicrobial effect against six species of gas-forming coliforms isolated from the colicky infants [96].

Given the likely link between the early bacterial pattern and later health status reported, a very early administration of probiotics when the gut microbiota is not fully established is of great interest and we have focused this review on this approach. Many attempts of early probiotic supplementation have been made for a long time, and numerous studies related to the use of infant formula supplemented with probiotics strains have been recently published [39]. This early use is reported to have some beneficial effects in terms of prevention of late development of some diseases. Administration is often given soon after birth, and the duration is variable according to the study, but often prolonged over several weeks or months. Lastly, dosages varied, ranging from 10^6 to ~10^9 CFU/mL or/g. The most frequently studied probiotic strains were *Bifidobacterium animalis* subsp *lactis, B longum, Lactobacillus rhamnosus, L reuteri, L johnsonii* and *Streptococcus thermophilus,* used alone or in combination.

Some studies have included the effects of such supplementation on growth. However, no significant effects have been shown on growth, but without any negative results [39]. Likewise, no reduction of gastrointestinal or respiratory infections, or reduction of antibiotic use have been reported, but a limited number of studies investigated such effect, avoiding to drawn final conclusions. Moreover, one difficulty to assess the health-promoting effects lies in the fact that the probiotics properties are strain-dependent and the use of different strains could explain the discrepancies between the observed effects. Second, mechanism(s) of action of the probiotics is not always well-established. Probiotics can have health-promoting effects related to their interaction with the gut microbiota, the barrier functions and the immune system. In particular, probiotic supplementations were shown to impact the intestinal maturation as reported with *Bifidobacterium lactis* supplementation of preterm infants which induced the maturation of the intestinal IgAs response [97]. Likewise, in fullterm neonates an infant formula containing two strains of probiotics allowed the preservation of high SIgA levels at 6 months compared to the control group [98]. Furthermore, such supplementation was suggested to have a synergistic effect on gut humoral immunity at 12 months of age, since it has shown that significant higher level of total IgM, IgA, and IgG titers was detected in infants who had been breastfed exclusively for at least 3 months and supplemented with probiotics compared with those breastfed receiving placebo [99]. Probiotic strains can also improve the intestinal barrier functions by inducing mucin production. Besides, they can interact directly with intestinal bacteria through secretion of bioactive factors preventing changes in tight junction proteins during inflammation [100].

The prevention of allergy through such early administration of probiotics is appealing. Though evidence of their effect is conflicting, their administration to infants at high risk for atopy and/or to their mothers seems to be effective for preventing infants from developing atopic disease [101,102]. Four studies investigated probiotic supplementation begun during pregnancy. Administration of *Lactobacillus* GG to the mother during pregnancy and breast-feeding appears to be a safe and effective method for enhancing the immunoprotective potential of breast milk and preventing atopic eczema in the infant [103,104], with a protective effect up to 7 years [105]. However, this preventive effect was not confirmed in a similar study by Kopp *et al*, may be due to differences in the study populations [106]. *L reuteri* supplementation in infants with a family history of allergic disease did not confirm a preventive effect against infant eczema but found a decreased prevalence of IgE-associated eczema during the second year [107]. Infants receiving *L rhamnosus* had a significantly lower risk of eczema than infants receiving placebo, but this was not the case for *B animalis* subsp *lactis* and there was no significant effect of these two strains on atopy [108]. Other trials consisting of supplementation with various probiotics strains only in infants from birth to 6 months of life did not find any reduction of the risk of atopic disease in high-risk infants [109-111]. Discrepancies between the observed effects could be linked to the various probiotics strains used. Indeed, the mechanism of their action could be through the maturation of the immune system, as suggested by the study of Roze *et al* where low levels of IgAs in the control group has been associated with atopy [98].

These data led the Nutrition Committee of ESPGHAN to conclude that there is too much uncertainty to draw reliable conclusions [39], confirmed through a recent review [112]. However, the Cochrane Database of Systematic Reviews claimed that there is a possible role a probiotics intervention in prevention of atopic dermatitis [113]. These promising results associated to the fact that the impact on the immune system has been shown to be strain-dependant [114] highlighting the importance of the choice of the probiotic strain argue for further studies in this field.

Identifying through animal studies and clinical studies a possible link between gut microbiota and obesity [69,84,86] may offer promising strategies through the gut modulation to prevent obesity. The intestinal microbiota may contribute to the development of inflammation and insulin resistance leading to overweight or obesity, either by its role in the regulation of energy homeostasis and fat storage or by the chronic inflammation it could induce, or both [21,115]. Reducing the susceptibility to obesity by early probiotics intervention would be a useful adjunct in strategies to alleviate the huge burden of childhood obesity which can be a risk factor for later diseases such as type 2 diabetes, hypertension and coronary heart disease [116]. The findings of early differences in microbiota of infants who later become overweight or obese [69] argues for an early intervention. Likewise, differences in obese and non obese children has been found [117,118]

Up to now, only one study on the effects on obesity of early probiotics supplementation has been conducted [119]. Pregnant women (n=159) were randomized and double-blinded to receive *L rhamnosus* or placebo 4 weeks before expected delivery; the intervention extending for 6 months postnatally. Anthropometric measurements were taken over 10 years. This perinatal probiotic administration appeared to moderate the initial phase of excessive weight gain, especially among children who later became overweight, but not the second phase of excessive weight gain, the impact being most pronounced at the age of 4 years. The effect of intervention was also shown as a tendency to reduce the birth-weight-adjusted mean body mass index at the age of 4 years. Another controlled trial has been performed but on children between 12 and 15 of age over a 12-week period [120]. The probiotics used was *L salivarius* and the objective was to investigate the effect of the probiotics supplementation on markers of inflammation and metabolic syndrome, showing no beneficial effects on these markers. This may be highlights again the usefulness of an early intervention before the onset of the clinical and/or biological signs.

6. Probiotics in preterm neonates

6.1. Gut bacterial establishment in preterm neonates

The current more obvious interest of probiotics use in neonates is very likely for preterm infants. In fact, preterm infants, and particularly those who are born at a low or very low gestational age and/or birth weight experience a delayed and abnormal pattern of gut colonization, particularly with regard to bifidobacteria and lactobacilli, normally dominant in healthy full term infants. The first studies on the gut bacterial colonization in preterm infants, based on culture methods and performed in the 80s, described a delayed colonization by many of the bacteria found in healthy fullterm infants [121-123]. However, more recent studies reported a greater delay either by culture [124-126] or culture-independent methods [50,124,126-130]. Recently, the use of a pyrosequencing-based method confirmed this aberrant pattern in low and very low birth weight infants [52].

The predominant facultative bacterial species in the fecal microbiota of preterm infants undergoing intensive care are staphylococci. Enterobacteria (mainly *Klebsiella* sp and *Enterobacter* sp) and enterococci are slightly delayed. Clostridia are the most common anaerobes during the first weeks of life, often the dominant anaerobic microbiota [124,126,131]. In contrast, *Bacteroides* and in particular bifidobacteria – known for their potential beneficial effects – seldom colonize preterm infants by contrast with fullterm infants [50,54,124]. Moreover, gestational age appears a major factor influencing their establishment [50,54]. Finally, the hospital environment can influence the bacterial pattern [131].

This bacterial establishment is the expression of colonization from the environment rather from maternal origin. A combination of more frequent birth through cesarean section, large antibiotic use, delayed initiation of enteral feedings, and exposure to the unusual microorganisms that populate the neonatal intensive care units may explain this abnormal pattern of colonization.

This impaired intestinal colonization may predispose preterm infants to diseases. Indeed, they are at high risk to acquire recurrent bacterial infections during their first weeks of life.

Both the permanent exposure to microorganisms due to frequent invasive procedures and the immaturity of the newborn immune system are responsible for the increased susceptibility to severe nosocomial infections. Early-onset sepsis remain an important cause among very preterm infants [132], thought to be due – at least partly – to the gut microbiota, Gram negative bacilli being the most frequent bacteria encountered in sepsis by contrast with fullterm infants [132]. Recent studies have demonstrated the origin of gut bacteria in these infections [133,134]. Besides, necrotizing enterocolitis (NEC) remains an important cause of morbidity and mortality among very preterm infants. Despite many investigations, its pathogenesis remains unclear [135]. The hypothesis that intestinal microbes are necessary for the development of NEC is supported by several lines of evidence [136]. No specific bacteria or bacterial pattern has been causally associated with the development of NEC although bacterial colonization is recognized as an important factor [137-139]. Implication of bacteria is thought to be due to fermentation of non-hydrolyzed lactose, a consequence of the immaturity of the intestinal lactasic equipment in preterm infants [140-142]. The genus *Clostridium* seems to be important in the pathogenesis of NEC [139,143,144], but other genera could be involved [51,130,145]. A decrease in microbial diversity [130] or an increase in enterococci and *Citrobacter* gene sequences in NEC infants has been observed [51].

Lastly, the very abnormal pattern observed particularly in VLBW infants could lead to an abnormal maturation of the functions of the intestinal ecosystem. Indeed, it could be a factor to develop late-onset disease such as allergy, obesity, such as suggested with a higher risk of allergy in infants born with a very low birth weight (VLBW)[63].

6.2. Probiotics in preterm neonates

Feeding oral probiotic bacteria may be therefore an effective way to change the abnormal pattern of colonization of preterm infants, and to have the potential to prevent the occurrence of gastrointestinal disorders in preterm infants. A relatively small number of trials have studied the effects of probiotics in those preterm infants. However, numerous meta-analyses or reviews (with a higher number than clinical trials, highlighting the great interest in this approach) have shown the potential benefits of such supplementation, leading to a significant and somewhat impressive reduction of all-cause mortality and NEC by more than half [146-148]. As for an example, the metaanalyse from the Cochrane Collaboration included 16 studies with 1371 infants treated with probiotics and 1376 controls [146]. Various probiotic strains have been used, i.e. lactobacilli, bifidobacteria or a combination of 2 or 3 strains. The most frequent *Lactobacillus* used was LGG. For bifidobacteria, *breve* and *longum* were the most frequent species administered. One study used *Saccharomyces boulardii.* Conclusions of this metaanalyse are concordant with other ones, with a significant decrease in the incidence of severe NEC (stage II or more) and of mortality. As highlighted for other applications, the effect is certainly strain-dependent with studies that did not found any beneficial supplementation regarding the incidence of NEC [149].

Other beneficial effects have been reported as a shortened time to full feeds. By contrast, if there is a trend toward a reduction of nosocomial sepsis, it does not reach the significance.

These beneficial effects are less obvious in extremely preterm infants, born with a very low birthweight (1000g or less, VLBW infants) [146]. This could be related with the fact that the probability to be colonized by probiotic strains diminished with decreasing birth weight [126]. Hence, in this latter study the improvement of gastrointestinal tolerance to enteral feeding was only reported in infants born with a birthweight >1000g. As infants weighting 1000g or less received antibiotic treatment more frequently, and had more frequent interruptions of enteral feeding than did infants weighing more than 1,000g, these findings suggest that these factors could prevent gut colonization by the probiotic strains, and, consequently, the capacity of probiotics to enhance intestinal function in extremely low birth weight infants [126].

Conclusions of the numerous reviews and metaanalyses strongly suggest that the use of probiotics in preterm infants could prevent tens of thousands of deaths annually. Hence, some authors recommend that it is time to change practice and to adopt the use of probiotics as a standard care in preterm infants [146,150]. However, controversies have emerged because there are yet too many unknowns about probiotics use [151,152]. One aspect concerns the safety although no negative effects have been reported even in long term follow-up [153]. However, data on this latter aspect are very scarce. Infrequent, systemic translocation of probiotics has been reported [38,154] raising some concerns about this side effect in the high-risk groups of low and very low birth weight infants who are characterized by high intestinal permeability, making this potential powerful tool a double-edge weapon. Increased incidence of NEC following probiotic administration has been observed in a preterm piglet model, may be related to the specific strain, dose, and the very immature gut immune system.[155]. A study in a pediatric unit even reported a trend toward an increase in nosocomial throughout a probiotic supplementation [156] although a routinary supplementation of VLBW infants with a probiotics strains over a 6-year period was safe [157].

To conclude, although there is encouraging data for the use of probiotics in particular in terms of NEC prevention, it may be reasonable to stand back from a routine use of probiotics in preterm infants. As suggested by several authors, probiotics supplementation should be a local decision [158-161]. Several questions have been raised. What is the interest of probiotic supplementation in units with low incidence of NEC? What are the mechanisms of action, which are not elucidated, in particular due to the lack of gut microbiota analyses in most of the studies? What are the beneficial effects apart reduction of incidence and severity of NEC, in particular concerning sepsis, since some results are promising, but large clinical trials are needed, as the ongoing study in Australia and New Zealand [162]. What is the safety of the various strains? Which product(s) should be administered, at what dose, when, and for how long [163]? Lastly, no general recommendation can be done currently for the special group of the VLBW infants regarding the lack of benefits of probiotics supplementation [146,160]. Further studies are thus recommended in this target population.

Lastly, no study had investigated the potential beneficial long-term effect of an early probiotics supplementation in terms of reduction of the risk of late-onset disease linked to an early dysbiosis such allergy and obesity for instance.

The Committee on Nutrition of ESPGHAN concluded – in a commentary published in 2010 – that there is not enough available evidence for a routine use of probiotics in preterm infants [164]. However, faced to some evidence of benefits of probiotics in preterm infants, guidelines have been proposed aiming at optimizing their use, emphasing that "routine" use does not equate "blind" use of probiotics, and raising the necessity to continue research in this field to provide answers to the current gaps [159].

7. Conclusion

The notion of "gut health" has become more and more popular. Currently, it is recognized that the gut microbiota contributes to the host health not only by assuming digestion and absorption of nutriments, but also by maturation of the immune system, defense against infection, signaling to the brain…

This leads to not only study the gut microbiota communities in terms of pathogenic relationships, as it was done for several decades, but also to study the endogenous microbiota and to investigate microorganism-host interactions in the gut that are, in fact, commensal or even mutualistic. Hence, currently several disease, which clinical symptom can be late in the life, are linked to dysbiosis that often occurred in the early step of gut colonization.

We need to learn more about the composition and functions of the gut microbiota and to the concept of early modulation of this microbiota. Thus, we are currently at the beginning of the era of probiotics which aim at counteracting deleterious effect of microorganisms with probiotics instead of using vaccines and antibiotics. This new field of medical microbiology is appealing and fascinating.

The current review aimed at giving the rational of the use of probiotics for promotion of health and prevention of disease through their use early in life when the gut microbiota is not fully established.

Several applications are claimed among them, some are appealing such as prevention of allergy. However, up to now, there are not enough data to recommend their routine use. But the potential interest in this field argues to do further research to validate the current beneficial results observed.

The most clear potential interest of early probiotic supplementation lies in taking care of preterm neonates, who are often colonized by an aberrant microbiota leading to high risks of early or late-onset of disease. Probiotic supplementation has been demonstrated to have benefits in terms of prevention of NEC. However, too many questions remain unanswered to recommend their routine use. One major concern is the safety linked to the ingestion of live microorganisms by an immature host. Hence, once again further research is needed in this exiting field with potential of health benefits.

Author details

Marie-José Butel, Anne-Judith Waligora-Dupriet and Julio Aires
Intestinal ecosystem, probiotics, antibiotics (EA 4065), Paris Descartes University, Faculty of Pharmaceutical and Biological Sciences, Paris, France

8. References

[1] Dethlefsen L, McFall-Ngai M, Relman DA. An ecological and evolutionary perspective on human-microbe mutualism and disease. Nature 2007,449:811-818.

[2] Bik EM. Composition and function of the human-associated microbiota. Nutr Rev 2009,67 Suppl 2:S164-S171.

[3] Manson JM, Rauch M, Gilmore MS. The commensal microbiology of the gastrointestinal tract. Adv Exp Med Biol 2008,635:15-28.

[4] Zoetendal EG, Rajilic-Stojanovic M, De Vos WM. High-throughput diversity and functionality analysis of the gastrointestinal tract microbiota. Gut 2008,57:1605-1615.

[5] Fujimura KE, Slusher NA, Cabana MD, Lynch SV. Role of the gut microbiota in defining human health. Expert Rev Anti Infect Ther 2010,8:435-454.

[6] Sekirov I, Russell SL, Antunes LC, Finlay BB. Gut microbiota in health and disease. Physiol Rev 2010,90:859-904.

[7] Wong JM, de SR, Kendall CW, Emam A, Jenkins DJ. Colonic health: fermentation and short chain fatty acids. J Clin Gastroenterol 2006,40:235-243.

[8] Blachier F, Mariotti F, Huneau JF, Tome D. Effects of amino acid-derived luminal metabolites on the colonic epithelium and physiopathological consequences. Amino Acids 2007,33:547-562.

[9] Stecher B, Hardt WD. The role of microbiota in infectious disease. Trends Microbiol 2008,16:107-114.

[10] Round JL, Mazmanian SK. The gut microbiota shapes intestinal immune responses during health and disease. Nat Rev Immunol 2009,9:313-323.

[11] Sartor RB. Microbial influences in inflammatory bowel diseases. Gastroenterology 2008,134:577-594.

[12] Reiff C, Kelly D. Inflammatory bowel disease, gut bacteria and probiotic therapy. Int J Med Microbiol 2010,300:25-33.

[13] Collins SM, Denou E, Verdu EF, Bercik P. The putative role of the intestinal microbiota in the irritable bowel syndrome. Dig Liver Dis 2009,41:850-853.

[14] Penders J, Thijs C, van den Brandt PA, Kummeling I, Snijders B, Stelma F, Adams H, van RR, Stobberingh EE. Gut microbiota composition and development of atopic manifestations in infancy: the KOALA Birth Cohort Study. Gut 2007,56:661-667.

[15] Chassaing B, Darfeuille-Michaud A. The commensal microbiota and enteropathogens in the pathogenesis of inflammatory bowel diseases. Gastroenterology 2011,140:1720-1728.

[16] Nishikawa J, Kudo T, Sakata S, Benno Y, Sugiyama T. Diversity of mucosa-associated microbiota in active and inactive ulcerative colitis. Scand J Gastroenterol 2009,44:180-186.

[17] Ouwehand AC, Nermes M, Collado MC, Rautonen N, Salminen S, Isolauri E. Specific probiotics alleviate allergic rhinitis during the birch pollen season. World J Gastroenterol 2009,15:3261-3268.

[18] Odamaki T, Xiao JZ, Iwabuchi N, Sakamoto M, Takahashi N, Kondo S, Miyaji K, Iwatsuki K, Togashi H, Enomoto T, Benno Y. Influence of Bifidobacterium longum BB536 intake on faecal microbiota in individuals with Japanese cedar pollinosis during the pollen season. J Med Microbiol 2007,56:1301-1308.

[19] Dahlqvist G, Piessevaux H. Irritable bowel syndrome: the role of the intestinal microbiota, pathogenesis and therapeutic targets. Acta Gastroenterol Belg 2011,74:375-380.

[20] Ley RE, Turnbaugh PJ, Klein S, Gordon JI. Microbial ecology: human gut microbes associated with obesity. Nature 2006,444:1022-1023.

[21] Musso G, Gambino R, Cassader M. Obesity, diabetes, and gut microbiota: the hygiene hypothesis expanded? Diabetes Care 2010,33:2277-2284.

[22] Zhang H, DiBaise JK, Zuccolo A, Kudrna D, Braidotti M, Yu Y, Parameswaran P, Crowell MD, Wing R, Rittmann BE, Krajmalnik-Brown R. Human gut microbiota in obesity and after gastric bypass. Proc Natl Acad Sci U S A 2009,106:2365-2370.

[23] Wu X, Ma C, Han L, Nawaz M, Gao F, Zhang X, Yu P, Zhao C, Li L, Zhou A, Wang J, Moore JE, Millar BC, Xu J. Molecular characterisation of the faecal microbiota in patients with type II diabetes. Curr Microbiol 2010,61:69-78.

[24] Larsen N, Vogensen FK, van den Berg FW, Nielsen DS, Andreasen AS, Pedersen BK, Al-Soud WA, Sorensen SJ, Hansen LH, Jakobsen M. Gut microbiota in human adults with type 2 diabetes differs from non-diabetic adults. PLoS One 2010,5:e9085.

[25] Duncan SH, Lobley GE, Holtrop G, Ince J, Johnstone AM, Louis P, Flint HJ. Human colonic microbiota associated with diet, obesity and weight loss. Int J Obes (Lond) 2008,32:1720-1724.

[26] Brugman S, Klatter FA, Visser JT, Wildeboer-Veloo AC, Harmsen HJ, Rozing J, Bos NA. Antibiotic treatment partially protects against type 1 diabetes in the Bio-Breeding diabetes-prone rat. Is the gut flora involved in the development of type 1 diabetes? Diabetologia 2006,49:2105-2108.

[27] De La Cochetiere MF, Durand T, Lepage P, Bourreille A, Galmiche JP, Dore J. Resilience of the dominant human fecal microbiota upon short-course antibiotic challenge. J Clin Microbiol 2005,43:5588-5592.

[28] Jernberg C, Lofmark S, Edlund C, Jansson JK. Long-term impacts of antibiotic exposure on the human intestinal microbiota. Microbiology 2010,156:3216-3223.

[29] McFarland LV. Antibiotic-associated diarrhea: epidemiology, trends and treatment. Future Microbiol 2008,3:563-578.

[30] Metchnikoff E. The prolongation of life: optimistic studies. G.P. Putnam's Sons ed. New York and London: 1908.

[31] FAO/WHO. Health and nutritional porperties of probiotics in food includion powder milk with live lactic acid bacteria. 30[suppl 2], S23-S33. 2001. Argentina.

[32] FAO/WHO Working group. Guidelines for the evaluation of probiotics in food. 2002. London, 30 avril-1er mai.

[33] Williams NT. Probiotics. Am J Health Syst Pharm 2010,67:449-458.

[34] Deshpande G, Rao S, Patole S. Progress in the field of probiotics: year 2011. Curr Opin Gastroenterol 2011,27:13-18.

[35] Park J, Floch MH. Prebiotics, probiotics, and dietary fiber in gastrointestinal disease. Gastroenterol Clin North Am 2007,36:47-63.

[36] Girardin M, Seidman EG. Indications for the use of probiotics in gastrointestinal diseases. Dig Dis 2011,29:574-587.

[37] Roberfroid M. Prebiotics: the concept revisited. J Nutr 2007,137:830S-837S.

[38] Boyle RJ, Robins-Browne RM, Tang ML. Probiotic use in clinical practice: what are the risks? Am J Clin Nutr 2006,83:1256-1264.

[39] Braegger C, Chmielewska A, Decsi T, Kolacek S, Mihatsch W, Moreno L, Piescik M, Puntis J, Shamir R, Szajewska H, Turck D, van GJ. Supplementation of infant formula with probiotics and/or prebiotics: a systematic review and comment by the ESPGHAN committee on nutrition. J Pediatr Gastroenterol Nutr 2011,52:238-250.

[40] Agostoni C, Goulet O, Kolacek S, Koletzko B, Moreno L, Puntis J, Rigo J, Shamir R, Szajewska H, Turck D. Fermented infant formulae without live bacteria. J Pediatr Gastroenterol Nutr 2007,44:392-397.

[41] Menard S, Candalh C, Ben AM, Rakotobe S, Gaboriau-Routhiau V, Cerf-Bensussan N, Heyman M. Stimulation of immunity without alteration of oral tolerance in mice fed with heat-treated fermented infant formula. J Pediatr Gastroenterol Nutr 2006,43:451-458.

[42] Hoarau C, Lagaraine C, Martin L, Velge-Roussel F, Lebranchu Y. Supernatant of Bifidobacterium breve induces dendritic cell maturation, activation, and survival through a Toll-like receptor 2 pathway. J Allergy Clin Immunol 2006,117:696-702.

[43] Adlerberth I, Wold AE. Establishment of the gut microbiota in Western infants. Acta Paediatr 2009,98:229-238.

[44] Campeotto F, Waligora-Dupriet AJ, Doucet-Populaire F, Kalach N, Dupont C, Butel MJ. [Establishment of the intestinal microflora in neonates]. Gastroenterol Clin Biol 2007,31:533-542.

[45] Martin R, Jimenez E, Heilig H, Fernandez L, Marin ML, Zoetendal EG, Rodriguez JM. Isolation of bifidobacteria from breast milk and assessment of the bifidobacterial population by PCR-DGGE and qRTi-PCR. Appl Environ Microbiol 2009,75:965-969.

[46] Solis G, de los Reyes-Gavilan CG, Fernandez N, Margolles A, Gueimonde M. Establishment and development of lactic acid bacteria and bifidobacteria microbiota in breast-milk and the infant gut. Anaerobe 2010,16:307-310.

[47] Palmer C, Bik EM, Digiulio DB, Relman DA, Brown PO. Development of the Human Infant Intestinal Microbiota. PLoS Biol 2007,5:e177.

[48] Biasucci G, Benenati B, Morelli L, Bessi E, Boehm G. Cesarean delivery may affect the early biodiversity of intestinal bacteria. J Nutr 2008,138:1796S-1800S.

[49] Huurre A, Kalliomaki M, Rautava S, Rinne M, Salminen S, Isolauri E. Mode of delivery - effects on gut microbiota and humoral immunity. Neonatology 2008,93:236-240.

[50] Jacquot A, Neveu D, Aujoulat F, Mercier G, Marchandin H, Jumas-Bilak E, Picaud JC. Dynamics and Clinical Evolution of Bacterial Gut Microflora in Extremely Premature Patients. J Pediatr 2010,158:390-396.

[51] Mshvildadze M, Neu J, Shuster J, Theriaque D, Li N, Mai V. Intestinal microbial ecology in premature infants assessed with non-culture-based techniques. J Pediatr 2010,156:20-25.

[52] Chang JY, Shin SM, Chun J, Lee JH, Seo JK. Pyrosequencing-based molecular monitoring of the intestinal bacterial colonization in preterm infants. J Pediatr Gastroenterol Nutr 2011,53:512-519.

[53] LaTuga MS, Ellis JC, Cotton CM, Goldberg RN, Wynn JL, Jackson RB, Seed PC. Beyond bacteria: a study of the enteric microbial consortium in extremely low birth weight infants. PLoS One 2011,6:e27858.

[54] Butel MJ, Suau A, Campeotto F, Magne F, Aires J, Ferraris L, Kalach N, Leroux B, Dupont C. Conditions of bifidobacterial colonization in preterm infants: a prospective analysis. J Pediatr Gastroenterol Nutr 2007,44:577-582.

[55] Moore DC, Elsas PX, Maximiano ES, Elsas MI. Impact of diet on the immunological microenvironment of the pregnant uterus and its relationship to allergic disease in the offspring--a review of the recent literature. Sao Paulo Med J 2006,124:298-303.

[56] Okada H, Kuhn C, Feillet H, Bach JF. The 'hygiene hypothesis' for autoimmune and allergic diseases: an update. Clin Exp Immunol 2010,160:1-9.

[57] Protonotariou E, Malamitsi-Puchner A, Rizos D, Papagianni B, Moira E, Sarandakou A, Botsis D. Age-related differentiations of Th1/Th2 cytokines in newborn infants. Mediators Inflamm 2004,13:89-92.

[58] Grönlund MM, Lehtonen OP, Eerola E, Kero P. Fecal microflora in healthy infants born by different methods of delivery: permanent changes in intestinal flora after cesarean delivery. J Pediat Gastroenterol Nutr 1999,28:19-25.

[59] Grönlund MM, Arvilommi H, Kero P, Lehtonen OP, Isolauri E. Importance of intestinal colonisation in the maturation of humoral immunity in early infancy: a prospective follow up study of healthy infants aged 0-6 months. Arch Dis Child Fetal Neonatal Ed 2000,83:F186-F192.

[60] Rautava S, Ruuskanen O, Ouwehand A, Salminen S, Isolauri E. The hygiene hypothesis of atopic disease--an extended version. J Pediatr Gastroenterol Nutr 2004,38:378-388.

[61] Kero J, Gissler M, Gronlund MM, Kero P, Koskinen P, Hemminki E, Isolauri E. Mode of delivery and asthma -- is there a connection? Pediatr Res 2002,52:6-11.

[62] Laubereau B, Filipiak-Pittroff B, von BA, Grubl A, Reinhardt D, Wichmann HE, Koletzko S. Caesarean section and gastrointestinal symptoms, atopic dermatitis, and sensitisation during the first year of life. Arch Dis Child 2004,89:993-997.

[63] Agosti M, Vegni C, Gangi S, Benedetti V, Marini A. Allergic manifestations in very low-birthweight infants: a 6-year follow-up. Acta Paediatr Suppl 2003,91:44-47.

[64] McKeever TM, Lewis SA, Smith C, Collins J, Heatlie H, Frischer M, Hubbard R. Early exposure to infections and antibiotics and the incidence of allergic disease: a birth cohort study with the West Midlands General Practice Research Database. J Allergy Clin Immunol 2002,109:43-50.

[65] Sjogren YM, Jenmalm MC, Bottcher MF, Bjorksten B, Sverremark-Ekstrom E. Altered early infant gut microbiota in children developing allergy up to 5 years of age. Clin Exp Allergy 2009,39:518-526.

[66] Bjorksten B, Sepp E, Julge K, Voor T, Mikelsaar M. Allergy development and the intestinal microflora during the first year of life. J Allergy Clin Immunol 2001,108:516-520.

[67] Sepp E, Julge K, Mikelsaar M, Bjorksten B. Intestinal microbiota and immunoglobulin E responses in 5-year-old Estonian children. Clin Exp Allergy 2005,35:1141-1146.

[68] Hong PY, Lee BW, Aw M, Shek LP, Yap GC, Chua KY, Liu WT. Comparative analysis of fecal microbiota in infants with and without eczema. PLoS One 2010,5:e9964.

[69] Kalliomaki M, Collado MC, Salminen S, Isolauri E. Early differences in fecal microbiota composition in children may predict overweight. Am J Clin Nutr 2008,87:534-538.

[70] De Cruz P, Prideaux L, Wagner J, Ng SC, McSweeney C, Kirkwood C, Morrison M, Kamm MA. Characterization of the gastrointestinal microbiota in health and inflammatory bowel disease. Inflamm Bowel Dis 2012,18:372-390.

[71] Fava F, Danese S. Intestinal microbiota in inflammatory bowel disease: friend of foe? World J Gastroenterol 2011,17:557-566.

[72] Conte MP, Schippa S, Zamboni I, Penta M, Chiarini F, Seganti L, Osborn J, Falconieri P, Borrelli O, Cucchiara S. Gut-associated bacterial microbiota in paediatric patients with inflammatory bowel disease. Gut 2006,55:1760-1767.

[73] Schippa S, Iebba V, Barbato M, Di NG, Totino V, Checchi MP, Longhi C, Maiella G, Cucchiara S, Conte MP. A distinctive 'microbial signature' in celiac pediatric patients. BMC Microbiol 2010,10:175.

[74] Schippa S, Conte MP, Borrelli O, Iebba V, Aleandri M, Seganti L, Longhi C, Chiarini F, Osborn J, Cucchiara S. Dominant genotypes in mucosa-associated Escherichia coli strains from pediatric patients with inflammatory bowel disease. Inflamm Bowel Dis 2009,15:661-672.

[75] Collado MC, Donat E, Ribes-Koninckx C, Calabuig M, Sanz Y. Specific duodenal and faecal bacterial groups associated with paediatric coeliac disease. J Clin Pathol 2009,62:264-269.

[76] Bolte ER. The role of cellular secretion in autism spectrum disorders: a unifying hypothesis. Med Hypotheses 2003,60:119-122.

[77] Finegold SM, Molitoris D, Song Y, Liu C, Vaisanen ML, Bolte E, McTeague M, Sandler R, Wexler H, Marlowe EM, Collins MD, Lawson PA, Summanen P, Baysallar M, Tomzynski TJ, Read E, Johnson E, Rolfe R, Nasir P, Shah H, Haake DA, Manning P, Kaul A. Gastrointestinal microflora studies in late-onset autism. Clin Infect Dis 2002,35:S6-S16.
[78] Song Y, Liu C, Finegold SM. Real-time PCR quantitation of clostridia in feces of autistic children. Appl Environ Microbiol 2004,70:6459-6465.
[79] Finegold SM, Downes J, Summanen PH. Microbiology of regressive autism. Anaerobe 2012,18:260-262.
[80] Thavagnanam S, Fleming J, Bromley A, Shields MD, Cardwell CR. A meta-analysis of the association between Caesarean section and childhood asthma. Clin Exp Allergy 2008,38:629-633.
[81] Bager P. Birth by caesarean section and wheezing, asthma, allergy, and intestinal disease. Clin Exp Allergy 2011,41:147-148.
[82] Bager P, Melbye M, Rostgaard K, Benn CS, Westergaard T. Mode of delivery and risk of allergic rhinitis and asthma. J Allergy Clin Immunol 2003,111:51-56.
[83] Decker E, Hornef M, Stockinger S. Cesarean delivery is associated with celiac disease but not inflammatory bowel disease in children. Gut Microbes 2011,2:91-98.
[84] Ajslev TA, Andersen CS, Gamborg M, Sorensen TI, Jess T. Childhood overweight after establishment of the gut microbiota: the role of delivery mode, pre-pregnancy weight and early administration of antibiotics. Int J Obes (Lond) 2011,35:522-529.
[85] Zhou L, He G, Zhang J, Xie R, Walker M, Wen SW. Risk factors of obesity in preschool children in an urban area in China. Eur J Pediatr 2011,170:1401-1406.
[86] Huh SY, Rifas-Shiman SL, Zera CA, Edwards JW, Oken E, Weiss ST, Gillman MW. Delivery by caesarean section and risk of obesity in preschool age children: a prospective cohort study. Arch Dis Child 2012 [Epub ahead of print].
[87] Cardwell CR, Stene LC, Joner G, Cinek O, Svensson J, Goldacre MJ, Parslow RC, Pozzilli P, Brigis G, Stoyanov D, Urbonaite B, Sipetic S, Schober E, Ionescu-Tirgoviste C, Devoti G, de Beaufort CE, Buschard K, Patterson CC. Caesarean section is associated with an increased risk of childhood-onset type 1 diabetes mellitus: a meta-analysis of observational studies. Diabetologia 2008,51:726-735.
[88] Davis JN, Whaley SE, Goran MI. Effects of breastfeeding and low sugar-sweetened beverage intake on obesity prevalence in Hispanic toddlers. Am J Clin Nutr 2012,95:3-8.
[89] Thomas DW, Greer FR. Probiotics and prebiotics in pediatrics. Pediatrics 2010,126:1217-1231.
[90] Hsieh MH, Versalovic J. The human microbiome and probiotics: implications for pediatrics. Curr Probl Pediatr Adolesc Health Care 2008,38:309-327.
[91] Allen SJ, Martinez EG, Gregorio GV, Dans LF. Probiotics for treating acute infectious diarrhoea. Cochrane Database Syst Rev 2010,CD003048.

[92] Szajewska H, Ruszczynski M, Radzikowski A. Probiotics in the prevention of antibiotic-associated diarrhea in children: A meta-analysis of randomized controlled trials. J Pediatr 2006,149:367-372.

[93] Johnston BC, Goldenberg JZ, Vandvik PO, Sun X, Guyatt GH. Probiotics for the prevention of pediatric antibiotic-associated diarrhea. Cochrane Database Syst Rev 2011,CD004827.

[94] Cohen-Silver J, Ratnapalan S. Management of infantile colic: a review. Clin Pediatr (Phila) 2009,48:14-17.

[95] Savino F, Pelle E, Palumeri E, Oggero R, Miniero R. *Lactobacillus reuteri* (American Type Culture Collection Strain 55730) versus simethicone in the treatment of infantile colic: a prospective randomized study. Pediatrics 2007,119:e124-e130.

[96] Savino F, Cordisco L, Tarasco V, Locatelli E, Di GD, Oggero R, Matteuzzi D. Antagonistic effect of *Lactobacillus* strains against gas-producing coliforms isolated from colicky infants. BMC Microbiol 2011,11:157.

[97] Mohan R, Koebnick C, Schildt J, Mueller M, Radke M, Blaut M. Effects of *Bifidobacterium lactis* supplementation on body weight, fecal pH, acetate, lactate, calprotectin and IgA in preterm infants. Pediatr Res 2008, 64:418-422.

[98] Roze JC, Barbarot S, Butel MJ, Kapel N, Waligora-Dupriet AJ, De M, I, Leblanc M, Godon N, Soulaines P, Darmaun D, Rivero M, Dupont C. An alpha-lactalbumin-enriched and symbiotic-supplemented v. a standard infant formula: a multicentre, double-blind, randomised trial. Br J Nutr 2012,107:1616-1622.

[99] Rinne M, Kalliomaki M, Arvilommi H, Salminen S, Isolauri E. Effect of probiotics and breastfeeding on the *Bifidobacterium* and *Lactobacillus/Enterococcus microbiota* and humoral immune responses. J Pediatr 2005,147:186-191.

[100] Sherman MP. New concepts of microbial translocation in the neonatal intestine: mechanisms and prevention. Clin Perinatol 2010,37:565-579.

[101] Betsi GI, Papadavid E, Falagas ME. Probiotics for the treatment or prevention of atopic dermatitis: a review of the evidence from randomized controlled trials. Am J Clin Dermatol 2008,9:93-103.

[102] Waligora-Dupriet AJ, Butel MJ. Microbiota and allergy: from dysbiosis to probiotics. In: Pereira C, editor. Allergic Diseases – Highlights in the Clinic, Mechanisms and Treatment. Rijeka: Intech; 2012. 413-434.

[103] Kalliomaki M, Kirjavainen P, Eerola E, Kero P, Salminen S, Isolauri E. Distinct patterns of neonatal gut microflora in infants in whom atopy was and was not developing. J Allergy Clin Immunol 2001,107:129-134.

[104] Kalliomaki M, Salminen S, Poussa T, Arvilommi H, Isolauri E. Probiotics and prevention of atopic disease: 4-year follow-up of a randomised placebo-controlled trial. Lancet 2003,361:1869-1871.

[105] Kalliomaki M, Salminen S, Poussa T, Isolauri E. Probiotics during the first 7 years of life: a cumulative risk reduction of eczema in a randomized, placebo-controlled trial. J Allergy Clin Immunol 2007,119:1019-1021.

[106] Kopp MV, Hennemuth I, Heinzmann A, Urbanek R. Randomized, double-blind, placebo-controlled trial of probiotics for primary prevention: no clinical effects of *Lactobacillus* GG supplementation. Pediatrics 2008,121:e850-e856.

[107] Abrahamsson TR, Jakobsson T, Bottcher MF, Fredrikson M, Jenmalm MC, Bjorksten B, Oldaeus G. Probiotics in prevention of IgE-associated eczema: a double-blind, randomized, placebo-controlled trial. J Allergy Clin Immunol 2007,119:1174-1180.

[108] Wickens K, Black PN, Stanley TV, Mitchell E, Fitzharris P, Tannock GW, Purdie G, Crane J. A differential effect of 2 probiotics in the prevention of eczema and atopy: a double-blind, randomized, placebo-controlled trial. J Allergy Clin Immunol 2008,122:788-794.

[109] Taylor AL, Dunstan JA, Prescott SL. Probiotic supplementation for the first 6 months of life fails to reduce the risk of atopic dermatitis and increases the risk of allergen sensitization in high-risk children: a randomized controlled trial. J Allergy Clin Immunol 2007,119:184-191.

[110] Taylor A, Hale J, Wiltschut J, Lehmann H, Dunstan JA, Prescott SL. Evaluation of the effects of probiotic supplementation from the neonatal period on innate immune development in infancy. Clin Exp Allergy 2006,36:1218-1226.

[111] Soh SE, Aw M, Gerez I, Chong YS, Rauff M, Ng YP, Wong HB, Pai N, Lee BW, Shek LP. Probiotic supplementation in the first 6 months of life in at risk Asian infants--effects on eczema and atopic sensitization at the age of 1 year. Clin Exp Allergy 2009,39:571-578.

[112] Szajewska H. Early nutritional strategies for preventing allergic disease. Isr Med Assoc J 2012,14:58-62.

[113] Boyle RJ, Bath-Hextall FJ, Leonardi-Bee J, Murrell DF, Tang ML. Probiotics for treating eczema. Cochrane Database Syst Rev 2008,CD006135.

[114] Menard O, Butel MJ, Gaboriau-Routhiau V, Waligora-Dupriet AJ. Gnotobiotic mouse immune response induced by *Bifidobacterium* sp. strains isolated from infants. Appl Environ Microbiol 2008,74:660-666.

[115] De Bandt JP, Waligora-Dupriet AJ, Butel MJ. Intestinal microbiota in inflammation and insulin resistance: relevance to humans. Curr Opin Clin Nutr Metab Care 2011,14:334-340.

[116] Park MH, Falconer C, Viner RM, Kinra S. The impact of childhood obesity on morbidity and mortality in adulthood: a systematic review. Obes Rev 2012 [Epub ahead of print].

[117] Nadal I, Santacruz A, Marcos A, Warnberg J, Garagorri M, Moreno LA, Martin-Matillas M, Campoy C, Marti A, Moleres A, Delgado M, Veiga OL, Garcia-Fuentes M, Redondo CG, Sanz Y. Shifts in *Clostridia, Bacteroides* and immunoglobulin-coating fecal bacteria associated with weight loss in obese adolescents. Int J Obes (Lond) 2009,33:758-767.

[118] Balamurugan R, George G, Kabeerdoss J, Hepsiba J, Chandragunasekaran AM, Ramakrishna BS. Quantitative differences in intestinal *Faecalibacterium prausnitzii* in obese Indian children. Br J Nutr 2010,103:335-338.

[119] Luoto R, Kalliomaki M, Laitinen K, Isolauri E. The impact of perinatal probiotic intervention on the development of overweight and obesity: follow-up study from birth to 10 years. Int J Obes (Lond) 2010,34:1531-1537.

[120] Gobel RJ, Larsen N, Jakobsen M, Molgaard C, Michaelsen KF. Probiotics to obese adolescents; RCT examining the effects on inflammation and metabolic syndrome. J Pediatr Gastroenterol Nutr 2012 [Epub ahead of print].

[121] Blakey JL, Lubitz L, Barnes GL, Bishop RF, Campbell NT, Gillam GL. Development of gut colonisation in pre-term neonates. J Med Microbiol 1982,15:519-529.

[122] Sakata H, Yoshioka H, Fujita K. Development of the intestinal flora in very low birth weight infants compared to normal full-term newborns. Eur J Pediatr 1985,144:186-190.

[123] Stark PL, Lee A. The bacterial colonization of the large bowel of pre-term low birth weight neonates. J Hyg Camb 1982,89:59-67.

[124] Campeotto F, Suau A, Kapel N, Magne F, Viallon V, Ferraris L, Waligora-Dupriet AJ, Soulaines P, Leroux B, Kalach N, Dupont C, Butel MJ. A fermented formula in preterm infants: clinical tolerance, gut microbiota, down regulation of fecal calprotectin, and up regulation of fecal secretory IgA. Br J Nutr 2011,105:1843-1851.

[125] Gewolb IH, Schwalbe RS, Taciak VL, Harrison TS, Panigrahi P. Stool microflora in extremely low birthweight infants. Arch Dis Child Fetal Neonatal Ed 1999,80:F167-F173.

[126] Rouge C, Piloquet H, Butel MJ, Berger B, Rochat F, Ferraris L, Des RC, Legrand A, De La Cochetiere MF, N'Guyen JM, Vodovar M, Voyer M, Darmaun D, Roze JC. Oral supplementation with probiotics in very-low-birth-weight preterm infants: a randomized, double-blind, placebo-controlled trial. Am J Clin Nutr 2009,89:1828-1835.

[127] Millar MR, Linton CJ, Cade A, Glancy D, Hall M, Jalal H. Application of 16S rRNA gene PCR to study bowel flora of preterm infants with and without necrotizing enterocolitis. J Clin Microbiol 1996,34:2506-2510.

[128] Roudiere L, Jacquot A, Marchandin H, Aujoulat F, Devine R, Zorgniotti I, Jean-Pierre H, Picaud JC, Jumas-Bilak E. Optimized PCR-Temporal Temperature Gel Electrophoresis compared to cultivation to assess diversity of gut microbiota in neonates. J Microbiol Methods 2009,79:156-165.

[129] Schwiertz A, Gruhl B, Lobnitz M, Michel P, Radke M, Blaut M. Development of the intestinal bacterial composition in hospitalized preterm infants in comparison with breast-fed, full-term infants. Pediatr Res 2003,54:393-399.

[130] Wang Y, Hoenig JD, Malin KJ, Qamar S, Petrof EO, Sun J, Antonopoulos DA, Chang EB, Claud EC. 16S rRNA gene-based analysis of fecal microbiota from preterm infants with and without necrotizing enterocolitis. ISME J 2009,3:944-954.

[131] Ferraris L, Butel MJ, Campeotto F, Vodovar M, Roze JC, Aires J. Clostridia in premature neonates' gut: incidence, antibiotic susceptibility, and perinatal determinants influencing colonization. PLoS One 2012,7:e30594.

[132] Stoll BJ, Hansen NI, Higgins RD, Fanaroff AA, Duara S, Goldberg R, Laptook A, Walsh M, Oh W, Hale E. Very low birth weight preterm infants with early onset neonatal sepsis: the predominance of gram-negative infections continues in the National Institute of Child Health and Human Development Neonatal Research Network, 2002-2003. Pediatr Infect Dis J 2005,24:635-639.

[133] Smith A, Saiman L, Zhou J, Della-Latta P, Jia H, Graham PL, III. Concordance of gastrointestinal tract colonization and subsequent bloodstream infections with gram-negative bacilli in very low birth weight infants in the neonatal intensive care unit. Pediatr Infect Dis J 2010,29:831-835.

[134] Das P, Singh AK, Pal T, Dasgupta S, Ramamurthy T, Basu S. Colonization of the gut with Gram-negative bacilli, its association with neonatal sepsis and its clinical relevance in a developing country. J Med Microbiol 2011,60:1651-1660.

[135] Obladen M. Necrotizing enterocolitis--150 years of fruitless search for the cause. Neonatology 2009,96:203-210.

[136] Morowitz MJ, Poroyko V, Caplan M, Alverdy J, Liu DC. Redefining the role of intestinal microbes in the pathogenesis of necrotizing enterocolitis. Pediatrics 2010,125:777-785.

[137] Waligora-Dupriet AJ, Dugay A, Auzeil N, Huerre M, Butel MJ. Evidence for clostridial implication in necrotizing enterocolitis through bacterial fermentation in a gnotobiotic quail model. Pediatr Res 2005,58:629-635.

[138] Lin PW, Nasr TR, Stoll BJ. Necrotizing enterocolitis: recent scientific advances in pathophysiology and prevention. Semin Perinatol 2008,32:70-82.

[139] Waligora-Dupriet AJ, Dugay A, Auzeil N, Nicolis I, Rabot S, Huerre MR, Butel MJ. Short-chain fatty acids and polyamines in the pathogenesis of necrotizing enterocolitis: Kinetics aspects in gnotobiotic quails. Anaerobe 2009,15:138-144.

[140] Kien CL. Colonic fermentation of carbohydrate in the premature infant : possible relevance to necrotizing enterocolitis. J Pediatr 1990,117:S52-S58.

[141] Lin J. Too much short chain fatty acids cause neonatal necrotizing enterocolitis. Med Hypotheses 2004,62:291-293.

[142] Szylit O, Maurage C, Gasqui P, Popot F, Favre A, Gold F, Borderon JC. Fecal short-chain fatty acids predict digestive disorders in premature infants. J Parent Enter Nutr 1998,22:136-141.

[143] Butel MJ, Roland N, Hibert A, Popot F, Favre A, Tessèdre AC, Bensaada M, Rimbault A, Szylit O. Clostridial pathogenicity in experimental necrotising enterocolitis in gnotobiotic quails and protective role of bifidobacteria. J Med Microbiol 1998,47:391-399.

[144] De La Cochetière MF, Piloquet H, Des Robert C, Darmaun D, Galmiche JP, Rozé JC. Early intestinal bacterial colonization and necrotizing enterocolitis in premature infants: the putative role of *Clostridium*. Pediatr Res 2004,56:1-5.

[145] Mai V, Young CM, Ukhanova M, Wang X, Sun Y, Casella G, Theriaque D, Li N, Sharma R, Hudak M, Neu J. Fecal microbiota in premature infants prior to necrotizing enterocolitis. PLoS One 2011,6:e20647.

[146] Alfaleh K, Anabrees J, Bassler D, Al-Kharfi T. Probiotics for prevention of necrotizing enterocolitis in preterm infants. Cochrane Database Syst Rev 2011,CD005496.

[147] Alfaleh K, Anabrees J, Bassler D. Probiotics reduce the risk of necrotizing enterocolitis in preterm infants: a meta-analysis. Neonatology 2010,97:93-99.

[148] Deshpande G, Rao S, Patole S, Bulsara M. Updated meta-analysis of probiotics for preventing necrotizing enterocolitis in preterm neonates. Pediatrics 2010,125:921-930.

[149] Luoto R, Matomaki J, Isolauri E, Lehtonen L. Incidence of necrotizing enterocolitis in very-low-birth-weight infants related to the use of *Lactobacillus* GG. Acta Paediatr 2010,99:1135-1138.

[150] Tarnow-Mordi WO, Wilkinson D, Trivedi A, Brok J. Probiotics reduce all-cause mortality and necrotizing enterocolitis: it is time to change practice. Pediatrics 2010,125:1068-1070.

[151] Neu J, Shuster J. Nonadministration of routine probiotics unethical--really? Pediatrics 2010,126:e740-e741.

[152] Soll RF. Probiotics: are we ready for routine use? Pediatrics 2010,125:1071-1072.

[153] Chou IC, Kuo HT, Chang JS, Wu SF, Chiu HY, Su BH, Lin HC. Lack of effects of oral probiotics on growth and neurodevelopmental outcomes in preterm very low birth weight infants. J Pediatr 2010,156:393-396.

[154] Ohishi A, Takahashi S, Ito Y, Ohishi Y, Tsukamoto K, Nanba Y, Ito N, Kakiuchi S, Saitoh A, Morotomi M, Nakamura T. *Bifidobacterium septicemia* associated with postoperative probiotic therapy in a neonate with omphalocele. J Pediatr 2010,156:679-681.

[155] Cilieborg MS, Thymann T, Siggers R, Boye M, Bering SB, Jensen BB, Sangild PT. The incidence of necrotizing enterocolitis is increased following probiotic administration to preterm pigs. J Nutr 2011, 141:223-230.

[156] Honeycutt TC, El KM, Wardrop RM, III, McNeal-Trice K, Honeycutt AL, Christy CG, Mistry K, Harris BD, Meliones JN, Kocis KC. Probiotic administration and the incidence of nosocomial infection in pediatric intensive care: a randomized placebo-controlled trial. Pediatr Crit Care Med 2007,8:452-458.

[157] Manzoni P, Lista G, Gallo E, Marangione P, Priolo C, Fontana P, Guardione R, Farina D. Routine *Lactobacillus rhamnosus* GG administration in VLBW infants: a retrospective, 6-year cohort study. Early Hum Dev 2011,87 Suppl 1:S35-S38.

[158] Neu J. Routine probiotics for premature infants: let's be careful! J Pediatr 2011,158:672-674.

[159] Deshpande GC, Rao SC, Keil AD, Patole SK. Evidence-based guidelines for use of probiotics in preterm neonates. BMC Med 2011,9:92.

[160] Mihatsch WA, Braegger CP, Decsi T, Kolacek S, Lanzinger H, Mayer B, Moreno LA, Pohlandt F, Puntis J, Shamir R, Stadtmuller U, Szajewska H, Turck D, van Goudoever JB. Critical systematic review of the level of evidence for routine use of probiotics for reduction of mortality and prevention of necrotizing enterocolitis and sepsis in preterm infants. Clin Nutr 2012,31:6-15.

[161] Mihatsch WA. What is the power of evidence recommending routine probiotics for necrotizing enterocolitis prevention in preterm infants? Curr Opin Clin Nutr Metab Care 2011,14:302-306.

[162] Garland SM, Tobin JM, Pirotta M, Tabrizi SN, Opie G, Donath S, Tang ML, Morley CJ, Hickey L, Ung L, Jacobs SE. The ProPrems trial: investigating the effects of probiotics on late onset sepsis in very preterm infants. BMC Infect Dis 2011,11:210.

[163] Szajewska H. Probiotics and prebiotics in preterm infants: Where are we? Where are we going? Early Hum Dev 2010,Suppl1:81-86.

[164] Agostoni C, Buonocore G, Carnielli VP, De CM, Darmaun D, Decsi T, Domellof M, Embleton ND, Fusch C, Genzel-Boroviczeny O, Goulet O, Kalhan SC, Kolacek S, Koletzko B, Lapillonne A, Mihatsch W, Moreno L, Neu J, Poindexter B, Puntis J, Putet G, Rigo J, Riskin A, Salle B, Sauer P, Shamir R, Szajewska H, Thureen P, Turck D, van Goudoever JB, Ziegler EE. Enteral nutrient supply for preterm infants: commentary from the European Society of Paediatric Gastroenterology, Hepatology and Nutrition Committee on Nutrition. J Pediatr Gastroenterol Nutr 2010,50:85-91.

Chapter 22

Probiotics and Mucosal Immune Response

Petar Nikolov

Additional information is available at the end of the chapter

http://dx.doi.org/10.5772/50042

1. Introduction

There is complex and ubiquitous interface between the probiotic and resident bacteria (human microbiota) at various mucosal sites and the mucosal immune system. The probiotic bacteria are normally exogenous and transient as the resident bacterial communities of the human body are relatively constant companions of the human body and the mucosal immune system. This interface may result in local and systemic immune responses thus contributing for the preservation of the biological individuality of the human macroorganism.

2. Human microbiota

The human microbiota is an aggregate of microorganisms that reside on the surface and in deep layers of skin, in the saliva and oral mucosa, in the conjunctiva, the urogenital, to some extend the respiratory and above all the gastrointestinal tract. They include mostly Bacteria, but also some Fungi and Archaea. All these body parts are offering a relatively stable habitat for the resident bacteria: constant nutrient influx, constant temperature, redox potential and humidity. The skin flora does not interact directly with the mucosal immune system so it would be excluded from the present book chapter.

2.1. Oral microbiota

The oral cavity shelters a very diverse, abundant and complex microbial community. Oral bacteria have developed mechanisms to sense their environment and evade or modify the host. Bacteria occupy the ecological niche provided by both the tooth surface and gingival epithelium. A varied microbial flora is found in the oral cavity, and Streptococcal anaerobes inhabit the gingival crevice. The oral flora is involved in dental caries and periodontal disease, which affect about 80 %. of the population in the Western world. Anaerobes in the oral flora are responsible for many of the brain, face, and lung infections that are frequently manifested by abscess formation. Oral bacteria include *Streptococci, Lactobacilli, Staphylococci, Corynebacteria* and various anaerobes in particular *Bacteroides*. The oral cavity of the new-born baby does not contain bacteria but rapidly becomes colonized with bacteria such as Streptococcus salivarius. With the appearance of the teeth during the first year colonization by *Streptococcus mutans* and *Streptococcus sanguinis* occurs as these organisms colonise the dental surface and gingiva. Other strains of streptococci adhere strongly to the gums and cheeks but not to the teeth. The gingival crevice area (supporting structures of the teeth) provides a habitat for a variety of anaerobic species. *Bacteroides* and *Spirochetes* colonize the mouth around puberty. However, a highly efficient innate host defense system constantly monitors the bacterial colonization and prevents bacterial invasion of human tissues. A dynamic equilibrium exists between dental plaque bacteria and the innate host defense system. [1, 2].

2.2. Respiratory microbiota

The nose, pharynx and trachea contain primarily those bacterial genera found in the normal oral cavity (for example, α-and β-hemolytic streptococci); however, anaerobes, *Staphylococci, Neisseriae* and *Diphtheroids* are also present. Potentially pathogenic organisms such as *Haemophilus, Mycoplasmas* and *Pneumococci* may also be found in the pharynx. Anaerobic organisms also are reported frequently. The upper respiratory tract is so often the site of initial colonization by pathogens (*Neisseria meningitides, C. diphtheriae, Bordetella pertussis,* etc.) and could be considered the first region of attack for such organisms. In contrast, the lower respiratory tract (small bronchi and alveoli) is usually sterile, because particles the size of bacteria do not readily reach it. If bacteria do reach these regions, they encounter host defense mechanisms, such as alveolar macrophages, that are not present in the pharynx [2].

2.3. Conjunctival microbiota

The conjunctiva harbors few or no organisms. *Haemophilus* and *Staphylococcus* are among the genera most often detected [2].

2.4. Urogenital microbiota

The urogenital flora is comprised mostly by the bacteria in the anterior urethra and the genital tract in women. In the anterior urethra of humans, *S. epidermidis*, enterococci, and diphtheroids are found frequently; *E. coli, Proteus,* and *Neisseria* (nonpathogenic species) are reported occasionally (10-30 %). The type of bacterial flora found in the vagina depends on the age, pH, and hormonal levels of the host. *Lactobacillus* spp. predominate in female infants (vaginal pH, approx. 5) during the first month of life. Glycogen secretion seems to cease from about I month of age to puberty. During this time, diphtheroids, *S. epidermidis*, streptococci, and *E. coli* predominate at a higher pH (approximately pH 7). At puberty, glycogen secretion resumes, the pH drops, and women acquire an adult flora in which *L. acidophilus, Corynebacteria, Peptostreptococci, Staphylococci, Streptococci* and *Bacteroides* predominate. After menopause, pH again rises, less glycogen is secreted, and the flora returns to that found in prepubescent females. Yeasts (*Torulopsis* and *Candida*) are occasionally found in the vagina (10-30 % of women); these sometimes increase and cause vaginitis [2].

2.5. Intestinal microbiota

The number of bacteria in the digestive system alone is at least as big as the number of the stars in our home galaxy – the Milky Way as it contains no less than 10^{11} stars [3], thus forming a specific bacterial microcosmos the human gut. The number of bacteria increases in a logarithmic progression along the digestive system: the stomach (10^1-10^3 colony-forming units per milliliter (cfu/ml)), duodenum (10^1-10^3 cfu/ml), distal small intestine (10^4-10^7 cfu/ml) and above all the colon (10^{11}-10^{12} cfu/ml). According to some authors the intestinal bacteria are forming the most densely populated ecosystem in the world [4]. The intestinal bacteria are really abundant when it comes to the various species and strains and their spatial distribution. The intestinal flora has a dynamic structure and is not isolated from the human host or the surrounding environment. There qualitative and quantitative variations in the gut flora depending on the diet, age, biotic and abiotic factors of the human environment, mucosal immune respose, presence or absence of organic disease of the host, intake of antibacterial medications, etc. The interface between the gut flora and the intestinal mucosal immune system is a perfect example for the interaction between the resident bacteria and the mucosal immune response. The gut flora is quite unique for each and every person and differs even in identical twins [5, 6]. The predominant bacterial genera and families inhabiting the human gut are presented on table 1 [4, 7-14]:

Location	Facultative anaerobes	Gram staining	Obligate anaerobes	Gram staining
Duodenum and Jejunum	*Lactobacillus* *Streptococccus* *Enterobacteriaceae*	+ + -	Solitary *Bacteroides*	-
Ileum	*Lactobacillus* *Streptococccus* *Enterococcus* *Enterobacteriaceae*	+ + + -	*Bacteroides* *Clostridia* *Veillonella*	- + -
Colon	*Lactobacillus* *Streptococccus* *Enterococcus* *Enterobacteriaceae*	+ + + -	*Bacteroides* *Bacillus* *Clostridium* *Fusobacterium* *Peptostreptococcus* *Bifidobacterium* *Eubacterium* *Ruminococcus*	- + + - + + + -

Table 1. Predominant bacterial genera and families inhabiting the human intestine.

The intestinal flora may be divided to resident and transient. The resident bacteria can colonize and multiply successfully in the human gut for continuous periods of time as the transient microbial species can only do so for limited periods of time. The resident bacteria are able to adhere to specific molecules of the host or other adhesive bacterial species. Most of the transient bacteria are unable to do so or can only do it for a short time. The transient bacteria are usually ingested trough the mouth and belong to various genera and species [15].

3. Probiotic bacteria

The probiotic bacteria belong to the transient species as their presence in the human body is always a result of exogenous intake. There are numerous definitions for probiotics and they all correct in a way of their own. The concept for probiotics is constantly evolving, but essentially designates that they are "Living microorganisms which favorably influence the health of the host by improving the indigenous microflora". This definition was given by R. Fuller back in 1989 [16] and is very distinct from the one of the World Health Organization given in the beginning of the 21st century – "Live microorganisms which when administered in adequate amounts confer a health benefit on the host" [17]. There are also many other definitions and they all speak of the "whats", the "whos" and the "whens" but none speaks of the "hows". So if one would wish to include the "hows" it may sound like "Living microorganisms which when administered in adequate amounts may change the balance and keep the human body move in the right direction…". It does not say "favorable" as probiotics also have side effects and still it does not speak enough of "hows" so it can't really become the universal definition for probiotics. The intake of probiotic bacteria can be reviewed not only from a therapeutic and immunological angle but also unraveled throught the prism of ecology and cognitive philosophy.

The probiotic bacteria exert the unique quality to change the balance in a balanced way. They way they work is quite complex and fall pretty much into the witty remark of Albert Einstein "Life is like riding a bicycle – in order to keep your balance, you must keep moving" [18]. Indeed probiotic bacteria are alive and keep moving so as the human body. So when we want to understand probiotics everything comes to the balance between the outer and the inner cosmos of humans mediated by their mucosal surfaces.

The majority of commercially available probiotic bacteria belong to the genera *Lactobacillus* and *Bifidobacterium* but also strains of *E. coli, Streptococcus, Enterococcus* and even *Bacillus, Oxalobacter,* etc. Some yeasts are also being used as probiotics – *Saccharomyces,* etc. All commercially available probiotic bacteria must exert 5 crucial technological and clinical properties (fig. 1).

All these properties are equally important but the positive effect is by all means the most significant one:

- **Origin**: bacteria descending from the human gastrointestinal tract (GIT) (preferably);
- **Safety**: probiotic bacteria should be non-pathogenic and sensitive to the most commonly used antibiotics;

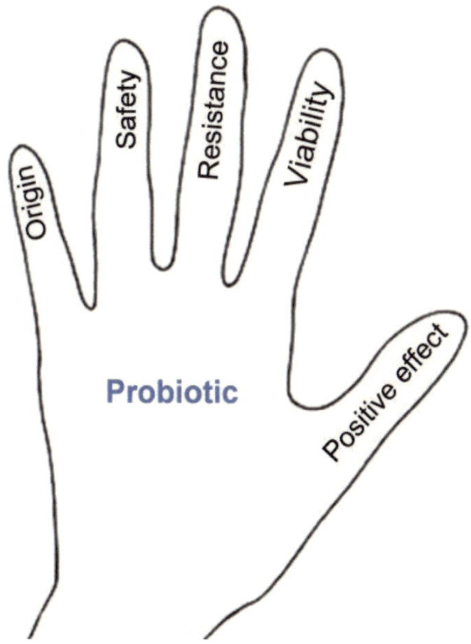

Figure 1. Main technological and clinical properties of the probiotic bacteria.

- **Resistance**: the bacterial strains should be able to survive the action of the stomach acid, the bile acids and the protease enzymes;
- **Viability**: these bacteria must survive the production process, proliferate in the small and/or large intestine, adhere to the gut epithelium and even colonize the small intestine and/or the colon for a finite time;
- **Positive effect**: their intake should be beneficial for health of the human macroorganism.

There is still conflicting evidence for the clinical efficacy of probiotic bacteria but yet they have been proven to be effective in infectious and antibiotic associated diarrhea [19, 20], urogenital infections [21, 22], immunologically mediated diseases such as inflammatory bowel disease (IBD) [23, 24] and atopic disease [25, 26], etc. Probiotic bacteria are being applied at various mucosal sites – orally, vaginally, as eye-drops, nasal sprays, etc. All mucosal sites are all connected in 3 different ways: anatomically, embryologically and most of all functionally.

4. Mucosal ecology

The intestinal flora is a specific blend of microorganisms, which have evolved and developed together with the macroorganism. These bacterial communities are highly variable and unique for all living persons. This is a result of time-limited migration of bacteria between humans in combination with their active interaction with the mucosal immune system, dietary and some genetic factors [27]. Human mucosal sites are classical habitats – they are normally populated by resident microorganisms. The human microbiota together with the mucosal surfaces of the human body form complex and dynamic ecosystems. All mucosal surfaces are directly exposed to the influence of environmental factors of the outer world – they are all located at the edge of the outer world and the inner cosmos of the human body. The edge effect in ecology is the effect of the juxtaposition or placing side by side of contrasting environments on an ecosystem. The highest diversity of species and the strongest influence of the living creatures over habitats are found on edges [28]. The abrupt changes in the microbial community and/or the habitat may alter the balance and alter the the delicate equilibrium between the resident flora and human host – the so called homeostasis. The exogenous introduction of probiotic bacteria is unique as in terms of ecology it can be considered both as an abiotic environmental factor and a biotic factor of the living matter. The mucosal surfaces with their indigenous microbial communities are also unique as they are the combining the role of a habitat and a part of a living organism at the same time. The probiotic bacteria may interact with the resident flora and the microorganism and alter the homeostasis. The probiotic bacteria however interact with the mucosal immune system like any other bacteria.

5. Intestinal homeostasis

In healthy individuals there is a tolerance towards the resident flora. Because of that tolerance normally there is no aggressive cellular or humoral immune response towards the indigenous flora. The tolerance towards the intestinal flora and numerous dietary compounds is called oral tolerance. The oral and other types of antigen specific tolerance are dependent also on the mucosal permeability and the antigen clearance of *lamina propria*. This delicate equilibrium may be disturbed in various ways and lead to the development of an active disease. An example of such a disease is the IBD, in which the local and systemic immune response are aiming for the resident intestinal bacteria. The mucosal immune system in IBD is trying to permanently eliminate the intestinal microbiota, thus leading to the development of a chronic inflammation [29]. The mucosal immune system plays a key role for the maintenance of the mucosal homeostasis.

6. Mucosal immune response

The complex and well-set interaction between the probiotic bacteria, the indigenous flora and the mucosal surfaces are all possible because of the mucosal immune system and particularly the mucosa associated lymphoid tissues (MALTs). The MALTs are dispersed aggregates of nonencapsulated organized lymphoid tissue within the mucosa, which are associated with local immune responses at mucosal surfaces. Human MALTs consist mainly of the lymphoid structures within the GIT, urogenital tract, respiratory tract, nasal and oral cavities, the salivary and lacrimal glands, the inner ear, the synovia and the lactating mammary glands. The three major regions of MALTs are the gut-associated lymphoid tissue (GALT), bronchus-associated lymphoid tissue (BALT) and nasal-associated lymphoid tissue (NALT) however, conjunctiva-associated lymphoid tissue (CALT), lacrimal duct-associated (LDALT), larynx-associated (LALT) and salivary duct-associated lymphoid tissue (DALT) have also been described [30-34]. The organization of the MALTs is similar to that of lymph nodes with variable numbers of follicles (B-cell area), interfollicular areas (T-cell area), and efferent lymphatics although afferent lymphatics are lacking. The overlying follicle associated epithelium is typically cuboidal with variable numbers of goblet cells and epithelial cells with either microvilli or numerous surface microfolds (M-cells). In addition, single lymphocytes can be observed within the epithelium, mucosa and *lamina propria*. All MALTs are morphologically similar although there are might be some differences in the percentage of T- and B-cells [35].

The GALT is typically organized into discrete lymphoid aggregates within the mucosa, submucosa and lamina propria of the small intestine called Peyer's patches (PP), the appendix, the mesenteric lymph nodes (MLN) and the solitary follicles. These aggregates are typically multiple lymphoid follicles with diffuse lymphatic tissue oriented towards the mucosa [36].

In the respiratory tract the NALT is the first site of contact for most airborne antigens and mostly presented by the tonsils and the adenoids at the entrance of the aerodigestive tract. The NALT bears certain similarities to the PP [34, 36].

The BALTs are organized aggregates of lymphocytes that are located within the bronchial submucosa. These aggregates are randomly distributed along the bronchial tract but are consistently present around the bifurcations of bronchi and bronchioli and always lie between an artery and a bronchus [34, 36].

The mucosal immune system has 3 main functions:

- protects the mucosa against pathogenic microorganisms;
- prevents the uptake of foreign proteins derived from ingested food, airborne matter and indigenous microbiota;
- prevents the development of potentially detrimental immune response to these antigens in case they reach the body interior – i.e. oral tolerance in the gut.

In contrast with the systemic immunity, which functions in a sterile milieu and often responds vigorously to "invaders", the MALT protects the structures that are replete with foreign matter. The MALT must economically select appropriate effector mechanisms and regulate their intensity to avoid bystander tissue damage.

All MALTs have two basic structures: organized and diffuse lymphoid tissue. In the GALT the organized tissues are mainly the PP, MLN and the appendix as the diffuse ones are the intraepithelial lymphocytes (IEL). [37, 38]. The other MALTs are similarly organized.

The mucosal immune response has 2 phases:

- inductive phase;
- effector phase.

Inductive phase

The antigen uptake in the intestinal mucosa (especially particular antigens) occurs either through the specialized sampling system represented by the M-cells overlying the PP or across normal epithelium overlying the *lamina propria*. The M-cells may transport various soluble antigens and even whole bacterial cells from the surface of the epithelium to the PP. Below the epithelium there are dendritic cells (DCs). The DCs perform phagocytosis of various antigens and present them to various immunocompetent cells in the mucosal immune system. The DCs may present the antigen to:

- T-lymphocytes in the PP;
- T-lymphocytes in the MLN – the antigen-loaded DCs may migrate from the PP through the afferent lymph vessels to the MLN and present the antigen there.

The cells, which present antigens are called antigen presenting cells (APC). Some MHC class II (+) enterocytes may also act as APC. The M-cells, DCs, PP and the MLN perform the antigen presentation and recognition, thus fulfilling the so called inductive phase of the immune response [39-41].

Effector phase

The diffuse lymphoid structures are mostly presented by the intraepithelial lymphocytes (IEL) – mature T-lymphocytes, and IgA producing plasma cells (activated B-cells). The T-lymphocytes are divided to CD4+ (helper or inducer) and CD8+ (suppressor or cytotoxic). In most cases the APC present the antigens to naïve CD4+ cells and activate them (fig. 2). The T-lymphocytes in lamina propria are predominantly CD4+, whereas the IEL are mostly CD8+. The activated CD4+ cells leave the organized lymphoid structures and using the lymphatic system reach the systemic circulation through the thoracic duct. The activated mucosal B-cells produce secretory IgA (sIgA), which is the principal mucosal immunoglobulin. Secretory IgA is a dimeric form of IgA and the two IgA molecules are binded by a joining chain. Secretory IgA inhibits the bacterial adhesion to the mucosa, carries out the lactoperoxidase and lactoferrin to the cell surface, takes part in the clearance of immune complexes and activates the alternative complement pathway. The IEL perform the effector phase of the immune response [37; 40].

The inductive and efector immune response are interdependent and sometimes overlapping.

The activated CD4+ may interact with other efector cells such as activated B-cells, CD8+ lymphocytes, etc. After priming, memory B- and T-cells migrate to other efector sites, followed by active proliferation, local induction of certain cytokines and production of secretory antibodies (IgA). The migration to other mucosal surfaces is called lymphocyte homing and it is possible because of the so called addressin receptors. By using the homing mechanism the lymphocytes sensitized in one part of the MALTs can reach all other mucosal sites [42]. About 80 % of the activated B-cells are found in the intestinal *lamina propria.* This is the main source of mucosal antibodies in MALTs [39; 43]. After priming, memory B- and T-cells migrate to effector sites, followed by active proliferation, local induction of certain cytokines and production of sIgA.

The intestinal epithelium and the GALT play a crucial role in the maintenance of the oral tolerance – antigen specific tolerance to orally ingested food and bacterial antigens [44]. All mucosal epithelial layers are a part of the innate immunity and serve as a first line of defense against numerous exogenous factors. The epithelial cells in the gut form a reliable and highly selective barrier between the intraluminal content and the body interior. The disruption of this barrier could lead to the development of an inflammatory response. This would be a result of the direct interaction between the GALT and the intraluminal antigens. This has been confirmed in animal models – the mice with genetically determined alterations of the intestinal permeability are developing intestinal inflammation [45, 46]. Normally there is a constant interaction between the intestinal epithelium and GALT thus making possible the existence of the oral tolerance [47].

There is a complex relationship between the intestinal immune system and the resident and transient intestinal microbiota and it is crucial for the epithelial cells and the mucosal immune system to distinguish between pathogenic and non-pathogenic agents. Intestinal epithelial cells and some enteroendocrine cells are capable of detecting bacterial antigens and initiating and regulating both innate and adaptive immune responses. Signals from bacteria can be transmitted to adjacent immune cells such as macrophages, dendritic cells and lymphocytes through molecules expressed on the epithelial cell surface – the so called pattern-recognitioning receptors (PRRs). There are numerous PRRs: major histo-compatibility complex I and II molecules and Toll-like receptors (TLRs). TLRs alert the immune system to the presence of highly conserved microbial antigens called pathogen-associated molecular patterns (PAMPs). They are present on most microorganisms. Examples of PAMPs include lipopolysaccharides (LPS), peptidoglycan, flagellin, and microbial nucleic acids [4, 48-50]. This is exactly how probiotic bacteria interact with the mucosal immune system – by their PAMPs.

There are at least ten types of human TLRs. In humans, TLRs are expressed in most tissues, including myelomonocytic cells, dendritic cells and endothelial and epithelial cells. Interaction of TLRs and PAMPs results in activation of a complex intracellular signaling cascade, up-regulation of inflammatory genes, production of pro- and anti-inflammatory inflammatory cytokines and interferons, and recruitment of myeloid cells. It also stimulates expression of co-stimulatory molecules required to induce an adaptive immune response of APC [4, 50]. The colonic epithelium expresses mostly TLR3 but also TLR4, TLR5, and TLR7 [51], while cervical and vaginal epithelial cells have a higher expression of TLR1, TLR2, TLR3, TLR5 and TLR6 [52]. TLR4 recognises LPS [53, 54], a constituent of the cell wall of Gram-negative bacteria, while TLR2 reacts with a wider spectrum of bacterial products such as lipoproteins, peptidoglycans and lipoteichoic acid found both in Gram-positive and Gram-negative bacteria [55, 56].

There is another family of membrane-bound receptors for detection of proteins and they are different from the TLRs. They are called NOD-like receptors or nucleotide-binding domain, leucine-rich repeat containing proteins (NLRs). The best characterised NLRs are NOD1 and NOD2. NRLs are located in the cytoplasm and are involved in the detection of bacterial PAMPs that enter the mammalian cell. NRLs are especially important in tissues where TLRs are expressed at low levels [57]. This is the case in the epithelial cells of the GIT where the cells are in constant contact with the microbiota, and the expression of TLRs must be down-regulated in order to avoid over-stimulation and permanent activation. However, if these intestinal epithelial cells get infected with invasive bacteria or bacteria interacting directly with the plasma membrane, they will come into contact with NLRs and will activate some certain defense mechanisms [58]. NLRs are also involved in sensing other endogenous warning signals which will result in the activation of inflammatory signalling pathways, such as nuclear factor-kappa B (NF-κB) and mitogen-activated protein kinases. Both NOD1 and NOD2 recognise peptidoglycan moieties found in bacteria. NOD1 can sense peptidoglycan moieties containing meso-diaminopimelic acid, which primarily are associated to gram-negative bacteria. NOD2 senses the muramyl dipeptide motif that can be found in a wider range of bacteria, including numerous probiotic bacteria [59, 60]. The ability of NRLs to regulate, for example, nuclear factor-kappa B (NF-κB) signalling and interleukin-1-beta (IL-1β) production, indicates that they are important for the pathogenesis of inflammatory human diseases, such as IBD and especially Crohn's disease.

NOD2 are expressed mostly by DCs, granulocytes, macrophages and Paneth cells, as the TNFα and IFNγ up-regulate the expression of NOD2 in epithelial cells in intestinal crypts [59, 61, 62]. The overall expression of NOD1 and NOD2 increases in inflammation [63, 64].

The microbiota alone can also predetermine the direction of this response with it's PAMPs and their interaction with human PRRs. The NLRs and TLRs play a crucial role in the regulation of the inflammatory response towards indigenous and transient microbiota. The synthesis of various pro- and anti-inflammatory cytokines and/or activation of NF-kB may alter the direction of the immune response – from inflammation to anergy.

The activation of the APC occurs after the binding of the PRRs with specific bacterial PAMPs. The types of PAMPs determine the selective activation of Th1, Th2, Th17 or Treg by the DCs (fig. 2).

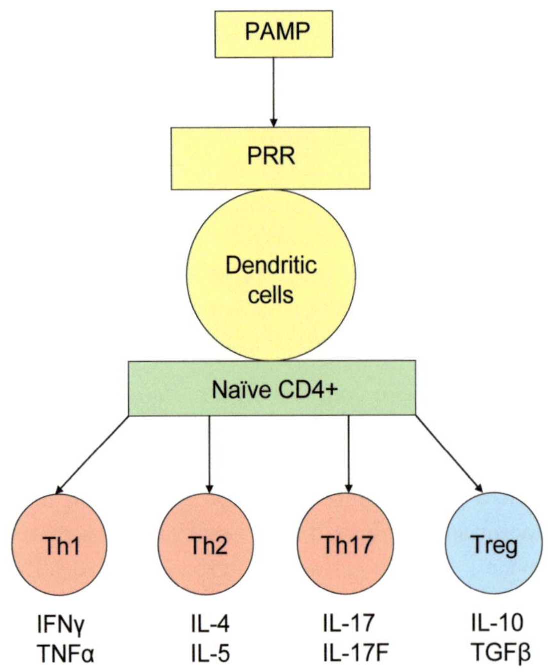

Figure 2. Interaction between the bacterial PAMPs, human PRRs, APCs, naïve CD4+ and activated CD4+ lymphocytes such as Th1, Th2, Th17 or Treg and their main cytokines.

The activated CD4+ lymphocytes may be divided in 2 groups:

- effector (Th1, Th2 and Th17);
- regulatory (Treg)

Effector CD4+ lymphocytes

- Th1-lymphocytes: they secrete IL-2, TNFα, IFNγ and GM-CSF. These lymphocytes take part mostly in the cell-mediated immune response, the normal functions of the macrophages and the delayed hypersensitivity reactions;
- Th2-lymphocytes: they secrete IL-4, IL-5, IL-6, IL-13 and mediate the humoral immune response, the synthesis of IgE and atopic disease;
- Th17-lymphocytes – some authors link them with the development of numerous autoimmune diseases. Their activation and functions are not fully studies and understood but they differ from the Th1- and Th2-lymphocytes. Their activation is mediated by TGF-β, IL-6, IL-21 and IL-23 but suppressed by IFNγ and IL-4. The Th17-lymphocytes secrete IL-17, IL-17F and IL-22.

Regulatory CD4+ lymphocytes

- Treg-lymphocytes: they secrete the anti-inflammatory IL-10 and TGFβ and mediate the intensity and the direction of the immune response. The animals with inborn deficiency of IL-10 and TGFβ develop acute enterocolitis with fatal consequences. This is a result of a paradoxical inflammatory response towards the resident intestinal flora [65-71];

There are parts of the indigenous microbiota that are less prone to induce inflammation, and there may even be bacterial genera with the ability to counteract inflammation. This seemingly inflammation-suppressing effect can be a result of different actions. The inflammation-suppressing fractions of the bacterial flora may be able to:

- counteract some of the inflammation-aggravating bacteria, which will decrease the inflammatory response;
- improve the barrier effect of the mucosa, which will inhibit the translocation of inflammation-inducing luminal contents into the body;
- directly interact with pro-inflammatory processes and cascades of the immune system.

All three actions may work simultaneously. Currently, the most studied inflammation-suppressing indigenous bacteria are certain species/strains of *Lactobacillus* and *Bifidobacterium*, and those are also the main bacteria used in the production of probiotics [72].

The inflammation alone can be a consequence of allergic reactions, infectious diseases and autoimmune diseases such as rheumatoid arthritis, diabetes type 1, multiple sclerosis and Crohn's disease, but a low-grade systemic inflammation also characterises the metabolic syndrome and the ageing human body. The long-term inflammation increases the risk for atherosclerosis, cancer, dementia and non-alcoholic fatty liver disease. Diabetes type 2 and obesity are also characterised by a low-grade inflammation but it is still unclear if the inflammation is the cause of the condition or just a result of it. The indigenous flora of the human body may trigger inflammation, and so favourable influence on the composition of the indigenous microbiota can be a strategy to mitigate inflammation. The use of probiotic bacteria can affect the composition of the resident flora, but probiotics may also have more direct effects on the immune system and the permeability of the mucosa. The better the barrier effect of the mucosa the smaller the risk of translocation of pro-inflammatory components originating from the mucosal microbiota [72].

7. Probiotics and mucosal immune response in clinical practice

The polarization of the immune response is the reason why the oral intake of probiotic bacteria has been proven to be effective in allergic inflammation – atopic dermatitis, vernal keratoconjunctivitis but also in inflammatory bowel disease [23, 24]; infectious and antibiotic induced diarrhea [19, 20], urogenital infections [21, 22], atopic disease [25, 26]. Probiotic-induced immune modulation at mucosal sites distant from the gut supports the 'hygiene theory' of allergy development [73]. The 'hygiene theory' links the recent increase in the prevalence of allergic disease with modern western lifestyle, through altered patterns of gut colonisation characterised by a skewing towards an IFN-γ mucosal cytokine response [74]. In addition some authors suggest that probiotics may have a place as adjunctive treatment in *H. pylori* infections and possibly in their prophylaxis [75].

Based on the clinical evidence we could assume that the effects of probiotic bacteria over the mucosal immune response may be divided into local and systemic. Indeed the efficacy of probiotic bacteria in atopic disease speaks of some systemic effect. Another perfect example for potential systemic efficacy are the immunological changes in breast milk, occurring after oral intake of *Lactobacillus bulgaricus* - "I. Bogdanov patent strain tumoronecroticance B-51" - ATCC 21815 [76]. According to the authors this is possible because of the functional entero-mammaric link and the functional redistribution of activated lymphocytes from the gut to the mammary gland and vice versa. In addition to this Dalmasso et al. [77] reported a novel biological property of probiotic bacteria: their capacity to affect immune cell redistribution by improving the competence of lymphatic endothelial cells to trap T lymphocytes.

The facilitation of oral tolerance and innocent bystander suppression by probiotic bacteria [78, 79] support the fact that particular probiotics not only drive protection against infection throughout the mucosal immune system, but also regulate the effector response. It is likely that different bacterial species operate through different mechanisms, indicating the importance of screening assays when identifying new isolates for clinical testing. It is suggested that a new term '*immunobiotics*', identifying those bacteria that promote health through activation of the mucosal immune apparatus, is a necessary evolutionary step as the foundation of our knowledge expand regarding the host–parasite relationships and their outcomes, as they relate to health and disease. Recognition of bacteria that promote mucosal T-cell function as '*immunobiotics*' moves probiotic biology forward by focusing on a mechanism of outcome, i.e. immunomodulation at distant mucosal sites. The human understanding of the interaction between the '*immunobiotic*' bacteria with the MALTs increases further and particular effector molecules and their receptor targets are being identified. A new focus in biotherapy can be expected to evolve. It still remains to convert predictable shifts in mucosal immunity into practical health gains for the benefits of immunobiotic therapy to be realised [74].

8. Conclusion

The Roman Emperor and Stoic Philosopher Marcus Aurelius has said "Constantly regard the universe as one living being, having one substance and one soul; and observe how all things have reference to one perception, the perception of this one living being; and how all things act with one movement; and how all things are the cooperating causes of all things which exist; observe too the continuous spinning of the thread and the contexture of the web." [80]. Indeed the probiotics, the resident flora and the mucosal immune system are extremely strongly related and act as a single equilibrium and should always be investigated and described together. There is a long way to go until we fully understand and manage to control the interaction between the probiotic bacteria and the mucosal immune system.

Author details

Petar Nikolov
Clinic of Gastroenterology, St. Ivan Rilsky University Hospital, Sofia, Bulgaria

Acknowledgement

This chapter was only possible because of the support from my family and the life lessons of my scientific mentor Prof. Zahariy Krastev.

9. References

[1] Rogers AH. Molecular Oral Microbiology. Norfolk: Caister Academic Press; 2008.

[2] Davis CP. Normal Flora. In: Baron S. (ed.) Medical Microbiology. 4th edition. Galveston (TX): University of Texas Medical Branch at Galveston; 1996.

[3] Encyclopedia Britannica. Astronomy: Milky Way Galaxy. http://www.britannica.com/EBchecked/topic/382567/Milky-Way-Galaxy (accesed 5 May 2012).

[4] O'Hara AM, Shanahan F. The gut flora as a forgotten organ. EMBO Reports 2006; 7, 688–693.

[5] Simon GL, Gorbach SL. Intestinal flora in health and disease. Gastroenterology 1984; 86, 174-193.

[6] Zoetendal EG, Akkermans ADL, Akkermans-van VWM et al. The Host Genotype Affects the Bacterial Community in the Human Gastronintestinal Tract. Microbial Ecology in Health and Disease 2001; 13, 129-134.

[7] Collignon A, Butel MJ. Establishment and composition of the gut microflora. In: Rambaud JC, Buts JP, Cortier G, Flourie B. (eds.) Gut microflora. Digestive physiology and pathology. 1st Edn. Montrouge: John Libbey Eurotext; 2006. 19-35.

[8] Marteau P, Pochart P, Doré J et al. Comparative study of bacterial groups within the human cecal and fecal microbiota.". Appl Environ Microbiol 2001; 67, 4939-4942.

[9] Isolauri E, Salminen S, Ouwehand AC. Probiotics. Best Practice & Research Clinical Gastroenterology 2004; 18, 299-313.

[10] Ouwehand AC, Vesterlund S. Health aspects of probiotics. IDrugs 2003; 6, 573-580.

[11] Bourlioux P, Koletzko B, Guarner F, Braesco V. The intestine and its microflora are partners for the protection of the host: report on the Danone Symposium "The Intelligent Intestine". Am J Clin Nutr 2003; 78, 675-683.

[12] Cummings JH, Englyst HN. Fermentation in the human large intestine and the available substrates. Am J Clin Nutr 1987; 45, 1243-1255.

[13] Silvester KR, Englyst HN, Cummings JH. Ileal recovery of starch from whole diets containing resistant starch measured in vitro and fermentation of ileal effluent. Am J Clin Nutr 1995; 62, 403-411.

[14] Guarner F, Malagelada JR. Gut flora in health and disease. Lancet 2003; 361, 512–519.

[15] Adlerberth I, Cerquetti M, Poilane I, Wold A, Collignon A. Mechanisms of Colonisation and Colonisation Resistance of the Digestive Tract Part 1: Bacteria/host Interactions. Microbial Ecol Health Dis 2000; 12, 223-239.

[16] Fuller R. Probiotics in man and animals. Journal of Applied Bacteriology 1989; 66, 365-378.

[17] World Health Organization. Food and Agriculture Organization of the United Nations, Health and Nutritional Properties of Probiotics in Food including Powder Milk with Live Lactic Acid Bacteria: http://www.who.int/foodsafety/publications/fs_management/en/probiotics.pdf (accesed on 5 May 2012)

[18] Albert Einstein. Letter to his son Eduard from 5 February 1930. In: Isaacson W (ed.). Einstein: His Life and Universe. New York: Simon & Schuster; 2007. p367.

[19] Alvarez-Olmos MI, Oberhelman RA. Probiotic Agents and Infectious Diseases: A Modern Perspective on a Traditional Therapy. Clinical Infectious Diseases 2001; 32(11) 1567-1576.

[20] D'Souza AL, Rajkumar C., Cooke J, Bulpitt CJ. Probiotics in prevention of antibiotic associated diarrhoea: meta-analysis. BMJ 2002; 324, 1361-1364.

[21] Reid G, Bruce AW, Fraser N, Heinemann C, Owen J, Henning B. Oral probiotics can resolve urogenital infections. FEMS Immunology & Medical Microbiology 2001; 30(1) 49–52.

[22] Reid G. Probiotics for Urogenital Health. Nutrition in Clinical Care 2002; 5(1) 3-8.

[23] Kruis W, Frič P, Pokrotnieks J, Lukáš M, Fixa B, Kaščák M, Kamm MA, Weismueller J, Beglinger C, Stolte M, Wolff C, Schulze J. Maintaining remission of ulcerative colitis with the probiotic Escherichia coli Nissle 1917 is as effective as with standard mesalazine. Gut 2004;53, 1617-1623.

[24] Gionchetti P, Rizzello F, Morselli C, Poggioli G, Tambasco R, Calabrese C, Brigidi P, Vitali B, Straforini G, Campieri M. High-Dose Probiotics for the Treatment of Active Pouchitis. Diseases of the Colon & Rectum 2007; 50(12) 2075-2084.

[25] Kalliomäki M, Salminen S, Arvilommi H, Kero P, Koskinen P, Isolauri E. Probiotics in primary prevention of atopic disease: a randomised placebo-controlled trial. Lancet 2001; 357(9262) 1076-9.

[26] Kalliomäki M, Salminen S, Poussa T, Arvilommi H, Isolauri E. Probiotics and prevention of atopic disease: 4-year follow-up of a randomised placebo-controlled trial. Lancet 2003; 361(9372) 1869-71.

[27] Dethlefsen L, McFall-Ngai M, Relman DA. An ecological and evolutionary perspective on human-microbe mutualism and disease. Nature 2007; 449, 811-818.

[28] Encyclopedia Britannica. Ecology: Edge effect. http://www.britannica.com/EBchecked/topic/179088/edge-effect (accessed on 7 May 2012).

[29] Haller D, Jobin C. Interaction Between Resident Luminal Bacteria and the Host: Can a Healthy Relationship Turn Sour? J Ped Gastroenterology & Nutr 2004; 38, 123-136.

[30] Bienenstock J, Befus AD. Mucosal Immunology. Immunology 1980; 41, 249-270.

[31] Brandtzaeg P, Nilssen DE, Rognum TO, Thrane PS. Ontogeny of the mucosal immune system and IgA deficiency. Gastroenterol Clin North Am 1991; 22, 397-439.

[32] Brandtzaeg P. Homing of mucosal immune cells – a possible connection between intestinal and articular inflammation. Alimentary Pharmacol Therapeutics 1997; 11, S24-S37.

[33] Gleeson M, Cripps AW, Clancy RL. Modifiers of the human mucosal immune system. Immunol Cell Biol 1995; 73: 397-404.

[34] Cesta MF. Normal Structure, Function, and Histology of Mucosa-Associated Lymphoid Tissue. Toxicol Pathol August 2006; 34(5) 599-608.

[35] Haley PJ. Species differences in the structure and function of the immune system. Toxicology 2003; 188, 49–71.

[36] Elmore SA. Enhanced Histopathology of Mucosa-Associated Lymphoid Tissue. Toxicol Pathol 2006; 34(5) 687-696.

[37] Mowat A, Viney JL. The anatomical basis of intestinal immunity. Immunol Rev 1997; 156, 145 - 166.

[38] Brandtzaeg P. Development and basic mechanisms of human gut immunity. Nutr Rev 1998; 56, S5-18.

[39] Simecka JW. Mucosal immunity of the gastrointestinal tract and oral tolerance. Advanced Drug Delivery Reviews 1998; 34, 235-259.

[40] Mowat A. Anatomical basis of tolerance and immunity to intestinal antigens. Nature Reviews Immunology 2003; 3, 331-341.

[41] Neutra MN, Kraehenbuhl JP. Transepithelial transport and mucosal defense: The role of M-cells. Trends in Cell Biol 1992; 2, 134-138.

[42] Picker LJ, Butcher EC. Physiological and Molecular Mechanisms of Lymphocyte Homing. Annu Rev Immunol 1992; 10, 561-591.

[43] Brandtzaeg P, Halstensen TS, Kett K, Krajci P, Kvale D, Rognum TO, Scott H, Sollid LM. Immunobiology and immunopathology of human gut mucosa: humoral immunity and intraepithelial lymphocytes. Gasteroenterology 1989; 97, 1562-1584.

[44] Yu Y, Sitaraman S, Gewirtz AT. Intestinal epithelial cell regulation of mucosal inflammation. Immunol Res 2004; 29, 55-68.

[45] Panwala CM, Jones JC, Viney JL. A novel model of inflammatory bowel disease: mice deficient for the multiple drug resistance gene, mdr1a, spontaneously develop colitis. J Immunol 1998; 161, 5733-5744.

[46] Hermiston ML, Gordon JI. Inflammatory bowel disease and adenomas in mice expressing a dominant negative N-cadherin. Science 1995; 270, 1203-1207.

[47] Mennechet FJ, Kasper LH, Rachinel N, Li W, Vandewalle A, Buzoni-Gatel D. Lamina propria CD4+ T lymphocytes synergize with murine intestinal epithelial cells to enhance proinflammatory response against an intracellular pathogen. J Immunol 2002; 168, 2988-2996.

[48] Cario E, Brown D, McKee M, Lynch-Devaney K, Gerken G, Podolsky DK. Commensal associated molecular patterns induce selective toll-like receptor-trafficking from apical membrane to cytoplasmic compartments in polarized intestinal epithelium. Am J Pathol 2002; 160, 165–173.

[49] Hershberg RM, Mayer LF. Antigen processing and presentation by intestinal epithelial cells—polarity and complexity. Immunol Today 2000; 21, 123–128.

[50] Testro AG, Visvanathan K. Toll-like receptors and their role in gastrointestinal disease. J Gastroenterol Hepatol 2009; 24, 943–954.

[51] Zarember KA, Godowski PJ. Tissue expression of human Tolllike receptors and differential regulation of Toll-like receptor mRNAs in leukocytes in response to microbes, their products, and cytokines. J Immunol 2002; 168, 554–561.

[52] Fichorova RN, Cronin AO Lien E, Anderson DJ, Ingalls RR. Response to Neisseria gonorrhoeae by cervicovaginal epithelial cells occurs in the absence of toll-like receptor 4-mediated signalling. J Immunol 2002; 168, 2424–2432.

[53] Poltorak A, He X, Smirnova I, Liu MY, Van Huffel C, Du X, Birdwell D, Alejos E, Silva M, Galanos C, et al. Defective LPS signaling in C3H/HeJ and C57BL/10ScCr mice: Mutations in Tlr4 gene. Science 1998; 282, 2085–2088.

[54] Qureshi ST, Lariviere L, Leveque G, Clermont S, Moore KJ, Gros P, Malo D. Endotoxin-tolerant mice have mutations in Toll-like receptor 4 (Tlr4). J Exp Med 1999; 189, 615–625.

[55] Schwandner R, Dziarski R, Wesche H, Rothe M, Kirschning CJ. Peptidoglycan- and lipoteichoic acid-induced cell activation is mediated by Toll-like receptor 2. J Biol Chem 1999; 274, 17406–17409.

[56] Takeuchi O, Kaufmann A, Grote K, Kawai T, Hoshino K, Morr M, Mühlradt PF, Akira S. Cutting edge: Preferentially the R-stereoisomer of the mycoplasmal lipopeptide macrophage-activating lipopeptide-2 activates immune cells through a toll-like receptor 2- and MyD88-dependent signaling pathway. J Immunol 2000; 164, 554–557.

[57] Philpott DJ, Girardin SE, Sansonetti PJ. Innate immune responses of epithelial cells following infection with bacterial pathogens. Curr. Opin. Immunol. 2001; 13, 410–416.

[58] Girardin SE, Tournebize R, Mavris M, Page AL, Li X, Stark GR, Bertin J, DiStefano PS, Yaniv M, Sansonetti PJ, et al. CARD4/Nod1 mediates NF-kappaB and JNK activation by invasive Shigella flexneri. EMBO Rep 2001; 2, 736–742.
[59] Girardin SE, Boneca IG, Carneiro LA, Antignac A, Jéhanno M, Viala J, Tedin K, Taha MK, Labigne A, Zähringer U, et al. Nod1 detects a unique muropeptide from gram-negative bacterial peptidoglycan. Science 2003; 300, 1584–1587.
[60] Hasegawa M, Yang K, Hashimoto M, Park JH, Kim YG, Fujimoto Y, Nuñez G, Fukase K, Inohara N. Differential release and distribution of Nod1 and Nod2 immunostimulatory molecules among bacterial species and environments. J Biol Chem 2006; 281, 29054–29063.
[61] Gutierrez O, Pipaon C, Inohara N. Induction of Nod2 in myelomonocytic and intestinal epithelial cells via nuclear factor-kB activation. J Biol Chem 2002; 277, 41701-41705.
[62] Kobayashi KS, Chamaillard M, Ogura Y, Henegariu O, Inohara N, Nuñez G, Flavell RA. Nod2-dependent regulation of innate and adaptive immunity in the intestinal tract. Science 2005; 307, 731-734.
[63] Berrebi D, Maudinas R, Hugot JP, Chamaillard3, Chareyre F, Lagausie PDe, Yang C, Desreumaux P, Giovannini M, Cézard J-P, Zouali H, Emilie D, Peuchmaur M. Card15 gene overexpression in mononuclear and epithelial cells of the inflamed Crohn's disease colon. Gut 2003; 52, 840-846.
[64] Rosenstiel P, Fantini M, Bräutigam K. TNF-[alpha] and IFN-[gamma] regulate the expression of the NOD2 (CARD15) gene in human intestinal epithelial cells. Gastroenterology 2003; 124, 1001-1009
[65] Mosmann T, Coffman R. Different patterns of lymphokine secretion lead to different functional properties. Ann Rev Immunol 1989; 7, 145-173.
[66] Fitch F, McKisic M, Landcki D, Gajewski T. Differential regulation of murine T-lymphocyte subsets. Ann Rev Immunol 1993; 11, 29-48.
[67] Groux H, Powrie F. Regulatory T-cells and inflamatory bowel disease. Immunol Today 1999; 20, 442-446.
[68] Harrington LE, Hatton RD, Mangan PR, Turner H, Murphy TL, Murphy KM. Interleukin 17-producing CD4+ effector T cells develop via a lineage distinct from the T helper type 1 and 2 lineages. Nat Immunol 2005; 6, 1123-1132.
[69] Dong C. TH17 cells in development: an updated view of their molecular identity and genetic programming. Nature Rev Immunol 2008; 8, 337–348.
[70] Manel N, Unutmaz D, Littman DR. The differentiation of human T(H)-17 cells requires transforming growth factor-beta and induction of the nuclear receptor RORgammat. Nature Immunol 2008; 9, 641–649.
[71] Korn T, Bettelli E, Oukka M, Kuchroo VK. IL-17 and Th17 Cells. Annu Rev Immunol 2009; 27, 485-517.
[72] Hakansson A, Molin G. Gut Microbiota and Inflammation. Nutrients 2011; 3, 637-682.
[73] Sly P, Holt P. Etiological factors of atopic disease in the respiratory tract. Mucosal Immunol 1999; 7, 13–14.
[74] Clancy R. Immunobiotics and the probiotic evolution. FEMS Immunology & Medical Microbiology 2003; 38(1) 9–12.

[75] Hamilton-Miller JMT. The role of probiotics in the treatment and prevention of Helicobacter pylori infection. International Journal of Antimicrobial Agents 2003; 22(4) 360–366.

[76] Nikolov P, Baleva M. The Alteration of secretory IgA in human breast milk and stool samples after the intake of a probiotic – report of 2 cases. Centr Eur J Med 2012; 7(1) 25-29.

[77] Dalmasso G, Cottrez F, Imbert V, Lagadec P, Peyron J-F, Rampal P, Czerucka D, Groux H. Saccharomyces boulardii inhibits inflammatory bowel disease by trapping T cells in mesenteric lymph nodes. Gastroenterology. 2006; 131, 1812–1825.

[78] Sudo N, Sawamura S, Tanaka K, Aiba Y, Kubo C, Koga Y. The requirement of intestinal bacterial flora for the development of an IgE production system fully susceptible to oral tolerance induction. J Immunol 1997; 159, 1739–1745.

[79] Kano H, Kaneko T, Kaminogawa S. Oral intake of Lactobacillus delbrueckii subsp. bulgaricus OLL1073R–1 prevents collagen-induced arthritis in mice. J Food Prot 2002; 65, 153–160.

[80] Aurelius M. Book 4. In: Aurelius M. (ed.) The Meditations of Marcus Aurelius. Stilwell: Digireads.com Publishing; 2005. p21. Available from http://books.google.co.uk/books?id=UFP1CIhPSKAC&pg=PA17&source=gbs_toc_r&cad=4#v=onepage&q&f=false (accessed 7 May 2012)

Section 3

Probiotics in Biotechnological Aspects

Chapter 23

Encapsulation Technology to Protect Probiotic Bacteria

María Chávarri, Izaskun Marañón and María Carmen Villarán

Additional information is available at the end of the chapter

http://dx.doi.org/10.5772/50046

1. Introduction

Probiotic bacteria are used in production of functional foods and pharmaceutical products. They play an important role in promoting and maintaining human health. In order, to produce health benefits probiotic strains should be present in a viable form at a suitable level during the product is shelf life until consumption and maintain high viability throughout the gastrointestinal tract. Many reports indicated that there is poor survival of probiotic bacteria in products containing free probiotic cells [1]. Providing probiotic living cells with a physical barrier to resist adverse environmental conditions is therefore an approach currently receiving considerable interest [2].

The encapsulation techniques for protection of bacterial cells have resulted in greatly enhanced viability of these microorganisms in food products as well as in the gastrointestinal tract. Encapsulation is a process to entrap active agents within a carrier material and it is a useful tool to improve living cells into foods, to protect [3, 4, 5, 6, 7], to extend their storage life and to convert them into a powder form for convenient use [8, 9, 10, 11]. In addition, encapsulation can promote controlled release and optimize delivery to the site of action, thereby potentiating the efficacy of the respective probiotic strain. This process can also prevent these microorganisms from multiplying in food that would otherwise change their sensory characteristics. Otherwise, materials used for design of protective shell of encapsulates must be food-grade, biodegradable and able to form a barrier between the internal phase and its surroundings.

2. Probiotics

2.1. Definition

Probiotics are defined as live microorganisms which, when administered in adequate amounts, confer health benefits to the host [12], including inhibition of pathogenic growth, maintenance of health promoting gut microflora, stimulation of immune system, relieving constipation, absorption of calcium, synthesis of vitamins and antimicrobial agents, and predigestion of proteins [13]. Several health benefits have been proved for specific probiotic bacteria, and recommendations for probiotic use to promote health have been published [14].

The term "probiotic" includes a large range of microorganisms, mainly bacteria but also yeasts. Because they can stay alive until the intestine and provide beneficial effects on the host health, lactic acid bacteria (LAB), non-lactic acid bacteria and yeasts can be considered as probiotics. LAB are the most important probiotic known to have beneficial effects on the human gastro-intestinal (GI) tract [15].

The effects of probiotics are strain-specific [16, 17, 18] and that is the reason why it is important to specify the genus and the species of probiotic bacteria when proclaiming health benefits. Each species covers various strains with varied benefits for health. The probiotic health benefits may be due to the production of acid and/or bacteriocins, competition with pathogens and an enhancement of the immune system [19]. Dose levels of probiotics depend on the considered strain [20], but 10^6–10^7 CFU/g of product per day is generally accepted [21].

2.2. Health benefits

There is evidence that probiotics have the potential to be beneficial for our health [22]. Multiple reports have described their health benefits on gastrointestinal infections, antimicrobial activity, improvement in lactose metabolism, reduction in serum cholesterol, immune system stimulation, antimutagenic properties, anti-carcinogenic properties, anti-diarrheal properties, improvement in inflammatory bowel disease and suppression of Helicobacter pylori infection by addition of selected strains to food products [23, 24, 25, 26, 27, 28, 29, 30, 31, 32, 33].

The beneficial effects of probiotic microorganisms appear when they arrive in the intestinal medium, viable and in high enough number, after surviving the above mentioned harsh conditions [34]. The minimum number of probiotic cells (cfu/g) in the product at the moment of consumption that is necessary for the fruition of beneficial pharmaceutical (preventive or therapeutic) effects of probiotics has been suggested to be represented by the minimum of bio-value (MBV) index [35]. According to the International Dairy Federation (IDF) recommendation, this index should be $\geq 10^7$ cfu/g up to the date of minimum durability [36]. Also, various recommendations have been presented by different researchers such as $>10^6$ cfu/g by all probiotics in yogurt [37, 38] and $>10^7$ cfu/g in the case of bifidobacteria [39]. Apart from the MBV index, daily intake (DI) of each food product is also determinable for their probiotic effectiveness. The minimum amount of the latter index has been recommended as approximately 10^9 viable cells per day [35, 38, 40].The type of culture media used for the enumeration of probiotic bacteria is also an important factor for determination of their viability, as the cell recovery rate of various media are different [35, 41].

Most existing probiotics have been isolated from the human gut microbiota. This microbiota plays an important role in human health, not only due to its participation in the digestion process, but also for the function it plays in the development of the gut and the immune system [42]. The mechanisms of action of probiotic bacteria are thought to result from modification of the composition of the endogenous intestinal microbiota and its metabolic activity, prevention of overgrowth and colonization of pathogens and stimulation of the immune system [43]. With regard to pathogen exclusion, probiotic bacteria can produce antibacterial substances (such as bacteriocins and hydrogen peroxide), acids (that reduce the pH of the intestine), block adhesion sites and be competitive for nutrients [44].

Recent studies have shown differences in the composition of the gut microbiota of healthy subjects [45], underlining the difficulties in defining the normal microbiota at microbial species level. Moreover, studies suggest that some specific changes in gut microbiota composition are associated with different diseases [46, 47]. This was confirmed by the comparison of the microbiome from healthy individuals with those of diseased individuals, allowing the identification of microbiota imbalance in human diseases such as inflammatory bowel disease or obesity [48, 49].

3. Encapsulation of probiotic living cells

Encapsulation is often mentioned as a way to protect bacteria against severe environmental factors [50, 51].The goal of encapsulation is to create a micro-environment in which the bacteria will survive during processing and storage and released at appropriate sites (e.g. small intestine) in the digestive tract. The benefits of encapsulation to protect probiotics against low gastric pH have been shown in numerous reports [50] and similarly for liquid-based products such as dairy products [21, 52].

Encapsulation refers to a physicochemical or mechanical process to entrap a substance in a material in order to produce particles with diameters of a few nanometres to a few millimetres. So, the capsules are small particles that contain an active agent or core material surrounded by a coating or shell. Encapsulation shell materials include a variety of polymers, carbohydrates, fats and waxes, depending of the core material to be protected, and this aspect will be discussed below in the this section.

The protection of bioactive compounds, as vitamins, antioxidants, proteins, and lipids may be achieved using several encapsulation technologies for the production of functional foods with enhanced functionality and stability. Encapsulation technologies can be used in many applications in food industry such as controlling oxidative reaction, masking flavours, colours and odours, providing sustained and controlled release, extending shelf life, etc. In the probiotic particular case, these need to be protected during the time from processing to consumption of a food product. The principal factors against them need to be protected are:

- Processing conditions (temperature, oxidation, shear, etc.)
- Desiccation (for dry food products)
- Storage conditions (packaging and environment: moisture, oxygen, temperature, etc.)

- Degradation in the gastrointestinal tract (low pH in stomach and bile salts in the small intestine).

Encapsulation technology is based on packaging of bioactive compounds in mili-, micro- or nano-scaled particles which isolate them and control their release upon applying specific conditions. The coating or shell of sealed capsules needs to be semipermeable, thin but strong to support the environmental conditions maintaining cells alive, but it can be designed to release the probiotic cell in a specific area of the human body. The scientific references related with probiotic encapsulation stress the degradation in the gastrointestinal tract, more than the processing conditions and the coating material usually employed can withstand acidic conditions in the stomach and bile salts form the pancreas after consumption. In this way, the protection of the biological integrity of probiotic bacteria is achieved during gastro-duodenal transit, achieving a high concentration of viable cells to the jejunum and the ileum.

The selection of the best encapsulation technology for probiotics needs to consider numerous aspects in order to guarantee the survival of bacteria during the encapsulation process, in storage conditions and consumption, as well as the controlled release in the specific desired area of gut. So, there are two important problematic issues considering probiotic encapsulation: the size of probiotics which exclude the nanoencapsulation technologies and the difficulties to keep them alive.

In this section the most common techniques used for microencapsulation of probiotics will be presented (Sect. 3.1), as well as the most usual microcapsule coating or shell materials (Sect. 3.2) and some marketing considerations for their application in food products (Sect. 3.3).

3.1. Main techniques for microencapsulation of probiotics

3.1.1. Spray-drying

Spray-drying is a commonly used technique for food ingredients production because it is a well-established technique suitable for large-scale, industrial applications. The first spray dryer was constructed in 1878 and, thus, it is a relatively old technique compared with competing technologies [53]. This technique is probably the most economic and effective drying method in industry, used for the first time to encapsulate a flavour in the 1930s. However, it is not so useful for the industrial production of encapsulated probiotics for food use, because of low survival rate during drying of the bacteria and low stability upon storage.

Drying is an encapsulation technique which is used when the active ingredient is dissolve in the encapsulating agent, forming an emulsion or a suspension. The solvent is commonly a hydrocolloid such as gelatine, vegetable gum, modified starch, dextrin, or non-gelling protein. The solution that is obtained is dried, providing a barrier to oxygen and aggressive agents [54].

In the spray-drying process a liquid mixture is atomized in a vessel with a single-fluid nozzle, a two-fluid nozzle or spinning wheel (depending of the type of spray dryer in use) and the solvent is then evaporated by contacting with hot air or other gas. Most of spray dryers used in food industry are concurrent in design, i.e. product enters the dryer flowing in the same direction as the drying air. The objective is to obtain a very rapid drying and to avoid that the temperature of the material dried exceeds the exit air temperature of the dryer (Figure 1).

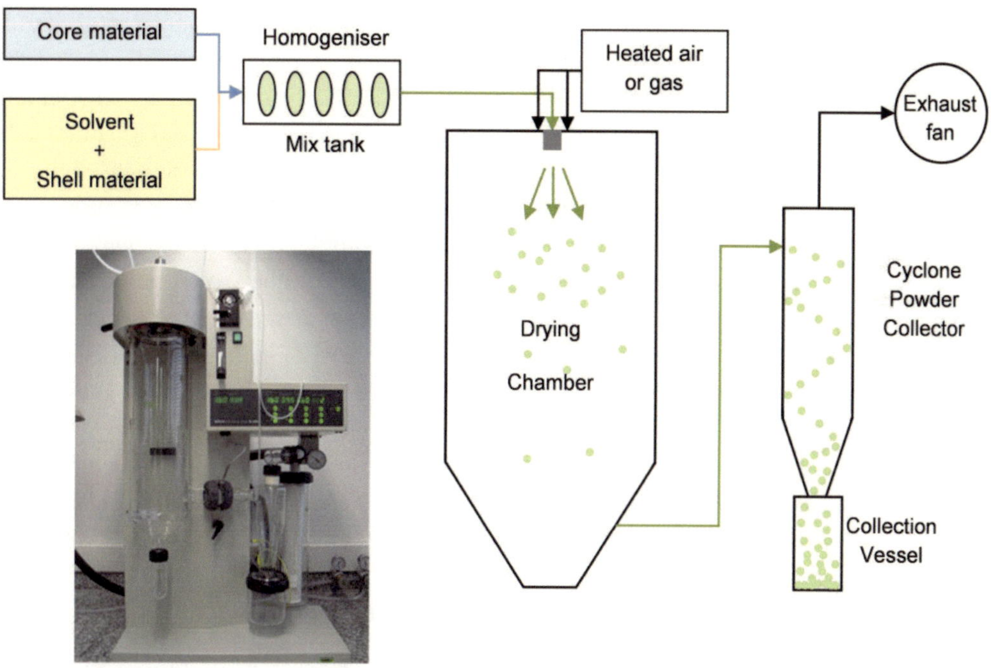

Figure 1. Schematic diagram of a spray-dry encapsulation process and image of a Mini Spray Dryer B-290 (BÜCHI), available at TECNALIA.

But also in a concurrent design, the conventional procedure requires to expose cells to high temperature and osmotic stresses due to dehydration witch results in relatively high viability and activity losses immediately after spraying and most likely also affects storage stability. However, some strains survive better than others. And parameters as drying temperature and time and shell material have also an important effect.

Using gelatinised modified starch as a carrier material, O'Riordan obtained good results in *Bifidobacterium* cells encapsulation with an inlet temperature of 100 ºC and oulet temperature of 45 ºC. Inlet temperatures of above 60 °C resulted in poor drying and the sticky product often accumulated in the cyclone. Higher inlet temperatures (>120 °C) resulted in higher outlet temperatures (>60 °C) and significantly reduced the viability of encapsulated [55]. The logarithmic number of probiotics decreases linearly with outlet air temperature of the spray-drier (in the range of 50 ºC - 80 ºC) [56]. So, the optimal outlet air temperature might be as low as possible, enough to assure the drying of the product and to avoid the sticky effect. Alternatively, a second draying step might be applied, using a fluid bed or a vacuum oven, for example, due to the optimal survival of probiotics is achieved with low water activity.

The successful spray drying of *Lactobacillus* and *Bifidobacterium* have previously reported for a number of different strains, including *L. paracasei* [57, 58], *Lactobacillus curvatus* [59], *L. acidophilus* [60], *L. rhamnosus* [61] and *Bifdobacterium ruminantium* [8]. Specifically, Favaro-Trindade and Grosso [6] used spray drying to encapsulate *B. lactis* and *L. acidophilus* in the enteric polymer cellulose acetate phthalate enriched with the fructooligosaccharide Raftilose1 (a prebiotic). In this work, the process was also appropriate, especially for *B. lactis* (Bb-12), since for entry temperature of 130 ºC and exit of 75 ºC, the counts in the powder and dispersion (feed) were similar; however, the *L. acidophilus* population showed a reduction of two log cycles. The atomization process and encapsulant agent cellulose acetate phthalate were effective in protecting these micro-organisms in acidic medium (hydrochloric acid solutions pH 1 and 2) during incubation for up to 2 h. In another study, *B. longum* B6 and *B. infantis* were encapsulated by spray drying, with gelatin, soluble starch, milk and gum arabic as encapsulating agents. Bifidobacteria in the encapsulated form showed a small reduction in their populations when exposed to acidic media and bile solutions when compared with those exposed in the free form. Among the encapsulants tested, gelatin and soluble starch were the most effective in providing protection to the micro-organisms in acidic medium and milk was the least effective [9]. Desmond and collaborators [57] encapsulated *L. acidophilus* in β-cyclodextrin and gum arabic. They used the spray drying process, in which entry and exit temperatures of 170 ºC and 90–85 ºC respectively, and observed a reduction of 2 log cycles in the microbial population. However, the microencapsulation process extended the shelf-life of the culture.

On the other hand, the most typical materials used as carrier in probiotic bacteria encapsulation are proteins and/or carbohidrates, which may be in the glassy state at storage temperatures to minimize molecular mobility and thus degradation. The presence of some prebiotics in the encapsulating material show higher count after spray drying for *Bifidobacterium*, depending of the physical properties of the prebiotic compound selected (thermoprotector effect, crystalinity, etc.) [62, 63] and a similar effect occurs for *Lactobacillus* bacteria [61, 64]. Some researchers have proposed the addition of thermo-protectants as inputs before drying with the intention of improving the resistance to the process and stability during storage [65]. In the case of Rodríguez-Huezo and collaborators [63] used a prebiotic as encapsulant (*'aguamiel'*) and a mixture of polymers composed of concentrated whey protein, 'goma mesquista' and maltodextrin. It is important to mention that not all the compound employees were efficient protectors. In fact, Ross and collaborators [66] reported that neither inulin nor polydextrose enhanced probiotic viability of spray-dried probiotics. In another study, it was also observed that when quercetin was added together with probiotics, the microencapsulation yields and survival rates were lower than for the micro-organism without quercetin [67]. A lot of other studies have employed of spray-drying technology to encapsulate probiotic cells, as noted in the table 1.

Probiotic strain	Materials	References
L. paracasei	Skim milk	[68]
Bifidobacterium PL1	Starch	[8]
L. acidophilus *B. lactis*	Cellulose acetate phthalate	[4]
L. bulgaricus	Alginate	[69]
L. paracasei	Milk based medium Gum acacia	[57]
B. breve *B. longum*	Whey protein	[70]
B. longum *B. infantis*	Gelatin soluble starch gum Arabica skim milk	[71]
B. lactis B1 *L. acidophilus LAC4*	Pectin and casein	[11]
L. rhamnosus GG	Whey protein-maltodextrin; Whey protein-inulin	[64]
B. longum *B. infantis*	Gelatin Starch Skim milk Gum arabic	[9, 72]

Table 1. Examples of encapsulated probiotic bacteria by Spray-drying Technology.

In summary, spray-drying technology offers high production rates at relatively low operating costs and resulting powders are stable and easily applicable [73]. However, most probiotic strains do not survive well the high temperatures and dehydratation during the spray-drying process. Loss of viability is principally caused by cytoplasmatic membrane damage although the cell wall, ribosomes and DNA are also affected at higher temperatures [74]. It was reported that the stationary phase cultures are more resistant to heat compare to cells in exponential growth phase [61].One approach used by a number of researchers to improve probiotic survival is the addition of protectants to the media prior to drying. For example, the incorporation of thermoprotectants, such as trehalose [75], non-fat milk solids and/ or adnitol [76], growth promoting factors including various probiotic/prebiotic combinations [77] and granular starch [78] have been shown to improve culture viability during drying and storage [79, 80].

Microencapsulation by spray-drying is a well-established process that can produce large amounts of material. Nevertheless, this economical and effective technology for protecting materials is rarely considered for cell immobilization because of the high mortality resulting from simultaneous dehydration and thermal inactivation of microorganisms.

3.1.2. Spray-cooling

This process is similar to spray-drying described before in relation with the production of small droplets. The principal difference in the spray-cooling process is the carrier material and the working conditions related with him. In the case, a molten matrix with low melting point is used to encapsulate the bacteria and the mixture is injected in a cold air current to enable the solidification of the carrier material.

It is interesting because the capsules produced in this way are generally not soluble in water. However, due the thermal conditions of the process, the spray-cooling is used rarely for probiotics encapsulation. As example of successful development, the patent US 5,292,657 [81] present the spray-cooling of probiotics in molten lipid atomized by a rotary disk in a cooling chamber. In any case, the contact time of the probiotics with the melt carrier material should remain very sort.

3.1.3. Fluid-bed agglomeration and coating

The fluid-bed technology evolved from a series of inventions patented by Dr. Wurster and colleagues at the University of Wisconsin Alumni Research Foundation (WARF) between 1957 and 1966 [82, 83, 84, 85]. These patents are based on the use of fluidising air to provide a uniform circulation of particles past an atomising nozzle. This nozzle is used to atomize a selected coating material (a melt product or an aqueous solution) which solidifies in a low temperature or by solvent evaporation. A proper circulation of the particles is recognised as the key to assure that all particles in the fluid-bed achieve a uniform coating. The most commonly used techniques are referred to as the bottom-spray (Wurster) fluid-bed process and the top-spray fluid-bed process (Figure 2); however, variations such as tangential-spray are also practised.

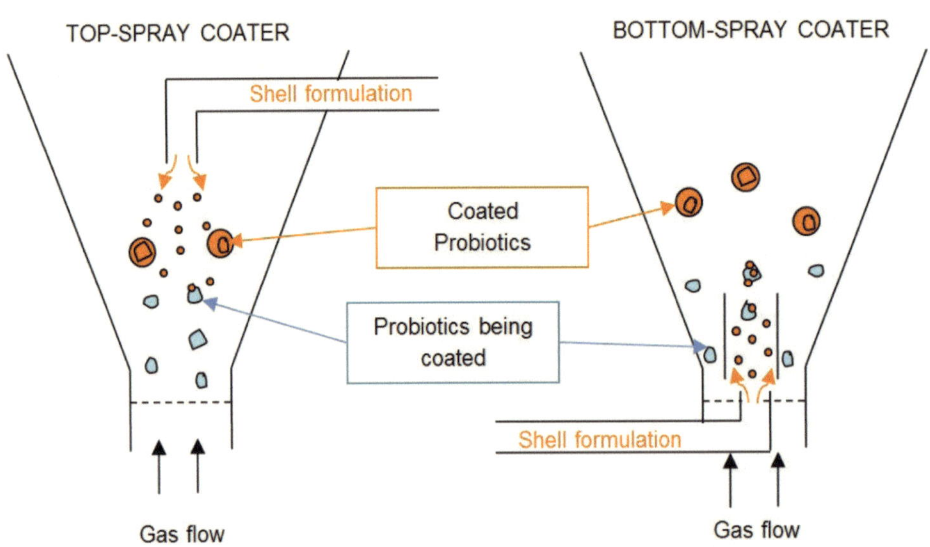

Figure 2. Schematic Diagrams of two types of the most commonly used fluid-bed coaters.

The top-spray fluid-bed coater is characterized by placement of nozzle above a fluidising bed and spraying down ware into the circulating flow of particles. This technique is useful for agglomeration or granulation. As particles flow is spray direction countercurrent, collisions involving wet particles are more probable and these collisions agglomerate particles. Bur the particles agglomerate become heavier and have less fluidization, so this phenomenon selectively agglomerates smaller particles and promotes agglomerate uniformity.

Placement of the nozzle at the bottom of a fluid bed provides the most uniform film on small particles and minimises agglomeration of such particles in the coating process compared with any other coating technique. This uniform coating is achieved because particles move further apart as they pass through the atomised spray from the nozzle and into an expansion region of the apparatus. This configuration allows the fluidising air to solidify or evaporate coating materials onto particles prior to contact between particles. A partition (centre tube) is used in Wurster fluid-bed coating to control the cyclic flow of particles in the process better than with de air distribution plate alone (Figure 3).

Figure 3. Expansion chamber for a bottom-spray (Wurster) fluid-bed process and detail of air distribution plate (from Glatt available at TECNALIA).

The most common coating material used for probiotics is lipid based, but proteins or carbohydrates can also be used [86]. This technique is among all, probably the most applicable technique for the coating of probiotics in industrial productions since it is possible to achieve large batch volumes and high throughputs. As example, Lallemand commercialize Probiocap™, and these particles are made in a fluid bed coating of freeze-dried probiotics with low melting lipids [87].

Specifically, Koo and collaborators [88], reported that *L. bulgaricus* loaded in chitosan-coated alginate microparticles showed higher storage stability than free cell culture. Later, Lee and researchers [69] showed that the microencapsulation in alginate microparticules coating with chitosan offers an effective way of delivering viable bacterial cells to the colon and maintaining their survival during refrigerated storage.

Fluidized-bed drying was recently investigated by Stummer and collaborators [89] as method for dehydration of *Enterococus faecium*. This study concludes to use fluidized-bed technology as a feasible alternative for the dehydration of probiotic bacteria by layering the cells on spherical pellets testing different protective agents as glucose, maltodextrin, skim milk, trehalose or sucrose, preferably skim milk or sucrose. According with the described procedure, it is possible to combine two manufacturing steps: (1) cell-dehydration preserving the optima cell properties and (2) the processing into suitable solid formulation with appropriate physical properties (the spherical pellets improve the flowability for filling capsules or dosing in different formulations.)

3.1.4. Freeze and vacuum-drying

Freeze-drying is also named lyophilisation. This drying technique is a dehydration process which works by freezing the product and then reducing the surrounding pressure to allow the frozen water to sublimate directly from the solid phase to the gas phase. The process is performed by freezing probiotics in the presence of carrier material at low temperatures, followed by sublimation of the water under vacuum. One of the most important advantages is the water phase transition and oxidation are avoided. In order to improve the probiotic activity upon freeze-drying and also stabilize them during storage, it is frequent the addition of cryoprotectans.

One of the most important aspects to decide is the choice of the optimal ending water content. This decision have to be a compromise between the highest survival rate after drying (higher survival rate with higher water content) and the lowest inactivation upon storage (better at low water activity, but not necessarily 0% of water content). According with King and collaborators [90], the loses in survival rates of freeze-dried probiotic bacteria under vacuum may be explained with a first-order kinetic and the rate constants can be described by an Arrhenius equation. But this equation might be affected by other factors as phase transition, atmosphere and water content.

In any case, the lyophilisation or freeze-drying is a very expensive technology, significantly more than spray-drying [56], even if it is probably most often used to dry probiotics. However, most of freeze-drying process only provide stability upon storage and not or limited during consumption. Because of that, this technique is used as a second step of encapsulation process. The freeze-drying is useful to dry probiotics previously encapsulated by other different techniques, as emulsion [91] or entrapment in gel microspheres [92]. In this way it is possible to improve the stability in the gastrointestinal tract and optimize the beneficial effect of probiotic consumption.

The Vacuum-drying is a similar process as freeze-drying, but it takes place at 0 - 40 ºC for 30 min to a few hours. The advantages of this process are that the product is not frozen, so the energy consumption and the related economic impact are reduced. In the product point of view, the freezing damage is avoided.

3.1.5. Emulsion-based techniques

An emulsion is the dispersion of two immiscible liquids in the presence of a stabilizing compound or emulsifier. When the core phase is aqueous this is termed a water-in-oil emulsion (w/o) while a hydrophobic core phase is termed an oil-in-water emulsion (o/w). Emulsions are simply produced by the addition of the core phase to a vigorously stirred excess of the second phase that contains, if it is necessary, the emulsifier (Figure 4). Nevertheless, even if the technique readily scalable, it produce capsules with an extremely large size distributions. Because of this limitation, there are several industrial efforts to achieve a narrow particle size distribution controlling the stirring and homogenization of the mixture.

There are also double emulsions, such water-in-oil-in-water (w/o/w). The technique is a modification of the basic technique in which an emulsion is made in of an aqueous solution in a hydrophobic wall polymer. This emulsion is the poured with vigorous agitation, into an aqueous solution containing stabilizer. The loading capacity of the hydrophobic core is limited by the solubility and diffusion to the stabilizer solution. The principal application of this technology is in pharmaceutical formulations.

Entrapment of probiotic bacteria in emulsion droplets has been suggested as a means of enhancing the viability of microorganism cells under the harsh conditions of the stomach and intestine. For example, Hou and collaborators [93] reported that entrapment of cells of lactic bacteria (*Lactobacillus delbrueckii* ssp. *bulgaricus*) in the droplets of reconstituted sesame oil body emulsions increased approximately 10^4 times their survival rate compared to free cells when subjected to simulated GI tract conditions.

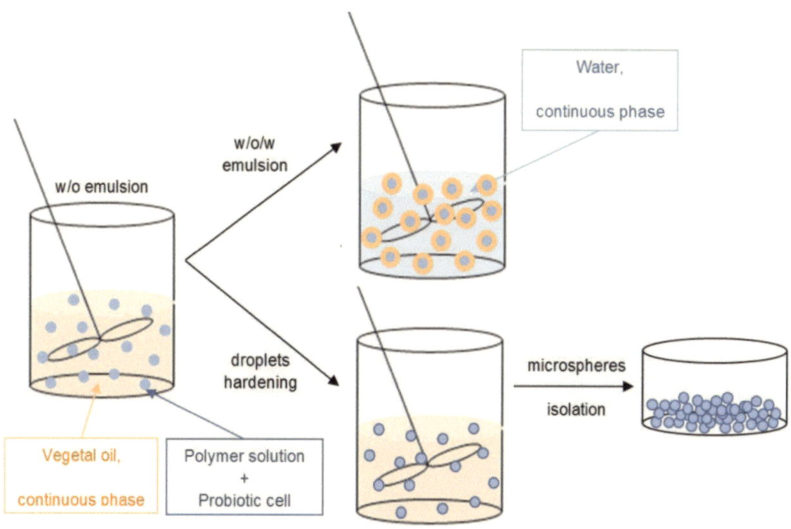

Figure 4. Probiotic cell encapsulation by water-in-oil and water-in-oil-in-water emulsions.

Nevertheless, Mantzouridou and collaborators [94] have presented an study investigating the effect of cell entrapment inside the oil droplets on viable cell count over storage and under GI simulating conditions, according to the type of emulsifier used: egg yolk, gum arabic/xanthan mixture or whey protein isolate. The study was performed with *Lactobacillus paracasei* and their entrapment in the oil phase of protein-stabilized emulsions protected the cells when exposed to GI tract enzymes, provided that the emulsions were freshly prepared. Following, however, treatment of aged for up to 4 weeks emulsions under conditions simulating those of the human GI environment, the microorganism did not survive in satisfactory numbers. The probiotic cells survived in larger numbers in aged emulsions when the cells were initially dispersed in the aqueous phase of a yolk-stabilized dressing-type emulsion and their ability to survive enzymatic attack was further enhanced by inulin incorporation.

Probiotic strain	Materials	References
B. bifidum	κ-Carrageenan	[96]
L. lactis spp. lactis	κ-Carrageenan	[97]
L. acidophilus *B. infantis*	Alginate/starch	[98]
L. acidophilus *B. bifidum*	Alginate	[99, 100]
B. longum	κ-Carrageenan	[101]
L. acidophilus	Alginate-starch	[102]
B. adolescentis	Alginate	[103
B. longum	κ-Carrageenan	[104
L. acidophilus *B. infantis*	Alginate/starch	[105]
B. breve	Milk fat and whey protein	[70]
L. acidophilus *L. casei* *L. rhamnosus* *B. infantis*	Alginate	[106]
L. acidophilus *B. bifidum*	Alginate/starch	[52]
L. bulgaricus	Alginate	[107]
L. casei *B. lactis*	Alginate	[108]
L. acidophilus *B. infantis*	Alginate/starch	[109]
L. acidophilus *B. lactis*		[110]
L. casei *B. lactis*	Ca-alginate	[111]
B. bifidum *B. infantis*	Alginate	[112]
L. rhamnosus *L. salivarius* *B. longum* *L. plantarum* *L. acidophilus* *L. paracasei* *B. lactis*		[113]
L. acidophilus *B. infantis*	Alginate/starch	[109]
L. acidophilus *B. lactis*		[110]
L. delbrueckii ssp. bulgaris		[93]
B. pseudolongum	Cellulose acetate phthalate	[114]
L. casei ssp. casei	κ-Carrageenan/locust bean gum	[115]
B. longum *Lactococcus lactis*	κ-Carrageenan/locust bean gum	[116]
B. lactis	Gelatin	[117]

Table 2. Examples of encapsulated probiotic bacteria by Emulsification Technology.

Lactobacillus rhamnosus has been encapsulated in a w/o/w emulsion. According to Pimentel-González and collaborators [95] the survival of the entrapped *L.rhamnosus* in the inner water phase of the double emulsion increased significantly under low pH and bile salt conditions in an *in vitro* trial, meanwhile the viability and survival of control cells decrease significantly under the same conditions.

In the table 2 details probiotic strains and carrier materials that have employed some researchers in the emulsification technology.

The emulsion methods produce capsules sized from a few micrometres to 1mm, approximately, but with a high dispersion compared to other techniques, as extrusion ones. Moreover, even if the emulsion techniques described before are easily scalable, these techniques have an important disadvantage to be applied in an industrial process because are batch processes. Nevertheless, it exist another promising technique different to the turbine used. The static mixers are small devices placed in a tube consisting in static obstacles or diversions where the two immiscible fluids are pumped [118, 119]. This system improves the size distribution, reduce shear and allows keeping the aseptic conditions because it might be a closed system (Figure 5). For example, nowadays this technology is used in dairy industry for viscous products, as admixing fruit pieces or cultures to yoghurt or to process ice cream or curds.

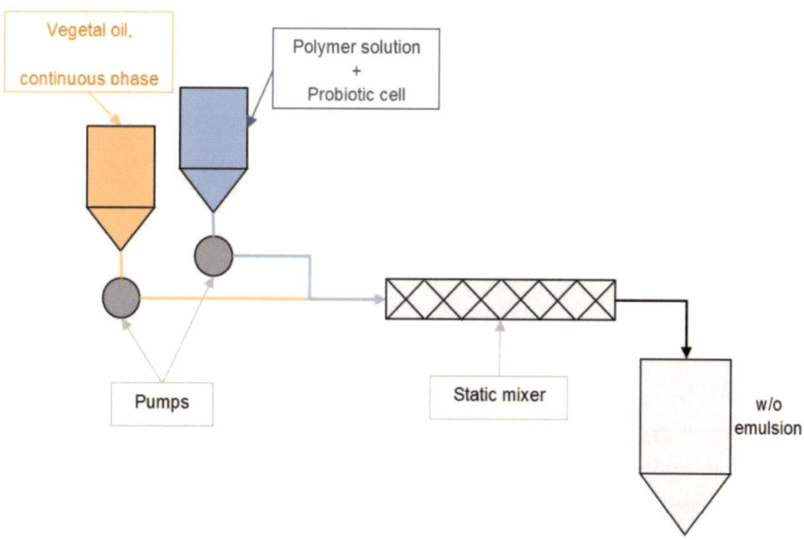

Figure 5. Schematic diagram of a static mixer system to make emulsions.

3.1.6. Coacervation

This process involves la precipitation of a polymer or several polymers by phase separation: simple or complex coacervation, respectively. Simple coacervation is based on "salting out" of one polymer by addition of agents as salts, that have higher affinity to water than the polymer. It is essentially a dehydration process whereby separation of the liquid phase results in the solid particles or oil droplets (starting in an emulsion process) becoming coated and eventually hardened into microcapsules. With regard to complex coacervation, it is a process whereby a polyelectrolyte complex is formed. This process requires the mixing of two colloids at a pH at which both polymers are oppositely charged (i.e. gelatine (+) and arabic gum (-)), leading to phase separation and formation of enclosed solid particles or liquid droplets.

The complex coacervation is one of the most important techniques used for flavour microencapsulation. But it is not the only use of this technique and the complex coacervation is also suitable for probiotic bacteria microencapsulation. And the most frequent medium used might be a water-in-oil emulsion [120].

Oliveira and collaborators [121] encapsulated *B. lactis* (BI 01) and *L. acidophilus* (LAC 4) through complex coacervation using a casein/pectin complex as the wall material. To ensure higher stability, the coacervated material was atomized. The process used and the wall material were efficient in protecting the microorganisms under study against the spray drying process and simulated gastric juice; however, microencapsulated *B. lactis* lost its viability before the end of the storage time. Specifically, microencapsulated *L. acidophilus* maintained its viability for a longer storage (120 days) at 7 and 37 ºC, *B. lactis* lost viability quickly.

Advantages of coacervation, compared with other methods for the encapsulation of probiotics, are a relatively simple low-cost process (which does not necessarily use high temperatures or organic solvents) and allow the incorporation of a large amount of micro-organisms in relation to the encapsulant. However, the scale-up of coacervation is difficult, since it is a batch process that yields coacervate in an aqueous solution. Therefore, to extend its shelf-life, an additional drying process should be applied, which can be harmful to cells.

3.1.7. Extrusion techniques to encapsulate in microspheres

The methods of bioencapsulation in microspheres include two principal steps: (1) the internal phase containing the probiotic bacteria is dispersed in small drops a then (2) these drops will solidify by gelation or formation of a membrane in their surface. Before this section, there are described emulsion systems and coacervation as different methods to obtain these drops and even the membrane formation, but also extrusion technology is useful in order to produce probiotic encapsulation in microspheres. There are different technologies available for this purpose and the selection of the best one is related with different aspects as desired size, acceptable dispersion size, production scale and the maximum shear that the probiotic cells can support.

When a liquid is pumped to go through a nozzle, first this is extruded as individual drops. Increasing enough the flow rate, the drop is transformed in a continuous jet and this continuous jet has to be broken in small droplets. So, the extrusion methods could be divided in two groups, dropwise and jet breakage (Figure 6), and the limit between them is established according to the minimum jet speed according to this equation (eq. 1):

$$v_{j,min} = 2\left(\sigma \middle/ d_j \right)^{0,5} \qquad \begin{aligned} v_{j,min} &= \text{minimum jet speed } (\mathrm{m/s}) \\ \sigma &= \text{surface tension } (\mathrm{N/m}) \\ \rho &= \text{liquid flowing density } (\mathrm{kg/m^3}) \\ d_j &= \text{jet diameter } (\mathrm{m}) \end{aligned} \tag{1}$$

Regardless of the selected technique, the liquid obtained drops have to be solidifying by gelation or external membrane formation (Figure 6). The resulting hydrogel beads are very porous and a polymeric coating is usually applied in order to assure a better retention of the encapsulated probiotic bacteria.

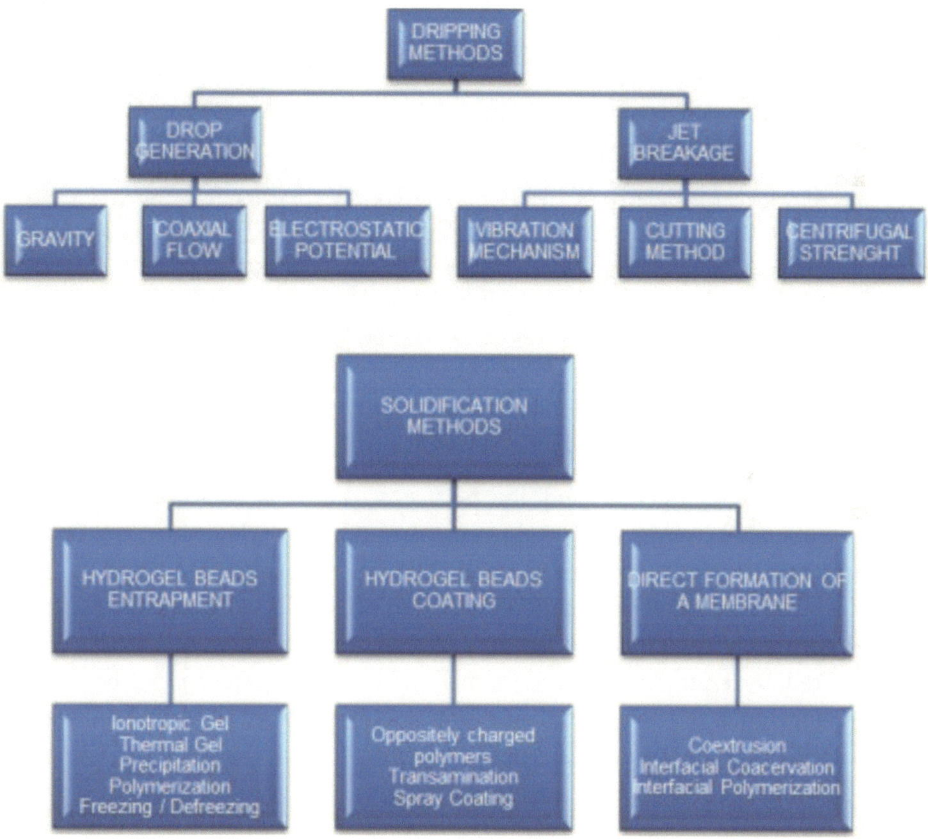

Figure 6. Classification of methods to make and solidify drops

- **Dripping by gravity**

This method is the simplest dripping method to make individual drops, but the size of the droplet will be determined by his weight and surface tension, as well as the nozzle perimeter. The typical diameter of a drop made by this technique is higher than 2 mm. Moreover, the flow is around several millilitres by hour and the method is not interesting for an industrial application. For example, in the Figure 7 is showed a cell immobilization process carried out at TECNALIA using the method of dripping by gravity. The nozzle diameter is 160 μm and the final size of hydrogel bead (after solidification in a Calcium Chloride solution) is 2,4±0,15 mm.

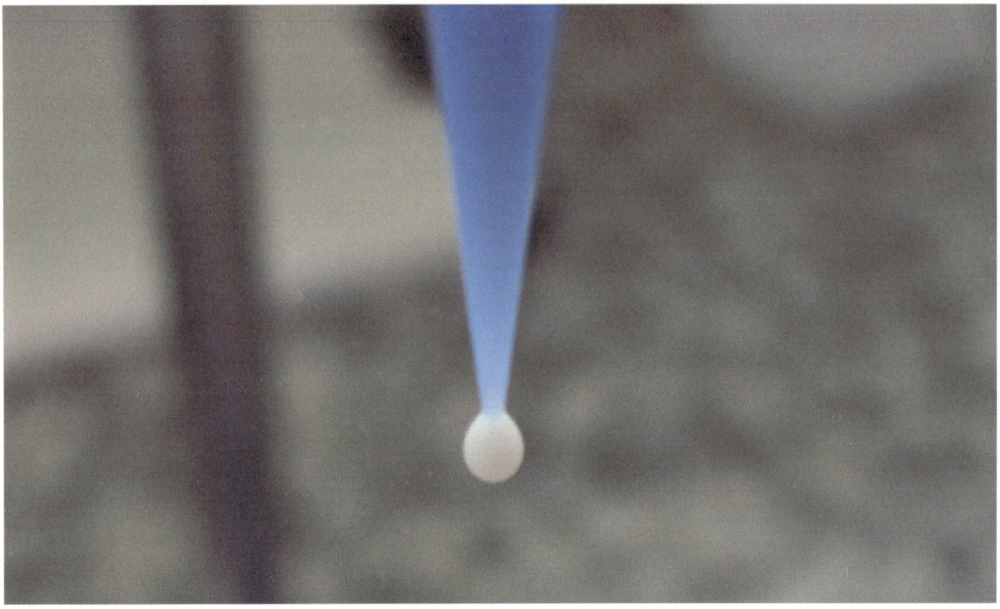

Figure 7. Cell encapsulation in an alginate matrix. Drop generation by gravity using a 160 μm nozzle.

- **Air o liquid coaxial flow and submerged nozzles**

Applying a coaxial air flow around the extrusion nozzle it is possible to reduce the microsphere diameter between a few micrometres and 1 mm. However, the flow rate is limited, less than 30 mL/h to avoid a continuous jet formation. The air flow might be replaced for a liquid one: with a suitable selection of the liquid flow the control of the surface tension is improved. Drops produced in air are generated as aerosols, while the drops produced, for example, in water are made as emulsions. The aerosol beads could be solidified using ionic gelation or hot air. The beads recovered as emulsion are usually extracted or the water is evaporated.

The Spanish enterprise Ingeniatrics Tecnologías has patent an owner Flow Focusing® technology, valid to work with air and liquid flow, and also an user-friendly bioencapsulation device for biotechnological research and clinical microbiology able to encapsulate high molecular weight compounds, microorganisms and cells in homogeneous particles of predictable and controllable size based on Flow Focusing® technology named Cellena® distributed by Biomedal (Figure 8).

Nevertheless, despite all the advantages, due to the mentioned low flow rate, this technique is not used in an industrial scale and also in a laboratory scale it is being replaced for the jet breakage techniques stated below.

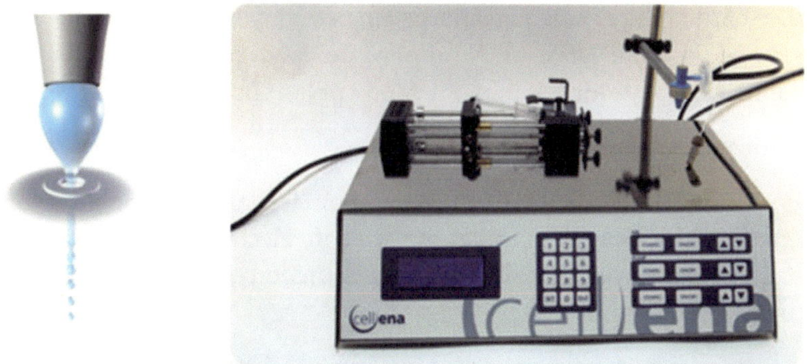

Figure 8. Flow-Focusing technology to make droplets and Cellena® equipment from Ingeniatrics Tecnologías.

The submerged nozzles usually are static, but they can be also rotating or vibrating to improve the droplet generation, but are always immersed in a carrier fluid. An example of the former consist of a static cup immersed in a water-immiscible oil such as mineral oil or vegetal oil and a concentric nozzle as is schematically showed in the Figure 9. Each droplet consist of core material being encapsulatd totally surrounded by a finite film of aqueous polymer solution, as gelatine, for example. The carrier fluid, a warm oil phase that cools after droplet formation, gels this polymer solution thereby forming gel beads with a continuous core/shell structure. The smaller diameter using this technique is typically around 1 mm.

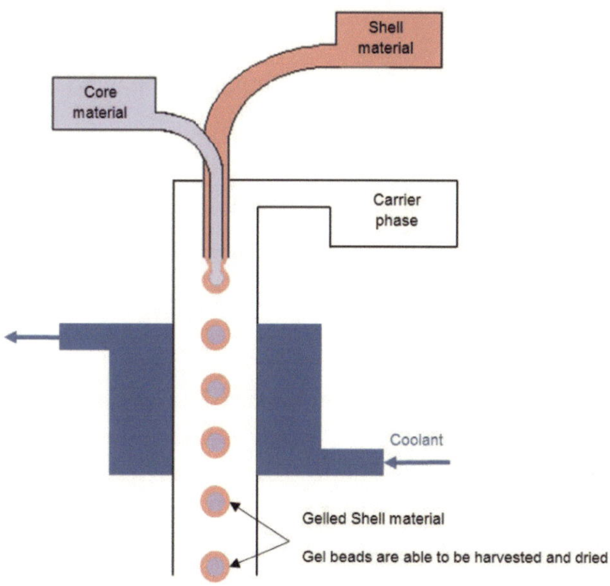

Figure 9. Schematic diagram of a submerged two-fluid static nozzle.

An example of this technology is provided by Morishita Jintan Co. Ltd in Japan These capsules are composed of three layers: a core freeze-dried probiotic bacteria in solid fat, with an intermediate hard fat layer and a gelatin-pectin outer layer [122]. However, the size of the capsules produced is quite large to be applied in food products (1.8-6.5 mm) and the technique is quite expensive for use in many food applications.

- **Electrostatic potential**

This technique is the last one of drop generation techniques. The droplet generation improves replacing the dragging forces by a high electrostatic potential between the capillary nozzle and the harvester solution. The electric forces help the gravity force in front of the surface tension.

Even if the capsules size is appropriated and the size distribution is narrow enough, this technique is more expensive than other extrusion ones and it is not fast enough to be scaled.

- **Vibration technology for jet break-up**

Applying a vibration on a laminar jet for controlled break-up into monodisperse microcapsules is one among different extrusion technologies for encapsulation of probiotic bacteria. The vibration technology is based on the principle that a laminar liquid jet breaks up into equally sized droplets by a superimposed vibration (Figure 10). The instability of liquid jets was theoretically analysed for Lord Rayleigh [123]. He showed that the frequency for maximum instability is related to the velocity of the jet and the nozzle diameter (eq. 2 and eq. 3).

$$f_{opt} = {}^{v_J}\!/_{opt} \qquad \begin{aligned} f_{opt} &= \text{optimal frequency } (\text{Hz}) \\ v_J &= \text{jet velocity } (\text{m/s}) \\ \lambda_{opt} &= \text{optimal wavelength } (\text{m}) \end{aligned} \tag{2}$$

$${}_{opt} = \pi\sqrt{2}d_N\sqrt{1+\frac{3}{\sqrt{\rho\sigma d_N}}} \qquad \begin{aligned} d_N &= \text{nozzle diameter } (\text{m}) \\ \eta &= \text{dynamic viscosity } (\text{kg/m s}) \\ \rho &= \text{density } (\text{kg/m}^3) \\ \sigma &= \text{Surface tension } (\text{N/m}) \end{aligned} \tag{3}$$

Using this technology, it is possible to obtain monodisperse droplets which size can be freely chosen in a certain range depending on the nozzle diameter and the frequency of the sinusoidal force applied (eq. 4). The droplets made are harvested in an accurate hardening bath. To avoid large size distributions due to coalescence effects during the flight and the hitting phase at the surface of hardening solution the use of a dispersion unit with an electrostatic dispersion unit is essential (Figure 10).

$$d_D = \sqrt[3]{1.5 d_{N\,opt}^2} \qquad \begin{aligned} d_D &= \text{droplet diameter } (\text{m}) \\ d_N &= \text{nozzle diameter } (\text{m}) \\ \lambda_{opt} &= \text{optimal wavelength } (\text{m}) \end{aligned} \tag{4}$$

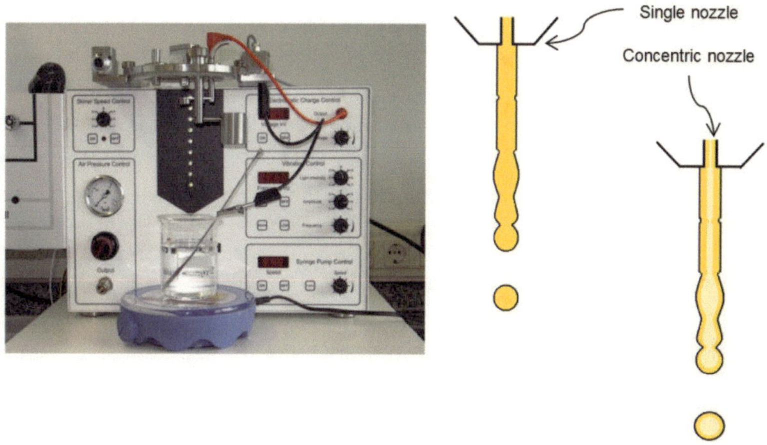

Figure 10. Image of Inotech Encapsulator IE-50R and schematic diagram of jet destabilization and breakage for single and concentric nozzles.

The Encapsulator BIOTECH (the updated version of IE-50R) from EncapBioSystems and Spherisator form BRACE GmbH are two different devices labels to produce microencapsuled probiotic bacteria using the vibration technology for jet breakage. The principal advantages of this technology are the low size dispersion (5-10%), a high flow rate (0.1-2 L/h) and is able to work in sterile conditions. The possibility of working with a wide range of materials (hot melt products, hydrogels, etc.) is also an important aspect to be considered, as well as the design with also concentric nozzles in the lab scale devices and with this kind of nozzles it is possible to produce capsules with a defined core region (solid or liquid) surrounded by a continuous shell layer. On the other side, the principal disadvantage of this technology is the limit in the viscosity for the liquid to be extruded.

But may be one of the most important advantage of the vibration devices commercialized is that the scale up of this technology is relatively "simple" and it consist in the multiplication of the number of nozzles, developing multinozzle devices. The only challenge is that each nozzle of a multinozzle plant must operate in similar production conditions: equal frequency and amplitude, and equal flow rate. In this way, the scale up is direct from the lab to a pilot or industrial scale.

- **JetCutter technology**

The bead production by JetCutter (from geniaLab) is achieved cutting a jet into cylindrical segments by a rotating micrometric cutting tool. The droplet generation is based on a mechanical impact of the cutting wire on the liquid jet. Some techniques as emulsion, simple dropping, electrostatic-enhanced dropping, vibration technique or rotating disc and nozzle techniques have in common that the fluids have to be low in viscosity, and not all of them may be used for large-scale applications. On the contrary, the JetCutter technique is especially capable of processing medium and highly viscous fluids up to viscosities of several thousand mPas.

For bead production by the JetCutter the fluid is pressed with a high velocity out of a nozzle as a solid jet. Directly underneath the nozzle the jet is cut into cylindrical segments by a rotating cutting tool made of small wires fixed in a holder (Figure 11). Driven by the surface tension the cylindrical segments form spherical beads while falling further down, where they finally can be harvested. The size of beads can be adjusted within a range between approximately 200 μm up to several millimetres, adjusting parameters as nozzle diameter, flow rate, number of cutting wires and the rotating speed of cutting tool.

Bead generation by a JetCutter device is achieved by the cutting wires, which cut the liquid jet coming out of the nozzle. But in each cut the wire produce a cutting loss. The device is designed to recover these losses, but it is important to minimize de lost volume selecting a smaller diameter of the cutting wire and angle of inclination of the cutting tool with regard to the jet (Figure 11). According with Pruesse and Vorlop [124], a suitable model of the cutting process might help to operator in the parameters selection. One of the most important parameters is the ratio of the velocities of the fluid and cutting wire, necessary to determinate the proper inclination angle (eq. 5), but the fluid velocity is also related with the bead size (eq. 6), while the diameter of the nozzle and wire determine the volume of cutting loses (eq. 7).

$$\propto = \arcsin\left({u_{fluid}}/{u_{wire}} \right) \qquad (5)$$

α = inclination angle
u_{fluid} = velocity of the fluid
u_{wire} = velocity of the cutting wire

$$d_{bead} = \sqrt[3]{\frac{3}{2} \bullet D^2 \bullet \left(\frac{u_{fluid}}{n \bullet z} - d_{wire} \right)} \qquad (6)$$

d_{bead} = bead diameter
D = nozzle diameter
d_{wire} = cutting wire diameter

$$V_{loss} = \frac{\pi \bullet D^2}{4} \bullet d_{wire} \qquad (7)$$

n = number of rotations
z = number of cutting wires
V_{loss} = Volume of the overall loss

Regarding the advantages of the JetCutter technology, besides the capacity for work with medium and highly viscous fluids, there are the narrow bead size dispersion and the wide range of possible sizes, as well as the high flow rate (approx. 0.1-5 L/h).

To scale up the JetCutter technology there are two ways. First, a multi-nozzle device can be used, in which nozzles are strategically distributed in the perimeter of the cutting tool. The second way is the increase of the cutting frequency, but this approach needs also a higher velocity of the jet and a too high speed of the beads might cause problems, as coalescence or deformation in the collection bath entrance. In order to overcome this problem, the droplets can be pre-gelled prior entering the collection bath using, for example, a tunnel equipped with nozzles spraying the hardening solution or refrigerating the falling beads.

The extrusion technique is the most popular microencapsulation or immobilization technique for micro-organisms that uses a gentle operation which causes no damage to probiotic cells and gives a high probiotic viability [21]. This technology does not involve deleterious solvents and can be done under aerobic and anaerobic conditions. The most important disadvantage of this method is that it is difficult to use in large scale productions due to the slow formation of the microbeads [15]. Various polymers can be used to obtain capsules by this method, but the most used agents are alginate, κ-carrageenan and whey proteins [125].

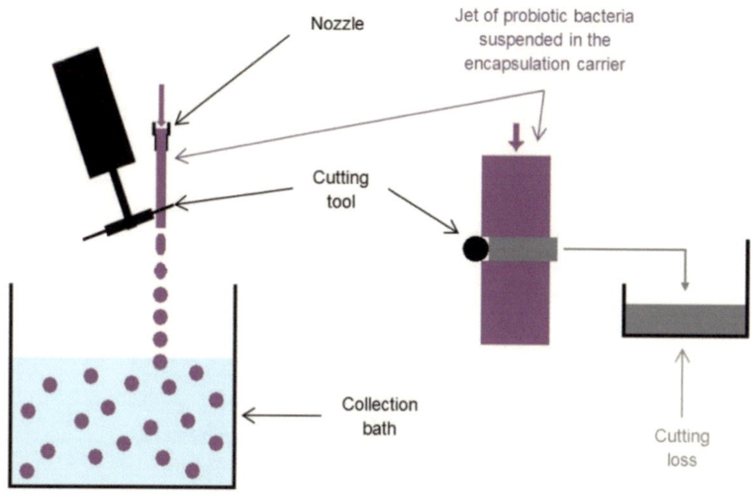

Figure 11. Schematic diagram of the JetCutter technology and representation of fluid losses due to the cutting wire impact.

Figure 12. Examples of two bioencapsulation process carried out at TECNALIA changing the nozzle diameter, cutting tool and inclination angle to obtain different bead size necessaries for several applications.

There are many studies with the extrusion techniques for probiotic protection and stabilization. In 2002, Shah and Ravula [126] encapsulated *Bifidobacterium* and *Lactobacillus acidophilus* in calcium alginate in frozen fermented milk-based dessert, and, in general, the survival of bacteria cells was improved by encapsulation. Some studies employed to encapsulate Bifidobacteria alginate alone and a mixture with other compounds and observed more resistant to the acidic medium than the free cells [5, 112, 127]. A similar result was observed by Chávarri and collaborators [67], where chitosan was used as coating material to improve the stability of alginate beads with probiotics. In this study, with extrusion technique, they showed an effective means of maintaining survival under simulated human gastrointestinal conditions. In the table 3 details probiotic strains and carrier materials that have employed some researchers in the extrusion technology.

Probiotic strain	Materials	References
L. casei		[128]
B. infantis	Gellan/xantan gum	[129]
L. acidophilus	Raftilose, raftiline and Storch	[130]
L. acidophilus *B. infantis*	Ca-alginate	[131]
L. acidophilus	Ca-alginate Chitosan	[132]
L. acidophilus *B. bifidum* *L. casei*	Alginate-chitosan	[133]
L. acidophilus	Alginate-chitosan	[134]
L. acidophilus *B. lactis*		[135]
L. acidophilus *B. lactis*		[136]
L. casei	Alginate/pectin	[137]
L. bulgaricus *S. thermophilus*	Ca-alginate	[138]
L. acidophilus *B. bifidum*	Alginate	[99, 100]
L. lactis	Ca-alginate	[139]
L. acidophilus	κ-Carrageenan	[140]
L. gasseri *B. bifidum*	Alginate-chitosan	[67]
B. lactis	Gellan/Xanthan gum	[141]
L. reuteri	Alginate	[142, 143]
L. rhamnosus	Whey protein	[144]
L. acidophilus	Ca-alginate	[145]
L. acidophilus	Ca-alginate	[140]
L. acidophilus *B. bifidum*	Alginate	[99, 100]
L. lactis		[139]
B. longum	Alginate	[146]
L. acidophilus *B. longum*	Alginate	[126]
B. bifidum *B. infantis*	Alginate	[147]
Lactococcus lactis ssp. cremoris	Alginate	[148]
L. rhamnosus GG	Whey protein	[150]
L. plantarum *L. rhamnosus*	Gellan gum Xanthan gum Pullulan gum Jamilan	[150]
L. acidophilus *B. lactis*	Alginate	[151]

Table 3. Examples of encapsulated probiotic bacteria by Extrusion Technology.

3.1.8. Adhesion to starch granules

Starch is unique among carbohydrates because it occurs naturally as discrete particles called granules. Their size depends on the starch origin ranging from 1 to 100 μm. They are rather dense and insoluble, and hydrate only slightly in water at room temperature. The granular structure is irreversibly lost when the granules are heated in water about 80 ºC, and heat and mechanical energy are necessaries to totally dissolve the granules.

Usually starches are partially or totally dissolved before they are used in food application, for example to be used as texturizing. Starch hydrolysates or chemically modified starches are used as microencapsulation matrices for lipophilic flavours [152, 153]. Partially hydrolysed and crosslinked starch granules were suggested to be suitable carriers for various functional food components [154]. To hydrolyse the starch granules, the use of amylases is the preferred way and corn starch seems to be the most suitable starch for his purpose.

Some probiotic bacteria were shown to be able to adhere to starch and a few investigations about the utilisation of starch granules to protect these bacteria were reported.

3.1.9. Compression coating

This technique involves compressing dried bacteria powder into a core tablet or pellet and the compressing coating material around the core to form the final compact (Figure 13). The compression coating has received a renewed interest for probiotic bacteria encapsulation used with gel-forming polymers in order to improve the stabilization of lyophilized bacteria during storage [155]. The viability in process of the bacteria is affected by the compression pressure and to improve the storage survival the coating material has a significant effect.

Due of the size of the final product obtained by compression coating, this technique is used for pharmaceutical and nutraceutical compounds development, but not for food ingredients obtention.

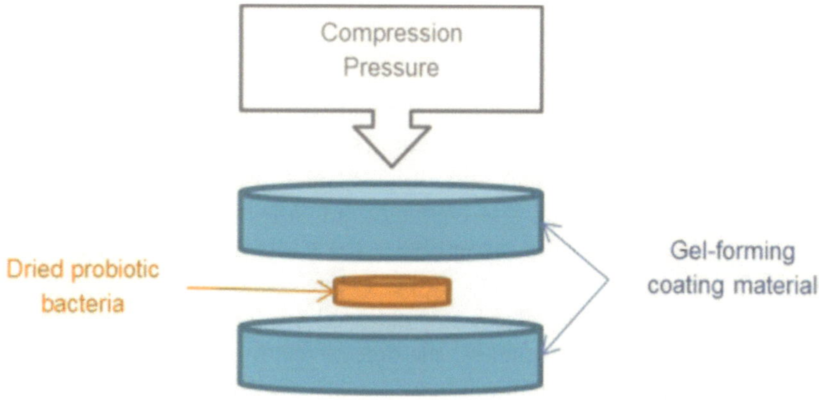

Figure 13. Schematic diagram of compression coating of probiotics.

3.2. Shell or carrier encapsulation materials

Microcapsules should be water-insoluble to maintain their structural integrity in the food matrix and in the gastrointestinal tract. The materials are used alone or in combination to form a monolayer. In this last case, coating the microcapsule with the double membrane can avoid their exposure to oxygen during storage and can enhance the resistance of the cells to acidic conditions and higher bile salt concentrations.

3.2.1. Ionic hydrogels

- **Alginate**

Alginate is surely the biopolymer most used and investigated for encapsulation. Alginates are natural occurring marine polysaccharides extracted from seaweed, but also they occur as capsular polysaccharides in some bacteria [156]. Being a natural polymer, alginic acids constitute a family of linear binary copolymers of 1-4 glycosidically linked α-L-guluronic acid (G) and its C-5 epimer β-D-mannuronic acid. (M). Alginates are the salts (or esters) of these polysaccharides. They are composed of several building blocks (100-3,000 units) liked together in a stiff and partly flexible chain. The relative amounts of the two uronic units and the sequential arrangements of them along the polymer chain vary widely, depending of the origin of the alginate: three types of blocks may be found: homopolymeric M-blocks (M-M), homopolymeric G-blocks (G-G) and heteropolymeric sequentially alternating MG-blocks (M-G). This composition and block structure are strongly related to the functional properties of alginate molecules within an encapsulation matrix.

Immobilisation or entrapment of probiotic bacteria in alginate it is possible due to it is a rapid, non-toxic and versatile method for cells. Dissolving alginate in water gives a viscous solution of which the viscosity will increase with the length of the macromolecule (number of monomeric units), and its solubility is also affected by the pH (at $pH < 3$ precipitate as alginic acid), the presence of counterions in water (alginate precipitates by crosslinking, gelling, with divalent ions such as Ca^{2+}, Ba^{2+}, Sr^{2+}...) and the sequential arrangements of the monomers (the flexibility of the alginate chains in solution increases in the order MG<MM<GG). The gelling occurs when a cation as Ca^{2+} take part in the interchain binding between G-bloks giving rise to a three-dimensional network (Figure 14).

The advantage of alginate is that easily form gel matrices around bacterial cells, it is safe to the body, they are cheap, mild process conditions (such as temperature) are needed for their performance, can be easily prepared and properly dissolve in the intestine and release entrapped cells. However, some disadvantages are attributed to alginate beads. For example, alginate microcapsules are susceptible to the acidic environment [136] which is not compatible for the resistance of the beads in the stomach conditions. Other disadvantage of alginate microparticle is that the microbeads obtained are very porous to protect the cells from its environment [157]. Nevertheless, the defects can be compensated by blending of alginate with other polymer compounds, coating the capsules by another compound or structural modification of the alginate by using different additives [21].

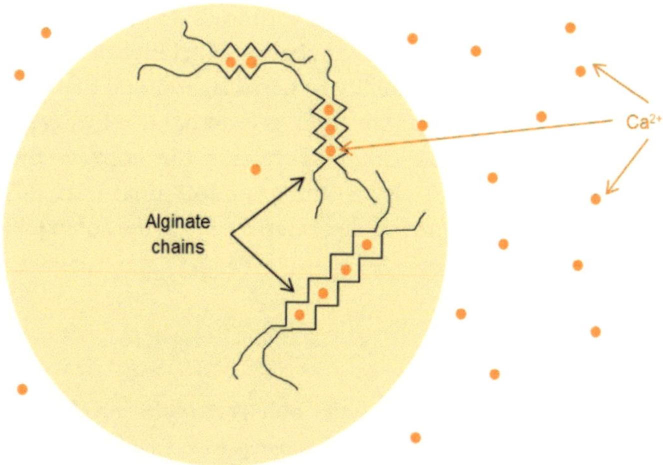

Figure 14. Gelation of an alginate bead when the Ca^{2+} gelling ions diffuse into the alginate-containing system.

- **Chitosan**

Chitosan is a deacetylated derivative of chitin, which is widely found in crustacean shells, fungi, insects and molluscs. This polymer is a linear polysaccharide, which can be considered as a copolymer consisting of randomly distributed β-(1,4) linked D-glucosamine and N-acetyl-D-glucosamine. The functional properties of chitosan are determined by the molecular weight, but also by the degree of acetylation (DA), which represents the proportion of N-acetyl-D-glucosamine units with respect to the total number of units [158]. Chitosan is soluble in acidic to neutral media, but solubility and viscosity of the solution is dependent on the length of chains and the DA.

As chitosan is a positively charged polymer, it forms ionic hydrogels by addition of anions such as pentasodium tripolyphosphate (TPP) and also by interaction with negatively charged polymers as alginate [67] or xanthan [159]. It is possible to obtain an hydrogel by precipitation in a basic medium or by chemical crosslinking with glutaraldehyde [160].

Figure 15. Chitosan microcapsules obtained by (left) TTP crosslinking using the IE-50R and size distribution of the particles, and (right) spray-dryer (TECNALIA).

Chitosan is biodegradable and biocompatible. Nevertheless, to be used in probiotic bacteria encapsulation it is necessary to consider the antibacterial activity of this polymer. Due the possibility of a negative impact in the viability of bacteria, and due that chitosan has a very good film-forming ability, chitosan is more used as external shell in capsules made with anionic polymers as alginate. This application of chitosan can improve the survival of the probiotic bacteria during storage and also in the gastrointestinal tract [67, 161, 162], and therefore, it is a good way of delivery of viable bacterial cells to the colon [67].

3.2.2. *Thermal hydrogels*

- **Gellan gum**

Gellan gum is a high molar mass anionic polyelectrolyte produced as an aerobic fermentation product by a pure culture of *Pseudomonas elodea* [163]. The chemical structure of gellan gum shows a tetrasaccharide repeating unit composed of one rhamnose, one glucoronic acid and two glucose units. It is possible to induce a thermo-reversible gelation upon cooling of gellan gum solutions and the gelation temperature will depend on the polymer concentration, ionic strength and type of counterions presents in the medium. The gels of gellan gums with low acyl content need the presence of divalent stabilizing cations [164].

Although gellan gum is able to generate gel-bead structure for microencapsulation, a disadvantage is that it is not used in this way for this purpose because of having a high gel-setting temperature (80-90°C for about 1 h) which results in heat injuries to the probiotic cells [129].

- **Xanthan**

Xanthan is a heteropolysaccharide with a primary structure consisting of repeated pentasaccharide units formed by two glucose units, two mannose units and one glucoronic acid unit. The polysaccharide is produced by fermentation of bacterium *Xanthomonas campestris* and posterior filtration or centrifugation. This polymer is soluble in cold water and hydrates rapidly. Even if xanthan is considered to be mainly non-gelling a mixture of both, xanthan and gellan gum has been used to encapsulate probiotic cells [19, 102] and contrary to alginate, the mixture presents high resistance towards acid conditions.

In contrary with alginate, mixture of xanthan-gellan is resistant to acidic conditions. Also, as opposed to from carrageenan which needs potassium ions for structural stabilization (it is harmful for the body in high concentrations), this gum can be stabilized with calcium ions [165, 166].

- **Carrageenan**

Carrageenans are a family of high molecular weight sulphated polysaccharides obtained from different species of marine red algae. The most frequently used is κ-carrageenan, opposite to λ- or τ-carrageenan. This polymer is largely used as thickening, gelling agent, texture enhancer or stabilizer on food, pharmaceutical and cosmetic formulations. His primary structure is based on an alternating disaccharide repeating unit of α-(1,3)-δ-galactose-4-sulphate and η-(1,4)-3,6-anhydro-δ-galactose.

κ-carrageenan requires high temperatures (60-90 ºC) for dissolution, especially when applied at high concentrations such as 2-5%. However, this material used for encapsulating probiotics requires a temperature comprised between 40 and 50 ºC at which the cells are added to the polymer solution. It forms thermoreversible gels by cooling in presence of K^+ as stabilizing ions. The gelation with divalent cations as Ca^{2+} or Cu^{2+} is also possible, but not so often used and the thermal gelation is the most common method [167, 168].

The κ-carrageenan beads for probiotic encapsulation can be produced using several technologies described in the extrusion as well as emulsion techniques.

The encapsulation of probiotic cells in κ-carrageenan beads keeps the bacteria in a viable state [96] but the produced gels are brittle and are not able to withstand stresses [19].

- **Gelatin**

Gelatin is a heterogeneous mixture of single or multi-stranded polypeptides, each with extended left-handed proline helix conformations and containing between 300 and 4,000 amino acid units. Gellatines generally have a characteristic primary structure determined by the parent collagen, because they are a irreversible hydrolysed form of collagen obtained from the skin, boiled crushed bones, connective tissues, organs and some intestines of animals. However they vary widely in their size and charge distribution and there are two types of gelatines depending on the treatment to obtain the gelatine: type-A gelatine is obtained from acid treated raw material and type-B gelatine is obtained from alkali treated one.

Gelatine is water-soluble, but the solutions have high viscosity and it forms a thermal hydrogel who melts to a liquid when heated and solidifies when cooled again. Gelatine gels exist over only a small temperature range, the upper limit being the melting point of the gel, which depends on gelatine grade and concentration (but is typically less than 35 °C) and the lower limit the freezing point at which ice crystallizes.

This material is useful to obtain beads using extrusion technologies or form a w/o emulsion by cooling, but to stabilize the gel the beads may need to be crosslinked using glutaraldehyde or salts of Chrome. In fact, it is largely used in complex coacervation technique combined with anionic polysaccharides such as arabic gum and others. The most important consideration is that both hydrocolloids have to be miscible at an appropriate pH to stabilize their charges and avoid the repulsion between similar charged groups.

3.2.3. Milk protein gel

Just like the gelatine, milk proteins are able to form gels in the suitable conditions. Proteins are chains of amino acid molecules connected by peptide bonds and there are may types of proteins due the high number of amino acids (22 units) and the different possibility of sequences. Among other proteins, milk proteins are very interesting as encapsulation material by their physic-chemical properties. There are two major categories of mil protein that are broadly defined by their chemical composition and physical properties. The caseins are proline-rich, open-structured rheomorphic proteins which have distinct hydrophobic and hydrophilic parts and 95% of caseins are naturally self-assembled into casein micelles. Whey proteins primarily include α-lactalbumin, β-lactoglobulin, immunoglobulins, and serum albumin, but also numerous minor proteins, but whey proteins are globular ones.

Milk proteins are natural vehicles for probiotics cells and owing to their structural and physico-chemical properties, they can be used as a delivery system [169]. For example, the proteins have excellent gelation properties and this specificity has been recently exploited by Heidebach and collaborators [170, 171] to encapsulate probiotic cells. The results of these studies are promising and using milk proteins is an interesting way because of their biocompatibility [169].

3.2.4. Starch

Starch is a polysaccharide composed by α-D-glucose units linked by glycosidic bonds, produced by all green plants. It consist of two constitutionally identical but architecturally different molecules: amylose and amylopectin. The amylose is the linear and helical chains of glucose polymer, while the amylopectin is the highly branched chains. The content of each fraction depends of the starch origin, but in general it contains around 20-30% amylose and 70-80% amylopectin.

Figure 16. Coloured maltodextrin microcapsules obtained by spray-drying (TECNALIA).

As it is described in the previous section, the probiotic bacteria can be encapsulated by adhesion to starch granules, but usually the starch is chemical or physically modified for different applications, even encapsulation as maltodextrins or cyclodextrins commonly used in combination with the spray-drying technology (Figure 16), fluid bed granulation, for examples.

Starch granule is an ideal surface for the adherence of the probiotics cells and the resistant starch (the starch which is not digested by pancreatic enzymes in the small intestine) can reach the colon where it is fermented [172]. Therefore, the resistant starch provides good enteric delivery characteristic that is a better release of the bacterial cells in the large intestine. Moreover, by its prebiotic functionality, resistant starch can be used by probiotic bacteria in the large intestine [173, 174].

Author details

María Chávarri, Izaskun Marañón and María Carmen Villarán
Bioprocesses & Preservation Area, Health Division, Tecnalia,Parque Tecnológico de Álava, Miñano (Álava), Spain

4. References

[1] De Vos, P., Faas, M. M., Spasojevic, M., & Sikkema, J. (2010) Encapsulation for preservation of functionality and targeted delivery of bioactive food components. International Dairy Journal, 20: 292-302.
[2] Kailasapathy, K. (2009). Encapsulation technologies for functional foods and nutraceutical product development. CAB Reviews: Perspectives in Agriculture, Veterinary Science, Nutrition and Natural Resources, 4(6).
[3] Favaro-Trindade C.S., Grosso C.R.F. (2000) The effect of the immobilization of *L. acidophilus* and *B. lactis* in alginate on their tolerance to gastrointestinal secretions. Michwissenschaft, 55: 496–9.
[4] Favaro-Trindade C.S., Grosso C.R.F. (2002) Microencapsulation of *L. acidophilus* (La-05) and *B. lactis* (Bb-12) and evaluation of their survival at the pH values of the stomach and in bile. Journal of Microencapsulation, 19: 485–494.
[5] Liserre A.M., Re M.I., Franco B.D.G.M. (2007) Microencapsulation of *Bifidobacterium animalis* subsp. *lactis* in modified alginate-chitosan beads and evaluation of survival in simulated gastrointestinal conditions. Food Biotechnology, 21: 1–16.
[6] Shima M., Matsuo T., Yamashita M., Adachi S. (2009) Protection of *Lactobacillus acidophilus* from bile salts in a model intestinal juice by incorporation into the inner-water phase of a W/O/W emulsion. Food Hydrocolloids, 23: 281–5.
[7] Thantsha M.S., Cloete T.E., Moolman F.S., Labuschagne P.W. (2009) Supercritical carbon dioxide interpolymer complexes improve survival of *B. longum* Bb-46 in simulated gastrointestinal fluids. International Journal of Food Microbiology, 129: 88–92.

[8] O'Riordan K., Andrews D., Buckle K., Conway P. (2001) Evaluation of microencapsulation of a *Bifidobacterium* strain with starch as an approach to prolonging viability during storage. Journal of Applied Microbiology, 91: 1059–66.

[9] Lian W., Hsiao H., Chou C. (2003) Viability of microencapsulated bifidobacteria in simulated gastric juice and bile solution. International Journal of Food Microbiology, 86: 293–301.

[10] Oliveira A.C., Moretti T.S., Boschini C., Baliero J.C.C., Freitas L.A.P., Freitas O. et al. (2007) Microencapsulation of *B. lactis* (BI 01) and *L. acidophilus* (LAC 4) by complex coacervation followed by spouted-bed drying. Drying Technology, 25: 1687–93.

[11] Oliveira A.C., Moretti T.S., Boschini C., Baliero J.C.C., Freitas O., Favaro-Trindade C.S. (2007) Stability of microencapsulated *B. lactis* (BI 01) and *L. acidophilus* (LAC 4) by complex coacervation followed by spray drying. Journal of Microencapsulation, 24: 685–93.

[12] FAO/WHO. Probiotics in Food. Health and Nutritional Properties and Guidelines for Evaluation, FAO Food and Nutrition Paper No. 85. World Health Organization and Food and Agriculture Organization of the United Nations, Rome; 2006.

[13] Rafter, J. (2003) Probiotics and colon cancer. Best Pract. Res. Clin. Gastroenterol. 17: 849-859.

[14] Floch, M. H. (2008) Editorial announcement. Journal of Clinical Gastroenterology, 42: 331.

[15] Burgain, J., Gaiani, C., Linder, M., & Scher, J. (2011) Encapsulation of probiotic living cells: From laboratory scale to industrial applications. *Journal of Food Engineering*, 104(4): 467-483.

[16] Luyer, M.D., Buurman, W.A., Hadfoune, M., Speelmans, G., Knol, J., Jacobs, J.A., Dejong, C.H.C., Vriesema, A.J.M., Greve, J.W.M. (2005) Strain specific effects of probiotics on gut barrier integrity following hemorrhagic shock. Infection and Immunity, 73 (6): 3689–3692.

[17] Canani, R.B., Cirillo, P., Terrin, G., Cesarano, L., Spagnuolo, M.I., De Vicenzo, A., Albano, F., Passariello, A., De Marco, G., Manguso, F., Guarino, A. (2007) Probiotics for treatment of acute diarrhea in children: a randomized clinical trial of five different preparations. British Medical Journal, 335 (7615): 340– 342.

[18] Kekkonen, R.A., Vasankari, T.J., Vuorimaa, T., Haahtela, T., Julkunen, I., Korpela, R. (2007). The effect of probiotics on respiratory infections and gastrointestinal symptoms during training in marathon runners. International Journal of Sport Nutrition and Exercise Metabolism, 17 (4): 352–363.

[19] Chen, M.J., Chen, K.N. (2007) Applications of probiotic encapsulation in dairy products. In: Lakkis, Jamileh M. (Ed.), Encapsulation and Controlled Release Technologies in Food Systems. Wiley-Blackwell, USA, 83–107.

[20] Sanders, M.E. (2008) Probiotics: definition, sources, selection and uses. Clinical Infectious Diseases, 46 (Suppl.): S58–S61.

[21] Krasaekoopt, W., Bhandari, B., Deeth, H. (2003) Evaluation of encapsulation techniques of probiotics for yoghurt. International Dairy Journal, 13 (1): 3–13.

[22] Wohlgemuth, S., Loh, G., Blaut, M. (2010) Recent developments and perspectives in the investigation of probiotic effects. International Journal of Medical Microbiology, 300 (1): 3–10.

[23] Gomes A.M.P., Malcata F.X. (1999) *Bifidobacterium* spp and *Lactobacillus acidophilus*: biological and theraputical revalant for use a probiotics. Trends Food Sci Technol., 10: 139-157.

[24] Kailasapathy, K., &Chin, J. (2000) Survival and therapeutic potential of probiotic organisms with reference to *Lactobacillus acidophilus* and *Bifidobacterium* spp. Immunology and Cell Biology, 78: 80-88.

[25] Agerholm-Larsen, L., Raben, A., Haulrik, N., Hansen, A.S., Manders, M., Astrup, A. (2000) Effect of 8 week intake of probiotic milk products on risk factors for cardiovascular diseases. Eur. J. Clin. Nutr., 54: 288–297.

[26] Gotcheva, V., Hristozova, E., Hrostozova, T., Guo, M., Roshkova, Z., Angelov, A. (2002) Assessment of potential probiotic properties of lactic acid bacteria and yeast strains. Food Biotechnol., 16: 211–225.

[27] Madureira, A. R., Pereira, C. I., Truszkowska, K., Gomes, A. M., Pintado, M. E., & Malcata, F. X. (2005) Survival of probiotic bacteria in a whey cheese vector submitted to environmental conditions prevailing in the gastrointestinal tract. International Dairy Journal, 15: 921-927.

[28] Nomoto, K. (2005) Review prevention of infections by probiotics. J. Biosci. Bioeng., 100: 583–592.

[29] Imasse, K., Tanaka, A., Tokunaga, K., Sugano, H., Ishida, H., Takahashi, S. (2007) *Lactobacillus reuteri* tablets suppress *Helicobacter pylori* infectionda double-blind randomised placebo-controlled cross-over clinical study Kansenshogaku zasshi. J. Jpn. Assoc. Infect. Dis., 81: 387–393.

[30] Jones, P. J., & Jew, S. (2007) Functional food development: concept to reality. Trends in Food Science and Technology, 18: 387-390.

[31] Sanders, M.E., Gibson, G., Gill, H.S. & Guarner, F. (2007) Probiotics: their potential to impact human health. CAST issue paper No. 36, October 2007. http://www.cast-science.org/publications.asp.

[32] Shah, N.P. (2007) Functional cultures and health benefits. Int. Dairy J., 17: 1262–1277.

[33] Chassard C., Grattepanche F., Lacroix C. (2011) "Chapter 4 Probiotics and Health Claims: Challenges for Tailoring their Efficacy." In: Kneifel W, Salminen S (Eds) Probiotics and Health Claims, Wiley-Blackwell, UK : 49-74.

[34] Gilliland S.E. (1989) A review of potential benefits to consumers. J Dairy Sci., 72: 2483-2494.

[35] Mortazavian, A.M., Sohrabvandi S. (2006) Probiotics and food Probiotic products. Eta Publication, Iran, (In Farsi).

[36] Ouwehand A.C., Salminen S.J. (1998) The health effects of cultured milk products with viable and non-viable bacteria. Int Dairy J., 8: 749-758.

[37] Robinson R.K. (1987) Survival of *Lactobacillus acidophilus* in fermented products-afrikanse. Suid A Frikaanse Tydskrif Vir Suiwelkunde, 19: 25-27.

[38] Kurman J.A., Rasic J.L. (1991) The health potential of products con- taining bifidobacteria. In: Therapeutic properties of functional milks, 117-158, R.K.

[39] Holcomb J.E., Frank J.F., Mc Gregor J.U. (1991) Viability of *Lactobacillus acidophilus* and *Bifidobacterium bifidum* in soft- serve frozen yogurt. Cult Dairy Product J., 26: 4-5.

[40] Shah N.P., Lankaputhra W.E.V., Britz M., Kyle W.S.A. (1995) Survival of *Lactobacillus acidophilus* and *Bifidobacterium longum* in commercial yoghurt during refrigerated storage. Int Dairy J., 5: 515-521.

[41] Mortazavian A.M., Sohrabvandi S., Reinheimer J.A. (2006) MRS- bile agar: Its suitability for the enumeration of mixed probiotic cultures in cultured dairy products. Milchwissenschaft, 62: 270-272.

[42] Round, J. L., & Mazmanian, S. K. (2009) The gut microbiota shapes intestinal immune responses during health and disease. Nature Reviews Immunology, 9: 313-323.

[43] Ouwehand, A.C., Salminen, S., Isolauri, E. (2002) Probiotics: an overview of beneficial effects. Antonie Van Leeuwenhoek, 82: 279–289.

[44] Modler, H. W., McKellar, R. C., A Yaguchi, M. (1990) Bifidobacteria and bifidogenic factors. Canadian Institute of Food Science and Technology Journal, 23: 29–41.

[45] Peso-Echarri, P., Martínez-Gracía, C., Ros-Berruezo, G., Vives, I., Ballesta, M., Solís, G., et al. (2011) Assessment of intestinal microbiota of full-term breast-fed infants from two different geographical locations. Early Human Development, 87: 511-513.

[46] Finegold, S. M., Dowd, S. E., Gontcharova, V., Liu, C., Henley, K. E., Wolcott, R. D., et al. (2010) Pyrosequencing study of fecal microflora of autistic and control children. Anaerobe, 16: 444-453.

[47] Kalliomäki, M., Collado, M. C., Salminen, S., & Isolauri, E. (2008) Early differences in fecal microbiota composition in children may predict overweight. American Journal of Clinical Nutrition, 87: 534-538.

[48] Qin, J., Li, R., Raes, J., Arumugam, M., Burgdorf, K. S., Manichanh, C., et al. (2010) A human gut microbial gene catalogue established by metagenomic sequencing. Nature, 464: 59-65.

[49] Turnbaugh, P. J., Hamady, M., Yatsunenko, T., Cantarel, B. L., Duncan, A., Ley, R. E., et al. (2009) A core gut microbiome in obese and lean twins. Nature, 457: 480-484.

[50] Anal, A. K., & Singh, H. (2007) Recent advances in microencapsulation of probiotics for industrial applications and targeted delivery. Trends in Food Science & Technology, 18(5): 240-251.

[51] Champagne, C. P., & Fustier, P. (2007) Microencapsulation for the improved delivery of bioactive compounds into foods. Current opinion in biotechnology, 18(2): 184-90.

[52] Kailasapathy, K. (2006) Survival of free and encapsulated probiotic bacteria and their effect on the sensory properties of yoghurt. LWT – Food Science and Technology, 39 (10): 1221–1227.

[53] Hayashi, H. (1989) Drying technologies of foods - their history and future. Drying Technology, 7(2): 315-369.

[54] Petrovic, T., Nedovic, V., Dimitrijevic-Brankovic, S., Bugarski, B., & Lacroix, C. (2007). Protection of probiotic microorganism by microencapsulation. CI&CEQ, 13(3): 169-174.

[55] O'Riordan, K., Andrews, D., Buckle, K., & Conway, P. (2001) Evaluation of microencapsulation of a *Bifidobacterium* strain with starch as an approach to prolonging viability during storage. Journal of applied microbiology, 91(6): 1059-66.

[56] Chávez, B.E. and A.M. Ledeboer. (2007) Drying of Probiotics: Optimization of Formulation and Process to Enhance Storage Survival. Drying Technology, 25(7-8): 1193-1201.

[57] Desmond, C., Ross, R. P., O'Callaghan, E., Fitzgerald, G., & Stanton, C. (2002) Improved survival of *Lactobacillus paracasei* NFBC 338 in spray-dried powders containing gum acacia. Journal of applied microbiology, 93(6): 1003-11.

[58] Gardiner, G., Ross, R.P., Collins, J.K., Fitzgerald, G., Stanton, C. (1998) Development of a probiotic Cheddar cheese containing human-derived *Lactobacillus paracasei* strains. Applied and Environmental Microbiology, 64 (6): 2192–2199.

[59] Mauriello, G., Aponte, M., Andolfi, R., Moschetti, G., & Villani, F. (1999) Spray-drying of bacteriocins-producing LAB. Journal of Food Protection, 62: 773-777.

[60] Prajapati, J. B., Shah, R. K., & Dave, J. M. (1987) Survival of *Lactobacillus acidophilus* in blended spray-dried acidophilus preparations. Australian Journal of Dairy Technology, 42: 17-21.

[61] Corcoran B.M., Ross R.P., Fitzgerald G.F. and Stanton C. (2004) Comparative survival of probiotic lactobacilli spray-dried in the presence of prebiotic substances. Journal of Applied Microbiology, 96: 1024–1039.

[62] Fritzen-Freire, C.B., et al. (2012) Microencapsulation of bifidobacteria by spray drying in the presence of prebiotics. Food Research International, 45(1): 306-312.

[63] Rodríguez-Huezo M.E., Durán-Lugo R., Prado-Barragán L.A., Cruz-Sosa F., Lobato-Calleros C., Alvarez-Ramírez J. et al. (2007) Pre-selection of protective colloids for enhanced viability of *Bifidobacterium bifidum* following spray-drying and storage, and evaluation of aguamiel as thermoprotective prebiotic. Food Research International, 40: 1299–306.

[64] Ying, D., et al. (2012) Enhanced survival of spray-dried microencapsulated *Lactobacillus rhamnosus* GG in the presence of glucose. Journal of Food Engineering, 109(3): 597-602.

[65] Anal A.K., Singh H. (2007) Recent development in microencapsulation of probiotics for industrial applications and targeted delivery. Trends in Food Science and Technology, 18: 240–51.

[66] Ross R.P., Desmond C., Fitzgerald G.F., Stanton C. (2005) Overcoming the technological hurdles in the development of probiotic foods. Journal of Applied Microbiology, 98: 410–7.

[67] Chávarri M., Marañón I., Ares R., Ibanez F.C., Marzo F., Villaran M.C. (2010) Microencapsulation of a probiotic and prebiotic in alginate-chitosan capsules improves survival in simulated gastro-intestinal conditions. International Journal of Food Microbiology, 142: 185–199.

[68] Gardiner, G.E., Bouchier, P., O' Sullivan, E., Kelly, J., Collins, J.K., Fitzgerald, G., Ross, R.P., Stanton, C. (2002) A spray-dried culture for probiotic cheddar cheese manufacture. International Dairy Journal, 12 (9): 749–756.

[69] Lee, J. S., Cha, D. S., & Park, H. J. (2004) Survival of freeze-dried *Lactobacillus bulgaricus* KFRI 673 in chitosan-coated calcium alginate microparticles. Journal of agricultural and food chemistry, 52(24): 7300-5.

[70] Picot, A., Lacroix, C. (2004) Encapsulation of Bifidobacteria in whey protein-based microcapsules and survival in stimulated gastrointestinal conditions and in yoghurt. International Dairy Journal, 14 (6): 505–515.

[71] Hung-Chi, H., Wen-Chian, L., Cheng, C.C. (2004) Effect of packaging conditions and temperature on viability of microencapsulated bifidobacteria during storage. Journal of the Science of Food and Agriculture, 84: 134-139.

[72] Lian, W. D., H. C. Hsiao and C. C. Chou. (2002) Survival of *Bifidobacteria* after spray-drying. International Journal of Food Microbiology, 74: 79–86.

[73] Petrovic, T., Nedovic, V., Dimitrijevic-Brankovic, S., Bugarski, B., & Lacroix, C. (2007) Protection of probiotic microorganism by microencapsulation. CI&CEQ, 13(3): 169-174.

[74] Teixeira, P., Castro, H., Mohácsi-Farkas, C., & Kirby, R. (2000) Identification of sites of injury in *Lactobacillus bulgaricus* during heat stress. Journal of Applied Microbiology, 62: 47-55.

[75] Conrad P.B., Miller D.P., Cielenski P.R., de Pablo J.J. (2000) Stabilization and preservation of *Lactobacillus acidophilus* in saccharide matrices. Cryobiology, 41: 17–24.

[76] Selmer-Olsen, E., Sorhaug, T., Birkeland, S. E., & Pehrson, R. (1999) Survival of *Lactobacillus helveticus* entrapped in Ca-alginate in relation to water content, storage and rehydration. Journal of Industrial Microbiology & Biotechnology, 23: 79-85.

[77] Desmond, C., Ross, R. P., O'Callaghan, E., Fitzgerald, G., & Stanton, C. (2002) Improved survival of *Lactobacillus paracasei* NFBC 338 in spray-dried powders containing gum acacia. Journal of applied microbiology, 93(6): 1003-11.

[78] Crittenden, R., Laitila, A., Forssell, P., Matto, J., Saarela, M., Mattila- Sandholm, T., et al. (2001) Adhesion of Bifidobacteria to granular starch and its implications in probiotic technologies. Applied and Environmental Microbiology, 67: 3469-3475.

[79] Picot, A., & Lacroix, C. (2003) Effects of micronization on viability and thermotolerance of probiotic freeze-dried cultures. International Dairy Journal, 13: 455-462.

[80] Picot, A., & Lacroix, C. (2003) Optimization of dynamic loop mixer operating conditions for production of o/w emulsion for cell microencapsulation. Lait, 83: 237-250.

[81] Rutherford, W.M.D.M., I.A., Allen, Jack E. (Boonville, IA), Schlameus, Herman W. (San Antonio, TX), Mangold, Donald J. (San Antonio, TX), Harlowe Jr., William W. (San Antonio, TX), Lebeda, Joseph R. (Urbandale, IA). (1994) Process for preparing rotary disc fatty acid microspheres of microorganisms, Pioneer Hi-Bred International, Inc. (Des Moines, IA): United States.

[82] Wurster, D.E. (1963) Granulating and coating process for uniform granules. Wisconsin Alumni Research Foundation: United States.

[83] Wurster, D.E. (1966) Particle coating process, Wisconsin Alumni Research Foundation: United States.

[84] Wurster, D.E., Battista, Joseph V., Lindlof, James A. (1965) Process for preparing agglomerates, Wisconsin, Alumni Res Found: United States.

[85] Lindlof, J.A., Wurster, Dale E. (1964) Apparatus for coating particles in a fluidized bed, Wisconsin, Alumni Res Found: United States.

[86] Champagne, C.P. and P. Fustier. (2007) Microencapsulation for the improved delivery of bioactive compounds into foods. Current Opinion in Biotechnology, 18(2): 184-190.

[87] Wu, W.-h.M., NY, Roe, William S. (Wurtsboro, NY), Gimino, Virgil G. (Highland Falls, NY), Seriburi, Vimon (Middletown, NY), Martin, David E. (Branchville, NJ), Knapp, Shaun E. (Middletown, NY). (2000) Low melt encapsulation with high laurate canola oil, Balchem Corporation (Slate Hill, NY): United States.

[88] Koo, S. M.; Cho, Y. H.; Huh, C. S.; Baek, Y. J.; Park, J. Y. (2001) Improvement of the stability of *Lactobacillus casei* YIT 9018 by microencapsulation using alginate and chitosan. J. Microbiol. Biotechnol., 11: 376-383.

[89] Stummer, S., et al. (2012) Fluidized-bed drying as a feasible method for dehydration of *Enterococcus faecium* M74. Journal of Food Engineering, 111(1): 156-165.

[90] King, V.A.-E., H.-J. Lin, and C.-F. Liu. (1998) Accelerated storage testing of freeze-dried and controlled low-temperature vacuum dehydrated *Lactobacillus acidophilus*. The Journal of General and Applied Microbiology, 44(2): 161-165.

[91] Heidebach, T., P. Först, and U. Kulozik. (2010) Influence of casein-based microencapsulation on freeze-drying and storage of probiotic cells. Journal of Food Engineering, 98(3): 309-316.

[92] Reid, A.A., et al. (2005) Microentrapment of probiotic bacteria in a Ca^{2+} -induced whey protein gel and effects on their viability in a dynamic gastro-intestinal model. Journal of Microencapsulation, 22(6): 603-619.

[93] Hou, R.C.W., et al. (2003) Increase of viability of entrapped cells of *Lactobacillus delbrueckii* ssp. *bulgaricus* in artificial sesame oil emulsion. Journal of Dairy Science, 86(2): 424-428.

[94] Mantzouridou, F., A. Spanou, and V. Kiosseoglou. (2012) An inulin-based dressing emulsion as a potential probiotic food carrier. Food Research International, 46(1): 260-269.

[95] Pimentel-González, D.J., et al. (2009) Encapsulation of *Lactobacillus rhamnosus* in double emulsions formulated with sweet whey as emulsifier and survival in simulated gastrointestinal conditions. Food Research International, 42(2): 292-297.

[96] Dinakar P., Mistry V.V. (1994) Growth and viability of *Bifidobacterium bifidum* in cheddar cheese. J Dairy Sci. 77: 2854-5864.

[97] Sodini, I., Boquien, C.Y., Corrieu, G., Lacroix, C. (1997) Use of an immobilized cell bioreactor for the continuous inoculation of milk in fresh cheese manufacturing. Journal of Industrial Microbiology and Biotechnology, 18 (1): 56–61.

[98] Godward, G., Kailasapathy, K. (2003) Viability and survival of free and encapsulated probiotic bacteria in Cheddar cheese. Milchwissenschaft, 58 (11–12): 624– 627.

[99] Özer, B., Uzun, Y.S., Kirmaci, H.A. (2008) Effect of microencapsulation on viability of *Lactobacillus acidophilus* LA-5 and *Bifidobacterium bifidum* BB-12 during Kasar cheese ripening. International Journal of Dairy Technology, 61 (3): 237–244.

[100] Özer, B., Kirmaci, H.A., Senel, E., Atamer, M., Hayaloglu, A. (2009) Improving the viability of *Bifidobacterium bifidum* BB-12 and *Lactobacillus acidophilus* LA-5 in white-brined cheese by microencapsulation. International Dairy Journal, 19 (1): 22–29.

[101] Adhikari, K., Mustapha, A., Grün, I.U. (2000) Viability of microencapsulated *bifidobacteria* in set yogurt during refrigerated storage. Journal of Dairy Science, 83 (9): 1946–1951.

[102] Sultana, K., Godward, G., Reynolds, N., Arumugaswamy, R., Peiris, P. (2000) Encapsulation of probiotic bacteria with alginate–starch and evaluation of survival in simulated gastrointestinal conditions and in yogurt. International Journal of Food Microbiology, 62 (1–2): 47–55.

[103] Truelstrup-Hansen, L., Allan-Wojotas, P.M., Jin, Y.L., Paulson, A.T. (2002) Survival of Ca-alginate microencapsulated *Bifidobacterium* spp. In milk and simulated gastrointestinal conditions. Food Microbiology, 19 (1): 35–45.

[104] Adhikari, K., Mustapha, A., Grün, I.U. (2003) Survival and metabolic activity of microencapsulated *Bifidobacterium* in stirred yogurt. Journal of Food Science, 68 (1): 275–280.

[105] Godward, G., Kailasapathy, K., 2003. Viability and survival of free, encapsulated and co-encapsulated probiotic bacteria in yoghurt. Milchwissenschaft, 58 (7–8): 396–399.

[106] Capela, P., Hay, T.K.C., Shah, N.P. (2006) Effect of cryoprotectants, probiotics and microencapsulation on survival of probiotic organisms in yoghurt and freeze- dried yoghurt. Food Research International, 39 (2): 203–211.

[107] Sheu, T.Y., Marshall, R.T. (1993) Microentrapment of *Lactobacilli* in calcium alginate gels. Journal of Food Science, 58: 557–561.

[108] Sheu, T.Y., Marshall, R.T., Heymann, H. (1993) Improving Survival of culture bacteria in frozen desserts by microentrapment. Journal of Dairy Science, 76 (7): 1902– 1907.

[109] Godward, G., Kailasapathy, K. (2003) Viability and survival of free, encapsulated and co-encapsulated probiotic bacteria in ice cream. Milchwissenschaft, 58 (3– 4): 161–164.

[110] Kailasapathy, K., Sultana, K. (2003) Survival and b-D-galactosidase activity of encapsulated and free *Lactobacillus acidophilus* and *Bifidobacterium lactis* in ice-cream. Australia}n Journal of Dairy Technology, 58 (3): 223–227.

[111] Homayouni, A., Azizi, A., Ehsani, M.R., Yarmand, M.S., Razavi, S.H. (2008) Effect of microencapsulation and resistant starch on the probiotic survival and sensory properties of synbiotic ice cream. Food Chemistry, 111 (1): 50–55.

[112] Khalil A.H., Mansour E.H. (1998) Alginate encapsulation bifidobacteria survival in mayonnaise. Journal of Food Science, 63: 702–5.

[113] Ding, W.K., Shah, N.P. (2008) Survival of free and microencapsulated probiotic bacteria in orange and apple juices. International Food Research Journal, 15 (2): 219–232.

[114] Rao A.V., Shiwnarin N., Maharij I. (1989) Survival of microencapsulated *Bifidobacterium pseudolongum* in simulated gastric and intestinal juices. Can Inst Food Sci Technol J., 22: 345-349.

[115] Arnauld J.P., Laroix C., Choplin L. (1992) Effect of agitation rate on cell release rate and metabolism during continues fermentation with entrapped growing *Lactobacillus casei* subsp. casei. Biotech Tech., 6: 265.

[116] Doleyres, Y., Fliss, I., & Lacroix, C. (2004) Continuous production of mixed lactic starters containing probiotics using immobilized cell technology. Biotechnology Progress, 20: 145-150.

[117] Annan N.T., Borza A.D., Truelstrup H. (2007) Encapsulation in alginate coated gelatin microspheres improves survival of the probiotic *bifidobacterium adolescentis*15703t during exposure to simulated gastro-intestinal conditions. J. Food Sci., 41(2): 184-193.

[118] Belyaeva, E., et al., New approach to the formulation of hydrogel beads by emulsification/thermal gelation using a static mixer. Chemical Engineering Science, 2004. 59(14): 2913-2920.

[119] Krstić, D.M., et al., Energy-saving potential of cross-flow ultrafiltration with inserted static mixer: Application to an oil-in-water emulsion. Separation and Purification Technology, 2007. 57(1): 134-139.

[120] Nag, A., K.-S. Han, and H. Singh. (2011) Microencapsulation of probiotic bacteria using pH-induced gelation of sodium caseinate and gellan gum. International Dairy Journal, 21(4): 247-253.

[121] Oliveira, A.C., Moretti, T.S., Boschini, C., Baliero, J.C., Freitas, O., Favaro-Trindade, C. S. (2007). Stability of microencapsulated *B. lactis* (BI 01) and *L. acidophilus* (LAC 4) by complex coacervation followed by spray drying. Journal of microencapsulation, 24(7): 673-681.

[122] Asada, M., et al. (2003) Capsules containing vital cells or tissues, M.J. Co, Editor.

[123] Lord Rayleigh, F.R.S., On the stability of jets. Proceedings London Mathematical Society, 1878. 10: p. 4-13

[124] Pruesse, U. and K.-D. Vorlop. (2004) The JetCutter Technology, in Fundamentals of Cell Immobilisation Biotechnology, V. Nedovic and R. Willaert, Editors. Kluwer Academic ublishers: Dordrecht, The Nederlands. p. 295-309.

[125] Rokka S., Rantamaki P. (2010) Protecting probiotic bacteria by microencapsulation: challenges for industrial applications. European Food Research and Technology, 231: 1–12.

[126] Shah N.P., Ravula R.R. (2000) Microencapsulation of probiotic bacteria and their survival in frozen fermented dairy desserts. Australian Journal of Dairy Technology, 55: 139–44.

[127] Cui J.H., Goh J.S., Kim P.H., Choi S.H., Lee B.J. (2000) Survival and stability of bifidobacteria loaded in alginate poly-l-lysine microparticles. International Journal of Pharmaceuticals, 210: 51–9.

[128] Lacroix C., Paquin C., Arnaud J.P. (1990) Batch fermentation with entrapped cells of *Lactobacillus casei*: optimization f the rheological properties of the entrapment gel matrix. Appl Microbiol Biotechnol., 32: 403-408.

[129] Sun W., Griffiths M.W. (2000) Survival of bifidobacteria in yogurt and simulate gastric juice following immobilization in gellan-xanthan beads. Int J Food Microbiol., 61: 17-25.

[130] Anjani, K., Iyer, C., Kailasapathy, K. (2004) Survival of co-encapsulated complementary probiotics and prebiotics in yoghurt. Milchwissenschaft, 59 (7–8): 396–399.

[131] Krasaekoopt, W., Bhandari, B., Deeth, H. (2004) Comparison of texture of yogurt made from conventionally treated milk and UHT milk fortified with low-heat skim milk powder. Journal of Food Science, 69 (6): E276–E280.

[132] Iyer, C., Kailasapathy, K. (2005) Effect of co-encapsulation of probiotics with prebiotics on increasing the viability of encapsulated bacteria under in vitro acidic and bile salt conditions and in yogurt. Journal of Food Science, 70 (1): M18–M23.

[133] Krasaekoopt, W., Bhandari, B., Deeth, H. (2006) Survival of probiotics encapsulated in chitosan-coated alginate beads in yoghurt from UHT- and conventionally treated milk during storage. LWT – Food Science and Technology, 39 (2): 177– 183.

[134] Urbanska, A.M., Bhathena, J., Prakash, S. (2007) Live encapsulated *Lactobacillus acidophilus* cells in yogurt for therapeutic oral delivery: preparation and in vitro analysis of alginate–chitosan microcapsules. Canadian Journal of Physiology and Pharmacology, 85 (9): 884–893.

[135] Kailasapathy, K., Harmstorf, I., Phillips, M. (2008) Survival of *Lactobacillus acidophilus* and *Bifidobacterium animalis* ssp. *Lactis* in stirred fruit yogurts. LWT – Food Science and Technology, 41 (7): 1317–1322.

[136] Mortazavian, A.M., Azizi, A., Ehsani, M.R., Razavi, S.H., Mousavi, S.M., Sohrabvandi, S., Reinheimer, J.A. (2008) Survival of encapsulated probiotic bacteria in Iranian yogurt drink (Doogh) after the product exposure to simulated gastrointestinal conditions. Milchwissenschaft, 63 (4): 427–429.

[137] Sandoval-Castilla, O., Lobato-Calleros, C., García-Galindo, H.S., Alvarez-Ramírez, J., Vernon-Carter, E.J. (2010) Textural properties of alginate–pectin beads and survivability of entrapped *Lb. Casei* in simulated gastrointestinal conditions and in yoghurt. Food Research International, 43 (1): 111–117.

[138] Prevost, H., Divies, C. (1987) Fresh fermented cheese production with continuous pre-fermented milk by a mixed culture of mesophilic lactic streptococci entrapped en Ca-alginate. Biotechnology Letters, 9 (11): 789–794.

[139] Prevost, H., Divies, C. (1992) Cream fermentation by a mixed culture of *Lactococci* entrapped in two-layer calcium alginate gel beads. Biotechnology Letters, 14 (7): 583–588.

[140] Tsen, J.H., Lin, Y.P., Huang, H.Y., An-Erl King, V. (2008) Studies on the fermentation of tomato juice by using κ-carrageenan immobilized *Lactobacillus acidophilus*. Journal of Food Processing and Preservation, 32 (2): 178–189.

[141] McMaster, L.D., Kokott, S.A., Slatter, P. (2005) Micro-encapsulation of *Bifidobacterium lactis* for incorporation into soft foods. World Journal of Microbiology and Biotechnology, 21 (5): 723–728.

[142] Muthukumarasamy, P., Holley, R.P. (2006) Microbiological and sensory quality of dry fermented sausages containing alginate-microencapsulated *Lactobacillus reuteri*. International Journal of Food Microbiology, 111 (2): 164–169.

[143] Muthukumarasamy, P., Holley, R.A. (2007) Survival of *Escherichia coli* O157:H7 in dry fermented sausages containing micro-encapsulated probiotic lactic acid bacteria. Food Microbiology, 24 (1): 82–88.

[144] Ainsley Reid, A., Champagne, C.P., Gardner, N., Fustier, P., Vuillemard, J.C. (2007) Survival in food systems of *Lactobacillus rhamnosus* R011 microentrapped in whey protein gel particles. Journal of Food Science, 72 (1): M31–M37.

[145] Ross, G.R., Gusils, C., Gonzalez, S.N. (2008) Microencapsulation of probiotic strains for swine feeding. Biological and Pharmaceutical Bulletin, 31 (11): 2121– 2125.

[146] Lee, K.Y., Heo, T.R. (2000) Survival of *Bifidobacterium longum* immobilized in calcium alginate beads in simulated gastric juices and bile salt solution. Applied and Environmental Microbiology, 66 (2): 869–973.

[147] Hussein, S.A., Kebary, K.M.K. (1999) Improving viability of bifidobacteria by microentrapment and their effect on some pathogenic bacteria in stirred yoghurt. Acta Aliment. 28: 113–131.

[148] Zhou, Y., Martins, E., Groboillot, A., Champagne, C.P., Neufeld, R.J. (1998) Spectrophotometric quantification of lactic bacteria in alginate and control of cell releasse with chitosan coating. J. Appl. Microbiol., 84: 342–348.

[149] Doherty, S. B., Auty, M. a., Stanton, C., Ross, R. P., Fitzgerald, G. F., & Brodkorb, A. (2012) Survival of entrapped *Lactobacillus rhamnosus GG* in whey protein micro-beads during simulated ex vivo gastro-intestinal transit. International Dairy Journal, *22*(1): 31-43.

[150] Jiménez-Pranteda, M. L., Poncelet, D., Náder-Macías, M. E., Arcos, A., Aguilera, M., Monteoliva-Sánchez, M., & Ramos-Cormenzana, A. (2012) Stability of lactobacilli encapsulated in various microbial polymers. Journal of Bioscience and Bioengineering, 113(2): 179-84.

[151] Moayednia, N., Ehsani, M. R., Emamdjomeh, Z., Asadi, M. M., & Mizani, M. (2010). A note on the effect of calcium alginate coating on quality of refrigerated strawberries. Irish Journal of Agricultural and Food Research, 49: 165-170.

[152] Jeon, Y.-J., et al. (2003) The suitability of barley and corn starches in their native and chemically modified forms for volatile meat flavor encapsulation. Food Research International, 36(4): 349-355.

[153] Murúa-Pagola, B., C.I. Beristain-Guevara, and F. Martínez-Bustos. (2009) Preparation of starch derivatives using reactive extrusion and evaluation of modified starches as shell materials for encapsulation of flavoring agents by spray drying. Journal of Food Engineering, 91(3): 380-386.

[154] Whisthler and R. L. (1991) Microporous granular starch matrix composition, Lafayette Applied Chemistry, Inc.

[155] Chan, E.S. and Z. Zhang. (2005) Bioencapsulation by compression coating of probiotic bacteria for their protection in an acidic medium. Process Biochemistry, 40(10): 3346-3351.

[156] Gudmund, S.-B., et al. (2006) Alginates, in Food Polysaccharides and Their Applications. CRC Press. p. 289-334.

[157] Gouin, S. (2004) Microencapsulation: industrial appraisal of existing technologies and trends. Trends in Food Science and Technology, 15 (7–8): 330–347.

[158] Chatelet, C., O. Damour, and Domard. A. (2001) Influence of the degree of acetylation on some biological properties of chitosan films. Biomaterials, 22(3): 261-268.

[159] Dumitriu, S. and Chornet E. (1998) Inclusion and release of proteins from polysaccharide-based polyion complexes. Advanced Drug Delivery Reviews, 31: 223-246.

[160] Jameela, S.R. and Jayakrishnan A. (1995) Glutaraldehyde cross-linked chitosan microspheres as a long acting biodegradable drug delivery vehicle: studies on the in vitro release of mitoxantrone and in vivo degradation of microspheres in rat muscle. Biomaterials, 16: 769-775.

[161] Capela, P. (2006) Use of cryoprotectants, prebiotics and microencapsulation of bacterial cells in improving the viability of probiotic organisms in freeze-dried yoghurt, in School of Molecular Science, Victoria University: Victoria (Australia). p. 158.

[162] Zhou, Y., et al. (1988) Spectrophotometric quantification of lactic bacteria in alginate and control of cell release with chitosan coating. Journal of Applied Microbiology, 84: 342-348.

[163] Jansson, P E, Lindberg B and Sandford P A. (1983) Structural studies of gellan gum, an extracellular polysaccharide elaborated by *Pseudomonas elodea.* Carbohydrate Res., 124: 135–139.

[164] Sworn G. (2000) Gellan gum. In: Phillips G.O., Williams P.A. (eds) Handbook of hydrocolloids. Woodhead Publishing Limited, Cambridge, England, pp 117–135.

[165] Klein, J., & Vorlop, D. K. (1985) Immobilization techniques: cells. In C. L. Cooney, & A. E. Humphrey (Eds.), Comprehensive biotechnology (pp. 542-550). Oxford, UK: Pergamon Press.

[166] Sanderson GR (1990). Gellan gum. In: Food Gels. pp. 201-233, P. Harris (Ed.).

[167] Mangione, M.R., Giacomazza, D., Bulone, D., Martorana, V., San Biagio, P.L. (2003) Thermoreversible gelation of n-Carrageenan: relation between conformational transition and aggregation, Biophysical Chemistry, 104: 95–105.

[168] Mangione, M.R., Giacomazza, D, Bulone, D, Martorana, V, Cavallaro, G, San Biagio, P.L. (2005) K+ and Na+ effects on the gelation properties of n-Carrageenan. Biophysical Chemistry, 113: 129– 135

[169] Livney, Y.D. (2010) Milk proteins as vehicles for bioactives. Current Opinion in Colloid and Interface Science, 15 (1–2): 73–83.

[170] Heidebach, T., Först, P., Kulozik, U. (2009) Microencapsulation of probiotic cells by means of rennet-gelation of milk proteins. Food Hydrocolloids, 23 (7): 1670– 1677.

[171] Heidebach, T., Först, P., Kulozik, U. (2009) Transglutaminase-induced caseinate gelation for the microencapsulation of probiotic cells. International Dairy Journal, 19 (2): 77–84.

[172] Kritchevsky D. (1995) Epidemiology of fiber, resistant starch and colorectal cancer. Eur J Cancer Prev., 4: 345-352.

[173] Haralampu S.G. (2000) Resistant starch-a review of the physical properties and biological impact of RS3. Carbohydrate Polymers., 41: 285-292.

[174] Thompson D.B. (2000) Strategies for the manufacture of resistant starch. Trends in Food Sci Technol., 11: 245-253.

Chapter 24

Different Methods of Probiotics Stabilization

Kamila Goderska

Additional information is available at the end of the chapter

http://dx.doi.org/10.5772/50313

1. Introduction

Starter cultures provide a basis in the production of fermented foods. Probiotics are the most important group of bacterial starter cultures. Commercial starter cultures were initially supplied in liquid form prior to the production of concentrated starter cultures. Progress in biotechnology later led to the application of concentrated starter cultures in frozen and freeze dried forms for direct incorporation into the food formulation. Application of frozen or freeze-dried starter cultures eliminates in –plant sub-culturing, reduces the costs associated with bulk culture preparation and lowers the risk of bacteriophage infection (Desmod et al. 2002).

Very low transportation and storage temperatures are the main commercial disadvantages of frozen starter cultures (Ghandi et al. 2012). Besides the risk of thawing, high transportation costs may limit the use of frozen starter cultures in distant areas or countries. Starters of probiotic bacteria are usually preserved by freeze thawing and lyophilization. In spite of being efficient methods, freezing and freeze drying have high manufacturing costs and energy consumption. For this reason, increasing attention has been paid on alternative dring processes such as spray drying, fluidized bed drying and vacuum drying.

Majority of vegetative forms of microorganisms are characterized by poor thermostability. They exhibit considerably high rates of dying and loss of activity as a result of thermal inactivation at the range of temperatures from 40 to 60°C. With regard to microbial biomass, there is certain critical water content (depending on the object property) which, when exceeded, results in dehydration inactivation. This can be attributed to the fact that in the case of vegetative forms of microorganisms water does not only provide environment for their life but it also acts as a substrate for biochemical reactions and its removal below a certain level prevents maintenance of metabolic functions and, consequently, leads to the death of cells. Among dehydration methods which allow maintaining viability of microbial biomass are: freeze-drying, sublimation drying, including fluidization drying using inert materials (carriers) and spray drying (Santivarangkna et al. 2008).

Freeze drying is therefore more convenient and easier as it does not require freezing conditions during distribution. Although freeze drying is the conventional drying technique used commercially by starter culture manufactures, it is lengthy and more expensive than other drying processes (Fonseca et al 2001, Ampatzoglou et al. 2010, Morgan et al. 2006). Many attempts have been made to develop alternative drying processes at lower cost and some authors have reported reasonable cell viability after drying (Tymczyszyn et al. 2008).

Spray drying is considered a good long-term preservation method for probiotic cultures. The spray drying of microorganisms dates to 1914 to the study of Rogers on dried lactic acid cultures. The concept of spray drying was first patented by Samuel Percy in 1872, and its industrial application in milk and detergent production began in the 1920s. The speed of drying and continuous production capability are very useful for drying large amounts of starter cultures. Since then, much research has been reported on the spray drying of bacteria without loss of cell activity in order to overcome the difficulties involved in handling and maintaining liquid stock cultures.

Spray drying is a unique process in which particles are formed at the same time as they are dried. It is a very suitable for the continuous production of dry solids in powder, granulate or agglomerate form liquid feed stocks as solutions, emulsions and pumpable suspensions. The end product of spray drying must comply with precise quality standards regarding particle size distribution, residual moisture, bulk density, and particle shape. In the spray drying process, dry granulated powders are produced from a slurry solution, by atomizing the wet product at high velocity and directing the spray of droplets into a flow of hot air e.g. 150-200°C. The atomized droplets have a very large surface area in the form of millions of micrometer-sized droplets (10-200μm), which results in a very short drying time when exposed to hot air in a drying chamber (Sunny-Roberts, Knorr 2009).

Spray drying involves atomization of a liquid feedstock into a spray of droplets and contacting the droplets with hot air in a drying chamber. The sprays are produced by rotary (wheel) or nozzle atomizers. Evaporation of moisture from the droplets and formation of dry particles proceed under controlled temperature and airflow conditions. Powder is discharged continuously from the drying chamber (Peighambardoust et al. 2011).

Spray drying is a common industrial and economic process for the preservation of microorganisms and for the preparation of starter cultures that are used to prepare lactic-fermented products. The survival of lactic acid bacteria is an important issue when spray drying is used for the preparation of microbial cultures. However biological activity of a lactic acid starter, which includes cell viability and physiological state, is a criterion for evaluating starter quality (Carvalho et al. 2004, Ananta et al. 2005).

It has been shown that both the water evaporation rate and the temperature of droplets containing microbial cells have a significant effect on their survival during spray drying.

Since it is not yet possible to quantify the changes occurring in the bacterial cells and their survival in situ when they are subjected to spray drying, single droplet drying is used instead. Single droplet drying, in which a single droplet is suspended in moving and conditioned air, provides the closest experimental resemblance to the spray drying environment. Single droplet drying can be conducted in various ways, for example (a) a single or a stream or streams of droplets could be allowed to fall under gravity in a tower-like dryer, (b) a droplet can be freely levitated using ultrasonic or aerodynamic fields, or (c) a droplet can be suspended on the tip of a fine glass filament. The first two method are not very popular as they are expensive and the heat and mass transfer rates in these environments are not close to the convective drying environment of spray drying. Li et al (2006) investigated the inactivation kinetics of two probiotic strains (*Bifidobacterium infantis* and *S. thermophilus*) in air temperature and relative humidity in the ranges of 70-100 °C and 3,7-0,5%, respectively, using single droplet drying in skim milk as a suspending medium. They reported that the inactivation mainly occurred at the early stage of the drying when the evaporation rate was high. The above studies do not offer a unanimous view whether the drying rate or the droplet temperature is the limiting factor of bacterial survival during drying.

Freeze drying is a preferred drying method for thermally sensitive bacteria as it keeps their survival at a reasonably high level. However, freeze drying is a batch process with a considerably long drying time. It is also expensive due to high energy requirements. For drying of starter cultures, spray drying can be a viable alternative if the survival can be raised to make it economically attractive. This is because spray drying is relatively inexpensive, energy efficient, high throughput and a hygienic process (Papapostolou et al. 2008, Carvalho et al. 2004).

In order to minimise cell death, the effects of drying parameters (inlet and outlet air temperatures, air flow rate, relative humidity, residence time, protective agents) on the survival and vitality of bacteria have to be understood to a considerable depth. The drying process causes damage to the cell wall and cellular components, especially cytoplasmic membrane and proteins, which results in the loss of survival. This cellular injury leads to cell inactivation and negatively impacts the productivity and characteristics of dried culture, and hence the cellular injury has to be minimised. Protective agents such as carbohydrates, proteins, amino acids, gums and skim milk are used to minimise the bacterial inactivation during drying. It is reported that low molecular weight carbohydrates such as sugars stabilize the membrane and protein chains of cellular macromolecules in dry state through hydrogen bonding in lieu of water when the water molecules are removed through desiccation. Protein are capable of forming relatively stable intracellular glasses, and by doing so, they can be more effective a protective materials for bacterial culture than sugars. It is reported that the combination of different protectans (e.g. mixtures of sugar and protein) can have synergistic effect on cell viability rather than acting individually. It has been shown that both the water evaporation rate and the temperature of droplets containing microbial cells have a significant effect on their survival during spray drying.

Encapsulation of probiotics is employed in order to increase the bacteria resistance to freezing and freezing drying of the food. In most of the studies the probiotic bacteria were entrapped in a gel matrix of biological nature materials such as alginate, κ-carrageenan, and gellan/xanthan (Semyonov et al. 2010, Kanmani et al. 2011). The core and wall solution was turned into drops of desired size by an extrusion method, employing an emulsion or by transfer from organic solvents. One problem in the probiotic entrapment approach is that the gel beads technologies stabilize the bacteria mostly in liquid products, and are difficult to scale up. To extend storage shelf-life t is convenient to convert the micro-capsules into a dry powder by employing techniques such as spray drying, freeze drying, and/or fluidized bed drying. The spray drying is an economic and effective technology, however, it causes high mortality as a results of simultaneous dehydratation, thermal and oxygen stresses imposed to bacteria during the drying process. Freeze drying is considered one of the most adequate methods for drying biological materials and sensitive foods. However, when this method was employed for drying probiotic bacteria and other cells, undesirable effects such leakage of the cell membrane due to changes in the physical state of membrane lipids or changes in he structure of sensitive proteins in the bacteria cell occur. Protective solutes such as cryoprotectans (saccharides and polyols) and other compatible solutes like adonitol, betaine, glycerol and skim milk were used to increase bacteria's viability and increase their survival during freeze-drying and subsequent storage. These studies lead to the conclusion that the effect of each protective agent on the viability of a specific lactic acid bacteria strain during or following the freeze-drying process have to be determined on a case-by-case basis (Heidebach et al. 2010, Krasaekoopt et al. 2003).

As mentioned above, dried probotic micro-capsules can be coated by an additional layer (shell) in order to protect the bacterial core from the acidic environment of the stomach and to avoid the deleterious effect of bile salts on the cell's membrane. This additional shell can help to release the bacterial core at a desired site in the GIT. In order to be further coated, bulk freezed powders are micronized to a narrow particle distribution. This process is complex, requires intensive energy, and decrease the viability of the dried cells.

The pharmaceutical industry utilized recently the spray freeze drying for pharmaceutical powders preparation. This method combines the narrow article size distribution of an extrusion device and the freeze-drying process to prepare a dry powder of desired particle size and of the narrow distribution. Spray freeze drying basic principle is to spray a solution containing dissolved/suspended material (e.g. protein) by an atomization nozzle into a cold vapor phase of a cryogenic liquid, such a liquid nitrogen, so the droplets may start freezing during their passage through the cold vapor phase, and completely freeze upon contact with the cryogenic liquid phase. The frozen droplets are then dried by lyophilization (Lian et al. 2002, Gardiner et al. 2002).

Spray freeze drying powders have a controlled size, larger specific surface area and a better porous character than spray-dried powders. The particles retain their spherical and porous morphology and can be further coated with an enteric food grade biological polymer which is designed to desintegrate at specific loci in the GIT.

Recently this method was further developed and the solution is sprayed under adequate pressure via a needle directly in liquid nitrogen. The cooling rates in the spray freezing section are dependent on many factors and thus are also very difficult to estimate. However it was claimed that maximum cooling rates by freezing in liquid nitrogen are the order of 300K/s, considered as upper boundary for the cooling rate. To the best of our knowledge the spray freeze drying method was not used yet to produce dry powder of probiotic cells.

Vacuum drying has been described to be the most promissory method to reserve sensible biological material because of its acceptable cost-effectiveness balance. However, the conditions of vacuum drying (time, temperature) must be optimized to allow the best bacterial recovery after dehydratation-rehydratation, avoiding cellular damages (Tymczyszyn et al. 2008).

It has been proposed that bacterial death results from the inactivation of critical sites in the cells. Membranes, nucleic acids and certain enzymes have been identified as cellular targets of damage caused by dehydratation. It has been reported that after dehydratation-rehydratation the microorganisms can be recovered even when the cellular membrane is damaged. In addition, it has also been observed that an increase in the absolute value of the zeta potential can be associated with an increase in the lag time. Changes in this parameter were correlated with a loss of the original orientation of the surface macromolecules and thus, the capacity to recover the surface properties after rehydratation. This indicates that there are other bacterial structural parameters besides the membrane integrity affecting the bacterial viability after dehydratation-rehydratation. In this sense, date obtained by Differential Scanning Calorimetry reveal that damage produced in membrane lipids, ribosomes and DNA are reversible, whereas damages produced in proteins are not.

When applying vacuum drying, it is important to consider that a thermal stress takes place in parallel to the hydric stress, probably inducing irreversible damages. For this reason, the exposure of microorganisms to high temperatures should be as short as possible and the correct choice of times and temperatures of dehydratation is crucial to achieve the best vacuum drying conditions.

The challenge of making vacuum drying a wide spread methodology for microorganisms' preservation is the difficulty of defining standardized conditions that allow the comparison of results obtained in different laboratories. The reason of the difficulty is that the times and temperatures for the dehydratation processes are related with the drying conditions (i.e.: exposure surface, pressure of the vacuum system, weight or volume of the sample, etc.), which in general are dependent on the equipment used. Therefore, to make results comparable, it becomes necessary to refer the experimental conditions, to a parameter that is independent to these experimental conditions, for example, the water activity of the sample after dehydratation in a given condition.

I consequence, considering that both time and temperatures of drying affect the final water activities of the samples, the definition of drying conditions in terms of the final water activity becomes important to define correlatable parameters with the state of dehydratation of the cells. This fact would help to attain the best conditions for the preservation processes.

Bifidobacteria benefit human health by improving the balance of intestinal microbiota and by strengthening mucosal defenses against pathogens. However, for probiotics to be therapeutically effective, it has been suggested that products should contain at least 6 log cfu/g of bacteria until the end of their shelf life. Although bifidobacteria are being increasingly recognized as probiotics that have advantageous properties, they are also fastidious, obligate anaerobes and, therefore, pose a technological challenge for the food industry. several factors have been claimed to affect the viability of bifidobacteria, including acidity, pH, time and temperature of storage, and oxygen content.

Within this context, microencapsulation of probiotic bacteria is currently drawing more and more attention for being a method to improve the stability of probiotic organisms in functional food products. Microencapsulation may improve the survival of these microorganisms, during both processing and storage, and also during passage through the human gastrointestinal tract. Spray drying is regarded as a microencapsulation method and it has been investigated as a means of stabilizing probiotic bacteria in a number of food matrices, most often composed of proteins, polysaccharides, sugars, and combination thereof. The survival rate of the culture during spray drying and subsequent storage depends upon a number of factors, which may include the species and strain of the culture, the drying conditions and also the use of encapsulating agents.

Reconstituted skim milk is an encapsulating agent that has shown a favorable effect on the improvement of cell survival during the spray drying process. Another approach to increase the viability of bifidobacteria is the use of prebiotics, which are nondigestible food ingredients that beneficially affect the host by selectively stimulating the growth and/or activity of bacteria in the colon. Inulin is a prebiotic whose degree of polymerization (DP) ranges between 10 to 60. It is extracted from chicory roots and consists of chains of fructose units. Oligofructose is obtained throught partial hydrolysis of inulin and therefore has a lower DP,, which range from 2 to 8. A mixture of oligofructose and inulin is known as oligofructose-enriched inulin. These prebiotics may potentially be exploited as carrier media for spray drying and may be useful for enhancing probiotic survival during processing. However, the use of different encapsulatin agents for production of microcapsules can result in different physical properties, depending on the structure and the characteristics of each agent. (Fritzen-Freire et al. 2012).

The study was conducted to evaluate the viability and the physical properties of *Bifidobacterium* BB-12 microencapsulated by spray drying partial replacement of reconstitutet skum milk (RSM), as encapsulating agent with the prebiotics inulin, oligofructose, and oligofructose-enriched inulin (at ratio of 1:1, 200g/ total concentration). The viable cell counts of the microcapsules were determined during storage for 180 days at 4°C and at -18°C. The partial replacement of RSM with inulin and the partial replacement of RSM with oligofructose-enriched inulin increased the initial count of bifidobacteria in the microcapsules. On the other hand, the microcapsules produced with oligofructose-enriched inulin and those produced with oligofructose showed better protection for the bifidobacterium during storage. The use of prebiotics did not affect the morphology of the microcapsules. However, the capsules produced with oligofructose showed a smaller particle size. The inclusion of prebiotics decreased the moisture content and water activity in the microcapsules. The microcapsules produced with inulin showed the lowest dissolution in water, while the microcapsules produced with oligofructose were the most hygroscopic. The total color difference of the microcapsules was not considered obvious to the human eye. The results of the thermoanalyses suggest an increase in the stability of the microcapsules produced with prebiotics. Finally, the results showed that the oligofructose-enriched inulin is the most appropriate prebiotic to be used as partial replacement of RSM to microcapsulate *Bifidobacterium* BB-12 by spray drying, with a great potentail as a functional ingredient to be applied in dairy foods. (Fritzen-Freire et al., 2012).

Ultrasonic vacuum spray dryer was used to produce a dry powder of highly viable probiotic cell. The drying was performed through two stages: vacuum spray drying of the solution followed by fluidized-bed drying of the powder. The embedding matrix was a combination of trehalose and maltodextrin. The effect of external and internal variables on cell survival during the drying process and storage were investigated. The hypothesis was that by minimizing the oxidative and thermal stresses in the drying stages, in addition to adequate formulation choice, the cell viability during the drying and storage will increase. It was concluded that during the drying process the faster the embedding matrix reaches a glassy state the higher was the probiotic survival. Evaluating water activity and moisture limit of the glassy matrix concluded that maltodextrin DE5 is a better encapsulating matrix than maltodextrin DE19. Combining trehalose to maltodextrin in the encapsulating matrix resulted in a significant increase in the survival up to 70.6±6.2%.

Higher temperatures used during spray drying may be detrimental to bacteria. However this is not the case for certain lactic acid bacteria. For example, similar survival rates were obtained on freeze-drying and spray-drying of concentrated cultures of *Lactobacillus bulgaricus*. cellular damage to probiotics may be reduced and viability preserved through control of drying parameters; specifically, by lowering the outlet temperature of spray dryers and the incorporation of appropriate carriers into the drying medium. The addition of sugars to the growth medium also influences the survival of dried probiotic preparations. The incorporation of glucose in formulations did not markedly influence the survival of probiotic during drying but had marked effects on *Lactobacillus* GG survival during subsequent long term storage. These results corroborate those of others who also found that although the survival of LGG during spray drying was not significantly affected when different media (reconstituted skim milk (RSM), RSM/polydextrose, RSM-Raftilose P95) were used, it did influenced survival of bacteria during long term storage. A similar result was reported, where the presence of sugars (fructose, trehalose or sucrose) or sugar alcohols (inositol, sorbitol) improved survival of *Lactobacillus plantarum* and *L. rhamnosus* during storage but not during the freeze-drying (Yang Ying et al. 2012).

The glucose-containing formulations in the study, improved storage stability of spray-dried LGG microcapsules stored under similar environmental conditions although the glass transition temperature of these formulations was depressed in comparison to those of formulations without glucose. It has been suggested that the incorporation of small sugars improves survival of bacteria during drying because of their ability to replace water that is removed from proteins/enzymes within the cells and reduce the membrane phase transition temperature. Results suggest that the effect of glucose is more significant during storage than during drying, even though glucose containing formulations did not maintain its glassy state at different storage conditions. The results of the work are in line with those of others which show that a glassy state during storage alone is not sufficient for stabilization of dried bacterial preparations.

Protectants which preserve the structural integrity of cell membranes, proteins and enzyme functions are required for improving viability during storage of dried probiotic preparations. These results suggest that a pre-requisite for LGG survival in the glassy state is the direct interactions between a low molecular weight sugar and cell components, which helps preserve cell functions during drying with subsequent beneficial effects on long term storage. Both the maintenance of a glassy state during storage and the incorporation of glucose or a low molecular weight sugar in the drying medium are required for optimal survival of probiotic powders during storage (Yang Ying et al. 2012).

The process for the formation of dry-encapsulated probiotics, using ultrasonic vacuum spray drying (UVSD), and microcapsule matrix composed of maltodextrin and trehalose were studied. The results of this study demonstrate thet using UVSD brought the matrix repidly to a glassy state and provided high survival of the probiotic cells- 3.3 x10^9 cfu/g dm, that was achieved with maltodextrin DE-trehalose (1:1) 20%g/100g matrix and 7.0 x10^9 cfu/g dm initial *L. paracasei* concentration. It was found that MD DE5 was a better encapsulation matrix than MD DE19, probably due to the fact that DE5 matrix maintained its glassy state at a higher a_w. The addition of trehalose increased the viability significantly during the drying and during storage of the dried powder. MD DE5-trehalose combination (1:1) resulted with the highest survival (70.6±6.2%). Evidently, further protection should be provided to the cells against oxidation, as storage in nitrogen was essential in order to gain storage stability. (Semyonov et al. 2011)

Improved production methods of starter cultures, which constitute the most important element of probiotic preparations, were investigated. The aim of the presented research was to analyse changes in the viability of *Lactobacillus. acidophilus* and *Bifidobacterium bifidum* after stabilization (spray drying, liophilization, fluidization drying) and storage in refrigerated conditions for 4 months. The highest numbers of live cells, up to the fourth month of storage in refrigerated conditions, of the order of 10^7 cfu/g preparation were recorded for the *B. bifidum* DSM 20239 bacteria in which the N-Tack starch for spray drying was applied. Fluidization drying of encapsulated bacteria allowed obtaining a preparation of the comparable number of live bacterial cells up to the fourth month of storage with those encapsulated bacteria, which were subjected to freeze-drying but the former process was much shorter. The highest survivability of the encapsulated *Lb. acidophilus* DSM 20079 and *B. bifidum* DSM 20239 cells subjected to freeze-drying was obtained using skimmed milk as the cryoprotective substance. Stabilisation of bacteria by microencapsulation can give a product easy to store and apply to produce dried food composition (Goderska, Czarnecki 2008) .

Author details

Kamila Goderska
Faculty of Food Science and Nutrition, Institute of Food Technology of Plant Origin, Department of Fermentation and Biosynthesis, University of Life Sciences in Poznań, Poland

2. References

Ampatzoglou A., Schurr B., Deepika G., Baipong S., Charalampopoulos D. Influence of fermentation on the acid tolerance and freeze drying survival of *Lactobacillus rhamnosus* GG. Biochemical Engineering Journal 2010, 52, 65-70

Ananta E., Volkert M., Knorr D. Cellular injuries and storage stability of spray-dried *Lactobacillus rhamnosus* GG. International dairy Journal 2005, 15, 399-409

Carvalho A.S., Silva J., Ho P., Teixeira P., Malcata F.X., Gibbs P. Relevant factors for the preparation of freeze-dried lactic acid bacteria. International Dairy Journal 2004, 14, 835-847

Desmond C., Stanton C., Fitzgerald G.F., Collins K., Ross R.P. Environmental adaptation of probiotic lactobacilli towards improvement of performance during spray drying. International Dairy Journal 2002, 12, 183-190

Fonseca F., Beal C., Corrieu G. Operating conditions that affect the resistance of lactic acid bacteria to freezing and frozen storage. Cryobiology, 2001, 43, 189-198

Fritzen-Freire C.B., Prudencio E.S., Amboni R.D.M.C., Pinto S.S., Negrao-Murakami A.N., Murakami F.S. Microencapsulation of bifidoabcteria by spray drying in the presence of prebiotics. Food Research International 2012, 45, 306-321

Gardiner G.E., Bouchier P., O'Sullivan E., Kelly J., Collins J.K., Fitzgerald G., Ross R.P., Stanton C. A spray-dried culture for probiotic Cheddar cheese manufacture. International Dairy Journal 2002, 12, 749-756

Ghandi A., Powell I., Chen X.D., Adhikari B. Drying kinetics and survival studies of dairy fermentation bacteria in convective air drying environment using single droplet drying. Journal of Food Engineering 2012, 110, 4,05-417

Goderska, Czarnecki Influence of microencapsulation and spray drying on the viability of *Lactobacillus* and *Bifidobacterium* strains. Polish Journal of Microbiology 57(2), 135-140

Kanmani P., KUmar R.S., Yuvaraj N., Paari K.A., Pattukumar V., Arul V. Effect of cryorpeservation and microencapsulation of lactic acid bacterium *Enterococcus faecium* MC13 for long-term storage. Biochemical Engineering Journal 2011, 58-59, 140-147

Krasaekoopt W., Bhandari B., Deeth H. Evaluation of encapsulation techniques of probiotics for yoghurt. International Dairy Journal 2003, 13, 3-13

Lian W.C., Hsiao H.C., Chou C.C. Survival of bifidobacteria after spray-drying. International Journal of Food Microbiology 2002, 74, 79-86

Morgan C.A., Herman N., White P.A., Vesey G. Preservation of micro-organisms by drying; A review. Journal of Microbiological Methods 2006, 66, 183-193

Papapostolou H., Bosnea L.A., Koutinas A.A., Kanellaki M. fermentation efficiency of thermally dried kefir. Bioresource technology 2008, 99, 6949-6956

Peighambardoust S.H., Tafti A.G., Hesari J. Application of spray dring for preservation of lactic acid starter cultures: a review. Trends in Food Sciences & Technology 2011, 22, 215-224

Santivarangkna C., Higl B., Foerst P. Protection mechanisms of sugars during different stages of preparation process of dried lactic acid starter cultures. Food Microbiology, 2008, 25, 429-441

Semyonov D., Ramon O., kaplun Z., Levin-Brener L., Gurevich N., Shimoni E. Microencapsulation of *Lactobacillus paracasei* by spray freeze drying. Food Research International 2010, 43, 193-202

Semyonov D., Ramon O., Shimoni E. Using ultrasonic vacuum spray dryer to porduce highly viable dry probiotics. LWT-Food Science and Technology 2011, 44, 1844-1852

Sunny-Roberts E.O., Knorr D. The protective effect of monosodium glutamate on survival of *Lactobacillus rhamnosus* GG and *Lactobacillus rhamnosus* E-97800 (E8000) strains during spray-drying and storage in trehalose-containing powders. International dairy Journal 2009, 19, 209-214

Tymczyszyn E.E., Diaz R., Pataro A., Sandonato N., Gomez-Zavaglia A., Disalvo E.A. Critical water activity for the preservation of *Lactobacillus bulgaricus* by vacuum drying. International Journal of Food Microbiology 2008, 128, 342-347

Ying D.Y., Sun J., Sanguansri L., Weerakkody R., Augustin M.A., Enhanced survival of spray-dried microencapsulated *Lactobacillus rhamnosus* GG in the presence of glucose. Journal of Food Engineering, 2012, 109, 597-602

Zbicinski I., Delag A., Strumillo C., Adamiec J. Advanced experimental analysis of drying kinetics in spray drying. Chemical Engineering Journal 2002, 86, 207-216

Chapter 25

Probiotic Food Products Classes, Types, and Processing

Saddam S. Awaisheh

Additional information is available at the end of the chapter

http://dx.doi.org/10.5772/51267

1. Introduction

Probiotic as a term is a relatively new word meaning "for life" and it is currently used to describe a group of bacteria when administered in sufficient quantity, confer beneficial effects for humans and animals [1]. The concept of probiotic bacteria is very old, and is associated with the consumption of fermented foods by human beings, for thousands of years. Since ancient times, man has made and eaten probiotic foods. The earliest types of probiotic food were cheeses and milks made by lactic acid bacterial (LAB) and fungal fermentation, and leavened bread fermented by yeasts fermentation [2]. Fermented food's health benefit has also been long known. Hippocrates and other scientists in the early ages had observed that some disorders of the digestive system could be cured by fermented milk, also, Plinius, the Roman historian, stated that fermented milk products can be used for treating gastroenteritis [3].

In the modern ages, the concern to understand the importance and mechanisms of action of probiotic bacteria to exert their beneficial effects has been raised. In the early 1900s, the Russian microbiologist Ilya Mechinikov, Nobel Prize laureate, attributed the good health and longevity of Bulgarian peoples to their high consumption of fermented probiotic foods. He not only identified the health-giving bacteria used to ferment these foods, he also concluded that the general human being's health is function of the balance between beneficial "good" probiotic bacteria and disease-causing "bad" bacteria in human gut [4]. At this time Henry Tissier, a French pediatrician, observed that children with diarrhea had in their stools a low number of bacteria characterized by a peculiar, Y shaped morphology, and these "bifid" bacteria were abundant in healthy children. Also, Tissier found that these bifidobacteria are dominant in the gut flora of breast-fed babies. The isolated bacterium named *Bacillus bifidus*, and was later renamed to the genus *Bifidobacterium*. Accordingly, he suggested that these bacteria could be administered to patients with diarrhea to help restore a healthy gut flora [2,3]. This claimed effect was due to bifidobacteria displacement of proteolytic bacteria causing the disease. The works of Metchnikoff and Tissier were the first scientific suggestions about the probiotic use of bacteria. However, In 1917, during sever shigellosis outbreak, the German professor Alfred Nissle isolated a nonpathogenic strain of *Escherichia coli* from the feces of a soldier who did not develop enterocolitis. Disorders of the intestinal tract were frequently treated with viable nonpathogenic bacteria to change or replace the intestinal microbiota. The *E. coli* strain Nissle 1917 is one of the few examples of a non-LAB probiotic. It was till 1960s, when the word "probiotic" was first proposed to describe substances produced by microorganisms and promote the growth of other microorganisms [5]. In 1989, Fuller, in order to point out the microbial nature of probiotics, redefined the word as "A live microbial feed supplement which beneficially affects the host animal by improving its intestinal balance" [6,7]. Another definition was proposed by [6] "a viable mono or mixed culture of bacteria which, when applied to animal or man, beneficially affects the host by improving the properties of the indigenous flora". A more recent, but probably not the last definition is "live microorganisms, which when consumed in adequate amounts, confer a health effects on the host beyond inherent basic nutrition [1,7].

As investigations continued in the probiotic field, its concept has been expanded to include bacteria from intestinal origin beside those bacteria isolated from fermented dairy products [8]. Nowadays, probiotic bacteria are available in a variety of food products, dietary supplements [9] and drugs [10]. Food products containing are almost dairy products – fluid milk and yogurt – due to the historical association of LAB with fermented milk. The most frequently used bacteria in these products include the *Lactobacillus* and *Bifidobacterium* species. Recently, new types of food products containing probiotic bacteria started to be introduced into the markets, including nondairy products, such as chocolate, cereals, beverages, fruits and vegetables products. In the near future wide range of nontraditional food products containing probiotic bacteria are expected to be introduced into the markets, as the researches in probiotic products development continue in both scientific and commercial centers around the world.

2. Safety of probiotic bacteria

Safety considerations of probiotic bacteria are of high importance, as most probiotic bacteria are marketed in foodstuffs or feed supplements. The safety of these microbes has been confirmed through a long experience of safe use in food as starter cultures [11-13]. Bacteria such as *Lactobacillus, Leuconostoc,* and *Pediococcus* species have long been involved in food processing throughout human history, and the ingestion of foods containing live dead bacteria, and metabolites of these bacteria has taken place for many centuries [14]. Generally, LAB are classified as generally recognized as safe (GRAS), and there were no reports of any harmful effects from the consumption of these bacteria through the long history of their use in the processing of many foods (i.e. fermented dairy, fermented vegetables ...etc.) [15]. In an epidemiological study of lactobacilli bacteremia case reports,

[16] concluded that the increased usage of probiotic products of lactobacilli did not cause any increase in incidence or frequency of bacteremia in Finland. However, it was found that under certain conditions, some lactobacilli strains have been associated with adverse effects, such as rare cases of bacteremia [12]. Ecologically, bifidobacteria are the predominant bacteria in the intestinal tract of breast-fed infants and are believed to contribute to the good health of infants. Until now, the safety of the bifidobacteria has not been questioned, as the reports of a harmful effect of these microbes on the host are very rare.

The concern of probiotic bacteria safety has been raised with the more recent use of intestinal isolates of bacteria delivered in high numbers to severely ill patients. Use of probiotic bacteria in ill persons is restricted to the strains and indications with proven efficacy. A multidisciplinary approach is necessary to assess the toxicological, immunological, gastroenterological, pathological, infectivity, the intrinsic properties of the microbes, virulence factors comprising metabolic activity, and microbiological effects of probiotic strains [1,17]. Conventional toxicology and safety evaluation is not sufficient, since a probiotic is meant to survive and/or grow in human colon in order to benefit humans. Several methods have been developed for evaluation the safety of LAB through the use of in vitro studies, animal studies, and human clinical studies [14]. Also, proposed studies on intrinsic properties and interactions between the host and probiotic bacteria can be used as means to assess the safety of probiotic bacteria [17,18]. Evaluation of the acute, sub-acute and chronic toxicity of ingestion of extremely large quantities of probiotic bacteria should be carried out for all potential strains. Such assessment may not be necessary for strains with established documented use.

Thus, safety considerations of probiotic bacteria should include:

1. Antibiotic resistance profiles.
2. Infectivity in immune-compromised animal models
3. Toxin production: probiotic bacteria must be tested for toxin production. One possible scheme for testing toxin production has been recommended by the EU Scientific Committee on Animal Nutrition.
4. Hemolytic activity.
5. Metabolic activities (D-lactate, bile salt de-conjugation).
6. Genetic and pathological side effects.
7. Epidemiological surveillance of adverse incidents in consumers (post market).

2.1. Antibiotics resistance profiles of probiotic bacteria

Most bacteria, including LAB and probiotic bacteria are resistant to some antibiotics. This resistance may be related to chromosomal, transposon or plasmid located genes [19]. However, data available on situations in which these genetic elements could be transferred is not sufficient, and whether the situation could arise to become a clinical problem is unknown yet. There is a concern over the use of probiotic bacteria that contain specific drug resistance genes in foods. Probiotic bacteria contain transferable drug resistance genes should not be used for human. So, there is an urgent need for the development of standardized methodology for the assessment of drug resistance profiles in lactobacilli and bifidobacteria. Due to the relevance of this problem, it has been suggested that further research is needed to assess the antibiotic resistance of these bacteria. When dealing with the selection of probiotic strains, it is recommended that probiotic bacteria should not harbor transferable genes encoding resistance to clinically used drugs. Also, research is needed concerning the antibiotic resistance of lactobacilli and bifidobacteria and the potential for transferring genetic elements to other intestinal and/or food borne bacteria. For example, some strains of *Enterococcus* display probiotic properties, but it was found that *Enterococcus* is emerging as an important cause of nosocomial infections and isolates are increasingly vancomycin resistant. Accordingly, *Enterococcus* is not recommended as a probiotic for human use [14].

3. Regulatory issues of probiotic products

As the global probiotic markets are expanding rapidly, the harmonization of national and international regulations and guidelines are becoming extremely important for evaluating the efficacy and safety of probiotic bacteria. Hence, there would always be a possibility of spurious and ineffective probiotic products with false claims being marketed, it becomes important that these products are standardized and fulfill essential prerequisite before being marketed. So far, there is no international harmonization of probiotic product regulations. Depending on the intended use of a probiotic, whether as a food/food ingredient, a dietary supplement, and/or a drug, regulatory requirements differ greatly among different countries [20]. For most countries, if a probiotic is to be used as a drug, then it must undergo the regulatory process as a drug, which is similar to that of any new therapeutic agent. The probiotic drug safety and efficacy for its intended use must be evaluated and approved before marketing. But, if a probiotic is to be used as a dietary supplement, it is considered as foods, and then these products do not need any evaluation or approval before being marketed. However, there is an urgent need for harmonization of these regulatory standards on probiotic bacteria at the international level to ensure the safety and efficacy of probiotic products for their effective utilization in different countries around the world. However, for most countries, probiotic bacteria are regulated under food and dietary supplements because most are taken orally as foods. These are differentiated from drugs in a number of ways, especially with respect to claims. Drug claims include efficacy in the treatment, mitigation or cure of a disease, whereas foods, feed additives and dietary supplements can only make general health claims, such as structure/function claim [21,22]. A 'health claim' is defined as "a statement, which characterizes the effect relationship of any substance to a disease or health-related condition, and these should be based upon well-established scientific evidences from national or international public health bodies. Examples include 'protects against cancer'. A structure/function claim is defined as "a statement of nutritional support that affects the structure or functioning of the human body, or characterizes the mechanism to maintain such structure or function. For example 'supports the immune system' [23,24]. No therapeutic claim or disease-prevention is known to have been approved by the United States, EU, or Canada [23].

3.1. FAO/WHO approach

The Joint FAO/WHO Expert Consultation on Evaluation of Health and Nutritional Properties of Probiotic bacteria has developed and proposed guidelines for evaluating probiotic bacteria in food that could lead to the harmonization of regulations and standards of probiotic bacteria health claims [1,25]. The recommended guidelines included: 1) using a combination of phenotypic and genotypic tests to identify the genus and species of the probiotic strain, as clinical evidences suggested that the health benefits of probiotic bacteria may be strain specific, 2) in vitro testing to delineate the mechanism of the probiotic effect, and 3) substantiation of the clinical health benefit of probiotic agents with human trials. In addition, the manufacturer should take on the responsibility (albeit not required by law) of providing guidance to consumers or clinicians about the type and extent of safety assessments that have been conducted on its products. According to The Joint FAO/WHO Expert Consultation recommendations, even though, that in most countries, only general health claims are allowed on probiotic foods, it is recommended that specific health claims may be allowed on probiotic foods, where sufficient scientific evidence is available. Such specific health claims should be permitted on the label and promotional material.

3.2. United states approach

In the USA, depending on how probiotic bacteria are intended to be used, they may be regulated as a dietary supplement and/or a biological agent. Biological agents require pre-market evaluation of the safety, purity and potency, as well as efficacy for approval by FDA, whereas, dietary supplements do not [25]. According to FDA, the determining factor as to whether a probiotic is a dietary supplement is whether it has been used as a food. A probiotic used for diagnosis, cure, mitigate, treat, or prevent disease is considered as a drug and/or a biological product. FDA's Center for Biologics Evaluation and Research (CBER) regulates probiotic products when used for clinical indications. CBER's Office of Vaccines Research and Review has regulatory jurisdiction over most probiotic products for clinical use [26]. Nevertheless, most probiotic bacteria are regulated as dietary supplements, which were regulated in 1994 by FDA via the Dietary Supplement and Health Education Act (DSHEA). According to DSHEA, probiotic dietary supplements may have a structure/function claim. It is the manufacturer responsibility to notify the FDA before marketing any probiotic product, and determine that the dietary supplements that it manufactures or distributes are safe, and that any claims made about them are substantiated by adequate evidence to show that they are not false or misleading. The manufacturer must also state on the label that the dietary supplement product is not intended to 'diagnose, treat, cure or prevent any disease' because only a drug can legally make such a claim. Unlike Canada and some European countries, the United States has no governmental standards for probiotics. As most probiotic bacteria are claimed to be GRAS, they are not subjected to any specific standards [22]. Currently there are no functional foods are regulated or marketed in the USA, and this is partly because there is no internationally accepted definition of a functional food [24]. The International Food information Council has suggested that functional foods be defined as foods that provide health benefits beyond basic nutrition [24,25].

3.3. European approach

As in the international level, the different European countries have different national regulations for probiotics. For example in Germany, France, and Italy, the probiotic bacteria in capsule, tablet or powder form have the pharmaceutical products status, whereas, in Denmark, Finland Netherlands, and Sweden, same probiotic products are regulated as food and/or food supplements [24,28]. Food supplements do not require authorities' notification or registration before marketing. In the EU, probiotic bacteria are legally regulated either as 1) foods; for examples, yogurts, dairy drinks, fermented fish, meats & vegetables, and cheeses; 2) food supplements; for examples, tablets, pills, powders, capsules, liquid concentrates in vials, and soft gels; or 3) novel foods. Novel foods are defined as foods/food a ingredient that does not have a significant history of human consumption within the EU countries prior to 15th May 1997 (97/258/EC). According to the novel food regulations, if a probiotic does not have a history of safe use, safety and quality guidelines are laid down. To date, probiotic bacteria for human foods are not governed under specific EU regulatory frame works. Novel Food regulation EU 258/97 is to relevant probiotic in some specific cases. There is therefore a considerable need for harmonization of European legislation on probiotic bacteria considered as food supplements. In contrast with the situation in the USA, even though, that the level of awareness and acceptance of probiotic bacteria in Europe is advanced, neither a legal definition nor specific regulations governing functional foods exist. However, according to Food Supplements Directive 2002/46/EC "food supplements' are defined as" foodstuffs the purpose of which is to supplement the normal diet and which are concentrated sources of nutrients or other substances with a nutritional or physiological effect, alone or in combination, marketed in dose form, namely forms such as capsules, pastilles, tablets, pills and other similar forms, sachets of powder, ampoules of liquids, drop dispensing bottles, and other similar forms of liquids and powders designed to be taken in measured small unit quantities". Even though, that this directive has generally been elaborated for vitamins and minerals, also, it already states that specific regulations must be laid down for nutrients other than vitamins and minerals. Given that probiotic bacteria fall within this definition of food supplements as used in this Directive, regulation of this kind can help to guarantee the safety and quality of the probiotic products [24].

3.4. Japanese approach

Japan is the only country that have legally defined and regulated functional foods, including probiotics, under the "Foods for Specific Health Use" (FOSHU) system by the Japanese Ministry of Health and Welfare [27]. The FOSHU system allows several health claims for probiotic bacteria include: 1) colonizes the intestines alive, 2) Increases the intestinal beneficial bacteria, 3) Inhibits harmful bacteria, 4) Maintains the balance of the intestinal flora, 5) Maintains the intestines good health, and 6) Promotes the maintenance of a good intestinal environment [28]. As a result, several probiotic products had received FOSHU approval in Japan [27]. FOSHU system requires the approval of the specific health claim prior to use, and this approval should be based on documented scientific evidences. FOSHU approved products are labeled for the specific health claim. In addition to approved FOSHU foods, many unapproved functional foods are available in Japan. These unapproved foods cannot carry an associated health claim but rely instead on consumer awareness of the probable health benefits of the ingredients [28].

3.5. Canadian approach

Health Canada (HC) and Natural Health Product Directorate (NHPD) which became a law in 2004 are the responsible regulators for food label and health claims in Canada [24,30]. Natural Health Products (NHPs) are considered as a subset of drugs under the Food and Drugs Act, and require assessment and licensing before being marketed. NHPs must be substantiated by sufficient evidence of safety and efficacy under recommended conditions of use, and must be manufactured under Good Manufacturing Practices. For HC/ NHDP, a probiotic is limited to nonpathogenic microorganisms, and is defined "as mono or mixed culture of live micro-organisms that benefit the microbiota indigenous to humans". Foods such as yogurt that contain "microbes" are controlled by the Food Products Directorate of HC. As with other food products regulated by HC, probiotic bacteria can carry a structure/function claim, a risk reduction claim, or a treatment claim. The amount and quality of the data to be supplied depend on the claim that is sought. The HC/NHPD regulations concerning probiotic bacteria have requirements related to toxicity and safety [23,31]. It is suggested to use a multidisciplinary approach to assess the pathological, genetic, toxicological, immunological, gastroenterological, and microbiological safety aspects of probiotic strains. Probiotic products in either capsule or liquid form as nutraceuticals' or as functional foods can be found in the marketplace in Canada today. It is not known how many petitions HC has received from companies related to probiotics. However, since its inception in 2004, HC/NHPD has not issued an approved health claim for any probiotic product [30].

4. Labeling requirements

Appropriate labeling and health claims are a pre-requisite for the consumer to make an informed choice. In addition to the general labeling requirements under the food laws of each country, necessary information should also be stated on the label [23,39]. Even though, that currently in most countries, only general health claims are labeled on foods containing probiotics, it is also recommended that specific health claims be allowed relating to the use of probiotics, where sufficient scientific evidence is available [22,25]. For example, the claim that a probiotic 'reduces the incidence and severity of rotavirus diarrhea in infants' would be more informative to the consumer than a general claim that probiotic bacteria' improve gut health'. Such specific health claims should be permitted on the label and promotional material. Also, it is the responsibility of the product manufacturer that an independent third party review by scientific experts in the field be conducted to establish that health claims are truthfully and not misleading labeled [20].

Hence, the following information must be displayed on the label:

1. Genus, species and strain: To clarify the identity of a probiotic present in food the microbial species must be stated on the label. Genus, species and strain designation should follow the standard international nomenclature. If the selection process has been undertaken, the identity of the strain should also be included since all probiotic effects are strain specific. Strain designation should not mislead consumers about the functionality of the strain.
2. Minimum viable numbers of each probiotic strain at the end of shelf life: The number of probiotic bacteria in food products should be clearly enumerated in order to include them on the label. The label should state the viable concentration of each probiotic present at the end of shelf life. The minimum efficacy level for each probiotic strain that to be maintained till the end of shelf life of product should be scientifically proven.
3. The serving size that delivers the effective dose of probiotic bacteria related to the health claim.
4. An accurate description of the physiological effect, as far as is allowable by law with the required scientific evidence.
5. Proper storage conditions including the temperature at which the product should be stored.
6. Corporate contact details for consumer information.
7. Safety in the conditions of recommended use.
8. Label information must not mislead the consumers to understand that consumption of the food, ingredient or nutrient of such food, can treat, relieve, cure or prevent a disease.

5. Probiotic food products

5.1. Dairy probiotic products

Dairy foods, fermented and non-fermented, have played important roles in the diet of humans worldwide for thousands of years. Since the observations of Mechinikov, in the early 1900s, there has been an increasing interest in the benefits of certain microorganisms; i.e. LAB and probiotic gut flora, and their effect on human general health, body functions, and life longevity. Currently hundreds of probiotic dairy products are manufactured and consumed around the world; typical examples include pasteurized milk, ice-cream, fermented milks, cheeses and baby milk powder [31-35]. The overall pattern of consumption of all types of probiotic dairy products is steadily expanding in the majority of countries in the world. The beneficial health claims are the main reasons behind the popularity and high consuming rates of these products in different communities. Milk is an excellent medium to carry or generate live and active cultured dairy products. The buffering capacity of milk helps to improve the survival of probiotic flora in the GI tract [35]. However, fermented foods remain the main vehicle to deliver probiotic bacteria [9,36]. Among the fermented milk products, yoghurt is by far the most popular and important vehicle for the delivery of probiotic bacteria [32,37]. Fermented dairy foods are well suited to promoting the positive health image of probiotic bacteria for several reasons: 1) fermented dairy foods already have a positive health image; 2) consumers have the fact that fermented foods contain living microorganisms (starter cultures); and 3) probiotic bacteria used as starter organisms combine the positive images of fermentation and probiotic cultures [38]. In probiotic fermented dairy products, viability of most of probiotic strains are affected as a result of antagonistic interaction between starter cultures and probiotic strains, as well as acid production in these cultured products [31,39]. As a result to these factors, a new trend in producing probiotic non-fermented dairy products has emerged. Wide range of probiotic non-fermented dairy products are produced and marketed by far, such as cheese, ice-cream, and fresh milk [31,33,40].

5.1.1. Fermented milks and yogurt (bio-yoghurt) probiotic products

For the maximum probiotic bacteria viability and optimal therapeutic effects, different types of food products were proposed as a carrier for probiotic bacteria by which consumers can take large amounts of viable probiotic cells. Yogurt, as a fermented milk product, is one of the most popular food carriers for the delivery of probiotic. Yogurt has long been recognized as a product with many desirable effects for consumers, and it is also important that most consumers consider yogurt to be 'healthy', add to that incorporation of probiotic bacteria, such as *L. acidophilus* and *B. bifidum*, into yogurt may add extra nutritional-physiological values [37,38]. Different types of yogurt and yogurt like products are manufactured around the world with different textures, including; natural-set yogurt, stirred yogurt, and drink yogurt, and these products differ greatly in their content of nonfat solids: 16–18%, 13–14%, and 11–12%, respectively [39].

Yogurt is a fermented milk product that has been prepared traditionally by allowing milk to ferment at 42–45°C. Modern yogurt production is a well-controlled process that utilizes ingredients of milk, milk powder, sugar, fruit, flavors, colorings, emulsifiers, stabilizers, and standard pure cultures of LAB (*Streptococcus thermophilus* and *L. bulgaricus*) to conduct the fermentation process. *S. thermophilus* and *L. bulgaricus* exhibit a symbiotic relationship during fermentation process of yogurt, with the ratio between the species changing constantly. The pH of commercial yogurt is usually in the range of 3.7–4.3 [38]. Recently new yogurt products, known as "Bio-Yogurt", have been manufactured by incorporating live probiotic strains in addition to the standard cultures, *S. thermophilus* and *L. bulgaricus,* into yogurt, since the recent discoveries in several aspects of bioscience support the hypothesis that, beyond nutrition, diet may modulate various functions in the body [32]. The Bio-Yogurt products have been formulated with different types of probiotic strains; mainly species of *Lactobacillus* and *Bifidobacteria*; include *L. acidophilus; L. casei; L. gasseri; L. rhamnosus; L. reuteri; B. bifidum; B. animalis; B. infantis; and B. longum* [32,34,35,41-43] Therefore, Bio-Yogurt is a yogurt that contains live probiotic cultures, the presence of which may give rise to claimed beneficial health effects. Different types of Bio-Yogurts are produced by far, including, plain, stirred, flavored, and fruits added Bio-Yogurts.

For the production of Bio-Yogurt, similar processing procedures of traditional yogurt are applied with the addition of live probiotic starter cultures. Heat treated, homogenized milk with increased protein content (3.6–3.8%) is inoculated with the standard starter cultures at 45°C and incubated for 3.5-5h [32]. The most common procedures of incorporation probiotic bacteria to Bio-Yogurt include: (1) addition of probiotic bacteria together with standard starter cultures; (2) two-step fermentation, which includes the fermentation of milk first with probiotic cultures to achieve high levels of viable cells, and then addition of standard starter cultures to complete fermentation; (3) two batches fermentation, in which two separate batches of pasteurized milk are fermented, one with probiotic cultures and the other with standard starter cultures, and then the two batches are mixed together; (4) the use of a probiotic alone as a starter culture. In this situation, the time of fermentation is generally higher than regular yogurt production using non-probiotic starter cultures [32,44]. However, the use of the probiotic bacteria alone in the production of yogurt was not sufficient to produce high quality product, where the pH values and the final characteristics (pH values 4.9-5.5, with poor curd formation) of yogurt manufactured by using probiotic species of Lactobacilli and Bifidobacteria, were unsatisfactory [32]. Probiotic bacteria generally tend to exhibit weak growth and acid production in milk, which invariably leads to long fermentation times and poor quality product. This may be due to the sensitive character of the microorganisms in these Bio-products, which adds to the usual difficulties encountered with novel food production (i.e. unusual palatability and consequent limited consumer acceptability) [45]. The poor quality and sensorial characteristics of Bio-Yogurt products are important challenges in probiotic industry [38]. To overcome the problem of the poor quality, two-step fermentation with mixed cultures of the probiotic bacteria and standard starter cultures was suggested. The use of the mixed cultures in the two-step fermentation resulted in yogurt with better acceptability and sensorial quality, and these include longer time for probiotic species to grow and multiply with making use of the traditional cultures to impart the traditional and favorable organoleptic characteristics [45]. Also, it is important to consider the effect of probiotic bacteria addition on the product sensorial characteristics, since metabolites produced by probiotic bacteria may lead to undesirable sensorial effects [46].

Different levels of probiotic bacteria in Bio-Yogurts have been recommended and specified, in order to exert the claimed health effects and considered as probiotic products. The National Yogurt Association (NYA) of the United States specifies that 10^8 cfu/mL of lactic acid bacteria at the time of manufacture, are required to use the NYA'Live and Active Culture' logo on the products containers [47]. In Japan, the Fermented Milks and Lactic Acid Bacteria Beverages Association has specified a minimum of 10^7 cfu/mL of bifidobacteria to be present in fresh dairy products as a standard [48]. Therefore, maintaining the probiotic bacteria viability and survivability during products manufacturing and storage is a very crucial factor for effective probiotic products. Different factors have been found to affect probiotic bacteria viability in Bio-Yogurt products, include, pH, oxygen residues, product composition, storage temperature, antagonistic activity among probiotic strains and with standard starter cultures. For example, survival of L. acidophilus is affected by the low pH of the yogurt [43], also, the addition of any ingredients, such as fruits or fruits constituents, that lower pH in yogurt may contribute to reduce the survivability of *L. acidophilus* [34]. Rapid loss of viability of *B. animalis subsp. lactis* was reported with increasing percentage of fruit pulp added into yogurt base [49]. Yogurt with high fat content inhibited probiotic cultures, particularly *B. bifidum* BBI [39]. Also, as the probiotic bacteria are oxygen sensitive, oxygen residues in yogurt has an inhibition effect on probiotic bacteria viability [45].

Bio-Yoghurts supplementation with different substances has showed variable effects on probiotic bacteria viability. The supplementation of Bio-Yogurt with ascorbic acid improved the viability of *L. acidophilus* in yoghurts [45]. Oxygen scavenging effect of ascorbic acid is one of the possible mechanisms that may help to improve the viability of probiotic bacteria. Moreover, due to their buffering capacity, the addition of whey protein may enhance the viability of some probiotic bacteria, especially in yogurts with added fruit pulp. Also, the incorporation of prebiotics (indigestible carbohydrates, such as fructooligosaccharides and inulin) [40], and neutraceuticals combination (isoflavones, phytosterols and omega-3-fatty acids) [28, 35] in yoghurt formulations seemed to stimulate the viability and activity of probiotic bacteria. Generally, prebiotics selectively stimulate the growth and activity of probiotic bacteria [20]. It was reported that incubation period, incubation temperature and storage time of yogurts affect probiotic bacteria viability [60]. On the other hand, as a result to oxygen incorporation into yogurts during stirring fruit pulp into yogurt base, stirred-yogurts have lower probiotic bacteria viability levels compared to plain-yogurts. Also, addition of cysteine at 250 and 500 mg/L to yogurt was associated with higher viability of L. acidophilus during manufacture and storage while viability of bifidobacteria was adversely affected by the same levels in different starter cultures, whereas, at level of 50 mg/L bifidobacteria demonstrated better viability. However, in mixed cultures Bio-Yogurt products, antagonistic and symbiotic interactions among probiotic cultures and between probiotic and standard starter cultures are very important factors affecting probiotic bacteria viability. The probiotic cultures must be compatible with each other and with the standard starter cultures, since these micro-organisms could produce inhibitory substances that damage each other and affect probiotic bacteria viability [40,51]. Different pattern of interactions have been demonstrated among different probiotic strains, include, strong, weak, and lack of inhibition [40]. Establishment of suitable combinations of mixed probiotic cultures, to guarantee the maximum probiotic bacteria viability and avoid any inhibitory effect during yogurt manufacturing and storage, requires the assessment of the pattern and the extent of interactions among the probiotic strains and the probiotic strains with the standard starter cultures.

5.1.2. Ice-cream and frozen probiotic products

Ice-cream is a frozen dairy product, consists of a mixture of components, include, milk, flavoring, sweeteners, stabilizers, and emulsifiers agents [52]. Several ice-cream related products, such as plain ice-cream, reduced fat, low fat, nonfat, fruit, and nut ice-creams, puddings, variegated, mousse, sherbet, frozen yoghurt, besides other frozen products are manufactured and marketed around the world [53]. Smoothness and softness are among the important physical criteria of ice-cream, and these criteria are conferred by vigorous agitation during freezing to incorporate air into frozen product [54]. Ice-cream is a highly appreciated product by people belonging to all age groups, include children, adults, and the elderly public, and by all social levels. Also, the ice-cream low acidity results in increased consumer acceptance, especially by those who prefer mild products.

During the last few decades, new type of the ice-cream products have been introduced to the markets, these products were developed by incorporating probiotic cultures into ice-cream products. The incorporation of probiotic cultures into ice-cream resulted in adding value to the ice-cream product and being considered as a functional product, in addition to being a rich food from the nutritional point of view, containing dairy based material, vitamins and minerals in its composition [33,52]. As a result to the composition/structure, manufacturing procedures, and storage conditions, ice-cream and frozen dairy desserts demonstrated great potential for use as vehicles for probiotic cultures. The ice-cream freezing storage temperature and low risk of temperature abuse during storage has leaded to higher viability of probiotic bacteria [54,56]. The ice-cream composition, which includes milk proteins, fat and lactose, as well as other compounds, make ice-cream a good vehicle for probiotic cultures. Also, ice-cream relatively high pH values (5.5 to 6.5) lead to an increased survival of the probiotic bacteria upon storage. Several studies showed the suitability of ice-creams as a vehicle for probiotic bacteria [33, 53].

The general steps involved in probiotic ice-cream manufacturing are: mixing the ingredients involved (milk, milk powder, sugar, emulsifiers, stabilizers); pasteurizing; cooling to a temperature of around 37–40°C, for the soured ice-cream, the freeze-dried starter cultures (usually yoghurt cultures) and the probiotic cultures is added; subsequent fermentation to a pH of 4.8–5.7, or the addition of a previously fermented inoculums containing both types of lactic cultures; cooling and keeping the mixture at 4°C for 24h for the maturation. Ice-cream mix is produced at this point. The mix is subsequently beaten/frozen, in order to produce the final product, which is packaged and maintained frozen throughout transport, commercial distribution, and storage for consumption. During all these steps after freezing, the temperature of the frozen product should be strictly controlled [33,53].

During probiotic ice-cream development, the ultimate aim of processes optimization is to enhance and maintain the probiotic survivability, so as to guarantee the product functional efficacy [55,56]. This includes the consideration of all the challenges involved in the production of conventional ice-cream. These challenges include: the ingredients microstructure and colloidal properties and/or components used in the formulation; the control of the ice crystallization; the choice of appropriate stabilizers; control of the fat destabilization and the emulsifier functionality [53,55]. Also, the incorporation of probiotic bacteria into an ice-cream products must not affect the product quality criteria, including physico–chemical parameters, such as the melting rate, and the sensory features, which must to be the same or even better than a conventional ice-cream. Ice-cream beaten, commonly known as overrun, is a process by which the air is incorporated into the product. Overrun is an intrinsic and compulsory step in the ice-cream processing, as it has a crucial impact on the physical properties and sensory acceptance of the ice-cream product, including, body lightness and the formation of a smooth structure, influencing characteristics such as the melt down and hardness properties. In fact, too little air gives the ice-cream a heavy, soggy body while too much air brings a fluffy body [57]. Therefore, overrun is a parameter that should be monitored in ice-cream formulations [58]. The overrun step, as a result to oxygen incorporation into the product, seems to affect the survival of probiotic cultures during processing and storage [33]. However, there is limited information about the effect of the overrun levels adopted during the processing of ice-cream on the survival of probiotic bacteria as well as the sensory acceptance of this kind of product. Recent reports indicated that higher overrun levels negatively influenced probiotic cultures; therefore it was recommended that lower overrun levels should be adopted during the manufacture of ice-cream in order to maintain its probiotic viability through the shelf life [56].

A decrease in the viability of some probiotic species during manufacturing and freezing of probiotic ice-cream was reported as a result to cells damage by freezing and thawing, mechanical stresses of mixing and overrunning during manufacturing and thereby exert a negative effect on functional efficacy of probiotic bacteria in frozen products after ingestion [57]. Addition of inulin and oligofructose demonstrated higher viability of *L. acidophilus* and B. lactis in ice-cream due to prebiotic effect. It is also found that viability of these probiotic bacteria may vary depending on the sugar levels of ice-cream [59,60]. In probiotic ice-cream development, great attention should be given to the other ingredients that are used in the product formulation, especially fruit pulp/juice, which give the product the final flavor. Fruits or their derivatives with a pronounced acidic character should be avoided in ice-creams containing probiotic cultures, since this attribute could influence their sensory acceptance and also decreased the viability of the cultures [61] as its addition decrease pH values. One of the strategies to ensure probiotic bacteria survivability in acidic products is to select acid resistant strains. A recent study suggests the addition of chemical compounds with buffering capacity – carbonate, and citrate salts – at acceptable levels before or during the incubation, in order to eliminate acidic stress [33]. However, fruits and/or flavorings additives with mild and low acidity ought to be used in ice-cream.

5.1.3. Cheese probiotic products

Cheese is the generic name for a group of fermented and non-fermented milk-based dairy products produced and consumed throughout the world in a great diversity of flavors, textures, and forms [62,63]. An essential part of the cheese making process is the curd formation, which involves the conversion of liquid milk into a solid mass that contains casein and fat of the milk. This is achieved by the addition of rennet or acid production by cheese starter cultures to coagulate the casein gel. Curd formation in rennet set cases is carried out through the action of chymosin on the k-casein steric stabilizing layer of the casein micelle. In cheese making, curd formation is usually followed by several processes such as pressing, salting and ripening. Many cheeses, known as ripened cheeses, need an additional time to ripen under controlled environmental conditions to achieve their own sensory features, particularly flavor and aroma [64]. All cheeses, whether rennet or acid set, can be classified as soft, semi-soft (semi-hard), hard, or very hard cheeses according to moisture contents [63].

As a result to the cultural aspects and the technologies involved with fermented milks and yogurt production, include, relatively short fermentation time, low pH values, oxygen residues, and antagonistic activity of yogurt starter cultures against probiotic bacteria, these cultured products may not be the optimal food carriers for probiotic bacteria to human, as this evidenced by poor probiotic bacteria viability in commercial yogurts [43,51]. In this case, cheese provides a valuable alternative as a food vehicle for probiotic delivery. Cheese high protein content provides probiotic bacteria with a good buffering protection against the high acidic condition in the gastrointestinal tract, and thus enhances probiotic bacteria survival throughout the gastric transit. Moreover, the dense matrix and relatively high fat content of cheese may offer additional protection to probiotic bacteria in the stomach. Also, the relatively high pH values and lack of antagonistic effects of starter cultures, in rennet set cheese may exert optimal conditions to maintain probiotic bacteria viability during cheese making and storage [31]. Accordingly, several soft, semi soft (semi hard), and hard probiotic cheese products have been developed and marketed in the last few years. Jordanian probiotic soft cheese was developed from goat's milk using L. acidophilus and L. reuteri [31]. Cheddar-like cheese was produced by using *B. infantis* [65]; whereas, cheddar cheese was produced by using *L. acidophilus, L. casei, L. paracasei and Bifidobacterium* spp. [66]. Also, probiotic bacteria of *Bifidobacterium, L. acidophilus* and *L. casei;* and *L. paracasei* A13 were used to produce Argentinian Fresco Cheese, respectively [39,67]. Moreover, it was shown that cheddar cheese is a good carrier to deliver Enterococcus faecium into the gastrointestinal tract of human [68]. Viability of probiotic bacteria during cheese processing and storage is the major challenge associated with the development of probiotic cheese. Probiotic bacteria should be technologically suitable for the incorporation into cheese products so that to retain both viability and functional efficacy during processing on a commercial scale and throughout consumption [69]. Furthermore, from a food processing perspective, it is desirable that such strains are suitable for large-scale industrial cheese production and withstand the processing conditions [70]. With regard to the development of probiotic cheese, this means that such strains should be grown to high cell level before addition into the cheese and/or be able to maintain viability during the manufacturing and/or ripening step [31,64]. In addition, a probiotic cheese should have the same sensory and nutritional qualities as the conventional cheese; the addition of probiotic cultures should not cause any loss in cheese quality. In this context, the level of proteolysis and lipolysis must be the same or even better than cheese which does not have probiotic bacteria [31,66].

Most of the probiotic cheeses have been developed by the addition of probiotic bacteria into cultured cheese [67,71]. In such products, viability of most probiotic strains was affected due to the antagonistic interaction between cheese starter cultures and probiotic bacteria, as well as acid production in these cultured products [64,69]. Compared to cultured type cheeses and due to its manufacturing process, fresh soft cheese seems to be ideally suited to serve as a carrier for probiotic bacteria as it is an un-ripened cheese, during storage it is submitted to refrigeration temperatures, and its shelf life is rather limited [31]. Fresh soft cheese is a semi-hard cheese and is manufactured in the Middle East and along the shores of the Mediterranean Sea [62]. Most of the soft cheeses are usually made by addition of rennet enzymes to pasteurized milk with no addition of starter cultures. Its pH is almost the same of original milk pH (6.3- 6.5). Moreover, soft cheese is very popular in many parts of world, because of its soft texture and favorable organoleptical characteristics [31]. As a result of these characteristics, soft cheese represents a promising vehicle to deliver probiotic to human. A number of scientific papers reporting the development of fresh cheeses containing recognized and potentially probiotic cultures have been published, which described suitable viable counts and a positive influence on the texture and sensorial properties of these cheeses [31,67]. Method of addition of probiotic bacteria into cheese has a crucial effect in the probiotic viability and functional efficacy during cheese processing and storage. There are two options for the addition of probiotic bacteria during cheese processing. First, probiotic bacteria can be added before the fermentation, together with the starter culture; second, after fermentation. In the first option, the optimal initial inoculum of probiotic to be added and the amount of probiotic which are lost in the whey during its drainage must be evaluated according to the process. In the second option, cheese must be cooled directly after probiotic addition, as metabolic activities of starters and probiotic bacteria are drastically controlled and reduced at these low temperatures. However, other methods for the addition of probiotic bacteria in a semi-hard cheese are the freeze-drying and spray-drying methods. These methods enhanced probiotic viability during cheese processing and storage via the protecting probiotic bacteria against different undesirable conditions encountered cheese processing [72].

Even though there is no specified level of probiotic bacteria in foods that would guarantee the biological activity, but it is increasingly recommended to ingest 10^8 CFU/day [1]. Having in mind that portions of around 100 g of cheese are usually consumed daily, populations of about 10^{6-7} CFU/g lead to an ingestion of 10^{8-9} CFU/daily portion. Addition of prebiotic substances was one of the valuable measures taken to maintain and enhance probiotic viability in cheese products. For example, addition of oligofructose and/or inulin to petit-suisse cheese enhanced the viability of both *L. acidophilus* and *B. animalis* subsp. *Lactis*, while addition of eucalyptus honey reduced the viability level of both probiotic bacteria in the same cheese. The low oligosaccharides content of honey may lead to poor growth and viability reported [73]. Moreover, inulin helps to improve the growth and viability of various probiotic species in a number of different products [50,73].

Also, probiotic bacteria used in food products, such as *Lactobacillus* and *Bifidobacterium* species are: oxygen sensitive or anaerobic; and acid and bile sensitive in nature [74]. Hence, the presence of oxygen, acid and bile may represent a threat for their survival. Several techniques have been applied to enhance and maintain the viability of probiotic bacteria under harsh conditions typical in cultured dairy products and cheeses, including the selection of probiotic strains tolerant to oxygen, acid and bile, the addition of amino acids and peptides [75]. Another strategy for enhancing bacterial tolerance to stress such as temperature, pH or bile salts is a prior exposure to sub-lethal levels of the given stresses. Stress responses may be used to enhance the survival of probiotic bacteria in stressful conditions and to improve their technological properties [76,77]. Moreover, another alternative for protecting probiotic bacteria to oxygen stress is the use of selected strains of *S. thermophilus* with high oxygen consumption rate as starter for the production of cheeses [75]. Salting of the curd, by immersing it in brine or rubbing salt on the surface is a common step in the manufacture of several varieties of cheeses. In several types of cheeses, specially ripened types, salt is added for preservative and sensorial purposes. However, this slat has an inhibition effect on the growth and the viability of probiotic bacteria in cheese [51]. It is well established that salt level is drastically reduce probiotic viability, especially when salt level is higher than 4% [78]. Therefore, processing of cheeses with high salt content should be optimized to minimize this inhibition effect of slat. Another option is to find ways to protect the probiotic bacteria from the hostile environment. One alternative is micro-encapsulation or cell incubation under sub-lethal conditions [79].

The packaging system is another important factor that is affecting probiotic viability and stability, especially during cheese storage stage. In general, probiotic dairy foods, including cheese, are packaged in plastics films which have different levels of permeability to oxygen. This becomes an important factor because most of the probiotic strains used in food are either oxygen sensitive anaerobic in nature. Therefore, oxygen low permeability plastic films should be used to pack these functional products; alternatively, the practice of adopting other alternatives, such as the use of vacuum packaging can be followed [80].

5.1.4. *Kefir*

Kefir is a traditional popular beverage consumed for thousands of years in the Central Asia and Middle East countries. It originates in the Caucasus Mountains in Central Asia. Kefir can be considered as natural probiotic fermented milk. It is an acidic-alcoholic fermented milk product, with uniform creamy consistency and a slight sour taste. Milk is fermented with kefir grains, small cluster of micro-organisms held together by a polysaccharide matrix named kefiran, and/or starter cultures prepared from grains [81]. Kefir grains look like pieces of coral or small clumps of cauliflower, which contain a complex mixture of lactic acid bacteria; Lactobacillus, Lactococcus, and Leuconostoc; acetic acid bacteria and yeast mixture [82]. Kefir grains usually contain lactose-fermenting yeasts; *Kluyveromyces lactis, K. marxianus* and *Torula kefir*; as well as non-lactose-fermenting yeasts *Saccharomyces cerevisiae* [81]. Yeasts are important in kefir fermentation because of the production of ethanol and carbon dioxide. *L. kefiri* is the dominant LAB in kefir, comprising about 80% of the LAB flora. The other 20% of the LAB flora in kefir comprises: *L. paracasei subsp. paracasei, L. acidophilus, L. bulgaricus, L. plantarum,* and *L. kefiranofaciens* [83].

The chemical and nutritional composition of kefir is variable and depends on the source and the fat content of milk, the composition of the grains or cultures and the technological process of kefir [84]. Kefir contains vitamins, minerals and essential amino acids that help the body with healing and maintenance functions and also contains easily digestible complete proteins. Kefir is rich in vitamins B1, B12, folic acid, vitamin K, and biotin, as well as calcium, magnesium, and phosphorus, beside essential amino acids such as tryptophan [83,84]. The benefits of consuming kefir in the diet are numerous. Kefir has frequently been claimed to be effective in improving several health and disease conditions, include cancer treatment, intestinal disorders, and promote bowel movement, constipation, flatulence, lactose intolerance [85]. Also, kefir antibacterial, anti-tumor, immunological, and hypocholesterolemic effects have been studied recently, and many reports indicated the efficacy of kefir products in possessing such effects [94,96-97].

Kefir beverages can be made from any type of milk; include, cow, goat or sheep, but commonly used is cow milk. Several substrates are produced in kefir aerobic fermentation includes lactic acid, acetic acid, CO_2 alcohol (ethanol) and aromatic compounds. These substrates provide kefir with its unique sensorial characteristics: fizzy, acid taste, tart and refreshing flavors [83]. There are several methods of kefir production. The traditional and industrial processes are the commonly used methods. The traditional method of making kefir involves the direct addition of kefir grains into milk. The raw milk is boiled and cooled to 20-25°C and inoculated with 2-10% (average of 5%) kefir grain. After 18-24h of fermentation, at 20-25°C, the grains are separated from the milk by filtering with a sieve. Grains can be dried at room temperature and kept at cold temperature to be used in the next inoculation. Kefir milk is cooled before consumption [81, 83]. In the industrial process of kefir, different methods with the same principle are usually applied to produce kefir. The first step is milk homogenization to 8% dry matter, and heating at 90-95°C for 5-10 minutes. Then cooling at 18-24°C and inoculate with 2-8% kefir grains and /or kefir starters in tanks. Fermentation time is 18 to 24h. The coagulum is pumped and distributed in bottles. After maturing either at 12-14°C or 3-10°C for 24h, kefir is stored at 4°C [81,83].

5.2. Non-dairy probiotic products

As mentioned earlier, dairy products are the main food carriers for probiotic bacteria to human. Limitations of these products such as the presence of allergens, high lactose and cholesterol contents, and the requirement for cold storage facilities have created the need to look for new probiotic product lines based on non-dairy substrates [88, 98]. Furthermore, the increase in the consumer vegetarianism throughout the developed countries generated an increasing demand for the vegetarian probiotic products, as well as the demand for new foods and tastes have initiated a trend in non-dairy probiotic product development [88, 89]. Accordingly, several ranges of non-dairy probiotic products have been developed and marketed in the last two decades. The market available non-dairy probiotic products include: fruits and vegetable, juices, non-dairy beverages, cereal based products, chocolate based products, meat…etc [88, 90-93].

Any new non-dairy probiotic food products should fulfill the consumer's expectancy and demands for the products that are pleasant and healthy; accordingly, the development process would be increasingly challenging [90, 95]. According to [94], new product development is a constant challenge for both scientific and applied research, and it has been observed that food design is essentially a problem of optimization to generate the best formulation. For this purpose, industries need to determine the basic formulation for each product, and the optimum levels of each ingredient to obtain the best sensorial and physicochemical criteria, chemical stability and shelf life, and reasonable price.

5.2.1. Fruits and vegetables probiotic products

Fruits and vegetables are considered healthy foods, as they contain several beneficial nutrients, such as minerals, vitamins, dietary fibers, and antioxidants. Unlike dairy products, fruits and vegetables lack allergens, lactose, and cholesterol, which adversely affect certain segments of the population [96]. Moreover, recent technologies advances have made alterations to some structural characteristics of fruits and vegetables matrices by modifying food components in a controlled way such as pH modification, and fortification of culture media, that might make fruits and vegetables ideal substrates for probiotic bacteria delivery to human [97] Accordingly, several type of probiotic fruits and vegetables products have been developed and marketed, such as fruits and vegetables juices, dried fruits, fermented vegetables, and vegetarian deserts [88,96-98].

As result to their pleasant taste and flavor, as well as acceptability by all age and economic groups, fruit and vegetables juices became one of the most studied, developed and consumed probiotic fruit and vegetable products [96,99,100]. Therefore, it is believed that there is a great potential in developing a new generation of non-dairy probiotic products through successful candidates that are chilled fruit juices and fermented vegetable juices [99,100]. Wide range of probiotic strains, mainly species of *Lactobacillus* and *Bifidobacteria,* such as *L. acidophilus, L. casei, L. paracasei, L. rhamnosus* GG, *L. plantarum, L. fermentum,* and *B. bifidum* have been widely used in the development of many fruit and vegetable products, specially juice products, include orange, pineapple, cranberry, cashew apple, tomato, cabbage, beet and carrot juices. These products have been tested for the suitability as carrier for probiotic bacteria, and the sensory acceptability by the consumer [96,99-101]. In the industrial scale, probiotic bacteria have been incorporated directly and in cell free form into these products. This practice was accompanied with the direct exposure of probiotic bacteria to the acidic conditions of juices and to other unfavorable process conditions, and consequently loose viability. Therefore, a special direct liquid inoculation system, that allows food producers to add the probiotic bacteria directly to the finished product, such as the innovated technology of Tetra Pak's aseptic dosing machine Flex Dos that allows the bacteria to be added to liquids just before they are filled into the cartons, is recommended to overcome the problems of direct inoculation [89]. This innovation is expected to significantly boost the market for the probiotic beverages, which have so far been restricted by the delicate nature of the ingredient and concerns over the contamination. Another challenge encountered the development and marketing of probiotic juices is the juices flavor and aroma. For example, unpleasant perfumery and dairy aromas, as well as sour and savory flavors have been observed in juices inoculated with *L. plantarum* It has been suggested that the perceptible off-flavors of probiotic orange juice, that often contribute to consumer dissatisfaction, may be masked by adding 10% (v/v) of tropical fruit juices [99].

However, variable patterns of probiotic bacteria viability have been demonstrated in fruit and vegetable juices. It was observed that probiotic's viability in different juices depends on the strains used, the characteristics of the substrate, the oxygen content and the final acidity of the product [45]. For example, when species of *Lactobacillus* and *Bifidobacterium* were added to orange, pineapple and cranberry juices, great differences were observed regarding the acid resistance, and all the strains survived for longer period in orange and pineapple juice compared to cranberry [96]. However, the micro-encapsulation technologies have been successfully applied using various matrices, such as agar, calcium pectate gel, chemically modified chitosan beads and alginates, to provide a physical barrier against unfavorable conditions to protect the probiotic cells from the damage caused by the external environment [100,102]. Vacuum impregnation is another technology applied to improve probiotic bacteria viability in fruit and vegetables products [103]. Using this technology, viability of *L. casei* was improved and sustained in dried apple slices for two months upon storage at room temperature. In this study, dried apple slices were immersed in probiotic cultures grown in liquid, usually natural juices, followed by applying a vacuum pressure of 50 mbar for 10 min, and then atmospheric pressure was restored leaving samples under the liquid for an additional 10 min period [97]. Moreover, fresh apple slices supplemented with *L. rhamnosus* GG was reported to represent a good vehicle for probiotic bacteria, as the probiotic bacteria maintained viability for 10 days at 2-4◦C [104]. Also, fermented table olive represents a potential carrier for delivery of L. paracasei IMPC 2.1 [91].

5.2.2. Cereals and soya probiotic products

Even though, that cereal nutritional quality, compared to milk and meat, is inferior because of their lower protein content, deficiency of certain essential amino acids (lysine), low starch availability, anti-nutrients substances (phytic acid, tannins and polyphenols) and the coarse nature of the grains, cereal grains are still considered as one of the most important food sources of protein, carbohydrates, vitamins, minerals and fiber for large segments of people all over the world [90]. Furthermore, cereal grains are good source of non-digestible carbohydrates that besides promoting several beneficial physiological effects can act as prebiotics that selectively stimulate the growth of *Lactobacilli* and *Bifidobacteria* in the colon [95]. Whole grains are also sources of many beneficial phytochemicals, including phytoestrogens, phenolic compounds, antioxidants, phytic acid and sterols [105].

Usually cereals are consumed either in a fresh or fermented states. There are a wide variety of traditional non-dairy fermented beverages produced around the world, most of them are non-alcoholic cereal beverages [101]. Even though, the non-dairy fermented cereal products have long been created throughout history for human nutrition, it just recently that probiotic characteristics of microorganisms involved in cereal foods fermentation have been evaluated. Examples of the traditional non-dairy cereal- based fermented beverages include, Boza, Tarhana, Kishk, Chicha, Kisra, Kenkey…etc. [89].

Several studies were carried out to develop probiotic cereal products, especially beverage type. The development of cereal based probiotic products requires the evaluation of the suitability of cereals as growth medium for probiotic bacteria. Probiotic bacteria, especially the strains of *Lactobacillus* and *Bifidobacteria,* have been recognized as complex microorganisms with high nutritional requirement, such as fermentable carbohydrates, amino acids, B vitamins, nucleic acids and minerals [74]. As mentioned earlier, cereals are good source for proteins, carbohydrates, vitamins, and minerals, beside their prebiotic content. These constituents may make cereals a suitable medium for probiotic bacteria growth. Beside that, fermented cereals, as a result to the fermentation process, may have more available nutrients for probiotic bacteria growth, such as improved protein quality and level of lysine, some amino acids may be synthesized, decreased level of carbohydrates as well as some non-digestible poly and oligosaccharides, and increased availability of group B vitamins, optimum pH conditions for enzymatic degradation of phytate and release minerals such as manganese (which is an important growth factor of LAB), iron, zinc and calcium [90]. Therefore fermentation of cereals may represent a cheap way to obtain a rich substrate that sustains the growth of probiotic bacteria. However, in the fermented cereal-based probiotic products, the antimicrobial activity of the LAB of the fermented cereals against added probiotic bacteria must also be considered and evaluated [92].

Several studies have been conducted to evaluate the suitability of different cereal grains to enhance probiotic bacteria growth and maintain their viability [88,92,108]. The oat-based, non-dairy products have been shown to enhance the survival of the probiotic strains *L. reuteri, L. acidophilus* and *B. bifidum*, all of human origin, upon storage at 6°C up to 30 days. These products were fermented by the three strains with and without the commercial yogurt culture. Products fermented in presence of yogurt culture showed lower probiotic bacteria viability compared to product fermented with probiotic bacteria solely. Yosa, a new probiotic oat-based fermented food, similar to flavored yogurt or porridge, contains LAB and bifidobacteria [90]. Yosa is considered as a healthy food due to its content of oat fiber and probiotic LAB, which in combination with the effect of b-glucane might reduce cholesterol and the effect of LAB in maintaining and improving the environment in the intestinal balance of the consumer. Maize, one of the most important sources of food for millions of people, particularly in Latin America and Africa. A maize porridge made of maize flour and barley malt, with high energy density and low viscosity, was fermented with four probiotic strains *L. reuteri, L. acidophilus* (2 strains) and *L. rhamnosus GG*. All strains exhibited a strong growth upon fermentation and storage [88], suggesting that maize porridge supplemented with barely malt is a good medium for probiotic growth. Also, and as a result to the desirable fruity flavors of fermented maize foods, probiotic fermented maize products could have a good world-wide acceptance. Rice is the major cereal in Asia, and its products could be an economical and beneficial medium to develop probiotic foods. The growth of four probiotic bacteria (*L. acidophilus, L. pentosus, L. plantarum* and *L. fermentum*) was found to be higher in germinated rough rice powder (5%, w/w) mixed with water than in only rice powder with added NaCl. Germinated rice grains found to have an increased content of reducing sugars, total protein and vitamins, mainly B vitamin, which is a very important element required for the growth of *L. plantarum* [74].

Soybean, the most important legume in the traditional Asian diet, is rich in high-quality protein. The products of soybean play an important role in the prevention of chronic diseases such as menopausal disorder, cancer, atherosclerosis, and osteoporosis [107]. Experiments studying the survival of probiotics indicate that soy products, include, soymilk, soy-based yogurt, vegetarian frozen desert, fermented soy tempeh, and soy cheese, are a good substrate for the growth of probiotic bacteria [88,92,106,109]. Soy yogurts were prepared with a yogurt starter in conjunction with either the probiotic bacteria *L. johnsonii, L. rhamnosus* GG or human derived *Bifidobacteria*. Probiotic frozen vegetarian soy deserts were developed with the incorporation of *L. acidophilus, L. rhamnosus, L. paracasei* ssp. *paracasei, Saccharomyces boulardii* and *B. lactis* [108]. The neutral pH of the frozen soy dessert improved the probiotic survivability since some probiotic organisms are susceptible to inactivation when stored in acidic conditions [31]. Moreover, it was reported that soymilk fermentation with probiotic bacteria (strains of *Lactobacillus* and *Bifidobacteria*) increased the antioxidative activities of the fermented soymilk, and this further increases the potential of developing a probiotic diet adjunct with probiotic fermented soymilk [88]. Recently, a new probiotic soy based cheese was developed on the basis of Chinese sufu [106]. The soy cheese was made from soymilk fermented with soy cheese bacterial starter cultures and *L. rhamnosus*. The probiotic strain showed good growing pattern during soy cheese fermentation, and good survivability upon storage

5.2.3. Meat probiotic products

Meat is a highly nutritious food with a high degree of nutrients bioavailability and consumers have a high degree of preference for its taste, flavor, and texture. Meat had shown an excellent vehicle for probiotics as a result to meat composition and structure. Furthermore, meat was found to have a protection effect on LAB against the lethal action of bile [109]. One of the most studied and processed probiotic meat products is the dry fermented sausages without heating. Beside the high nutritional value, the characteristics of this type of meat product make it an ideal food matrix for probiotic delivery to human, as, it is a fermented product so the addition of probiotic bacteria will not alter the product sensorial characteristics, also, it is not heat treated, and so the viability of probiotic bacteria will not be reduced. These fermented products are prepared from seasoned, raw meat that is stuffed in casings and is allowed to ferment and mature by LAB starter cultures. The currently commonly employed LAB strains in meat starter cultures include *L. casei, L. curvatus, L. pentosus, L. plantarum, L. sakei, Pediococcus acidilactici* and *P. pentosaceus* [110]. The incorporation of microorganisms that have probiotic criteria is receiving increasing interest. However, few reports so far are available concerning the incorporation of probiotic bacteria into dry fermented sausages. *L. gasseri* JCM1131 has been demonstrated to be useful as a potential probiotic strain for application in meat fermentation and improving its safety

[111]. The efficacy of *L. rhamnosus* FERM P-15120 and *L. paracasei* subsp. *paracasei* FERM P-15121 has also been reported, as potential probiotics in meat products [112]. A mixed culture of the traditional starter culture and a potential probiotic culture of *L. casei* LC-01 or *B. lactis* Bb-12 have been successfully employed in sausage production [113].

The importance of using probiotic bacteria from the meat dominant strains supports the demand for higher numbers of viable cells at the time of consumption, which is a prerequisite for the probiotic to insure beneficial effects on the host. Furthermore, the use of a probiotic starter culture would prove superior in providing more safety, taste and health benefits, as compared to the traditional cultures [114]. LAB strains, include *L. acidophilus, L. crispatus, L. amylovorus, L. gallinarum, L. gasseri,* and *L. johnsonii,* were found to be suitable for meat fermentation and to enhance product safety [111]. Also, it has been reported that the selection of *L. plantarum* and *L. pentosus* isolated from Scandinavian-type fermented sausage as a promising probiotic meat starter cultures [121]. Moreover, *L. plantarum* and *L. curvatus* strains isolated from Greek dry-fermented sausages were resistant to 0.3% bile salts [116].

Various studies have shown that probiotic organisms survive poorly in fermented foods [117]. Nonetheless, probiotic organisms may be encapsulated by the sausage matrix consisting of meat and fat. Alginate-microencapsulated probiotics (*L. reuteri* and *B. longum*) may be an option in the formulation of fermented meat products such as sausages with viable health-promoting bacteria; nevertheless, their inhibitory action against some pathogen organisms could be reduced [118]. *B. longum* and *L. reuteri* encapsulated in Alginate were a suitable option for this purpose. Recently, *B. longum* was successfully protected in-vivo and in-vitro by encapsulation in innovated encapsulation material of succinylated β-lactoglobulin tablets [119].

5.2.4. Chocolate probiotic products

Chocolate is one of the most popular products all over the world, due to its delicious taste and flavor, high nutritious energy, fast metabolism and good digestibility. The presence of cocoa butter, milk and milk based materials, as well as sugar in its composition can be the warranty of an appropriate ingestion of proteins, carbohydrates, fats, minerals and vitamins [120]. Chocolate in its original form has long been known to lift mood, increase mental activity, to control appetite, and improve heart health. However, the high sugar content of conventional brands has raised concerns that their consumption is contributing to the current obesity epidemic, to osteoporosis development in older women, and the raising diabetes incidence in the Western industrialized nations. Nowadays, one of the most important trends in chocolate manufacturing is originated by the consumers' demand of functional or health-promoting chocolate, i.e., chocolate that not only do not adversely consumer health, but also remedy or prevent illnesses such as heart disease, osteoporosis, cancer, diabetes…etc. [121,122] Chocolate itself is a functional food, as it contains sufficient polyphenolic antioxidants and flavonoids compounds. These beneficial compounds in chocolate have been attributed to chocolate health beneficial effects. However, it is now possible for manufacturers to create functional foods by fortifying and enhancing their products to give them added health benefits have never been possible before, by incorporating probiotic bacteria to chocolate products [120] Developing a Probiotic chocolate product that is affordable and also nutritional for many more people is a challenge. The application of probiotic bacteria into chocolate could offer a good alternative to common dairy products, and allow broadening the health claims of chocolate based food products. Indeed, recent market research on functional food has shown that, in relation to chocolate, digestive health was one of the most important drivers of consumer acceptance [122,123].

The development of probiotic-containing chocolate involves a good understanding of the selected probiotic strains, the chocolate manufacturing process and the different critical points of the process for probiotic survival, as well as the application of specific protective technology [123]. Viability of probiotic bacteria in a product at the point of consumption is an important consideration for the efficacy, as they have to survive during the processing and shelf life of food and supplements, transit through high acidic conditions of the stomach and enzymes and bile salts in the small intestine [95]. Moreover, the sensorial acceptability of the product from the consumer is another limiting factor that determines the success of the product [124]. A few numbers of attempts were made to develop probiotic chocolate products so far. Recently, a chocolate mousse was developed by using probiotic and prebiotic ingredients. Probiotic and synbiotic chocolate mousses were supplemented with *L. paracasei* subsp. *paracasei* LBC 82, solely or together with the prebiotic ingredient inulin [122] It was shown that the chocolate mousse was an excellent vehicle for the delivery of *L. paracasei*, as it enhanced probiotic bacteria growth and viability during chocolate mousse processing and shelf life, and the prebiotic ingredient inulin did not interfere in its viability, as well as the addition of the probiotic microorganism and of the prebiotic ingredient did not interfere in the sensorial preference of the product. Moreover, another chocolate product was evaluated to support the growth and survivability of *L. rhamnosus* IMC 501 and *L. paracasei* IMC 502 mixed 1:1 (SYNBIO). The survival and viability of probiotics were determined during the product processing and shelf-life. The values of viable probiotic bacteria showed that this product could represent an ideal vehicle for probiotic bacteria to human [123].Furthermore, a chocolate product has been evaluated as a potential protective carrier for oral delivery of a microencapsulated mixture of *L. helveticus* CNCM I-1722 and *B. longum* CNCMI-3470 [124], the data in this study indicated that the coating of the probiotics in chocolate is an excellent solution to protect them from environmental stress conditions and for optimal delivery.

Author details

Saddam S. Awaisheh
Associate Professor of Food Science, Department of Food Sciences & Nutrition,
Al-Balqa Applied University

6. References

[1] Food and Agriculture Organization of the United Nations and World Health Organization. Report of a Joint FAO/WHO Working group on Drafting Guidelines for the Evaluation of Probiotics in Food, London, Ontario, Canada 2002. Available at: ftp://ftp.fao.org/es/esn/food/wgreport2.pdf. Accessed 15 April 2012.

[2] Kopp-Hoolihan L. Prophylactic and Therapeutic Uses of Probiotics: A Review. Journal of American Dietary Association 2001;101(2) 229-241.

[3] Suvarna VC, Boby UV. Probiotics in Human Health: A Current Assessment. Current Science 2005;88 1744-1748.

[4] Metchnikoff E. The prolongation of Life. London, UK: William Heinemann, 1907.

[5] Lilley DM, Stillwell RH. Probiotics: Growth promoting factors produced by microorganisms. Food Science 1965;147 747–748.

[6] Huis in't Veld JHJ, Havenaar R. Selections Criteria and Application of Probiotic Microorganisms in Man and animal. Microecology Therapy 1997;26 43–58.

[7] Guarner F, Schaafsma GJ. Probiotics. International Journal of Food Microbiology 1998;39 237-238.

[8] Zeng XQ, Pan DD, Guo YX. The Probiotic Properties of *Lactobacillus buchneri* P2. Journal of Applied Microbiology 2010;108 2059-2066.

[9] Parvez S, Malik KA, Kang KA, Kim HY. Probiotics and Their Fermented Food Products Are Beneficial For Health-Review. Journal of Applied Microbiology 2006;100 1171–1185.

[10] Sanders ME, Morelli L, Tompkins TA. Spore Formers as Human Probiotics: *Bacillus, Sporolactobacillus,* and *Brevibacillus*. Comprehensive Review in Food Science and Food Safety 2003;2 101-110.

[11] Naidu AS, Bidlack WR, Clemens RA. Probiotic Spectra of Lactic Acid Bacteria. Critical Reviews in Food Science and Nutrition 1999;39 113-126.

[12] Saxelin M, Rautelin H, Salminen S, Makela PH. The Safety of Commercial Products with Viable *Lactobacillus* Strains. Infectious Diseases in Clinical Practice 1996;5 331-335.

[13] Sanders ME. Probiotics. Food Technology 1999;53(11) 67-77.

[14] Aguirre M, Collins MD. Lactic Acid Bacteria and Human Clinical Infection. Journal of Applied Bacteriology 1993;75 95–107.

[15] Gilliland SE. Health and Nutritional Benefits from Lactic Acid Bacteria. FEMS Microbiology Reviews 1990;87 175-188.

[16] Salminen MK, Järvinen A, Saxelin M, Tynkkynen S, Rautelin H, Valtonen V. Increasing Consumption of *Lactobacillus* GG as A Probiotic and the Incidence of Lactobacilli Bacteraemia in Finland. Clinical Microbiology and Infection 2001;7(Suppl 1) 802-808.

[17] Holzapfel WH, Haberer P, Snel J, Schillinger U, Huis in't Veld JH. Overview of Gut Flora and Probiotics. International Journal of Food Microbiology 1998;41 85-101.

[18] Ishibashi N, Yamazaki S. Probiotics and Safety. American Journal of Clinical Nutrition 2001;73(2 Suppl) 465S-470S.

[19] Adams MR, Marteau P. On the Safety of Lactic Acid Bacteria from Food. International Journal of Food Microbiology 1995;27 263–264.

[20] Salminen S, Bouley C, Boutron-Ruault MC, Cummings JH, Franck A, Gibson GR, Isolauri E, Moreau MC, Roberfroid M, Rowland I. Functional Food Science and Gastrointestinal Physiology and Function. British Journal of Nutrition 1998;80 S147-S171.

[21] Reid G, Zalai C, Gardiner G. Urogenital Lactobacilli Probiotics, Reliability, and Regulatory Issues. Journal Dairy Science 2001;84(E. Suppl.) E164-E169.

[22] Pineiro M, Stanton C. Probiotic Bacteria: Legislative Framework—Requirements to Evidence Basis. Journal of Nutrition 2007;137 850S–853S.

[23] Vanderhoof JA, Young R. Probiotics in the United States. Clinical Infectious Diseases 2008;46 S67–72.

[24] Sanders ME, Tompkins T, Heimbach JT, Kolida S. Weight of Evidence Needed to Substantiate a Health Effect for Probiotics and Prebiotics, Regulatory Considerations in Canada, EU, and US. European Journal of Nutrition 2004;44 303–310.

[25] Venugopalan V, Shriner KA, Wong-Beringer A. Regulatory Oversight and Safety of Probiotic Use. Emerging Infectious Diseases 2010;16(11) 1661-1665.

[26] Food and Agriculture Organization of the United Nations and World Health Organization. Report of a Joint FAO/WHO. Expert Consultation on Evaluation of Health and Nutritional Properties Of Probiotics In Food Including Powder Milk With Live Lactic Acid Bacteria, Córdoba, Argentina. 1-4 October 2001. Available at:ftp://ftp.fao.org/docrep/fao/meeting/009/y6398e.pdf. Accessed 14 April 2012.

[27] Dietary Supplement Health and Education Act 1994; Pub L. No.103-417.

[28] Sanders ME, Huis in't Veld J. Bringing a Probiotic-Containing Functional Food to the Market: Microbiological, Product, Regulatory and Labeling Issues. Antoine Van Leeuwenhoek 1999;76 293–315.

[29] Yamada K, Sato-Mito N, Nagata J, Umegaki K. Health Claim Evidence: Requirements in Japan. Journal of Nutrition 2008;138 1192S–1198S.

[30] Health Canada. Evidence for safety and efficacy of finished natural health products. Section additional requirements for probiotics-7.2.1 safety considerations 2004. Available at: http://hc-sc.gc.ca/dhp-mps/prodnatur/legislation/docs/efe-paie_e.html#72. Accessed 10 March 2012.

[31] Health Canada. List of licensed natural health products. 2007. Available at: http://www.hc-sc.gc.ca/dhp-mps/prodnatur/applications/licen-prod/lists/listapprnhp-listeapprpsn_e.html. Accessed 10 March 2012.

[32] Awaisheh SS. Development of Probiotic Soft Cheese Manufactured Using Goat's Milk With the Addition of Thyme. MilchWissenSchaft 2011;66(1) 51-54.

[33] Awaisheh SS, Hadaddin MS, Robinson RK. Incorporation of Selected Nutraceuticals and Probiotic Bacteria Into Fermented Dairy Product. International Dairy Technology 2005;10 1189-1195.

[34] Cruz AG, Antunes AE, Sousa ALOP, Faria JAF, Saad SMI. Ice-Cream as a Probiotic Food Carrier. Food Research International 2009;42 1233-1239.

[35] Kailasapathy K, Phillips HM. Survival of *Lactobacillus acidophilus* and *Bifidobacterium animalis ssp. lactis* in Stirred Fruit Yogurts. LWT-Food Science Technology 2008;41 1317–1322.

[36] Tamime AY, Marshall V, Robinson RK. Microbiological and Technological Aspects of Milks Fermented by *Bifidobacteria*. Journal of Dairy Research 1995;62 151-187.

[37] Stanton C, Gardiner G, Meehan H, Collins K, Fitzgerald G, Lynch PB, Ross RB. Market Potential for Probiotics. *The* American Journal of Clinical Nutrition 2001;73(supplement) 476s-483s.

[38] Lourens-Hattingh A, Viljoen BC. Yogurt as Probiotic Carrier Food: A Review. International Dairy Journal 2001 ;11 1-17.

[39] Heller KJ. Probiotic Bacteria in Fermented Foods: Product Characteristics and Starter Organisms. Journal of Clinical Nutrition 2001;73(suppl) 374S-379S.

[40] Vinderola CG, Prosellon W, Ghiberto D, Reinheimer JA. Viability of Probiotic (*Bifidobacterium, Lactobacillus acidophilus* and *Lactobacillus casei*) and Non-Probiotic Microflora in Argentinian Fresco Cheese. Journal of Dairy Science 2000;83 1905–1911.

[41] Awaisheh SS, Al-Dmoor HM, Omar SS, Hawari A, Al-Rwaily MM. Impact of Selected Nutraceuticals on Viability of Probiotic Strains in Milk During Refrigerated Storage at 4°C for 15 Days. International Journal of Dairy Technology 2012;65(2) 268-273.

[42] Anderson J, Gilliland SE. Effect of Fermented Milk (Yogurt) Containing *Lactobacillus Acidophilus* L1 on Serum Cholesterol in Hypercholesterolemic Humans. Journal of American Collection of Nutrition 1999;18 43-50.

[43] Hekmat S, Soltani H, Reid G. Growth and Survival of *Lactobacillus reuteri* RC-14 and *Lactobacillus rhamnosus* GR-1 in Yogurt for Use as a Functional Food. Innovative Food Science Emerging Technology 2009;10 293–296.

[44] Kailasapathy K. Survival of Free and Encapsulated Probiotic Bacteria and Their Effect on the Sensory Properties of Yoghurt. LWT-Food Science and Technology 2006;39 1221-1227.

[45] Tamime AY, Saarela M, Korslund-Søndergaard A, Mistry VV. and Shah NP. Production and Maintenance of Viability of Probiotic Micro-Organisms in Dairy Products. In: Tamime AY. (ed.) Probiotic Dairy Products. Blackwell Publishing, Oxford, UK: 2005; P44-51.

[46] Dave RI, Shah NP. Effectiveness of Ascorbic Acid as an Oxygen Scavenger in Improving Viability of Probiotic Bacteria in Yoghurts Made with Commercial Starter Cultures. International Journal of Dairy Technology 1997;7 435-443.

[47] Ostlie H, Helland MH, Narvhu J. Growth and Metabolism of Probiotics in Fermented Milk. International Journal of Food Microbiology 2003;87 17-27.

[48] Ishibashi N., Shimamura S. Bifidobacteria: Research and Development in Japan. Journal of Food Technology 1993;47(6) 126, 129–134.

[49] Nighsowonger BD, Brashears MM, Gilliland SE. Viability of *Lactobacillus acidophilus* and *Lactobacillus casei* in Fermented Milk Products during Refrigerated Storage. Journal of Dairy Science 1996; 79:212-219.

[50] Capela P, Hay TK, Shah NP. Effect of Cryoprotectants Prebiotics and Microencapsulation on Survival of Probiotic Organisms in Yoghurt and Freeze Dried Yoghurt. Food Research International 2006;39 203-211.

[51] Vinderola CG, Mocchiutti, P, Reinheimer AJ. Interactions Among Lactic Acid Starter and Probiotic Bacteria Used for Fermented Dairy Products. Journal of Dairy Science 2002;85 721–729.

[52] Marshall RT, Goff HD, Hartel RW. Ice Cream. New York: Springer; 2003.

[53] Goff D. 65 Years of Ice-Cream Science. International Dairy Journal 2008;18(7) 754–758.

[54] Marshall RT, Arbuckle WS. Ice-Cream. New York: Chapman & Hall; 1996.

[55] Akin MS. Effects of Inulin and Different Sugar Levels on Viability of Probiotic Bacteria and the Physical and Sensory Characteristics of Probiotic Fermented Ice Cream. MilchWissenSchaft 2005;60(3) 297–300.

[56] Ferraz JL, Cruz AD, Cadena RS, Freitas MQ, Pinto UM, Carvalho CC, Faria JAF, Bolini HMA. Sensory Acceptance and Survival of Probiotic Bacteria in Ice Cream Produced with Different Overrun Levels. Journal of Food Science 2011;71(1) S24-S28.

[57] Alamprese C, Foschino R. Technology and Stability of Probiotic and Prebiotic Ice Creams. In: Shah NP. Cruz AG. Faria JAF. (Ed.) Probiotic and Prebiotic Foods: Technology, Stability and Benefits to Human Health. New York: Nova Publisher; 2011. P235–98.

[58] Sung KK, Goff HD. Effect of Solid Fat Content on Structure in Ice Creams Containing Palm Kernel Oil and High-Oleic Sunflower Oil. Journal of Food Science 2011;75 274-9.

[59] Akalin AS, Erisir D. Effects of Inulin and Oligofructose on the Rheological Characteristics and Probiotic Culture Survival in Low-Fat Probiotic Ice-cream. Journal of Food Science 2008;73 184-188.

[60] Akın MB, Akın MS, Kırmacı Z. Effects Of Inulin and Sugar Levels on The Viability of Yogurt and Probiotic Bacteria and the Physical and Sensory Characteristics in Probiotic Ice Cream. Food Chemistry 2007;104 93–99.

[61] Favaro-Trindade CS, De Carvalho-Balieiro JC, Dias PF, Sanino FA, Boschini C. Effects of Culture PH and Fat Concentration on Melting Rate and Sensory Characteristics of Probiotic Fermented Yellow Mombin (*Spondias mombin* L) Ice Creams. Food Science and Technology International 2007;13 285–291.

[62] Davis JG. Cheese Manufacturing Methods. Churchill, Livingstone London, 1976; vol.(3)891–896.

[63] Fox PF, Guinee TP, Cogan TP, McSweeney PLH. Fundamentals of Cheese Science. Gaithersburg: Aspen. 2000. p.1-9.

[64] Ong L, Shah NP. Probiotic Cheddar Cheese: Influence of Ripening Temperatures on Survival of Probiotic Microorganisms, Cheese Composition And Organic Acid Profiles. LWT-Food Science Technology 2009;42 1260–1268.

[65] Ross RP, Fitzgerald G, Collins K, Stanton C. Cheese Delivering Biocultures-Probiotic Cheese. The Australian Journal of Dairy Technology 2002;57 71-78.

[66] Ong L, Henriksson A. and Shah NP. Development of Probiotic Cheddar Cheese Containing *Lactobacillus acidophilus, Lactobacillus casei, Lactobacillus paracasei* and *Bifidobacterium* spp. and the Influence of These Bacteria on Proteolytic Patterns and Production of Organic Acid. International Dairy Journal 2006;16 446–456.

[67] Vinderola G, Prosello W, Molinari F, Ghiberto D, Reinheimer J. Growth of *Lactobacillus paracasei* A13 in Argentinian Probiotic Cheese and Its Impact on The Characteristics of The Product. International Journal of Food Microbiology 2009;135 171-174.

[68] Gillian E, Gardiner, Bouchier P, O'Sullivan E, Kelly J, Collins JK, Fitzgerald G, Ross RP, Stanton C. A Spray-Dried Culture for Probiotic Cheddar Cheese Manufacture. International Dairy Journal 2002;12 749–756.

[69] Fortin M-H, Champagne CP, St-Gelais D, Britten M, Fustier P, Lacroix M . Effect of Time of Inoculation, Starter Addition, Oxygen Level and Salting on The Viability of Probiotic Cultures During Cheddar Cheese Production. International Dairy Journal 2011; 21 75-82.

[70] Stanton C, Desmond C, Coakley M, Collins JK, Fitzgerald G, Ross RP. Challenges Facing Development of Probiotic-Containing Functional Foods. In: Farnworth ER. (ed.) Handbook of Fermented Functional Foods. Boca Raton; 2003. P50-79.

[71] Songisepp E, Kullisaar T, Hutt P, Elias P, Brilene T, Zilmer M, Mikelsaar M. A New Probiotic Cheese with Antioxidative and Antimicrobial Activity. Journal of Dairy Science 2004; 87 2017-2023.

[72] Bergamini CV, Hynes ER, Quiberoni A, Sua´rez VB, Zalazar CA. Probiotic Bacteria as Adjunct Starters: Influence of the Addition Methodology on Their Survival in A Semi-Hard Argentinean Cheese. Food Research International 2005; 38(5) 597-604.

[73] Cardarelli HR, Buriti FC, Castro IA, Saad SM . Inulin and Oligofructose Improve Sensory Quality and Increase the Probiotic Viable Count in Potentially Synbiotic Petit-Suisse Cheese. LWT-Food Science Technology 2008; 41(6) 1037-1046.

[74] Gomes MP, Malcata FX. *Bifidobacterium* spp. and *Lactobacillus acidophilus*: Biological, Biochemical, Technological and Therapeutical Properties Relevant for Use as Probiotics. Trends in Food Science and Technology 1999;10 139-157.

[75] Boylston TD, Vinderola CG, Ghoddusi HB, Reinheimer JA. Incorporation of Bifidobacteria into Cheeses: Challenges and Rewards. International Dairy Journal 2004; 14 375-387.

[76] Roy D. Technological Aspects Related to the Use of *Bifidobacteria* in Dairy Products 2005;85(1-2) 39-56.

[77] Saarela M, Rantala M, Hallamaa K, Nohynek L, Virkajarvi I, Matto J. Stationary-Phase Acid and Heat Treatments for Improvement of the Viability of Probiotic *Lactobacilli* and *Bifidobacteria*. Journal of Applied Microbiology 2004;96(6) 1205-1214.

[78] Gobbetti M, Corsetti A, Smacchi E, Zocchetti A, de Angelis A. Production of Crescenza Cheese by Incorporation of Bifidobacteria. Journal of Dairy Science 1998;81(1) 37-47.

[79] Ozer B, Kirmaci HA, Sxenel E, Atamer M, Hayaloglu A. Improving the Viability of *Bifidobacterium bifidum* Bb-12 and *Lactobacillus acidophilus* La-5 in White-Brined Cheese By Microencapsulation. International Dairy Journal 2009; 19(1) 22-29.

[80] Kasımoglu A, Goncuoglu M, Akgun S. Probiotic White Cheese with *Lactobacillus acidophilus*. International Dairy Journal 2004;14(12) 1067-1073.

[81] Farnworth ER. Kefirda complex probiotic. Food Science and Technology Bulletin Functional Foods 2005;2 1-17.

[82] Guzel-Seydim Z, Twyffels J, Seydim C, Greene K . Turkish Kefir and Kefir Grains: Microbial Enumeration and Electron Microscobic Observation. International Journal of Dairy Technology 2005;58 25–29.

[83] Chen TH, Wang SY, Chen KN, Liu JR, Chen MJ. Microbiological and Chemical Properties of Kefir Manufactured by Entrapped Microorganisms Isolated From Kefir Grains. Journal of Food Science 2009;92 3002-3013.

[84] Zubillaga M, Weill R, Postaire E, Goldman C, Caro R, Boccio J. Effect of Probiotics and Functional Foods and Their Use in Different Diseases. Nutrition Research 2001;21 569-579.

[85] Hosono A, Tanabe T, Otani H. Binding Property of Lactic Acid Bacteria Isolated From Kefir Milk With Mutagenic Amino Acid Pyrolyzates. MilchWissenSchaft 1990;45 647-651.

[86] Tamai Y, Yoshimitsu N, Watanabe Y, Kuwabara Y, Nagai S. Effects of Milk Fermented by Culturing with Various Lactic Acid Bacteria and A Yeast on Serum Cholesterol Level in Rats. Journal of Fermentation and Bioengineering 1996;81 181-182.

[87] Zacconi C, Parisi MG, Sarra PG , Dallavalle P, Bottazzi V. Competitive Exclusion of *Salmonella Kedougou* in Kefir Fed Chicks. Microbiological Alim. Nutrition 1995; 12 387-390.

[88] Heenan CN, Adams C, Hoskena RW, Fleet H. Survival and Sensory Acceptability of Probiotic Microorganisms in A Nonfermented Frozen Vegetarian Dessert. LWT-Food Science and Technology 2004;37 461-466.

[89] Prado FC, Parada JL, Pandey A, Soccol CR. Trends in Non-Dairy Probiotic Beverages. Food Research International 2008; 41 111–123.

[90] Blandino A, Al-Aseeri ME, Pandiella SS, Cantero D, Webb C. Cereal-Based Fermented Foods and Beverages: Review. Food Research International 2003;36 527–543.

[91] De Bellis P, Valerio F, Sisto A, Lonigro SL, Lavermico P. Probiotic Table Olives: Microbial Populations Adhering on Olive Surface in Fermentation Sets Inoculated with the Probiotic Strain *Lactobacillus Paracasei* IMPC2 in An Industrial Plant. International Journal of Food Microbiology 2010;140 6-13.

[92] Farnworth ER, Mainville I, Desjardins MP, Gardner N, Fliss I, Champagne C. Growth of Probiotic Bacteria and Bifidobacteria in A Soy Yogurt Formulation. International Journal of Food Microbiology 2007;116 174–181.

[93] Helland MH, Wicklund T, Narvhus JA. Growth and Metabolism of Selected Strains of Probiotic Bacteria, in Maize Porridge with Added Malted Barley. International Journal of Food Microbiology 2004;91 305– 313.

[94] Lavermicocca P. Highlights on New Food Research. Digestive and Liver Disease 2006;38(Suppl.2) S295-S299.

[95] Reid G. Probiotics and Prebiotics-Progress and Challenges. International Dairy Journal 2008;18 969-975.

[96] Sheehan VM, Ross P, Fitzgerald GF. Assessing the Acid Tolerance and the Technological Robustness of Probiotic Cultures for Fortification in Fruit Juices. Innovative Food Science and Emerging Technology 2007;8 279–284.

[97] Betoret N, Puente L, Dıaz MJ, Pagan MJ, Garcıa MJ, Gras ML, Martınez-Monzo J, Fito P. Development of Probiotic-Enriched Dried Fruits by Vacuum Impregnation. Journal of Food Engineering 2003;56 273-277.

[98] Granato D, Branco GF, Nazzaro F, Cruz AG, Faria JAF. Functional Foods and Nondairy Probiotic Food Development: Trends, Concepts, and Products. Comprehensive Reviews in Food Science and Food Safety 2010;9 292-302.

[99] Luckow T, Sheehan V, Fitzgerald G, Delahunty C. Exposure, Health Information and Flavored Masking Strategies for Improving the Sensory Quality of Probiotic juice. Appetite 2006;47 315-325.

[100] Yoon KY, Woodams EE, Hang YD. Production of Probiotic Cabbage Juice by Lactic Acid Bacteria. Bioresource Technologies 2006;97 1427-1430.

[101] Pereira ALF, Maciel TC, Rodrigues S. Probiotic Beverage From Cashew Apple Juice Fermented with *Lactobacillus casei*. Food Research International 2011;44 1276-1283.

[102] Nedovic V, Kalusevic A, Manojlovic V, Levic S, Bugarski B. An Overview of Encapsulation Technologies for Food Applications. Proceeding in Food Science 2011; 1806-1815.

[103] Fito P, Chiralt A, Betoret N, Gras ML, Cháfer M, Martínez-Monzó J, Andrés A, Vidal D. Vacuum Impregnation and Osmotic Dehydration in Matrix Engineering. Application in Functional Fresh Food Development. Journal of Food Engineering 2001;49 175-183.

[104] Rößle C, Auty MAE, Brunton N, Gormley RT, Butler F. Evaluation of Fresh-Cut Apple Slices Enriched with Probiotic Bacteria. Innovative Food Science and Emerging Technologies 2010;11 203–209.

[105] Katina K, Liukkonen KH, Kaukovirta-Norja A, Adlercreutz H, Heinonen SM, Lampi AM, Pihlava JM, Poutanen K. Fermentation-Induced Changes in the Nutritional Value of Native or Germinated Rye. Journal of Cereal Science 2007;46 348-355.

[106] Rivera-Espinoza Y, Gallardo-Navarro Y. Review: Non-Dairy Probiotic Products. Food Microbiology 2010;27 1–11.

[107] Liu DM, Li L, Yang XQ, Liang SZ, Wang JS. Survivability of *Lactobacillus rhamnosus* During the Preparation of Soy Cheese. Food Technologies Biotechnology 2006;44 417–422.

[108] Martensson O, Osteb O, Holst O. The Effect of Yoghurt Culture on the Survival of Probiotic Bacteria in Oat-Based, Non-Dairy Products. Food Research International 2002;35 775–784.

[109] Wang YC, Yu RC, Yang HY, Chou CC. Antioxidatives Activities of Soymilk Fermented with Lactic Acid Bacteria. Food Microbiology 2006; 23 128–135.

[110] Ganzle M, Hertel C, van der Vossen J, Hammes W. Effect of Bacteriocin Producing *Lactobacilli* on the Survival of *Escherichia coli* and *Listeria* in A Dynamic Model of the Stomach and the Small Intestine. International Journal of Food Microbiology 1999;48 21–35.

[111] Arihara K, Ota H, Itoh M, Kondo Y, Sameshima T, Yamanaka H, Akimoto M, Kanai S, Miki T. *Lactobacillus acidophilus* Group Lactic Acid Bacteria Applied to Meat Fermentation. Journal of Food Science 1998;63 544–547.

[112] Sameshima T, Magome C, Takeshita K, Arihara K, Itoh M, Kondo Y. Effect of Intestinal *Lactobacillus* Starter Cultures on the Behaviour of *Staphylococcus aureus* in Fermented Sausage. International Journal of Food Microbiology 1998;41 1–7.

[113] Hammes WP, Hertel C. New Developments in Meat Starter Cultures. Meat Science 1998;49 125-138.

[114] Amor MS, Mayo B. Selection Criteria for Lactic Acid Bacteria to Be Used As Functional Starter Cultures in Dry Sausage Production: An Update. Meat Science 2007;76 138-146.

[115] Klingberg TD, Axelsson L, Naterstad K, Elsser D, Bude BB. Identification of Potential Starter Cultures for Scandinavian-Type Fermented Sausages. International Journal of Food Microbiology 2005;105 419-431.

[116] Papamanoli E, Tzanetakis N, Litopoulou-Tzanetaki E, Kotzekidou P. Characterization of Lactic Acid Bacteria Isolated From a Greek Dry-Fermented Sausage in Respect of Their Technological and Probiotic Properties. Meat Science 2003;65 859-867.

[117] Lucke FK. Utilization of Microbes to Process and Preserve Meat. Meat Science 2000;56 105–115.

[118] Muthukumarasamy P, Holley RA. Survival of *Escherichia coli* O157:H7 in Dry Fermented Sausages Containing Micro-Encapsulated Pobiotic Lactic Acid Bacteria. Food Microbiology 2007;24 82–88.

[119] Poulin JF, Caillard R, Subirade M. β-Lactoglobulin Tablets as A Suitable Vehicle for Protection and Intestinal Delivery of Probiotic Bacteria. International Journal of Pharmaceutics 2010;405 47–54.

[120] Kris-Ethertona PM, Keenb CL. Evidence that the Antioxidant Flavonoids in Tea and Cocoa are Beneficial for Cardiovascular Health. Current Opinion in Lipidology 2002;13 41-49.

[121] Egan BM, Laken MA, Donovan JL, Woolson RF. Brief Review: Does Dark Chocolate Have a Role in the Prevention and Management of Hypertension? Commentary on the Evidence. Hypertension. 2010.

[122] Aragon-Alegro LC, Alegro JHA, Cardarelli HR, Chiu MC, Saad SMI. Potentially Probiotic and Symbiotic Chocolate Mousse. LWT-Food Science and Technology 2007;40 669–675.

[123] Coman MM, Cecchini C, Verdenelli MC, Silvi S, Orpianesi C, Cresci A. Functional Foods as Carriers for SYNBIO, A Probiotic Bacteria Combination. International Journal of Food Microbiology 2012; doi:10.1016/j.ijfoodmicro.2012.06.003. (in press).

[124] Possemiers S, Marzorati M, Verstraete W, Van de Wiele T. Bacteria and chocolate: A successful combination for probiotic delivery. International Journal of Food Microbiology. 2010; 141:17-103.

Chapter 26

Biotechnological Aspects in the Selection of the Probiotic Capacity of Strains

Andrea Carolina Aguirre Rodríguez and Jorge Hernán Moreno Cardozo

Additional information is available at the end of the chapter

http://dx.doi.org/10.5772/50050

1. Introduction

Several genera of bacteria and yeast have been reported as probiotics. The most used are of the genus *Lactobacillus, Bifidobacteruin and Saccharomyces.* Although the benefits of its use have been widely reported, the selection of probiotic strains with effective capacity has been a complex process that must take into account efficacy and safety conditions. In this way, the selection of strains can be divided into three stages:

1. Selection and characterization of strains
2. Capacity Assessment In vitro probiotic
3. Capacity Assessment In vivo probiotic

Strain selection includes sources of screening, identification, assessing growth conditions of biomass such as growth kinetics, substrates, pH and temperature allowing calculation appropriate kinetic parameters for comparing strains in order to establish the feasibility of industrial scale production. Also take into account the conditions of preservation and maintenance of microorganisms in stock collections to ensure genetic and metabolic stability of selected strains [1].

Some of the effects reported *in vitro* probiotics are the production of enzymes, vitamins and amino acids, adherence capacity, the antagonistic effect on pathogenic microorganisms, tolerance to bile salts, production of bacteriocins, resistance to gastric juices, the reduction of cholesterol levels and immune system stimulation among others. In general, probiotic characteristics depend on many aspects that usually does not have a single strain, it is often necessary to include characteristics of several strains in a single product.

A probiotic is a preparation or a product that contains viable microorganisms in sufficient numbers, which alter the microflora (by implantation or colonization) in a compartment of the host provoking beneficial effects to that host's health [2]. In general, the probiotic characteristics depend on multiple aspects, which are generally not specific to a single strain. Some of the probiotic effects reported *in vitro* are the production of enzymes, vitamins, and amino acids, the capacity of adherence, the antagonistic effect on pathogenic microorganisms, tolerance to bile salts, production of bacteriocins, resistance to gastric juices, reduction of cholesterol levels, and stimulation of the immune system among others [3].

Lactic acid bacteria (LAB) belong to a group of bacteria that ferment sugars like glucose and lactose to produce lactic acid. This is important because it generates a decrease of pH and, hence, the inhibition of pathogenic and alteration microorganisms. Within this group, the existence of aerobic and anaerobic microorganisms and facultative anaerobes is recognized. The most representative LAB genre that have been used as probiotics are: *Lactobacillus, Leuconostoc, Streptococcus, Bifidobacterium,* and *Pediococcus* [4].

Lactic acid bacteria have effects that have been widely reported like the capacity to produce bacteriocins, which have antimicrobial activity against pathogens *like Listeria monocytogenes, Escherichia coli, and Salmonella* among others [5].

Likewise, the role of Lactic acid bacteria has been evaluated in food allergies, specifically in milk proteins where it has been suggested that probiotics have immunoregulatory characteristics in pathologies where the immune system [6], is implied like atopic dermatitis [6,7], genitourinary tract infections [9,10], colon cancer prevention, and reduction of colonization by *Helycobacter pylori* among others [11,12].

Probiotics, especially those contained in fermented milk, play a very important role in the prevention and treatment of diarrhea, given that they produce local intestinal and systemic effects that aid in preventing and reducing post-antibiotic therapy intestinal infections.

Several mechanisms exist by which a microorganism presents interaction against others; basically, three forms exist:

1. Competition for space,
2. Competition for nutrients,
3. Production of antimicrobial compounds attributed to the accumulation of products of fermentation processes like lactic acid, hydrogen peroxide, and bacteriocins.

Regarding yeasts, the probiotic capacity of *S. cerevisiae var boulardii* has been broadly studied; however, little is known about its action mechanism, given that research has focused on other microorganisms of greater use, mainly those from the group of the lactic acid bacteria previously mentioned [13] . This yeast has been reported as a supplement in the diets of monogastric animals like poultry, indicating that its use as a probiotic reduces some enteropathogens, produces favorable changes in the intestinal mucosa, and improves the productive behavior with rations low in protein [14]. It has also been recognized for promoting growth, increasing the production of vitamin B, helping in weight gain, improving the digestion of some foods, stimulating the immune system, improving the assimilation of nutrients, and correcting the microbial population balance.

In evaluating the probiotic capacity of strains it is important to verify their tolerance to the conditions of the gastrointestinal tract, recreating the intestinal conditions in *in-vitro* tests; thereafter, the effect should be evaluated *in vivo* [15].

2. Selection and screening of strains

A reliable probiotic product requires correct identification of the bacterial species used and a statement on the label of the species actually present. This is important because quite often the identity of the microorganisms recovered does not always correspond to the information indicated on the product label [16].

The first step for the selection of a strain with probiotic capacity is the determination of its taxonomic classification, which can give an indication of the origin, habitat, and physiology of the strain. The classification of probiotics is based on comparing the highly conserved molecules, *i.e.*, genes encoding ribosomal RNA (rRNA). Main progress in molecular biology methods has permitted sequencing the 16s and 23s rRNA subunits and, consequently, the generation of data bases of sequences of desired probiotic strains. Additionally, strains currently closely related have been distinguished by using methods based on molecular biology like plasmid profile, restriction enzyme analysis, ribotyping, random amplified DNA, and pulsed electrophoresis [17].

Once the taxonomic identification has taken place, a screening process is carried out by evaluating some physiological aspects or criteria like: [16]

- Fermentation of carbohydrates and enzymatic activity
- Adhesion to intestinal sites or areas that leads to colonization and favors equilibrium of intestinal microbiota, aids in intestinal permeability, inflammation relief, and strengthening of the barrier.
- Production of metabolites with antimicrobial activity and/or with effects at epithelial level, which help to strengthen the barrier and regulate bowel movements.
- Production of cytokines that reduce the risk of developing inflammation and generate a protection against deviations in the intestinal immune response.
- Evaluation of the link to specific toxins like mycotoxins, cyanotoxins, heavy metals, and other diet and water contaminants. This leads to the protection of the intestinal integrity and reduction of the risk of induced deviations.
- Characterization of the quorum sensing, based on detection and reaction against deviations in the diversity of intestinal tract microbiota, which favors the equilibrium of intestinal microbiota and immune response.
- Safety properties like production of anti-inflammatory cytokines contrary to pro-inflammatory cytokines and absence of antibiotic resistance genes.
- Tolerance to gastrointestinal conditions like stability at acidic pH, and tolerance to bile.

Additionally, in 2003, the FAO established some desirable key criteria for the selection of probiotics like: [12,18,19]

- Safety criteria: origin, pathogenicity, and infectivity, virulence factors (toxicity, metabolic activity, and intrinsic properties)
- Technological criteria: genetically stable strains, long-term viability of processing and storage, good sensory properties, phage resistance, and large-scale production.
- Functional criterion: tolerance to gastric juices and acids, tolerance to bile, adhesion to the surface of the intestinal mucosa, and effects on health validated and documented.
- Desired physiological criteria: immunomodulation, antagonistic activity to gastrointestinal pathogens, anti-mutagenic and anti-carcinogenic properties, and metabolism of cholesterol and lactose.

3. Conservation of strains

Freeze – drying is commonly used for the long – term preservation and storage of microorganisms in stock collections as well as for the production of starter cultures for the food industry. The choice of an appropriate suspending medium is of primary importance to increase the survival rate of the lactic acid bacteria (LAB) and yeasts during and after freeze – drying although the success of the process also depends on several factors such as growth phase, extent of drying, rehydratation, suspension medium, cruoprotectors, and so forth. During freezing or freeze – drying, cellular damage may occur, resulting in a mixed population containing unharmed cells and dead cells as well as those sublethally injured. Damage may not lead directly to death since in a suitable environment the injured cells may repair and regain normal functions.

LAB and yeasts can also be preserved for short – term storage. The techniques may be:

3.1. Short term storage

For daily or weekly use. Rich undefined media such as MRS broth (polypeptone 10g; meat extract 10g; yeast extract 10g; glucose 20g, ammonium citrate 2g; sodium acetate 5g; MgSO47H20 0,2g; MnSO44H2O 0,05g; KH2PO4 2g; Tween 80 1mL; the pH is adjusted to 6.4 ± 0.2 before autoclaving) [20] LAPTg broth (yeast extract 10g; universal peptone 10g; tryptone 16g; glucose 10g; Tween 80 1m; the pH is adjusted to 6.6 before autoclaving), [21] M17 broth (phytone peptone 5g; polypeptone 5g; yeast extract 5g; beef extract 2.5g; lactose 5g; acorbic acid 0.5g; β – disodium glycerophosphate 19g; MgSO47H20 1mL; the pH is adjusted to 7.1 before autoclaving) [22], or Elliker broth (tryptone 20g; yeast extract 5g; gelation 2.5g; dextrose 5g; lactose 5g; sucrose 5g; sodium chloride 4g; sodium acetate 1.5g; ascorbic acid 0.5g; the pH is adjusted to 6.8 before autoclaving) [23] are commonly used for LAB. For the storage of yeasts, rich undefined media such as YPG (yeast extract 10g; peptone 20g; glucose 20g), YGC (yeast extract 5g; glucose 20g; chloramphenicol 0.1g).

3.2. Storage on liquid medium

Tubes of any of the broth media, as described previously. Inoculum: bacterial cells, grown for 16 h in any of the media described to approximately 10^8 – 10^9 CFU/mL or McFarland´s tube No. 3

3.3. Long – term storage

Where inmediate acces is less important, but maintenance of the characteristics of the species and the strains is the primary objective.

3.4. Lyophilization

Cultures grown in any of the cultures media describe previously, for 16 h (overnight) at 37°C. In the case of thermophilic species the optimum incubation temperature may be in the range 39 – 41°C.

- Prepare outer vials by placing a small amount of silica gel granules (6 – 16 mesh) in the vial to cover about half of the bottom. Add a small cotton wad to cushion the inner vial and heat at 100° C overnight. The silica gel should be dark blue after heating; this serves as a moisture indicator during storage. Place vials in a dry box (<10% relative humidity) to cool.
- Aseptically, mix equal amount of inoculum (washed) and suspending medium in a sterile tube or bottle.
- Inoculation of the inner vial: Six drops of the mixture (0.2mL) are transferred to the bottom of each vial with a sterile Pasteur pipet.
- Replace the cotton plug and trim it so the cotton is even with the rim of the vial. Place de inner vial in a pan, in racks, or in boxes in a freezer at – 60°C to – 70°C and let the sample freeze for 1 – 2 h.
- Chamber – type freeze – dryer: The plates of the freeze dryers should be frozen as well. Let the condenser cool at – 60°C to – 70°C about 30 – 45 min and then place the frozen inner vials on the plates. Evacuate the system to below 30 µmHg.
- Start the process in the afternoon and allow to run about 18 h. The system is monitored by a thermistor vacuum gage. When the vacuum sensor is placed between the product and the condenser, it will show an increase in pressure as drying occurs. However, when drying is complete, the pressure should return to below 30 µmHg.
- When the cycle is complete, close the vacuum line between the chamber containing the plates with the dried samples and the condenser. Open the valve on the inlet port to admit air, allowing pressure in the cabinet to reach atmospheric pressure.
- Insert the inner vials into the outer vials. Tamp at ¼ inch plug of glass fiber paper above the cotton – plugged inner vial. Heat the outer vial in an air/gas torch, rotating the vial and keeping the flame just above the glass fiber paper until the glass begins to constrict. Pull the top of the vial slowly with forceps until the constriction is a narrow capillary tube. Cools the vials in a dry cabinet.
- Attach each vial to a port of a manifold. Each port has a single – holed rubber stopper that fits the open end of the vial. Evacuate the system to less than 50 µmHg. Seal the vials at the capillary using a double – flame air/gas torch.
- Store vials at 2 – 8°C. To open the vials, heat the tip of the outer vial in a flame, then squirt a few drops of water on the hot tip to crack the glass. Strike with a file or pencil to remove the tip. Remove the fiber paper insulation and the inner vial. Use forceps to

gently remove the cotton plug and rehydratate with 0.3 – 0.4mL of appropriate broth medium. When suspended, transfer the content to 5 – 6mL of broth and incubate ate the selected temperature for 16 – 18 h.

3.5. Freezing

- Inoculum: washed bacterial cells obtained by centrifugation of cultures grown for 16 h in any of the media described, and takedn to half of the initial volume (approximately $10^8 - 10^9$ CFU/mL or McFarland´s tube No. 3) with sterile distilled water.
- Inoculation into NFM: Harvest and wash once by centrifugation the cells rom a 10mL overnight culture. Resuspend the cell pellet into 1 – 2 mL 10% NFM supplemented with 1% (w/v) glucose, 0.5% (w/v) yeast extract, and 10% (v/v) glycerol (final concentration) and store in a domestic freezer (- 20°C to – 30°C) or even better, at – 60°C to – 70°C.
- Inoculation into glycerol solution: Take an aliquot of the washed pellet and make up to a glycerol concentration of 15 – 50%.
- Transfer the mixture of the sterile cryovials, freeze, and store as described previously.
- Routine transfers are made by scraping a little of the culture from the surface of the frozen medium and transferring to fresh medium.
- Survival is for several years, cultures stored at – 70°C surviving longer than those kept at – 20°C.
- For thawing, place the cryovials at room temperature or in water bath at 37°C and inoculate tubes containing 5 – 10mL of the proper liquid medium. Incubate the tubes at the selected temperature for 16 – 18 h. Make at least two or three transfers in fresh medium before using.

3.6. Storage under liquid nitrogen

- Inoculum: bacterial cells, grown at the selected temperature for 16 h in any of the liquid media described, to a cell density (approximately $10^8 - 10^9$ CFU/mL or McFarland´s tube No. 3).
- Mix equal quantities of inoculum (washed previously) and the glycerol 95% (v/v) solution (or other cryoprotectant) in a sterile tube, so that the final concentration of glycerol is 10% (v/v). Transfer 1mL of the mixture to each of the ampules.
- Freeze the preparations in a domestic freezer or cooling bath, to – 30°C, at a rate of about 5°C/min and allow to dehydratate for 2h.
- Transfer the frozen ampules, without thawing, to the liquid – nitrogen refrigerator.
- Maintain the level of liquid nitrogen to where the ampules are completely submerged.
- Cultures are revived by rapid thawing in a water bath at 37°C.

4. Culture media for biomass production

One of the biotechnological aspects of biomass production implies the design or selection of the culture medium. For the selection phase of strains, commercial culture media may be used that favor growth of the biomass and rapid development of the exponential phase of the microorganism being evaluated. For said purpose, conditions must be established for the bioreactor operation, such as: temperature, oxygenation, agitation, volume and ideal carbon source to reach high concentrations of biomass (10^{12}- 10^{14}).

After standardizing the production process in the commercial culture medium, evaluation of economic substrates must be carried out in the greater-scale production phase. For the production of yeasts with probiotic capacity, substrates have been evaluated with sugar cane molasses, which contributes necessary nutrients for growth and production of the strain under study [15]. These molasses have compounds that favor development of biomass like high contents of carbohydrates (sucrose, glucose, and fructose), proteins, fats, calcium, phosphorus, amino acids, and vitamins among others. Sugar cane molasses can be satisfactorily used as substrate; however [24], analyzed that for more demanding microorganisms it is necessary to supplement with certain free amino acids or ammonium sulfate that serve as a source of nitrogen and suggested controlling pH for the media with sugar cane molasses become excellent substrates for microbial fermentations.

5. Growth kinetics

5.1. Production of inoculum

This stage seeks to diminish the adaptation phase of the microorganism in fermentation. For this, initially, an enriched culture medium must be prepared for the microorganism sought to be evaluated; for lactic acid bacteria an MRS broth [20] is used and for yeasts an YGC broth. Thereafter, the contents of a vial are added onto an agar plate, and then this is incubated at the necessary temperature and time for the growth of the characteristic colonies. During this stage of the process the macro and microscopic characteristics of the strain are evaluated. Then, a cell suspension in saline solution 0.85% (p/v) is conducted until obtaining a concentration corresponding to an absorbance of 0.5 to 540 nm for LAB or 620 nm for yeasts. This suspension is added to the culture medium and it is incubated at the optimal growth temperature of the microorganism with constant agitation at 150 rpm, during 12 hours [15].

6. Discontinuous fermentation

Discontinuous fermentation seeks to produce a high concentration of microorganisms in exponential phase; this must be quantified through specific techniques like spectrophotometry and dry weight or plate counts; likewise, the consumption of the substrate must be quantified during the time of fermentation. A volume corresponding to 10% (v/v) of inoculum must be added to the sterile culture medium. The conditions of the culture must be kept at 150 rpm, 30 ºC during a maximum of 20 hours. Samplings are made every two hours to determine the concentration of biomass and concentration of residual substrate. Once the culture conditions have been established, discontinuous fermentations will be carried out at bioreactor scale [15].

The culture in the bioreactor must keep the same conditions of inoculation preparation, agitation, aeration, temperature, and time established during the previous stage.

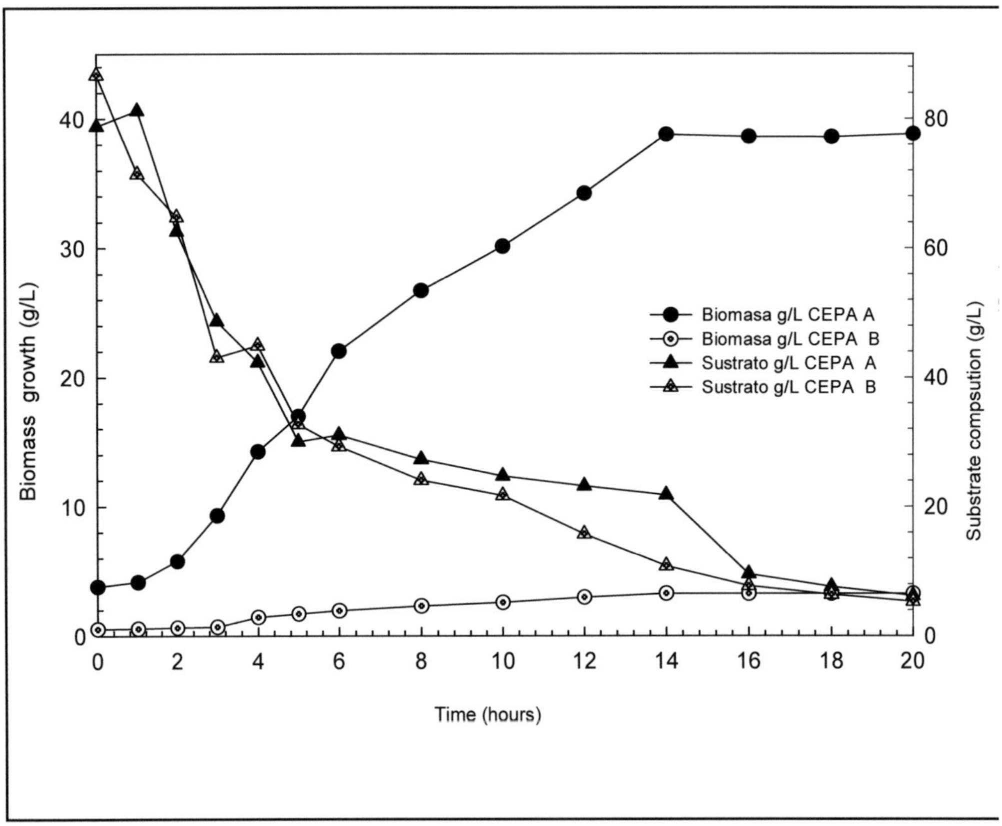

Figure 1. shows the growth kinetics results obtained by Ortiz *et al.*, at bioreactor level with a concentration of 20% (p/v) of sugar cane molasses in which is noted increased concentration of *S. cerevisiae* biomass (strain A), during 14 hours, compared to the control strain (Strain B) at Erlenmeyer level [15].

The data obtained are generally evaluated with the calculation of kinetic parameters that permit comparing the behavior of strains and operating conditions. The biomass concentration obtained along the fermentation process are logarithmically transformed according to formula 1 with the prior elaboration of the biomass pattern curve (g/L).

$$LN(\frac{X}{X_{0)}}) \tag{1}$$

Where:

X_0 represents the biomass (g/L) at the time 0 of the process (once inoculated).

X represents the biomass (g/L) during each of the hours of the process.

In addition, kinetic parameters are calculated like biomass/substrate yield Y(x/s) (g/g) (Formula 2); specific growth rate, *mx* (h-1) (Formula 3); and time of duplication, *td* (h) (Formula 4),

$$Y_{\left(x/_{s}\right)} = \frac{dx}{ds} \tag{2}$$

$$\mu_{(x)} = \frac{1}{x}\frac{dx}{dt} \tag{3}$$

$$td = \frac{Ln(2)}{\mu_{(x)}} \tag{4}$$

7. *in-vitro* tests to evaluate probiotic capacity

7.1. Tolerance to bile salts

Resistance to bile salts is a mechanism involving membrane proteins bound to ATP, which permit efficiently transporting bile acids. The presence of vesicles in yeasts has been found, similar to those found in mammals that can internalize salts, for their later degradation through catabolic enzymes [15,21].

Another mechanism by which yeast is resistant to high concentrations of bile salts is the accumulation of polyols and glycerol as elements to regulate cell osmotic pressure with the external environment.

To evaluate tolerance to bile salts, an adequate culture medium is prepared for the microorganism to be evaluated and it is supplemented with bile salts (Bile Oxgall Difco®) to obtain different concentrations (0.05, 0.1, 0.15, 0.2, 0.25, and 0.3% (p/v)). Thereafter, it is inoculated with a previously obtained suspension of the microorganism equivalent to 10^8

cells/ml. The samples are incubated under ideal conditions for each microorganism. Upon completing the incubation period, the biomass is quantified via the plate count technique [15,25].

7.2. Tolerance to ph

Tolerance to pH may be due to two types of Na^+/H^+ antiporters in yeast; Nha1p, found in the plasma membrane and Nhx1p, which is located in the pre-vacuolar/endosomal compartment. These proteins catalyze the exchange of monovalent cations (Na^+ or K^+) and H^+ through the membranes, so that they regulate the concentrations of cations and pH at organelle and cytoplasmic levels [26,27]. Another of the possible regulation mechanisms is an ATPase located in the cytoplasmic membrane; it can create an electrochemical proton gradient that leads to the secondary transport of solutes and which is implied in keeping pH close to neutral [28]. The capacity to withstand pH ranges and concentrations of bile salts was demonstrated in combination with the capacity to grow at 37 °C, ensuring that these were selection criteria to evaluate the probiotic potential of strains [29].

Tolerance to pH was assessed adjusting the culture medium to different pH ranges (2.0, 2.5, 3.0, 3.5, 4.0, 4.5, and 5.0) with concentrated HCl. Each tube was inoculated with a suspension of the microorganism to be evaluated at a previously obtained concentration of 10^8 cells/mL. The samples were incubated at ideal conditions for each microorganism. Once done with the incubation period, plate counts were carried out via the plate count technique [25].

7.3. Determination of resistance to gastric juices

Another test that shows the probiotic capacity of a strain is resistance to gastric juices. The gastric juice secreted has a pH ~2.0 and a concentration of salts ~ 0.5% (p/v) along with catabolic enzymes [30].

Tolerance to gastric juices was evaluated by preparing artificial gastric juice, for which NaCl (2 g/l) and pepsin (3.2 g/l) were added, adjusting to final pH from 2.0 - 2.3 with concentrated HCl. As control, artificial gastric juice was adjusted to neutral pH 6.5 – 7.0 with NaOH *5N*. Sterilization was conducted through filtration with 0.22-μm membrane. The artificial gastric juice and the control were inoculated with a suspension of the microorganism at a concentration of 10^8 cells/ml; these were incubated at 30 ºC, taking samples at different times (0, 1, 2, 3, 4, and 24 hours). Plate counts were carried out in each sampling [25].

7.4. Reduction of cholesterol in the presence of bile salts

Cholesterol reduction is a desired characteristic, given that for humans the condition of hypercholesterolemia or increased levels of cholesterol in blood is considered the greatest risk for the development of heart disease; and in animals lower presence of cholesterol generates high-quality meats and of great demand, given that they are fat free. The administration of probiotics has demonstrated that they can notably reduce cholesterol levels [31,32].

Cholesterol does not destabilize or precipitate in the medium due to its conjugation with bile salts, which is why it is possible that the microorganisms assimilate the cholesterol present in the medium to incorporate it to its cell membrane. Studies suggest that yeasts exposed to culture medium enriched with cholesterol were more difficult to lyse after being subjected to sonication than yeasts that did not grow in the medium enriched with cholesterol, which indicates a possible morphological change in the wall or in the cell membrane, given that upon the microorganism incorporating this sterol onto its structure, it becomes more resistant to cell lysis, compared to those not incorporating it [29].

It is important to add bile salts to the culture medium with added cholesterol to extract samples to elaborate the pattern curve. This is because the bile salts are present in the organism during activities of lipid emulsion, solubilization, and absorption in the intestine [33]. To evaluate the reduction of cholesterol, a culture medium was prepared supplemented with bile salts (Bile Oxgall Difco®). Thereafter, 224.2 µg/ml of Lipids Cholesterol Rich (Sigma ®) were added. This medium was inoculated with 1% (v/v) of the suspension of the microorganism to be evaluated at a concentration of 10^8 cells/mL. The mixture was incubated for 12 hours at the adequate temperature according to the microorganism.

To evaluate the reduction of cholesterol, a culture medium was prepared supplemented with bile salts (Bile Oxgall Difco®). Thereafter, 224.2 µg/ml of Lipids Cholesterol Rich (Sigma®) were added. This medium was inoculated with 1% (v/v) of the suspension of the microorganism to be evaluated at a concentration of 10^8 cells/mL. The mixture was incubated for 12 hours at the adequate temperature according to the microorganism.

The medium was centrifuged at 8000 x g for 15 minutes, 3 ml of ethanol at 95% (v/v) were added to the supernatant, followed by 2 ml of potassium hydroxide at 50% (v/v). Afterwards, the samples were heated to 60 ºC for 10 minutes, then 5 ml of hexane and 3 ml of distilled water were added agitating in vortex after adding each component. From the aqueous phase (hexane layer) 2.5 ml were transferred onto a tube; this was evaporated in a furnace at 60 ºC. The residue formed was resuspended in 4 ml of *0 – phthalaldehyde*. After remaining at rest at room temperature for 10 minutes, 2 ml of concentrated sulfuric acid were added. Finally, absorbance at 550 nm was measured against the target reagent with prior elaboration of a pattern curve with a concentrated solution of 130 µg of cholesterol/mL.

To evaluate the reduction of cholesterol, a culture medium was prepared supplemented with bile salts (Bile Oxgall Difco®). Thereafter, 224.2 µg/ml of Lipids Cholesterol

Rich (Sigma®) were added. This medium was inoculated with 1% (v/v) of the suspension of the microorganism to be evaluated at a concentration of 10^8 cells/mL. The mixture was incubated for 12 hours at the adequate temperature according to the microorganism.

The medium was centrifuged at 8000 g for 15 minutes, 3 ml of ethanol at 95% (v/v) were added to the supernatant, followed by 2 ml of potassium hydroxide at 50% (v/v). Afterwards, the samples were heated to 60 ºC for 10 minutes, then 5 ml of hexane and 3 ml of distilled water were added agitating in vortex after adding each component. From the aqueous phase (hexane layer) 2.5 ml were transferred onto a tube; this was evaporated in a furnace at 60 ºC. The residue formed was resuspended in 4 ml of *0 – phthalaldehyde*. After remaining at rest at room temperature for 10 minutes, 2 ml of concentrated sulfuric acid were added. Finally, absorbance at 550 nm was measured against the target reagent with prior elaboration of a pattern curve with a concentrated solution of 130 μg of cholesterol/mL [25].

7.5. Adherence test

One of the important criteria for a probiotic strain is the ability to adhere to the mucous surface of the gastrointestinal tract, given that "*in vivo*" probiotic microorganisms adhere to enterocytes avoiding possible strains from effecting cell adherence as pathogenicity mechanism. Exclusion through the competition for adhesion sites and for substrate is one of the action mechanisms of yeasts used as probiotics.

Cells can be used from the Caco-2 cell line from adenocarcinoma of human colon, which develops characteristics of mature enterocytes and provides a uniform population of differentiated cells, which can be used under conditions defined to quantify adhering microorganisms. According to the study in which adherence tests were conducted of Caco-2 cells with several strains of *Lactobacilli,* it was determined that strains presenting an adherence count below 40 microorganisms in the 20 fields counted at random were considered as non-adhering, between 41 and 100 microorganisms as adhering and over 100 microorganisms as strongly adhering [34].

The Caco-2 cell line must be grown at 37 ºC in an environment with 5% CO_2 by using the minimum essential medium (GIBCO ®) until observing a monolayer. Then, the cells were washed three times with sterile PBS (pH 7.0 ± 0.2). A total of 5 ml of culture was taken from the strains previously grown at culture conditions; then, they were centrifuged and washed with sterile PBS (pH 7.0 ± 0.2) and resuspended in minimum essential medium.

The Caco-2 cells were inoculated with 0.8 ml of the culture of the previously treated microorganism. The mixture was incubated at 37 °C during 90 minutes in an environment with 5% CO_2, and then four washes were carried out with sterile PBS (pH 7.0 ± 0.2). This was followed by Wright's staining, which was observed in the inverted microscope counting the number of microorganisms adhered to the Caco-2 cells in 20 random microscopic fields. Adherence capacity is expressed as the number of microorganisms adhered to 100 Caco-2 cells [34].

Figure 2 and 3 shows the behavior of two strains that adhered to the Caco-2 cell line, where it is observed that the strain of study isolated from sugar cane molasses (strain A) had greater adhesion than the control strain (strain B) [15].

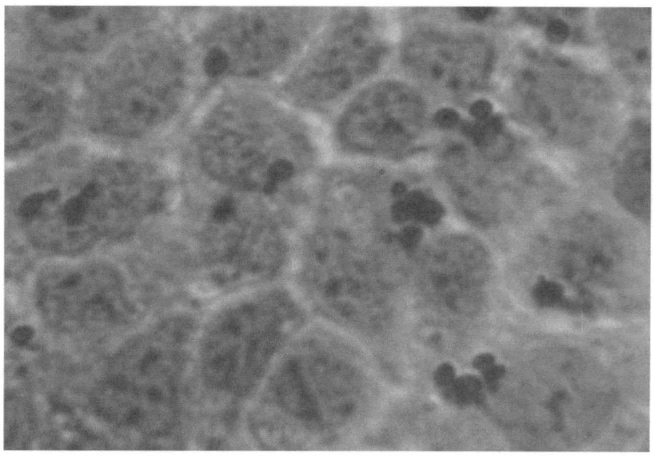

Figure 2. Inverted microscope adherence analysis of *Saccharomyces cerevisiae* (strain A) to Caco-2 cells with Wright's staining.

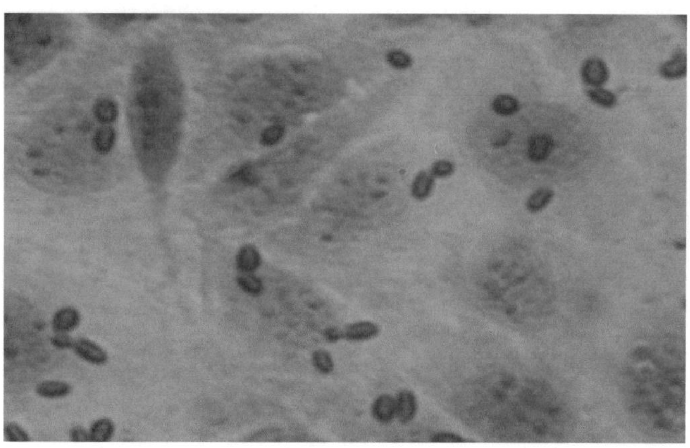

Figure 3. Inverted microscope adherence analysis of *S. cerevisiae* var. *boulardii* (strain B) to Caco-2 cells with Wright's staining.

Author details

Andrea Carolina Aguirre Rodríguez and Jorge Hernán Moreno Cardozo
Department of Microbiology, Pontificia Universidad Javeriana, Colombia

8. References

[1] Rubio A, Hernandez C, Aguirre A, Poutou R. (2008)In vitro preliminary identification of probiotic propieties of S. cerevisiae strains. MVZ Córdoba 200;13(1):1157-69.

[2] Schrezenmeier JyDV, M. (2001)Probiotics, prebiotics and symbiotics -approching a definition. Probiotics, prebiotics and symbiotics -approching a definition. 1;73:361-4.

[3] Lee Y, Salminen S. (2009). Handbook of Probiotics and Prebiotics. Second Edition ed. A John Wiley & Sons I, Publication, Editor. New Jersey.

[4] Ortega M MA, Aranceta J, Mateos J, Requejo A, Sierra L.(2002)Alimentos Funcionales-Probioticos. Panamericana EM, editor. Madrid.

[5] Vallejo M, V E, Horiszny C, E M. (2009). Inhibición de *Escherichia coli* O157:H7 por cepas *Lactobacillus* aisladas de queso ovino. Analecta Veterinaria. 29(1):15-9.

[6] Cross M SL, Gils. (2001) Anti-Allergy Properties of Fermented Foods: An Important Inmunoregulatory Mechanism of Lactic Acid Bacteria? Intern Immunopharm;1:891-901.

[7] Majamaa H, Isolauri E. (1997). Probiotic: A novel approach in the management of food allergy. J Allergy Clin Inmunol. 99:179-85.

[8] Murch S. (2001). Toll of Allergy Reduced by Probiotics. Lancet;357:1057-9.

[9] Hooton M. (2001). Recurrent Urinary Trct Infection in Women. Int J Antimicr Agents. 17:259-68.

[10] Reid G. (2000) Probiotic Agents to Protect the Urogenital Trac Against Infection. Am J Clin Nutr;73(Suppl):1682-7.

[11] Vandenplas Y, Brunser O, H S. (2009). *Saccharomyces boulardii* in childhood. Eur J Pedriatr;168:253-65.

[12] Shah N. (2007) Funcional Cultures and Health Benefits. International Dairy Journal2007;17:1262-77.

[13] Guslandi M, Giollo P, Testoni P. (2004) A pilot trial of *Saccharomyces boulardii* in ulcerative colitis. *European Journal of Gastroenterology and Hepatology*;15:697-8.

[14] Agarwal N, Kamra D, Chaudhary L, Sahoo A, Pathak N. (2000) Selection of *Saccharomyces cerevisiae* strains for use as a microbial feed additive. *Letters in Applied Microbiology*;31:270-3.

[15] Ortiz A, Reuto J, Fajardo E, Sarmiento S, Aguirre A, Arbelaez G, *et al.* In Vitro Preliminary Identification of Probiotic Propierties of S. cerevisiae Strains. Rev MVZ Cordoba 2008;13(1):1169-2008.

[16] Guei monde M, Salminen S. (2006) New methods for selecting and evaluating probiotics. Digestive and Liver Disease;38(2):242-7.

[17] Vasiljevic T, Shah N. 2008. Probiotics-From Metchnikoff to bioactives. International Dairy Journal;18:714-28.

[18] Guidelines for the Evaluation of Probiotics in Food, (2002).

[19] Morell M. (2007). In vitro assessment of probiotic bacteria: From survival to functionality. International Dairy Journal;17:1278-83.

[20] De man J, Rogosa M, Sharpe M. A (1960) Medium for the Cultivation of lactobacilli. J Appl Bacteriol;23:130-5.

[21] Raibaud P, Coulet M, Galpin J, Mocquot G. (1961) Stidies on the bacterial flora of the alimentary tract of pigs. JAppl Bacteriol1961;24:285-91.

[22] Terzaghi B, Sandine W. (1981). Bacteriophage production Following Exposure of Lactic Streptococci to Ultraviolet Radiation. J Appl Bacteriol;24:285-91.

[23] Elliker P, Anderson A, Hannesson G. (1956). An Agar medium for Lactic Acid Streptococci and lactobacilli. J Dairy Sci;39:1611-2.

[24] Brandeberg T, Gustafsson L, C F. (2006). The impact of severe nitrogen limitation and microaerobic conditions on extended continuous cultivations of *Saccharomyces cerevisiae* with cell recirculation. *Enzyme and Microbial Technology*;40:585-93.

[25] Spencer JyR, A.2001. Food Microbiology Protocols. New Jersey, USA: Humana Press Inc.

[26] Mitsui K, Yasui, H., Nakamura, N. y Kanazawa, H (2005). Oligomerization of the *Saccharomyces cerevisiae* Na+/H+ antiporter Nha1p: Implications for its antiporter activity. *Biochimica et Biophysica Acta*;1720:185-96.

[27] Ohgaki R, Nakamura, N., Mitsui, K. y Kanazawa, H. (2005). Characterization of the ion transport activity of the budding yeast Na^+/H^+ antiporter, Nha1p, using isolated secretory vesicles. *Biochimica et Biophysica Acta*;1712:185-96.

[28] Thomas K, Hynes S, Ingledew M. (2002) Influence of Medium Buffering Capacity on Inhibition of *Saccharomyces cerevisiae* Growth by Acetic and Lactic Acids. Environmental Microbiology;68:1616-23.

[29] Psomas E, Fletouris D, Litopoulou E, Tzanetakis N. (2003). Assimilation of Cholesterol by Yeast Strains Isolated from Infant Feces and Feta Cheese. Journal of Dairy Science;86:3416-22.

[30] Martins F, Ferreira, F., Penna, F. Rosa, C., Drummond, R., Neves, M. y Nicol, J.(2005). Estúdio do potencial probiótico de linhagens de *Saccharomyces cerevisiae* a través de testes in Vitro. Revista de Biologia e Ciencias da terra;5:1-13.

[31] Khani S, Hosseini HM, Taheri M, Nourani MR, Fooladi AAI. (2012). Probiotics as an alternative strategy for prevention and treatment of human diseases: A review. Inflammation and Allergy - Drug Targets;11(2):79-89.

[32] Kumar R, Grover S, Batish VK. (2012) Bile Salt Hydrolase (Bsh) Activity Screening of Lactobacilli: In Vitro Selection of Indigenous Lactobacillus Strains with Potential Bile

Salt Hydrolysing and Cholesterol-Lowering Ability. Probiotics and Antimicrobial Proteins:1-11.

[33] Begley M, Hill, C., Cormac, G. y Gahan, M. (2006). Bile salt hydrolase activity in probiotics. Applied and Enviromental Microbiology;73(3):1729-38.

[34] Jacobsen C, Rosenfeldt V, Hayford A, Moller P, Michaelsen K, Paerregaard A, *et al.* (1999) Screening of Probiotic Activities of Forty-Seven Strains of *Lactobacillus* spp. by *In Vitro* Techniques and Evaluation of the Colonization Ability of Five Selected Strains in Humans. Applied and Enviromental Microbiology;16:4949-56.

Section 4

Aquaculture

Chapter 27

Probiotics in Larvae and Juvenile Whiteleg Shrimp *Litopenaeus vannamei*

I.E. Luis-Villaseñor, A.I. Campa-Córdova and F.J. Ascencio-Valle

Additional information is available at the end of the chapter

http://dx.doi.org/10.5772/50123

1. Introduction

In penaeid shrimps, *Vibrio* spp. is the main cause of bacterial diseases, such as *V. parahaemolyticus, V. alginolyticus, V. harveyi* (Garriques and Arevalo, 1995) and *V. penaeicida* (Aguirre-Guzmán and Ascencio-Valle, 2001). Possible mode of infection consists of three basic steps: (i) the bacterium penetrates the host cuticle or exoskeleton wound by means of chemotactic motility; (2) within the host tissues the bacterium deploys iron-sequestering systems; e.g., sidero-phores, to "steal" iron from the host; and (3) the bacterium eventually damages the organisms by means of extracellular products, e.g. hemolysins and proteases (Thompson et al., 2004). Containing high loads of either *Vibrio parahaemolyticus* or *V. harveyi* induced the rounding up and detachment of epithelial cells from the basal lamina of the midgut trunk. Epithelial cell detachment of epithelial was not seen in the presence of non-pathogenic bacteria (probiotics) (Chen et al., 2000; Martin et al., 2004). Pathogens like *Vibrio* spp., which cause detachment of the epithelium in the midgut trunk, can affect high mortality in shrimp by eliminating 2 layers that protect the shrimp from infections: the epithelium and the peritrophic membrane it secretes. In addition, loss of the epithelium may affect the regulation of water and ion outtake into the body (Mykles 1977, Neufeld and Cameron 1994).

Prevention and control of diseases had led to increase the use of antibiotics developing drug resistant bacteria, which are difficult to control and eradicate. An alternative to antibiotic treatment is the use of probiotics or beneficial bacteria that control pathogens. Probiotics are generally defined as viable microorganisms that, when to human or animals, beneficially affect the health of the host by improving the indigenous microbial balance (Fuller, 1989; Havenaar et al., 1992). Generally, probiotic strains have been isolated from indigenous and exogenous microbiota of aquatic animals (Vine et al., 2004). Probiotics may protect their host from pathogens by producing metabolites that inhibit the colonization or growth of other microorganisms or by competing with them for resources such as nutrients or space (Vine et al. 2004). Studies of probiotics to improve growth or survival in crustacean larvae are scarce. Recently, methods for improving water quality of hatcheries and application of probiotics has gained momentum (Balcázar et al., 2007a; Gómez et al., 2008; Guo et al., 2009; Van Hai et al., 2009). Daily administration of probiotics based on *Bacillus* spp. during hatchery and farming stages leads to higher feed conversion ratios, improved specific growth rates, and higher final shrimp biomass than controls (Guo et al., 2009; Liu et al., 2009a). Metamorphosis improved with administration of the probiotic *B. fusiformis* (Guo, et al., 2009). Zhou et al. (2009) found that *B. coagulans* SC8168, as a water additive at certain concentrations, significantly increased survival and some digestive enzyme activities of shrimp larvae. *Bacillus* spp. possesses adhesion abilities, produce bacteriocins, and provide immunostimulation (Ravi et al., 2007).

The criteria of probiotic selection to be used in aquaculture systems has been discussed by some authors. Nguyen et al. (2007) suggest that the beneficial effect of the probiotics on the host has been wrongly attributed to what is found during *in vitro* observations, that *in vivo* physiology might be different from in vitro metabolic processes. Development of suitable probiotics is not a simple task and requires full-scale trials, as well as development of appropriate monitoring tools and controlled production (Decamp et al., 2008). *In vitro* and *in vivo* studies are needed to demonstrate antagonisms to pathogens and their effect on survival and growth of the host. The main purpose of using probiotics is to maintain or reestablish a favorable relationship between friendly and pathogenic microorganisms that constitute the flora of intestinal or skin mucus of aquatic animals. Since, successful probiotic is expected to have a few specific properties in order to certify a beneficial effect (Ali, 2000).

Bacteria present in the aquatic environment influence the composition of the gut microbiota and vice versa. The genus present in the intestinal tract generally seems to be those from the environment or the diet that can survive and multiply in the intestinal tract (Cahill, 1990). Therefore, probiotic strains have been isolated from indigenous and exogenous microbiota of aquatic animals. Gram-negative facultative anaerobic bacteria such as *Vibrio* and *Pseudomonas* constitute the predominant indigenous microbiota of a variety of species of marine animals (Onarheim *et al.*, 1994). On the other hand, the indigenous microbiota of freshwater animals tends to be dominated by members of the genera *Aeromonas, Plesiomonas,* representatives of the family *Enterobacteriaceae,* and obligate anaerobic bacteria of the genera *Bacteroides, Fusubacterium,* and *Eubacterium* (Sakata 1990). Lactic acid producing bacteria, which are prevalent in the mammal or bird gut, are generally sub-dominant in fishes and are represented essentially by the genus *Carnobacterium* (Ringo & Vadstein 1998). Ideally, microbial probiotics should have a beneficial effect and not cause any harm to the host. Therefore, all strains have to be non-pathogenic and non-toxic in order to avoid undesirable side-effects when administrated to aquatic animals (Chukeatirote, 2002).

Some research and products talk about the multifactorial action of the probiotics (Gomez et al., 2007; Tuohy et al., 2003) on aquatic animals. However, the multifactorial effect is not agreed with evidence or is overestimate. Sometimes, this type of publicity about the potential of those products really affects the perspective of real probiotic designed for aquaculture industry.

Different modes of action or properties are desire on the potential probiotic like antagonism to pathogens (Ringo and Vadstein, 1998), ability of cells to produce metabolities (like vitamins) and enzymes (Ali, 2000), colonization or adhesion properties (Olsson *et al.*, 1992), enhance the immune systems (Perdigon et al., 1995) and others. On the other hand, a criterion to discard potential harmful bacteria is the ability to produce toxins that induce lysis of host cells (Zamora-Rodríguez, 2003)

Various mechanisms have been proposed to explain their beneficial effects, including competition for adhesion sites, competition for nutrients, enzymatic contribution to digestion, improved water quality, and stimulation of the host immune response (Kumar Sahu et al., 2008). Selection of probiotics in aquaculture enterprises is usually based on results of tests showing antagonism toward the pathogens, an ability to survive and colonize the intestine, and a capacity to increase an immune response in the host. Adhesion of probiotic microorganisms to the intestinal mucus is considered important for many of the observed probiotic health effects (Ouwehand et al., 2000). Adhesion is regarded a prerequisite for colonization (Alander et al., 1999).

The composition of the bacterial community in an aquaculture environment has a strong influence on the internal bacterial flora of farmed animals, which is vital for their nutrition, immunity and disease resistance (Luo et al. 2006). The intestinal microbiota of aquatic organisms in culture is an important factor in maintaining the healthy, either by preventing pathogen colonization, degradation of food, production of antimicrobial compounds, producing nutrients and maintaining normal mucosal immune (Escobar-Briones et al., 2006). The interest in investigating the intestinal microbiota is based on the need for a better understanding of how probiotics can influence the bacterial composition. Another important function was to emerge in recent years suggesting that the effect of the commensal microbiota influence processes such as lipid metabolism and development of the host immune response. The inter-relationship between the microbiota and the host are clearly important in relation to health and the imbalance between these systems results in disease development. Several studies listed the benefits or these probiotics to culture organisms, however, few works that the type of modulation is performed to the intestinal microbiota and its effects on health of the host organism. The interest to investigated the intestinal microbiota is based on the need for a better understanding of how probiotics can influence the bacterial composition. Such studies have been widely performed in vertebrates (Brikbeck, 2005; Austin, 2006; Escobar-Briones et al., 2006; McKellep Bakke, 2007), but in invertebrates is very limited. The intestinal microbiota of aquatic organisms has shown a high dependence of bacterial colonization during early development, environmental conditions and change in diet (Ringo et al., 1995, 2006; Ringo and Birkbeck, 1999; Olafsen, 2001). For that to know the impact that probiotics in the modulation of intestinal microbita should be studied. We investigated the effect of *Bacillus* probiotics was showed trait inhibitory to *Vibrio* and ability to adhere and grow, on intestinal mucus on the survival and rate of development of whiteleg shrimp *L. vannamei* larvae to understand mechanisms of how endemic *Bacillus* probiotic strains improve the health of larvae. Moreover, analyzed the composition of bacterial communities in the juvenile shrimp L. vannamei know the impact that probiotics in the modulation of intestinal microbiota.

2. Antagonism test

Antagonism in the world of bacteria is a highly prevalent phenomenon: one bacterium species suppresses the development or inhibits the growth of other microorganisms (Egorov, 2004). A common way to select probiotic is to perform *in vitro* antagonism test. *Bacillus* spp. produce polypeptides (bacitracin, gramicidin S, polymyxin, and tyrothricin) that are active against a broad range of Gram positive and Gram negative bacteria, which also explains the inhibitory effect on pathogenic *Vibrio* (Drablos et al., 1999; Morikawa et al., 1992; Perez et al., 1993). The antagonism of *Bacillus* is due mainly to the production of antimicrobial proteins and antibiotics as well as chemical compounds synthesized by secondary metabolism pathways (Hu et al., 2010), competition for essential nutrients and adhesion sites. We scrutinized their ability to inhibit the growth of *Vibrio* species utilized the two-layer method described by Dopazo et al. (1988) (Figure 1), shows that only two isolates *Bacillus tequilensis* and *B. amyloliquefaciens* (YC5-2 and YC2-a) inhibited growth of *V. campbelli* (CAIM 333) and *V. vulnificus* (CAIM 157).

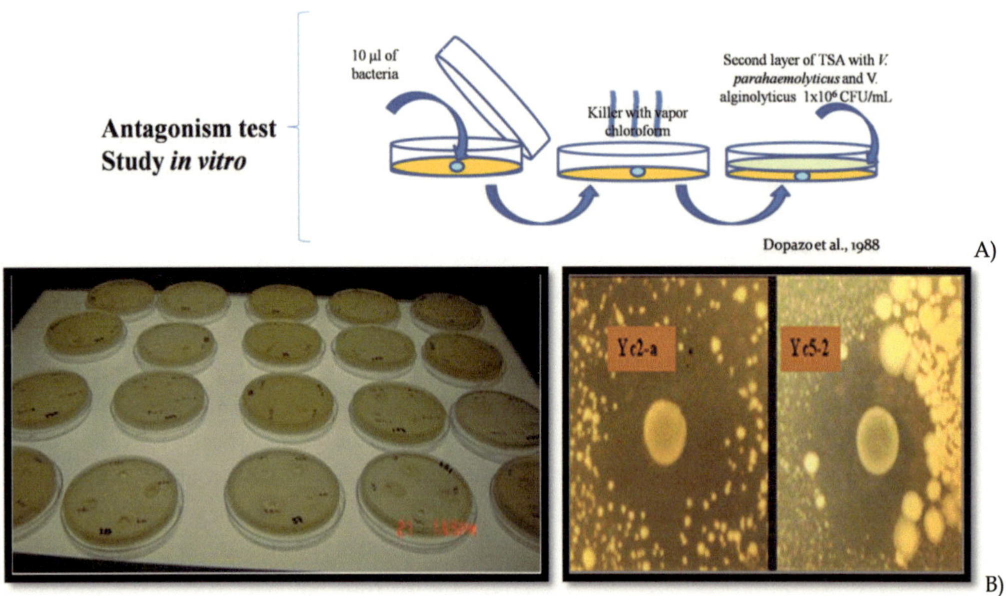

Figure 1. A) Schematic from Antagonism test utilized the two-layer method described by Dopazo et al. (1988). B) Zone inhibition obtained by *Bacillus amyloliquefaciens* (strain YC2-a) and *Bacillus tequilensis* (strain YC5-2) against *Vibrios parahaemolyticus*.

The well diffusion test (Balcázar et al., 2007) showed that 24-h cultures of inactivated isolates YC5-2 (*Bacillus tequilensis*), *YC2-a (B. amyloliquefaciens)* YC3-b (*B. endophyticus*) and C2-2 (*B. endophyticus*) were able to inhibit *V. parahaemolyticus* (CAIM 170) and *V. harveyi* (CAIM 1793). *V. alginolyticus* (CAIM 57) showed sensitivity but no inhibition to these probiotic strains (Luis-Villaseñor et al., 2011) (Figure 1, Table 1). *Bacillus* strains isolated from shrimp inhibited vibriosis by a well-diffusion method. The antagonism test showed that probiotic strains were able to inhibit pathogenic strains of *V. harveyi* (CAIM 1793), *V. parahaemolyticus* (CAIM 170), *V. campbelli* (CAIM 333), *V. alginolyticus* (CAIM 57), and *V. vulnificus* (CAIM 157). Similar results were obtained by Balcazar et al. (2007a), where *B. subtilis* UTM 126 was able to inhibit *V. parahaemolyticus* PS-107. Nakayama et al. (2009) found that cell-free supernatant from *B. subtilis, B. licheniformis*, and *B. megaterium* inhibited growth of one *V. harveyi* strain for 24 h. Decamp et al. (2008) administered *B. subtilis* and *B. licheniformis* to larval *L. vannamei* and *Penaeus monodon* and this inhibited growth of *Vibrio* strains and increased the survival rate of the shrimp.

Isolate	Gram	Hemolytic activity		Inhibition zone (mm)				
		Erythrocytes	Hemocytes	*V. parahaemolyticus* CAIM 170	*V. harveyi* CAIM 1793	*V. campbelli* CAIM 333	*V. vulnificus* CAIM 157	*V. alginolitycus* CAIM 57
YC5-2**	+	γ	NR	17.5±0.7	11±1.8	5±1.4	18±1.4	*
YC2-a**	+	γ	NR	13.5±1.0	12±3.0	9±1.4	6.5±0.2	*
C2-2	+	γ	NR	21.5±1.1	11.5±2.1	NR	NR	*
YC3-B	+	γ	NR	13.5±2.1	11±2.1	NR	NR	*
YC1-A	+	α	4.5±0.7	16.5±2.1	8.85±0.5	9±0.5	21.9±1.6	*
YC3-C	+	α	4.5±1.4	17.5±0.7	10±1.4	9.1±0.1	18.7±1.1	*
YC3-A	+	α	3.5±0.7	13.45±1.1	8±1.4	8.15±1.6	18.1±1.8	*
YC2-B	+	β	8.5±0.7	8.5±2.1	13±1	NR	4.6±0.8	*
YC3-D	+	β	8.7±0.3	9.5±0.7	11±1	NR	10.75±0.4	*

** = Inhibitory effect for the two-layer method (Dopazo et al. 1988). γ = Growth, but not hemolysis. NR = Negative to the test.

Table 1. Test of antagonism of probiotics isolates against pathogenic *Vibrio* strains. * = Bacteriostatic effect.

3. Hemolytic activity of *Bacillus* strains

The principal purpose of the use of probiotics in to produce a proper relationship between useful microorganism and the pathogenic microflora and their environment. Probiotics should be of animal-species origin, this criteria is based on ecological reasons, and takes into consideration the original habitat of the selected bacterial (in intestinal flora) (Farzanfar, 2006). One of the most important features of a probiotic is that it does not harm the host (Kesarcodi-Watson et al., 2008). Some *Bacillus* spp. produce hemolysins, which could be a health risk to the host (Liu et al., 2009b). Bernheimer and Grushoff (1967) demonstrated that *B. cereus, B. alvei, B. laterosporus, B. subtilis* contained streptolysin and lysins. To measure hemolytic activity of the various *Bacillus* strains on erythrocytes, nine isolated *Bacillus* probiotic strains were inoculated by streaking on plates containing blood-based agar supplemented with 5% (w/v) human sterile blood and 3% (w/v) NaCl. Plates were incubated at 37 °C for 24 h and results were determined, as described by Koneman et al. (2001), as: α-hemolysis (slight destruction of hemocytes and erythrocytes with a green zone around the bacterial colonies); β-hemolysis (hemolysin that causes a clean hemolysis zone around the bacterial colonies); and γ- hemolysis (without any change in the agar around the bacterial colonies.

Hemolytic activity in shrimp hemocytes was tested, as described by Chin-I et al. (2000). Briefly, a 1-mL syringe was rinsed with EDTA buffer (450 mmol L^{-1} NaCl, 10 mmol L^{-1}5 KCl, 10 mmol L^{-1} EDTANa2, and 10 mmol L^{-1} HEPES at pH 7.3). After disinfecting the surface of the shrimp weighing ~20 g with 70% ethanol; hemolymph was drawn with a sterile needle from between the fifth pair of pereiopods; 1 mL hemolymph was immediately transferred to a sterilized tube containing 0.2 mL EDTA buffer and stained with 133 μL 3% (w/v) Rose Bengal dye (#R4507, Sigma St. Louis, MO) dissolved in EDTA buffer with gentle shaking to achieve complete mixing. Aseptically, 1 mL of the stained hemolymph preparation was added to 15 mL sterile basal agar medium containing (10 g L^{-1} Bacto 12 peptone (#211677, Difco), 5 g L^{-1} HCl, and 15 g L^{-1} Bacto agar (#214050, Difco) at pH 6.8) cooled to 45–50 °C, followed by gentle mixing and poured into Petri dishes. Shrimp blood agar plates with a rose red color were considered satisfactory because of the homogenously distributed stained hemocytes. When the hemocytes were destroyed by hemolytic bacteria, a clear zone (4 mm) appeared around the colonies. Four *Bacillus* strains isolated from the gut of adult *L. vannamei* (YC2-a, *B. amyloliquefaciens*; YC3-b, *B. endophyticus*; YC5-2, *B. tequilensis* and C2-2; *B. endophyticus*) exhibited type γ hemolytic activity(without any change in the agar around the bacterial colonies), three *Bacillus* strains (*B. licheniformis* strains YC1-a, YC3-a, and YC3-c) exhibited type α hemolytic activity (slight destruction of hemocytes around the bacterial colonies), and two *Bacillus* strains (YC3-d and YC2-b) having type β hemolytic activity (destruction of hemocytes, showed a clean zone around the bacterial colonies) (Luis-Villaseñor et al., 2011).

4. Mucus adhesion assay and bacterial growth in mucus

The intestinal epithelium is a natural barrier of the gastrointestinal tract providing defense against extrinsic invasions. The resident microflora, especially the beneficial ones, plays a crucial role in maintaining the host healthiness in numerous ways including; preserving the niche balance of intestinal microflora, reducing the colonization and invasion of pathogens, retaining the epithelial integrity and promoting immune function (Ouwehand et al., 1999; Herich and Levkut 2002). The strains with the highest adhesion ability have the greatest effect on host healthiness and performance (Majamaa et al., 1995; Shornikova et al., 1997; Kirjavainen et al., 1998; Ouwehand et al., 1999). Mucus composition varies from site to site. Among its major components is a group of high molecular weight glycoproteins called mucins. Depending upon the location, mucus may also contain various electrolytes, sloughed epithelial cells, plasma proteins, immunoglobulins, lysozime, bacteria and their products, digested food material, digestive enzymes, epithelial cell membrane glycoproteins, and other components (Gibbons, 1982). The suggested functional properties of mucins are: Lubrication of epithelial surfaces; Diffusion barrier to nutrients, drugs, ions, toxins, and macromolecules; binding of bacteria, virus, parasites; Detoxification by heavy metal binding; Protection of mucosa against proteases; Interaction with immune surveillance system, and Interaction of membrane mucins with microfilaments (actins) (Forstner and Forstner, 1989).

The protective role of mucosal surfaces against potentially harmful substances such as acids, digestive enzyme, food lectins, toxins, bacterial and others infectious agents (Forstner and Forstner, 1989). The cell wall of Gram-positive bacteria is made up of a think, multilayered peptidoglycan sacculus (also called murein) containing teichoic acids, proteins and polysaccharides (Vinderola et al., 2004). Mucin and cell surface carbohydrate are usually considered to be highly hydrophilic, although like other oligosaccharides, they can probably adopt amphipathic configurations (Sundari et al., 1991) to present a hydrophobic surface for interactions with some bacterial structures (Forstner and Forstner, 1994).

The ability to adhere to the intestinal mucus in considered one of the main criteria in the selection of potential probiotics as adhesion prolongs their permanence in the intestine and thus allows them to exert healthful effect (Apostolou et al., 2001).

During characterization of potential probiotics, we scrutinized their ability to adhere and colonize the intestine of shrimp. The dot-blot assays described in the present report is based on the formation of a complex between adhesion promoting compounds from the cell surface of the bacteria and the enzymatically labeled receptor in gastrointestinal mucus, followed by the visualization of bound components on a solid phase matrix (Rojas et al., 2002). Seven strains (YC2-a, YC3-b, YC5-2, C2-2, YC1-a, YC3-a and YC3-C) adhered to porcine gastric and crudes shrimp mucus (Fig 2). The seven isolates were able to grow in the mucus 24 h after inoculation; after 48 h viable cell counts were tower. These strains were examined for their ability to grown shrimp intestinal mucus. Sterility of mucus was confirmed on specific media. The number of viable cells decreased by ~50% at 48 h; strains22 YC5-2, YC3-a, YC3-c, YC1-a, and YC2-a had viable cell counts between 18×106 UFC mL^{-1} and 10×109 UFC mL^{-1} at 24 h, which decreased to between 1.3×106 UFC mL^{-1} and 0.126×106 UFC mL^{-1} at 48 h; however, abundant free spores were observed in five strains with epifluorescence microscopy (Table 2). Strains YC3-b and C2-2 had viable cell counts between 1.87×106 UFC mL and 4.14×106 UFC mL at 24 h, showing a decrease at 48 h with viable bacteria remaining about 0.18×106 UFC mL^{-1} for both strains. Similar studies reported that strains of *Bacillus* spp. able to grow in water and colonize the digestive tract of shrimp. This ability is related to competitive exclusion. However, in vitro activity assays cannot be used to predict a possible in vivo effect (Balcázar et al., 2006).

	CFU mL^{-1}	
	Time (h)	
Bacterial Strains	24	48
YC3-B	1.87×10^6	0.18×10^6
C2-2	4.14×10^6	0.18×10^6
YC5-2	$>10 \times 10^9$	0.126×10^6
YC2-a	18×10^6	1.3×10^6
YC3-A	$>10 \times 10^9$	0.27×10^6
YC3-C	$>10 \times 10^9$	0.84×10^6
YC1-A	$>10 \times 10^9$	0.54×10^6
Control	0	0

Table 2. Growth of bacterial in mucus of shrimp *Litopenaeus vannamei*

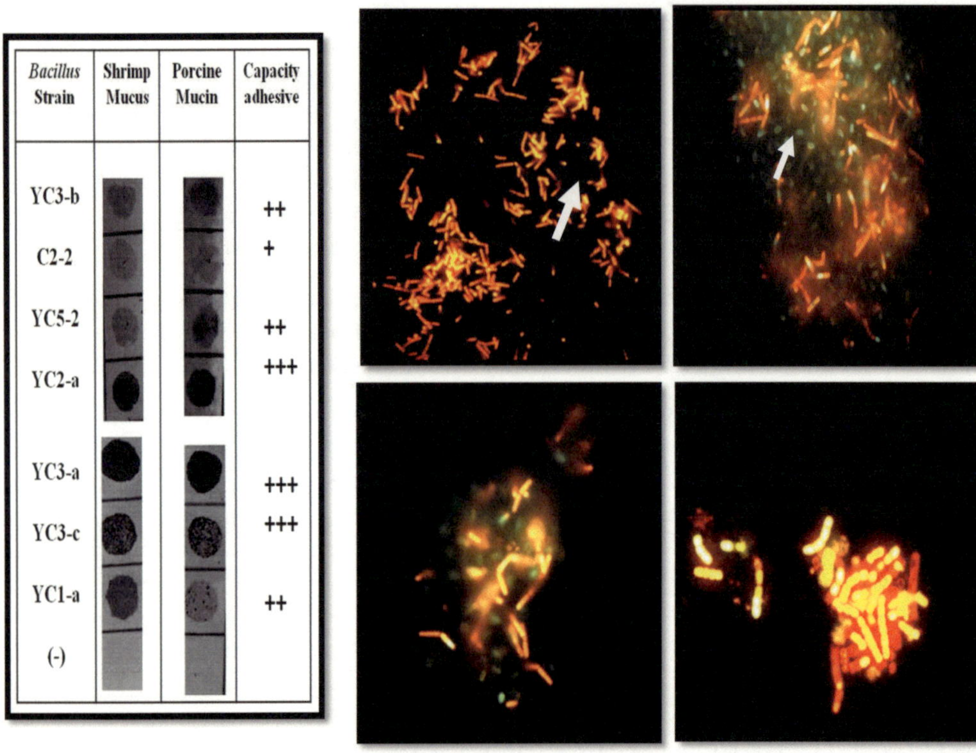

Figure 2. A) Testing of adhesion of bacterial isolates to shrimp mucus and mucin by the Dot-blot method,(-): negative control (Buffer Hepes-Hanks) Capacity: weak adhesion(+), moderate adhesion (++), strong adhesion (+++). B) Acridine orange staining of *Bacillus* spp. Adhered to mucus of shrimp observed by fluorescent microscope.

The presence of *Bacillus* species, whether as spores or vegetative cells, within the gut could arise from ingestion of bacteria associated with soil. However, a more unified theory is now emerging in which *Bacillus* species exist in an endosymbiotic relationship with their host, being able temporarily to survive and proliferate within the GIT. In some cases though, the endosymbiont has evolved further into a pathogen, exploiting the gut as its primary portal of entry to the host (*B. anthracis*) or as the site for synthesis of enterotoxins (*B. cereus, B. thuringiensis*) (Jensen et al., 2005).

5. Larval culture

Previous studies showed that inoculation with a probiotic strain during cultivation of larval *L. vannamei* (nauplii stage V) prevented colonization by a pathogenic strain, because the probiotic succeeds in colonizing the gut of the larvae (Zherdmant et al., 1997; Gómez-Gil et al., 2000). In this study, the effects of the probiotic strains cultured alone or mixed in the larval culture were evaluated. *Bacillus* strains were tested on larval shrimp using a daily concentration of 1×10^5 CFU mL^{-1}, starting each bioassay at nauplii V and a density of 225 nauplii L^{-1}. Inoculations of four natural, commercial products and antibiotic oxytetracycline were added directly to the water. Larvae inoculated with potential probiotic isolates at a density of 1×105 CFU mL^{-1} had significantly better survival than the control. The highest larval survival, compared to the control (4.9%) was inoculated with isolate YC5-2 (67.3%) and the commercial probiotic Alibio™ (57.4%). The low survival of the control shrimp (5%) in the second trial reinforced the view that probiotics are highly effective for increasing survival of larvae. Srinivas et al. (2010) showed that traditional practices (large exchange of water, application of disinfectants and antimicrobials, or both) are required to successfully complete the larval cycle; hence, the low survival rate in our control group in our bioassay was expected.

The larvae were sampled to determine the effect of the potential probiotics on larval development and rate of development, using the index of development (ID) described by Villegas and Kanazawa (1979):

$$ID = (\Sigma[i\ n_i]) / n,$$

Where i is the absolute value attributed to each larval stage (3 = ZIII; 4 = MI, 5 = MII; 6 = MIII, and 7 = PL1), n_i is the total number of larvae at stage I, and n is the number of organisms measured.

A mix of two strains induced the highest rate of development (7.00), followed by Alibio™ (6.35). Highest larval survival occurred with single-strain treatments, but the highest rate of larval development was obtained with the *Bacillus* mix. The onset of exogenous feeding by larvae of penaeid shrimp is a critical phase in survival, growth, and development because the larval gut is exposed to microbes at the transition from nauplii 5 to zoea I (Jones et al., 1997). In our study, *Bacillus tequilensis* (strain YC5-2), *B. endophyticus* (strains C2-2 and YC3-b), and *B. amyloliquefaciens* (strain YC2- a) significantly increased development of larvae (Luis-Villaseñor et al., 2011). Using probiotics, modification of bacterial communities in tank water improves cultivation of larval crustaceans (Balcazar et al.,2007b; Garriques and Arevalo, 1995; Gómez et al., 2008; Guo et al., 2009; Nogami and Maeda, 1992) and bivalves Douillet and Langdon (1993, 1994); Riquelme et al., 1996, 1997, 2001). Our study advances previous work demonstrating that probiotics maintain a balanced and natural bacterial community that improves production of shrimp larvae, which is also reflected in the rate of development, as demonstrated in our two bioassays with *Bacillus* spp.

Figure 3. Larvae shrimps of *Litopenaeus vannamei* in stage Zoea III and Mysis I.

Decamp et al. (2008) administered *B. subtilis* and *B. licheniformis* to larval *L. vannamei* and *Penaeus monodon* and this inhibited growth of *Vibrio* strains and increased the survival rate of the shrimp. Inhibitory effects of *Bacillus* are attributed to various causes: alterations of the pH in growth medium, use of essential nutrients, and production of volatile compounds (Chaurasia et al., 2005; Gullian et al., 2004; Yilmaz et al., 2006).

6. Modulation of microbiota

Intestinal bacteria thrive in a stable, nutrient rich environment but serve beneficial function to the host including energy salvage of otherwise indigestible complex carbohydrates, vitamin and micronutrient synthesis, activation of immune response, development and competitive exclusion of pathogenic microorganisms (Neish et al., 2010). It is clear that bacterial species of the gut can influence the health and robustness of the host. One of the problems associated with evaluating *Bacillus* products (or indeed any probiotic product) for aquaculture is determining whether the observed effect is due to the action of the bacterium on the host gut or due to an indirect effect on water quality or antagonism of external pathogens . Regardless, sufficient evidence suggests that adding *Bacillus* as spores or vegetative cells to rearing ponds has a beneficial effect. It is important to know the origin of the probiotic strain in order to increase the probability of survives and colonize the gastrointestinal tract of the host (Vine et al., 2004). The interest in investigating the intestinal microbiota is based on the need for a better understanding of how probiotics can influence the bacterial composition. For instance, Oxley et al., 2002, examined the bacterial flora of healthy wild and reared *P. mergulensis* shrimp and found a high abundance of *Vibrio*, the authors also found that the bacterial floras of wild and reared penaeid shrimp are similar and suggested that shrimp may influence and/or select the composition of their gut microbiota. To study the intestinal microbiota composition, culture-dependent methods are considered inadequate because more those 99% of all bacteria cannot yet be cultivated (Amann et al. 1995). Composition of the aquatic bacterial community in ponds has a strong influence on the internal bacterial flora of farmed marine animals, which is vital for their nutrition, immunity, and disease resistance (Luo et al., 2006). At the same time, it also impacts, and is impacted by, the bacterial communities in the nearby marine environments that receive aquacultural effluents (Guo & Xu 1994). Intestinal microbiota of cultivated aquatic organisms is an important factor in maintaining health, either by preventing colonization by pathogens, decomposition of food, production of antimicrobial compounds, releasing nutrients, and maintaining normal mucosal immunity (Escobar-Briones et al., 2006).

Single Strain Conformation Polymorphism (SSCP) is based on sequence-specific separation of polymerase chain reaction (PCR)-derived rRNA gene amplicons in polyacrylamide gels is used to study the diversity of microbes based on the sequence difference of PCR products of 16S rDNA gene amplified from different microbes (Dohrmann and Tebbe, 2004). Our interest in intestinal microbiota is based on the need for understanding how probiotics influence bacterial composition. Similar studies have been performed in vertebrates (Brikbeck et al., 2005; Austin, 2006; Escobar-Briones et al., 2006; Bakke-McKellep, 2007; He et al., 2009;Nayak, 2010; Tapia-Paniagua et al., 2010). In invertebrates, studies are limited; they include Pacific white shrimp *Litopenaeus vannamei* (Johnson et al., 2008), Kuruma shrimp

Marsupenaeus japonicus (Liu et al., 2010), European lobster *Homarus gammarus* L. (Daniels 2 et al., 2010), and Chinese shrimp *Fenneropenaeus chinensis* (Liu et al., 2011).

We used probiotic strains of *Bacillus* that are antagonistic to pathogenic strains of Vibrio, are not harmful to juvenile shrimp, and adhere to and grow on intestinal mucosa, which is an important factor in colonizing or at least remaining for a moderate amount of time in the shrimp gut (Luis-Villaseñor et al., 2011).In our study, SSCP analysis using universal primers targeting the V4 and V5 regions of the 16S rRNA gene were used to visualize the bacterial diversity and identify the dominant intestinal bacterial in juvenile shrimp *L. vannamei* (Fig. 4). Tanks were stocked with 21 shrimp (8± 0.1 g each), and inoculated daily with one of the following treatments:

1. *Bacillus* mix at a density of 0.1×10^6 CFU mL^{-1}.
2. Commercial probiotic Alibio®at $1{\times}10^6$ CFU mL^{-1}.
3. Control: Juvenile *L. vannamei* without probiotics.

Each treatment and control was performed in quintuplicate and each replicate was represented by one tank.

A total of 119 bands from four SSCP gels were registered, sequenced, and identified. Analysis of the SSCP fingerprints showed that the composition of the intestinal microbiota of juvenile *L. vannamei* exposed to a *Bacillus* mix was modified. The shrimp treated with *Bacillus* mix showed higher bacterial diversity than the control groups. Liu et al. (2010) reported that the addition of *Bacillus* spp. in feed of the shrimp *Marsupenaeus japonicus* increased individual variation and the total diversity of bacterial species.

A comparison of the patterns obtained from shrimp gut samples inoculated with probiotics at 5 days showed uniformity in the composition of the microbiota and clustering with high similarity of 71.3% and71.21% for *Bacillus* mix and Alibio, respectively. However, both exhibited a lower similarity that control group by 23.7% (Fig. 5a).

The dendrogram analysis at day 10 showed that SSCP pattern in samples from shrimp treated with *Bacillus* mix were clustered into one group was 62.3% for M1-M2 and 82.8% for M4-M5, whereas shrimps treated with Alibio were clustered into a different one had similarity of 72.7% (A1-A5). Results were heterogeneous in the Control group, with similarity of 50.6% for C1-C4 and 84.6% for C2-A4 (Fig. 5b). Similarity at day 15 had the highest homogeneity between treatments: 86.9% for the *Bacillus* mix treatments (M1-M3) and 93.2% (M2-M4) and 87.6% for the Alibio treatments (A1-A3) and 93.9% (A1-A5) (Fig. 6a). Similar banding patterns occurred at day 20, reaching 89.9% to 98.5%. Variation in the communities with eachtreatment group did not vary greatly (Fig. 6b).

In our study, most of the OTUs identified by SSCP gels treated with the probiotics belong to phylogenic groups class α- andγ-proteobacteria, flavobacteria, shingobacteria, and fusobacteria, compared with other species of invertebrates, where the microbiota were represented by class α-, γ-, and ε-proteobacteria in fleshy prawn *Fenneropenaeus chinensis* (Lui et al., 2011),by fusobacteria and γ-proteobacteria in giant tiger prawn *Penaeus monodon*

(Chaiyapechara et al., 2011), and by derribacteres, mollicutes, γ- and ε-proteobacteria, small fractions of firmicutes, cytophaga-flavobacter-bacteroides, verricomicrobiae, β- and δ-proteobacteria in vent shrimp (Durand et al., 2010).Furthermore, the gut content of shrimps inoculated with the *Bacillus* mix and Alibio had higher bacterial diversity, compared with the controls, supported by the total number of OTU´s.

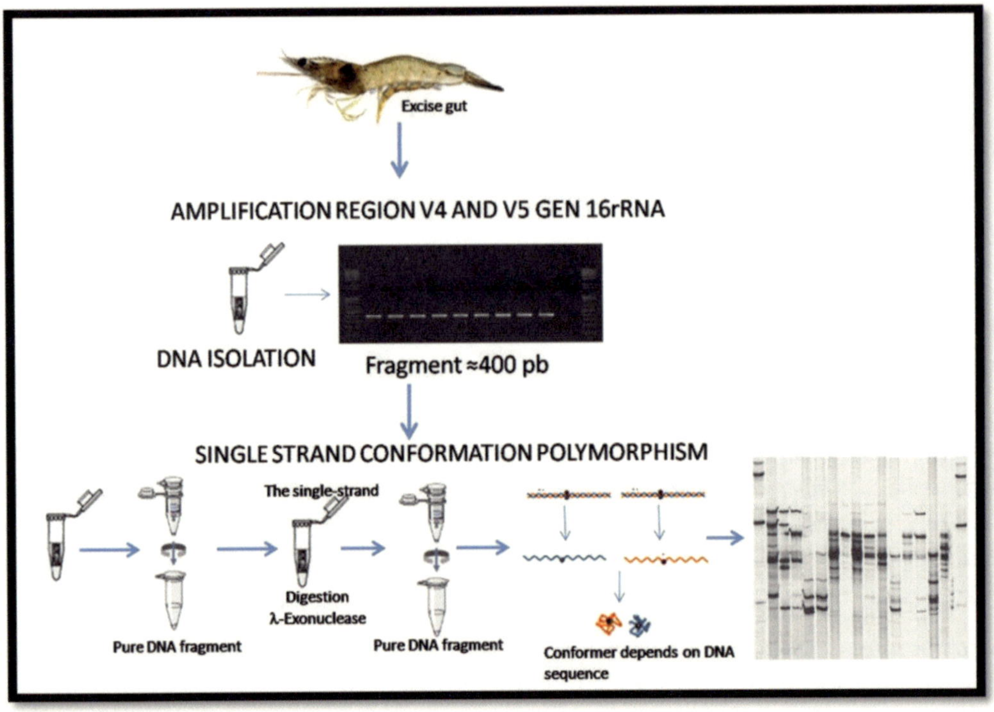

Figure 4. Schematic illustrating the process of Single strand conformation polymorphism (SSCP).

The intestinal bacterial community shows a similar dominance of α-proteobacteria and flavobacteria at all times in shrimp treated with probiotics. The resident community included *Maribius salinus* and *Donghicola eburneus* (α-proteobacteria) and *Wandonia haliotis* (flavobacteria) in all treatments. Dominance of γ-proteobacteria occurs in the intestinal community of other crustaceans, including *Fenneropenaeus chinensis* (Liu et al., 2011), ornate rock lobster *Panulirus ornatus* (Payne et al., 2007), *Rimicaris exoculata* (Durand et al., 2009), European lobster *Homarus gammarus* L. (Daniels et al., 2010), and *Penaeus monodon* (Chaiyapechara et al., 2011).

Sequence analysis showed that at day 5, intestines of the shrimp were dominated by phylogenetic groups flavobacteria and α-proteobacteria., At day 15, the *Bacillus* mix treatment had small populations of α-proteobacteria and flavobacteria,the Alibio treatment led to the appearance of sphingobacteria and fusobacteria. At day 20, α- and γ-proteobacteria, sphingobacteria, and flavobacteria were present, with few variations between treatments.

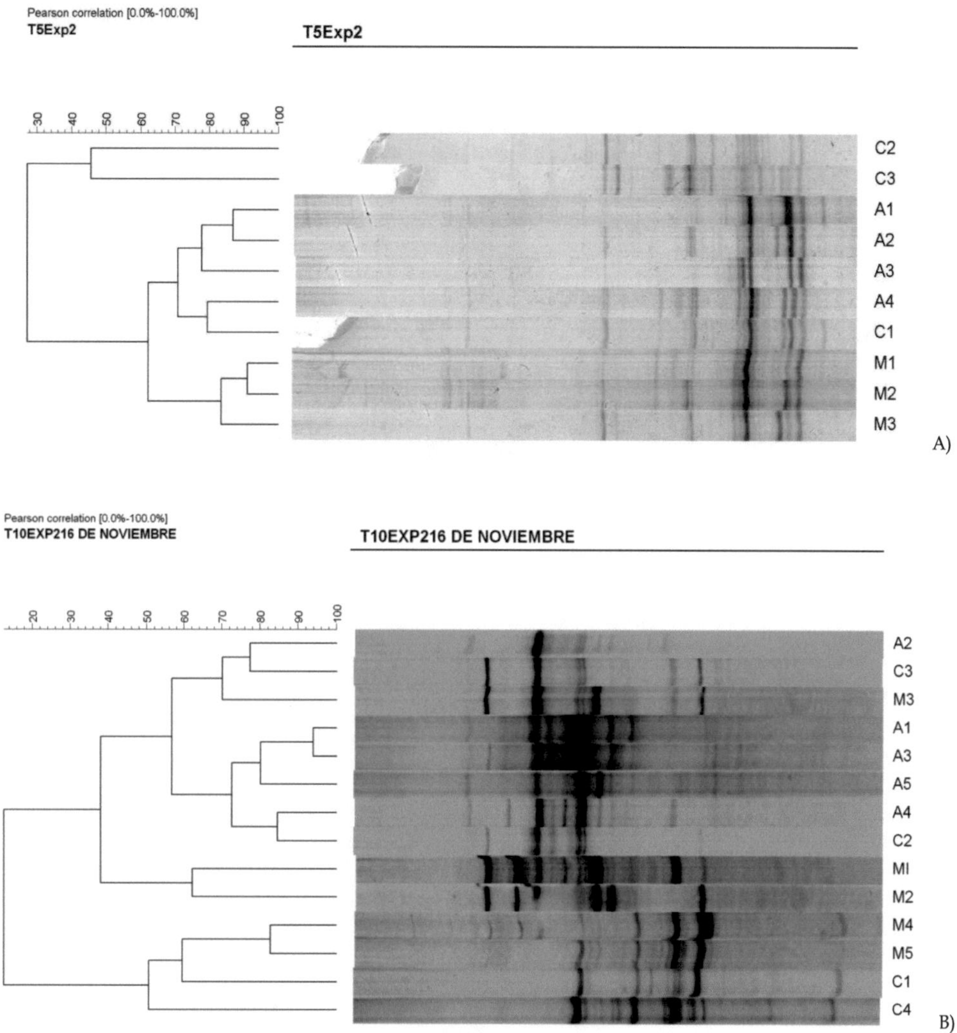

Figure 5. Dendrogram illustrating the relationship (percent similarity) between bacterial communities in gut of shrimp at 5 d (a) and 10 d (b) inoculated with probiotics; M1–M5 (*Bacillus*mix), A1–A5 (commercial probiotic), C1–C4 (without probiotics). Scale of dendrogram show similarity percent of clusters. The dendrogram was calculated with UPGMA and Pearson correlation.

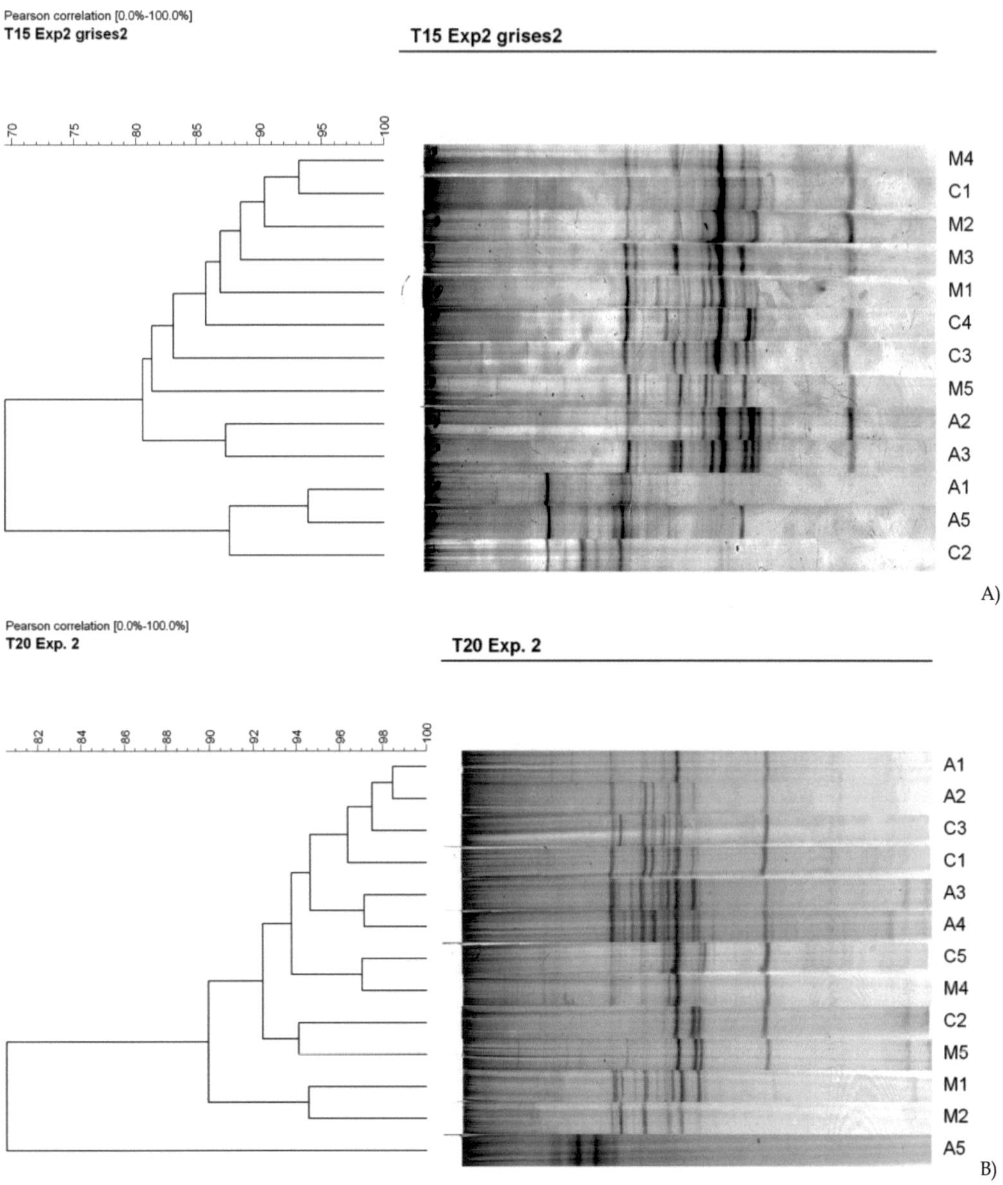

Figure 6. Dendrogram illustrating the relationship (percent similarity) between bacterial communities in shrimp gut at 15 d (a) and 20 d (b) inoculated with probiotics; M1–M5 (*Bacillus*mix), A1–A5 (commercial probiotic), C1–C4 (without probiotics). Scale of dendrogram showed similarity percent of clusters. The dendrogram was calculated with UPGMA and Pearson correlation.

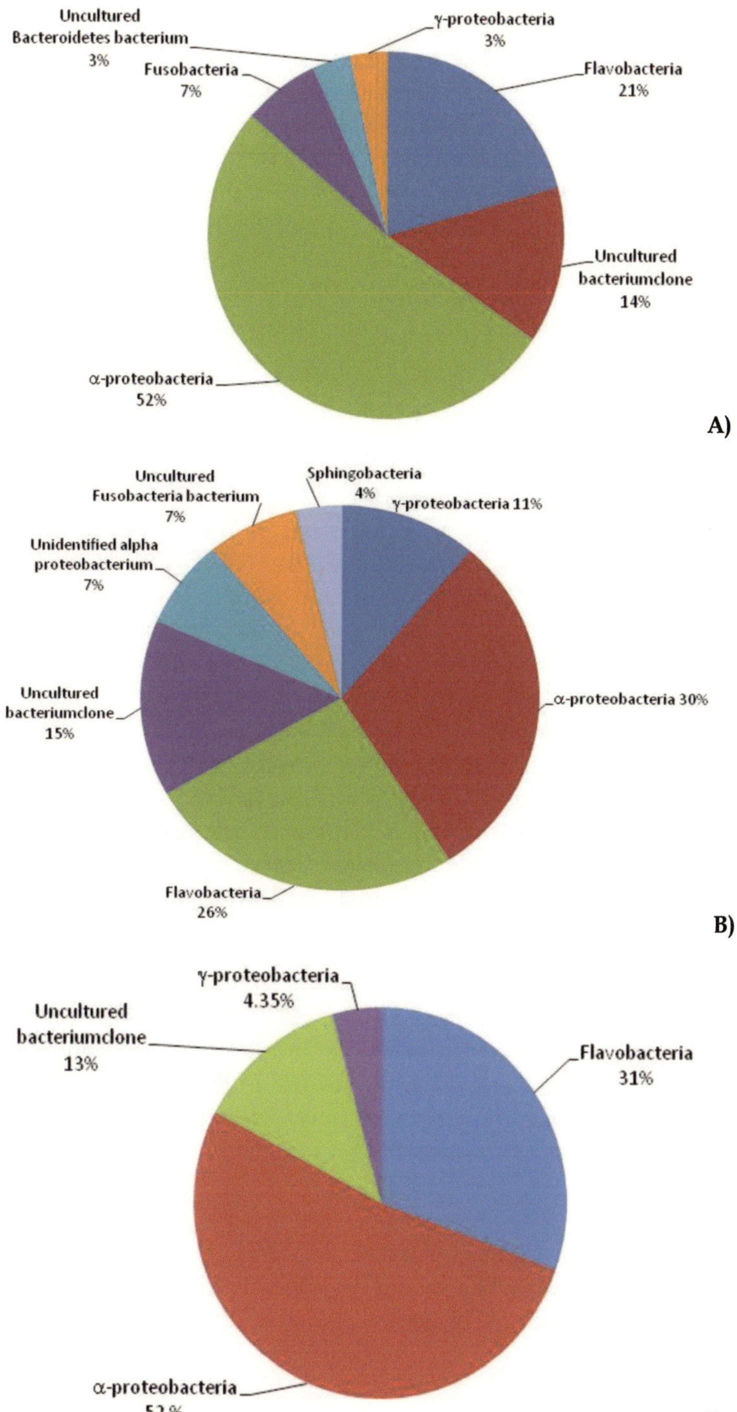

Figure 7. Composition of intestinal bacterial community of individual *L. vannamei* inoculated with probiotics *Bacillus* mix (M5-M20), Alibio (A5-A20), and Control (C5-C20) based on 16S rRNA.

Dempsey et al., (1989) suggest that only one or two phylogenic groups dominate the shrimp gut and have very low diversity. The most common genera of gut microbiota in aquatic invertebrates are *Vibrio, Pseudomonas, Flavobacterium, Micrococcus,* and *Aeromonas* (Harris, 1993). These reports of gut communities in shrimp were based mainly on culture dependent microbiological techniques. Comparisons with molecular techniques indicate that 10–50% of population is cultivable (Holzapfel et al., 1988; Wilson et al., 1996). Since the SSCP monitors the predominant bacteria in a sample, bands representing *Bacillus* probionts were not detected because the density of probiotic strains was $<0.1 \times 10^6$CFU mL^{-1}. Smalla et al., (2007) reported that DGGE and SSCP can contribute to the generation of the same bands, hence, leading to an underestimate of diversity. Likewise, Muyzer et al., (2003) shows that DGGE can only detect 1–2% of the microbial population representing the dominant species present in microbial communities.

7. Conclusion

Bacillus spp. exposed to *L. vannamei* increased survival, and development in larvae, and modulated the intestinal microbiota in juvenile shrimp. This study demonstrated that the management the properly combinations of selected *Bacillus* isolates are a good option to improve health, rate of development, and survival in shrimp. The isolates we tested were antagonistic to pathogenic strains of *Vibrio* and were not harmful to the larvae. Their ability to adhere and grow in intestinal mucosa is an important factor in colonizing or at least remaining for short time periods in the gut of shrimp. More rapid development also occurred when the larvae were treated with mixtures of Bacillus strains. Treatment Mix-2 increased survival and larval development, compared to the control group. Similar results were found by Guo et al. (2009), where *B. fusiformis* increased survival and accelerated metamorphosis of *P. monodon* and *L. vannamei* larvae. This study demonstrated that management that combines properly selected *Bacillus* isolates are a good option in larviculture to improve health, rate of development, and rate of survival of whiteleg shrimp.

In summary, analysis of SSCP fingerprints demonstrated that the composition of the intestinal microbiota of shrimp inoculated with the *Bacillus* mix was distinctly different from the control group. The *Bacillus* mix significantly reduced species diversity and richness and increased similarity of the microbial communities within the probiotic replicates, reducing diversity compared to the control, predominantly consisting of α-and γ-proteobacteria, fusobacteria, sphingobacteria, and flavobacteria.

Author details

I.E. Luis-Villaseñor, A.I. Campa-Córdova and F.J. Ascencio-Valle
Centro de Investigaciones Biológicas del Noroeste S.C., México

8. References

Aguirre-Guzmán, G. Ascencio-Valle, F. 2001. Infectious diseases in shrimp species with aquaculture potential. Recent Research of Developmental Microbiology 4: 333-348.

Alander, M., Satokari, R., Korpela, R. 1999. Persistence of colonisation of human colonic mucosa by a probiotic strains, *Lactobacillus rhamnosus* GG after oral consumption. Applied and Enviromental Microbiology 65:351-354.

Ali, A. 2000. Probiotic in fish farming-Evaluation of a candidate bacterial mixture. Sveriges Lantbruks Universitet. Umea, Senegal.

Amann, R. I., Ludwig, W., Schleifer, K.H., 1995. Phylogenetic identification and in situ detection of individual microbial cells without cultivation. Microbiol Rev 59, 143-169.

Austin B, 2006. The bacterial microflora of fish, revised. Scientific World J 6:931–945

Apostolou, E., Kirjavainen, P.V., Saxelin, M., Rautelin, H., Valtonen, V., Salminen, S., Ouwehand, A.C. 2001. Good adhesión properties of probiotics: a potential risk for bacteremia?. FEMS Immunology and Medicine Mocrobiology 31: 35-39.

Baake–McKellep AM, Penn MH, Salas S, Refstie S, Sperstad S, Landsverk T&Krogdahl A. 2007. Effects of dietary soybean meal, insulin and oxytetracycline on intestinal microbiota and epithelial cell stress, apoptosis and proliferation in the teleost Atlantic Salmon (Salmosalar L.). Brit J Nutr 97:699–713

Balcázar, J.L., Rojas-Luna, T., 2007a. Inhibitory activity of probiotic Bacillus subtilis UTM 126 against vibrio species confers protection against vibriosis in juvenile shrimp (Litopenaeus vannamei). Curr. Microbiol. 55, 409–412.

Balcázar, J.L., Ruíz-Zarzuela, I., Cunningham, D., Vendrell, D., Múzquiz, J.L., 2006. The role of probiotics in aquaculture. Vet. Microbiol. 114, 173–186.

Bernheimer, A.W., Grushoff, P. 1967. Extracellular Hemolysins of Aerobic Sporogenic Bacilli. J. Bacteriol. 93 (5), 1541-1543.

Birkbeck TH & Ringo E, 2005. Pathogenesis and the gastrointestinal tract of growing fish. In: Microbial Ecology in Growing Animals (ed. By W. Holzapfel& P. Naughton Biology in Growing Animals Series (ed by S.G. Pierzynowski& R. Zabielski). Elsevier 208–234

Cahill M.M. 1990. Bacterial flora of fishes: a review. Microbial Ecology19, 21-41.

Chaiyapechara S, Rungrassamee W, Suriyachay I, Bacterial community associated with the intestinal tract of *P. monodon* in commercial farms. Microb ecol (2011), doi: 10.1007/s00248-011-9936-2

Chaurasia, B., Pandey, A., Palni, L.M.S., Trivedi, P., Kumar, B., Colvin, N., 2005. Diffusible and volatile compounds produced by an antagonistic *Bacillus subtilis* strain cause structural deformations in pathogenic fungi in vitro. Microbiol. Res. 160, 75–81.

Chen, F.R., Liu, P.C., Lee, K.K. 2000. Lethal attribute of serine protease secreted by Vibrio alginolyticus strains in Kuruma Prawn Penaeus japonicas. Zool Natur for Sch 55:94-99.

Chukeatirote E. 2002. Potential use of probiotics. Song J. Sci. Tech. 25: 275-282.

Daniels CL, Merrifield DL, Boothroyd DP, Davies SJ, Factor JR & Arnold KE, 2010 Effect of dietary *Bacillus* spp. and mannan oligosaccharides (MOS) on European lobster (*HomarusgammarusL.*) larvae growth performance, gut morphology and gut microbiology and gut microbiota. Aquaculture 304: 49–57

Decamp, O., Moriarty, D.J.W., Lavens, P., 2008. Probiotics for shrimp larviculture: review of field data from Asia and Latin America. Aquac. Res. 39, 334-338.

Dempsey AC, Kitting CI & Rosson RA, 1989. Bacterial variability individual panaeid shrimp digestive tracts. Crustaceana 56: 267–278

Dohrmann, A. B., Tebbe, C.C., 2004. Microbial community analysis by PCR-single-strand conformation polymorphism (PCR-SSCP), PCR-single-strand conformation polymorphism (PCR-SSCP), p. 809–838. *In* G. A. Kowalchuk, F. J. de Bruijn, I. M. Head, A. D. Akkermans, and J. D. van Elsas (ed.), Molecular microbial ecology manual. Kluwer Academic Publishers, Dordrecht, The Netherlands.

Dopazo, C., Lemos, M., Lodeiros, C., Barja, J., Toranzo, A., 1988. Inhibitory activity of antibiotic-producing bacteria against fish pathogens. J. Appl. Bacteriol. 65, 97–101.

Douillet, P.A., Langdon, C.J., 1993. Effect of marine bacteria on the culture ofaxenic oyster *Crassostrea gigas*. Biol. Bull. 184, 36-51.

Douillet, P.A., Langdon, C.J., 1994. Use of a probiotic for the culture of larvae of the Pacific oyster (*Crassostrea gigas* Thunberg). Aquaculture 119, 25-40.

Drablos, F., Nicholson, D., Ronning, M., 1999. EXAFS study of zinc coordination in Bacitracin A. Biochim, Biophys, Acta 1431, 433–442.

Durand L, Zbinden M, Cueff-Gauchard V, Duperron S, Roussell EG, Shillito B & Cambon–Bonavita MA, 2009. Microbial diversity associated with the hydrothermal shrimp Rimicarisexoculata gut and occurrence of a resident microbial community. FEMS Microbiol Ecol 71: 291–303

Egorov, N.S. 2004. Fundamentals of the theory of antibiotics: A manual (in Russian), Moscow. In: Semenov, A., Sgibnev, A., Cherkasov, S.V., Bukharin, O. V. 2007. Bacterial regulation of antagonistic activity of bacteria. Bulletin of Experimental Biology and Medicine 144(5): 702-705.

Escobar-Briones, L., Olvera-Novoa, M.A., Puerto-Castillo, C. 2006. Avances sobre ecología microbiana del tracto digestivo de la tilapia y sus potenciales implicaciones. VIII Simposium internacional de Nutrición Acuícola. 15-17 de Noviembre. Universidad Autónoma de Nuevo León, Monterrey, Nuevo León, México.

Farzanfar, A., 2006. The use of probiotics in shrimp aquaculture. FEMS Immunol Med Microbiol 48, 149-158.

Forstner, G., Forstner, J., Fahim, R. 1989. Small intestinal mucin: polymerization and the link glycopeptides. Soc Exp Biol 259-271.

Forstner, J.F., Forstner G.G. 1994. Gastrointestinal mucus, in physiology of the gastrointestinal tract. L.R. Johnson, Ed. Raven Press: New York 1255-1281.

Fuller, R. 1989. Probiotics in man and animals. Journal of Applied Bacteriology 66:265-378.

Gatesoupe, F.J. (1999) The use of probiotics in aquaculture. Aquaculture 180, 147–165.

Garriques, D., Arevalo, G., 1995. An evaluation of the production and use of a live bacterial isolate to manipulate the microbial flora in the commercial production of Penaeus vannamei postlarvae in Ecuador. In: Browdy, C.L., Hopkins, J.S. (Eds.), Swimming through Troubled Water. : Proceedings of the Special Session on Shrimp Farming, Aquaculture 1995. World Aquaculture Society, Baton Rouge, LA, pp. 53–59.

Gibbons, R. J., 1982. Review and discussion of role of mucus in mucosal defense, in Recent Advances in Mucosal Immunity, L. A. H. W. Strober, and K.W.Sell, Editor. Raven Press: New York. p. 343-351.

Gómez R, Geovanny D, Balcazar JL, Shen MA, 2007. Probiotics as control agents in Aquaculture. J. Ocean Univ. China. 6: 76-79.

Gómez, R., Geovanny, D., Shen, M.A., 2008. Influence of probiotics on the growth and digestive enzyme activity of white Pacific shrimp (Litopenaeus vannamei). Ocean Coast Sea Res. 7, 215–218.

Gómez-Gil, B., Roque, A., Turnbull, J.F., 2000. The use and selection of probiotic bacteria for use in the culture of larval aquatic organisms. Aquaculture 191, 259–270.

Gullian, M., Thompson, F., Rodríguez, J., 2004. Selection of probiotic bacteria and study of their immunostimulatory effect in *Penaeus vannamei*. Aquaculture 233, 1–14.

Guo, J.J., Liu, K.F., Cheng, S.H., Chang, C.I., Lay, J.J., Hsu, Y.O., Yang, J.Y., Chen, T.I., 2009. Selection of probiotic bacteria for use in shrimp larviculture. Aquac. Res. 40, 609–618.

Guo P & Xu M, 1994 The bacterial variation in the water environment of cultured prawn pond. Oceano et Limno Sinica 25:625–629.

Harris J, Seiderer L & Lucas M, 1991. Gut microflora of two salt marsh detritivore thalassinid prawns, Upogebia africana and Callianas safraussi. Microb Ecol 21: 277–296

Havennar, R., ten Brink, B., Huis in't Veld, J.H.J. 1992. Selection of strains for probiotics use. In Probiotics, the Scientific Basis ed. Fuller, R. pp. 209-224. London: Chapman and Hall.

Herich, R., Levkut, M. 2002. Lactic acid bacteria, probiotics and immune system. Vet Med 47:169-180.

Hu, H. Q., Li, X.S., He, H. 2010. Characterization of an antimicrobial material from a newly isolated Bacillus amyloliquefaciens from mangrove for biocontrol of Capsicum bacterial wilt. Biological control 54, 359-365.

Jensen, G.B., Hansen, B.M., Eilenberg, J. and Mahillon, J. 2003. The hidden lifestyles of Bacillus cereus and relatives. Environ. Microbiol. 5, 631–640.

Jones, D.A., Yule, A.B., Holland, D.L., 1997. Larval nutrition. In: D'Abramo, L.R., Conclin, D.E., Akiyama, D.M. (Eds.), Crustacean Nutrition. Advances in World Aquaculture. World Aquaculture Society, Baton Rouge, LA.

Johnson CN, Barnes S, Ogle J &Grimes JD. 2008. Microbial Community Analysis of Water, Foregut, and Hindgur during Growth of Pacific White Shrimp, *Litopenaeus vannamei*, in Closed-System Aquaculture. J World Aquacult Soc 39:251–257.

Kesarcodi-Watson A., Kaspar H., Lategan M.J., Gibson L., 2008. Probiotic in aquaculture The need, principles and mechanisms of action and screening process. Aquaculture 274, 1-14.

Kirjavainen, P.V., Ouwehand, A.C., Isolauri, E., Salminen, S.J. 1998. The ability of probiotic bacteria to bind to human intestinal mucus. FEMS Microbiol Lett 167:85-189.

Liu CH, Chiu CS, Ho PL & Wang SW. 2009. Improvement in the growth performance of white shrimp, *Litopenaeus vannamei*, by a protease-producing probiotic, *Bacillus subtilis* E20, fronnatto. J Appl Microbiol 107:1031–1041.

Liu H, Liu M, Wang B, Jiang K, Jiang S, Sun S & Wang L (2010) PCR-DGGE analysis of intestinal bacteria and effect of *Bacillus* spp. On intestinal microbial diversity in kuruma shrimp (*Marsupenaeusjaponicus*). Chin J Oceanol Limn 28:808–814

Liu H, Wang L, Liu M, Wang B, JiangK, Ma S & Li Q, 2011. The intestinal microbial diversity in Chinese shrimp (*Fenneropenaeuschinensis*) as determined by PCR-DGGE and clone library analyses. Aquaculture 317:32–36

Luis-Villaseñor, I.E., Macias-Rodriguez, M.E., Gomez-Gil, B., Ascencio-Valle, F., Campa-Cordova, A.I. Beneficial effects of four *Bacillus* strains on the larval cultivation of *Litopenaeus vannamei*, Aquaculture (2011), doi: 10.1016/j.aquaculture.2011.08.036

Luo, P., Hu, C., Xie, Z., Zhang, L., Ren, C., Xu, Y., 2006. PCR-DGGE analysis of bacterial community composition in brackish water Litopenaeus vannamei culture system. J. Trop Oceanogr 25(2):49–53.

Majamaa, H., Isolauri, E., Saxelin, M., Vesikari, T. 1995. Lactici acid bacteria in the treatment of acute rotavirus gastroenteritis. J Pediatr Gastroenterol Nutr 20: 333-338.

Martin, G.G., Rubin, N., Swanson, E. 2004. *Vibrio parahaemolyticus and Vibrio Harvey* cause detachment of the epithelium from the midgut trunk of the penaeid shrimp Sicyonia ingenti. Diseases of Aquatic Organisms 60:21–29.

Morikawa, M., Ito, M., Imanaka, T., 1992. Isolation of a new surfactin producer *Bacillus pumilus* A-1, and cloning and nucleotide sequence of the regulator gene, psf-1. J. Ferment. Bioeng. 74, 255–261.

Mykles D.L. 1977. The ultrastructure of the posterior midgut caecum of *Pachygrapsus crassipies* (Decapoda, Brachyura) adapted to low salinity. Tissue cell 9:681–691.

Nakayama, T., Lu, H., Nomura, N., 2009. Inhibitory effects of *Bacillus* probionts on growth and toxin production of Vibrio harveyi pathogens of shrimp. Lett. Appl. Microbial. 49, 679–684.

Nayak, S.K. 2010. Probiotics and immunity: A fish perspective. Fish and Shellfish Immonology 29: 2-14.

Neufeld D.S., Cameron, J.N. 1994. Mechanism of the net uptake of water in moulting blue crags (*Callinectes sapidus*) acclimated to high and low salinities. J. Exp Biol 188:11–23.

Nguyen, T.N.T., Dierckens, K., Sorgeloos, P., Bossier, P., 2007. A review of the functionality of probiotic in the larviculture food chain. Marine Biotechnol. 10, 1-12.

Nogami K. Maeda M. 1992. Bacteria as biocontrol agents for rearing larvae of the crab *Portunus tribeculatus*. Canadian Journal of Fisheries and Aquatic Sciences 49, 2373-2376.

Olsson JC, Westerdahk A, Conway PL, Kjelleberg S 1992. Intestinal colonization potential of turbot (*Scophthalmus maximus*) and dab (*Limanda limanda*) associated bacteria with inhibitory effects against *Vibrio anguillarum.* App. Env. Microbio. 58: 551-556.

Onarheim, A. M., Wilik, R., Burghardt, J. Stackebrandt, E. 1994. Characterization and identi®cation of two *Vibrio* species

indigenous to the intestine of ®sh in cold sea water; description of *Vibrio iliopiscarius* sp. nov. *Syst Appl Microbiol* 17, 370±379.

Oxley, A.P., Shipton, W., Owens, L., McKay, D. 2002. Bacterial flora from the gut of the wild and cultured banana prawn, Penaeus merguienesis. J Appl Microbiol 93:214-223.

Ouwehand, A.C., Kirjavainen, P.V., Shortt, C., Salminen, S. 1999. Probiotics: mechanisms and established effects. Int Dairy J 9:43-52.

Ouwehand, A.C., Tölkkö, S., Kulmala, J., Salminen, S., Salminen, E. 2000. Adhesion of inactivated probiotic strains to intestinal mucus. Letters in Applied Microbiology 31:82-86.

Payne MS, Hall MR, Sly L & Bourne DG (2007) Microbial diversity within early-stage cultured Panulirusornatusphyllosomas. Appl Environ Microbiol 73:1940–1951

Perdigon G, Alvarez S, Rachid M, Agüero G, Gobbato N 1995. Probiotic bacteria for humans: clinical systems for evaluation of effectiveness: immune system stimulation by probiotics. J. Dairy Sc. 78: 1597-1606.

Perez, C., Suarez, C., Castro, G.R., 1993. Antimicrobial activity determined in strains of Bacillus circulans cluster. Folia Microbiol. 38, 25–28.

Ravi, A.V., Musthafa, K.S., Jegathammbal, G., Kathiresan, K., Pandian, S.K., 2007. Screening and evaluation of probiotics as a biocontrol agent against pathogenic Vibrios in marine aquaculture. Lett. Appl. Microbiol. 45, 219-223.

Ringo E. & O Vadstein. 1998. Colonization of *Vibrio pelagius* and *Aeromonas caviae* in early developing turbot (*Scophtalmus maximus*) larvae. Journal of Applied Microbiology. 84: 227 - 233.

Riquelme, C., Hayashida, G., Araya, R., Uchida, A., Satomi, M., Ishida, Y., 1996. Isolation of a native bacterial strain from the scallop *Argopecten purpuratus* with inhibitory effects against pathogenic *Vibrios*. J. Shellfish Res. 15, 369–374.

Riquelme, C., Araya, R., Vergara, N., Rojas, A., Guaita, M., Candia, M., 1997. Potential probiotic strains in the culture of the Chilean scallop *Argopecten purpu*ratus (Lamarck, 1819). Aquaculture 154, 17-26.

Riquelme, C., Jorquera, M.A., Rojas, A.I., Avendano, R.E., Reyes, N., 2001. Addition of inhibitor producing bacteria to mass cultures of *Argopecten purpuratus* larvae (Lamarck, 1819). Aquaculture 192, 111-119.

Rojas, M. Conway, P.L., 2001. A Dot-Blot assay for adhesive components relative to probiotics. Methods Enzymol. 336, 389-402.

Sakata T. 1990. Microflora in the digestive tract of fish and shellfish. In: Microbiology in Poecilotherms (ed. By R. Lesel), pp.171-176. Elsevier,Amsterdam, theNetherlands.

Shornikova, A.V., Casas, I.A., Isolauri, E., Vesikari, T. 1997. Lactobacillus reuteri as a therapeutic agent in acute diarrhoea in Young children. J. Pediatr Gastroenteril Nutr 24: 399-404.

Smalla K, Oros-Sichler M, Milling A, Heuer H, Baumgarte S, Becker R, Neuber G, Kropf S, Ulrich A, Tebbe CC (2007) Bacterial diversity of soils assessed by DGGE, T-RFLP and SSCP fingerprints of PCR-amplified 16S rRNA gene fragments: Do the different methods provide similar results?. J Microbiol Meth 69:470–479.

Srinivas, S. P., Abdulaziz, A., Natamai, S. J., Prabhakaran, P., Balachandran, S., Radhakrishnan, P., Rosamma, P., Ambat, M., Bright, I.S.S., 2009. *Penaeus monodon* larvae can be protected from *Vibrio harveyi* infection by pre-emptive treatment of a rearing system with antagonistic or non-antagonistic bacterial probiotics. Aquac. Res. 1-14.

Sundari, C.S., Raman, B., Balasubramanian, D., 1991. Hydrophobic surfaces in oligosaccharides; linear dextrins are amphiphilic chains. Biochim Biophys Acta 1065:35-41.

Tapia-Paniagua, S.T., Chabrillo´ n, M., Dı´az-Rosales, P., de la Banda, I.G., Lobo, C., Balebona, M.C. and Morin˜igo, M.A. (2010) Intestinal microbiota diversity of the flat fish Solea senegalensis (Kaup, 1858) following probiotic administration. Microb Ecol 60, 310–319.

Thompson, F.L., Lida, T., Swings, J. 2004. Biodiversity of Vibrios. Microbiology and molecular biology reviews. 68(3):403-431.

Tuohy, K.M., Probert, H.M., Smejkal, C.W., Gibson, G.R. 2003. Using probiotics and prebiotics to improve gut health. Drug Discovery Today 8 (15):692-700.

Van Hai, N., Buller, N., Fotedar, R., 2009. Effects of probiotics (Pseudomonas synxantha and Pseudomonas aeruginosa) on
the growth, survival and immune parameters of juvenile western king prawns (Penaeus latisulcatus Kishinouye, 1896). Aquac. Res. 40, 590–602.

Villegas, C.T. Kanazawa, A., 1979. Relationship between diet composition and growth of zoeal and mysis stages of *Penaeus japonicus* (Bate). Fish Res. J. Philippines 4, 32–40.

Vinderola, C.G., Medici, M., Perdigón G. 2004. Relationship between interaction sites in the gut, hydrophobicity, mucosal immunomodulating capacites and cell wall protein profiles in indigenous and exogenous bacteria. Journal of Applied Microbiology 96:230-243.

Vine, N.G., Leukes, W.D., Kaiser, H., Baxter, J., Hecht, T., 2004. Competition for attachment of aquaculture candidate probiotic and pathogenic bacteria on fish intestinal mucus. J. Fish Dis. 27, 319–326.

Wilson K. Blitchington R. 1996. Human colonic biota studied by ribosomal DNA sequence analysis. Appl Environ Microb 62:2273–2278.

Yilmaz, M., Soran, H., Beyatli, Y., 2006. Antimicrobial activities of some *Bacillus* spp. strains isolated from the soil. Microbiol. Res. 161, 127–131.

Zamora-Rodríguez L.M. 2003. Aislamiento, identificación y conservación de cultivos de bacterias lácticas antagonistas de microbiota contaminante de sangre de matadero. Doctoral Dissertation, Universidad de Girona, España a, 259pp.

Zherdmant, M.T., San Miguel, L., Serrano, J., Donoso, E., Miahle, E., 1997. Estudio y utilización de probióticos en el Ecuador. Panorama Acuícola 2, 28.

Zhou, X., Wang, Y. Li, W., 2009. Effect of probiotic on larvae shrimp (*Penaeus vannamei*) based on water quality, survival rate and digestive enzyme activities. Aquaculture 287, 349-353.

Chapter 28

Probiotic Biofilms

Mariella Rivas and Carlos Riquelme

Additional information is available at the end of the chapter

http://dx.doi.org/10.5772/50124

1. Introduction

Microalgae are in global scale primary producers, they are involved in all marine and fresh waters ecosystems. The growth of microalgae is correlated directly with the chlorophyll *a* concentration, and the bacterial population, and both variables are tightly related with the number of planktonic cells [1, 2]. However, there are numerous studies completed at date about microalgae, often the associated communities of bacteria have not been considered. Recently it has been evidenced that there is not only a positive correlation between bacteria and microalgae concentration but there is also a positive correlation between the extracellular polymeric substances (EPS), which is bigger in bacteria-microalgae mixed cultures than in microalgae axenic cultures [3]. These bacterial communities play a critical role in modulating the population dynamic and the algal metabolism. The kinds of interactions between algae and symbiotic bacteria under photoautotrophic conditions may involve mutualism and commensalism [4]. The role of bacteria is important because they act as a source of inorganic nutrients, feeding, and in viral lysis in algal growth control, physiology, and events of cellular differentiation [5, 6]. Bacteria in microalgal phycosphere stimulate algal growth creating a favorable environment [figure 1; 7], regenerating organic and inorganic nutrients [8, 9], or producing growing factors, including trace metals, vitamins, phytohormones and chelates [10, 11]. Nevertheless, in some described cases microbiota can inhibit algal growth. Algaecide bacteria are investigated as a one of the key biological agents in the abrupt end of microalgae blooms [12]. Algaecide bacteria attack and kill directly the microalgae or produce special compounds to lyse these cells [13, 14, 15]. Other non-algaecide bacteria can inhibit the microalgal growth changing the microenvironment of the microalgae [16] or by competing with the microalgae for nutrients [17, 18].

Other described processes that occur between bacteria and microalgae involve various ecological relationships such as competence, parasitism and other important microbiological processes [19]. Thereby, the microalgae can inhibit and/or induce the bacterial growth due to the production of organic exudates or toxic metabolites. Inversely, the bacteria can produce stimulating or inhibiting effects in microalgae through the production or absence of nutrients and/or stimulating or inhibiting substances which affect microalgae [20, 21, 22]. Delucca and McCracken (1977) [23] suggest that the interactions bacteria-algae are not randomly but highly specific. There are numerous data which report that the extracellular products from algae are capable to stimulate the growth of bacterial strains [21, 22] through the excretion of carbohydrates, organic acids, nitrogenous substances and vitamins [24]. Some studies in natural ecosystems have determined that organic substances derived from phytoplankton are used by bacteria as a substrate for growing. However, microalgae also inhibit bacterial growth by production of organic exudates or toxic metabolites. There are several reports suggesting a synergistic action between microalgae and its bacterial flora associated [figure 2; 25].

Most part of microbial life develops in biofilm form, either in surface or aggregates. In this ecosystem, bacteria and microalgae are the predominant components and they are the basis of the trophic chain and of the organic matter recirculation. A biofilm is a microbial consortium associated with EPS and other molecules attached to a submerged surface. The formation of a biofilm begins with the accumulation of organic molecules over a submerged surface, this physicochemical event occurs in a few seconds or minutes after the immersion of any surface in a liquid. Few hours later of the establishment of a macromolecular film, the bacterial colonization starts [26].

A mature biofilm is capable to maintain the concentrations of ammonium and phosphate present in the surrounding medium at low levels. Thompson et al. (2002) [27] determined that the decline of the ammonium concentrations is related with the increase of the chlorophyll *a* in biofilms, determining that the ammonium was absorbed mainly by the microalgae to produce new biomass. In Thompson et al. (2002) [27] experiments, most of the ammonium ingest in biofilm occurs at 10-15 days after the beginning of the experiment, when the chlorophyll *a* concentration reaches 5 μgcm^{-2}. In this case, the microalgae community is dominated by pennates diatoms (*Amphora, Campylopyxis, Navícula, Sinedra, Hantschia* and *Cylindrotheca*) and filamentous cyanobacteria (*Oscillatoria* and *Spirulina*). The fact that a biofilm effectively absorbs or transforms the ammonium present in the water column has important applications as probiotic for health of cultivable species such as juveniles of mollusks and crustaceans, including *Farfantepenaeus paulensis*, due to that shrimps tolerate high nitrate (>15000 μM) and nitrite (>1000 μM) concentrations [28], but ammonium in high concentrations is lethal, and can inhibit seriously the ingestion of food and growth [29, 30].

Mainly, the use of bacteria-microalgae biofilms would be applicable to tanks of intensive cultures in which there are a great accumulation of dissolved nitrogen, especially ammonium, as a result of addition of food and excretion of organisms maintained in high density, being one of the most important problems in intensive culture of shrimp and other mollusks, affecting the ingestion of food, growth and survival [28, 30]. One alternative to maintain a high water quality is the biological treatment, based in the use of pre-colonized filters by microorganisms that absorb the excess of nutrients from water. A similar process occurs in nature, where biofilms associated with a matrix of EPS attached are responsible of many biogeochemical cycles in aquatic ecosystems, especially the one of the nitrogen [31]. The eutrophication process accelerates if the main form of nitrogen inputted in the ecosystem is ammonium. This happens due to that the primary producers use less energy to incorporate this source of N into the amino acids and proteins, while the nitrate form must be transformed inside the cells to ammonium, with a higher cost of energy. Therefore, autotrophic cells grow faster in presence of ammonium forms than nitrate [27]. Thus, the presence of biofilms could reduce the eutrophication in the water mass that receives the effluents of aquaculture rich in ammonium through the absorption of this.

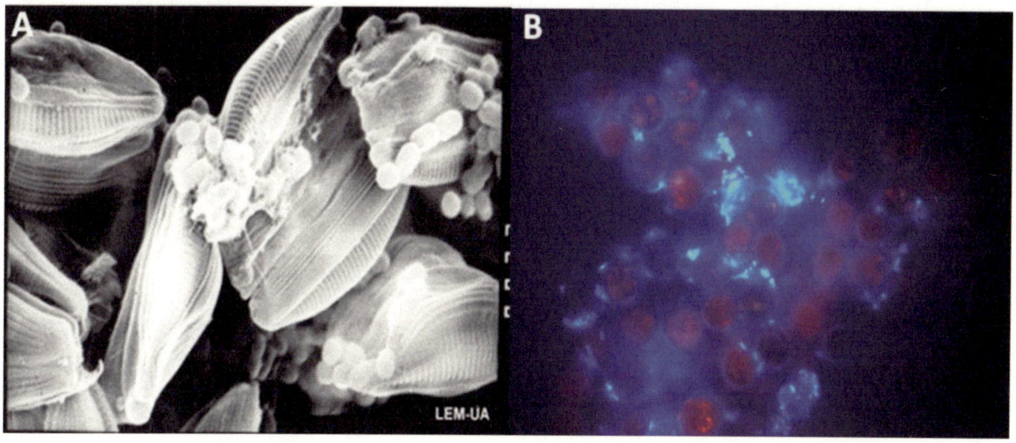

Figure 1. A, Biofilm from bacteria *Alteromonas* sp. and microalga *Navicula incerta*. B, Biofilm from microalga *Botryococcus braunii* and bacteria *Rhizobium* sp.

Nevertheless, a point to consider is that the biofilms have been thoughtful as reservoirs of pathogens bacteria, like *Vibrio harveyi*, which can affect crustacean's cultures such as shrimp. Pathogens bacteria present in biofilms are difficult to eliminate through the use of antibiotics, due to the hardness of the access of these molecules into the biofilms [32]. However, the results of Thompson et al. (2002) [27] indicate that the ingestion and transformation of nitrogen by the biofilm may help to reduce the occurrence of pathogens bacteria, due to that this microorganisms normally are present in situations where nitrogenous compounds are extremely high [33]. On the other way, lots of microalgae present in biofilms are capable to produce antibiotics that prevent the growing of pathogens bacteria [34, 35]. Protozoa that inhabit biofilms could also control abundance of pathogenic bacteria through the grazing [36]. Avila-Villa et al. (2011) [37] evaluate the presence of pathogen bacteria in microalgae, determining that species of these kind of bacteria such as NHPB (necrotizing hepatic pancreatitis bacteria) don′t attach to the surface of any microalgae and besides, they don′t survive in presence of these species, confirming the production of antibiotic substances by these microalgae species [38]. Respect to the benthic microalgae *Navicula* sp., this can easily form biofilms, and some bacteria thrive there using the exudates of the microalgae and the excreted extracellular products (carbohydrated substances and with nitrogen, organic acids and lipids) as a source of nutrients [39]. Besides, it has been documented that predominant bacteria linked to biofilms of algae are γ-proteobacteria and α-proteobacteria [40]. Thus it is possible, on the contrary to the expected effect, that the elimination of a biofilm could increase the risk to develop pathogenic bacteria. Also, is important to note that biofilms are considered an important source of food for cultivable species such as *Daphnia* [41], Nile tilapia [42] and carpa [43]. Despite the low protein content measured in biofilms, the microorganisms in there can provide essential elements such as; polyunsaturated fatty acids, sterols, amino acids, vitamins and carotenoids [36]. Thus, the biofilm probably contribute to the increment of weight and total biomass of juvenile of crustaceans like *F. paulensis* [27]. On the other hand, biofilms are essential in crustacean's cultures too like fresh water crab *Cherax quadricarinatus,* and also another kind of cultures, the presence of biofilms impact directly in water quality of cultures, increasing survival almost in 100% when they are feed with biofilms and also there is an increment in the growth of juveniles [44]. Different species of cultured crustaceans have improved their growth or survival when biofilms are used as a food source [27, 45, 46]. Moreover, water quality in culture systems is remarkably improved by the use of the biofilm [27, 47].

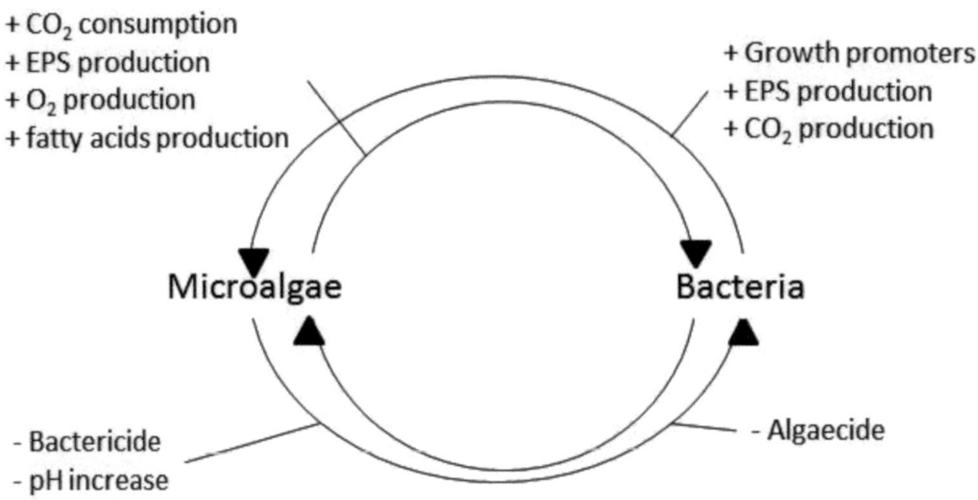

Figure 2. Interactions between microalgae and bacteria.

2. Probiotic role

Aquaculture is an important economic activity worldwide, in an attempt to improve the production of organisms it has been used a great quantity of antibiotics in an indiscriminate way for diseases control. Due to this, nowadays its use is questioned because the bacterial resistance generated and for the tons of antibiotics released to the biosphere during the last 60 years [48]. Recently, as an alternative for improve the growth of the cultured organisms, disease control and to improve the immune system it has been proposed the use of probiotics [49, 50, 51, 52]. The term "probiotic" is defined as "live microorganisms administered in appropriated quantities as food or food supplement that have benefic effects in the intestinal microbiological equilibrium of the host" [53]. The benefits for the host consist in to optimize the degradation and absorption of the food, favoring the autochthonous microbiota balance [49] reducing the pathogenic load [50]. According to the literature, most of the probiotics proposed as agents of biological control in aquaculture are bacteria from genus *Vibrio* and *Bacillus* [50].

In natural habitats, most bacteria are associated to algae and can have both effects in the algal growth, beneficial or deleterious. The interaction between algae and bacteria are complex and include competition for resources [54], production of antimicrobial agents [55, 56], stress protection through the production of extracellular polymeric substances, and the junction of metals or transformation through the production of exudates [57]. The algal cells can associate with a range of bacterial communities [58, 59] and this association vary from to share the general habitat, to a colonization of bacteria in the algal surface (epiphytic biofilm) and the endophytic association of bacteria inside de algal cells. There are reports that show that the presence of a large number and diversity of bacteria associated with algal cultures enhances the growth of algal species [table 1; 60]. This increase in growth rate suggests that the relationship between algae and bacteria in these cultures is beneficial to algae. Grossart et al. (2006) [59] also found that the cell density of *Skeletonema costatum* in the exponential phase of growth was significantly higher in the presence of bacteria. The ability of bacteria to increase algal growth depends on the growth phase of algae in which is added [59]. It has been determined that the cell densities of *Thalassiosira rotula* remain higher when is exposed to bacteria in the exponential phase of growth, but if is exposed in the stationary phase, the algal cell densities decrease rapidly. The response of the algae will then depend on the species of bacteria and the medium in which the algae obtain their nutrients and vitamins [5, 61]. It has been observed that bacteria specifically isolated from the surface of marine diatoms have a greater positive effect on algal growth than those isolated from the ocean [54], suggesting that the spatial relationships between bacteria and algae can be important. Rier and Stevenson (2002) [62] suggest that bacteria tend to be effective competitors for resources because they have (i) rapid growth rate, (ii) a ratio of volume per surface area larger (iii) rapid rates of phosphorus intake. In the oligotrophic conditions of the open sea the algae-bacteria relationship is consolidated because the concentration of the non-algal dissolved organic matter is very low and bacteria prefer carbon derived from algae as an energy source. This was verified in laboratory bioassays in which dissolved organic matter decreases rapidly when bacteria are present, demonstrating that they have a rapid dissolution and decomposition of organic matter [59].

There are many studies reporting the growth promoter effect on microalgae by bacteria (table 1). Induction of bacterial growth in specific cultures has been reported for a few species of microalgae such as *Chlorella vulgaris, C. sorokiniana* and *B. braunii*, and growth promoter bacterial strains are mainly of *Azospirillum spp* and a *Rhizobium sp.* [63, 64, 65, 66, 67]. Induction of growth in plants used in agriculture through the use of plant growth promoter bacteria (PGPB) [68] is an established fact, involving the use of different mechanisms between plants and bacteria, in which the final product of these many associations is to improve a characteristic of the plant, usually depending on the uses of the plant for human consumption [69]. On the other hand, induction of aquatic microalgae by bacteria, although it was discovered decades ago, is an emerging field in which the majority of studies have been performed in recent years [65, 70,71]. The main interest in this artificial association between algae and bacteria is due to obtaining a community associated with better characteristics than the microalgae alone [73] for applications such as removal of contaminants from wastewater [8], or use as food [74] or as a probiotic. The mechanisms by which growth-promoter bacteria in plants (PGBP) [68] affect the growth of plants vary widely. PGPB directly affect the metabolism of plants giving substances that are usually of low availability. These bacteria are capable of fixing atmospheric nitrogen, solubilize phosphorus and iron, and produce plant hormones such as; auxins, giggerelins, cytokinins, ethylene, nitrite and nitric oxide. Additionally, they improve stress tolerance in plants (drought, high salinity, metal toxicity and the presence of pesticides). One or more of these mechanisms may contribute to increase the growth and development of plants, higher than normal in standard culture conditions [69, 75]. Most PGPB are *Bacillus* spp. that work by diseases control [76], however some species of *Bacillus* promote the absence of disease by stimulating the immune system [77]. Possible interactions between *Bacillus* spp. with microalgae are unknown. Thereby, *Azospirillum* is one of the few genera of bacteria known to promote the growth of microalgae (Microalgae growth promoter bacteria, MGPB) [65]. *Azospirillum* is the most studied PGPB in agriculture [77]. Its habitat is the rhizosphere, N_2-fixing bacteria that is very versatile in its nitrogen transformations. In addition to fix N_2 under microaerobic conditions, act as denitrifying under anaerobic or microaerobic conditions, and can assimilate NH_4^+, NO_3^-, o NO_2^- and acts as a general PGPB for many species of plants, including the microalgae *Chlorella* [65]. *Azospirillum* spp. significantly alters the metabolism of microalgae, mainly producing indole-3-acetic acid (IAA) [78] and increasing the nitrogen cycle enzymes in these algae [73]. Although several studies described that inoculation of marine phytoplankton and freshwater bacteria sometimes increase their productivity [74], these studies are descriptive and exploratory and there is no mechanism described or demonstrated by which the phenomenon occurs. Despite the induction of microalgal growth by bacteria, not all interactions are positives; interaction of *C. vulgaris* with their associated bacteria *Phyllobacterium myrsinacearum* induces culture senescence [65, 79]. In a study by Hernández et al. (2009) [66] was employed the PGPB *Bacillus pumilus* Es4, originally isolated from the rhizosphere. This PGPB fix atmospheric nitrogen, produce IAA in vitro in the presence of tryptophan, besides to efficiently produce siderophores and increase growth in a cactus for long periods of time. *B. pumilus* Es4 also induces the growth of the microalga *C. vulgaris* acting as a MGPB, but this occurs only in the absence of nitrogen. *Chlorella* spp. is able to grow without nitrogen by a limited period of time, using ammonium that can be produced and recycled within the organism by a variety of metabolic pathways, such as photorespiration, phenylpropanoid metabolism, use of compounds of nitrogen transport, and amino acids catabolism [66, 80]. In this regard, *Chlorella* growth in the absence of other microorganisms can be explained by the differential activity of the enzyme glutamate dehydrogenase. This enzyme serves as a bond between the nitrogen and carbon metabolism due to its ability to assimilate ammonium to glutamate or to deaminate the glutamate to 2-oxoglutarate and ammonium under stress conditions [80, 81]; thus, the ammonium may be re-absorbed by *Chlorella* and used to a limited growth.

De Bashan and Bashan (2008) [78], proposed and studied a model of microalgae and bacteria immobilized in alginate to analyze and evaluate their possible interactions. In their study described the following sequence of events occurring during the interaction between the two microorganisms. Randomly immobilization of *Chlorella* spp. occurs first with a PGPB strain within a matrix and nutrients are in the surrounding medium that diffuses freely. In a given time (from 6 to 48 hours), depending on the bacteria-microalgae combination, both microorganisms are in the same cavity of the sphere, mainly in the periphery [79]. Here the bacteria secrete indole-3-acetic acid (IAA) and other undefined signal-molecules, possibly near the microalgal cells. At this stage, the activity microalgal enzyme (glutamine synthetase and glutamate dehydrogenase) does not increase. In the next phase of interaction, after 48 h occurs the increment of the enzymatic activity, production of photosynthetic pigments, and nitrogen and phosphorus intake. It also occurs releasing of oxygen as a byproduct of photosynthesis [for review see 65]. The most notable effect is the increasing by 2 to 3% on growth of microalgae with PGPB on those without PGPB [65]. This model proposed by Bashan and Bashan (2008) [78] has been evaluated in various combinations of microalgae-PGPB demonstrating the induction of growth in *C. sorokiniana* and *B. pumilus*, and others *C. vulgaris* and *A. brasilense* Sp6 [table 1; 78]. At cell and culture level there is an increase in the absorption of ammonium. The addition of exogenous tryptophan (precursor of the phytohormone IAA and the main mechanism by which *Azospirillum* affects the growth of *Chlorella* [64]) also induces a significant increase in the growth of microalgae. It also increases the activity of glutamate dehydrogenase, a key enzyme in ammonium assimilation in plants. Other PGPB such as *B. pumilus* and other microalgae, such as *C. sorokiniana* have been tested successfully (table 1). These options create opportunities for many combinations of microalgae and PGPB. Similarly, different alginates and derivatives from many macroalgae are commercially available [72] and to design the necessary combination and entrapment schemes. Because the immobilization of microorganisms is commonly used with other polymers [83], this model is not restricted to alginates, but each polymer has its advantages and disadvantages to be studied in future studies.

The EPS (a heterogeneous mixture of polysaccharides, proteins, nucleic acids, lipids and humic acids [84]) have a key role in biofilms, recently defined as a stabilization mechanism in mixed biofilms of bacteria and microalgae and present in a significantly higher percentage only when microalgae are associated with bacteria [3]. Furthermore, EPS are also important for the recycling of trace metals in aquatic systems, favoring metal binding to bacterial and algal agglomerates, and colloidal material/EPS, allowing the removal from surface waters and large particles [57]. Bacterial colonization is superior in stressed algal cells more than in healthy algal cells [54], which can be related to the release of organic material from the cell after cell lysis as part of a process of senescence, or under conditions of induced stress, such as exposure to contaminant metals [60]. The inability to detect visually bacteria from axenic cultures may be due to a very close association of the bacteria in the algal phycosphere or in the cell wall, or bacteria are in endophytic form in the algal cell, making it impossible to remove the bacteria from the algae using physical techniques. What's more, it appears that algal species benefit from the presence of bacteria, increasing their growth rate [60, 67]. The production of exudates of communities in bacteria/microalgae mixed biofilm increase in exposure to metals [85]. These exudates may be produced from algae or bacteria, but they are used as a mechanism of survival and resistance to stress for entire biofilm [60].

Type of study	Microalga species	Bacterial strain (s)	Reference (s)
Growth promotion	*Oscillatoria* sp.	*Pseudomonas* sp., *Xanthomonas* sp., *Flavobacterium* sp.	23
Growth promotion (dry wt, cell no., colony size, cell size)	*Asterionella gracilis*	*Pseudomonas* sp., *Vibrio* sp.	20
Antibacterial activity	*Chattonella marina*	*Pseudomonas*	20
Growth promotion	*Asterionella gracilis*	*Flavobacterium* NAST	20
Antibacterial activity	*Skeletonema costatum*	*Vibrio* sp., *Listonella anguillarum*, *Vibrio fisheri*	108
Growth promotion	*Isochrysis galbana*	*Vibrio* sp. C33, *Pseudomonas* sp. 11, *Arthrobacter* sp. 77	22
Antibacterial activity	*Tetraselmis suecica*	*Listonella anguillarum, V. alginolyticus, V. salmonicida, V. vulnificus, Vibrio* sp.	34
Growth promotion (dry wt, cell no., colony size, cell size)	*C. vulgaris*	*A. brasilense* Cd. Sp6, Sp245; A. lipoferum JA4	65, 70
Delayed senescence	*C. vulgaris*	*A. brasilense* Cd; P. myrsinacearum	79
Population control	*C. vulgaris*	*A. brasilense* Cd; P. myrsinacearum	59, 79
Lipids	*C. vulgaris*	*A. brasilense* Cd	126
Modification of fatty acids	*C. vulgaris*	*A. brasilense* Cd	126
Cell-cell interactions	*C. vulgaris*	*A. brasilense* Cd	126
Mitigation of heat and intense sunlight	*C. Sorokiniana*	*A. brasilense* Cd	126
Population dynamics	*C. vulgaris*	*A. brasilense* Cd	63
Mitigation of tryptophan inhibition	*C. vulgaris*	*A. brasilense* Cd	63
Mitigation of pH inhibition	*C. vulgaris*	*A. brasilense* Cd	8
Photosynthetic pigments	*C. vulgaris*	*A. brasilense* Cd, *Phyllobacterium myrsinacearum, B. pumilus*	8, 66, 72, 105, 126
Nutrient starvation	*C. Sorokiniana*	*A. brasilense* Cd	70
Enzymes in the nitrogen cycle	*C. vulgaris*	*A. brasilense* Cd	70
Hormones	*C. Sorokiniana*	*A. brasilense* Cd; *B. pumilus*	66, 70
Absorption of nitrogen and phosphorus	*C. vulgaris, C. Sorokiniana*	*A. brasilense* Cd, Sp6, Sp245; FAJ0009, SpM7918; *A. lipoferum* JA4, JA4::ngfp15	73
Growth promotion	*Botryococcus braunii*	*Rhizobium* sp.	67

Table 1. Studies of paired microalga-bacteria interactions.

3. Induction of larval settlement

Benthic diatoms present in the biofilm plays an important role in the marine ecosystem not only serve as food for advanced stages of development of marine invertebrate larvae [86], but also with bacteria and other microorganisms, form an attractive site for larval settlement in the process of metamorphosis [87]. There are numerous studies which have determined the characteristics that make a substrate optimal for larval settlement, and which are the effects of various biofilms in controlling larval settlement events [87, 88, 89, 90]. In the natural environment, the development of a biofilm formed by diatoms and other organisms is preceded by primary colonization of bacteria [91] aided by the EPS which act as "glue" and work at the cellular and molecular level to establish a strong and irreversible binding to a given substrate [92]. This succession of microorganisms often precedes the subsequent stages in a substrate, in which the macroorganisms eventually begin to be dominant [26].

Avendaño-Herrera and Riquelme (2007) [87] showed how optimize the production of a biofilm formed by the diatom *Navicula veneta* and a bacterium of the genus *Halomonas* sp., proposed model for the use in the induction of larval settlement. When the strain of *Halomonas* spp. was added to the diatom occurs an acceleration of growth of *N. veneta* [87], this occurs only when adding live bacteria, indicating the requirement of precursors of extracellular products excreted by the bacteria. Without the presence of *Halomonas* the microalgal biomass obtained is 65% lower. Is important to note that the diatom-bacteria biofilm can be used efficiently to provide food for species such as, abalone or scallop juvenile stages, and/or to colonize substrates that are used for adhesion, favoring larval settlement and reducing production time in macroorganisms cultures [93]. In addition, phytoplankton cultures are widely used in the aquaculture industry for a variety of purposes; these cultures are described as "green water" because they contain high levels of phytoplankton species such as *Nannochloropsis* sp. and *Chlorella* sp. The "green water" is added to the tanks with fish larvae and to enrich zooplankton, and provide a direct and indirect nutrition for the larvae. Moreover, the "green water" reduces water clarity, minimizing larval exposure to light, which acts as a stressor [94]. According to this, the presence of phytoplankton improves water quality by reducing the ammonium ion concentrations and increasing concentrations of dissolved oxygen through photosynthesis. Notably, phytoplankton also produces antibacterial substances that can prevent disease outbreaks [95, 96, 97, 98]. Among these, important are some members of the *Roseobacter* clade (Alphaproteobacteria) such as *Phaeobacter* and *Ruegeria* that suppress the growth of the fish pathogen *Vibrio anguillarum* by producing tropodithietic acid (TDA) [98, 99, 100, 101]. Also the abundance of bacteria from *Roseobacter* clade is highly correlated with phytoplankton blooms [102].

4. Chemical signals in bacteria-microalgae biofilms

According to the study of Sharifah and Eguchi (2011) [94] there is synergy and beneficial contribution by using bacteria belonging to the *Roseobacter* clade together with phytoplankton like *N. oculata*. In their study they used approximately between 11.4 to 13.2%

of bacteria in indoor cultures of *N. oculata*. These levels are comparable to the concentration of bacteria in coastal sea water (<1-25%) [102, 103]. Most of the cultivable bacteria in the *Roseobacter* clade corresponding to the genera *Phaeobacter, Silicibacter, Sulfitobacter, Roseobacter* and *Roseovarius*, which have potentially probiotic properties [99, 100, 102]. When these species are adding with phytoplankton to the tanks with fish larvae increased larval survival [95, 96, 97, 104] for growth inhibition of pathogenic bacteria. This process could be mediated by at least two possible mechanisms. The first one involves the preferential entry of nutrients or competition for nutrients, by bacteria. The second one, and more complex, involves a direct interaction between phytoplankton and microbes such as phytoplankton and pathogenic bacteria, probiotic bacteria and pathogenic bacteria, and phytoplankton-probiotic bacteria and pathogenic bacteria. Regarding the first mechanism, competition for entry of nutrients, the abundance of the *Roseobacter* clade in the coastal sea is correlated with the release of organic substances from natural phytoplankton blooms such as dimethylsulfoniopropionate (DMSP) and amino acids [105, 106]. In turn *N. oculata* may also excrete some substances similar to DMSP or amino acids that support more optimally bacterial growth of the clade [94]. Referring to the second mechanism described above, involving complex interactions, there is no direct inhibition of fish pathogens by phytoplankton, in contrast to other studies [107, 108]. As there is no difference in the viability of *V. anguillarum* by using probiotic bacteria it was concluded that there is no direct inhibition on the viability of *V. anguillarum*. In contrast, a study of the diatom *Skeletonema costatum* and the macroalgae *Ulva clathrata*, they produce organic compounds that inhibit the growth of *V. anguillarum* directly [107, 108].

From this point of view, the *Roseobacter* clade is beneficial and acts as a probiotic to induce the spread of scallop [109] and larvae of turbot [110] by removing fish pathogens. Other studies show that bacterial cell density of the clade in the range of 10^6-10^9 $CFUml^{-1}$ is needed to reduce pathogenic bacterial population by 10% [94]. Added to this, the static conditions favor culture biofilm formation by allowing bacteria of the genera *Phaeobacter, Silicibacter, Sulfitobacter, Roseobacter, Pseudoalteromonas* and *Roseovarius* produce tropodithietic acid (TDA), antibacterial compound produced by *Phaeobacter* spp., *Silicibacter* sp. and *Ruegeria* sp. [100, 111]. Static culture conditions and the presence of a brown pigment are indicators of the production of TDA [100]. However, in the study of Sharifah and Eguchi (2011) [94] *Roseobacter* clade members produced different antibacterial compounds to TDA, and the cultures were incubated under agitation and did not produce brown pigment. Interestingly, the previous study demonstrated that agitated *Roseobacter* cultures are able to eliminate *V. anguillarum* only in the presence of substances excreted from phytoplankton, and none of these species belongs to *Phaeobacter* sp. previously described [101]. The Inhibitory activity *of Sulfitobacter* sp., *Thalassobius* sp., *Rhodobacter* sp. and *Antarctobacter* sp., is significantly affected by the thermostable substances excreted by *N. oculata* [94]. Microalgae *N. oculata, N. granulata, N. oceanica* and *N. salina* produce putrescine, a thermostable polyamine [112]. Moreover, *N. oculata* CCMP525 produces signaling molecules like low molecular weight *n*-acyl-homoserine lactones which are produced by bacteria to the communication system cell-to-cell regulating gene expression [quorum sensing; 113]. The analogues of *n*-acyl-homoserine lactones are thermostable. These compounds can be secreted by *N. oculata* and act as signaling molecules for communication with *Sulfitobacter* sp. RO3 resulting in growth inhibition of *V. anguillarum*. These results demonstrate that phytoplankton cultures used as "green water" for the production of fish larvae have a key role in enhancing the inhibitory effect of *Roseobacter* clade against *V. anguillarum*. A similar inhibitory effect was also observed in *Chlorella* sp., other marine microalgae used in aquaculture [94].

5. Other applications

Immobilization of microorganisms on polymers because the production of different products and environmental and agricultural applications is well known and have increased in the last two decades [93, 114, 115]. The immobilization of microalgae is a common approach for many applications of bioremediation [66]. Immobilization in several substances provides to the microorganisms several advantages over free-living microorganisms. These advantages include: (i) a continuous source of nutrients without competition with other microorganisms [116] and (ii) protection against environmental stress [66, 117], bacteriophages, toxins, and UV irradiation [118]. A recently developed treatment for tertiary domestic wastewaters uses the green microalga *Chlorella spp.* and the plant growth promoter bacteria (PGPB) *Azospirillum brasilense*, both bound and immobilized in alginate beads [116]. Each unit in this technological model, a single polymer sphere, contains within cavities that serve as matrix for the folding of microalgae and bacteria [66, 78, 119]. Additionally, the entrapment of microorganisms may also be within the solid matrix polymer of the polymeric sphere. In some cases, microbial cells are on the surface or partially in or out of the gel matrix. During the formation of alginate spheres the number of organisms is higher outside than inside. However, this approach can be used in aquaculture as a feeding method for growing mollusks such as *Haliotis rufescens* [120].

The algae are the organisms most commonly used to assess metal contamination and bioavailability in aquatic systems, are highly sensitive to heavy metals such as Cu, Fe and Cd in environmentally relevant concentrations. Algae are primary producers and affect nutrient cycling in marine and fresh water ecosystems, and in aquaculture [121]. As such, the algae are considered ecologically significant organisms and the ideal candidates for ecotoxicological studies. However, algae are rarely isolated in the environment, but are part of complex planktonic communities and biofilms. The alteration of community structure may influence the overall function (e.g. respiration, photosynthesis) and community sensitivity to toxicants. Although the tests of toxicity for single-species used in microalgae are highly sensitive and reproducible, they do not have a realistic environment. Interactions between algae and associated bacteria, in plankton or in biofilms, may alter algal sensitivity to pollutants. Recent research has attempted to develop multi-species algal test in the evaluation of metals based on toxicity [122, 123]. These studies explored the toxicological response of individual algal species when they are exposed in combination with one or other species of algae.

Bacteria can have both positive and negative effects on algae in polluted environments. For example, the tolerance of the green macroalga *Enteromorpha compressa* to copper in a coastal environment in Chile attributed to an epiphytic bacterial community colonizing the surface [1]. Bacterial biofilms can mediate metal toxicity to the host organism by limiting the diffusion of toxins, protective effects of high concentrations of extracellular polymeric substances, protective effects of stored nutrients trapped, and effects due to a larger surface area (less toxic per cell). While the effects of metals in biofilms are widely reported [85, 124, 125], there are few studies on the effects of metal toxicity to algae biofilms.

6. Conclusions

Since the first studies of bacteria-microalgae interactions decades ago, it has been elucidate and discovered several events in which the close connection between these two heterotrophs and autotrophs components is evidenced. Showing that the coupling of microalgae-bacteria produces changes in the excreted compounds in the surrounding environment, that affects positively or negatively to others organisms.

Most of the interactions are strongly regulated by chemical signals. Although it has been described lots of phenomena in positive and negative interactions in biofilms, there are a few investigations that explore the chemical and molecular nature of chemical compounds involved in this interactions which are produced by microorganisms, this is why in the future will be required to deepen in the study of mechanisms involved in the growth of mixture biofilms.

The use of this biofilms in nature can be easily developed in the laboratory; they can be used increasing and affecting some specific compounds which are useful for a third organism of commercial interest. As well, in phenomena like larval settlement, induction of growth and increment of biomass rich in lipids has revealed a great potential probiotic use, particularly in aquatic industry which require more attention to the involved mechanisms in the action of this beneficial biofilms. These uses will allow us to get a better understanding of the role of these microbial consortiums in nature, and also a biotechnological orientation could be spread for the production of these beneficial biofilms in a stable and standard form.

Author details

Mariella Rivas
Centro Científico Tecnológico para la Minería CICITEM., Antofagasta, Chile

Carlos Riquelme
Laboratorio de Ecología Microbiana,
Centro de Bioinnovación, Universidad de Antofagasta, Antofagasta, Chile

7. References

[1] Riquelme C, Araya R, Vergara N, Rojas A, Guaita M, Candia M (1997) Potential probiotic strain in the culture of the Chilean scallop *Argopecten purpuratus* (Lamarrck, 1819). Aquaculture 154, 17-26.

[2] Abarzua M, Basualto S, Urrutia H (1995) Relación entre la abundancia y la biomasa de fitoplancton y bacterioplancton heterotrofico en aguas superficiales del Golfo de Arauco, Chile. Investigaciones Marinas Valparaíso 23, 67-74.

[3] Lubarsky HV, Hubas C, Chocholek M, Larson F, Manz W, Paterson DM, Gerbersdorf SU (2010) The stabilisation potential of individual and mixed assemblages of natural bacteria and microalgae. PLOS One 5(11), e13794.

[4] Watanabe K, Takihana N, Aoyagi H, Hanada S, Watanabe Y, Ohmura N, Saiki H, Tanaka HI (2005) Symbiotic association in *Chlorella* culture. FEMS Microbiol Ecol 51, 187-196

[5] Grossart HP, Simon M (2007) Interactions of planktonic algae and bacteria: effects on algal growth and organic matter dynamics. Aquat Microb Ecol 47, 163–176.

[6] Matsuo Y, Imagawa H, Nishizawa M, Shizuri Y (2005) Isolation of an algal morphogenesis inducer from a marine bacterium. Science 307,1598.

[7] Mouget J-L, Dakhama A, Lavoie MC, de la Noüe J (1995) Algal growth enhancement by bacteria: is consumption of photosynthetic oxygen involved? FEMS Microbiol Ecol 18, 35– 43.

[8] Hernandez J-P, de-Bashan LE, Bashan Y (2006) Starvation enhances phosphorus removal from wastewater by the microalga *Chlorella* spp. coimmobilized with *Azospirillum brasilense*. Enzyme Microb Tech 38, 190-198.

[9] Jiang L, et al. (2007) Quantitative studies on phosphorus transference occurring between *Microcystis aeruginosa* and its attached bacterium (*Pseudomonas* sp.). Hydrobiologia 581, 161–165.

[10] Amin SA, et al. (2009) Photolysis of iron-siderophore chelates promotes bacterial-algal mutualism. Proc Natl Acad Sci USA 106, 17071-17076.

[11] Croft MT, Lawrence AD, Raux-Deery E, Warren MJ, Smith AG (2005) Algae acquire vitamin B12 trough a symbiotic relationship with bacteria. Nature 438(3), 90-93.

[12] Adachi M, Fukami K, Kondo R, Nishijima T (2002) Identification of marine algicidal *Flavobacterium* sp. 5 N-3 using multiple probes and whole-cell hybridization. Fisheries Sci 68, 713-720.

[13] Amaro AM, Fuentes MS, Ogalde SR, Venegas JA, Suarez-Isla BA (2005) Identification and characterization of potentially algal-lytic marine bacteria strongly associated with the toxic dinoflagellate *Alexandrium catenella*. J Eukaryot Microbiol 52, 191-200.

[14] Simon N, Biegala IC, Smith EA, Vaulot D (2002) Kinetics of attachment of potentially toxic bacteria to *Alexandrium tamarense*. Aquat Microb Ecol 28, 249 –256.

[15] Su JQ, et al. (2007) Isolation and characterization of a marine algicidal bacterium against the toxic dinoflagellate *Alexandrium tamarense*. Harmful Algae 6, 799–810.

[16] Ferrier M, Martin J, Rooney-Varga J (2002) Stimulation of *Alexandrium fundyense* growth by bacterial assemblages from the Bay of Fundy. J Appl Microbiol 92, 706 –716.

[17] Otsuka S, Abe Y, Fukui R, Nishiyama M, Sendoo K (2008) Presence of previously undescribed bacterial taxa in non-axenic *Chlorella* cultures. J Gen Appl Microbiol 54, 187–193.

[18] Wang H, Haywood D, Laughinghouse IV, Matthew A, Chen F, Willliams E, Place AR, Zmora O, Zohar Y, Zheng T, Hilla RT (2012) Novel Bacterial Isolate from Permian Groundwater, Capable of Aggregating Potential Biofuel-Producing Microalga *Nannochloropsis oceanica* IMET1. Appl Environ Microbiol 78, 1445-1453.

[19] Salvensen I, Reitan KI, Skjermo J, Oie J (2000) Microbial environment in marine larviculture: Impact of algal growth rates on the bacterial load in six microalgae. Aquaculture International 8, 275-287.

[20] Riquelme CE, Fukami K, Ishida Y (1988) Effects of bacteria on the growth of a marine diatom, *Asterionella glacialis*. Bull Jpn Soc Microb Ecol 3, 29–34.

[21] Munro PD, Barbour A, Birkbeck TH (1995) Comparison of the growth and survival and larval turbot in the absence of cultivable bacteria with dose in the presence of *Vibrio anguillarum, Vibrio alginolyticus* or a marine *Aeromonas* sp. Applied and Environmental Microbiology 61, 4425-4428.

[22] Avendaño R, Riquelme C (1999) Establishment of mixed-culture probiotics and microalgae as food for bivalve larvae. Aquaculture Research 30, 893-900.

[23] Delucca R, McCraken M (1977) Observations on interactions between naturally collected bacteria and several species of algae. Hydrobiologia 55, 71-75.

[24] Lodeiros C, Campos Y, Marin N (1991) Production de antibióticos por la flora bacteriana asociada a monocultivos microalgales de utilidad en acuacultura. Sociedad de Ciencias Naturales 136, 213-223.

[25] Riquelme C.E., R.E. Avendaño-Herrera. 2003. Interacción bacteria-microalga en el ambiente marino y uso potencial en acuicultura. *Revista Chilena Historia Natural* 76(4), 725-736.

[26] Wahl M (1989) Marine epibiosis I. Fouling and antifouling: some basic aspects. Mar Ecol Prog Ser 58, 175–189.

[27] Thompson FL, Abreu PC, Wasielesky W (2001) Importance of biofilm for water quality and nourishment in intensive shrimp culture. Aquaculture 203, 263-278.

[28] Cavalli RO, Wasielesky JW, Franco CS, Miranda FC (1996) Evaluation of the short-term toxicity of ammonia, nitrite and nitrate to *Penaeus paulensis* ŽCRUSTACEA, DECAPODA. Broodstock Arq Biol Tecnol 39 (3), 567–575.

[29] Miranda FKC (1997) Efeito da amonia na sobrevivencia e crescimento de juvenis de camarao-rosa *Penaeus paulensis* Perez-Farfante, 1967 CRUSTACEA: DECAPODA. Effect of ammonium on the survival and growth of the pink-shrimp *Penaeus paulensis* Perez-Farfante, 1967 CRUSTACEA: DECAPODA. juveniles. MSc Thesis, Fundacao Universidade do Rio Grande, Rio Grande, RS, Brazil, 122 pp.

[30] Wasielesky JW, Marchiori MA, Santos MH (1994) Efeito da amonia no crescimento de pos-larvas do camarao rosa, *Penaeus paulensis,* Perez Farfante, 1967 Decapoda Penaeidae. Effect of ammonium on the growth of the pink-shrimp *Penaeus paulensis,* Perez Farfante, 1967 Decapoda: Penaeidae. post larvaex *Nauplius* 2, 99–105.

[31] Meyer-Reil M (1994) Microbial life in sedimentary biofilms—the challenge to microbial ecologists. Review Mar Ecol Prog Ser 112, 303–311.

[32] Costerton JW, Stewart PS, Greenberg EP (1999) Bacterial biofilms: a common cause of persistent infections. Science 284, 1318–1322.

[33] Austin B, Austin D (1999) Bacterial Fish Pathogens: Disease of Farmed and Wild Fish. 3rd edn. Springer, Chichester, 457 pp.

[34] Austin B, Day JG (1990) Inhibition of prawn pathogenic *Vibrio* spp. by a commercial spray-dried preparation of *Tetraselmis suecica*. Aquaculture 90, 389–392.

[35] Alabi AO, Cob ZC, Jones DA, Latchford JW (1999) Influence of algal exudates and bacteria on growth and survival of white shrimp larvae fed entirely on microencapsulated diets. Aquacult Int 7 (Ž3), 137–158.

[36] Thompson FL, Abreu PC, Cavalli RO (1999) The use of microorganisms as food source for *Penaeus paulensis* larvae. Aquaculture 174, 139–153.

[37] Avila-Villa LA, Martínez-Porchas M, Gollas-Galván T, López-Elías JA, Mercado L, Murguia-López A, Mendoza-Cano F, Hernández-López J (2011) Evaluation of different microalgae species *Artemia* (*Artemia franciscana*) as possible vectors of necrotizing hepatopancreatitis bacteria. Aquaculture 318, 273-276.

[38] Sánchez-Saavedra MP, Licea-Navarro A, Berlández-Sarabia J 2010. Evaluation of the antibacterial activity of different species of phytoplankton. Revista de Biología Marina y Oceanografía 45, 531–536.

[39] Caroppo C, Stabili L, Cavallo RA, 2003. Diatoms and bacteria diversity: stury of their relationships in the Southern Adriatic Sea. Mediterranean Marine Science 4, 73–82.

[40] Dovretsov S 2010. Marine biofilms. Chapter 9. In: Durr, S., Thomason, J.C. (Eds.), Biofouling. Wiley-Blackwell, Singapore, p. 405.

[41] Langis R, Proulx D, Noue J, Couture P 1988. Effects of bacterial biofilm on intensive *Daphnia* culture. Aquacult. Eng. 7, 21–38.

[42] Shrestha MK, Knud-Hansen CF 1994. Increasing attached microorganism biomass as a management strategy for Nile Tilapia *Oreochromis niloticus*. production. Aquacult. Eng. 13, 101–108.

[43] Ramesh MR, Shankar KM, Mohan CV, Varghese TJ 1999. Comparison of three plant substrates for enhancing carp growth through bacterial biofilm. Aquacult. Eng. 19, 119–131.

[44] Viau VE, Ostera JM, Tolivia A, Ballester ELC, Abreu PC, Rodríguez EM (2012) Contribution of biofilm to water quality, survival and growth of juveniles of the freshwater crayfish *Cherax quadricarinatus* (Decapoda, Parastacidae). Aquaculture 324-325, 70-78.

[45] Abreu PC, Ballester EC, Odebrecht C, Wasielesky JW, Cavalli RO, Graneli W, Anesio AM (2007) Importance of biofilm as food source for shrimp (Farfantepenaeus paulensis) evaluated by stable isotopes (d13C and d15N). Journal of Experimental Marine Biology and Ecology 347, 88-96.

[46] Ballester EC, Wasielesky JW, Cavalli RO, Abreu PC (2007) Nursery of the pink shrimp *Farfantepenaeus paulensis* in cages with artificial substrates: biofilm composition and shrimp performance. Aquaculture 269, 355-362.

[47] Holl CM, Otoshi C, Unabia CR (2011) Nitrifying biofilms critical for water quality in intensive shrimp RAS. Global Aquaculture Advocate 14, 38-39.

[48] Scientific committee on animal nutrition (SCAN) (2003) Opinion of the scientific committee on animal nutrition on the criteria for assessing the safety of microorganisms resistant to antibiotics of human clinical and veterinary importance. European Commission Health and Consumer Protection Directorate General. Available at: http://europa.eu/food/fs/sc/scan/out108.en.pdf/

[49] Verschuere L, Rombaut G, Sorgeloos P, Verstraete (2000) Probiotic bacteria as biological control agents in aquaculture. Microbiology and Molecular Biology Reviews 64, 655-671.

[50] Balcazar JL, De Blas L, Ruiz-Zarzuela L, Cunningham D, Vendrell D, Muzquiz JL (2006) The role of probiotics in aquaculture. Veterinary Microbiology 114, 173-186.

[51] Wang YB, Han JZ (2007) The role of probiotic cell wall hydrophobicity in bioremediation of aquaculture. Aquaculture 269, 349-354.

[52] Wang YB, Li JR, Lin J (2008) Probiotics in aquaculture: challenges and outlook. Aquaculture 281 (1-4), 1-4.

[53] Sihag RC, Sharma P (2012) Probiotics: The new ecofriendly alternative measures of disease control for sustainableaquaculture. Journal of fisheries and aquatic science. DOI:10.3923/jfas.

[54] Grossart HP (1999) Interactions between marine bacteria and axenic diatoms (*Cylindrotheca fusiformis, Nitzschia laevis* and *Thalassiosira weissflogii*) incubated under various conditions in the lab. Aquat Microb Ecol 19, 1–11.

[55] Fukami K, Nishijima T, Ishida Y (1997) Stimulative and inhibitory effects of bacteria on the growth of microalgae. Hydrobiologia 358, 185–191.

[56] Gross E (2003) Allelopathy of aquatic autotrophs. Crit Rev Plant Sci 22, 313–339.

[57] Koukal B, Rossé P, Reinhardt A, Ferrari B, Wilkinson KJ, Loizeau JL, Dominik J (2007) Effect of *Pseudokirchneriella subcapitata* (Chlorophyceae) exudates on metal toxicity and colloid aggregation. Water Res 41, 63–70.

[58] Schäfer H, Abbas B, Witte H, Muyzer G (2002) Genetic diversity of 'satellite' bacteria present in cultures of marine diatoms. FEMS Microbiol Ecol 42, 25–35.

[59] Grossart HP, Czub G, Simon M (2006) Algae–bacteria interactions and their effects on aggregation and organic matter flux in the sea. Environ Microbiol 8, 1074–1084.

[60] Levy JL, Stauber JL, Wakelin SA, Jolley DF (2009) The effect of bacteria on the sensitivity of microalgae to copper in laboratory bioassays. Chemosphere 74, 1266-1274.

[61] Gurung TB, Urabe J, Nakanishi M (1999) Regulation of the relationship between phytoplankton *Scenedesmus acutus* and heterotrophic bacteria by the balance of light and nutrients. Aquat Microb Ecol 17, 27-35.

[62] Rier ST, Stevenson RJ (2002) Effects of light, dissolved organic carbon, and inorganic nutrients on the relationship between algae and heterotrophic bacteria in stream periphyton. Hydrobiologia 489, 179–184.

[63] de-Bashan LE, Antoun H, Bashan Y (2005) Cultivation factors and population size control uptake of nitrogen by the microalgae *Chlorella vulgaris* when interacting with the microalgae growth-promoting bacterium *Azospirillum brasilense*. FEMS Microbiol Ecol 54, 197–203.

[64] de-Bashan LE, Antoun H, Bashan Y (2008b) Involvement of indole- 3-acetic-acid produced by the growth-promoting bacterium *Azospirillum* spp. in promoting growth of *Chlorella vulgaris*. J Phycol 44, 938–947.

[65] Gonzalez LE, Bashan Y (2000) Increased growth of the microalga *Chlorella vulgaris* when coimmobilized and cocultured in alginate beads with the plant-growth-promoting bacterium *Azospirillum brasilense*. Appl Environ Microbiol 66, 1527–1531.

[66] Hernandez JP, de-Bashan LE, Rodriguez J, Rodriguez Y, Bashan Y (2009) Growth promotion of the freshwater microalga *Chlorella vulgaris* by the nitrogen-fixing, plant growth-promoting bacterium *Bacillus pumilus* from arid zone soils. European Journal of soil Biology 45, 88-93.

[67] Rivas MO, Vargas P, Riquelme CE (2010) Interactions of *Botryococcus braunii* cultures with bacterial biofilms.Microbial Ecology 60(3), 628-635.

[68] Bashan Y, Holguin G (1998) Proposal for the division of plant growth-promoting rhizobacteria into two classifications: biocontrol–PGPB (plant growth-promoting bacteria) and PGPB. Soil Biol Biochem 30, 1225-1228.

[69] Bashan Y, de-Bashan LE (2005) Bacteria/plant growth-promotion, in: D. Hillel (Ed.), Encyclopedia of Soils in the Environment, Vol. 1, Elsevier, Oxford, UK pp. 103–115.

[70] de-Bashan LE, Trejo A, Huss VAR, Hernandez JP, Bashan Y (2008a) *Chlorella sorokiniana* UTEX 2805, a heat and intense, sunlight-tolerant microalga with potential for removing ammonium from wastewater. Bioresour Technol 99, 4980–4989.

[71] de Bashan LE, Bashan Y (2003) Microalgae growth-promoting bacteria: a novel approach in wastewater treatment. Rev Colomb Biotechnol 5, 85-90.

[72] Yabur R, Bashan Y, Hernandez-Carmona G (2007) Alginate from the macroalgae *Sargassum sinicola* as a novel source for microbial immobilization material in wastewater treatment and plant growth promotion. J Appl Phycol 19, 43–53.

[73] de-Bashan LE, Magallon P, Antoun H, Bashan Y (2008c) Role of glutamate dehydrogenase and glutamine synthetase in *Chlorella vulgaris* during assimilation of ammonium when jointly immobilized with the microalgae growth-promoting bacterium *Azospirillum brasilense*. J Phycol 44, 1188–1196.

[74] Garg SK, Bhatnagar A (1999) Effect of *Azospirillum* and *Azotobacter* inoculation on pond productivity and fish growth under fresh water conditions. Indian J Microbiol 39, 227–233.

[75] Rodriguez H, Fraga R, Gonzalez T, Bashan Y (2006) Genetics of phosphate solubilization and its potential applications for improving plant growth-promoting bacteria. Plant Soil 287, 15–21.

[76] Kloepper JW, Ryu CM, Zhang S (2004) Induced systemic resistance and promotion of plant growth by *Bacillus* spp. Phytopathology 94, 1259–1266.

[77] Bashan Y, Moreno M, Troyo E (2000) Growth promotion of the seawater-irrigated oilseed halophyte *Salicornia bigelovii* inoculated with mangrove rhizosphere bacteria and halotolerant *Azospirillum* spp. Biol Fertil Soils 32, 265–272.

[78] de-Bashan LE, Bashan Y (2008) Joint immobilization of plant growth-promoting bacteria and green microalgae in alginate beads as an experimental model for studying plant-bacterium interactions. Applied and Environmental Microbiology 74(21), 6797–6802.

[79] Lebsky VK, Gonzalez-Bashan LE, Bashan Y (2001) Ultrastructure of coimmobilization of the microalga *Chlorella vulgaris* with the plant growth-promoting bacterium *Azospirillum brasilense* and with its natural associative bacterium *Phyllobacterium myrsinacearum* in alginate beads. Can J Microbiol 47, 1–8.

[80] Dubois F, Terce-Laforgue T, Gonzalez-Moro MB, Estavillo JM, Sangwan R, Gallais A, Hirel B (2003) Glutamate dehydrogenase in plants: is there a new story for an old enzyme? Plant Physiol Biochem 41, 565–576.

[81] Robinson SA, Slade AP, Fox GG, Phillips R, Ratcliffe G, Stewart GR (1991) The role of glutamate dehydrogenase in plant nitrogen metabolism, Plant Physiol 95, 509–516.

[82] Gacesa P (1998) Bacterial alginate biosynthesis – recent progress and future prospects.Microbiology 144, 1133-1143.

[83] O'Reilly AM, Scott JA (1995) Defined coimmobilization of mixed microorganism cultures. Enzyme Microbiology Technology 17, 636-646.

[84] Flemming HC, Wingender J (2001) Relevance of microbial extracelular polymeric substances (EPSs)- Part I: Structural and ecological aspects. Water Science and Technology 43, 1-8.

[85] Garcia-Meza JV, Barranguet C, Admiraal W (2005) Biofilm formation by algae as a mechanism for surviving on mine tailings. Environ Toxicol Chem 24, 573–581.

[86] Takami H, Kawamura T, Yamashita Y (1997) Contribution of diatoms as food sources for post-larval abalone *Haliotis discus hannai* on a crustose coralline alga. Molluscan Res 18, 143–151.

[87] Avendaño-Herrera R, Riquelme C (2007) Production of a diatom-bacteria biofilm in a photobioreactor for aquaculture applications. Aquaculture Engineering 36, 97-104.

[88] Pearce CM, Bourget E (1996) Settlement of larvae of the giant scallop, *Placopecten magellanicus* (Gmelin), on various artificial and natural substrata under hatchery-type conditions. Aquaculture 141, 201-221.

[89] Harvey M, Bourget E, Gagne N (1997) Spat settlement of the giant scallop, *Placopecten magellanicus* (Gmelin 1791), and other bivalve species on artificial filamentous collector coated with chitinous material. Aquaculture 148, 277–298.

[90] Daume S, Brand-Gardner S, Woelkerling W (1999) Settlement of abalone larvae (*Haliotis laevigata* Donovan) in response to non-geniculate coralline red algae (Corallinales, Rhodophyta). J Exp Mar Biol Ecol 234, 125–143.

[91] Allison DG, Gilberg P (1992) Bacterial biofilms. Sci Prog 76, 301-321.

[92] Wetherbee R, Lind JL, Burke J, Quatrano R (1998) The first kiss: establishment and control of initial adhesion by raphid diatoms. Journal of Phycology 34, 9-15.

[93] Lebeau T, Robert MR (2003) Diatom cultivation and biotechnologically relevant products. Part I. Cultivation at various scales. Appl Microbiol Biotechnol 60, 612–623.

[94] Sharifah EN, Eguchi M (2011) The phytoplankton *Nannochloropsis oculata* enhances the ability of *Roseobacter* clade bacteria to inhibit the growth of fish pathogen *Vibrio anguillarum*. PLOS One 6(10), e26756.

[95] Muller-Feuga A (2000) The role of microalgae in aquaculture: situation and trends. J Appl Phycol 12, 527–534.

[96] Oie G, Makridis P, Reitan KI, Olsen Y (1997) Protein and carbon utilization of rotifers (*Brachionus plicatilis*) in first feeding of turbot larvae (*Scophthalmus maximus* L.). Aquaculture 153, 103–122.

[97] Palmer PJ, Burke MJ, Palmer CJ, Burke JB (2007) Developments in controlled green-water larval culture technologies for estuarine fishes in Queensland, Australia and elsewhere. Aquaculture 272, 1–21.

[98] Prado S, Montes J, Romalde JL, Barja JL (2009) Inhibitory activity of *Phaeobacter* strains against aquaculture pathogenic bacteria. Int Microbiol 12, 107–114.

[99] Porsby CH, Nielsen KF, Gram L (2008) *Phaeobacter* and *Ruegeria* species of the Roseobacter clade colonize separate niches in a Danish turbot (*Scophthalmus maximus*)-rearing farm and antagonize *Vibrio anguillarum* under different growth conditions. Appl Environ Microbiol 74, 7356–7364.

[100] Bruhn JB, Gram L, Belas R (2007) Production of antibacterial compounds and biofilm formation by *Roseobacter* species are influenced by culture conditions. Appl Environ Microbiol 73, 442–450.

[101] D'Alvise PW, Melchiorsen J, Porsby CH, Nielsen KF, Gram L (2010) Inactivation of *Vibrio anguillarum* by attached and planktonic Roseobacter cells. Appl Environ Microbiol 76, 2366–2370.

[102] Buchan A, Gonza′lez JM, Moran MA (2005) Overview of the marine *Roseobacter* lineage. Appl Environ Microbiol 71, 5665–5677. 103. Brinkhoff T, Giebel HA, Simon M (2008) Diversity, ecology, and genomics of the *Roseobacter* clade: a short overview. Arch Microbiol 189, 531–539.

[103] Nakase G, Eguchi M (2007) Analysis of bacterial communities in *Nannochloropsis* sp. cultures used for larval fish production. Fish Sci 73, 543–549.

[104] Gonzalez-Bashan LE, Lebsky V, Hernandez JP, Bustillos JJ, Bashan Y (2000) Changes in the metabolism of the microalgae *Chlorella vulgaris* when coimmobilized in alginate with the nitrogen-fixing *Phyllobacterium myrsinacearum*. Can J Microbiol 46, 653–659.

[105] Zubkov MV, Fuchs BM, Archer SD, Kiene RP, Amann R, et al. (2001) Linking the composition of bacterioplankton to rapid turnover of dissolved dimethylsulphoniopropionate in an algal bloom in the North Sea. Environ Microbiol 3, 304–311.

[106] Lu K, Lin W, Liu J (2008) The characteristics of nutrient removal and inhibitory effect of *Ulva clathrata* on *Vibrio anguillarum* 65. J Appl Phycol 20, 1061–1068.

[107] Naviner M, Bergé JP, Durand P, Le Bris H (1999) Antibacterial activity of marine diatom *Skeletonema costatum* against aquacultural pathogens. Aquaculture 174, 15-24.

[108] Ruiz-Ponte C, Samain JF, Sánchez JL, Nicolas JL (1999) The benefit of a Roseobacter species on the survival of scallop larvae. Mar Biotechnol 1, 52-59.

[109] Planas M, Peréz-Lorenzo M, Hjelm M, Gram L, Fiksdal IU, et al. (2006) Probiotic effect in vivo of *Roseobacter* strain 27-4 against *Vibrio* (*Listonella*) *anguillarum* infections in turbot (*Scophthalmus maximus* L.) larvae. Aquaculture 255, 323–333.

[110] Holmström C, Kjelleberg S (1999) Marine *Pseudoalteromonas* species are associated with higher organisms and produce biologically active extracellular agents. FEMS Microbiol Ecol 30, 285–293.

[111] Hamana K (2008) Cellular polyamines of phototrophs and heterotrophs belonging to the lower eukaryotic phyla Cercoza, Euglenozoa, Heterokonta and Metamonada. J Gen Appl Microbiol 54, 135-140.

[112] Natrah FMI, Kenmegne MM, Wiyoto W, Sorgeloos P, Bossier P, et al. (2011) Effects of micro-algae commonly used in aquaculture on acyl-homoserine lactones quorum sensing. Aquaculture 317, 53–57.

[113] Moreno-Garrido I (2008) Microalgae immobilization: current techniques and uses. Bioresource Technol 99, 3949–3964.

[114] Subashchandrabose SR, Ramakrishnan B, Megharaj M, Venkateswarlu K, Naidu R (2011) Consortia of cyanobacteria/microalgae and bacteria: Biotechnological potential. Biotechnology Advances 29, 896-907.

[115] de-Bashan LE, Hernández JP, Morey T, Bashan Y (2004) Microalgae growth-promoting bacteria as "helpers" for microalgae a novel approach for removing ammonium and phosphorus from municipal wastewater. Water Res 38, 466-474.

[116] Moreno-Garrido I, Codd GA, Gadd GM, Lubian LM (2002) Acumulación de Cu y Zn por células microalgales marinas *Nannochloropsis gaditana* (Eustigmatophyceae) inmovilizadas en alginato de calcio. Cienc Mar 28, 107–119.

[117] Zohar-Perez C, Chernin L, Chet I, Nussinovitch A (2003) Structure of dried cellular alginate matrix containing fillers provides extra protection for microorganisms against UVC radiation. Radiat Res 160, 198–204.

[118] Covarrubias SA, de-Bashan LE, Moreno M, Bashan Y (2012) Alginate beads provide a benefical physical barrier against native microorganisms in wastewater treated with immobilized bacteria and microalgae. Appl Microbiol Biotechnol 93, 2669-2680.

[119] Silva-Aciares FR, Carvajal PO, Mejías CA, Riquelme CE (2011) Use of macroalgae supplemented with probiotics in the *Haliotis rufescens* (Swainson, 1822) culture in Northern Chile. Aquaculture Research 42, 953-961.

[120] Azam F, Malfatti F (2007) Microbial structuring of marine systems. Nature Rev Microbiol 5, 782-791.

[121] Franklin NM, Stauber JL, Lim RP (2004) Development of multispecies algal bioassays using flow cytometry. Environ Toxicol Chem 23, 1452–1462.

[122] Yu Y, Kong F, Wang M, Qian L, Shi X (2007) Determination of short-term copper toxicity in a multiseries microalgal population using flow cytometry. Ecotox Environ Safety 66, 49–56.

[123] Barranguet C, Van den Ende FP, Rutgers M, Breure AM, Greijdanus M, Sinke JJ, Admiraal W (2003) Copper-induced modifications of the trophic relations in riverine algal-bacterial biofilms. Environ Toxicol Chem 22, 1340-1349.

[124] Massieux B, Boivin MEY, Van den Ende FP, Langenskiöld J, Marvan P, Barranguet C, Admiraal W, Lannbroek HJ, Zwart G (2004) Analysis of structural and physiological profiles to assess the effects of Cu on biofilm microbial communities. Appl Environ Microbiol 70, 4512–4521.

[125] de-Bashan LE, Moreno M, Hernandez JP, Bashan Y (2002) Removal of ammonium and phosphorus ions from synthetic wastewater by the microalgae *Chlorella vulgaris* coimmnobilized in alginate beads with the microalgae growth-promoting bacterium *Azospirillum brasilense*. Water Res 36, 2941-2948.